Chitosan, Chitosan Derivatives and Their Applications

Chitosan, Chitosan Derivatives and Their Applications

Editor

Agnieszka Ewa Wiącek

Basel • Beijing • Wuhan • Barcelona • Belgrade • Novi Sad • Cluj • Manchester

Editor
Agnieszka Ewa Wiącek
Department of Interfacial
Phenomena
Maria Curie-Skłodowska
University
Lublin
Poland

Editorial Office
MDPI
St. Alban-Anlage 66
4052 Basel, Switzerland

This is a reprint of articles from the Special Issue published online in the open access journal *Molecules* (ISSN 1420-3049) (available at: www.mdpi.com/journal/molecules/special_issues/chitosan_chemistry).

For citation purposes, cite each article independently as indicated on the article page online and as indicated below:

Lastname, A.A.; Lastname, B.B. Article Title. *Journal Name* **Year**, *Volume Number*, Page Range.

ISBN 978-3-7258-0254-8 (Hbk)
ISBN 978-3-7258-0253-1 (PDF)
doi.org/10.3390/books978-3-7258-0253-1

Contents

About the Editor

Agnieszka Ewa Wiącek

Prof. Agnieszka Ewa Wiącek studied chemistry in 1989-1994 at Maria Curie-Sklodowska University in Lublin (Poland) and was employed as a research assistant at the Department of Physical Chemistry in 1994. She received a D.Sc. degree in chemistry in September 2000, with a PhD thesis discussing the effect of natural stabilizers on oil/water emulsion stability. In 2013, she received a Dr. habilitation degree in physical chemistry, producing a dissertation titled "Effect of the selected biologically-active substances, mainly phospholipids and (phospho)lipases on the interfacial properties of dispersed systems". In 2018, she secured an associate professor position. She published more than 90 scientific papers (including 15 monoauthor's paper) which were cited more than 1800 times according to GoogleScolar. Additionally she is the author or co-author of approximately 200 conference presentations. Between 1997 and 2004, she was an examiner of Recruitment Commission of Chemistry Faculty in Lublin and outer cities. She was promoter of more than 50 master of science theses (M. Sc.) or licentiate theses (B. Sc.) and 2 doctoral theses. She participated in seven scientific projects, including international projects.

Preface

"Chitosan, Chitosan Derivatives and Their Applications" is devoted to the latest advances in research on chitosan-based systems and focuses on highlighting scientific discoveries carried out in leading laboratories around the world. This topic is a developing field of science, but a very promising one in terms of practical applications. The manuscripts included in this book focus on all aspects of the production, modification, enzymology and application of chitin/chitosan-based systems. The reprint is divided into two parts, each of which begins with an overview article. The first half of the collected articles concerns with the application potential in agricultural and nutritional sciences. The second part covers medical and pharmaceutical applications. At the end of the presented series, several articles devoted to the synthesis and characterization of chitosan-based materials were collected. All mentioned articles (21 in total) represent a high scientific level and the latest trends in research on chitin, chitosan and their derivatives. I wish you a pleasant read.

Agnieszka Ewa Wiącek
Editor

molecules

MDPI

Review

Strategies for the Preparation of Chitosan Derivatives for Antimicrobial, Drug Delivery, and Agricultural Applications: A Review

Rajeev Shrestha [1,*], Anusree Thenissery [1], Rahul Khupse [2] and Gireesh Rajashekara [1,*]

1 Center for Food Animal Health, Department of Animal Sciences, The Ohio State University, Wooster, OH 44691, USA; thenissery.1@buckeyemail.osu.edu

2 College of Pharmacy, University of Findlay, Findlay, OH 45840, USA; khupse@gmail.com

* Correspondence: shrestha.144@osu.edu (R.S.); rajashekara.2@osu.edu (G.R.)

Abstract: Chitosan has received much attention for its role in designing and developing novel derivatives as well as its applications across a broad spectrum of biological and physiological activities, owing to its desirable characteristics such as being biodegradable, being a biopolymer, and its overall eco-friendliness. The main objective of this review is to explore the recent chemical modifications of chitosan that have been achieved through various synthetic methods. These chitosan derivatives are categorized based on their synthetic pathways or the presence of common functional groups, which include alkylated, acylated, Schiff base, quaternary ammonia, guanidine, and heterocyclic rings. We have also described the recent applications of chitosan and its derivatives, along with nanomaterials, their mechanisms, and prospective challenges, especially in areas such as antimicrobial activities, targeted drug delivery for various diseases, and plant agricultural domains. The accumulation of these recent findings has the potential to offer insight not only into innovative approaches for the preparation of chitosan derivatives but also into their diverse applications. These insights may spark novel ideas for drug development or drug carriers, particularly in the antimicrobial, medicinal, and plant agricultural fields.

Keywords: chitosan (CS) derivatives; antimicrobial activity; drug delivery; infectious diseases; plant agriculture

Citation: Shrestha, R.; Thenissery, A.; Khupse, R.; Rajashekara, G. Strategies for the Preparation of Chitosan Derivatives for Antimicrobial, Drug Delivery, and Agricultural Applications: A Review. *Molecules* **2023**, *28*, 7659. https://doi.org/10.3390/molecules28227659

Academic Editor: Agnieszka Ewa Wiącek

Received: 28 September 2023
Revised: 14 November 2023
Accepted: 15 November 2023
Published: 18 November 2023

1. Introduction

Over the last few decades, natural polymers, including biopolymers, polyesters, and proteins, have displayed diverse applications in the fields of agriculture, pharmaceuticals, food safety, and the cosmetic industry due to their biodegradability, eco-friendliness, compatibility, and cost-effectiveness [1–3]. As a result, natural polymers have gained significantly increased interest in creating novel drugs, owing to their distinct and intricate chemical composition, which is responsible for their wide and diverse range of biological effects [4,5]. Among these, chitosan (CS) stands out as a widely utilized polysaccharide that has been extensively studied across various scientific disciplines [6–8]. CS is composed of linear polysaccharides with varying amounts of ($\beta 1 \rightarrow 4$) linked residues of *N*-acetyl-2 amino-2-deoxy-D-glucose (glucosamine, GlcN) and 2-amino-2-deoxy-D-glucose (*N*-acetyl-glucosamine, GlcNAc) residues distributed inside the polymer. It is derived from chitin, polyβ-[1,4]-*N*-acetyl-d-glucosamine) (or poly(*N*-acetyl-d-glucosamine) via deacetylation in the presence of sodium hydroxide or enzymes (Scheme 1) [9–11]. Chitin is found in crustaceans, including crabs, shrimp, shellfish, and lobsters, and its extraction primarily involves demineralization, deproteinization, and discoloration [12,13]. Its hydrophilic and reactive properties are limited by the presence of acetyl groups (CH_3CO^-) attached to the amino groups of GlcNAc. These acetyl groups function not only as barriers for water molecules, but also as protective shields, which reduce the availability of the amino groups for reactions [14]. For the biological and physical activity of CS, the degree of deacetylation (DD) is the pivotal factor, which represents the proportion of β-1,4-D-glucosamine repeating units in the polysaccharides during the deacetylation process originating from chitin [9–11,14]. A higher DD denotes an increased presence of free amino groups, which can form ammonium (NH_4^+) groups within a pH range of 5–6, resulting in increased water solubility and the enhancement of the biological impact of CS [15]. Furthermore, a lower molecular weight (MW) is another factor that decreases with a higher DD, which impacts the bioactivity of CS [15]. CS with a lower MW has demonstrated antibacterial, antioxidant, and anti-tumor properties due to its increased water solubility, ease of permeability to cells, and binding efficacy to DNA or RNA [16,17].

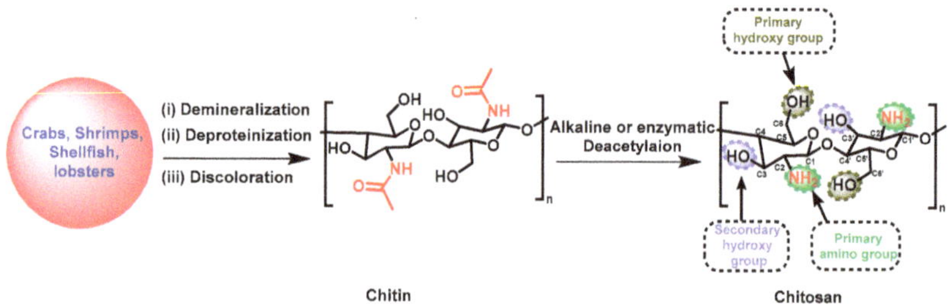

Scheme 1. Synthetic strategy for the preparation of CS.

CS is a cationic biopolymer agent that can interact with negatively charged substances such as lipids, fats, ions, cholesterol, proteins, and other organic and inorganic molecules [18]. CS was first identified in 1811 in mushrooms. Extensive research has revealed that CS significantly exhibits antimicrobial, antiviral, anticancer, non-toxic, water treatment, biocompatible, and biodegradable properties [19,20]. However, its solubility in acidic conditions (pH less than 6) limits certain bioactivities and applications. To address these issues and to enhance the bioactive and physiochemical effectiveness under neutral and physiological conditions, the modification of CS is a promising alternative since it has three active functional groups: C2-NH_2, C3-OH, and C6-OH (Scheme 1) [21]. The nucleophilic substitution reactions involving the amino and hydroxyl groups, followed by multi-step reactions, have been studied, including *N*-alkylation/amination, acetylation, cross-linking, oxidation, grafting, and sulfation [22,23]. Synthetic reactions are accomplished through chemical or enzymatic catalysis, polymerization, or hydrolysis conditions to achieve the desired product [24]. The CS derivatives, which vary in terms of their degree of substitution (DS), a crucial factor in the structure–activity relationship (SAR), demonstrate unique physiochemical characteristics. Novel scaffolds with these characteristics have enhanced biocompatibility, bioactivity, biodegradability, and non-toxicity, while retaining the inherent pharmacological effects of their antibacterial, anticancer, and antiviral properties [25]. Moreover, CS derivatives have also been widely utilized in the field of nanomaterials due to their immense potential as carriers for targeted drugs, adjuvants, or vaccines, facilitating the treatment of various diseases including cancer, diabetes, Alzheimer's, cardiovascular disorders, and inflammatory conditions [26,27]. These nanomaterials have shown superior outcomes compared to conventional drug formulations.

Herein, we explore the recent progress of CS derivatives, focusing on modifications to their functional groups conjugated to the polymer skeleton as well as substitutions in the active sites, including C2-NH_2, C3-OH, and C6-OH. Importantly, we discuss the synthesis process of CS derivatives, categorizing them into various chemical groups based on their synthetic pathway or functional characteristics, which included alkylated, acylated, Schiff base, quaternary ammonia, guanidine, and heterocyclic CS derivatives. Moreover, this review also comprehensively covers the diverse applications of CS and its derivatives as nanomaterials in the antimicrobial, agricultural, and drug delivery fields, including the elucidation of their mechanisms of action and the factors that contribute to superior outcomes compared to conventional approaches.

2. Global Research Trends of Chitosan

The global research trends surrounding CS underscore its significant scientific and technological value. As projected by Precedence Research, the CS market is expected to surpass US $29 billion by 2030, reflecting its substantial importance across various sectors [28]. These trends are driven by several key factors. Notably, there is a growing demand for CS in various end-user industries, an increasing emphasis on global water treatment activities, significant advancements in healthcare and the medical sector in industrialized countries, and a heightened awareness of public health concerns, including rising obesity rates. The Asia–Pacific region has assumed a prominent position, dominating the CS market share [28]. Over the last two decades, CS has ascended to a leading role in research involving biomedicine, nanomedicine, and cell and tissue engineering [29]. The increase in research endeavors is well noted by the heightened number of patents, particularly in the biomedical sector, where its gelling properties enable viscosity and density adjustments under specific physiochemical conditions to enhance biocompatibility [30] (Table 1). Furthermore, the applications of CS underscore these global research trends. CS finds utility in diverse fields, which have extended constant research efforts in drug delivery, dentistry, ophthalmology, wound dressings (some marketed CS-based wound dressings: HemCon® Bandage PRO, HemCon ChitoFlex® PRO, Chitopack C, ChiGel, Tegaderm, Tegasorb, and Traumastat), cell encapsulation, bioimaging, tissue engineering, food packaging, gelling and coating, abiotic stress prevention in plants, enhanced water availability in crops, controlled-release fertilizers, dye-sensitized solar cells, wastewater and sludge treatment, and metal extraction [31,32]. This multifaceted use highlights CS's pivotal role in shaping global research trends across healthcare, agriculture, and various industries on a worldwide scale, making it a central focus in the realm of biopolymers.

Table 1. Patents on chitosan-based materials in antimicrobial, drug delivery, and agricultural applications.

Patent Area	Application	Final Form	References
Biomedical/Antimicrobial	As a teat sealant wherein CS hydrogels accelerate involution and prevent infection of mammary gland	Hydrogels	[33]
Biomedical/Antimicrobial	An adsorbent biodegradable wound dressing	Films	[34]
Biomedical/Antimicrobial	CS coating in metallic nanoparticles as an anti-biofilm agent on medically implantable articles	Nanoparticles/powder	[35]
Biomedical/Antimicrobial	For coating *Listeria iuanuii*, to enhance negative surface charge and thus immune response	CS oligosaccharide	[36]
Biomedical/Antimicrobial	Antimicrobial wound dressing material with quaternized CS	Quaternized CS sponges	[37]
Biomedical/Antimicrobial	Formulation of CS with metallic nanoparticles for treating skin lesions/infections	Gel matrix	[38]
Biomedical/Drug delivery	CS hydrogel used for targeted drug delivery	Hydrogels	[39]
Biomedical/Drug delivery	Low-toxicity and high-affinity CS derivative for drug delivery	Quaternized, acylated CS derivative	[40]
Biomedical/Drug delivery	Delivery of IL17RC protein that function by improving nasal adsorption	Nanoparticles	[41]
Biomedical/Drug delivery	CS–nucleic acid polyplexes for delivery of nucleic acid encoding IL-2	Nanoparticles	[42]
Biomedical/Drug delivery	Treatment of transected peripheral nerve injuries	CS hydrogel	[43]
Agricultural	Gel microspheres of CS with manganese and prothioconazole for use against pests	Gel microspheres	[44]
Agricultural	Antifungal activity in plants	CS derivative	[45]
Agricultural	Pesticides that comprise CS and berberine	Powder, suspension	[46]
Agricultural	Liquid fertilizer with CS oligosaccharides and amino acids	Liquid	[47]

3. Preparation Methods of Chitosan Derivatives

Owing to the efficient biological and physiochemical properties of CS derivatives, various pathways of CS modification have been explored. The methods involve single- to multi-step reactions, including direct modification, catalytic or enzymatic reactions, chemical grafting, and cross-linking reactions under various conditions [23]. In terms of the chemical reactivity of CS, 2-NH_2 displays more reactivity compared to the OH group. Therefore, to achieve the selective functionalization of the OH group, NH protection by using the phthaloyl group has been employed [48–50]. However, OH protection groups, including triphenylmethyl [51], trimethylsilyl [52], and tertiary-butyl dimethylsilyl [53,54], have also been utilized to achieve the specific functionalization of the NH group, resulting in a high DS in the products.

Although CS is soluble in acidic water through the binding process of H^+ with the N atom of amino group, CS exhibits a high degree of crystallinity due to the intermolecular and intramolecular hydrogen bonding, leading to its near insolubility in normal water [55]. Therefore, the key benefit of CS modification is the enhancement of the water solubility of CS in high pH ranges by introducing hydrophilic groups like amino, carbonyl, carboxyl, hydroxyl, or sulfhydryl groups into the CS [55–59]. These functional groups are capable of not only destroying the original hydrogen bonding and crystallinity of CS but also creating new hydrogen bonds with water, thereby increasing the solubility of the product. Consequently, the modified CS is often employed to improve the solubility of the product, demonstrating the synergetic efficacy of attached functional groups for antimicrobial, antifungal, and other biological activities [23,24]. The researchers have employed CS with a wide range of MWs (ranging from 50 to 1400 kDa) and DDs (ranging from 70 to 95%) in their experiments to produce CS derivatives with various ranges of DSs of the functional group. A higher DS leads to better water solubility and higher biological potentials [29,60].

3.1. Alkylation, Acylation, and Schiff Base-Based Chitosan Derivatives

The alkylation of CS entails the hydrophobic modification of this material. The resulting hydrophobic traits yield specific benefits, especially the restriction of liquid interchange between the external surroundings and wound tissue. This property creates a protective barrier on the wound surface that impedes the infiltration of airborne bacteria and upholds gas exchange, ultimately leading to rapid wound healing [61]. The alkylation of CS (**2a**) is conducted by employing various alkyl or allyl halides such as bromoethane, bromohexane, bromododecane, bromohexadecane, chlorobutane, and allyl bromide in the presence of a strong base like sodium hydroxide (Scheme 2) [62–64]. However, the Nie group reported that the Michael addition reaction of CS with hydroxyethylacryl in acidic media led to alkylated CS (**2b**) [65]. On the other hand, CS has been acylated with various acyl chlorides in the presence of NaOH, with the initial step involving the dissolution of CS in acetic acid with water (**2c**) [66,67]. The acylation reaction has the potential to disrupt both the intramolecular and intermolecular H-bonding within the CS, leading to a reduction in its crystallinity and enhancing water solubility [68]. In recent studies, researchers have explored the reaction of CS with pre-prepared acyl chlorides or acetic acid containing diverse heterocycles like benzothiazole, thiadiazole, thymine, or pyrazoles to produce the acylated CS bearing the heterocycles (**2d, 2e**) [69–76]. Similarly, Zhang's group developed CS-bearing quinolinyl urea derivatives (**2f**). The reaction was accomplished by initiating a reaction between methyl chloroformate and CS, followed by a subsequent reaction with 2-aminoquinoline [77]. Specifically, the incorporation of heterocyclic groups into CS, with substitution degrees of 57.4% [75] and 47% [76], resulted in enhanced antimicrobial activities.

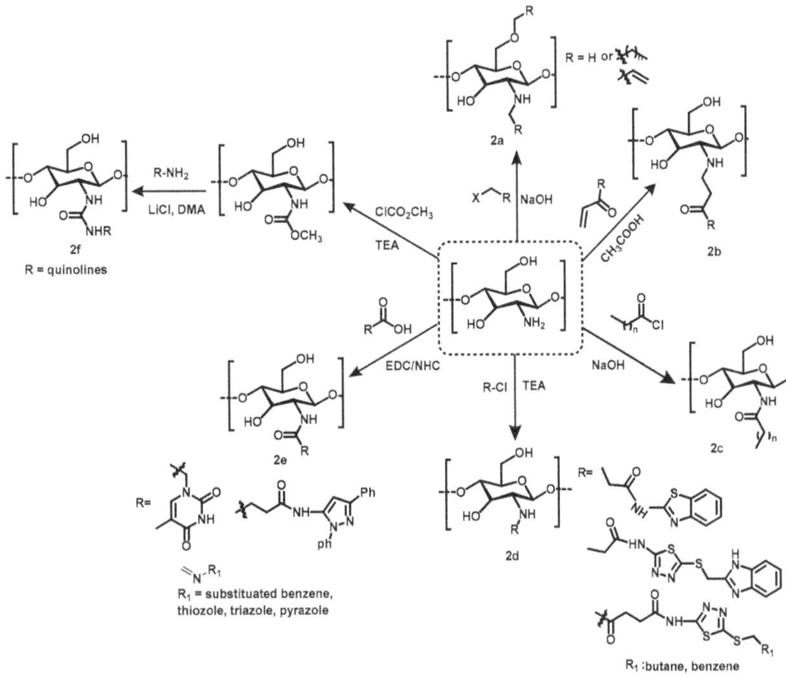

Scheme 2. Synthesis scheme of chitosan derivatives through alkylation and acylation processes.

In the course of the modification of CS derivatives by using CS (MW: 100–300 kDa, DD: 74–95%), the preparation of Schiff bases emerges as the most straightforward method. This approach includes the reaction of an amine group with benzaldehydes through an intermediate carbinolamine (Scheme 3). This method not only facilitates the introduction of new groups, such as heterocycles, but also enables subsequent alkylation after reduction [78,79]. In a plethora of studies, the attachment of heterocycles such as thiophene, furan, pyrazole, indoles, quinolone, benzothiazole, oxazole, and imidazole-bearing Schiff base CS derivatives (**2g**; DS: 1.15–56%) was extensively explored in acetic acid as a mild condition [80–84]. Incorporating heteroatoms into CS as Schiff base resulted in a hydrophobic characteristic; nevertheless, the presence of pyrazole and thiophene groups exhibited water solubility for CS [80–84]. Additionally, the reduction of Schiff bases with sodium borohydride has been widely employed as an alternative alkylation strategy, leading to the attachment of new functional groups or heterocycles (DS: 36–81%) (**2h**) [85]. Moreover, Khrustalev et al. revealed innovative CS derivatives with novel heterocyclic structures (**2i**), forming complexes with platinum metals [86]. Herein, CS of various MWs (viscosity average MW: 3.7×10^4 D to 17.8×10^4 D) was employed to introduce nitrile derivatives of CS (**2i'**) in the range of DS, 17% to 59%. This synthesized platinum complex was not soluble in water; however, the 1,2,4-oxadiazoline CS derivatives prepared from nitrones, a functional group consisting of *N*-oxide and imine, were soluble in water and showed a high antibacterial activity coupled with low toxicity [86].

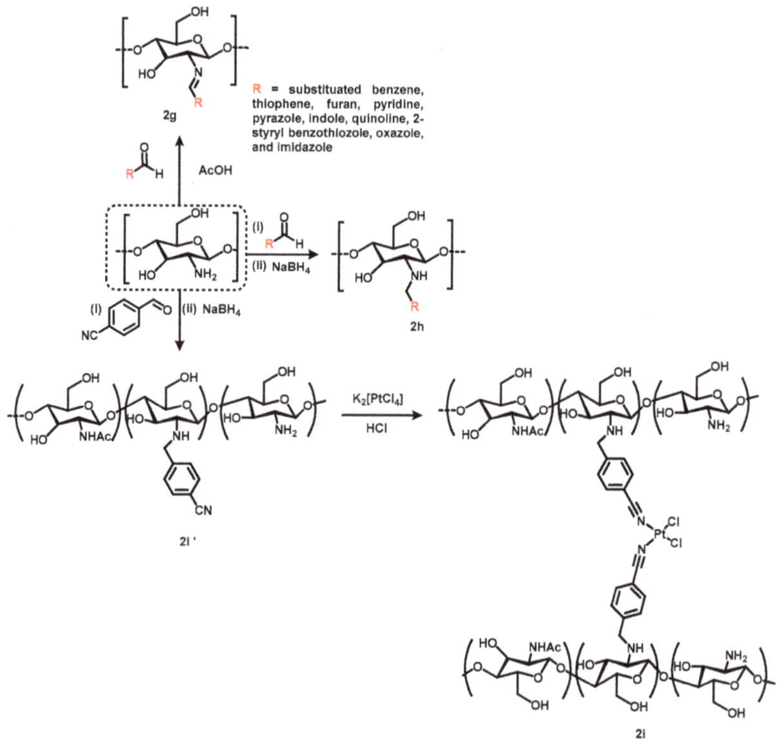

Scheme 3. Synthesis of chitosan derivatives bearing Schiff bases.

3.2. *Quaternary Ammonium Chitosan Derivatives*

Quaternary ammonium CS derivatives are the modified form of CS salt bearing a positively charged nitrogen atom, covalently attached with alkyl or aryl groups. The hydrophilic quaternary ammonium group enhances the water solubility of CS by introducing a quaternary ammonium salt group. The salt increases the electrostatic charge and weakens intramolecular hydrogen bonds, resulting in a high level of water solubility [87]. This quaternary ammonium salt improves not only the hydrophilicity of the final CS derivatives but also enhances its antimicrobial properties, making it valuable in diverse applications, including wound dressings, drug carriers, and other biomedical fields [88]. In addition, the polymeric quaternary ammonium CS bearing heterocyclic compounds has attracted a significant antimicrobial agent due to the synergetic effects of CS, quaternary ammonia salts, and heterocyclic compounds [89]. Commonly, the quaternary ammonium salt, *N*-trimethyl CS (TMC, **3a**), is prepared via the methylation of CS (MW: 5–200 kDa; DD: 73–97%) in the presence of sodium iodide and iodomethane under alkaline conditions (Scheme 4) [90–99]. However, there is the possibility of *O*-methylation, which can be reduced by using either an acidic or heterogeneous medium. The inclusion of heterocyclic and aromatic moieties shows promising prospects for drug development [91]. By taking advantage of this, TMC has been further subjected to derivatization, enabling the incorporation of heterocyclic and aromatic groups. In this respect, Guo et al. prepared the intermediate 6-bromo-6-deoxy-*N*-trimethyl quaternary ammonium CS through the bromination of *N*-bromobutanimide (NBS) on TMC. Subsequently, it was further reacted with coumarins with electron-donating or electron-withdrawing groups under basic conditions to yield coumarins bearing TMC (DS: 69–82%) (**3b**) [92]. In another study of their group, 6-bromo-6-deoxy-*N*-trimethyl quaternary ammonium CS was further investigated by treating it with ethylenediamine and mono- or bi-substituted polyhydroxyl benzaldehyde in two steps, leading to the formation of phenol or polyphenol-containing trimethyl quaternary ammonium CS (**3c**) [93].

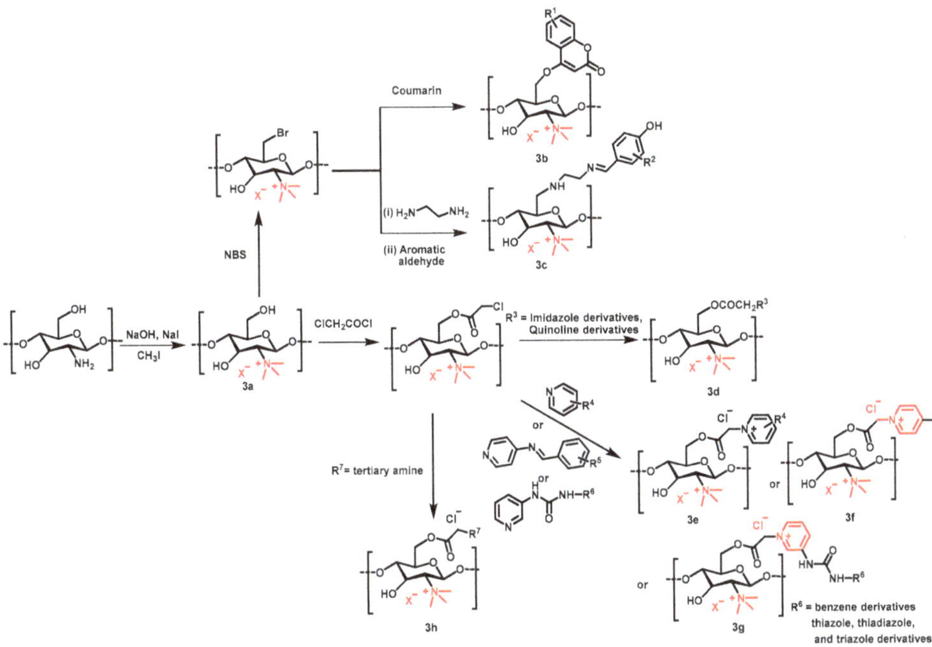

Scheme 4. Synthesis of quaternary ammonium salt, *N*-trimethyl chitosan derivatives. The red color represents the functional group of quaternary ammonium salt.

In the meantime, an alternative approach has been introduced to prepare *O*-chloroacetyl quaternary ammonium CS, which has been further employed in chemical modifications involving heterocyclic compounds due to its high effectiveness in nucleophilic substitution reactions (Scheme 4). For example, compounds such as quinoline (DS: 24.34–98.7%), imidazole (DS: 12.7–65.5%), amino pyridines (DS: 72–82%), pyridines, substituted pyridines bearing Schiff bases (DS: 21.5–58.3%) or urea derivatives (DS: 21–91%), and tertiary amines (DS: 45.3–84.3%) were subjected to reactions with *O*-chloroacetyl quaternary ammonium CS, resulting in the formation of double quaternary ammonium CS salt derivatives (**3d–3h**) [96–99]. The advantages of double-quaternary ammonium salt include making CS more cationic and highly soluble in water.

While developing the novel derivatives of quaternary ammonium CS, the water-soluble (3-Chloro-2-hydroxypropyl) trimethy/triethyl ammonium chloride (CHTAC) or glycidyl trimethyl ammonium chloride serves as an excellent quaternary ammonium source, which has been employed in nucleophilic substitution reactions, either C2-NH_2 or C6-OH, of CS (Scheme 5) [100–108]. In recent years, *n*-alkyl quaternary ammonium CS (**3i**) and 2-hydroxypropyl trimethyl ammonium chloride CS (**3j′**) were synthesized through the reaction of *n*-alkyl CS or CS with a quaternary ammonium source, CHTAC. The *n*-alkyl CS was initially obtained through the alkylation of CS with *n*-butylaldehyde using the Schiff alkali method, followed by reduction using sodium borohydride [100]. The optimal synthesis condition for a higher DS (70–90%) of **3i** was optimized in various temperatures, molar ratios of chemicals, and reaction times. While the extended *n*-alkyl chain notably reduces the intermolecular hydrogen bonds in CS, it alone does not lead to complete dissolution in water. However, when combined with quaternary ammonium salt, it significantly enhances water solubility [100]. On the other hand, **3j′** was further dialyzed with carboxymethyl insulin [101] and other acetate derivatives against deionized water [102], followed by lyophilization to obtain the desired quaternary ammonium CS derivatives (**3j**; DS: 17.5–40.9% and 46–81%). In the meantime, the Zheng group also explored imidazole-based quaternary ammonium CS derivatives (DS: 36%) through multistep nucleophilic substitution reactions [103]. The synthesis process comprised a sequential generation of intermediary compounds, *N*-chlorobutanol CS and *N*-(1-carboxybutyl-4-(3-methyl-imidazole) CS chloride, through 4-chlorobutanol chloride and *N*-methylimidazole. The ultimate product (**3k**) was obtained through the reaction between CHTAC and *N*-(1-carboxymethyl-4-(3-methyl-imidazole)) CS chloride. Likewise, hydroxypropyl trimethyl ammonium chloride CS (HACC) has demonstrated itself as an excellent water-soluble quaternary CS derivative, which is commonly prepared from the reaction of CS and glycidyl trimethyl ammonium chloride in deionized water [104,105]. Furthermore, the introduction of thioctate [106] and Schiff base [107] functionalities onto HACC was accomplished through treatments with α-lipoic acid and 4-diethylaminobenzaldehyde to give the new derivatives **3l** (DS: 51–86%) and **3m**. In an alternate approach to the aforementioned procedure, the researchers initially established a Schiff base of CS, followed by a reaction with glycidyl trimethyl ammonium chloride, yielding *O*-quaternary ammonium *N*-benzylidene CS. To expand the scope of chemical modifications, *O*-quaternary ammonium *N*-acyl thiourea CS (**3n**; DS: 81–93%) was synthesized through the incorporation of ammonium thiocyanate and chloroacetyl chloride into the reaction sequence [108].

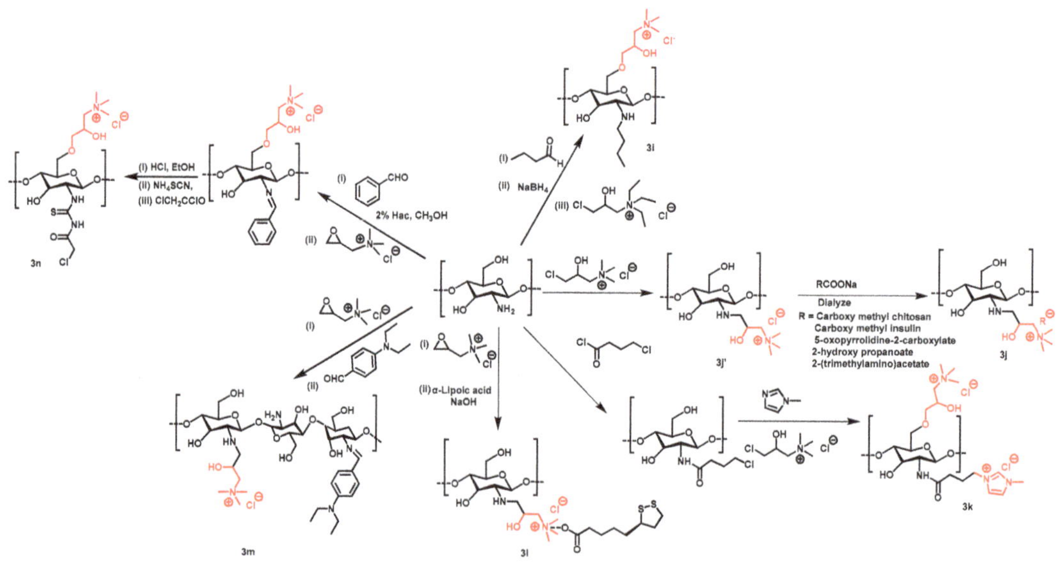

Scheme 5. Synthesis of chitosan quaternary ammonium derivatives through a quaternary ammonium source. The red color represents the functional group of quaternary ammonium salt.

Another category of quaternary ammonium salt is the pyridinium salt group, which acts as a cationic surfactant, is soluble in water [87,109], and exhibits noteworthy antimicrobial and antifungal activities [110,111]. Pyridine functions as a nucleophile, capable of replacing the alkyl halide to synthesize the quaternary pyridinium compound [112]. Typically, CS undergoes modification through a reaction with either the halide group or the pyridine group, enabling subsequent functionalization with pyridine salts (Scheme 6) [113]. Huang et al. grafted pyridine to *N*-chlorobutyryl CS, resulting in the formation of pyridinium CS salt (**3o**, DS: 38%) [114]. *N*-chlorobutyryl CS was prepared from the reaction of CS and 4-chlorobutyryl chloride via acylation. Recently, Tan et al. synthesized CS-bearing pyridinium salt along with quaternary ammonium salt (DS: 35–97%) (**3p**) through a treatment between CS and narcotic acid, followed by a reaction with bromopropyl trialkyl ammonium bromides [115]. In another study, Guo et al. also introduced CS derivatives that contain pyridine-bearing Schiff bases (**3q**, DS: 22.7–50%) [116] and heterocyclic scaffolds along with thiourea (**3r**, **3s**; DS: 21–91%, 18–88%) [98,117]. In contrast with these conventional approaches, the Zincke reaction has been explored for the direct transformation of primary amines into *N*-aryl pyridinium salt [118]. This process typically involves the use of *N*-(2,4-dinitrophenyl)-pyridinium chloride, commonly known as Zincke salt, as the reagent, which follows an SN(ANRORC) mechanism involving nucleophilic substitution addition, ring opening, and subsequent ring closure [119]. In this respect, the CS amino group at the C2 position was directly converted to CS pyridinium salt (**3t**) via the Zincke reaction [120,121].

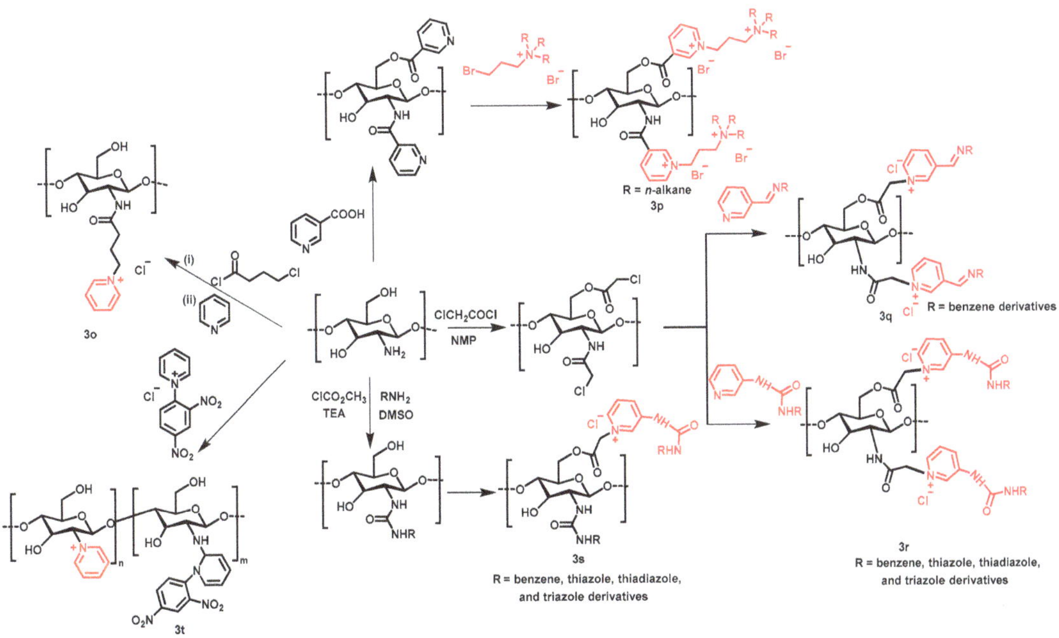

Scheme 6. Synthesis of chitosan derivatives bearing pyridinium salt. The red color represents the functional group of pyridinium salt.

The application of click chemistry has demonstrated significant utility in introducing quaternary ammonium salts into CS molecules, primarily due to its notable benefits of high efficiency and selectivity [122]. For instance, the initial step involved the development of *N*-trimethyl ammonium CS, which also served as a C2-NH_2 protection group. Then, the reaction was followed by either azidation [123] or *O*-acetylation [124,125] to proceed with the cycloaddition reaction of alkyne-azides, resulting in the integration of water soluble triazole substituents with quaternary ammonium CS (**2u**, **2v**; DS: 72–83%, 28–92%, 53–92%), which exhibited antifungal and antioxidant properties (Scheme 7).

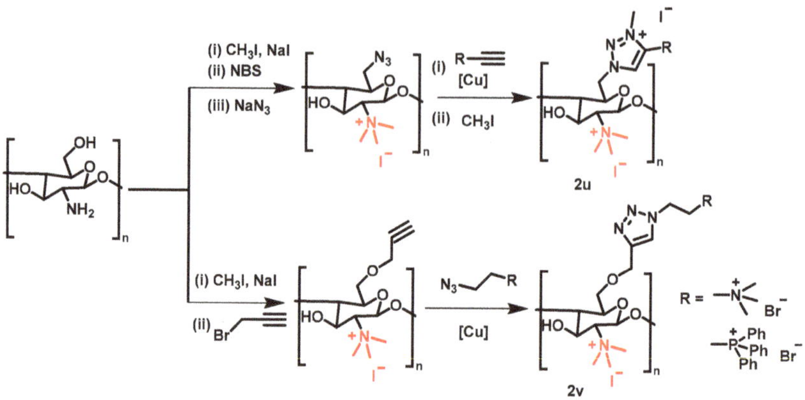

Scheme 7. Synthesis of quaternary ammonium chitosan derivatives followed by click chemistry. The red color represents the functional group of quaternary ammonium salt.

3.3. Guanidine Chitosan Derivatives

Guanidines have demonstrated a significant interest as a valuable functional group incorporated with multiple drugs for various therapeutic applications and biological activities, such as antimicrobial, antifungal, antimalarial, and others [126]. From a chemical perspective, guanidines function as the nitrogenous analog of carbonic acid, behaving as strong organic bases with higher pKa values. However, by introducing the appropriate substituents, such as CS, onto the nitrogen atom, the basicity of guanidine can be adjusted, and it is highly soluble in neutral aqueous solutions by disrupting hydrogen bonding and hydrophobic effects in polymers, resulting in the remarkable versatility of guanidine motif-bearing scaffolds [55,127]. Water-soluble CS-bearing guanidines (**4a**, **4b**; DS: 8.4–14.2%, 14–30%) are commonly synthesized through the reaction of cyanimide or aminoiminomethanesulfonic acid with CS in the presence of hydrochloric acid (Scheme 8) [128,129]. In the alternative pathway, the microwave-assisted grafting of guanidine oligomers with CS was also investigated, and the thus-formed CS guanidine (**4c**) in various DSs was utilized for the production of hygiene paper products [130]. Moreover, the main composition of the peptide, L-arginine, or L-arginine Schiff base, was also coupled with CS in a basic medium to develop biomimetic CS derivatives (**4d**; DS: 6–61.5%) for the improvement of the permeable capabilities of cell-penetrating peptides, wound healing, and antimicrobial and antioxidant activities [131–135].

Furthermore, CS–guanidine derivatives incorporating additional moieties such as poly (3-hydroxy butyrate) (**4e**) or quaternized carboxymethyl (**4f**; DS: 16–73%) were also synthesized (Scheme 8) [136,137]. In this process, cyanoguanidine or cyanimide functioned as the source of the guanidine. However, the reactions involved multiple sequential steps of the nucleophilic reaction of poly(3-hydroxybutyrate) or epichlorohydrin. Another approach for synthesizing guanidinylated CS derivatives involved the utilization of protective groups like tertiarybutyl dimethylsilyl (TBDMS) and tertiarybutyloxycarbonyl (Boc). These protective groups effectively overcome the limitation of selective modification of the amino group with a high degree of substitution. In the course of the reaction, the TBDMS group was initially introduced to the hydroxyl group of CS possessing a free amino group, enabling the successful synthesis of a 100% guanidinium substitution of the amino group, thereby yielding the desired guanidinium CS derivatives (**4g**, **4h**; DS: 13–100%) [138]. In the exploration of diverse derivatives, the Mingshan Li group functionalized the CS nanofiber membrane with the polyhexamethylene guanidine group (**4i**). The synthesis process included the oxidation of CS in the presence of sodium periodate ($NaIO_4$), electrospinning with polyvinyl alcohol (PVA), and functionalization with polyhexamethylene guanidine [139]. Thus, the formed nanofiber membrane exhibited not only antibacterial activities but also high adsorption capacities of Cu (II) and Congo red.

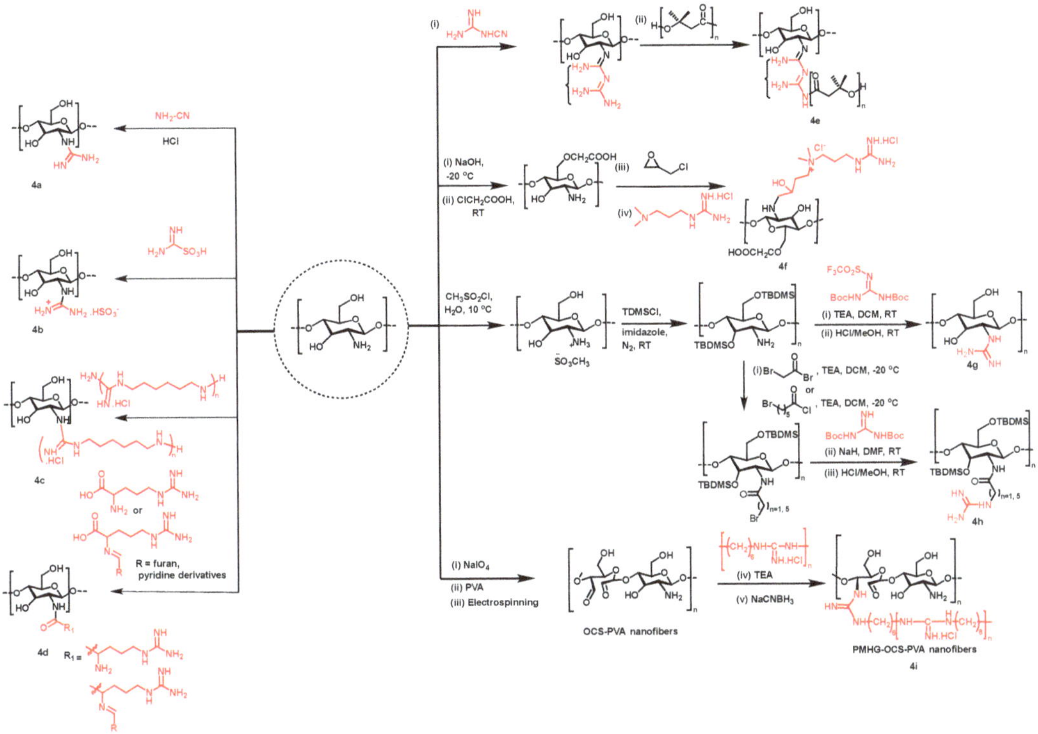

Scheme 8. Synthetic strategy for guanidine chitosan derivatives. The red color represents the functional group of guanidine.

In contrast to the aforementioned method, Peter Hesemann et al. employed a different approach to produce an *N*-guanidinium CS derivative. This was carried out in the presence of scandium (III) triflate and acetic acid, followed by a subsequent sol–gel reaction utilizing 3-(trihydroxysilyl)-1-propanesulphonic acid to synthesize a guanidinium CS silica hybrid material (**4j**) (Scheme 9) [140]. This material demonstrated remarkable effectiveness as an adsorbent for removing dyes from wastewater. On the other hand, the same group investigated the reaction between CS and carbodiimides in ionic liquid media in the presence of 1-butyl-3-methylimidazolium chloride (BMIM Cl) to produce *N*-guanidinium CS (*N*,*N*′-dicyclohexyl) chloride (**4k**) and *N*-guanidinium CS (*N*-(3-dimethylaminopropyl)-*N*′-ethyl hydrochloride) chloride (**4l**), showing significant amicrobial properties [141]. Similarly, the Sedghi group devised a semi-conductive water-soluble biguanidine/polyaniline composite (**4m**; DS: 18.6%) through the reaction of CS with cyanoguanidine followed by aniline incorporation (Scheme 10). Subsequently, the composite was further introduced into a self-healing waterborne polyurethane system, with shape-memory effect properties suitable for bone tissue engineering [142].

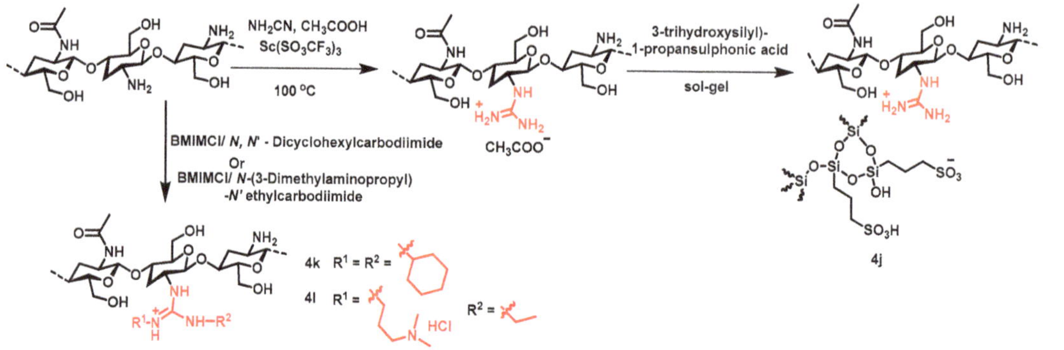

Scheme 9. Synthetic strategy of guanidine chitosan salt derivatives. The red color represents the functional group of guanidine.

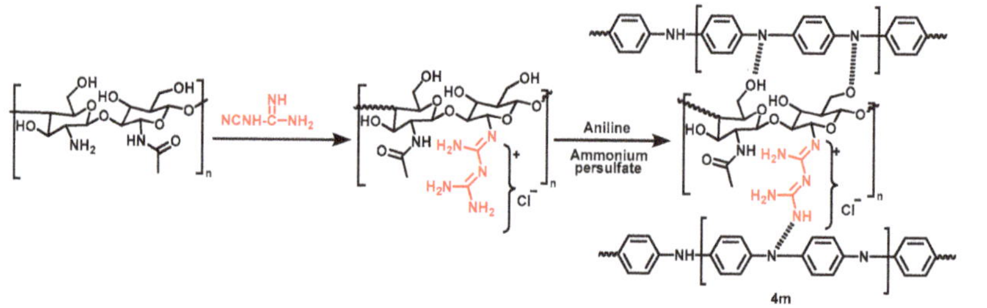

Scheme 10. Synthetic strategy of biguanidine/polyaniline composite. The red color represents the functional group of guanidine.

3.4. Heterocyclic Chitosan Derivatives via Click Chemistry or N-Functionalized Reaction

Copper-mediated click chemistry has emerged as a versatile approach for synthesizing triazole moieties that involve the cycloaddition reaction between an azide and terminal alkyne, allowing for the incorporation of diverse bioactive functional groups into biomacromolecule backbones [143,144]. Furthermore, this strategy serves as a stable antimicrobial pharmacophore, displaying its dual significance [145]. In this regard, CS has been widely functionalized with heterocyclic scaffolds, such as substituted triazole groups formed through click reactions, and demonstrated significant antimicrobial and antifungal activities [49,124,146–151]. The primary advantage of incorporating triazole, a polar moiety, is its excellent water solubility, which imparts hydrophilic properties to the resulting CS derivatives bearing triazole groups [146–151]. CS has been modified by introducing an azido group or alkyne group into the 2-*N* or 6-*O* position for the click reaction (Scheme 11). The Li group introduced the azido group at both active positions of CS through the reaction of chloroacetyl chloride, followed by sodium azide, and subsequently reacted with prop-2-yn-1-yl nicotinate in the presence of copper sulfate to achieve antifungal (1,2,3-triazol-4-yl)methyl nicotinate CS (**5a**) [151]. The phthalolyl group was utilized for the protection of the 2-NH_2 group, facilitated the 6-*O* position for azidation through stepwise bromination and azidation reactions, and was followed by the 3+2 cycloaddition reaction to provide diverse CS derivatives with triazoles (**5b**) [49,152]. Alternatively, another approach involved azidation through a nucleophilic NH_2 reaction with compounds like 1-azido-3-chloro-2-propanol or 2(azidomethyl) oxiran.

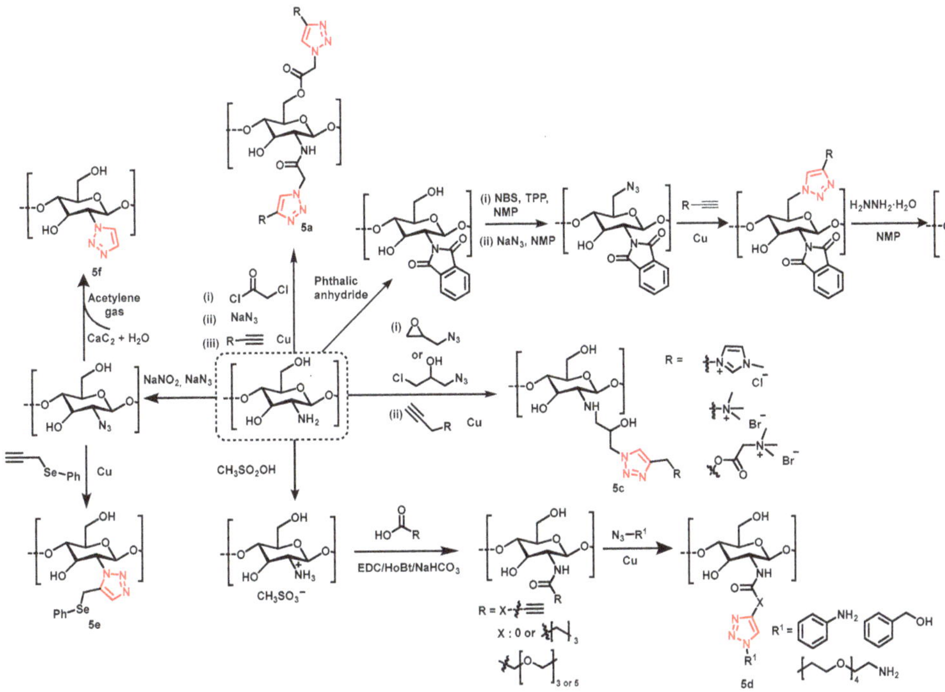

Scheme 11. Synthetic strategy of heterocyclic chitosan derivatives through click chemistry. The red color represents the newly constructed triazole moiety.

Subsequent reactions with substituted acetylene in the presence of a Cu catalyst also led to the triazole-functionalized CS (**5c**) [153,154]. In addition, alkylation on the 2-*N* position through CS mesylate salt [155] or direct *N*-azidation in the presence of sodium nitrite and sodium azide [156,157] also enabled the click reactions to synthesize CS triazoles (**5d**, **5e**, **5f**).

In a further study of CS-based heterocycles, the researchers developed cross-linked CS-based hydrogels by using Diels–Alder reactions (Scheme 12). Furan-modified CS, which was obtained via the reaction of furfural with CS, was then subjected to the Diels–Alder click reaction with poly(ethylene)glycol-maleimide derivatives or pre-synthesized maleimide-functionalized CS in acidic conditions to prepare the hydrogels (**5g**, **5h**; DS: 21–31%) with the capability of controlled drug release [158,159]. Moreover, the researchers have also constructed CS azalactone gel (**5i**) or genepin-crosslinked CS hydrogel (**5j**) through the direct substitution of oxygen with the nitrogen of the NH_2 group of CS (Scheme 13) [160,161]. In another approach for attaching the heterocycles to CS, the direct reaction of isobenzofurandione or pyrano-quinolinedione derivatives with CS led to the CS-heterocycles **5k–5n** (DS: 7–18%) and **5o** (DS: 34.79–92.05%) through either the direct interaction with the NH_2 group or the ring opening of benzofurandiones [162,163].

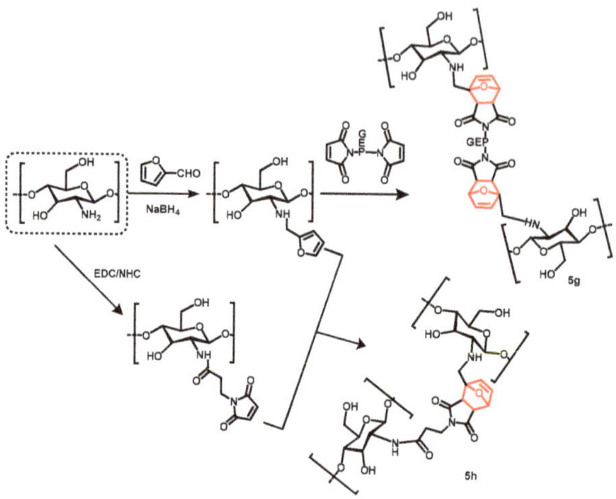

Scheme 12. Synthetic strategy of furan-modified chitosan derivatives through Diels–Alder click reaction. The red color represents the creation of new bonds through Diels–Alder click reaction.

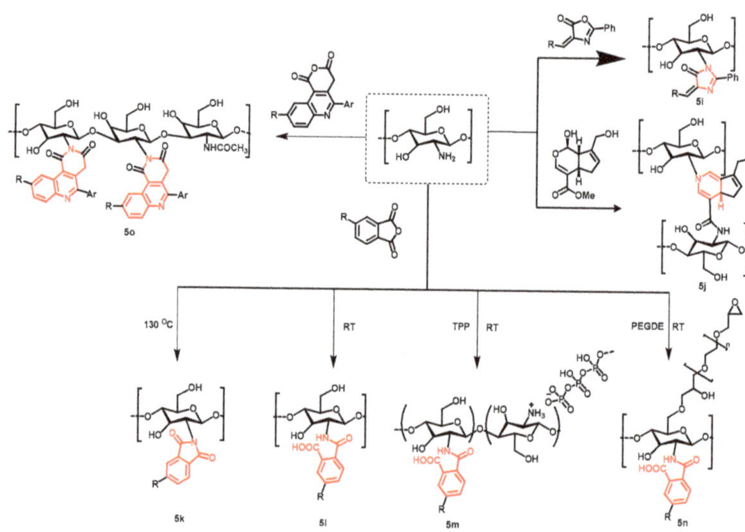

Scheme 13. Synthetic strategy of heterocyclic chitosan derivatives through *N*-functionalized reaction. The red color represents the addition of new heterocyclic moiety through direct *N*-functionalization reaction.

4. Applications of Chitosan Derivatives

4.1. Antimicrobial Actions of Chitosan

CS has garnered extensive attention across a wide spectrum of applications, ranging from its role as an antimicrobial agent to its utility as a drug carrier. Numerous research investigations have sought to unveil CS's antimicrobial attributes against a diverse array of microorganisms, including bacteria, fungi, yeast, and algae, primarily due to its biocompatible, biodegradable, and non-toxic properties [20,164]. The antimicrobial effects of CS are rooted in its physicochemical characteristics, which encompass the degree of deacetylation, its structural advantage in possessing reactive hydroxyl groups at the C-3 and C-6 positions, environmental conditions, and the specific microorganism in question [164]. The native CS, chitin, exhibits limited antimicrobial efficacy due to the presence of acetylated repeating units, which restrain the exposure of free amino groups. Furthermore, it is highly hydrophobic in nature because of the strong intra- and intermolecular hydrogen bonding [165]. On the other hand, CS-bearing repeating units of β-(1 → 4)-2-amino-2-deoxy-β-D-glucose offer primary amine functionality, typically achieved through the deacetylation process. A higher DD in the CS during the deacetylation process signifies an increased presence of free amino groups. These amino groups can interact with microbial cell surfaces through electrostatic attraction. Additionally, CS's solubility in acidic water allows it to form cationic polyelectrolytes, facilitating its easy interaction with microbial cells and leading to cell lysis and cell death [166]. High-molecular-weight and low-molecular-weight CS exhibit distinct actions against microbes. High-molecular-weight CS's antimicrobial activity primarily resides in the extracellular compartment, as it cannot penetrate the cell wall. Here, it operates by altering cell permeability, resulting in a reduced nutrient uptake from the extracellular milieu and a diminished availability of essential metals due to chelation. In contrast, low-molecular-weight CS displays the capability to exert both extracellular and intracellular effects. This translates to alterations in mitochondrial function and consequential impacts on RNA and protein synthesis [167,168].

CS has undergone extensive testing against both Gram-positive and Gram-negative bacteria for antibacterial activity (Figure 1, Table 2). However, the precise mechanism of action remains incompletely understood. Most studies underscore CS's polycationic nature as a key contributor to its antibacterial efficacy [169]. Furthermore, its interactions with the bacterial cell wall are also dependent on the pH of the environment. When the environmental pH falls below CS's pKa, electrostatic interactions occur between the polycationic CS and the anionic components of the bacterial cell wall. These components include lipopolysaccharides in the Gram-negative cell wall and certain proteins on the cell surface of bacteria. These interactions lead to the exertion of antibacterial effects [22,169]. Interestingly, the extent of electrostatic interactions is largely influenced by the number of amino groups linked to C-2 on the CS backbone. A higher number of amino groups results in more positively charged sites, enhancing the interaction with the carboxyl-negative charges on the bacterial cell wall [170].

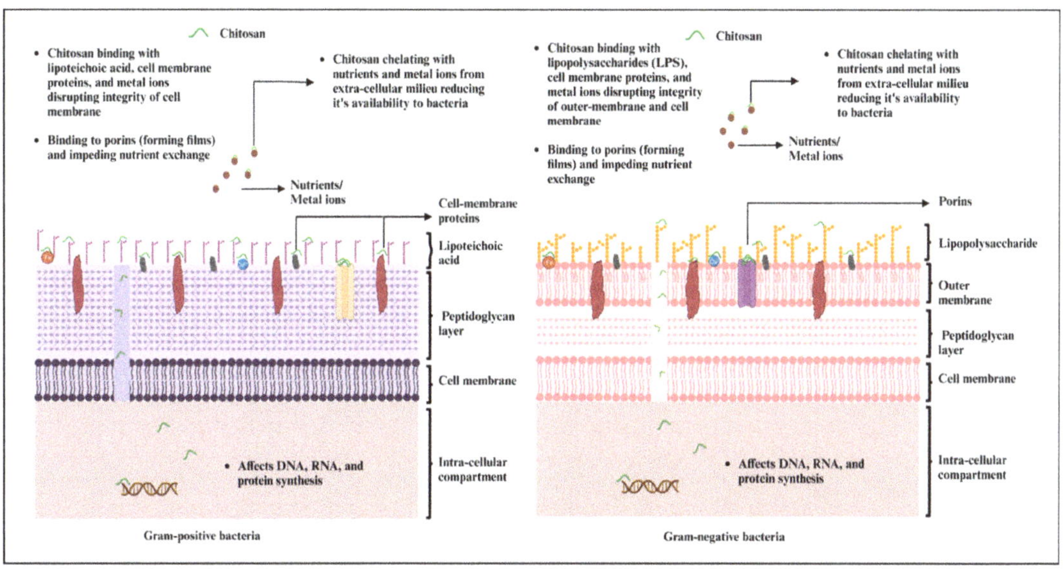

Figure 1. Mechanism of antimicrobial effects of chitosan and chitosan-based materials against Gram-positive and Gram-negative bacteria (created with BioRender.com, accessed on 27 September 2023).

Conversely, when the environmental pH exceeds the pKa, electrostatic interactions weaken, and chelating and hydrophobic effects come into play for antibacterial activity [169,170]. When chelation effects outweigh electrostatic forces, metal ions like Fe^{2+}, Cu^{2+}, and Zn^{2+} present on the bacterial surface can be chelated by CS's amino groups [164,169,171]. Many of these divalent cations play a vital role in stabilizing bacterial cell membranes through their interactions with phosphate groups in teichoic acids and lipopolysaccharides (LPSs) in Gram-positive and Gram-negative bacteria, respectively [172,173]. Variations in susceptibility to CS exist between Gram-positive and Gram-negative bacteria due to marked differences in their cell wall compositions. Gram-positive bacterial cell walls consist of peptidoglycan, teichoic acids, and lipoteichoic acids [174,175]. These components contribute to high-density negative charges within the cell wall, hindering ion movement across the membrane. Positively charged CS can interact with these negatively charged teichoic acids in peptidoglycans, leading to cell membrane damage and the release of intracellular constituents [22,176]. Studies have also demonstrated that interactions with CS can result in peptidoglycan hydrolysis, intensifying electrostatic interactions, and enhancing antibacterial activity [177].

In Gram-negative bacteria, CS targets negatively charged lipopolysaccharides, causing the disruption of the outer membrane, the penetration of the cell membrane, and ultimately, bacterial death [178,179]. High-molecular-weight CS can form a dense polymer film on the bacterial cell surface, obstructing porins and impeding nutrient exchange, which ultimately leads to bacterial cell death [178,180]. Evidence of this is observed in scanning electron microscope studies, which revealed vesicle-like structures on the outer membrane of CS-treated *Escherichia coli* (*E. coli*) and *Salmonella typhimurium* (*S. typhimurium*) [181]. On the contrary, low-molecular-weight CS can penetrate bacterial cells, affecting DNA, RNA, and protein synthesis [20]. Studies have shown that when *E. coli* has been exposed to oleoyl-CS nanoparticles, the nanoparticles bind to DNA and RNA and influence the electrophoretic mobility of both nucleic acids [171]. CS and its derivatives have been extensively explored for their antimicrobial properties against various pathogens, offering diverse mechanisms of action [182]. Thioether CS oligosaccharides and CS–silver nanocomposites have been explored for their synergistic antimicrobial effects against multiple Gram-positive and Gram-negative bacteria [183,184]. Furthermore, another study reported the increased antibacterial activity of *N*-(2-ethylamino)-CS and *N*-2(2,6-diaminohexanamide)-CS polymers, attributed to the presence of amino groups and their hydrophobic interactions [185]. Similar studies have introduced CS Schiff bases with aromatic substitutions, which exhibit enhanced interactions with bacterial cell components [186]. CS Schiff bases were developed to potentially interact with specific components in both Gram-positive and Gram-negative bacteria, exerting antibacterial action [187]. The synthesis and antimicrobial properties of *N*-guanidinium CS derivatives were also investigated, highlighting their permeability-enhancing effects through adsorption and intracellular binding [141].

CS nanoparticles, when loaded with antibiotics, peptides, and proteins, have displayed promising results in possessing antibacterial effects and disrupting biofilm formation by bacteria. For instance, ceftriaxone sodium encapsulated within CS nanoparticles has exhibited heightened intracellular antibacterial efficacy against *S. typhimurium* [188]. Cefazolin-loaded CS nanoparticles have demonstrated enhanced antibacterial activity, particularly against multi-drug-resistant Gram-negative bacteria [189]. Meanwhile, tetracycline-loaded *O*-carboxymethyl CS nanoparticles have shown the potential to augment the action of antibiotics against *Staphylococcus Aureus* (*S. aureus*) infections [190]. Similarly, CS-based systems have been widely used for assessing their potential as carriers for probiotics or beneficial bacteria that can be used for controlling pathogenic bacterial infections. In one such study, our group demonstrated that CS–alginate nanoparticles carrying *E. coli* Nissle effectively inhibited *Campylobacter jejuni* growth in vitro [191].

Within the realm of CS metal nanocomposites, silver-based variants have received substantial attention. Several studies have revealed their effectiveness against a spectrum of bacteria, including *S. aureus*, *E. coli*, and *S. typhimurium* [190,192,193]. Furthermore,

CS-based systems have been evaluated for their impact on biofilms. Anti-biofilm properties have been explored using preparations such as CS–streptomycin conjugates and CS–gold nanoparticles against various Gram-positive and Gram-negative bacteria. These studies have revealed that CS conjugation enhances the drug's ability to disrupt biofilms by improving its penetration and contact with the bacterial surface [194]. Due to these antimicrobial properties, CS-based systems have undergone extensive investigation for their potential applications in wound healing, drug delivery carriers, and various biomedical applications.

The utilization of CS as an antimicrobial agent presents a dynamic landscape of potential research avenues and practical applications. Nanoparticle engineering is poised to harness CS's inherent properties for the development of finely tuned nanoparticles, enabling precise, targeted antibacterial delivery with a focus on enhanced efficacy. Synergy exploration within combination therapies involving CS and other antimicrobial agents unveils an exciting frontier in the fight against antibiotic-resistant pathogens. In-depth investigations into the intricate molecular interactions governing CS's biofilm disruption capabilities hold significant promise for novel strategies for combating persistent biofilm-associated infections. Concurrently, rigorous assessments of CS's biocompatibility and toxicity profiles are vital, ensuring the safe integration of CS-based therapies into clinical practice. The vast potential of CS derivatives including CS oligomers and quaternized CS beckons, as they offer tailored solutions to specific antibacterial challenges. Elucidating the precise mechanisms underpinning CS's antibacterial action, encompassing cell membrane interactions and resistance mechanisms, as well as understanding the host immune responses to CS, not only advances fundamental knowledge but also guides the rational design of more potent therapies in the future.

Table 2. Application of chitosan and its derivative for antibacterial activities.

CS Derivative/Preparations	Bacteria	Assay	Effect/Mechanism of Action	References
CS with benzoimidazolyl-thiadiazole	*S. aureus, B. subtilis, E. coli, P. aeruginosa, C. albicans*	Agar well diffusion method	Presence of polar groups, sulfur, and nitrogen, increases the solubility and electrostatic attraction between polymer and bacterial cell wall and increases cell death	[70]
Thiadiazole CS derivative	*E. coli, P. aeruginosa, B. subtilis, S. aureus*	Agar well diffusion method	Increase solubility due to cationic groups contributing to improved antibacterial activity	[71]
1,3,4-thiadiazole modified CS	*E. coli, P. aeruginosa, B. subtilis, S. aureus*	Agar well diffusion method	Increase solubility due to hydrophilic thiadiazole derivatives	[72]
Thymine-modified CS	*P. aeruginosa, A. baumannii, S. aureus, E. coli*, MRSA	96-well plate microdilution method, scanning electron microscopy	Increase ζ-potential of the positively charged thiamine-modified CS derivatives; higher positive electric density; enhance electrostatic interaction with the negatively charged bacterial membrane	[74]
CS linked with diphenyl pyrazole with succinic anhydride	*B. subtilis, S. aureus, P. aeruginosa*	Agar well diffusion method	Polycationic nature of CS and presence of amino group in pyrazole; penetration of CS into nucleus, blocking binding sites of RNA to DNA, inhibiting synthesis of cell wall proteins	[75]
CS coupled with 4-((5, 5-dimethyl-3-oxocyclohex-1-en-1-yl) amino) benzene-sulfonamide (CS Schiff base)	*E. coli, S. aureus*	Visible spectroscopy analysis, MIC assay	Increase hydrophobicity of CS Schiff base, improving interaction with peptidoglycan	[78]
CS with heteroaryl pyrazole derivatives (CS Schiff base)	*E. coli, K. pneumonia, S. aureus, S. mutans*	Agar well diffusion method	Functional modification with pyrazole ring bearing pyridyl moiety enhanced antibacterial effect	[80]

Table 2. *Cont.*

CS Derivative/Preparations	Bacteria	Assay	Effect/Mechanism of Action	References
CS with formyl pyrazole derivatives (CS/pyrazole Schiff base)	*S. aureus, B. cereus, E. coli*	Agar well diffusion method	Presence of furan ring and nitro groups; enhance entry of CS to nucleus and interaction with RNA/DNA	[81]
CS with indole-3-carboxaldehyde and 4-dimethylaminobenzaldehyde (phenolic CS Schiff bases)	*S. aureus, B. cereus, E. coli, P. aeruginosa, S.* spp.	Agar well diffusion method	Interaction with bacterial cell membrane and disruption of cell wall integrity	[82]
Chitooligosaccharide-niacin acid conjugate	*S. aureus, E. coli, V. harveyi*	Broth dilution assay	Increase lipophilicity, hydrophobic interaction of the trialkyl chain, increased interaction of quaternary ammonium salts and lipid structure of bacterial cell membrane	[115]
Pyridine-4-aldehyde Schiff bases grafted chloracetyl CS oligosaccharide derivatives	*S. aureus, E. coli*	Agar well diffusion method	Increase positive charge of the derivative and interaction with bacterial cell membrane	[116]
L-arginine Schiff bases acylated CS derivatives	*B. cinerea, S. aureus, E. coli*	Plate colony counting method	Free positive charge and guanidine carried by CS combines with the negatively charged components of the bacterial cell wall	[135]
N-guanidinium CS acetate, *N*-guanidinium CS chloride	*E. coli, P. aeruginosa, S. aureus, B. subtilis*	Turbidimetric method	Cationic groups enhance the permeability via adsorption followed binding to intracellular constituents	[141]
Chitotriazolan (poly(β(1-4)-2-(1H-1,2,3-triazol-1-yl)-2-deoxy-D-glucose))	*S. aureus, E. coli*	96-well plate microdilution method	Loss of integrity to bacterial cell wall	[146]
Calcium–CS–triazole nanocomplex	*E. coli, B. subtilis*	Agar well diffusion method	Enhance antibacterial activity contributed by triazole moiety	[157]
CS hydrogel containing hydroxypropyl methylcellulose (HPMC)	*S. aureus, P. aeruginosa*	Biofilm assay/ confocal scanning laser microscopy	Increase adhesiveness and penetration into biofilm and disruption	[182]
Thioether CS oligosaccharide (CS oligosaccharide coupled with 3-bromopropene and tiopronin)	*S. aureus, B. subtilis, L. monocytogenes, E. coli, P. aeruginosa*	MIC and MBC determination, cytotoxicity evaluation, antioxidant activity assays	Enhance antimicrobial capability due to carboxyl groups possessing positive charges, which bind to negatively charged cell membranes. Increased iron chelation from tiopronin.	[183]
Biogenic CS–silver nanocomposite	*S. aureus, P. aeruginosa, S.* spp.*, E.* spp.*, S.* spp.*, Shigella* spp.	Well diffusion technique for antimicrobial susceptibility	Synergistic effect of CS and silver; greater adsorption onto the surface of bacterial cells and can easily penetrate the bacterial cell wall causing cell death	[184]
N-(2-ethylamino)-CS and *N*-2(2,6-diaminohexanamide)-CS polymers	*S. aureus*	Bacterial growth inhibition, cytocompatibility studies	Presence of the amino groups: hydrophobic interaction with the bacterial wall is responsible for the enhanced activity	[185]
CS Schiff base (CS with 2,4,6-trimethoxybenzaldehyde)	*S. aureus, E. coli, P. aeruginosa, K. pneumonia, S. haemolyticus*	Turbidity assay	Enhance interaction of CS with peptidoglycan and plasma membrane due to aromatic substitution	[186]
CS coupled with cyclohexanone and 2-*N*-methyl pyrrolidone (CS Schiff base)	*S. aureus, E. coli, P. aeruginosa, B. cereus*	Agar well diffusion assay, MIC determination, bactericidal studies	Interaction with teichoic and lipoteichoic acid in Gram-positive bacteria and with O-specific side chain of LPS of Gram-negative bacteria	[187]

Table 2. *Cont.*

CS Derivative/Preparations	Bacteria	Assay	Effect/Mechanism of Action	References
CS with immobilized ZnO nanoparticles	*E. coli, S. aureus*	Agar well diffusion method	Antibacterial activities	[195]
Low-molecular-weight CS	*E. coli DH5α, P. putida* F1, *L. lactis* IO-1, *B. subtilis* 168	Zeta potential measurement, contact angle measurement	Alteration of cell surface charge, cell surface hydrophobicity	[196]
CS nanoparticles	*P.* spp.	Disk diffusion test, antibiofilm assay	Compromise structural integrity of the biofilms, heterogeneous destruction of the biofilm matrix	[197]
Thiazolium CS (Quaternary CS)	*L. innocua, S. epidermidis, S. aureus*, MRSA	Broth dilution method	Combine antibacterial effect of thiazolium and CS	[198]
CS covalently bounded to isocyanate terminated quaternary ammonium salt and sulfopropylbetaine	*E. coli, S.aureus*	Real-time turbidity monitoring via automated turbidimetric system	NH_2 group of the derivative binds with Mg^{2+} and Ca^{2+} in bacterial outer membrane	[199]

Note: *S. aureus*: *Staphylococcus aureus*, *B. subtilis*: *Bacillus subtilis*, *E. coli*: *Escherichia coli*, *P. aeruginosa*: *Pseudomonas aeruginosa*, *C. albicans*: *Candida albicans*, *A. baumannii*: *Acinetobacter baumannii*, *MRSA*: *Methicillin-resistant Staphylococcus aureus*, *K. pneumonia*: *Klebsiella pneumonia*, *S. mutans*: *Streptococcus mutans*, *B. cereus*: *Bacillus cereus*, *S.* spp.: *Salmonella* spp., *V. harveyi*: *Vibrio harveyi*, *B. cinerea*: *Botrytis cinerea*, *L. monocytogenes*: *Listeria monocytogenes*, *E.* spp.: *Enterococcus* spp., *S. haemolyticus*: *Staphylococcus haemolyticus*, *P. putida* F1: *Pseudomonas putida* F1, *E. coli DH5α*: *Escherichia coli DH5α*, *L. lactis* IO-1: *Lactococcus lactis* IO-1, *B. subtilis* 168: *Bacillus subtilis* 168, *L. innocua*: *Listeria innocua*, *S. epidermidis*: *Staphylococcus epidermidis*.

4.2. Chitosan Derivatives as Drug or Vaccine Delivery System

Controlled drug or vaccine release is highly emphasized as a strategy for controlling disease pathologies, achieved through the targeted accumulation of therapeutic compounds at the site of the ailment or the utilization of inventive therapeutic agents to control infectious pathogens [200,201]. Drug encapsulation and dosage reduction emerge as the most convenient methods for achieving controlled drug release [202]. CS and its derivative-based nanoparticles and films have received considerable attention as highly suitable candidates for drug or antigen carriers due to their mucoadhesive properties, pH sensitivity, and biocompatibility. The release of drugs/antigens from CS nanoparticles in biological fluids occurs through diffusion, erosion, or swelling mechanisms [203] (Figure 2). Additionally, their distinct antimicrobial activities collaboratively synergize with drugs to enhance their antimicrobial effectiveness [204].

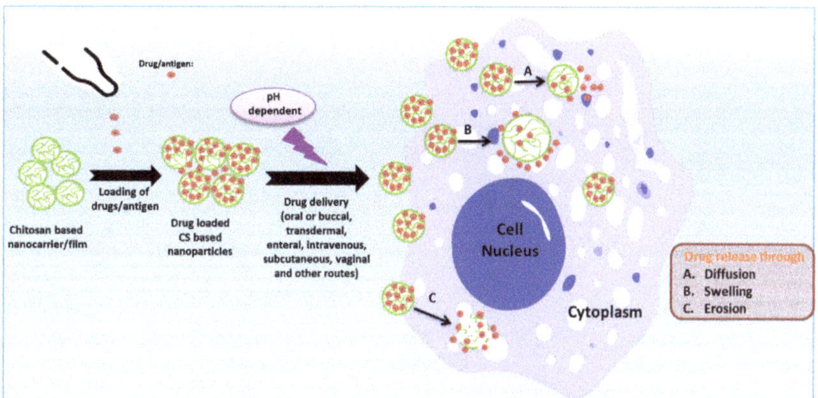

Figure 2. The mechanisms of release of drugs/antigens from chitosan nanoparticles (created with Microsoft Power Point 2010 and Chemdraw 19.1).

The recent applications of CS-based nanomaterials as drug carriers against infectious diseases are summarized in Table 3 [205–218]. For example, the nasal delivery of hesperidin-loaded CS nanoparticles exhibited enhanced cellular uptake within inflammatory microenvironments and effectively suppressed cytokine storm syndrome and acute lung injury [205]. Meanwhile, CS biguanidine nanoparticles displayed remarkable inhibition against various *Mycobacterium tuberculosis* strains, including sensitive, multidrug-resistant (MDR), and extensively drug-resistant (XDR) types, surpassing the performance of the standard anti-tuberculosis drug, isoniazid [206]. Sawant et al. demonstrated that tuberculosis prevention could be achieved through the pulmonary delivery of rifampicin with improved solubility when encapsulated within octanoyl CS nanoparticles [207]. Additionally, CS-based materials proved effective in delivering various anti-infectious drugs and antibiotics, including carvacrol [208], doripenem [209], chlorhexidine [210], ciprofloxacin [211], dolutegravir [212], endolysin Cpl-1 [213], and amphotericin B [214]. These materials successfully inhibited the growth of pathogens such as *Salmonella enterica*, *Pseudomonas aeruginosa*, *Staphylococcus aureus*, *Staphylococcus epidermidis*, *Escherichia coli*, human immunodeficiency virus HIV, and fungal species. In the expanded application, the eradication of waterborne pathogens such as rotavirus and *Shigella* spp. was also investigated through CS-wrapped carbon nanotubes [215].

Table 3. Applications of CS-based nano/micro particles as drug/vaccine carriers for infectious pathogens.

CS Derivative/Nanomaterials	Preparation	Interference	References
Hesperidin-loaded CS nanoparticle	Emulsification and evaporation methods	• Nasal delivery of the anti-inflammatory HPD compound to inflammatory lungs; • Enhance cellular uptake in the inflammatory microenvironment compared with free HPD; • Suppress the cytokine storm syndrome and acute lung injury/acute respiratory distress syndrome in a murine model of inflammatory lung disease.	[205]
CS biguanidine nanoparticle	One pot, green, ionic gelation method	• Inhibitory activity against the growth of three different *Mycobacterium tuberculosis* pathogens (sensitive, MDR, and XDR) compared with isoniazid drug (standard anti-tuberculosis drug).	[206]
Rifampicin-loaded octanoyl CS nanoparticle	Double emulsion solvent evaporation technique	• Pulmonary delivery; • Optimization of drug release, solubility, and pulmonary release of rifampicin through this carrier; • Aim at improving the interventions for tuberculosis.	[207]
Carvacrol-loaded bovine serum albumin (BSA) CS nanoparticle	Ionic gelation method	• Mucosal delivery; • Ensure delayed release and limited degradation in the gastric conditions; • Effective eradication of *Salmonella enterica* serovar *Typhi*.	[208]
Doripenem-loaded CS micro-particles	Ionic gelation, spray-drying method	• Effective pulmonary delivery of doripenem; • Significant effect on reducing the infection by *P. aeruginosa*.	[209]
Chlorhexidine-infused CS	One-pot method	• Epidermal delivery; • Show improved antibacterial activity against *S. aureus* and *S. epidermidis*.	[210]
Ciprofloxacin-loaded CS/poly (2-ethyl-2-oxazoline)	Solvent evaporation method	• Antimicrobial properties against *S. aureus* and *E. coli*; • Mucoadhesive vaginal drug delivery system for sheep.	[211]
CS nanoparticle loaded with dolutegravir	Spray-drying method	• High drug loading and releasing activity; • Anti-HIV activity.	[212]

Table 3. *Cont.*

CS Derivative/Nanomaterials	Preparation	Interference	References
CS nanoparticle loaded with phage endolysin Cpl-1	Ionic gelation method	• Mucoadhesive drug delivery; • Increased therapeutic applicability of endolysins.	[213]
Magnetic CS loaded with amphotericin B	Chemical coprecipitation method followed by surface coating	• Stable and controlled delivery of amphotericin B; • Interfere with biofilm formation and fungicidal effect.	[214]
CS-wrapped carbon nano tubes	Cross-linking method of CS and CNTs	• Separation, identification, and eradication of rotavirus and *Shigella*, waterborne pathogens in water samples of diverse sources.	[215]
CS/lecithin nanoparticles loaded with antibiotics	Central rotatable composite design method	• Epidermal delivery; • Increase the accumulation of antibiotics in the epidermis with higher retention ratio compared to the commercial formulations; • Test against *E. coli* and *S. aureus*.	[216]
CS–hyaluronic acid-based nanoparticle containing ceftazidime	Ionic gelation method	• Mucoadhesive drug delivery; • Increase interaction with ocular surface and drug residence time; • Prolong drug release profile.	[217]
CS nanoparticle containing LL-37	Ionotropic gelation method	• Increase half-life and prolonged LL37 biological activity; • Inhibition of high biofilm formation by Methicillin-resistant *S. aureus* when compared to LL37 alone.	[218]
Inactivated PR8 influenza virus-loaded CS alginate and trimethyl CS nanoparticles	Ionic gelation method	• Immunoadjuvant potential after nasal immunization; • Exhibited higher IgG2a/IgG1 ratio as criteria for Th1-type immune response; • Efficient intranasal antigen delivery system for nasal vaccines.	[219]
Inactivated Dengue virus (DENV)-loaded trimethyl CS nanoparticle	Ionotropic gelation method	• Highly immunogenic, inducing greater levels of antibodies (total IgG, IgG1, IgG2a and neutralizing antibodies) and T cells (activated $CD4^+$ and $CD8^+$ T cells) against DENV-2 compared to soluble DENV-2 immunogens.	[220]
CS-based nanoparticle vaccine loaded with *Salmonella Enteritidis* outer membrane proteins and flagellin proteins	Ionic gelation method	• Mucosal delivery; • Induce specific immune responses against the bacteria and reduced cecal colonization in birds; • Almost 65.9% birds which were vaccinated had cleared the infection.	[221]
E. coli Nissle 1917 microencapsulated CS–alginate	Ionic gelation method	• Reduce *Campylobacter jejuni* colonization in infected chickens; • Improve the intestinal morphology, increase immune responses of treated chickens via likely activation of the Th1, Th2, and Th17 pathways.	[222]
Mast cell activator C48/80 with CS nanoparticle	Mixing of C48/80 compound with CS, lyophilization	• Exhibit adsorption of model antigen C48/80 and biocompatible onto the nanoparticle; • Demonstrate first time for the hepatitis B surface antigen loaded Chi-C48/80 nanoparticle as nasal vaccination strategy.	[223]
Newcastle disease virus (NDV/La Sota) and infectious bronchitis virus (IBV/H120)-loaded quaternized CS (-2-hydroxypropyl trimethyl ammonium chloride CS (*N*-2-HACC) and *N*, *O*-carboxymethyl CS) nanoparticles	Polyelectrolyte composite method	• Newcastle disease and Infectious Bronchitis elicit immune response in chickens after intranasal administration; • Inducement of higher titers of IgG and IgA antibodies; • Increase the proliferation of lymphocytes and induce higher levels of cytokines, including IL-2, IL-4, and IFN-γ.	[224]

Table 3. *Cont.*

CS Derivative/Nanomaterials	Preparation	Interference	References
CS nanoparticles encapsulating influenza H9N2 HA2 and M2e mRNA molecules	Anionic gelation method	• Delivery of immunogenic antigens to antigen presenting cells; • Effective penetration of the mucosal barrier to reach immune initiation sites; • Efficient release of mRNA molecules into antigen presenting cells; • Higher systemic IgG, mucosal IgA antibody response.	[225]

Note: HPD: hesperidin, MDR: multidrug-resistant, XDR: extensively drug-resistant, *P. aeruginosa*: *Pseudomonas aeruginosa*, *S. aureus*: *Staphylococcus aureus*, *S. epidermidis*: *Staphylococcus epidermidis*, *E. coli*: *Escherichia coli*.

In highlighting the effort in the encapsulation process, the inactivated virus, RNA, proteins, and probiotics have also been capsulated in the CS derivatives, which serve as a nanovaccine delivery system to increase the immune response against the pathogens [219–225]. As an instance, Tafaghodi et al. encapsulated the inactivated PR8 influenza virus within CS alginate and trimethyl CS nanoparticles which exhibited immunoadjuvant capabilities after nasal immunization, thereby stimulating a Th1-type immune response [219]. Inactivated dengue virus (DENV)-loaded CS nanoparticles displayed their high immunogenicity, inducing elevated levels of antibodies and T cells against DENV-2 in comparison with soluble DENV-2 immunogens [220]. CS-based nanoparticles, containing *Salmonella Enteritidis* outer membrane proteins and flagellin proteins, emerged as a mucosal vaccine, inducing specific immune responses and reducing cecal colonization in avian hosts [221]. Our group also demonstrated the microencapsulation of probiotic *E. coli* Nissle 1917 within the CS–alginate, which reduced *Campylobacter jejuni* colonization in infected chickens [222].

Due to their cationic nature and high endocytosis potential, CS and its derivatives are not only utilized for drug and vaccine delivery for infectious diseases but also find applications in addressing various other conditions, such as cancer, gastrointestinal disorders, pulmonary ailments, Alzheimer's disease, and diabetes [226–239]. Table 4 provides a summary of drug delivery systems designed to treat a range of diseases utilizing various forms of CS and its derivatives. CS-based nanoparticles enhance the drug-targeting ability and improve the blood retention time through desorption, swelling, and erosion. For instance, a PEGylated CS nanoparticle was employed to deliver a chemotherapeutic agent such as doxorubicin (DOX) in combination with breast cancer-specific monoclonal antibodies to treat breast cancer [226]. Thiolated CS nanoparticles delivered vincristine by binding to leaky prostate tumor cells, stimulating the receptors of these cells, and facilitating their internalization through endocytosis [227]. In another approach to the drug delivery system, the formulation of CS-based films and patches was employed in the ocular and transdermal delivery of the drugs brimonidine tartrate (BT) and tizanidine, respectively [231,232]. The CS-BT film significantly improved the transport of BT across the cornea, offering a potential alternative eye drop in the future, whereas the transdermal delivery of tizanidine facilitated efficient drug permeation through the skin, maintaining a therapeutic concentration. Moreover, CS and its derivatives have been fabricated through the crosslinking and polymerization methods to prepare hydrogels that have demonstrated a controlled, painless, and biocompatible transdermal release of therapeutic ingredients [233]. In an effort to find suitable carriers, CS-based micelles and liposomes have emerged as innovative tools to address the challenges associated with drug administration, especially the issues of low water solubility and limited drug permeability [234–238]. For example, the amphiphilic colloidal structures, especially CS-based micelles, including the CS-bearing thiourea group, incorporated with gold nanorods and carboxymethyl CS with a vitamin E succinate nanomicellar system (*O*-CMCTS-VES), effectively facilitated the release of drugs like paclitaxel and DOX for antitumor activities [234,235]. In addition, CS derivatives were employed to encapsulate liposomes, which possess minimal toxicity and cell-like membranes, along with drugs like resveratrol, aptamer-functionalized 5-fluorouracil, photosensitizer HPPH, and hypoxia-activated prodrug TH-302, enhancing the efficient targeted delivery of these drugs [236–238].

Table 4. Application of chitosan and its functionalized nanoparticles for drug delivery system.

Form	Drug Carrier System	Drug or Enzymes	Preparation	Interference	References
Nanoparticle	Anti-hMAM and anti-HER2 PEGylated CS nanoparticle	Doxorubicin (DOX)	Mixing the DOX in the PEG-CS solution, cross-linking with TPP.	• Cytotoxic against MCF-7 cancer cells (breast cancer).	[226]
	Hyaluronic acid-coated thiolated CS nano particle	Vincristine	Ionic gelation process	• In vitro anticancer activity against cancerous prostate cells.	[227]
	Human serum albumin (HSA)-coated trimethyl CS	IR780 (I) or bufalin (B)	Ionic gelation process and electrostatic absorption	• Inhibition of cell proliferation and mitochondrion activity of metastatic 4T1 breast cancer cells; • Enhance the tumor-targeting and penetrating capability for photo thermal therapy, efficient prevention of lung metastasis of breast cancer.	[228]
	O-carboxymethyl CS nanoparticle	Glipizide	Ionotropic gelation method	• Significant impact in insulin-resistant/type 2 diabetes.	[229]
	CS nanoparticle	Sitagliptin	Ionic gelation method	• Increase symptomatic relief of Alzheimer's disease by increasing the level of sitagliptin in the brain.	[230]
Film	CS-based film	Brimonidine tartrate (BT)	Dissolution method	• High corneal permeability with fast drug release for potential ocular drug delivery.	[231]
Patches	CS, thiolated CS patches with flaxseed oil and coriander oil	Tizanidine	The solvent casting method	• In vitro transdermal drug release and drug permeation, skin irritation; • In vivo application, pharmacokinetics analysis, and stability studies.	[232]
Hydrogel	Carboxymethyl CS–silk fibroin peptide/oxidized pullulan (CMCS-SFP/OPL) hydrogel-based microneedle	Salvia miltiorrhiza	Polydimethylsiloxane (PDMS) mold	• Direct drug release from HFM-1; • Alternative transdermal delivery of water-soluble Salvia miltiorrhiza.	[233]
Micelles	CS thiourea with gold nanorods	Paclitaxel	Ring opening polymerization, dialysis, gold-thiolate complex formation	• Light-sensitive drug delivery vehicle; • Cytotoxicity against MCF7 cells.	[234]
	O-CMCTS-VES micelles nanoparticles	Doxorubicin	Dehydrative condensation of the carboxyl group of the carboxymethyl CS and the amino group of the vitamin E succinate, self-assembled nano-micelles	• Efficient cellular uptake by HePG2 cancer cells (human liver cancer) and inhibition of tumor (62.7%).	[235]

Table 4. *Cont.*

Form	Drug Carrier System	Drug or Enzymes	Preparation	Interference	References
Liposome	Trimethylated CS-coated flexible liposomes	Resveratrol	Solvent injection method, electrostatic adsorption	• Topical drug (Resveratrol) delivery to reduce blue-light-induced retinal damage; • Modulation for the mitochondrial membrane potential and protection of ARPE-19 cells from damage by H_2O_2.	[236]
	Aptamer-functionalized liposomal coated with calcium alginate/CS/PEC	5-Fluorouracil (AFL5-FU)	Optimized thin film method, freeze–thaw process	• Oral formulation for colon cancer therapy; • Cytotoxicity against on the colon cancer cell line, HT-29 (IC_{50}; 66.53 3M).	[237]
	CS oligosaccharide (CO) nanoparticle based on liposomes	Photosensitizer HPPH and the hypoxia-activated prodrug TH302 (CO-HPPH-TH302/Lipo)	O/W emulsification method, mixing the lipids, drugs, and CO-OA	• Activation for PDT and chemotherapy by HPPH and TH302; • Inhibition of the growth of CD44-overexpressing tumors with minimal damage to normal tissues by CD44+-targeted liposomes.	[238]

Note: Anti-hMAM: anti-human mammaglobin, Anti-HER2: anti-human epidermal growth factor, HFM-1: Helicase for Meiosis 1, O-CMCTS-VES: carboxymethyl CS/vitamin E succinate nano-micellar System, PEC: photoelectrolyte complexes, HPPH: 2-[1-hexyloxyethyl]-2-devinyl pyropheophorbide-a (Photochlor), TH302: Evofosfamide, CO: CS oligosaccharide, OA: oleic acid, PDT: photodynamic therapy.

The drug-release characteristics of CS nanoparticles exhibit intricate responsiveness to pH alterations, rendering them a versatile tool for precision drug delivery. Research indicates that alterations in pH levels induce significant shifts in CS nanoparticle size and drug molecule release. In a study incorporating 5-fluorouracil (5-FU) within CS nanoparticles, it was observed that as the pH transitions from 3 to 5, there is a substantial increase in particle size, resulting in a continuous and rapid release of 5-FU. Conversely, when the pH exceeds 5, a notable reduction in particle size is evident, particularly at pH 7.4. This pH-responsive behavior can be attributed to the protonation of primary amino groups within the CS chain, leading to heightened electric density and increased repulsion forces between interconnected CS chains. Consequently, the surface density of protonated amino groups and the degree of protonization are adaptably responsive to shifts in solution pH [240]. CS nanoparticles undergo a reversible process of swelling and contraction, contingent upon the environmental pH, resulting in particle sizes fluctuating from approximately 450 nm to 150 nm [241]. The pH-sensitive surface charge reversal also facilitates enhanced cellular uptake and endolysosomal escape, extending blood circulation time, mitigating side effects, and optimizing drug delivery efficiency [242]. Such adaptability holds significant promise for targeted drug delivery, aligning with the varying pH levels found within the human body, which include gastric juice with a low pH of 1–1.5, the physiological environment exhibiting a pH of 7.4, the tumor extracellular microenvironment typically with a pH 6.5, and the pH within endosomes and lysosomes ranging from 5.0–6.2 to 4.0–5.0, respectively [242,243]. Moreover, studies highlight the enhancement of responsive properties through the integration of pH-sensitive molecules with CS, further streamlining drug delivery in response to environmental cues [244]. Additionally, investigations demonstrate that under acidic conditions (pH 3.5), there is a swift release of model drugs, while at slightly alkaline conditions (pH 7.4), the release rate is attenuated [245].

Despite the encouraging implications of CS and its derivatives as a drug or vaccine carrier system, they suffer from specific limitations. These limitations involve factors such as a restricted capacity to carry larger-sized drugs or vaccines, issues related to stability, the occurrence of burst drug release upon administration, limitations in targeting specific diseases, and the potential for an immune response in certain individuals. Nevertheless, ongoing research on the development of novel drug carriers that are more biodegradable, non-invasive, and have a high drug-loading capacity is actively employed to overcome these constraints. In the future, there is potential for designing CS derivative nanoparticles in a computational way that may not only be capable of their targeting efficacy but also have the ability to interact with proteins and signaling pathways. This system could potentially reduce the drug dosage requirements with minimal side effects and improve the patient's condition.

4.3. Chitosan Derivatives in Plant Agriculture

CS and its derivatives have gained significant interest in the agriculture field, especially for their role in enhancing plant growth, stimulating root and shoot development, increasing crop yield, and supporting environmentally friendly farming practices (Figure 3) [246]. The cationic property of CS derivatives enables effective interaction with the negatively charged components of the bacterial cell membrane, resulting in bactericidal or bacteriostatic effects [247]. On the other hand, the chelating characteristic of CS also establishes it as a tremendous antifungal agent [248]. Similarly, it has also demonstrated its role as an elicitor by activating the defensive genes in plant disease control [249].

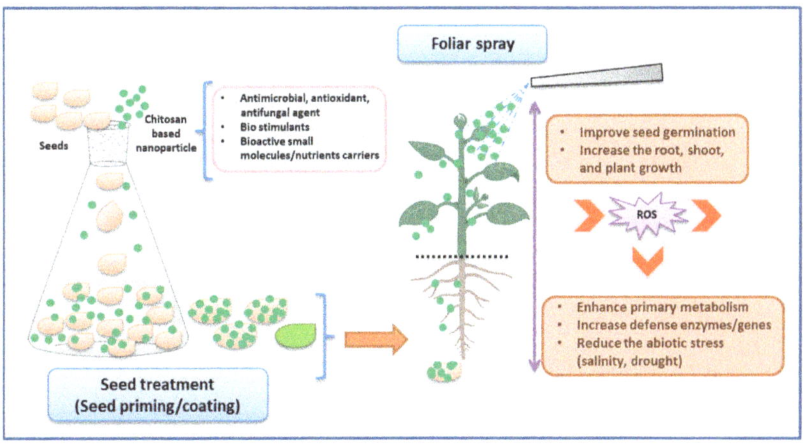

Figure 3. Schematic representation of the application of chitosan-based nanoparticles in plant agriculture (created with Microsoft PowerPoint 2010, Chemdraw 19.1, and BioRender.com, accessed on 27 September 2023).

Considering the encouraging advantages, CS derivatives have been utilized in a wide range of crops, including vegetables, flowers, fruits, cereals, and medicinal plants, in various forms such as seed treatment, foliar spraying, coating fruits or vegetables, and delivering nutrients through nanomaterials [250] (Table 5). For example, Udayashankar group investigated the CS exogenous seed priming of cucumbers seeds, resulting in the inducement of phytohormone production and an increment of resistance against cucumber powdery mildew disease [251]. Recent studies demonstrated the use of CS through seed priming, seed soaking, or seedling application in mung bean [252], *Platycodon grandiflorus* [253], and *Carum copticum* L. [254] for reducing abiotic stress and increasing photosynthesis and antioxidant activity, resulting in efficient plant growth. Similarly, plant diseases such as root rot caused by *Fusarium solani* in fenugreek were also managed with seed treatment [255]. CS derivatives such as CS–ferulic acid, CS–caffeic acid conjugate, and carboxymethyl CS were explored for seed treatment or seed coating in cucumbers [256] or *Prunus davidiana* [257]. As a result, they either increased the antioxidant activity or enhanced the biomass, photosynthetic capacity, and absorption of phosphorus and potassium to improve plant growth.

In further applications, CS nanoparticles have been significantly introduced in seed treatment [258]. For instance, Jogaiach et al. explored CS-derived nanoparticles in tomato seed treatment, which led to a remarkable increase in immune responses by accumulating defense enzymes and lignin to eradicate bacterial wilt disease [259]. CS incorporated with organic compounds or metals, such as garlic essential oil or silver-loaded nanoparticles, demonstrated its application in seed treatment on barley or chilies, which reduced the antifungal activities in plant growth [260,261].

CS and its derivatives, along with nanoparticles, were also employed as foliar sprays to reduce plant diseases, functioning as elicitors, plant protectors, or biostimulants [262,263]. For example, the foliar spraying of CS or its nanoparticles for tomatoes elicited an increment in shoot biomass and flower numbers, with reduced interaction with arbuscular mycorrhizal fungi (AMF) [264]. CS foliar spraying significantly reduced drought stress in bermudagrass by enhancing the turf quality, chlorophyll content, and leaf relative water content, and decreasing the level of electrolyte leakage, malonaldehyde, and hydrogen peroxide content [265]. In another study, the spray of CS derivatives such as CS lactate on *Ocimum basilicum* L. and *Melissa officinalis* L. accumulated the valuable phenolic group rosmarinic acid (RA) and increased the shoot biomass [266]. This showed the capability of herbal plants to produce functional foods. In addition, by using the CS–selenium nanoparticle, Skalicky et al. explored foliar spray with bitter melons, alleviating salt stress by enhancing antioxidant enzyme activity, elevating proline concentration, maintaining relative water content and K^+, and reducing malonaldehyde and hydrogen peroxide levels, as well as sodium accumulation in plant tissues [267].

Despite the plethora of studies of CS applications demonstrating remarkable outcomes in enhancing plant growth, disease resistance, and stress tolerance, there always remains a requirement to determine the optimal techniques, dosage, and timing for its most effective use [246,268]. Through ongoing research, technological innovation, and eco-friendly techniques, CS has the capability to revolutionize harmless agricultural methods, which could pave the way for a more sustainable agricultural future.

Table 5. Chitosan and its derivatives for applications in plant agriculture.

CS or Its Derivatives	Form	Plant	Interference	References
CS	Seed priming	Cucumber	• Upregulation of phytohormones; • Resistance against cucumber powdery mildew disease.	[251]
CS	Seed priming	Mung bean	• Alleviate water stress, increase antioxidant system, improve growth.	[252]
CS	Seed soaking	*P. grandiflorus*	• Enhances plant growth, photosynthesis, resistance, yield, and quality.	[253]
CS	Seedling and callus culture	*C. copticum* L.	• Improve plant tolerance to oxidative stress under salinity; • Elevation of essential oils and antioxidant activity.	[254]
CS	Seed treatment	Fenugreek	• Controlling the root rot disease by reducing *F. solani;* • Increase seed germination.	[255]
CS–ferulic acid and CS–caffeic acid conjugate	Seed treatment	Cucumber	• Increase plant growth; • Antioxidant activity.	[256]
Carboxymethyl CS	Seed coating	*P. davidiana*	• Increase the biomass and photosynthetic capacity; • Promote the absorption of phosphorus and potassium; • Decrease nitrogen uptake.	[257]

Table 5. *Cont.*

CS or Its Derivatives	Form	Plant	Interference	References
CS nanoparticle	Seed treatment	Tomato	• Enhancing the immune response by accumulating the defense enzyme and lignin.	[259]
Garlic essential oil-loaded CS nanoparticle	Seedling	Barley	• Antifungal activity; • Alternative to tebuconazole as seed-dressing agent.	[260]
CS silver nanocomposites	Seed treatment	Chili	• Antifungal activity against anthracnose pathogen.	[261]
CS and its nanoparticles	Foliar spray	Tomato	• Enhance plant growth and flowering.	[264]
CS	Foliar spray	Bermudagrass	• Improve drought tolerance.	[265]
CS lactate	Foliar spray	*O. basilicum* L. and *M. officinalis* L.	• Enhance the accumulation of valuable phytochemicals.	[266]
CS–selenium nanoparticle	Foliar spray	Bitter melon	• Increased salinity tolerance in bitter melon plants.	[267]

Note: *P. grandifloras*: *Platycodon grandifloras*, *C. copticum* L.: *Carum copticum* L., *P. davidiana*: *Prunus davidiana*, *O. basilicum* L.: *Ocimum basilicum* L., *M. officinalis* L.: *Melissa officinalis* L.

5. Conclusions and Future Outlooks

CS is considered a promising polysaccharide renowned for its extensive array of potential applications in the fields of biology, medicine, water treatment, the food industry, cosmetics, and biodegradable packaging. Certain constraints persist, such as a lower degree of deacetylation and elevated molecular weight, which directly result in poor water solubility in neutral aqueous solutions. The modification of CS through different strategic pathways is one perspective that not only improves the water solubility but also enhances the bioactivity of the final product. However, the CS-bearing long chain or hydrophobic scaffolds may diminish the water solubility; nevertheless, they can still be effectively incorporated into films, fibers, or polymeric materials to enhance the hydrophobicity and stability of the materials [269]. This review highlighted the recent methodologies for synthesizing CS derivatives and their notable uses in antimicrobial functions, nanomaterials employed in drug delivery systems, and applications within the agricultural sector. Recent progress in the development of CS derivatives has been achieved through various methods, such as forming Schiff bases with heterocyclic compounds, generating quaternary ammonium salts and guanidines, employing click chemistry, and using them in direct *N*-functionalization reactions. There are also critical challenges, including separation, purity, yield, and stability, to developing new CS-bearing compounds.

Exploring CS's use as an antimicrobial agent unmasks a dynamic array of research possibilities, especially in the medical and agricultural fields. CS demonstrates significant bioactivity in combating both animal and plant pathogens. The research on nanoparticle engineering involving CS and its derivatives as drug carriers has significantly demonstrated the effective delivery of drugs or vaccines to target diseases in the pharmaceutical field. Antimicrobial dressings based on CS and its derivatives have received FDA approval [270,271]. However, new applications as drug carriers are still awaiting approval due to insufficient research on their mutagenicity and genotoxicity. Simple and mild preparation of CS-based nanoparticles is always encouraged for ideal drug delivery systems due to their water solubility. In addition, researchers have observed that treating seeds or foliar sprays by applying CS derivatives or nanomaterials demonstrated effective results in enhancing plant growth, increasing crop yields, and advancing food security. This approach not only reduces the need for pesticides and fertilizers but also contributes to environmental pollution reduction and promotes the sustainability of entire agricultural systems. Nevertheless, the detailed molecular mechanisms by which they influence seeds and entire plants remain unclear, underscoring the need for further research into their genomic, proteomic, and metabolic effects.

Looking ahead, the development of novel CS derivatives through the incorporation of biologically active heterocyclic groups using recent techniques and methodologies involving different reaction mechanisms like CO, CN, or CH activation reactions is a highly promising prospect. However, research into the structure–activity relationship is an important factor in developing bioactive derivatives, along with a high degree of substitution of the functional groups. Researchers are consistently encouraged to explore novel reactions in conjugation with the structure–activity relationship. The exact mechanism of modified CS derivatives against antimicrobial activities has not been thoroughly explored; nevertheless, these effects are often generalized with unmodified CS. Advanced research regarding the mechanism of antimicrobial activities due to the specific functional group of CS derivatives is needed. Continuous research and development efforts are needed to enhance the water solubility, mitigate toxicity, and bolster the immune responses and functioning mechanisms of CS-based derivatives for applications in human drug delivery and agrochemicals. Moreover, contemporary technologies like computational and artificial intelligence can be used to serve not only in the design and synthesis of innovative CS derivatives and nanoparticles, but also as a means to approach the most realistic mechanisms for antimicrobial effects, drug delivery systems, and plant growth enhancement.

Author Contributions: G.R., R.S. and R.K. were responsible for the conception of this review work. R.S. and A.T. conducted the literature review, manuscript writing, and created the image. G.R., R.S., A.T. and R.K. reviewed and edited the manuscript. All authors have read and agreed to the published version of the manuscript.

Funding: The research in the Rajashekara laboratory was supported by the U.S. Department of Agriculture (USDA) National Institute for Food and Agriculture (NIFA) (grants number 2015-68004-23131 and 2020-6701-31401), and by the United States Department of Agriculture Specialty Crops Research Initiative project awards 2019-51181-30010 and 2022-51181-38242.

Institutional Review Board Statement: Not applicable.

Informed Consent Statement: Not applicable.

Data Availability Statement: Not applicable.

Conflicts of Interest: The authors declare no conflict of interest.

References

1. Paradowska-Stolarz, A.; Wieckiewicz, M.; Owczarek, A.; Wezgowiec, J. Natural Polymers for the Maintenance of Oral Health: Review of Recent Advances and Perspectives. *Int. J. Mol. Sci.* **2021**, *22*, 10337. [CrossRef]
2. Gupta, R.K.; Guha, P.; Srivastav, P.P. Natural Polymers in Bio-Degradable/Edible Film: A Review on Environmental Concerns, Cold Plasma Technology and Nanotechnology Application on Food Packaging—A Recent Trends. *Food Chem. Adv.* **2022**, *1*, 100135. [CrossRef]
3. Puertas-Bartolomé, M.; Mora-Boza, A.; García-Fernández, L. Emerging Biofabrication Techniques: A Review on Natural Polymers for Biomedical Applications. *Polymers* **2021**, *13*, 1209. [CrossRef] [PubMed]
4. Tong, X.; Pan, W.; Su, T.; Zhang, M.; Dong, W.; Qi, X. Recent Advances in Natural Polymer-Based Drug Delivery Systems. *React. Funct. Polym.* **2020**, *148*, 104501. [CrossRef]
5. Vasile, C.; Pamfil, D.; Stoleru, E.; Baican, M. New Developments in Medical Applications of Hybrid Hydrogels Containing Natural Polymers. *Molecules* **2020**, *25*, 1539. [CrossRef]
6. Hamedi, H.; Moradi, S.; Hudson, S.M.; Tonelli, A.E.; King, M.W. Chitosan Based Bioadhesives for Biomedical Applications: A Review. *Carbohydr. Polym.* **2022**, *282*, 119100. [CrossRef]
7. Fan, P.; Zeng, Y.; Zaldivar-Silva, D.; Agüero, L.; Wang, S. Chitosan-Based Hemostatic Hydrogels: The Concept, Mechanism, Application, and Prospects. *Molecules* **2023**, *28*, 1473. [CrossRef] [PubMed]
8. Ressler, A. Chitosan-Based Biomaterials for Bone Tissue Engineering Applications: A Short Review. *Polymers* **2022**, *14*, 3430. [CrossRef] [PubMed]
9. Mohammed, M.H.; Williams, P.A.; Tverezovskaya, O. Extraction of Chitin from Prawn Shells and Conversion to Low Molecular Mass Chitosan. *Food Hydrocoll.* **2013**, *31*, 166–171. [CrossRef]
10. Harmsen, R.A.G.; Tuveng, T.R.; Antonsen, S.G.; Eijsink, V.G.H.; Sørlie, M. Can We Make Chitosan by Enzymatic Deacetylation of Chitin? *Molecules* **2019**, *24*, 3862. [CrossRef]

11. Kaczmarek, M.B.; Struszczyk-Swita, K.; Li, X.; Szczęsna-Antczak, M.; Daroch, M. Enzymatic Modifications of Chitin, Chitosan, and Chitooligosaccharides. *Front. Bioeng. Biotechnol.* **2019**, *7*, 243. [CrossRef] [PubMed]
12. No, H.K.; Meyers, S.P.; Lee, K.S. Isolation and Characterization of Chitin from Crawfish Shell Waste. *J. Agric. Food Chem.* **1989**, *37*, 575–579. [CrossRef]
13. Arbia, W.; Arbia, L.; Adour, L.; Amrane, A. Chitin extraction from crustacean shells using biological methods—A review. *Food Technol. Biotechnol.* **2013**, *51*, 12–25.
14. Pillai, C.K.S.; Paul, W.; Sharma, C.P. Chitin and Chitosan Polymers: Chemistry, Solubility and Fiber Formation. *Prog. Polym. Sci.* **2009**, *34*, 641–678. [CrossRef]
15. Li, X.; Xia, W. Effects of concentration, degree of deacetylation and molecular weight on emulsifying properties of chitosan. *Int. J. Biol. Macromol.* **2011**, *48*, 768–772. [CrossRef]
16. Kou, S.G.; Peters, L.M.; Mucalo, M.R. Chitosan: A review of sources and preparation methods. *Int. J. Biol. Macromol.* **2021**, *169*, 85–94. [CrossRef]
17. Kou, S.G.; Peters, L.; Mucalo, M. Chitosan: A review of molecular structure, bioactivities and interactions with the human body and micro-organisms. *Carbohydr. Polym.* **2022**, *282*, 119132. [CrossRef]
18. Zargar, V.; Asghari, M.; Dashti, A. A Review on Chitin and Chitosan Polymers: Structure, Chemistry, Solubility, Derivatives, and Applications. *ChemBioEng Rev.* **2015**, *2*, 204–226. [CrossRef]
19. Kim, S. Competitive Biological Activities of Chitosan and Its Derivatives: Antimicrobial, Antioxidant, Anticancer, and Anti-Inflammatory Activities. *Int. J. Polym. Sci.* **2018**, *2018*, 1708172. [CrossRef]
20. Ke, C.-L.; Deng, F.-S.; Chuang, C.-Y.; Lin, C.-H. Antimicrobial Actions and Applications of Chitosan. *Polymers* **2021**, *13*, 904. [CrossRef] [PubMed]
21. Negm, N.A.; Hefni, H.H.H.; Abd-Elaal, A.A.A.; Badr, E.A.; Abou Kana, M.T.H. Advancement on Modification of Chitosan Biopolymer and Its Potential Applications. *Int. J. Biol. Macromol.* **2020**, *152*, 681–702. [CrossRef] [PubMed]
22. Sahariah, P.; Másson, M. Antimicrobial Chitosan and Chitosan Derivatives: A Review of the Structure–Activity Relationship. *Biomacromolecules* **2017**, *18*, 3846–3868. [CrossRef]
23. Argüelles-Monal, W.; Lizardi-Mendoza, J.; Fernández-Quiroz, D.; Recillas-Mota, M.; Montiel-Herrera, M. Chitosan Derivatives: Introducing New Functionalities with a Controlled Molecular Architecture for Innovative Materials. *Polymers* **2018**, *10*, 342. [CrossRef] [PubMed]
24. Cohen, E.; Poverenov, E. Hydrophilic Chitosan Derivatives: Synthesis and Applications. *Chem. Eur. J.* **2022**, *28*, e202202156. [CrossRef]
25. Boominathan, T.; Sivaramakrishna, A. Recent Advances in the Synthesis, Properties, and Applications of Modified Chitosan Derivatives: Challenges and Opportunities. *Top. Curr. Chem.* **2021**, *379*, 19. [CrossRef]
26. Frigaard, J.; Jensen, J.L.; Galtung, H.K.; Hiorth, M. The Potential of Chitosan in Nanomedicine: An Overview of the Cytotoxicity of Chitosan Based Nanoparticles. *Front. Pharmacol.* **2022**, *13*, 880377. [CrossRef] [PubMed]
27. Naskar, S.; Kuotsu, K.; Sharma, S. Chitosan-Based Nanoparticles as Drug Delivery Systems: A Review on Two Decades of Research. *J. Drug Target.* **2019**, *27*, 379–393. [CrossRef]
28. Precedence Research. *Chitosan Market Size to Surpass around US$ 29 Billion by 2030*; GlobeNewswire: Los Angeles, CA, USA, 2021.
29. Wang, W.; Meng, Q.; Li, Q.; Liu, J.; Zhou, M.; Jin, Z.; Zhao, K. Chitosan derivatives and their application in biomedicine. *Int. J. Mol. Sci.* **2020**, *21*, 487. [CrossRef]
30. Kertmen, A.; Dziedzic, I.; Ehrlich, H. Patentology of chitinous biomaterials. Part II: Chitosan. *Carbohydr. Polym.* **2023**, *301*, 120224. [CrossRef]
31. Thambiliyagodage, C.; Jayanetti, M.; Mendis, A.; Ekanayake, G.; Liyanaarachchi, H.; Vigneswaran, S. Recent Advances in Chitosan-Based Applications-A Review. *Materials* **2023**, *16*, 2073. [CrossRef]
32. Elangwe, C.N.; Morozkina, S.N.; Olekhnovich, R.O.; Krasichkov, A.; Polyakova, V.O.; Uspenskaya, M.V. A Review on Chitosan and Cellulose Hydrogels for Wound Dressings. *Polymers* **2022**, *14*, 5163. [CrossRef]
33. Lacasse, P.; Lanctôt, S.; Fustier, P.; Bégin, A.; Taherian, A.R.; Bisakowski, B. Chitosan hydrogels for accelerating involution and preventing infection of the mammary gland at drying-off. U.S. Patent No. 10, 828, 319, 10 November 2020.
34. Armendariz, I.O.; Gomez, L.E.V.; Estrada, S.A.M. Biodegradable Film Made of Carboximethyl Chitosan and Mimosa Tenuifloraethanolic Extract for Wounds Treatment. Patent MX 2016007284 A, 3 June 2016. Available online: https://lens.org/157-946-545-478-401 (accessed on 4 December 2017).
35. Travan, A.; Borgogna, M.A.; Parravicini, M.; De Lollis, A.; Siccardi, F. Biocompatible Compositions Comprising a Biocompatible Thickening Polymer and a Chitosan Derivative. U.S. Patent No. 17/597,280, 6 October 2022.
36. Chuan, W.; Xiang, Z.; Lin, S.; Sijing, L.; Mingjuan, J.; Huan, H.; Yongyu, L.; Ting, L.; Yuzhen, Z. Chitosan Oligosaccharide Coated Listeria Iuanuii and Preparation Method and Application Thereof. Patent CN 109248312 A, 23 March 2021. Available online: https://lens.org/019-931-768-976-410 (accessed on 22 January 2019).
37. Weiwei, W.; Wenshuai, L.; Pingsheng, H.; Huijuan, S.; Deling, K. Double Quaternized Chitosan Derivative, Synthetic Method Thereof, Composite Sponge Biological Dressing Containing Double Quaternized Chitosan Derivative and Application Thereof. Patent CN 111690078 A, 20 September 2022. Available online: https://lens.org/158-377-650-080-779 (accessed on 22 September 2020).
38. Triviño, G.C. Chitosan Gels (A) Containing Metal Nanoparticles of Copper, Silver and Antibiotics (Ciprofloxacin, Cefotaxime, Gentamicin and Cloxacillin). U.S. Patent No. 16/973,000, 21 October 2021.

39. Alcantar, N.A.; Falahat, R.; Wiranowska, M.; Toomey, R.G.U.S. Enhanced targeted drug delivery system via chitosan hydrogel and chlorotoxin. U.S. Patent No. 9522114B1, 20 December 2016.
40. Aleksandrovna, Z.A.; Vladimirovna, K.M.; Tsyrendorzhievna, S.B.; Viktorovna, S.E.; Petrovich, V.V. Method of Producing Chitosan Derivatives for Visualizing Cell Membranes and Creating Drug Delivery Systems with High Mucoadhesion. Patent RU 2697872 C1, 21 August 2019. Available online: https://lens.org/079-602-098-619-834 (accessed on 27 September 2023).
41. Jianhua, Z.; Liangxia, W.; Sheng, G.; Ling, C.; Chaoying, W.; Yongli, L.; Wanting, X.; Yumei, M. Chitosan nanoparticle CS-IL-17RC and preparation method thereof. CN 107519481 A, 27 August 2017. Available online: https://lens.org/105-221-674-703-471 (accessed on 29 December 2017).
42. Panicker, R.K.G.; Veilleux, D.; Lian, P.; Tam, N.C.M.; Fleet, C.; Cheung, A.; Dauphinee, S.; Chen, X.; Jose, L.O.R.A. Chitosan Polyplex-Based Localized Expression of Il-12 Alone or in Combination with Type-I Ifn Inducers for Treatment of Mucosal Cancers. U.S. Patent No. 17/438,922, 24 November 2022.
43. Pillay, V.; Choonara, Y.E.; Kumar, P.; Du Toit, L.C.; Ramburrun, P. Biodegradable Implant. U.S. Patent No. US10478527B2, 19 November 2019.
44. Qiliang, H.; Lidong, C.; Chunli, X.; Chong, C.; Pengyue, Z.; Fengmin, L.; Yongquan, Z. Method for Preparing Prothioconazole Sustained-Release Gel Particles by Carboxymethyl Chitosan-Metal Ion Crosslinking Method. Patent CN 111887244 A, 15 July 2020. Available online: https://lens.org/095-820-269-282-291 (accessed on 6 November 2020).
45. Arne, T.; Bjugan, A.B.; Linda, H.; Morten, S.; Hafizur, R.; Line, N.A.; Vincent, E. Composition Comprising Chitosan and a Fungicide. U.S. Patent No. 10172353B2, 8 January 2019.
46. Jiajia, C. Berberine and Chitosan Containing Pesticide Composition. Patent CN 110024801 A, 27 May 2019. Available online: https://lens.org/196-874-974-995-823 (accessed on 19 July 2019).
47. Zhiping, P.; Jichuan, H.; Yuting, T.; Zhijun, L.; Linxiang, Y.; Xuena, W. Method for Preparing Liquid Fertilizer Containing Amino Acid and Chitosan by Using Shrimp-Crab Residual Materials. Patent CN 108456033 A, 2 February 2018. Available online: https://lens.org/040-149-136-430-541 (accessed on 28 August 2018).
48. Chen, Y.; Li, J.; Li, Q.; Shen, Y.; Ge, Z.; Zhang, W.; Chen, S. Enhanced Water-Solubility, Antibacterial Activity and Biocompatibility upon Introducing Sulfobetaine and Quaternary Ammonium to Chitosan. *Carbohydr. Polym.* **2016**, *143*, 246–253. [CrossRef]
49. Ifuku, S.; Matsumoto, C.; Wada, M.; Morimoto, M.; Saimoto, H. Preparation of Highly Regioselective Amphiprotic Chitosan Derivative via "Click Chemistry". *Int. J. Biol. Macromol.* **2013**, *52*, 72–76. [CrossRef]
50. Kurita, K.; Ikeda, H.; Yoshida, Y.; Shimojoh, M.; Harata, M. Chemoselective Protection of the Amino Groups of Chitosan by Controlled Phthaloylation: Facile Preparation of a Precursor Useful for Chemical Modifications. *Biomacromolecules* **2002**, *3*, 1–4. [CrossRef]
51. Holappa, J.; Nevalainen, T.; Soininen, P.; Másson, M.; Järvinen, T. Synthesis of Novel Quaternary Chitosan Derivatives via N-Chloroacyl-6-O-Triphenylmethylchitosans. *Biomacromolecules* **2006**, *7*, 407–410. [CrossRef]
52. Kurita, K.; Hirakawa, M.; Kikuchi, S.; Yamanaka, H.; Yang, J. Trimethylsilylation of Chitosan and Some Properties of the Product. *Carbohydr. Polym.* **2004**, *56*, 333–337. [CrossRef]
53. Rathinam, S.; Ólafsdóttir, S.; Jónsdóttir, S.; Hjálmarsdóttir, M.Á.; Másson, M. Selective Synthesis of N,N,N-Trimethylated Chitosan Derivatives at Different Degree of Substitution and Investigation of Structure-Activity Relationship for Activity against P. Aeruginosa and MRSA. *Int. J. Biol. Macromol.* **2020**, *160*, 548–557. [CrossRef] [PubMed]
54. Song, W.; Gaware, V.S.; Rúnarsson, Ö.V.; Másson, M.; Mano, J.F. Functionalized Superhydrophobic Biomimetic Chitosan-Based Films. *Carbohydr. Polym.* **2010**, *81*, 140–144. [CrossRef]
55. Sogias, I.A.; Khutoryanskiy, V.V.; Williams, A.C. Exploring the factors affecting the solubility of chitosan in water. *Macromol. Chem. Phys.* **2010**, *211*, 426–433. [CrossRef]
56. Panda, P.K.; Yang, J.M.; Chang, Y.H.; Su, W.W. Modification of different molecular weights of chitosan by p-Coumaric acid: Preparation, characterization and effect of molecular weight on its water solubility and antioxidant property. *Int. J. Biol. Macromol.* **2019**, *136*, 661–667. [CrossRef] [PubMed]
57. Wu, Q.X.; Lin, D.Q.; Yao, S.J. Design of chitosan and its water soluble derivatives-based drug carriers with polyelectrolyte complexes. *Mar. Drugs* **2014**, *12*, 6236–6253. [CrossRef]
58. Panda, P.K.; Dash, P.; Chang, Y.H.; Yang, J.M. Improvement of chitosan water solubility by fumaric acid modification. *Mater. Lett.* **2022**, *316*, 132046. [CrossRef]
59. Wu, Q.X.; Lin, D.Q.; Yao, S.J. Fabrication and formation studies on single-walled CA/NaCS-WSC microcapsules. *Mater. Sci. Eng. C* **2016**, *59*, 909–915. [CrossRef]
60. Wang, W.; Xue, C.; Mao, X. Chitosan: Structural modification, biological activity and application. *Int. J. Biol. Macromol.* **2020**, *164*, 4532–4546. [CrossRef] [PubMed]
61. Vilčnik, A.; Jerman, I.; Šurca Vuk, A.; Koželj, M.; Orel, B.; Tomšič, B.; Simončič, B.; Kovač, J. Structural Properties and Antibacterial Effects of Hydrophobic and Oleophobic Sol−Gel Coatings for Cotton Fabrics. *Langmuir* **2009**, *25*, 5869–5880. [CrossRef]
62. Pei, L.; Cai, Z.; Shang, S.; Song, Z. Synthesis and Antibacterial Activity of Alkylated Chitosan under Basic Ionic Liquid Conditions. *J. Appl. Polym. Sci.* **2014**, *131*, 40052–40059. [CrossRef]
63. Sun, X.; Li, J.; Shao, K.; Su, C.; Bi, S.; Mu, Y.; Zhang, K.; Cao, Z.; Wang, X.; Chen, X.; et al. A Composite Sponge Based on Alkylated Chitosan and Diatom-Biosilica for Rapid Hemostasis. *Int. J. Biol. Macromol.* **2021**, *182*, 2097–2107. [CrossRef]

64. Akopova, T.A.; Demina, T.S.; Cherkaev, G.V.; Khavpachev, M.A.; Bardakova, K.N.; Grachev, A.V.; Vladimirov, L.V.; Zelenetskii, A.N.; Timashev, P.S. Solvent-Free Synthesis and Characterization of Allyl Chitosan Derivatives. *RSC Adv.* **2019**, *9*, 20968–20975. [CrossRef]
65. Ma, G.; Yang, D.; Zhou, Y.; Xiao, M.; Kennedy, J.F.; Nie, J. Preparation and Characterization of Water-Soluble N-Alkylated Chitosan. *Carbohydr. Polym.* **2008**, *74*, 121–126. [CrossRef]
66. Naidoo, S.; Nomadolo, N.; Matshe, W.M.R.; Cele, Z.; Balogun, M. Exploring the Potential of N-Acylated Chitosan for the Removal of Toxic Pollutants from Wastewater. *IOP Conf. Ser. Mater. Sci. Eng.* **2019**, *655*, 012047. [CrossRef]
67. Reis, B.; Gerlach, N.; Steinbach, C.; Haro Carrasco, K.; Oelmann, M.; Schwarz, S.; Müller, M.; Schwarz, D. A Complementary and Revised View on the N-Acylation of Chitosan with Hexanoyl Chloride. *Mar. Drugs* **2021**, *19*, 385. [CrossRef] [PubMed]
68. Champagne, L.M. *The Synthesis of Water Soluble N-Acyl Chitosan Derivatives for Characterization as Antibacterial Agents*; Louisiana State University and Agricultural & Mechanical College: Baton Rouge, LA, USA, 2008.
69. Gamal, A.; Ibrahim, A.G.; Eliwa, E.M.; El-Zomrawy, A.H.; El-Bahy, S.M. Synthesis and Characterization of a Novel Benzothiazole Functionalized Chitosan and Its Use for Effective Adsorption of Cu(II). *Int. J. Biol. Macromol.* **2021**, *183*, 1283–1292. [CrossRef] [PubMed]
70. Mohamed, A.E.; Elgammal, W.E.; Eid, A.M.; Dawaba, A.M.; Ibrahim, A.G.; Fouda, A.; Hassan, S.M. Synthesis and Characterization of New Functionalized Chitosan and Its Antimicrobial and In-Vitro Release Behavior from Topical Gel. *Int. J. Biol. Macromol.* **2022**, *207*, 242–253. [CrossRef]
71. Ibrahim, A.G.; Fouda, A.; Elgammal, W.E.; Eid, A.M.; Elsenety, M.M.; Mohamed, A.E.; Hassan, S.M. New Thiadiazole Modified Chitosan Derivative to Control the Growth of Human Pathogenic Microbes and Cancer Cell Lines. *Sci. Rep.* **2022**, *12*, 21423. [CrossRef] [PubMed]
72. Mohamed, A.E.; Elgammal, W.E.; Dawaba, A.M.; Ibrahim, A.G.; Fouda, A.; Hassan, S.M. A Novel 1,3,4-Thiadiazole Modified Chitosan: Synthesis, Characterization, Antimicrobial Activity, and Release Study from Film Dressings. *Appl. Biol. Chem.* **2022**, *65*, 54. [CrossRef]
73. Ibrahim, A.G.; Elgammal, W.E.; Hashem, A.H.; Mohamed, A.E.; Awad, M.A.; Hassan, S.M. Development of a Chitosan Derivative Bearing the Thiadiazole Moiety and Evaluation of Its Antifungal and Larvicidal Efficacy. *Polym. Bull.* **2023**, 1–33. [CrossRef]
74. Deng, P.; Chen, J.; Yao, L.; Zhang, P.; Zhou, J. Thymine-Modified Chitosan with Broad-Spectrum Antimicrobial Activities for Wound Healing. *Carbohydr. Polym.* **2021**, *257*, 117630. [CrossRef]
75. Hamodin, A.G.; Elgammal, W.E.; Eid, A.M.; Ibrahim, A.G. Synthesis, Characterization, and Biological Evaluation of New Chitosan Derivative Bearing Diphenyl Pyrazole Moiety. *Int. J. Biol. Macromol.* **2023**, *243*, 125180. [CrossRef]
76. Mi, Y.; Li, Q.; Miao, Q.; Tan, W.; Zhang, J.; Guo, Z. Enhanced Antifungal and Antioxidant Activities of New Chitosan Derivatives Modified with Schiff Base Bearing Benzenoid/Heterocyclic Moieties. *Int. J. Biol. Macromol.* **2022**, *208*, 586–595. [CrossRef]
77. Wang, L.; Liu, X.; Tan, W.; Li, Q.; Guo, Z.; Zhang, J. Preparation and Antioxidant Activity of Novel Chitosan Oligosaccharide Quinolinyl Urea Derivatives. *Carbohydr. Res.* **2022**, *521*, 108678. [CrossRef]
78. Omer, A.M.; Eltaweil, A.S.; El-Fakharany, E.M.; Abd El-Monaem, E.M.; Ismail, M.M.F.; Mohy-Eldin, M.S.; Ayoup, M.S. Novel Cytocompatible Chitosan Schiff Base Derivative as a Potent Antibacterial, Antidiabetic, and Anticancer Agent. *Arab. J. Sci. Eng.* **2023**, *48*, 7587–7601. [CrossRef]
79. Antony, R.; Arun, T.; Manickam, S.T.D. A Review on Applications of Chitosan-Based Schiff Bases. *Int. J. Biol. Macromol.* **2019**, *129*, 615–633. [CrossRef] [PubMed]
80. Hamed, A.A.; Abdelhamid, I.A.; Saad, G.R.; Elkady, N.A.; Elsabee, M.Z. Synthesis, Characterization and Antimicrobial Activity of a Novel Chitosan Schiff Bases Based on Heterocyclic Moieties. *Int. J. Biol. Macromol.* **2020**, *153*, 492–501. [CrossRef]
81. El-Gharably, A.A.; Kenawy, E.-R.S.; Safaan, A.A.; Aboamna, S.A.; Mahmoud, Y.A.-G.; Mandour, H.S.A. Synthesis, Characterization and Application of Chitosan Conjugated Heterocyclic Compounds. *J. Polym. Res.* **2022**, *29*, 141. [CrossRef]
82. Hassan, M.A.; Omer, A.M.; Abbas, E.; Baset, W.M.A.; Tamer, T.M. Preparation, Physicochemical Characterization and Antimicrobial Activities of Novel Two Phenolic Chitosan Schiff Base Derivatives. *Sci. Rep.* **2018**, *8*, 11416. [CrossRef] [PubMed]
83. Manchaiah, A.S.; Badalamoole, V. Novel Heterocyclic Chitosan-based Schiff Base: Evaluation as Adsorbent for Removal of Methyl Orange from Aqueous Solution. *Water Environ. J.* **2020**, *34*, 364–373. [CrossRef]
84. El-Naggar, M.M.; Haneen, D.S.A.; Mehany, A.B.M.; Khalil, M.T. New Synthetic Chitosan Hybrids Bearing Some Heterocyclic Moieties with Potential Activity as Anticancer and Apoptosis Inducers. *Int. J. Biol. Macromol.* **2020**, *150*, 1323–1330. [CrossRef]
85. Paula, H.C.B.; Silva, R.B.C.; Santos, C.M.; Dantas, F.D.S.; De Paula, R.C.M.; De Lima, L.R.M.; De Oliveira, E.F.; Figueiredo, E.A.T.; Dias, F.G.B. Eco-Friendly Synthesis of an Alkyl Chitosan Derivative. *Int. J. Biol. Macromol.* **2020**, *163*, 1591–1598. [CrossRef]
86. Kritchenkov, A.S.; Egorov, A.R.; Artemjev, A.A.; Kritchenkov, I.S.; Volkova, O.V.; Kiprushkina, E.I.; Zabodalova, L.A.; Suchkova, E.P.; Yagafarov, N.Z.; Tskhovrebov, A.G.; et al. Novel Heterocyclic Chitosan Derivatives and Their Derived Nanoparticles: Catalytic and Antibacterial Properties. *Int. J. Biol. Macromol.* **2020**, *149*, 682–692. [CrossRef]
87. Wang, L.; Zhang, J.; Liu, X.; Tan, W.; Li, Q.; Guo, Z. Preparation of Novel Chitosan Oligosaccharide Quaternary Ammonium Derivatives Bearing Quinoline with Antioxidative and Antibacterial Activities. *Starch-Stärke* **2023**, *75*, 2300001. [CrossRef]
88. Andreica, B.-I.; Cheng, X.; Marin, L. Quaternary Ammonium Salts of Chitosan. A Critical Overview on the Synthesis and Properties Generated by Quaternization. *Eur. Polym. J.* **2020**, *139*, 110016. [CrossRef]
89. Wang, Y.; Yang, Z.; Hu, H.; Wu, J.; Finšgar, M. Indolizine Quaternary Ammonium Salt Inhibitors: The Inhibition and Anti-Corrosion Mechanism of New Dimer Derivatives from Ethyl Acetate Quinolinium Bromide and n-Butyl Quinolinium Bromide. *Colloids Surf. Physicochem. Eng. Asp.* **2022**, *651*, 129649. [CrossRef]

90. Muzzarelli, R.A.A.; Tanfani, F. The N-Permethylation of Chitosan and the Preparation of N-Trimethyl Chitosan Iodide. *Carbohydr. Polym.* **1985**, *5*, 297–307. [CrossRef]
91. Ali, I.; Nadeem Lone, M.; Al-Othman, Z.A.; Al-Warthan, A.; Marsin Sanagi, M. Heterocyclic Scaffolds: Centrality in Anticancer Drug Development. *Curr. Drug Targets* **2015**, *16*, 711–734. [CrossRef]
92. Li, Q.; Wei, L.; Zhang, J.; Gu, G.; Guo, Z. Significantly Enhanced Antioxidant Activity of Chitosan through Chemical Modification with Coumarins. *Polym. Chem.* **2019**, *10*, 1480–1488. [CrossRef]
93. Li, Q.; Mi, Y.; Tan, W.; Guo, Z. Highly Efficient Free Radical-Scavenging Property of Phenolic-Functionalized Chitosan Derivatives: Chemical Modification and Activity Assessment. *Int. J. Biol. Macromol.* **2020**, *164*, 4279–4288. [CrossRef]
94. Wei, L.; Li, Q.; Chen, Y.; Zhang, J.; Mi, Y.; Dong, F.; Lei, C.; Guo, Z. Enhanced Antioxidant and Antifungal Activity of Chitosan Derivatives Bearing 6-O-Imidazole-Based Quaternary Ammonium Salts. *Carbohydr. Polym.* **2019**, *206*, 493–503. [CrossRef]
95. Li, Q.; Zhang, C.; Tan, W.; Gu, G.; Guo, Z. Novel Amino-Pyridine Functionalized Chitosan Quaternary Ammonium Derivatives: Design, Synthesis, and Antioxidant Activity. *Molecules* **2017**, *22*, 156. [CrossRef]
96. Wei, L.; Tan, W.; Wang, G.; Li, Q.; Dong, F.; Guo, Z. The Antioxidant and Antifungal Activity of Chitosan Derivatives Bearing Schiff Bases and Quaternary Ammonium Salts. *Carbohydr. Polym.* **2019**, *226*, 115256. [CrossRef]
97. Zhang, J.; Tan, W.; Li, Q.; Dong, F.; Guo, Z. Synthesis and Characterization of N,N,N-Trimethyl-O-(Ureidopyridinium)Acetyl Chitosan Derivatives with Antioxidant and Antifungal Activities. *Mar. Drugs* **2020**, *18*, 163. [CrossRef] [PubMed]
98. Zhang, J.; Mi, Y.; Sun, X.; Chen, Y.; Miao, Q.; Tan, W.; Li, Q.; Dong, F.; Guo, Z. Improved Antioxidant and Antifungal Activity of Chitosan Derivatives Bearing Urea Groups. *Starch-Stärke* **2020**, *72*, 1900205. [CrossRef]
99. Wei, L.; Mi, Y.; Zhang, J.; Li, Q.; Dong, F.; Guo, Z. Evaluation of Quaternary Ammonium Chitosan Derivatives Differing in the Length of Alkyl Side-Chain: Synthesis and Antifungal Activity. *Int. J. Biol. Macromol.* **2019**, *129*, 1127–1132. [CrossRef] [PubMed]
100. Jiang, Y.; Zhang, Y.; Zhang, H.; Zhu, R.; Hu, Y. Synthesis of N-alkylated Quaternary Ammonium Chitosan and Its Long-term Antibacterial Finish for Rabbit Hair Fabric. *Polym. Adv. Technol.* **2022**, *33*, 314–325. [CrossRef]
101. Mi, Y.; Zhang, J.; Han, X.; Tan, W.; Miao, Q.; Cui, J.; Li, Q.; Guo, Z. Modification of Carboxymethyl Inulin with Heterocyclic Compounds: Synthesis, Characterization, Antioxidant and Antifungal Activities. *Int. J. Biol. Macromol.* **2021**, *181*, 572–581. [CrossRef]
102. Mi, Y.; Miao, Q.; Cui, J.; Tan, W.; Guo, Z. Novel 2-Hydroxypropyltrimethyl Ammonium Chitosan Derivatives: Synthesis, Characterization, Moisture Absorption and Retention Properties. *Molecules* **2021**, *26*, 4238. [CrossRef] [PubMed]
103. Li, W.; Duan, Y.; Huang, J.; Zheng, Q. Synthesis, Antioxidant and Cathepsin D Inhibition Activity of Quaternary Ammonium Chitosan Derivatives. *Carbohydr. Polym.* **2016**, *136*, 884–891. [CrossRef] [PubMed]
104. Spinelli, V.A.; Laranjeira, M.C.M.; Fávere, V.T. Preparation and Characterization of Quaternary Chitosan Salt: Adsorption Equilibrium of Chromium(VI) Ion. *React. Funct. Polym.* **2004**, *61*, 347–352. [CrossRef]
105. Sangeetha, Y.; Meenakshi, S.; SairamSundaram, C. Corrosion Mitigation of N-(2-Hydroxy-3-Trimethyl Ammonium)Propyl Chitosan Chloride as Inhibitor on Mild Steel. *Int. J. Biol. Macromol.* **2015**, *72*, 1244–1249. [CrossRef]
106. Tan, W.; Lin, C.; Zhang, J.; Li, Q.; Guo, Z. Synthesis of Hydroxypropyltrimethyl Ammonium Chitosan Derivatives Bearing Thioctate and the Potential for Antioxidant Application. *Molecules* **2022**, *27*, 2682. [CrossRef]
107. Li, P.; He, L.; Li, X.; Liu, X.; Sun, M. Corrosion Inhibition Effect of N-(4-Diethylaminobenzyl) Quaternary Ammonium Chitosan for X80 Pipeline Steel in Hydrochloric Acid Solution. *Int. J. Electrochem. Sci.* **2021**, *16*, 150931. [CrossRef]
108. Li, Z.; Yang, F.; Yang, R. Synthesis and Characterization of Chitosan Derivatives with Dual-Antibacterial Functional Groups. *Int. J. Biol. Macromol.* **2015**, *75*, 378–387. [CrossRef] [PubMed]
109. Kratky, M.; Vinsova, J. Antimycobacterial activity of quaternary pyridinium salts and pyridinium N-oxides-review. *Curr. Pharm. Des.* **2013**, *19*, 1343–1355. [CrossRef] [PubMed]
110. Sowmiah, S.; Esperança, J.M.S.S.; Rebelo, L.P.N.; Afonso, C.A.M. Pyridinium Salts: From Synthesis to Reactivity and Applications. *Org. Chem. Front.* **2018**, *5*, 453–493. [CrossRef]
111. Wei, L.; Li, Q.; Tan, W.; Dong, F.; Luan, F.; Guo, Z. Synthesis, Characterization, and the Antioxidant Activity of Double Quaternized Chitosan Derivatives. *Molecules* **2017**, *22*, 501. [CrossRef] [PubMed]
112. Oyervides-Muñoz, E.; Avérous, L.; Sosa-Santillán, G.D.J.; Pollet, E.; Pérez-Aguilar, N.V.; Rojas-Caldera, C.M.; Fuentes-Avilés, J.G.; García-Astrain, C. EDC-Mediated Grafting of Quaternary Ammonium Salts onto Chitosan for Antibacterial and Thermal Properties Improvement. *Macromol. Chem. Phys.* **2019**, *220*, 1800530. [CrossRef]
113. Franconetti, A.; Contreras-Bernal, L.; Prado-Gotor, R.; Cabrera-Escribano, F. Synthesis of Hyperpolarizable Biomaterials at Molecular Level Based on Pyridinium–Chitosan Complexes. *RSC Adv.* **2015**, *5*, 74274–74283. [CrossRef]
114. Jia, R.; Duan, Y.; Fang, Q.; Wang, X.; Huang, J. Pyridine-Grafted Chitosan Derivative as an Antifungal Agent. *Food Chem.* **2016**, *196*, 381–387. [CrossRef]
115. Lin, C.; Guo, Z.; Jiang, A.; Liang, X.; Tan, W. Cationic Chitooligosaccharide Derivatives Bearing Pyridinium and Trialkyl Ammonium: Preparation, Characterization and Antimicrobial Activities. *Polymers* **2022**, *15*, 14. [CrossRef]
116. Zhang, J.; Zhang, S.; Wang, L.; Tan, W.; Li, Q.; Guo, Z. The Antioxidant and Antibacterial Activities of the Pyridine-4-Aldehyde Schiff Bases Grafted Chloracetyl Chitosan Oligosaccharide Derivatives. *Starch-Stärke* **2023**, *75*, 2100268. [CrossRef]
117. Zhang, J.; Sun, X.; Chen, Y.; Mi, Y.; Tan, W.; Miao, Q.; Li, Q.; Dong, F.; Guo, Z. Preparation of 2,6-Diurea-Chitosan Oligosaccharide Derivatives for Efficient Antifungal and Antioxidant Activities. *Carbohydr. Polym.* **2020**, *234*, 115903. [CrossRef]

118. Zincke, T.; Heuser, G.; Möller, W.I. Ueber Dinitrophenylpyridiniumchlorid und Dessen Umwandlungsproducte. *Justus Liebigs Ann. Chem.* **1904**, *333*, 296–345. [CrossRef]
119. Cheng, W.-C.; Kurth, M.J. The Zincke Reaction. *A Review. Org. Prep. Proced. Int.* **2002**, *34*, 585–608. [CrossRef]
120. Venegas-García, D.J.; Steiger, B.G.K.; Wilson, L.D. A Pyridinium-Modified Chitosan-Based Adsorbent for Arsenic Removal via a Coagulation-like Methodology. *RSC Sustain.* **2023**, *1*, 1259–1269. [CrossRef]
121. Gonçalves, F.; Freitas, R. Modification of Chitosan by Zincke Reaction: Synthesis of a Novel Polycationic Chitosan-Pyridinium Derivative. *J. Braz. Chem. Soc.* **2019**, *30*, 2318–2323. [CrossRef]
122. Meng, X.; Edgar, K.J. "Click" Reactions in Polysaccharide Modification. *Prog. Polym. Sci.* **2016**, *53*, 52–85. [CrossRef]
123. Li, Q.; Sun, X.; Gu, G.; Guo, Z. Novel Water Soluble Chitosan Derivatives with 1,2,3-Triazolium and Their Free Radical-Scavenging Activity. *Mar. Drugs* **2018**, *16*, 107. [CrossRef]
124. Tan, W.; Zhang, J.; Mi, Y.; Dong, F.; Li, Q.; Guo, Z. Synthesis, Characterization, and Evaluation of Antifungal and Antioxidant Properties of Cationic Chitosan Derivative via Azide-Alkyne Click Reaction. *Int. J. Biol. Macromol.* **2018**, *120*, 318–324. [CrossRef]
125. Tan, W.; Zhang, J.; Mi, Y.; Dong, F.; Li, Q.; Guo, Z. Enhanced Antifungal Activity of Novel Cationic Chitosan Derivative Bearing Triphenylphosphonium Salt via Azide-Alkyne Click Reaction. *Int. J. Biol. Macromol.* **2020**, *165*, 1765–1772. [CrossRef]
126. Kim, S.-H.; Semenya, D.; Castagnolo, D. Antimicrobial Drugs Bearing Guanidine Moieties: A Review. *Eur. J. Med. Chem.* **2021**, *216*, 113293. [CrossRef]
127. Eckert-Maksić, M.; Glasovac, Z.; Trošelj, P.; Kütt, A.; Rodima, T.; Koppel, I.; Koppel, I.A. Basicity of Guanidines with Heteroalkyl Side Chains in Acetonitrile. *Eur. J. Org. Chem.* **2008**, *2008*, 5176–5184. [CrossRef]
128. Zhai, X.; Sun, P.; Luo, Y.; Ma, C.; Xu, J.; Liu, W. Guanidinylation: A Simple Way to Fabricate Cell Penetrating Peptide Analogue-Modified Chitosan Vector for Enhanced Gene Delivery: Guanidinylation. *J. Appl. Polym. Sci.* **2011**, *121*, 3569–3578. [CrossRef]
129. Hu, Y.; Du, Y.; Yang, J.; Kennedy, J.; Wang, X.; Wang, L. Synthesis, Characterization and Antibacterial Activity of Guanidinylated Chitosan. *Carbohydr. Polym.* **2007**, *67*, 66–72. [CrossRef]
130. Lahmer, R.A.; Williams, A.P.; Townsend, S.; Baker, S.; Jones, D.L. Antibacterial Action of Chitosan-Arginine against Escherichia Coli O157 in Chicken Juice. *Food Control* **2012**, *26*, 206–211. [CrossRef]
131. Lv, J.; Jiang, M.; Fang, Y.; Zhang, W.; Wu, M.; Zheng, F.; Lei, K.; Shang, L.; Zhao, Y. The Mimicked Cell Penetration Peptides of Chitosan Derivates with Low Molecular Weight for Transdermal Enhancement Studies. *Eur. Polym. J.* **2023**, *195*, 112197. [CrossRef]
132. Zhou, Y.; Liu, G.; Huang, H.; Wu, J. Advances and Impact of Arginine-Based Materials in Wound Healing. *J. Mater. Chem. B* **2021**, *9*, 6738–6750. [CrossRef] [PubMed]
133. Xiao, B.; Wan, Y.; Zhao, M.; Liu, Y.; Zhang, S. Preparation and Characterization of Antimicrobial Chitosan-N-Arginine with Different Degrees of Substitution. *Carbohydr. Polym.* **2011**, *83*, 144–150. [CrossRef]
134. Lv, H.-X.; Zhang, Z.-H.; Wang, X.-P.; Cheng, Q.-Q.; Wang, W.; Huang, X.-H.; Zhou, J.-P.; Zhang, Q.; Hou, L.-L.; Huo, W. A Biomimetic Chitosan Derivates: Preparation, Characterization and Transdermal Enhancement Studies of N-Arginine Chitosan. *Molecules* **2011**, *16*, 6778–6790. [CrossRef]
135. Cui, J.; Sun, Y.; Wang, L.; Miao, Q.; Tan, W.; Guo, Z. Preparation of L-Arginine Schiff Bases Modified Chitosan Derivatives and Their Antimicrobial and Antioxidant Properties. *Mar. Drugs* **2022**, *20*, 688. [CrossRef]
136. Salama, H.E.; Saad, G.R.; Sabaa, M.W. Synthesis, Characterization and Antimicrobial Activity of Biguanidinylated Chitosan-g-Poly[(R)-3-Hydroxybutyrate]. *Int. J. Biol. Macromol.* **2017**, *101*, 438–447. [CrossRef] [PubMed]
137. Sang, W.; Tang, Z.; He, M.-Y.; Hua, Y.-P.; Xu, Q. Synthesis and Preservative Application of Quaternized Carboxymethyl Chitosan Containing Guanidine Groups. *Int. J. Biol. Macromol.* **2015**, *75*, 489–494. [CrossRef]
138. Sahariah, P.; Óskarsson, B.M.; Hjálmarsdóttir, M.Á.; Másson, M. Synthesis of Guanidinylated Chitosan with the Aid of Multiple Protecting Groups and Investigation of Antibacterial Activity. *Carbohydr. Polym.* **2015**, *127*, 407–417. [CrossRef] [PubMed]
139. Chen, S.; Li, C.; Hou, T.; Cai, Y.; Liang, L.; Chen, L.; Li, M. Polyhexamethylene Guanidine Functionalized Chitosan Nanofiber Membrane with Superior Adsorption and Antibacterial Performances. *React. Funct. Polym.* **2019**, *145*, 104379. [CrossRef]
140. Salama, A.; Hesemann, P. New N-Guanidinium Chitosan/Silica Ionic Microhybrids as Efficient Adsorbent for Dye Removal from Waste Water. *Int. J. Biol. Macromol.* **2018**, *111*, 762–768. [CrossRef] [PubMed]
141. Salama, A.; Hasanin, M.; Hesemann, P. Synthesis and Antimicrobial Properties of New Chitosan Derivatives Containing Guanidinium Groups. *Carbohydr. Polym.* **2020**, *241*, 116363. [CrossRef] [PubMed]
142. Shaabani, A.; Sedghi, R. Preparation of Chitosan Biguanidine/PANI-Containing Self-Healing Semi-Conductive Waterborne Scaffolds for Bone Tissue Engineering. *Carbohydr. Polym.* **2021**, *264*, 118045. [CrossRef] [PubMed]
143. Agrahari, A.K.; Bose, P.; Jaiswal, M.K.; Rajkhowa, S.; Singh, A.S.; Hotha, S.; Mishra, N.; Tiwari, V.K. Cu(I)-Catalyzed Click Chemistry in Glycoscience and Their Diverse Applications. *Chem. Rev.* **2021**, *121*, 7638–7956. [CrossRef]
144. Badria, A. Click Chemistry: A Promising Tool for Building Hierarchical Structures. *Polymers* **2022**, *14*, 4077. [CrossRef] [PubMed]
145. Chen, Y.; Yu, L.; Zhang, B.; Feng, W.; Xu, M.; Gao, L.; Liu, N.; Wang, Q.; Huang, X.; Li, P.; et al. Design and Synthesis of Biocompatible, Hemocompatible, and Highly Selective Antimicrobial Cationic Peptidopolysaccharides via Click Chemistry. *Biomacromolecules* **2019**, *20*, 2230–2240. [CrossRef]
146. Rathinam, S.; Hjálmarsdóttir, M.Á.; Thygesen, M.B.; Másson, M. Chitotriazolan (Poly(β(1-4)-2-(1H-1,2,3-Triazol-1-Yl)-2-Deoxy-d-Glucose)) Derivatives: Synthesis, Characterization, and Evaluation of Antibacterial Activity. *Carbohydr. Polym.* **2021**, *267*, 118162. [CrossRef]

147. Chen, Y.; Wang, F.; Yun, D.; Guo, Y.; Ye, Y.; Wang, Y.; Tan, H. Preparation of a C_6 Quaternary Ammonium Chitosan Derivative through a Chitosan Schiff Base with Click Chemistry. *J. Appl. Polym. Sci.* **2013**, *129*, 3185–3191. [CrossRef]
148. Tirino, P.; Laurino, R.; Maglio, G.; Malinconico, M.; d'Ayala, G.G.; Laurienzo, P. Synthesis of Chitosan–PEO Hydrogels via Mesylation and Regioselective Cu(I)-Catalyzed Cycloaddition. *Carbohydr. Polym.* **2014**, *112*, 736–745. [CrossRef] [PubMed]
149. Koshiji, K.; Nonaka, Y.; Iwamura, M.; Dai, F.; Matsuoka, R.; Hasegawa, T. C6-Modifications on Chitosan to Develop Chitosan-Based Glycopolymers and Their Lectin-Affinities with Sigmoidal Binding Profiles. *Carbohydr. Polym.* **2016**, *137*, 277–286. [CrossRef] [PubMed]
150. Barbosa, M.; Vale, N.; Costa, F.M.T.A.; Martins, M.C.L.; Gomes, P. Tethering Antimicrobial Peptides onto Chitosan: Optimization of Azide-Alkyne "Click" Reaction Conditions. *Carbohydr. Polym.* **2017**, *165*, 384–393. [CrossRef]
151. Qin, Y.; Liu, S.; Xing, R.; Li, K.; Yu, H.; Li, P. Synthesis and Antifungal Evaluation of (1,2,3-Triazol-4-Yl)Methyl Nicotinate Chitosan. *Int. J. Biol. Macromol.* **2013**, *61*, 58–62. [CrossRef]
152. Tan, W.; Zhang, J.; Zhao, X.; Dong, F.; Li, Q.; Guo, Z. Synthesis and Antioxidant Action of Chitosan Derivatives with Amino-Containing Groups via Azide-Alkyne Click Reaction and N-Methylation. *Carbohydr. Polym.* **2018**, *199*, 583–592. [CrossRef]
153. Lunkov, A.; Shagdarova, B.; Lyalina, T.; Dubinnyi, M.A.; Karpova, N.; Lopatin, S.; Il'ina, A.; Varlamov, V. Simple Method for Ultrasound Assisted «click» Modification of Azido-Chitosan Derivatives by CuAAC. *Carbohydr. Polym.* **2022**, *282*, 119109. [CrossRef]
154. Kritchenkov, A.S.; Egorov, A.R.; Dysin, A.P.; Volkova, O.V.; Zabodalova, L.A.; Suchkova, E.P.; Kurliuk, A.V.; Shakola, T.V. Ultrasound-Assisted Cu(I)-Catalyzed Azide-Alkyne Click Cycloaddition as Polymer-Analogous Transformation in Chitosan Chemistry. High Antibacterial and Transfection Activity of Novel Triazol Betaine Chitosan Derivatives and Their Nanoparti-cles. *Int. J. Biol. Macromol.* **2019**, *137*, 592–603. [CrossRef]
155. Sahariah, P.; Másson, M. Efficient Synthesis of Chitosan Derivatives as Clickable Tools. *Eur. Polym. J.* **2022**, *166*, 111039. [CrossRef]
156. Nornberg, A.B.; De Aquino, T.F.B.; Martins, C.C.; Luchese, C.; Wilhelm, E.A.; Jacob, R.G.; Hartwig, D.; Fajardo, A.R. Organoselenium-Chitosan Derivative: Synthesis via "Click" Reaction, Characterization and Antioxidant Activity. *Int. J. Biol. Macromol.* **2021**, *191*, 19–26. [CrossRef]
157. Manisha D., P.; Chawla, R.; Dutta, P.K. "Click" Synthesized Non-Substituted Triazole Modified Chitosan from CaC2 as a Novel Antibacterial and Antioxidant Polymer. *J. Polym. Res.* **2022**, *29*, 179. [CrossRef]
158. Guaresti, O.; García–Astrain, C.; Palomares, T.; Alonso–Varona, A.; Eceiza, A.; Gabilondo, N. Synthesis and Characterization of a Biocompatible Chitosan–Based Hydrogel Cross–Linked via 'Click' Chemistry for Controlled Drug Release. *Int. J. Biol. Macromol.* **2017**, *102*, 1–9. [CrossRef] [PubMed]
159. Ruiz-Pardo, C.; Silva-Gutiérrez, L.; Lizardi-Mendoza, J.; López-Franco, Y.; Peniche-Covas, C.; Argüelles-Monal, W. Chitosan Hydrogels Based on the Diels–Alder Click Reaction: Rheological and Kinetic Study. *Polymers* **2022**, *14*, 1202. [CrossRef] [PubMed]
160. Azmy, E.A.M.; Hashem, H.E.; Mohamed, E.A.; Negm, N.A. Synthesis, Characterization, Swelling and Antimicrobial Efficacies of Chemically Modified Chitosan Biopolymer. *J. Mol. Liq.* **2019**, *284*, 748–754. [CrossRef]
161. Muzzarelli, R.; El Mehtedi, M.; Bottegoni, C.; Aquili, A.; Gigante, A. Genipin-Crosslinked Chitosan Gels and Scaffolds for Tissue Engineering and Regeneration of Cartilage and Bone. *Mar. Drugs* **2015**, *13*, 7314–7338. [CrossRef]
162. Kandile, N.G.; Mohamed, H.M. New Chitosan Derivatives Inspired on Heterocyclic Anhydride of Potential Bioactive for Medical Applications. *Int. J. Biol. Macromol.* **2021**, *182*, 1543–1553. [CrossRef]
163. Kandile, N.G.; Mohamed, M.I.; Zaky, H.T.; Nasr, A.S.; Ali, Y.G. Quinoline Anhydride Derivatives Cross-Linked Chitosan Hydrogels for Potential Use in Biomedical and Metal Ions Adsorption. *Polym. Bull.* **2022**, *79*, 2461–2486. [CrossRef]
164. Rabea, E.I.; Badawy, M.E.-T.; Stevens, C.V.; Smagghe, G.; Steurbaut, W. Chitosan as Antimicrobial Agent: Applications and Mode of Action. *Biomacromolecules* **2003**, *4*, 1457–1465. [CrossRef]
165. Zia, K.M.; Barikani, M.; Zuber, M.; Bhatti, I.A.; Sheikh, M.A. Molecular engineering of chitin based polyurethane elastomers. *Carbohy. Polym.* **2008**, *74*, 149–158. [CrossRef]
166. Sudarshan, N.R.; Hoover, D.G.; Knorr, D. Antibacterial action of chitosan. *Food Biotechnol.* **1992**, *6*, 257–272. [CrossRef]
167. Perinelli, D.R.; Fagioli, L.; Campana, R.; Lam, J.K.W.; Baffone, W.; Palmieri, G.F.; Casettari, L.; Bonacucina, G. Chitosan-Based Nanosystems and Their Exploited Antimicrobial Activity. *Eur. J. Pharm. Sci.* **2018**, *117*, 8–20. [CrossRef] [PubMed]
168. Yilmaz Atay, H. Antibacterial Activity of Chitosan-Based Systems. In *Functional Chitosan*; Jana, S., Jana, S., Eds.; Springer: Singapore, 2019; pp. 457–489. ISBN 9789811502620.
169. Kong, M.; Chen, X.G.; Xing, K.; Park, H.J. Antimicrobial Properties of Chitosan and Mode of Action: A State of the Art Review. *Int. J. Food Microbiol.* **2010**, *144*, 51–63. [CrossRef] [PubMed]
170. Jeon, Y. Antimicrobial Effect of Chitooligosaccharides Produced by Bioreactor. *Carbohydr. Polym.* **2001**, *44*, 71–76. [CrossRef]
171. Xing, K.; Chen, X.G.; Liu, C.S.; Cha, D.S.; Park, H.J. Oleoyl-Chitosan Nanoparticles Inhibits Escherichia Coli and Staphylococcus Aureus by Damaging the Cell Membrane and Putative Binding to Extracellular or Intracellular Targets. *Int. J. Food Microbiol.* **2009**, *132*, 127–133. [CrossRef] [PubMed]
172. Clifton, L.A.; Skoda, M.W.A.; Le Brun, A.P.; Ciesielski, F.; Kuzmenko, I.; Holt, S.A.; Lakey, J.H. Effect of Divalent Cation Removal on the Structure of Gram-Negative Bacterial Outer Membrane Models. *Langmuir* **2015**, *31*, 404–412. [CrossRef]
173. Hosseinnejad, M.; Jafari, S.M. Evaluation of Different Factors Affecting Antimicrobial Properties of Chitosan. *Int. J. Biol. Macromol.* **2016**, *85*, 467–475. [CrossRef]

174. Brown, S.; Santa Maria, J.P.; Walker, S. Wall Teichoic Acids of Gram-Positive Bacteria. *Annu. Rev. Microbiol.* **2013**, *67*, 313–336. [CrossRef] [PubMed]
175. Pasquina-Lemonche, L.; Burns, J.; Turner, R.D.; Kumar, S.; Tank, R.; Mullin, N.; Wilson, J.S.; Chakrabarti, B.; Bullough, P.A.; Foster, S.J.; et al. The Architecture of the Gram-Positive Bacterial Cell Wall. *Nature* **2020**, *582*, 294–297. [CrossRef]
176. Matica, M.A.; Aachmann, F.L.; Tøndervik, A.; Sletta, H.; Ostafe, V. Chitosan as a Wound Dressing Starting Material: Antimicrobial Properties and Mode of Action. *Int. J. Mol. Sci.* **2019**, *20*, 5889. [CrossRef]
177. Je, J.-Y.; Kim, S.-K. Chitosan Derivatives Killed Bacteria by Disrupting the Outer and Inner Membrane. *J. Agric. Food Chem.* **2006**, *54*, 6629–6633. [CrossRef]
178. Feng, P.; Luo, Y.; Ke, C.; Qiu, H.; Wang, W.; Zhu, Y.; Hou, R.; Xu, L.; Wu, S. Chitosan-Based Functional Materials for Skin Wound Repair: Mechanisms and Applications. *Front. Bioeng. Biotechnol.* **2021**, *9*, 650598. [CrossRef] [PubMed]
179. Gan, L.; Chen, S.; Jensen, G.J. Molecular Organization of Gram-Negative Peptidoglycan. *Proc. Natl. Acad. Sci. USA* **2008**, *105*, 18953–18957. [CrossRef]
180. Khalid, S.; Piggot, T.J.; Samsudin, F. Atomistic and Coarse Grain Simulations of the Cell Envelope of Gram-Negative Bacteria: What Have We Learned? *Acc. Chem. Res.* **2019**, *52*, 180–188. [CrossRef] [PubMed]
181. Helander, I.M.; Nurmiaho-Lassila, E.-L.; Ahvenainen, R.; Rhoades, J.; Roller, S. Chitosan Disrupts the Barrier Properties of the Outer Membrane of Gram-Negative Bacteria. *Int. J. Food Microbiol.* **2001**, *71*, 235–244. [CrossRef] [PubMed]
182. Chen, C.-P.; Hsieh, C.-M.; Tsai, T.; Yang, J.-C.; Chen, C.-T. Optimization and Evaluation of a Chitosan/Hydroxypropyl Methylcellulose Hydrogel Containing Toluidine Blue O for Antimicrobial Photodynamic Inactivation. *Int. J. Mol. Sci.* **2015**, *16*, 20859–20872. [CrossRef]
183. Xia, W.; Wei, X.-Y.; Xie, Y.-Y.; Zhou, T. A Novel Chitosan Oligosaccharide Derivative: Synthesis, Antioxidant and Antibacterial Properties. *Carbohydr. Polym.* **2022**, *291*, 119608. [CrossRef]
184. Arjunan, N.; Kumari, H.L.J.; Singaravelu, C.M.; Kandasamy, R.; Kandasamy, J. Physicochemical Investigations of Biogenic Chitosan-Silver Nanocomposite as Antimicrobial and Anticancer Agent. *Int. J. Biol. Macromol.* **2016**, *92*, 77–87. [CrossRef]
185. Rahayu, D.P.; De Mori, A.; Yusuf, R.; Draheim, R.; Lalatsa, A.; Roldo, M. Enhancing the Antibacterial Effect of Chitosan to Combat Orthopaedic Implant-Associated Infections. *Carbohydr. Polym.* **2022**, *289*, 119385. [CrossRef]
186. Tamer, T.M.; Zhou, H.; Hassan, M.A.; Abu-Serie, M.M.; Shityakov, S.; Elbayomi, S.M.; Mohy-Eldin, M.S.; Zhang, Y.; Cheang, T. Synthesis and Physicochemical Properties of an Aromatic Chitosan Derivative: In Vitro Antibacterial, Antioxidant, and Anticancer Evaluations, and in Silico Studies. *Int. J. Biol. Macromol.* **2023**, *240*, 124339. [CrossRef]
187. Hassan, M.A.; Tamer, T.M.; Omer, A.M.; Baset, W.M.A.; Abbas, E.; Mohy-Eldin, M.S. Therapeutic Potential of Two Formulated Novel Chitosan Derivatives with Prominent Antimicrobial Activities against Virulent Microorganisms and Safe Profiles toward Fibroblast Cells. *Int. J. Pharm.* **2023**, *634*, 122649. [CrossRef]
188. Zaki, N.M.; Hafez, M.M. Enhanced Antibacterial Effect of Ceftriaxone Sodium-Loaded Chitosan Nanoparticles Against Intracellular Salmonella Typhimurium. *AAPS PharmSciTech* **2012**, *13*, 411–421. [CrossRef] [PubMed]
189. Jamil, B.; Habib, H.; Abbasi, S.; Nasir, H.; Rahman, A.; Rehman, A.; Bokhari, H.; Imran, M. Cefazolin Loaded Chitosan Nanoparticles to Cure Multi Drug Resistant Gram-Negative Pathogens. *Carbohydr. Polym.* **2016**, *136*, 682–691. [CrossRef]
190. Maya, S.; Indulekha, S.; Sukhithasri, V.; Smitha, K.T.; Nair, S.V.; Jayakumar, R.; Biswas, R. Efficacy of Tetracycline Encapsulated O-Carboxymethyl Chitosan Nanoparticles against Intracellular Infections of Staphylococcus Aureus. *Int. J. Biol. Macromol.* **2012**, *51*, 392–399. [CrossRef]
191. Mawad, A.; Helmy, Y.A.; Shalkami, A.-G.; Kathayat, D.; Rajashekara, G.E. Coli Nissle Microencapsulation in Alginate-Chitosan Nanoparticles and Its Effect on Campylobacter Jejuni in Vitro. *Appl. Microbiol. Biotechnol.* **2018**, *102*, 10675–10690. [CrossRef] [PubMed]
192. Rhim, J.-W.; Hong, S.-I.; Park, H.-M.; Ng, P.K.W. Preparation and Characterization of Chitosan-Based Nanocomposite Films with Antimicrobial Activity. *J. Agric. Food Chem.* **2006**, *54*, 5814–5822. [CrossRef]
193. Kumar-Krishnan, S.; Prokhorov, E.; Hernández-Iturriaga, M.; Mota-Morales, J.D.; Vázquez-Lepe, M.; Kovalenko, Y.; Sanchez, I.C.; Luna-Bárcenas, G. Chitosan/Silver Nanocomposites: Synergistic Antibacterial Action of Silver Nanoparticles and Silver Ions. *Eur. Polym. J.* **2015**, *67*, 242–251. [CrossRef]
194. Mu, H.; Liu, Q.; Niu, H.; Sun, Y.; Duan, J. Gold Nanoparticles Make Chitosan–Streptomycin Conjugates Effective towards Gram-Negative Bacterial Biofilm. *RSC Adv.* **2016**, *6*, 8714–8721. [CrossRef]
195. P, M.R.; Muraleedaran, K.; Mujeeb, V.M.A. Applications of Chitosan Powder with in Situ Synthesized Nano ZnO Particles as an Antimicrobial Agent. *Int. J. Biol. Macromol.* **2015**, *77*, 266–272. [CrossRef] [PubMed]
196. Hanpanich, O.; Wongkongkatep, P.; Pongtharangkul, T.; Wongkongkatep, J. Turning Hydrophilic Bacteria into Biorenewable Hydrophobic Material with Potential Antimicrobial Activity via Interaction with Chitosan. *Bioresour. Technol.* **2017**, *230*, 97–102. [CrossRef]
197. Rivera Aguayo, P.; Bruna Larenas, T.; Alarcón Godoy, C.; Cayupe Rivas, B.; González-Casanova, J.; Rojas-Gómez, D.; Caro Fuentes, N. Antimicrobial and Antibiofilm Capacity of Chitosan Nanoparticles against Wild Type Strain of *Pseudomonas* sp. Isolated from Milk of Cows Diagnosed with Bovine Mastitis. *Antibiotics* **2020**, *9*, 551. [CrossRef]
198. Muñoz-Nuñez, C.; Cuervo-Rodríguez, R.; Echeverría, C.; Fernández-García, M.; Muñoz-Bonilla, A. Synthesis and Characterization of Thiazolium Chitosan Derivative with Enhanced Antimicrobial Properties and Its Use as Component of Chitosan Based Films. *Carbohydr. Polym.* **2023**, *302*, 120438. [CrossRef]

199. Zou, W.; Gu, J.; Li, J.; Wang, Y.; Chen, S. Tailorable Antibacterial and Cytotoxic Chitosan Derivatives by Introducing Quaternary Ammonium Salt and Sulfobetaine. *Int. J. Biol. Macromol.* **2022**, *218*, 992–1001. [CrossRef]
200. Alenzi, A.M.; Albalawi, S.A.; Alghamdi, S.G.; Albalawi, R.F.; Albalawi, H.S.; Qushawy, M. Review on Different Vesicular Drug Delivery Systems (VDDSs) and TheirApplications. *Recent Pat. Nanotechnol.* **2023**, *17*, 18–32. [CrossRef]
201. Pacheco, C.; Baião, A.; Ding, T.; Cui, W.; Sarmento, B. Recent Advances in Long-Acting Drug Delivery Systems for Anticancer Drug. *Adv. Drug Deliv. Rev.* **2023**, *194*, 114724. [CrossRef] [PubMed]
202. Lukova, P.; Katsarov, P. Contemporary Aspects of Designing Marine Polysaccharide Microparticles as Drug Carriers for Biomedical Application. *Pharmaceutics* **2023**, *15*, 2126. [CrossRef] [PubMed]
203. Lee, J.H.; Yeo, Y. Controlled drug release from pharmaceutical nanocarriers. *Chem. Eng. Sci.* **2015**, *125*, 75–84. [CrossRef]
204. Desai, N.; Rana, D.; Salave, S.; Gupta, R.; Patel, P.; Karunakaran, B.; Sharma, A.; Giri, J.; Benival, D.; Kommineni, N. Chitosan: A Potential Biopolymer in Drug Delivery and Biomedical Applications. *Pharmaceutics* **2023**, *15*, 1313. [CrossRef] [PubMed]
205. Jin, H.; Zhao, Z.; Lan, Q.; Zhou, H.; Mai, Z.; Wang, Y.; Ding, X.; Zhang, W.; Pi, J.; Evans, C.E.; et al. Nasal Delivery of Hesperidin/Chitosan Nanoparticles Suppresses Cytokine Storm Syndrome in a Mouse Model of Acute Lung Injury. *Front. Pharmacol.* **2021**, *11*, 592238. [CrossRef]
206. Abdallah, H.M.; Abu Elella, M.H.; Abdel-Aziz, M.M. One-Pot Green Synthesis of Chitosan Biguanidine Nanoparticles for Targeting *M. tuberculosis*. *Int. J. Biol. Macromol.* **2023**, *232*, 123394. [CrossRef]
207. Petkar, K.C.; Chavhan, S.; Kunda, N.; Saleem, I.; Somavarapu, S.; Taylor, K.M.G.; Sawant, K.K. Development of Novel Octanoyl Chitosan Nanoparticles for Improved Rifampicin Pulmonary Delivery: Optimization by Factorial Design. *AAPS PharmSciTech* **2018**, *19*, 1758–1772. [CrossRef]
208. Niaz, T.; Sarkar, A.; Mackie, A.; Imran, M. Impact of Albumin Corona on Mucoadhesion and Antimicrobial Activity of Carvacrol Loaded Chitosan Nano-Delivery Systems under Simulated Gastro-Intestinal Conditions. *Int. J. Biol. Macromol.* **2021**, *169*, 171–182. [CrossRef]
209. Yildiz-Peköz, A.; Akbal, O.; Tekarslan, S.H.; Sagirli, A.O.; Mulazimoglu, L.; Morina, D.; Cevher, E. Preparation and Characterization of Doripenem-Loaded Microparticles for Pulmonary Delivery. *J. Aerosol Med. Pulm. Drug Deliv.* **2018**, *31*, 347–357. [CrossRef]
210. Hemmingsen, L.M.; Julin, K.; Ahsan, L.; Basnet, P.; Johannessen, M.; Škalko-Basnet, N. Chitosomes-In-Chitosan Hydrogel for Acute Skin Injuries: Prevention and Infection Control. *Mar. Drugs* **2021**, *19*, 269. [CrossRef] [PubMed]
211. Abilova, G.K.; Kaldybekov, D.B.; Irmukhametova, G.S.; Kazybayeva, D.S.; Iskakbayeva, Z.A.; Kudaibergenov, S.E.; Khutoryanskiy, V.V. Chitosan/Poly(2-Ethyl-2-Oxazoline) Films with Ciprofloxacin for Application in Vaginal Drug Delivery. *Materials* **2020**, *13*, 1709. [CrossRef] [PubMed]
212. Priya Dharshini, K.; Fang, H.; Ramya Devi, D.; Yang, J.-X.; Luo, R.-H.; Zheng, Y.-T.; Brzeziński, M.; Vedha Hari, B.N. pH-Sensitive Chitosan Nanoparticles Loaded with Dolutegravir as Milk and Food Admixture for Paediatric Anti-HIV Therapy. *Carbohydr. Polym.* **2021**, *256*, 117440. [CrossRef] [PubMed]
213. Gondil, V.S.; Dube, T.; Panda, J.J.; Yennamalli, R.M.; Harjai, K.; Chhibber, S. Comprehensive Evaluation of Chitosan Nanoparticle Based Phage Lysin Delivery System; A Novel Approach to Counter *S. pneumoniae* Infections. *Int. J. Pharm.* **2020**, *573*, 118850. [CrossRef]
214. Zareshahrabadi, Z.; Khorram, M.; Pakshir, K.; Tamaddon, A.-M.; Jafari, M.; Nouraei, H.; Ardekani, N.T.; Amirzadeh, N.; Irajie, C.; Barzegar, A.; et al. Magnetic Chitosan Nanoparticles Loaded with Amphotericin B: Synthesis, Properties and Potentiation of Antifungal Activity against Common Human Pathogenic Fungal Strains. *Int. J. Biol. Macromol.* **2022**, *222*, 1619–1631. [CrossRef]
215. Pramanik, A.; Jones, S.; Gao, Y.; Sweet, C.; Begum, S.; Shukla, M.K.; Buchanan, J.P.; Moser, R.D.; Ray, P.C. A Bio-Conjugated Chitosan Wrapped CNT Based 3D Nanoporous Architecture for Separation and Inactivation of Rotavirus and Shigella Waterborne Pathogens. *J. Mater. Chem. B* **2017**, *5*, 9522–9531. [CrossRef]
216. Liu, L.; Ma, Q.; Wang, S.; Gao, Y.; Zhu, C.; Zhao, W.; Sun, W.; Ma, H.; Sun, Y. Efficient Epidermal Delivery of Antibiotics by Self-Assembled Lecithin/Chitosan Nanoparticles for Enhanced Therapy on Epidermal Bacterial Infections. *Int. J. Biol. Macromol.* **2022**, *218*, 568–579. [CrossRef]
217. Silva, M.; Calado, R.; Marto, J.; Bettencourt, A.; Almeida, A.; Gonçalves, L. Chitosan Nanoparticles as a Mucoadhesive Drug Delivery System for Ocular Administration. *Mar. Drugs* **2017**, *15*, 370. [CrossRef]
218. Rashki, S.; Safardoust-Hojaghan, H.; Mirzaei, H.; Abdulsahib, W.K.; Mahdi, M.A.; Salavati-Niasari, M.; Khaledi, A.; Khorshidi, A.; Mousavi, S.G.A. Delivery LL37 by Chitosan Nanoparticles for Enhanced Antibacterial and Antibiofilm Efficacy. *Carbohydr. Polym.* **2022**, *291*, 119634. [CrossRef] [PubMed]
219. Mosafer, J.; Sabbaghi, A.-H.; Badiee, A.; Dehghan, S.; Tafaghodi, M. Preparation, Characterization and In Vivo Evaluation of Alginate-Coated Chitosan and Trimethylchitosan Nanoparticles Loaded with PR8 Influenza Virus for Nasal Immunization. *Asian J. Pharm. Sci.* **2019**, *14*, 216–221. [CrossRef]
220. Jearanaiwitayakul, T.; Sunintaboon, P.; Chawengkittikul, R.; Limthongkul, J.; Midoeng, P.; Chaisuwirat, P.; Warit, S.; Ubol, S. Whole Inactivated Dengue Virus-Loaded Trimethyl Chitosan Nanoparticle-Based Vaccine: Immunogenic Properties in Ex Vivo and in Vivo Models. *Hum. Vaccines Immunother.* **2021**, *17*, 2793–2807. [CrossRef]
221. Acevedo-Villanueva, K.Y.; Lester, B.; Renu, S.; Han, Y.; Shanmugasundaram, R.; Gourapura, R.; Selvaraj, R. Efficacy of Chitosan-Based Nanoparticle Vaccine Administered to Broiler Birds Challenged with Salmonella. *PLoS ONE* **2020**, *15*, e0231998. [CrossRef] [PubMed]

222. Helmy, Y.A.; Closs, G.; Jung, K.; Kathayat, D.; Vlasova, A.; Rajashekara, G. Effect of Probiotic *E. coli* Nissle 1917 Supplementation on the Growth Performance, Immune Responses, Intestinal Morphology, and Gut Microbes of Campylobacter Jejuni Infected Chickens. *Infect. Immun.* **2022**, *90*, e00337-22. [CrossRef]
223. Bento, D.; Jesus, S.; Lebre, F.; Gonçalves, T.; Borges, O. Chitosan Plus Compound 48/80: Formulation and Preliminary Evaluation as a Hepatitis B Vaccine Adjuvant. *Pharmaceutics* **2019**, *11*, 72. [CrossRef] [PubMed]
224. Zhao, K.; Li, S.; Li, W.; Yu, L.; Duan, X.; Han, J.; Wang, X.; Jin, Z. Quaternized Chitosan Nanoparticles Loaded with the Combined Attenuated Live Vaccine against Newcastle Disease and Infectious Bronchitis Elicit Immune Response in Chicken after Intranasal Administration. *Drug Deliv.* **2017**, *24*, 1574–1586. [CrossRef]
225. Hajam, I.A.; Senevirathne, A.; Hewawaduge, C.; Kim, J.; Lee, J.H. Intranasally Administered Protein Coated Chitosan Nanoparticles Encapsulating Influenza H9N2 HA2 and M2e mRNA Molecules Elicit Protective Immunity against Avian Influenza Viruses in Chickens. *Vet. Res.* **2020**, *51*, 37. [CrossRef]
226. Helmi, O.; Elshishiny, F.; Mamdouh, W. Targeted Doxorubicin Delivery and Release within Breast Cancer Environment Using PEGylated Chitosan Nanoparticles Labeled with Monoclonal Antibodies. *Int. J. Biol. Macromol.* **2021**, *184*, 325–338. [CrossRef]
227. Naseer, F.; Ahmad, T.; Kousar, K.; Kakar, S.; Gul, R.; Anjum, S. Formulation of Surface-Functionalized Hyaluronic Acid-Coated Thiolated Chitosan Nano-Formulation for the Delivery of Vincristine in Prostate Cancer: A Multifunctional Targeted Drug Delivery Approach. *J. Drug Deliv. Sci. Technol.* **2022**, *74*, 103545. [CrossRef]
228. Hu, H.; Qi, Q.; Dong, Z.; Yu, X.; Mo, Y.; Luo, J.; Wang, Y.; Du, S.; Lu, Y. Albumin Coated Trimethyl Chitosan-Based Targeting Delivery Platform for Photothermal/Chemo-Synergistic Cancer Therapy. *Carbohydr. Polym.* **2020**, *241*, 116335. [CrossRef]
229. El-Dakroury, W.A.; Zewail, M.B.; Amin, M.M. Design, Optimization, and in-Vivo Performance of Glipizide-Loaded O-Carboxymethyl Chitosan Nanoparticles in Insulin Resistant/Type 2 Diabetic Rat Model. *J. Drug Deliv. Sci. Technol.* **2023**, *79*, 104040. [CrossRef]
230. Wilson, B.; Mohamed Alobaid, B.N.; Geetha, K.M.; Jenita, J.L. Chitosan Nanoparticles to Enhance Nasal Absorption and Brain Targeting of Sitagliptin to Treat Alzheimer's Disease. *J. Drug Deliv. Sci. Technol.* **2021**, *61*, 102176. [CrossRef]
231. Li, B.; Wang, J.; Gui, Q.; Yang, H. Drug-Loaded Chitosan Film Prepared via Facile Solution Casting and Air-Drying of Plain Water-Based Chitosan Solution for Ocular Drug Delivery. *Bioact. Mater.* **2020**, *5*, 577–583. [CrossRef]
232. Akhlaq, M.; Azad, A.K.; Fuloria, S.; Meenakshi, D.U.; Raza, S.; Safdar, M.; Nawaz, A.; Subramaniyan, V.; Sekar, M.; Sathasivam, K.V.; et al. Fabrication of Tizanidine Loaded Patches Using Flaxseed Oil and Coriander Oil as a Penetration Enhancer for Transdermal Delivery. *Polymers* **2021**, *13*, 4217. [CrossRef] [PubMed]
233. Wei, H.; Liu, S.; Tong, Z.; Chen, T.; Yang, M.; Guo, Y.; Sun, H.; Wu, Y.; Chu, Y.; Fan, L. Hydrogel-Based Microneedles of Chitosan Derivatives for Drug Delivery. *React. Funct. Polym.* **2022**, *172*, 105200. [CrossRef]
234. Pourjavadi, A.; Bagherifard, M.; Doroudian, M. Synthesis of Micelles Based on Chitosan Functionalized with Gold Nanorods as a Light Sensitive Drug Delivery Vehicle. *Int. J. Biol. Macromol.* **2020**, *149*, 809–818. [CrossRef]
235. Chen, X.; Gu, J.; Sun, L.; Li, W.; Guo, L.; Gu, Z.; Wang, L.; Zhang, Y.; Zhang, W.; Han, B.; et al. Efficient Drug Delivery and Anticancer Effect of Micelles Based on Vitamin E Succinate and Chitosan Derivatives. *Bioact. Mater.* **2021**, *6*, 3025–3035. [CrossRef] [PubMed]
236. Gu, H.; Chen, P.; Liu, X.; Lian, Y.; Xi, J.; Li, J.; Song, J.; Li, X. Trimethylated Chitosan-Coated Flexible Liposomes with Resveratrol for Topical Drug Delivery to Reduce Blue-Light-Induced Retinal Damage. *Int. J. Biol. Macromol.* **2023**, *252*, 126480. [CrossRef]
237. Khodarahmi, M.; Abbasi, H.; Kouchak, M.; Mahdavinia, M.; Handali, S.; Rahbar, N. Nanoencapsulation of Aptamer-Functionalized 5-Fluorouracil Liposomes Using Alginate/Chitosan Complex as a Novel Targeting Strategy for Colon-Specific Drug Delivery. *J. Drug Deliv. Sci. Technol.* **2022**, *71*, 103299. [CrossRef]
238. Ding, Y.; Yang, R.; Yu, W.; Hu, C.; Zhang, Z.; Liu, D.; An, Y.; Wang, X.; He, C.; Liu, P.; et al. Chitosan Oligosaccharide Decorated Liposomes Combined with TH302 for Photodynamic Therapy in Triple Negative Breast Cancer. *J. Nanobiotechnol.* **2021**, *19*, 147. [CrossRef] [PubMed]
239. Jafernik, K.; Ładniak, A.; Blicharska, E.; Czarnek, K.; Ekiert, H.; Wiącek, A.E.; Szopa, A. Chitosan-Based Nanoparticles as Effective Drug Delivery Systems-A review. *Molecules* **2023**, *28*, 1963. [CrossRef] [PubMed]
240. Aydin, R.S.T.; Pulat, M. 5-Fluorouracil encapsulated chitosan nanoparticles for pH-stimulated drug delivery: Evaluation of controlled release kinetics. *J. Nanomater.* **2012**, *2012*, 42. [CrossRef]
241. Lim, E.K.; Huh, Y.M.; Yang, J.; Lee, K.; Suh, J.S.; Haam, S. pH-triggered drug-releasing magnetic nanoparticles for cancer therapy guided by molecular imaging by MRI. *Adv. Mater.* **2011**, *23*, 2436–2442. [CrossRef] [PubMed]
242. Hu, X.; Wu, T.; Bao, Y.; Zhang, Z. Nanotechnology based therapeutic modality to boost anti-tumor immunity and collapse tumor defense. *J. Control. Release* **2017**, *256*, 26–45. [CrossRef]
243. Liu, H.; Liu, W.; Tan, Z.; Zeng, Z.; Yang, H.; Luo, S.; Wang, L.; Xi, T.; Xing, Y. Promoting Immune Efficacy of the Oral Helicobacter pylori Vaccine by HP55/PBCA Nanoparticles against the Gastrointestinal Environment. *Mol. Pharm.* **2018**, *15*, 3177–3186. [CrossRef]
244. Bhavsar, C.; Momin, M.; Gharat, S.; Omri, A. Functionalized and graft copolymers of chitosan and its pharmaceutical applications. *Expert Opin. Drug Deliv.* **2017**, *14*, 1189–1204. [CrossRef]
245. Gierszewska, M.; Ostrowska-Czubenko, J.; Chrzanowska, E. pH-responsive chitosan/alginate polyelectrolyte complex membranes reinforced by tripolyphosphate. *Eur. Polym. J.* **2018**, *101*, 282–290. [CrossRef]

246. Stasińska-Jakubas, M.; Hawrylak-Nowak, B. Protective, Biostimulating, and Eliciting Effects of Chitosan and Its Derivatives on Crop Plants. *Molecules* **2022**, *27*, 2801. [CrossRef]
247. Riseh, R.S.; Hassanisaadi, M.; Vatankhah, M.; Babaki, S.A.; Barka, E.A. Chitosan as a Potential Natural Compound to Manage Plant Diseases. *Int. J. Biol. Macromol.* **2022**, *220*, 998–1009. [CrossRef]
248. García-Carrasco, M.; Valdez-Baro, O.; Cabanillas-Bojórquez, L.A.; Bernal-Millán, M.J.; Rivera-Salas, M.M.; Gutiérrez-Grijalva, E.P.; Heredia, J.B. Potential Agricultural Uses of Micro/Nano Encapsulated Chitosan: A Review. *Macromol* **2023**, *3*, 614–635. [CrossRef]
249. Doares, S.H.; Syrovets, T.; Weiler, E.W.; Ryan, C.A. Oligogalacturonides and Chitosan Activate Plant Defensive Genes through the Octadecanoid Pathway. *Proc. Natl. Acad. Sci. USA* **1995**, *92*, 4095–4098. [CrossRef] [PubMed]
250. Sun, W.; Shahrajabian, M.H.; Petropoulos, S.A.; Shahrajabian, N. Developing Sustainable Agriculture Systems in Medicinal and Aromatic Plant Production by Using Chitosan and Chitin-Based Biostimulants. *Plants* **2023**, *12*, 2469. [CrossRef] [PubMed]
251. Jogaiah, S.; Satapute, P.; De Britto, S.; Konappa, N.; Udayashankar, A.C. Exogenous Priming of Chitosan Induces Upregulation of Phytohormones and Resistance against Cucumber Powdery Mildew Disease Is Correlated with Localized Biosynthesis of Defense Enzymes. *Int. J. Biol. Macromol.* **2020**, *162*, 1825–1838. [CrossRef]
252. Hidangmayum, A.; Dwivedi, P. Effect of Chitosan Seed Priming on Mungbean Seedlings Subjected to Different Levels of Water Potential. *Acta Physiol. Plant.* **2023**, *45*, 6. [CrossRef]
253. Liu, H.; Zheng, Z.; Han, X.; Zhang, C.; Li, H.; Wu, M. Chitosan Soaking Improves Seed Germination of Platycodon Grandiflorus and Enhances Its Growth, Photosynthesis, Resistance, Yield, and Quality. *Horticulturae* **2022**, *8*, 943. [CrossRef]
254. Razavizadeh, R.; Adabavazeh, F.; Komatsu, S. Chitosan Effects on the Elevation of Essential Oils and Antioxidant Activity of Carum copticum L. Seedlings and Callus Cultures under in Vitro Salt Stress. *J. Plant Biochem. Biotechnol.* **2020**, *29*, 473–483. [CrossRef]
255. Ghule, M.R.; Ramteke, P.K.; Ramteke, S.D.; Kodre, P.S.; Langote, A.; Gaikwad, A.V.; Holkar, S.K.; Jambhekar, H. Impact of Chitosan Seed Treatment of Fenugreek for Management of Root Rot Disease Caused by Fusarium Solani under In Vitro and In Vivo Conditions. *3 Biotech* **2021**, *11*, 290. [CrossRef]
256. Nedved, E.L.; Kalatskaja, J.N.; Ovchinnikov, I.A.; Rybinskaya, E.I.; Kraskouski, A.N.; Nikalaichuk, V.V.; Hileuskaya, K.S.; Kulikouskaya, V.I.; Agabekov, V.E.; Laman, N.A. Growth Parameters and Antioxidant Activity in Cucumber Seedlings with the Application of Chitosan and Hydroxycinnamic Acids Conjugates under Salt Stress. *Appl. Biochem. Microbiol.* **2022**, *58*, 69–76. [CrossRef]
257. Xu, D.; Li, H.; Lin, L.; Liao, M.; Deng, Q.; Wang, J.; Lv, X.; Deng, H.; Liang, D.; Xia, H. Effects of Carboxymethyl Chitosan on the Growth and Nutrient Uptake in Prunus Davidiana Seedlings. *Physiol. Mol. Biol. Plants* **2020**, *26*, 661–668. [CrossRef]
258. Hoang, N.H.; Le Thanh, T.; Sangpueak, R.; Treekoon, J.; Saengchan, C.; Thepbandit, W.; Papathoti, N.K.; Kamkaew, A.; Buensanteai, N. Chitosan Nanoparticles-Based Ionic Gelation Method: A Promising Candidate for Plant Disease Management. *Polymers* **2022**, *14*, 662. [CrossRef] [PubMed]
259. Narasimhamurthy, K.; Udayashankar, A.C.; De Britto, S.; Lavanya, S.N.; Abdelrahman, M.; Soumya, K.; Shetty, H.S.; Srinivas, C.; Jogaiah, S. Chitosan and Chitosan-Derived Nanoparticles Modulate Enhanced Immune Response in Tomato against Bacterial Wilt Disease. *Int. J. Biol. Macromol.* **2022**, *220*, 223–237. [CrossRef] [PubMed]
260. Mondéjar-López, M.; Rubio-Moraga, A.; López-Jimenez, A.J.; García Martínez, J.C.; Ahrazem, O.; Gómez-Gómez, L.; Niza, E. Chitosan Nanoparticles Loaded with Garlic Essential Oil: A New Alternative to Tebuconazole as Seed Dressing Agent. *Carbohydr. Polym.* **2022**, *277*, 118815. [CrossRef] [PubMed]
261. Gowda, S.; Sriram, S. Green Synthesis of Chitosan Silver Nanocomposites and Their Antifungal Activity against Colletotrichum Truncatum Causing Anthracnose in Chillies. *Plant Nano Biol.* **2023**, *5*, 100041. [CrossRef]
262. Sathiyabama, M.; Manikandan, A. Foliar Application of Chitosan Nanoparticle Improves Yield, Mineral Content and Boost Innate Immunity in Finger Millet Plants. *Carbohydr. Polym.* **2021**, *258*, 117691. [CrossRef]
263. Chouhan, D.; Mandal, P. Applications of Chitosan and Chitosan Based Metallic Nanoparticles in Agrosciences-A Review. *Int. J. Biol. Macromol.* **2021**, *166*, 1554–1569. [CrossRef]
264. El Amerany, F.; Meddich, A.; Wahbi, S.; Porzel, A.; Taourirte, M.; Rhazi, M.; Hause, B. Foliar Application of Chitosan Increases Tomato Growth and Influences Mycorrhization and Expression of Endochitinase-Encoding Genes. *Int. J. Mol. Sci.* **2020**, *21*, 535. [CrossRef]
265. Hao, T.; Yang, Z.; Liang, J.; Yu, J.; Liu, J. Foliar Application of Carnosine and Chitosan Improving Drought Tolerance in Bermudagrass. *Agronomy* **2023**, *13*, 442. [CrossRef]
266. Hawrylak-Nowak, B.; Dresler, S.; Rubinowska, K.; Matraszek-Gawron, R. Eliciting Effect of Foliar Application of Chitosan Lactate on the Phytochemical Properties of *Ocimum basilicum* L. and *Melissa officinalis* L. *Food Chem.* **2021**, *342*, 128358. [CrossRef] [PubMed]
267. Sheikhalipour, M.; Esmaielpour, B.; Behnamian, M.; Gohari, G.; Giglou, M.T.; Vachova, P.; Rastogi, A.; Brestic, M.; Skalicky, M. Chitosan–Selenium Nanoparticle (Cs–Se NP) Foliar Spray Alleviates Salt Stress in Bitter Melon. *Nanomaterials* **2021**, *11*, 684. [CrossRef]
268. Ji, H.; Wang, J.; Chen, F.; Fan, N.; Wang, X.; Xiao, Z.; Wang, Z. Meta-Analysis of Chitosan-Mediated Effects on Plant Defense against Oxidative Stress. *Sci. Total Environ.* **2022**, *851*, 158212. [CrossRef]
269. Zhang, Z.; Jin, F.; Wu, Z.; Jin, J.; Li, F.; Wang, Y.; Wang, Z.; Tang, S.; Wu, C.; Wang, Y. O-acylation of chitosan nanofibers by short-chain and long-chain fatty acids. *Carbohydr. Polym.* **2017**, *177*, 203–209. [CrossRef] [PubMed]

270. Wedmore, I.; McManus, J.G.; Pusateri, A.E.; Holcomb, J.B. A special report on the chitosan-based hemostatic dressing: Experience in current combat operations. *J. Trauma Acute Care Surg.* **2006**, *60*, 655–658. [CrossRef] [PubMed]
271. Malmquist, J.P.; Clemens, S.C.; Oien, H.J.; Wilson, S.L. Hemostasis of oral surgery wounds with the HemCon Dental Dressing. *J. Oral Maxillofac. Surg.* **2008**, *66*, 1177–1183. [CrossRef] [PubMed]

molecules MDPI

Review

Protective, Biostimulating, and Eliciting Effects of Chitosan and Its Derivatives on Crop Plants

Maria Stasińska-Jakubas and Barbara Hawrylak-Nowak *

Department of Botany and Plant Physiology, Faculty of Environmental Biology, University of Life Sciences in Lublin, Akademicka 15, 20-950 Lublin, Poland; maria.jakubas@up.lublin.pl
* Correspondence: barbara.nowak@up.lublin.pl

Abstract: Chitosan is a biodegradable and biocompatible polysaccharide obtained by partial deacetylation of chitin. This polymer has been gaining increasing popularity due to its natural origin, favorable physicochemical properties, and multidirectional bioactivity. In agriculture, the greatest hopes are raised by the possibility of using chitosan as a biostimulant, a plant protection product, an elicitor, or an agent to increase the storage stability of plant raw materials. The most important properties of chitosan include induction of plant defense mechanisms and regulation of metabolic processes. Additionally, it has antifungal, antibacterial, antiviral, and antioxidant activity. The effectiveness of chitosan interactions is determined by its origin, deacetylation degree and acetylation pattern, molecular weight, type of chemical modifications, pH, concentration, and solubility. There is a need to conduct research on alternative sources of chitosan, extraction methods, optimization of physicochemical properties, and commercial implementation of scientific progress outcomes in this field. Moreover, studies are necessary to assess the bioactivity and toxicity of chitosan nanoparticles and chitosan conjugates with other substances and to evaluate the consequences of the large-scale use thereof. This review presents the unique properties of chitosan and its derivatives that have the greatest importance for plant production and yield quality as well as the benefits and limitations of their application.

Keywords: chitosan; biostimulants; biotic elicitor; polycationic polymers; secondary metabolites

Citation: Stasińska-Jakubas, M.; Hawrylak-Nowak, B. Protective, Biostimulating, and Eliciting Effects of Chitosan and Its Derivatives on Crop Plants. *Molecules* **2022**, *27*, 2801. https://doi.org/10.3390/molecules27092801

Academic Editor: Agnieszka Ewa Wiącek

Received: 6 April 2022
Accepted: 26 April 2022
Published: 28 April 2022

1. Introduction

Chitosan is a biopolymer obtained from chitin, which is the second most common natural polysaccharide after cellulose [1,2]. The discovery and first research on this compound date back to the 19th century, when the links between chemistry, botany, and medicine were discerned. Chitin was probably discovered in 1799 by English scientist A. Hachett, who extracted the compound from shells of marine invertebrates and described it as "a material with particular resistance to ordinary chemicals". However, he did not conduct further research on this compound. For this reason, it is believed that chitin, originally called fungine, was first isolated from fungi by French researcher H. Braconnot and subsequently described by Swiss chemist A. Hoffman in his doctoral thesis. Almost 20 years later, the same compound was isolated from insect cuticle by A. Odier and named chitin (Greek: chitōn—tunic, coat). In turn, chitosan was discovered accidentally in 1859 by French physiologist C. Rouget during the production of natural soap. The process of boiling water-dissolved chitin with the addition of concentrated potassium hydroxide resulted in the deacetylation of chitin in the alkaline solution yielding chitosan. Intensive research on the structure and properties of chitin and chitosan was carried out in the following years. The modified form of chitin was eventually named "chitosan" by German chemist and physiologist F. Hoppe-Seiler in 1894; however, the chemical structure of this compound was determined only in the mid-20th century [3–7].

Currently, the term chitosan denotes a group of biopolymer substances obtained in the process of chemical or enzymatic deacetylation of chitin with different (but not lower than 50%) deacetylation degrees. Both chitosan and chitin are linear copolymers consisting of D-glucosamine and N-acetyl-D-glucosamine linked together by β-1,4-glycosidic bonds. The main difference between these polymers is their deacetylation degree (DD) expressed as a percentage and defined as the ratio of the number of amino groups ($–NH_2$) to the total number of acetylamino groups ($–NHCOCH_3$) present in chitin [1,4,8]. In addition, the solubility in dilute acids is a practical criterion for discriminating chitosan from chitin. The extraction of chitin begins with demineralization aimed at dissolving the calcium carbonate by acid treatment followed by deproteinization which uses, depending on the method (biological or chemical), enzymes or alkaline solutions. In order to obtain pure and colorless chitin, the last step is decolorization (Figure 1). The deacetylation process results in the removal of the acetyl groups ($–CH_3CO$) from chitin and substitution of reactive $–NH_2$ groups, thus resulting in the formation of chitosan, whose deacetylation degree depends primarily on the duration of the reaction, the temperature (90–120 °C), and the concentration of the aqueous NaOH solution (40–50%) used in the process. Although it occurs naturally in the cell walls of some fungi, bacteria, and algae as well as insect cuticles, chitosan for industrial and laboratory use is obtained from chitin derived mainly from the shells of marine invertebrates (shrimps, crabs, lobsters, and krill), which are wastes of marine food processing industry [1,8–10]. Until recently, the use of marine resources was considered one of the most effective methods for the extraction of chitin and the production of chitosan. An additional advantage was the possibility of disposal of a huge amount of waste generated by the marine food sector. However, some disadvantages and threats posed by this practice have been noticed in recent years, e.g., dependence on the fish industry, environmental hazards, and allergenicity of the final product, which depends on its purity. According to Ravindranathan et al. [11], chitosan thoroughly purified from impurity proteins and endotoxins is low-allergenic. However, the global chitin and chitosan market is developing very dynamically, and it has become important to look for alternative sources of these polymers for commercial and sustainable production. Currently, an increase in the number of studies on the production and applications of chitosan from various insect species can be observed. The primary cause of the interest is the speed of insect reproduction and the possibility of year-round breeding. Additionally, the results of many studies show that insect chitosan has even more favorable properties than that derived from shellfish. Recently, there have been attempts to produce chitosan from chitin derived from cell walls of fungi, which are the second-largest source of chitin after marine invertebrates [10,12–15], and biological methods are increasingly being used to obtain these polymers [2].

Chitosan is a compound with a number of physicochemical properties determining its suitability to be used in medicine, pharmacy, environmental protection, and agriculture as well as food, cosmetic, textile, and paper industries. These properties include nontoxicity; biodegradability; biocompatibility; hydrophilicity; film-forming properties; high sorption capacity; and high affinity for metals, proteins, and dyes. A summary of the extraction and applications of chitosan is presented in Figure 1. However, chitosan also exhibits many physicochemical properties that impede its wider use in some areas. The main limitation is its solubility, which largely depends on the isolation conditions, degree of deacetylation, type of acetyl group distribution, and molecular weight. There are many types of chitosan which are well soluble in both alkaline and neutral media, but in most cases, the best solvents for this group of polysaccharides are some acids. Additionally, chitosan swells strongly in an acidic environment and is characterized by poor mechanical and chemical strength. Therefore, chitosan is subjected to various types of physical and chemical modifications primarily aimed at improvement of its solubility and other physicochemical properties, e.g., mechanical strength, thus expanding the spectrum of its potential applications [8,13,16–19]. Three basic groups of chitosan modification techniques are used, i.e., methods leading to an increase in the molecular weight or extension of the chitosan chain; generation of chitosan derivatives through substitution reactions; and methods of physical, chemical, or enzymatic depolymerization. Chitosan derivatives obtained through modifications are an important object in research on their application in biotechnology, agro-technology, and medicine as well as food, pharmaceutical, and textile industries [20,21].

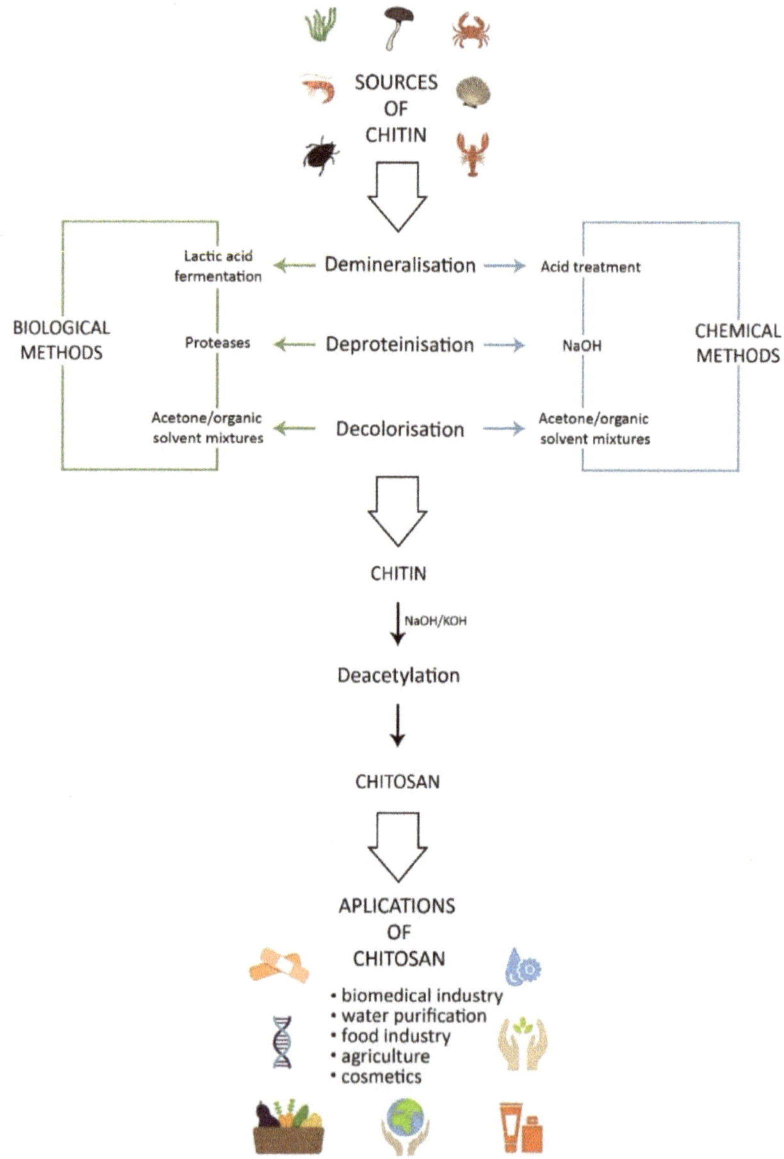

Figure 1. Extraction, preparation, and applications of chitosan and chitosan derivatives.

2. Biological Activity of Chitosan

Chitosan is a biopolymer with a wide potential for application in plant production mainly due to its biocompatibility and biodegradability as well as high biological activity. Although it is not part of the structure of plant tissues, chitosan exerts a considerable impact on plant growth and development. It initiates and modulates various types of physiological reactions, e.g., defense and immune responses in plants, and regulates metabolic processes [22]. Chitosan also exhibits strong antibacterial, antifungal, and antiviral properties. The first investigations of its antimicrobial activity were reported in 1979 by Allan and Hadwiger, whose publication contributed to the growing interest in the potential application of chitosan in the agricultural industry. However, the practical use of this compound in plant production became possible with the wider availability of industrial amounts of chitosan. The compound has been used for protecting plants against pathogens and fighting diseases affecting plants during the vegetation and postharvest period only since the 1990s [23].

The mechanism of the antimicrobial activity of chitosan has not been fully elucidated yet, but many literature reports indicate that it is most probably related to its polycationic nature. The interactions of chitosan molecules with negatively charged molecules on the surface of microbial cells lead to agglutination of their cell walls, loss of intracellular components, and cell death. Another hypothesis assumes that chitosan limits the entry of pathogens into plant cells by reduction in stomatal opening. In turn, the antifungal activity of chitosan results from its ability to bind and inhibit the synthesis of toxins and stimulate induced systemic resistance and production of many secondary fungistatic metabolites, e.g., phytoalexins, abscisic acid, methyl jasmonate, and phenolic compounds, and various enzymes such as hypha-degrading chitinase and β-glucanase. Additionally, microscopic studies have shown that chitosan induces clear morphological changes in fungal cells at various stages of development. This polymer also reinforces plant cell walls through lignification and constitutes a mechanical barrier limiting plant contact with adverse external factors [4,24–29]. Consequently, the effectiveness of chitosan action against microorganisms is associated with its bidirectional action: control of the presence of pathogens and induction of plant defense reactions [30].

The high biological activity of chitosan in the induction of plant immune response may be related to the processes taking place in plant cell walls. Acid pectins present in the cell walls bind calcium and form chain dimers at higher concentrations of the element. Cationic chitosan can interact with negatively charged pectin and pectin dimers, thus influencing their supramolecular conformation, which induces a specific alarm signal informing plant cells about the degradation of cell walls and the presence of pathogens. The plant response to chitosan–pectin dimer complexes is considerably stronger than that in the case of separately interacting components [31].

Chitosan has been shown to have antioxidant activity consisting in the neutralization of such reactive oxygen species (ROS) as superoxide anion radicals, free hydroxyl radicals, and hydrogen peroxide (H_2O_2). It is also involved in interactions of hydroxyl and amine groups with metal ions resulting in metal chelation, adsorption, and ion exchange [27,32]. However, research results have indicated that chitosan can also induce the synthesis of H_2O_2 in plant cells as a signal molecule in defense reactions to stress and increase the activity of superoxide dismutase (SOD), peroxidases (POX), and catalase (CAT), i.e., enzymes involved in the direct neutralization of ROS [33]. As reported by Hawrylak-Nowak et al. [34], spraying plants with a chitosan lactate solution increased CAT and guaiacol peroxidase (GPOX) activity in *Ocimum basilicum* and ascorbate peroxidase (APX) in *Melissa officinalis*. Similarly, foliar application of chitosan nanoparticles increased the activity of CAT, APX, and glutathione reductase (GR) in *Catharanthus roseus* exposed to salt stress [35]. In turn, Quitadamo et al. [36] reported an important role of chitosan as a regulator of antioxidant responses to salinity in *Triticum durum*. It was observed that the use of this polymer neutralized the harmful effects of stress through a reduction in the content of superoxide radicals, H_2O_2, and malondialdehyde and via enhancement of CAT activity. In turn, the addition of chitosan as an elicitor in *Psammosilene tunicoides* hair root cultures was found to induce the production of nitric oxide and increase the activity of ROS-scavenging enzymes [37]. A similar effect was observed in *Curcuma longa* [38], where chitosan contributed to an increase in the activity of POX and polyphenol oxidase (PPO). Additionally, the results of research on the induction of oxidative responses by salicylic acid, chitosan, and exogenous H_2O_2 in *Capsicum annuum* plants demonstrated that, in comparison with other substances, the foliar application of chitosan induced the lowest H_2O_2 accumulation in the leaves of this species [39]. The antioxidant activity of chitosan is also one of the criteria of its suitability to be used for enhancement of the storage stability of raw materials, which is described in detail in the next section. For example, in a study conducted by Wang and Gao [40], treatment of strawberries with chitosan induced a number of defense reactions; i.e., it increased the activity of CAT, GPOX, glutathione peroxidase (GSH-POX), dehydroascorbate reductase (DHAR), and monodehydroascorbate reductase (MDHAR).

The biological activity of chitosan and the viscosity of its solutions, and thus the effectiveness of its use, are determined by a number of different factors, e.g., the origin, type of modifications, polymerization and deacetylation degrees, pH, positive charge, concentration, solubility, and chelation capacity [26,41]. Recent studies also indicate the importance of the pattern of chitosan acetylation, which may be an important molecular carrier of information [42].

The effect of chitosan depends on the plant species, developmental stage, and physiological condition. In turn, the antimicrobial activity of this polymer depends on the type of microorganisms. However, molecular weight is the most important parameter with a large impact on the level of biological activity of chitosan [26,33]. For example, it was observed in an experiment carried out by Kulikov et al. [43] that a decrease in the molecular weight of chitosan was accompanied by an increase in its inhibitory activity against mosaic virus infection in *Phaseolus vulgaris*. Animal-derived chitosan is a high-molecular-weight polymer. Moreover, it exhibits antimicrobial activity only in an acidic environment, in which it is soluble. Therefore, it is important to develop technologies for modification and adaptation of chitosan in order to elicit adequately targeted biological activity of this compound [9,44].

3. Application of Chitosan as a Biostimulant in Cultivation of Plants

Plant production is one of the most important elements of agriculture and the economy, which require continuous progress. The introduction of various novel technologies has contributed to the rapid increase in agricultural performance, but the search for ecological solutions increasing the efficiency of plant production has currently become essential [33]. Pro-ecological activities in this area are also enforced by the latest changes in agricultural policies, such as the European Green Deal, with one of the strategies aimed at a reduction in the use of plant protection products and the promotion of organic farming [45]. Hence, the interest in the use of natural-origin substances as stimulants of plant growth and development, the so-called biostimulants, has increased in recent years. The concept of biostimulants was proposed at the end of the 20th century but has not been clearly defined to date. Moreover, there are no adequate legal regulations for the systematization of available preparations or registration of new agents. However, the term "plant biostimulant" is assumed to denote any substance or formulation that is not a plant constituent, fertilizer, or pesticide but contains natural compounds (single or mixtures) or microorganisms. It is intended to be applied onto the whole plant, a part of a plant, or the rhizosphere in order to intensify natural physiological processes, increase plant resistance to stress, enhance the utilization of minerals, and improve the size and quality of crop yields [46–48].

Given its biological activity and film-forming properties, chitosan is widely used in treatments improving the propagation material, such as coating or encapsulation. It is most often used as an independent bioactive compound or a binding substance in combination with other agents stimulating plant growth and development and ensuring protection against pathogens. It is believed that coating the propagation material with a solution of this polymer mainly ensures adequate water and gas permeability, thus stimulating germination and development of seedlings [25,49]. It is also suggested that chitosan can activate hydrolytic enzymes required for the degradation and mobilization of storage substances, such as starch and protein [50]. A study conducted by Ruan and Xue [51] showed that coating *Oryza sativa* seeds with chitosan accelerated their germination and increased the tolerance of these plants to salt stress. In turn, experiments carried out on tubers of freesia [52] showed that the application of a 0.2% chitosan solution increased the biomass and the number of progeny tubers in this species. Subsequent studies [53] demonstrated a positive effect of coating *Ornithogalum saundersiae* bulbs with a 0.5% aqueous solution of oligochitosan with 5000 or 100,000 g $\times$ mol^{-1} molecular weight on plant yields and most of the analyzed biometric and physiological indicators. It was also found that different doses of chitosan stimulated seed germination and had a positive effect on the fresh weight of cucumber roots and shoots [54]. Similar results were reported by Zeng et al. [55]; i.e., treatment with a chitosan solution stimulated germination and increased yields in soybean.

Current literature reports indicate a high effectiveness of foliar or soil application of chitosan in the stimulation of plant growth. This may be related to its stimulating effect on the uptake of water and essential minerals and its impact on the osmotic pressure in cells [41]. It was reported that a solution of this polymer sprayed on strawberry plants exerted a beneficial effect on plant growth and fruit yield [56]. This finding was confirmed in a study conducted by Rahman et al. [57], where foliar treatment of strawberry with chitosan had a positive effect not only on the growth and yield of fruits but also on their chemical composition. Furthermore, the results reported by Poterańska et al. [58] indicated that foliar application of chitosan-containing agents had a positive effect on the weight of haskap berries. In turn, *Eustoma grandiflorum* plants grown in soil with the addition of chitosan flowered much earlier and produced a larger number and greater weight of flowers [59]. Equally positive results were reported by Chookhongkha et al. [60], who additionally indicated a variable effectiveness of chitosan depending on its molecular weight. It was found that 1% high-molecular-weight chitosan applied to soil had a beneficial effect on fruit and seed yields in *Capsicum annuum*.

The literature provides information on the potential use of chitosan in the fertilizer industry as an ingredient for the production of sustained (SRF) or controlled release (CRF) fertilizers. The interest in this polymer has mainly been aroused by its beneficial effect on soil and plants and some physicochemical features. Given its film-forming properties as well as high biocompatibility and biodegradability, chitosan can be used as a coating material regulating the rate of release of minerals into the soil solution; additionally, it can be used to solve the problem with the disposal of residues of coatings produced from non-biodegradable polymers. In the agrochemical industry, chitosan nanoparticles are used both to optimize the activity and efficiency of various types of formulations and to reduce their toxicity to the environment [23,61,62]. An example of such an application was shown in an experiment carried out by Abdel-Aziz et al. [63], in which the foliar application of a nanochitosan-NPK fertilizer contributed to an increase in wheat growth and yields compared to plants treated with traditional forms of nitrogen, phosphorus, and potassium.

4. Chitosan as a Plant Protection Agent

The presence of harmful organisms poses a serious threat to agriculture and many industries worldwide. According to data published in 2019 by the Food and Agriculture Organisation of the United Nations (FAO), agrophages cause approximately 20–40% of losses of the world's crops annually [64]. The use of chemical plant protection products raises increasing concerns related to their negative impact on the environment despite their effectiveness. Given the growing ecological and consumer awareness and changes in the agricultural sector, the search for natural and safe methods for the protection of crops against pathogens and pests has become relevant. Chitosan is commonly described in the literature as a stimulant of plant resistance inducing natural defense mechanisms, which may reduce the amounts of synthetic plant protection products used in cultivation [33]. With its film-forming properties, this polymer can also act as a physical barrier limiting the contact between plants and pathogenic microorganisms [65]. Moreover, due to their potent antimicrobial activity, chitosan-based products can be used separately or in combination with other agents as biocides applied as hydrogels for coating tubers, fruits, and seeds; as a solution for soil or foliar application during or after vegetation; or as medium supplements in hydroponic or tissue cultures [33]. Numerous studies have been carried out to assess the potential use of chitosan as a plant health-promoting agent. It was shown that soaking freesia tubers in a 0.2% solution of chitosan with different molecular weights (in the range of 2000–970,000 g $\times$ mol^{-1}) exerted a positive effect on their health [52]. In turn, the treatment of cucumber seeds with chitosan increased the resistance of seedlings to *Phytophthora capsici* in a concentration-dependent manner (0, 125, 250, and 500 ppm). The highest concentration ensured complete resistance of young cucumber plants to blight caused by the pathogen [54]. Treatment of soybeans with a chitosan solution was reported to reduce the presence of herbivorous insects [55]. Moreover, chitosan was found to induce an effective reduction in the viability of *Phomopsis viticola* spores [66]. Given its properties, it is also possible to use this polymer for the protection of herbal plants against pathogens. Szczeponek et al. [67] demonstrated the efficacy of a chitosan-based formulation against fungal species that infest lemon balm and peppermint most frequently. Chitosan can also be used to control pathogenic soil nematodes and increase plant resistance to these pests [41].

Chitosan is also one of the few active substances contained in formulations and fungicides approved for use in forestry for the protection of forest nurseries against pathogenic diseases [68]. As shown in an experiment carried out by Aleksandrowicz-Trzcińska et al. [69], foliar application of a chitosan-containing formulation increased the resistance of Scots pine to fungal infections. Other results of laboratory studies indicated that chitosan exerted a negative effect on growth and caused changes in the morphological and structural structure of *Cylindrocladium floridanum*, *Cylindrocarpon destructans*, *Fusarium acuminatum*, and *Fusarium oxysporum* fungi responsible for root rot in forest nurseries [70]. Moreover, there are a number of chitosan-based agents on the market for coating injuries and wounds in trees and shrubs.

5. Application of Chitosan in Storage

The main goal of storage is to reduce food losses by ensuring the longest possible shelf life, good quality of stored products, and the best possible protection of product stability before further distribution. An important role in this process is mainly played by the selection of optimal conditions and storage methods as well as the monitoring of such product quality parameters as the color, weight, firmness, content of bioactive compounds, and rate of ethylene and carbon dioxide production. However, in the case of low-processed raw materials, it is often necessary to employ various types of preservation methods. One of the options to preserve the quality of stored raw materials consists in a reduction in respiration and transpiration through the use of natural edible coatings. The potential of using chitosan in storage is associated with its physicochemical properties and biological activity. Given its natural origin, ability to form semipermeable coatings, and high antioxidant and antimicrobial activity, this polymer facilitates the maintenance and optimization of the postharvest stability of raw materials and food products and influences their chemical composition. Additionally, chitosan delays fruit ripening through the limitation of ethylene and carbon dioxide release. Therefore, it can be widely used in the production of biodegradable films, food packaging, fibers, gels, and films with high strength and flexibility. It can also be used as nanoparticles. Moreover, this polymer can be applied directly to fruits and vegetables as a single ingredient or in combination with other substances in order to create edible protective coatings with bacteriostatic and fungistatic properties [28,44,71]. Modifications achieved by combining chitosan with various types of substances offer many possibilities for the production of edible coatings and food packaging with required properties. For instance, conjugates of chitosan with phenolic (gallic and caffeic) acids produced durable films with adequate properties to serve as a barrier against water vapor and oxygen. They exhibited stronger antioxidant and antimicrobial activity than films based on traditional chitosan [72].

Numerous reports have demonstrated the suitability of chitosan for storage purposes, as its properties help to enhance the stability and quality of stored products. One of the examples is the research on the effect of chitosan coatings on the shelf life and quality of plum fruit. The results of the experiment showed that the use of coatings containing 2% chitosan had a significant effect on the maintenance of the color, firmness, and weight of fruit stored at low temperature [73]. Similar results were obtained in a study of guava fruit stored at low temperature; i.e., the use of 2% chitosan coatings exerted a positive effect on the quality, firmness, and weight of fruit [74]. It was also observed that chitosan coatings increased the antioxidant properties of apricot fruits and contributed to the longer maintenance of a high total content of phenolic compounds [75]. Moreover, another experiment showed that chitosan coatings increased the storage stability of longan fruits and had a positive effect on their quality, color, and weight during storage [76]. A similar effect was reported in a study of *Actinidia melanandra* fruits, whose storage stability increased after the application of gel coatings containing this polymer [77]. In an experiment conducted by Zhu et al. [78], the application of chitosan coatings delayed the maturation and degradation of mangoes.

A study conducted by Tayel et al. [79] demonstrated that coating lemon fruit with chitosan inhibited the occurrence of *Penicillium* fungi. Moreover, the application of a chitosan-based formulation during the potato vegetation period was found to exert a limiting effect on the presence of *Fusarium* spp. and *P. carotovorum* subsp. *carotovorum* causing dry and wet tuber rot during storage [80]. A similar protective effect against the presence of the bacterial pathogen *Acidovorax citrulli*, which causes fruit blotch, was achieved in the case of watermelon seedlings [81]. In turn, as reported by He et al. [82], the use of a chitosan spray before harvesting strawberries had a beneficial effect on the quality and shelf life of the fruit through maintenance of the content of sugars; vitamin C; and numerous secondary metabolites, e.g., total phenolic compounds, flavonoids, and anthocyanins. Furthermore, the use of different concentrations of chitosan with different molecular weights was shown to increase the resistance of tomato fruit to grey mold caused by *Botrytis cinerea*. This was associated with both the direct antifungal activity of chitosan and the induction of a biochemical defense response in the fruit, which resulted in increased accumulation of phenolic compounds and enhanced activity of PPO [83].

6. Chitosan and its Derivatives as Biotic Elicitors

The potential for application of chitosan and its derivatives in the elicitation process is related to the biological activity of this compound, which primarily involves the ability to stimulate natural plant defense mechanisms and increase plant resistance to stress. This is associated with various types of physiological and biochemical changes, such as oxidative stress; accumulation of H_2O_2; synthesis of secondary metabolites (polyphenolic compounds, phytoalexins, flavonoids, alkaloids), enzymes (chitinase, glucanase, protease), and growth inhibitors (abscisic acid, jasmonic acid, salicylic acid); and accumulation of lignin and callose. The effect of chitosan on plants is reflected in changes in the chromatin structure, the inhibition of H^+-ATPase activity in the cell membrane, the activation of MAP kinases, and an increase in the cytosolic Ca^{2+} concentration [23,44,84]. Elicitation is aimed at stimulation of the biosynthesis of secondary metabolites present in plants or induction of the formation of new substances. It can solve problems related to the insufficient amount of bioactive compounds produced by plants, which does not cover the current needs, and bring benefits by providing high-quality raw materials with increased content of health-enhancing compounds [85]. The effectiveness of chitosan in elicitation largely depends on its solubility; therefore, chitosan salts, such as lactate or acetate, are the most commonly used derivatives of this polymer [21,34].

Numerous research publications report the effectiveness of chitosan and its derivatives in the elicitation process. Studies on *Curcuma longa* [86] indicated that the foliar application of a 0.1% chitosan solution triggered defense responses in the plants and had a positive effect on both plant growth and accumulation of curcumin in their rhizomes. A stimulating effect of 0.1% and 0.2% chitosan solutions was also reported in a study of *Stevia rebaudiana*. The application of chitosan increased the biomass and concentration of phenolic compounds and rebaudioside A [87]. A field experiment conducted on *Origanum vulgare* ssp. *hirtum* demonstrated that the use of different concentrations of chitosan oligosaccharides (50,

200, 500, or 1000 ppm) exerted a beneficial effect on plant growth and the content of polyphenolic compounds [88].

The study conducted by Gerami et al. [89] suggested the possibility of the use of chitosan as an elicitor increasing *S. rebaudiana* tolerance to salinity and reducing its phytotoxic effects on these plants. Similar results were obtained in an experiment conducted by Safikhan et al. [90], in which chitosan mitigated the harmful effects of salt stress, stimulated growth, and had a positive effect on the physiological parameters of *Silybum marianum*. Furthermore, a study conducted in water deficit conditions showed that chitosan treatment of *Salvia officinalis* reduced the negative impact of drought stress. It also had a positive effect on the quantity and quality of essential oil, the total content of phenolic compounds and flavonoids, and the antioxidant properties of sage extracts [91]. A stimulating effect of a chitosan suspension and a chitosan solution in 1% acetic acid on the production of flavonoids in *Ononis arvensis* in in vitro conditions was demonstrated as well [92]. A chitosan solution in acetic acid used as an elicitor in *Mentha piperita* suspension cultures produced a significant increase in the accumulation of menthol [93]. In turn, the treatment of *M. piperita* with chitosan in greenhouse conditions increased the total content of phenolic compounds and flavonoids and enhanced the antioxidant activity of the extracts [94]. Chitosan also stimulated the production of triterpenoid saponins in *Psammosilene tunicoides* hair root cultures [37]. Moreover, the polymer was found to have a beneficial effect on biomass accumulation in cell cultures of three basil species: *Ocimum basilicum*, *O. sanctum*, and *O. gratissimum* [95]. Other studies carried out on *O. basilicum* indicated the effectiveness of chitosan elicitation in the stimulation of biomass growth, intensified accumulation of total phenolic and terpene compounds, elevation of rosmarinic acid and eugenol concentrations, and enhancement of antioxidant activity [96]. The foliar application of chitosan lactate stimulated the biosynthesis of rosmarinic acid, anthocyanins, and phenolic compounds in *O. basilicum* and *M. officinalis* raw materials [34].

The effects of the application of chitosan and its derivatives on the accumulation of secondary metabolites in selected plant species are shown in Table 1.

Table 1. Effects of application of chitosan and its derivatives on the level of secondary metabolites in selected plant species.

Plant Species	Plant Growth Conditions	Chitosan Form	Dose, Method, and Number of Chitosan Applications	Effect of Chitosan on the Level of Secondary Metabolites	Reference
Artemisia annua	laboratory conditions; hairy root cultures	chitosan	50, 100, or 150 mg L^{-1} of chitosan added to hairy root cultures	increased artemisinin production	Putalun et al. [97]
Curcuma longa	field conditions	chitosan	0.1% chitosan; foliar application; seven treatments	increased curcumin content	Sathiyabama et al. [38]
Dracocephalum kotschyi	glasshouse; mixture of peat, sandy soil, and perlite substrate	chitosan	100 or 400 mg L^{-1} of chitosan; triple foliar application	enhanced biosynthesis of total phenolic and flavonoid compounds, including rosmarinic acid and apigenin	Kahromi and Khara [98]
Catharanthus roseus	greenhouse; sandy soil	chitosan nanoparticles	1% chitosan nanoparticles; single foliar application in salinity stress conditions	increased alkaloid accumulation	Hassan et al. [35]

Table 1. *Cont.*

Plant Species	Plant Growth Conditions	Chitosan Form	Dose, Method, and Number of Chitosan Applications	Effect of Chitosan on the Level of Secondary Metabolites	Reference
Fragaria × annanasa	field conditions	chitosan	125, 250, 500, or 1000 ppm chitosan; foliar application; six treatments	increased amount of phenolic compounds, carotenoids, flavonoids, and anthocyanins in strawberry fruits	Rahman et al. [57]
Ginkgo biloba	laboratory conditions; callus cultures	chitosan	25, 50, 100, or 200 mg L^{-1} of chitosan added to MS medium	enhanced production of total flavonoids and total phenolic compounds	Elateeq et al. [99]
Iberis amara	laboratory conditions; cell suspension cultures	chitosan	50, 100, or 200 mg L^{-1} of chitosan	enhanced total phenolic compounds, flavonoid, flavonol, and anthocyanin contents	Taghizadeh et al. [100]
Isatis tinctoria	laboratory conditions; hairy root cultures	chitosan	50, 100, 150, 200, or 400 mg L^{-1} of chitosan; hairy root cultures elicited for 6–96 h	increased total flavonoid accumulation	Jiao et al. [101]
Melissa officinalis	phytotron room; soil substrate	chitosan lactate	100 or 500 mg L^{-1} of chitosan lactate; single foliar application	increased accumulation of rosmarinic acid, anthocyanins, and total phenolic compounds	Hawrylak-Nowak et al. [34]
Mentha piperita	greenhouse; soil phosphate	chitosan	50 or 100 µM chitosan; single foliar application	increased content of phenolic and flavonoid compounds	Salimgandomi and Shabrangi [94]
Ocimum basilicum	greenhouse; potting substrate irrigated with a fertilizer solution	chitosan	0.01%, 0.05%, 0.1%, 0.5% or 1% chitosan; seed soaking (30 min.)	increased content of total phenolic and terpenic compounds (rosmarinic acid, eugenol)	Kim et al. [96]
Origanum vulgare ssp. hirtum	field conditions	chitosan oligosaccharides	50, 200, 500, or 1000 ppm chitosan oligosaccharides; single foliar application	increased accumulation of total flavonoids and total polyphenolic compounds	Yin et al. [88]
Psammosilene tunicoides	laboratory conditions; hairy root cultures	chitosan	200 mg L^{-1} of chitosan; hairy roots elicited by chitosan for nine days	enhanced accumulation of total saponins, increased content of quillaic acid, gypsogenin, and gypsogenin-3-O-β-D-glucuronopyranoside	Qui et al. [37]
Salvia officinalis	field conditions	chitosan	0.25 or 0.50 g L^{-1} of chitosan; single foliar application in reduced irrigation conditions	increased amount of total phenolic and flavonoid content; enhanced production of α- and β-pinene, limonene, α- and β-thujone, camphor, and 1,8-cineole in the essential oil	Vosoughi et al. [91]

Table 1. *Cont.*

Plant Species	Plant Growth Conditions	Chitosan Form	Dose, Method, and Number of Chitosan Applications	Effect of Chitosan on the Level of Secondary Metabolites	Reference
Satureja isophylla	greenhouse; sandy soil	chitosan	0.2 or 0.4 g L^{-1} of chitosan; foliar application	increased amount of essential oil; increased concentrations of essential oil constituents (carvacrol, β-bisabolene)	Salehi et al. [102]
Stevia rebaudiana	greenhouse; perlite and peat substrate	chitosan	0.2, 0.4, or 0.6 g L^{-1} of chitosan; double foliar application in salinity stress conditions	increased content of steviol glycosides: stevioside and rebaudioside A	Gerami et al. [89]
Sylibum marianum	laboratory conditions; cell suspension cultures	chitosan	0.5, 1, 2.5, 5, 10, 25, or 50 mg L^{-1} of chitosan in MS medium	enhanced accumulation of total flavonoids, total phenolic compounds, and silymarin	Shah et al. [103]

7. Prospects for New Applications of Chitosan

Due to the wide availability and a number of its beneficial properties, an increase in the number of potential applications of chitosan and its derivatives has been observed in recent years. Nevertheless, despite its numerous advantages, the practical commercial-scale use of this polymer is still limited by some of its physicochemical features and is dependent on progress in the development of optimal methods for chemical modification and adaptation of the polymer for specific applications. To date, the greatest success in this regard has been achieved in nanotechnology, which is regarded as one of the five most groundbreaking fields of science and technology of the last century [6,104,105]. The achievements of this discipline facilitate a wide use of chitosan nanomaterials in various types of biomedical, cosmetic, food, ecological, and agrotechnical innovations [13,16]. The application of nanoparticles facilitates more effective utilization of the biological properties of chitosan.

One of the most important prospects for the use of chitosan in nanotechnology is the concept of sustainability of the agrochemical industry through the production of the so-called agronanochemicals. It mainly involves the preparation of chitosan nanoparticles and chitosan nanocomposites or encapsulation of active substances in chitosan-based nanocarriers. The preparations primarily exhibit high efficiency and bioavailability; hence, they can constitute a more economical environmentally friendly alternative limiting the amounts of currently available chemicals [17,104,106,107].

Chitosan may exert a positive effect on plant photosynthetic efficiency, ability to biosynthesize and accumulate chlorophyll, and nutrient uptake [27,106,108,109]. The available research results demonstrate that chitosan nanoparticles have a greater impact on the macronutrient uptake efficiency than chitosan oligomers. Experiments carried out in greenhouse conditions showed that chitosan nanoparticles contributed to an increase in photosynthetic efficiency; chlorophyll content; and nitrogen, phosphorus, and potassium absorption in *Coffea canephora* [110]. Similar results were reported by Sathiyabama and Manikandan [111], who found that the foliar application of nanochitosan stimulated growth and increased the mineral content in *Eleusine coracana*.

Various research results also indicate the possibility of using chitosan to increase plant tolerance to abiotic and biotic stress factors and mitigate their phytotoxic effects [35,107,109]. Many reports show the effectiveness of chitosan in increasing plant resistance to salinity, drought, or low-temperature stress [112,113]. It was shown that spraying with a solution containing 1% chitosan nanoparticles alleviated the effects of salt stress through the activation of defense mechanisms in *Catharanthus roseus* [35,109]. The foliar application of a solution containing chitosan nanoparticles (400 mg L^{-1}) to bananas enhanced their tolerance to low temperature and improved biometric parameters via a reduction in the oxidative stress intensity [113]. In turn, Kocięcka and Liberacki [112] reported the effectiveness of chitosan, its derivatives, and nanoparticles in the stimulation of the resistance of cereals to such abiotic stress factors as low temperature, drought, and salinity and in the enhancement of the health of plants and improvement of their quality.

This polymer can also be used in combination with other compounds not only to extend the spectrum of its activity but also to produce materials with more favorable mechanical, chemical, thermal, and barrier properties [17,107,114]. The use of chitosan in the production of nanocomposites may also increase the effectiveness of other chemical compounds or elements, e.g., zinc, iron, copper, and silver [107,114]. In addition to the stimulation of the efficiency of agrochemicals, these nanocomposites enhance the biosynthesis of secondary metabolites and may accelerate plant regeneration. An example of this type of use is the application of zinc encapsulated chitosan nanoparticles, which had a beneficial effect on the growth, health, and quality of maize kernels [115]. In turn, studies of *Capsicum annuum* tissue cultures showed that a composite of zinc oxide and chitosan nanoparticles contributed to a significant increase in its elicitation efficiency [116].

Chitosan nanoparticles can reduce the amount and frequency of application of plant protection products, ameliorate their negative effects, and help to monitor and remove their excess from soils and waters [23,117]. Due to its cationic nature and high affinity to metals and dyes, chitosan is also one of the preferred natural and relatively cheap adsorbents replacing the currently used active carbon, with similar sorption properties in water engineering [2,10,105,118]. In water remediation, similar to other applications, chitosan can be used alone or in combination with other substances, e.g., activated carbon or silicon dioxide. Nanocomposites made of chitosan and activated carbon, whose combination ensures a significant increase in their adsorption capacity, are one example [118,119].

The main applications of chitosan and its derivatives in plant production are summarized in Figure 2.

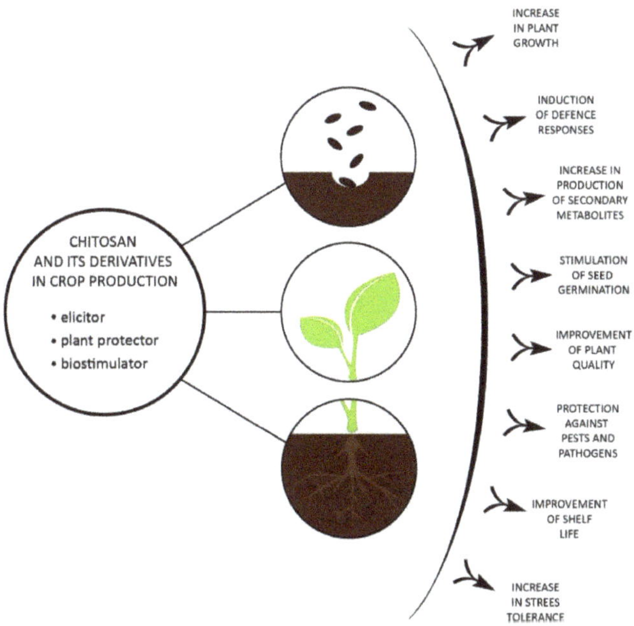

Figure 2. Application of chitosan and its derivatives in plant production.

8. Conclusions

Chitosan is an easily available environmentally friendly biopolymer with numerous favorable biological properties; hence, it has many applications in various fields. Additionally, it can be a potential solution to various ecological problems, especially in the production of plants by increasing their internal potential. Biodegradable and biocompatible chitosan and chitosan-based nanomaterials are becoming essential in agriculture due to their unique properties, such as biostimulating, eliciting, and antimicrobial activity and stimulation of plant growth and tolerance to environmental stresses. However, the use of other effective sources of chitosan, the methods for extraction and optimization of its physicochemical properties, and the practical implementation of laboratory results should be investigated comprehensively. Moreover, detailed studies on the potential toxicity of chitosan-based nanomaterials and the ecological consequences of their large-scale use are required. At present, these substances are not being widely used in agriculture, as the mechanisms of their biological activity in plants and action against pathogenic microorganisms have not been fully elucidated to date.

Author Contributions: Conceptualization, B.H.-N. and M.S.-J.; formal analysis, B.H.-N.; writing—original draft preparation, M.S.-J.; writing—review and editing, B.H.-N.; visualization, M.S.-J. All authors have read and agreed to the published version of the manuscript.

Funding: This research received no external funding. The APC (Article Processing Charges) was partially funded by the Doctoral School of the University of Life Sciences in Lublin (grant number SD/48/NB/2022).

Institutional Review Board Statement: Not applicable.

Informed Consent Statement: Not applicable.

Data Availability Statement: Not applicable.

Conflicts of Interest: The authors declare no conflict of interest.

References

1. Hidangmayum, A.; Dwivedi, P.; Katiyar, D.; Hemantaranjan, A. Application of chitosan on plant responses with special reference to abiotic stress. *Physiol. Mol. Biol. Plants* **2019**, *25*, 313–326. [CrossRef] [PubMed]
2. Muthu, M.; Gopal, J.; Chun, S.; Devadoss, A.J.P.; Hasan, N.; Sivanesan, I. Crustacean waste-derived chitosan: Antioxidant properties and future perspective. *Antioxidants* **2021**, *10*, 228. [CrossRef] [PubMed]
3. Khoushab, F.; Yamabhai, M. Chitin research revisited. *Mar. Drugs* **2010**, *8*, 1988–2012. [CrossRef]
4. Ferri, M.; Tassoni, A. Chitosan as elicitor of health beneficial secondary metabolites in in vitro plant cell cultures. In *Handbook of Chitosan Research and Applications*; Mackay, R.G., Tait, J.M., Eds.; Nova Science Publishers Inc.: Hauppauge, NY, USA, 2011; pp. 389–414.
5. Berezina, N. Production and application of chitin. *Phys. Sci. Rev.* **2016**, *1*, 20160048. [CrossRef]
6. Bakshi, P.S.; Selvakumar, D.; Kadirvelu, K.; Kumar, N.S. Chitosan as an environment friendly biomaterial–A review on recent modifications and applications. *Int. J. Biol. Macromol.* **2020**, *150*, 1072–1083. [CrossRef]
7. Iber, B.T.; Kasan, N.A.; Torsabo, D.; Omuwa, J.W. A review of various sources of chitin and chitosan in nature. *J. Renew. Mater.* **2022**, *10*, 1097–1123. [CrossRef]
8. From Chitin to Chitosan. Available online: http://www.glycopedia.eu/IMG/pdf/from_chitin_to_chitosan-2.pdf. (accessed on 17 November 2021).
9. Pandit, A.; Indurkar, A.; Deshpande, C.; Jain, R.; Dandekar, P. A systematic review of physical techniques for chitosan degradation. *Carbohydr. Polym. Technol. Appl.* **2021**, *2*, 100033. [CrossRef]
10. Ma, J.; Faqir, Y.; Tan, C.; Khaliq, G. Terrestrial insects as a promising source of chitosan and recent developments in its application for various industries. *Food Chem.* **2022**, *373*, 131407. [CrossRef]
11. Ravindranathan, S.; Koppolu, B.P.; Smith, S.G.; Zaharoff, D.A. Effect of chitosan properties on immunoreactivity. *Mar. Drugs* **2016**, *14*, 91. [CrossRef]
12. Hahn, T.; Tafi, E.; Paul, A.; Salvia, R.; Falabella, P.; Zibek, S. Current state of chitin purification and chitosan production from insects. *J. Chem. Technol. Biotechnol.* **2020**, *95*, 2775–2795. [CrossRef]
13. Santos, V.P.; Marques, N.S.S.; Maia, P.C.S.V.; Lima, M.A.B.; Franco, L.O.; Campos-Takaki, G.M. Seafood waste as attractive source of chitin and chitosan production and their applications. *Int. J. Mol. Sci.* **2020**, *21*, 4290. [CrossRef] [PubMed]

14. Mohan, K.; Ganesan, A.R.; Muralisankar, T.; Jayakumar, R.; Sathishkumar, P.; Uthayakumar, V.; Chandirasekar, R.; Revathi, N. Recent insights into the extraction, characterization, and bioactivities of chitin and chitosan from insects. *Trends Food Sci. Technol.* **2020**, *105*, 17–42. [CrossRef] [PubMed]
15. Shahrajabian, M.H.; Chaski, C.; Polyzos, N.; Tzortzakis, N.; Petropoulos, S. Sustainable agriculture systems in vegetable production using chitin and chitosan as plant biostimulants. *Biomolecules* **2021**, *11*, 819. [CrossRef] [PubMed]
16. Venkatraman, P. Chitosan: An antimicrobial polymer. *J. Undergrad. Mater. Res.* **2015**, *5*, 11–13. [CrossRef]
17. Azmana, M.; Mahmood, S.; Hilles, A.R.; Rahman, A.; Arifin, M.A.B.; Ahmed, S. A review on chitosan and chitosan-based bionanocomposites: Promising material for combatting global issues and its applications. *Int. J. Biol. Macromol.* **2021**, *185*, 832–848. [CrossRef] [PubMed]
18. Varum, K.M.; Ottoy, M.H.; Smidsrod, O. Water-solubility of partially N-acetylated chitosans as a function of pH: Effect of chemical composition and depolymerisation, *Carbohydr. Polym.* **1994**, *25*, 65–70. [CrossRef]
19. Blagodatskikh, I.V.; Kulikov, S.; Vyshivannaya, O.; Bezrodnykh, E.; Tikhonov, V. N-Reacetylated oligochitosan: pH dependence of self-assembly properties and antibacterial activity. *Biomacromolecules* **2017**, *18*, 1491–1498. [CrossRef]
20. Ji, J.; Wang, L.; Yu, H.; Chen, Y.; Zhao, Y.; Zhang, H.; Amer, W.A.; Sun, Y.; Huang, L.; Saleem, M. Chemical modifications of chitosan and its applications. *Polym. Plast. Technol. Eng.* **2014**, *53*, 1494–1505. [CrossRef]
21. Kowalczyk, D.; Kordowska-Wiater, M.; Nowak, J.; Baraniak, B. Characterization of films based on chitosan lactate and its blends with oxidized starch and gelatin. *Int. J. Biol. Macromol.* **2015**, *77*, 350–359. [CrossRef]
22. Placek, M.; Dobrowolska, A.; Wraga, K.; Zawadzińska, A.; Żurawik, P. The use of chitosan in cultivation, preservation and protection of horticultural plants. *Zesz. Probl. Post. Nauk Roln.* **2009**, *3–4*, 101–109. (In Polish)
23. Malerba, M.; Cerana, R. Chitosan effects on plant systems. *Int. J. Mol. Sci.* **2016**, *17*, 996. [CrossRef] [PubMed]
24. Bautista-Banos, S.; Hernandez-Lauzardo, A.N.; Velazquez-del Valle, M.G.; Hernandez-Lopez, M.; Ait Barka, E.; Bosquez-Molina, E.; Wilson, C.L. Chitosan as a potential natural compound to control pre- and postharvest diseases of horticultural commodities. *Crop Prot.* **2006**, *25*, 108–118. [CrossRef]
25. Hadwiger, L.A. Multiple effects of chitosan on plant systems: Solid science or hype. *Plant Sci.* **2013**, *208*, 42–49. [CrossRef] [PubMed]
26. Hosseinnejad, M.; Jafari, S.M. Evaluation of different factors affecting antimicrobial properties of chitosan. Review. *Int. J. Biol. Macromol.* **2016**, *85*, 467–475. [CrossRef] [PubMed]
27. Divya, K.; Jisha, M.S. Chitosan nanoparticles preparation and applications. *Environ. Chem. Lett.* **2018**, *16*, 101–112. [CrossRef]
28. Morin-Crini, N.; Lichtfouse, E.; Torri, G.; Crini, G. Applications of chitosan in food, pharmaceuticals, medicine, cosmetics, agriculture, textiles, pulp and paper, biotechnology and environmental chemistry. *Environ. Chem. Lett.* **2019**, *17*, 1667–1692. [CrossRef]
29. Wang, W.; Xue, C.; Mao, X. Chitosan: Structural modification, biological activity and application, Review. *Int. J. Biol. Macromol.* **2020**, *164*, 4532–4546. [CrossRef]
30. Hassan, O.; Chang, T. Chitosan for eco-friendly control of plant disease. *Asian J. Plant Pathol.* **2017**, *11*, 53–70. [CrossRef]
31. Cabrera, J.C.; Boland, A.; Cambier, P.; Frettinger, P.; Van Cutsem, P. Chitosan oligosaccharides modulate the supramolecular conformation and the biological activity of oligogalacturonides in *Arabidopsis*. *Glycobiology* **2010**, *20*, 775–786. [CrossRef]
32. Rajalakshmi, A.; Krithiga, N.; Jayachitra, A. Antioxidant activity of the chitosan extracted from shrimp exoskeleton. *Middle-East J. Sci. Res.* **2013**, *16*, 1446–1451. [CrossRef]
33. Katiyar, D.; Hemantaranjan, A.; Singh, B. Chitosan as a promising natural compound to enhance potential physiological responses in plant: A review. *Ind. J. Plant Physiol.* **2015**, *20*, 1–9. [CrossRef]
34. Hawrylak-Nowak, B.; Dresler, S.; Rubinowska, K.; Matraszek-Gawron, R. Eliciting effect of foliar application of chitosan lactate on the phytochemical properties of *Ocimum basilicum* L. and *Melissa officinalis* L. *Food Chem.* **2021**, *342*, 128358. [CrossRef] [PubMed]
35. Hassan, F.A.S.; Ali, E.; Gaber, A.; Fetouh, M.I.; Mazrou, R. Chitosan nanoparticles effectively combat salinity stress by enhancing antioxidant activity and alkaloid biosynthesis in *Catharanthus roseus* (L.) G. Don. *Plant Physiol. Biochem.* **2021**, *162*, 291–300. [CrossRef] [PubMed]
36. Quitadamo, F.; De Simone, V.; Beleggia, R.; Trono, D. Chitosan-induced activation of the antioxidant defense system counteracts the adverse effects of salinity in durum wheat. *Plants* **2021**, *10*, 1365. [CrossRef]
37. Qiu, H.; Su, L.; Wang, H.; Zhang, Z. Chitosan elicitation of saponin accumulation in *Psammosilene tunicoides* hairy roots by modulating antioxidant activity, nitric oxide production and differential gene expression. *Plant Physiol. Biochem.* **2021**, *166*, 115–127. [CrossRef]
38. Sathiyabama, M.; Bernstein, N.; Anusuya, S. Chitosan elicitation for increased curcumin production and stimulation of defence response in turmeric (*Curcuma longa* L.). *Ind. Crops Prod.* **2016**, *89*, 87–94. [CrossRef]
39. Mejía-Teniente, L.; Durán-Flores, D.; Chapa-Oliver, A.; Torres-Pacheco, I.; Cruz-Hernández, A.; González-Chavira, M.; Ocampo-Velázquez, R.; Guevara-González, R. Oxidative and molecular responses in *Capsicum annuum* L. after hydrogen peroxide, salicylic acid and chitosan foliar applications. *Int. J. Mol. Sci.* **2013**, *14*, 10178–10196. [CrossRef]
40. Wang, S.; Gao, H. Effect of chitosan-based edible coating on antioxidants antioxidant enzyme system and postharvest fruit quality of strawberries (*Fragaria × aranassa* Duch). *Food Sci. Technol.* **2013**, *52*, 71–79. [CrossRef]
41. Chakraborty, M.; Hasanuzzaman, M.; Rahman, M.; Khan, M.A.R.; Bhowmik, P.; Mahmud, N.U.; Tanveer, M.; Islam, T. Mechanism of plant growth promotion and disease suppression by chitosan biopolymer. *Agriculture* **2020**, *10*, 624. [CrossRef]

42. Basa, S.; Nampally, M.; Honorato, T.; Das, S.N.; Podile, A.R.; El Gueddari, N.E.; Moerschbacher, B.M. The pattern of acetylation defines the priming activity of chitosan tetramers. *J. Am. Chem. Soc.* **2020**, *142*, 1975–1986. [CrossRef]
43. Kulikov, S.N.; Chirkov, S.N.; Il'ina, A.V.; Lopatin, S.A.; Varlamov, V.P. Effect of the molecular weight of chitosan on its antiviral activity in plants. *Appl. Biochem. Microbiol.* **2006**, *42*, 200–203. [CrossRef]
44. Rabea, E.I.; Badawy, M.E.T.; Stevens, C.V.; Smagghe, G.; Steurbaut, W. Chitosan as antimicrobial agent: Applications and mode of action. *Biomacromolecules* **2003**, *4*, 1457–1465. [CrossRef] [PubMed]
45. A European Green Deal. Available online: https://ec.europa.eu/info/strategy/priorities-2019-2024/european-green-deal (accessed on 7 December 2021).
46. Dmytryk, A.; Rój, E.; Wilk, R.; Chojnacka, K.; Górecki, H. Effect of new biostimulators on the initial phase of plant growth. *Przem. Chem.* **2014**, *93*, 1020–1025. (In Polish)
47. Matyjaszczyk, E. The introduction of biostimulants on the Polish market. The present situation and legal requirements. *Przem. Chem.* **2015**, *10*, 1841–1844. (In Polish) [CrossRef]
48. Posmyk, M.M.; Szafrańska, K. Biostimulators: A new trend towards solving an old problem. *Front. Plant Sci.* **2016**, *7*, 748. [CrossRef]
49. Korbecka-Glinka, G.; Wiśniewska-Wrona, M.; Kopania, E. The use of natural polymers for treatments enhancing sowing material. *Polimery* **2021**, *66*, 11–20. [CrossRef]
50. Hameed, A.; Sheikh, M.; Hameed, A.; Farooq, T.; Basra, S.; Jamil, A. Chitosan priming enhances the seed germination, antioxidants, hydrolytic enzymes, soluble proteins and sugars in wheat seeds. *Agrochimica* **2013**, *67*, 32–46.
51. Ruan, S.; Xue, Q. Effects of chitosan coating on seed germination and salt-tolerance of seedling in hybrid rice (*Oryza sativa* L.). *Acta Agron. Sin.* **2002**, *28*, 803–808.
52. Salachna, P.; Bartkowiak, A.; Mazurkiewicz-Zapałowicz, K.; Placek, M. Assessment of the effect of chitosan on the yield and health of freesia tubers (*Freesia* Eckl. Ex klatt) cv. 'Versailles'. *Rocz. AR Pozn. CCCLXXXIII Ogrodn.* **2007**, *41*, 177–181. (In Polish)
53. Salachna, P. Use of chitosan derivatives to improve the growth of ornamentals. *Inż. Ekolog.* **2017**, *18*, 63–68. [CrossRef]
54. Zohara, F.; Surovy, M.Z.; Khatun, A.; Prince, M.F.R.K.; Akanda, A.M.; Rahman, M.; Islam, T. Chitosan biostimulant controls infection of cucumber by *Phytophthora capsici* through suppression of asexual reproduction of the pathogen. *Acta Agrobot.* **2019**, *72*, 1763. [CrossRef]
55. Zeng, D.; Luo, X.; Tu, R. Application of bioactive coatings based on chitosan for soybean seed protection. *Int. J. Carbohydr. Chem.* **2012**, *2012*, 104565. [CrossRef]
56. El-Miniawy, S.M.; Ragab, M.E.; Youssef, S.M.; Metwally, A.A. Response of strawberry plants to foliar spraying of chitosan. *Res. J. Agric. Biol. Sci.* **2013**, *9*, 366–372.
57. Rahman, M.; Mukta, J.A.; Sabir, A.A.; Gupta, D.R.; Mohi-Ud-Din, M.; Hasanuzzaman, M.; Miah, M.G.; Rahman, M.; Islam, M.T. Chitosan biopolymer promotes yield and stimulates accumulation of antioxidants in strawberry fruit. *PLoS ONE* **2018**, *13*, e0203769. [CrossRef] [PubMed]
58. Poterańska, N.; Mijowska, K.; Ochmian, I. The influence of foliar calcium fertilizers and bio-stimulators on bushes growth, yield and fruit quality of blue honeysuckle (*Lonicera caerulea* L.) Czarna cultivar. In *Research and Development of Young Scientists in Poland–Life Sciences*; Young Scientists: Szczecin, Poland, 2015; pp. 132–138. (In Polish)
59. Ochta, K.; Taniguchi, A.; Konishi, N.; Hosoki, T. Chitosan treatment affects plant growth and flower quality in *Eustoma grandiflorum*. *HortScience* **1999**, *34*, 233–234. [CrossRef]
60. Chookhongkha, N.; Miyagawa, S.; Jirakiattikul, Y.; Photchanachai, S. Chili growth and seed productivity as affected by chitosan. In Proceedings of the International Conference on Agriculture Technology and Food Sciences (ICATFS'2012), Manila, Philippines, 17–18 November 2012; pp. 146–149.
61. Lubkowski, K.; Grzmil, B. Controlled release fertilizers. *Pol. J. Chem. Technol.* **2007**, *9*, 81–84. [CrossRef]
62. Gumelar, M.D.; Hamzah, M.; Hidayat, A.S.; Saputra, D.A.; Idvan. Utilization of chitosan as coating material in making NPK slow release fertilize. *Macromol. Symp.* **2020**, *391*, 1900188. [CrossRef]
63. Abdel-Aziz, H.M.M.; Hasaneen, M.N.A.; Omer, A.M. Nano chitosan-NPK fertilizer enhances the growth and productivity of wheat plants grown in sandy soil. *Span. J. Agric. Res.* **2016**, *14*, e0902. [CrossRef]
64. New Standards to Curb the Global Spread of Plant Pests and Diseases. Available online: https://www.fao.org/news/story/en/item/1187738/icode/ (accessed on 17 December 2021).
65. Orzali, L.; Corsi, B.; Forni, C.; Riccioni, L. Chitozan in agriculture: A new challenge for managing plant disease. In *Biological Activities and Application of Marine Polysaccharide*; Shalaby, E.A., Ed.; IntechOpen: London, UK, 2016; pp. 17–36. [CrossRef]
66. Król, E. Influence of some chemicals on the viability of *Phomopsis viticola* Sacc. spores. *J. Plant Prot. Res.* **2005**, *45*, 195203.
67. Szczeponek, A.; Mazur, S.; Nawrocki, J. The usage of chitosan in protection of some peppermint and lemon balm pathogens. *Prog. Chem. Appl. Chitin Deriv.* **2006**, *Monograph XI*, 193–200.
68. Karmiłowicz, E. The use of fungicides in a protection of forest nurseries against fungal diseases in Poland. *Post. Ochr. Roślin* **2019**, *59*, 53–61. (In Polish)
69. Aleksandrowicz-Trzcińska, M.; Bogusiewicz, A.; Szkop, M.; Drozdowski, S. Effect of chitosan on disease control and growth of scots pine (*Pinus sylvestris* L.) in a forest nursery. *Forests* **2015**, *6*, 3165–3176. [CrossRef]
70. Laflamme, P.; Benhamou, N.; Bussires, G.; Dessureault, M. Different effect of chitosan on root rot fungal pathogens in forest nurseries. *Can. J. Bot.* **2000**, *77*, 1460–1468. [CrossRef]

71. Abd El-Hack, M.E.; El-Saadony, M.T.; Shafi, M.E.; Zabermawi, N.M.; Arif, M.; Batiha, G.E.; Khafaga, A.F.; Abd El-Hakim, Y.M.; Al-Sagheer, A.A. Antimicrobial and antioxidant properties of chitosan and its derivatives and their applications: A review. *Int. J. Biol. Macromol.* **2020**, *164*, 2726–2744. [CrossRef]
72. Wang, Y.; Du, H.; Xie, M.; Ma, G.; Yang, W.; Hu, Q.; Pei, F. Characterization of the physical properties and biological activity of chitosan films grafted with gallic acid and caffeic acid: A comparison study. *Food Packag. Shelf Life* **2019**, *22*, 100401. [CrossRef]
73. Kumar, P.; Sethi, S.; Sharma, R.R.; Srivastav, M.; Varghese, E. Effect of chitosan coating on postharvest life and quality of plum during storage at low temperature. *Sci. Hortic.* **2017**, *226*, 104–109. [CrossRef]
74. Hong, K.; Xie, J.; Zhang, L.; Sun, D.; Gong, D. Effects of chitosan coating on postharvest life and quality of guava (*Psidium guajava* L.) fruit during cold storage. *Sci. Hortic.* **2012**, *144*, 172–178. [CrossRef]
75. Ghasemnezhad, M.; Shiri, M.A.; Sanavi, M. Effect of chitosan coatings on some quality indices of apricot (*Prunus armeniaca* L.) during cold storage. *Casp. J. Env. Sci.* **2010**, *8*, 25–33.
76. Jiang, Y.; Li, Y. Effects of chitosan coating on postharvest life and quality of longan fruit. *Food Chem.* **2001**, *73*, 139–143. [CrossRef]
77. Kaya, M.; Cesoniene, L.; Daubaras, R.; Leskauskaite, D.; Zabulione, D. Chitosan coating of red kiwifruit (*Actinidia melanandra*) for extending of the shelf life. *Int. J. Biol. Macromol.* **2016**, *85*, 355–360. [CrossRef]
78. Zhu, X.; Wang, Q.; Cao, J.; Jiang, W. Effects of chitosan coating on postharvest quality of mango (*Mangifera indica* L. Cv. Tainong) fruits. *J. Food Process. Preserv.* **2008**, *32*, 770–784. [CrossRef]
79. Tayel, A.A.; Moussa, S.H.; Salem, M.F.; Mazrou, K.E.; El-Tras, W.F. Control of citrus molds using bioactive coatings incorporated with fungal chitosan/plant extracts composite. *J. Sci. Food Agric.* **2016**, *96*, 1306–1312. [CrossRef] [PubMed]
80. Kurzawińska, H.; Mazur, S. Effect of selected preparations used in vegetation period on occurrence of potato tubers rot during the storage. *Post. Ochr. Roślin* **2012**, *52*, 73–77. (In Polish)
81. Li, B.; Shi, Y.; Shan, C.; Zhou, Q.; Ibrahim, M.; Wang, Y.; Wu, G.; Li, H.; Xie, G.; Sun, G. Effect of chitosan solution on the inhibition of *Acidovorax citrulli* causing bacterial fruit blotch of watermelon. *J. Sci. Food Agric.* **2013**, *93*, 1010–1015. [CrossRef] [PubMed]
82. He, Y.; Bose, S.K.; Wang, W.; Jia, X.; Lu, H.; Yin, H. Pre-harvest treatment of chitosan oligosaccharides improved strawberry fruit quality. *Int. J. Mol. Sci.* **2018**, *19*, 2194. [CrossRef] [PubMed]
83. Badawy, M.E.I.; Rabea, E.I. Potential of the biopolymer chitosan with different molecular weights to control postharvest gray mold of tomato fruit. *Postharvest Biol. Technol.* **2009**, *51*, 110–117. [CrossRef]
84. El Hadrami, A.; Adam, L.R.; El Hadrami, I.; Daayf, F. Chitosan in plant protection. *Mar. Drugs* **2010**, *8*, 968–987. [CrossRef]
85. Gorelick, J.; Bernstein, N. Elicitation: An underutilized tool in the development of medicinal plants as a source of therapeutic secondary metabolites. *Adv. Agron.* **2014**, *124*, 201–230. [CrossRef]
86. Sathiyanarayanan, A.; Sathiyabama, M. Effect of chitosan on growth, yield and curcumin content in turmeric under field condition. *Biocatal. Agric. Biotechnol.* **2016**, *6*, 102–106. [CrossRef]
87. Mehregan, M.; Mehrafarin, A.; Labbafi, M.R.; Naghdi Badi, H.A. Effect of different concentrations of chitosan biostimulant on biochemical and morphophysiological traits of stevia plant (*Stevia rebaudiana* Bertoni). *J. Med. Plants* **2017**, *16*, 169–181.
88. Yin, H.; Frette, X.C.; Christensen, L.P.; Grevsen, K. Chitosan oligosaccharides promote the content of polyphenols in Greek oregano (*Origanum vulgare* ssp. *hirtum*). *J. Agric. Food Chem.* **2012**, *60*, 136–143. [CrossRef] [PubMed]
89. Gerami, M.; Majidian, P.; Ghorbanpour, A.; Alipour, Z. *Stevia rebaudiana* Bertoni responses to salt stress and chitosan elicitor. *Physiol. Mol. Biol. Plants* **2020**, *26*, 965–974. [CrossRef] [PubMed]
90. Safikhan, S.; Khoshbakht, K.; Chaichi, M.R.; Amini, A.; Motesharezadeh, B. Role of chitosan on the growth, physiological parameters and enzymatic activity of milk thistle (*Silybum marianum* (L.) Gaertn.) in a pot experiment. *J. Appl. Res. Med. Aromat. Plants* **2018**, *10*, 49–58. [CrossRef]
91. Vosoughi, N.; Gomarian, M.; Pirbalouti, A.G.; Khaghani, S.; Malekpoor, F. Essential oil composition and total phenolic, flavonoid contents, and antioxidant activity of sage (*Salvia officinalis* L.) extract under chitosan application and irrigation frequencies. *Ind. Crops Prod.* **2018**, *117*, 366–374. [CrossRef]
92. Tumova, L.; Backovska, M. Chitosan and the flavonoid production. *Herba Pol.* **1999**, *45*, 114–119.
93. Chang, J.H.; Shin, J.H.; Chung, I.S.; Lee, H.J. Improved menthol production from chitosan-elicited suspension culture of *Mentha piperita*. *Biotechnol. Lett.* **1998**, *20*, 1097–1099. [CrossRef]
94. Salimgandomi, S.; Shabrangy, A. The effect of chitosan on antioxidant activity and some secondary metabolites of *Mentha piperita* L. *J. Pharm. Health Sci.* **2016**, *4*, 135–142.
95. Mathew, R.; Sankar, P.D. Effect of methyl jasmonate and chitosan on growth characteristics of *Ocimum basilicum* L., *Ocimum sanctum* L. and *Ocimum gratissimum* L. cell suspension cultures. *Afr. J. Biotechnol.* **2012**, *11*, 4759–4766. [CrossRef]
96. Kim, H.; Chen, F.; Wang, X.; Rajapakse, N.C. Effect of chitosan on the biological properties of sweet basil (*Ocimum basilicum* L.). *J. Agric. Food Chem.* **2005**, *53*, 3696–3701. [CrossRef]
97. Putalun, W.; Luealon, W.; De-Eknamkul, W.; Tanaka, H.; Shoyama, Y. Improvement of artemisinin production by chitosan in hairy root cultures of *Artemisia annua* L. *Biotechnol. Lett.* **2007**, *29*, 1143–1146. [CrossRef]
98. Kahromi, S.; Khara, J. Chitosan stimulates secondary metabolite production and nutrient uptake in medicinal plant *Dracocephalum kotschy*. *J. Sci. Food Agric.* **2021**, *101*, 3898–3907. [CrossRef] [PubMed]
99. Elateeq, A.; Saad, Z.; Eissa, M.A.; Ullah, S. Effect of chitosan and light conditions on the production of callus biomass, total flavonoids and total phenolics in *Ginkgo biloba* L. *J. Agric. Res.* **2021**, *46*, 28–42. [CrossRef]

100. Taghizadeh, M.; Nekonan, M.; Setorki, M. Enhancement production of phenolic compounds in the cell suspension culture of *Iberis amara* L.: The effect of chitosan elicitation. *Plant Cell Tissue Organ Cult.* 2021; *submitted.* [CrossRef]
101. Jiao, J.; Gai, Q.-Y.; Wang, X.; Qin, Q.-P.; Wang, Z.-Y.; Liu, J.; Fu, Y.-J. Chitosan elicitation of *Isatis tinctoria* L. hairy root cultures for enhancing flavonoid productivity and gene expression and related antioxidant activity. *Ind. Crop. Prod.* **2018**, *124*, 28–35. [CrossRef]
102. Salehi, S.; Rezayatmand, Z.; Ghasemi Pirbalouti, A. The effect of foliar application of chitosan on yield and essential oil of savory (*Satureja isophylla* L.) under salt stress. *J. Herb. Drug* **2017**, *8*, 101–108. [CrossRef]
103. Shah, M.; Jan, H.; Drouet, S.; Tungmunnithum, D.; Shirazi, J.H.; Hano, C.; Abbasi, B.H. Chitosan elicitation impacts flavonolignan biosynthesis in *Silybum marianum* (L.) Gaertn. cell suspension and enhances antioxidant and anti-inflammatory activities of cell extracts. *Molecules* **2021**, *26*, 791. [CrossRef]
104. Sun, C.; Zeng, Z.; Cui, H.; Verheggen, F. Polymer-based nanoinsecticides: Current developments, environmental risks and future challenges. A review. *Biotechnol. Agron. Soc. Environ.* **2020**, *24*, 59–69. [CrossRef]
105. Sadiq, A.C.; Olasupo, A.; Ngah, W.S.W.; Rahim, N.Y.; Suah, F.B.M. A decade development in the application of chitosan-based materials for dye adsorption: A short review. *Int. J. Biol. Macromol.* **2021**, *191*, 1151–1163. [CrossRef]
106. Maluin, F.N.; Hussein, M.Z. Chitosan-based agronanochemicals as a sustainable alternative in crop protection. *Molecules* **2020**, *25*, 1611. [CrossRef]
107. Yu, J.; Wang, D.; Geetha, N.; Khawar, K.M.; Jogaiah, S.; Mujtaba, M. Current trends and challenges in the synthesis and applications of chitosan-based nanocomposites for plants: A review. *Carbohydr. Polym.* **2021**, *261*, 117904. [CrossRef]
108. Qavami, N.; Naghdi Badi, H.; Labbafi, M.R.; Mehregan, M.; Tavakoli, M.; Mehrafarin, A. Overview on chitosan as a valuable ingredient and biostimulant in pharmaceutical industries and agricultural products. *Trakia J. Sci.* **2017**, *15*, 83–91. [CrossRef]
109. Mukhtar Ahmed, K.B.; Khan, M.M.A.; Siddiqui, H.; Jahan, A. Chitosan and its oligosaccharides, a promising option for sustainable crop production–A review. *Carbohydr. Polym.* **2020**, *227*, 115331. [CrossRef] [PubMed]
110. Van, S.N.; Minh, H.D.; Anh, D.N. Study on chitosan nanoparticles on biophysical characteristics and growth of Robusta coffee in green house. *Biocatal. Agric. Biotechnol.* **2013**, *2*, 289–294. [CrossRef]
111. Sathiyabama, M.; Manikandan, A. Foliar application of chitosan nanoparticle improves yield, mineral content and boost innate immunity in finger millet plants. *Carbohydr. Polym.* **2021**, *258*, 117691. [CrossRef] [PubMed]
112. Kocięcka, J.; Liberacki, D. The potential of using chitosan on cereal crops in the face of climate change. *Plants* **2021**, *10*, 1160. [CrossRef] [PubMed]
113. Wang, A.; Li, J.; AL-Huqail, A.A.; AL-Harbi, M.S.; Ali, E.F.; Wang, J.; Ding, Z.; Rekaby, S.A.; Ghoneim, A.M.; Eissa, M.A. Mechanisms of chitosan nanoparticles in the regulation of cold stress resistance in banana plants. *Nanomaterials* **2021**, *11*, 2670. [CrossRef]
114. Silva, A.O.; Cunha, R.S.; Hotza, D.; Machado, R.A.F. Chitosan as a matrix of nanocomposites: A review on nanostructures, processes, properties, and applications. *Carbohydr. Polym.* **2021**, *272*, 118472. [CrossRef]
115. Choudhary, R.C.; Kumaraswamy, R.V.; Kumari, S.; Sharma, S.S. Zinc encapsulated chitosan nanoparticle to promote maize crop field. *Int. J. Biol. Macromol.* **2019**, *127*, 126–135. [CrossRef]
116. Asgari-Targhi, G.; Iranbakhsh, A.; Oraghi Ardebili, Z.; Hatami Tooski, A. Synthesis and characterization of chitosan encapsulated zinc oxide (ZnO) nanocomposite and its biological assessment in pepper (*Capsicum annuum*) as an elicitor for *in vitro* tissue culture applications. *Int. J. Biol. Macromol.* **2021**, *189*, 170–182. [CrossRef]
117. Kumar, S.; Chauhan, N.; Gopal, M.; Kumar, R.; Dilbaghi, N. Development and evaluation of alginate-chitosan nanocapsules for controlled release of acetamiprid. *Int. J. Biol. Macromol.* **2015**, *81*, 631–637. [CrossRef]
118. Yadav, M.; Goswami, P.; Paritosh, K.; Kumar, M.; Pareek, N.; Vivekanand, V. Seafood waste: A source for preparation of commercially employable chitin/chitosan materials. *Bioresour. Bioprocess* **2019**, *6*, 8. [CrossRef]
119. Silva Alves, D.C.; Healy, B.; Pinto, L.A.A.; Cadaval, T.R.S., Jr.; Breslin, C.B. Recent developments in chitosan-based adsorbents for the removal of pollutants from aqueous environments. *Molecules* **2021**, *6*, 594. [CrossRef] [PubMed]

molecules

MDPI

Review

Green-Chemical Strategies for Production of Tailor-Made Chitooligosaccharides with Enhanced Biological Activities

Reeba Thomas [1], Tamo Fukamizo [1,2] and Wipa Suginta [1,*]

1 School of Biomolecular Science and Engineering (BSE), Vidyasirimedhi Institute of Science and Technology (VISTEC), Payunai, Wangchan District, Rayong 21210, Thailand; reeba.t_s19@vistec.ac.th (R.T.); tamo0111fuka@gmail.com (T.F.)
2 Department of Advanced Bioscience, Kindai University, 3327-204 Nakamachi, Nara 631-8505, Japan
* Correspondence: wipa.s@vistec.ac.th

Abstract: Chitooligosaccharides (COSs) are b-1,4-linked homo-oligosaccharides of *N*-acetylglucosamine (GlcNAc) or glucosamine (GlcN), and also include hetero-oligosaccharides composed of GlcNAc and GlcN. These sugars are of practical importance because of their various biological activities, such as antimicrobial, anti-inflammatory, antioxidant and antitumor activities, as well as triggering the innate immunity in plants. The reported data on bioactivities of COSs used to contain some uncertainties or contradictions, because the experiments were conducted with poorly characterized COS mixtures. Recently, COSs have been satisfactorily characterized with respect to their structures, especially the degree of polymerization (DP) and degree of *N*-acetylation (DA); thus, the structure–bioactivity relationship of COSs has become more unambiguous. To date, various green-chemical strategies involving enzymatic synthesis of COSs with designed sequences and desired biological activities have been developed. The enzymatic strategies could involve transglycosylation or glycosynthase reactions using reducing end-activated sugars as the donor substrates and chitinase/chitosanase and their mutants as the biocatalysts. Site-specific chitin deacetylases were also proposed to be applicable for this purpose. Furthermore, to improve the yields of the COS products, metabolic engineering techniques could be applied. The above-mentioned approaches will provide the opportunity to produce tailor-made COSs, leading to the enhanced utilization of chitin biomass.

Keywords: chitooligosaccharides; biological activities; chitinase; chitosanase; transglycosylation; glycosynthase

Citation: Thomas, R.; Fukamizo, T.; Suginta, W. Green-Chemical Strategies for Production of Tailor-Made Chitooligosaccharides with Enhanced Biological Activities. *Molecules* **2023**, *28*, 6591. https://doi.org/10.3390/molecules28186591

Academic Editor: Agnieszka Ewa Wiącek

Received: 26 May 2023
Revised: 23 August 2023
Accepted: 30 August 2023
Published: 13 September 2023

1. Introduction

Chitin, a b-1,4-linked polysaccharide of *N*-acetylglucosamine (GlcNAc), is the second most abundant biomass on the earth, next to cellulose. Three forms of chitin are present; α-chitin, the most abundant and tightly compacted crystalline structure, is arranged in an anti-parallel fashion [1,2]. α-chitin forms are commonly found in arthropods including crabs, lobsters, and shrimp, and also in fungal cell walls [3]. β-chitin, in contrast to α-chitin, has weak hydrogen bonding arranged in parallel forms, resulting in weak intramolecular forces, and is distributed in diatoms, squid pens, and annelids [4]. The least form is γ-chitin, which contains a mixture of both parallel and anti-parallel forms [4]. De-*N*-acetylated derivatives of chitin, chitosan, are mainly found in fungal cell walls and exhibit two forms, Type I and Type II [5,6]. Modern solid-state NMR revealed the polymorphism of these polysaccharide structures [7,8]. Although a similar polysaccharide, cellulose (a glucan having b-1,4-glycosidic linkages), is most efficiently utilized as an essential material in our daily life, the utilization of chitin and chitosan is much less intensive. Many attempts have been made to enhance the utilization of these polysaccharides [9], but effective approaches still remain to be developed. In previous attempts at enhancing chitin utilization, chitin and chitosan were degraded to chitooligosaccharides (COS) by acid or enzymatic hydrolysis [10], and the oligosaccharides were characterized with respect to their antimicrobial, anti-inflammatory, antioxidant, immune-potentiating, and antitumor activities [11]. Nowadays, COSs are well known to exhibit such biological activities in vitro or in vivo. These activity data were obtained using COS mixtures, which included b-1,4-linked oligosaccharides, composed of *N*-acetylglucosamine (GlcNAc), glucosamine (GlcN), or both (Figure 1); hence, they are not fully characterized with respect to their chemical structures.

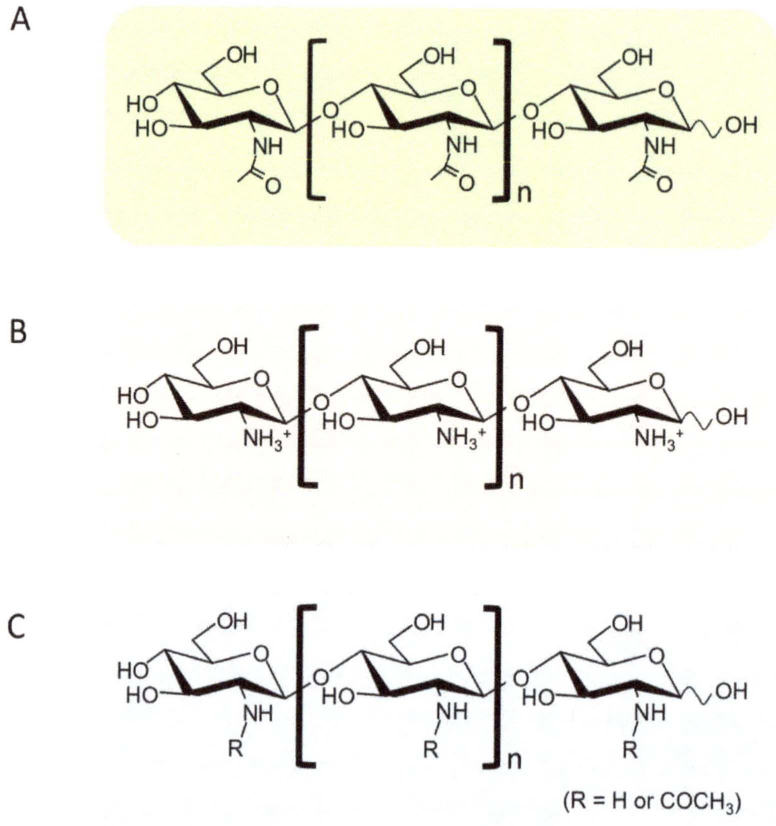

Figure 1. The molecular structures of COSs with different bioactivities. (**A**) Chitooligosaccharides $(GlcNAc)_n$, n = 1, 2, 3, ..., (**B**) chitosan oligosaccharides $(GlcN)_n$, n = 1, 2, 3, ..., (**C**) partially *N*-acetylated chitooligosaccharides, ······GlcNAc-GlcN······ n = 1, 2, 3, ...

However, recent developments in separation and analytical methods for carbohydrates have enabled exact determinations of the degrees of polymerization (DP) and of *N*-acetylation (DA), as well as sequences, which can be analyzed by NMR, IR, and mass spectrometry [12–15]. In fact, the biological activities of COS were found to be controlled by the three structural factors: DP, DA, and sequence. For example, the antioxidant and anti-inflammatory activities of COS were dependent on their DP, as well as on DA [16,17]. Similar dependencies were also observed for other biological activities, as described previously [11]. Furthermore, it was proposed that the specific sequences composed of GlcNAc and GlcN, so called "ChitoCode", were recognized by some proteins with significant functions in fungal interactions with other organisms [18]. Similar mechanisms of molecular recognition may exist in other interaction systems composed of living organisms that have chitin/chitosan as structural components. To obtain COSs with the desired biological activities, it has now become possible to design and synthesize COSs with specified sequences, so called tailor-made COSs. Here, we have made a comprehensive review of the biological activities of COSs and discussed their structure–function relationship. This review also underlines the green-chemical methods reported for the synthesis of pure COSs with designated sequences. These methods are considered promising approaches for the production of high-quality COSs with enhanced biological activities (Figure 2).

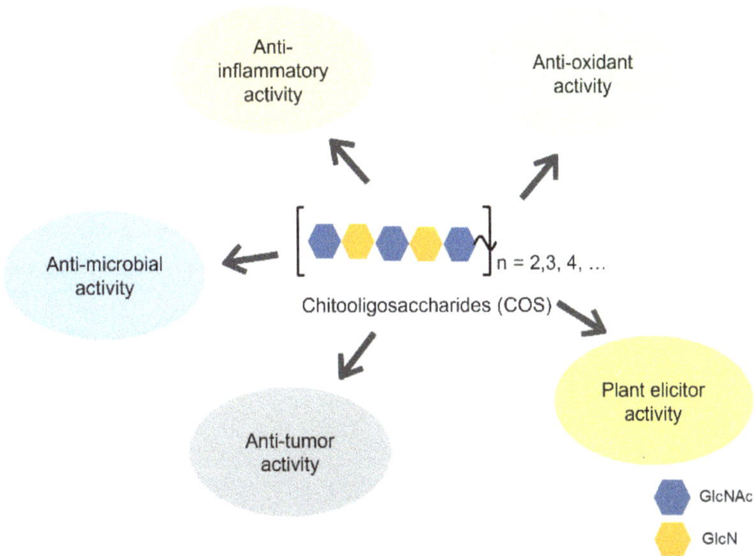

Figure 2. A summary of biological activity of COSs.

2. Biological Activities of COSs

2.1. Antimicrobial Activity

COSs have been recognized to exhibit antimicrobial activities, inhibiting the growth of a variety of phytopathogenic and animal-pathogenic fungi, bacteria, and viruses. The bactericidal activities of COSs with higher (5–10 kDa), medium (1–5 kDa), and lower molecular weights (<1 kDa) were evaluated by counting the colony formed on the medium, comparing to the control, and found to be 98, 62, and 51% toward *Escherichia coli* (Gram-negative); 47, 35, and 22% toward *Pseudomonas aeruginosa* (Gram-negative); 82, 57, and 23% toward *Staphylococcus epidermidis* (Gram-positive); and 98, 63, and 57% toward *Lactobacillus bulgaricus*, respectively [19]. COSs with higher molecular weights had greater antimicrobial effects than those with medium and lower ones. Higher DPs were also reported to be beneficial for exerting the antibacterial activities of COSs [20,21]. It is likely that the positive charges of COSs interact with negative charges on the bacterial cell surfaces, leading to inhibition of cell growth. In contrast, the shorter COSs (DA = 0.15; DP = 2, 3, and 4) and low molecular weight chitosan produced by a chitosanase from *Bacillus thuringiensis* B-387 from chitosan polysaccharide (DA, 0.15), of which the average MWs were in a range of 14–46 kDa, exhibited greater antibacterial activities toward *E. coli* in vitro (MIC, 70 $mg{\cdot}mL^{-1}$) in comparison with the initial polysaccharides (MIC, 120 $mg{\cdot}mL^{-1}$) [22]. Ganan et al. analyzed the antifungal activities of soluble and well-defined COSs. Determination of the antifungal activities of COS fractions with varying average DPs (17, 31, 54, and 62) indicated that fractions with intermediate DPs (31 and 54) had the strongest inhibitory effects; MICs toward *Candida guillermondii* were 78, 39, 39, and 1250 $mg{\cdot}mL^{-1}$, respectively [23]. Another report proposed that short-chain COSs exhibited high fungicidal activities toward the *Candida* species [24]. The contradictions in these findings may arise from differences in COS-preparation methods and the targeted microorganisms, as well as in experimental conditions. The antimicrobial activities may depend not only on the electrostatic interactions of free amino groups of COSs but also on the interaction of GlcNAc residues with the hydrophobic regions of microbial cell surfaces. A summary of the antimicrobial activities of COSs is presented in Table 1.

Table 1. A summary of antimicrobial activity of COSs.

MW or DP	Source, DA, or Sequence	Activity	Assay	Ref.
Higher MW (5–10 kDa), Medium MW (1–5 kDa), and Lower MW (<1 kDa)	From chitosan	Higher MW > Medium MW > Lower MW	Growth inhibition toward Gram-positive, Gram-negative, and lactic acid bacteria	[19]
Chitosan polymers and Chitosan oligomers	From chitosan	Polymers > Oligomers	Growth inhibition toward *Staphyllococcus aureus* Gram-positive and Gram-negative bacteria	[20,21]
Chitosan polymers and DPs of 2–4, and Lower MW chitosans	Chitosan with DA, 0.15 or 0.5	DP of 2–4 and Lower MW chitosans > Initial polymers	Growth inhibition toward fungi and bacteria	[22]
Chitosans with average DPs of 17, 31, 54, and 62	From chitosan (DA, 0.15)	Chitosans with DP of 31 and 54 > Chitosans with DP of 17 and 62	Growth inhibition toward yeast, fungi, including *Candida*	[23]
Chitosans (70 and 600 kDa) Chitosan oligomers	From chitosan	Oligomers > Chitosans	Growth inhibition toward *Candida*	[24]

2.2. *Anti-Inflammatory Activity*

Four COSs with different DAs (0, 0.12, 0.5, and 0.85) and the same DP distributions were employed for assays of anti-inflammatory activity [17]. The results revealed that COSs with a DA of 0.12 had greater anti-inflammatory activity than COSs with other DAs. Another study demonstrated that low-MW COSs with DP of 3–5 obtained from chitosan were found to be effective against allergic inflammation in vivo [25]. In support of the fact that low-MW COSs possess the highest anti-inflammatory activity, Pangestuti et al. reported that depolymerized products from chitosan (MW < 10 kDa) reduced the production of nitric oxide and prostaglandin E2 by inhibiting inducible nitric oxide synthase (iNOS) and cyclooxygenase expression [26]. The effects were clearly dependent on the MW of COSs. The nitric oxide productions in BV2 microglia cell were about 10, 15, 20, and 23 mM for the treatment with 1, 1–3, 3–5, and 5–10 kDa COSs, respectively, while it was 25 mM in the control. The prostaglandin E2 productions were about 0.8, 1.0, 1.2, and 1.4 $ng \cdot ml^{-1}$, while it was 1.9 $ng \cdot ml^{-1}$ in the control. The effect of COSs (MW < 1 kDa) on nitric oxide production was further studied by Wei et al. [27]. Pretreatment with this COS could inhibit the production of nitric oxide by suppressing iNOS expression in activated microglial cells. Sanchez et al. [28] prepared COSs composed of 42% fully de-*N*-acetylated and 54% mono-*N*-acetylated oligomers through a single enzymatic hydrolysis step. This combination of COSs demonstrated anti-inflammatory effects [28]. Taken together, COSs with lower DPs (3–5) and a lower DA significantly attenuated the inflammation. It is notable, however, that completely de-*N*-acetylated COSs (DA = 0) exhibited no effective anti-inflammatory activity. The anti-inflammatory effects of COSs are summarized in Table 2.

Table 2. A summary of anti-inflammatory bioactivity of COSs.

MW or DP	Source, DA, or Sequence	Activity	Assay	Ref.
COSs with similar DP distribution	Different DAs of 0, 0.12, 0.50, and 0.85	COSs with DA of 0.12 have the highest activity	Inhibition of LPS-induced inflammatory cytokine burst, down-regulating its mRNA expression, and reducing phosphorylation of IκBα	[17]
Chitosan oligomer mixture (DPs of 3–5)	DA of 0.0	Active toward allergic asthma inflammation in vivo	Inhibition of degranulation, cytokine generation in RBL-2H3 cells, and lung inflammation in allergic asthma model mice	[25]
Chitosan oligomers with MW < 1 kDa, 1–3 kDa, 3–5 kDa, and 5–10 kDa	From chitosan	The lower the MW, the higher the activity	Inhibition of iNOS and cyclooxygenase expressions	[26]
Chitosan oligomers with DP of 2–6 (weight percentages; 4, 16, 29, 37, and 14%, respectively)	From chitosan with DA < 0.05	Active toward N9 microglia cells	Inhibition of NO production by suppressing iNOS expression	[27]
42% fully de-*N*-acetylated/54% mono-*N*-acetylated oligomers (42/54) and 50% fully de-*N*—acetylated/27% mono-*N*-acetylated oligomers (50/27)	From chitosan	42/54 attenuated the inflammation both in vivo and in vitro, but 50/27 promoted the inflammatory response	Effects of COS preparations on inflammation in lipopolysaccharide-induced mice and in RAW264.7 macrophages	[28]

2.3. *Antioxidant Activity*

Fully *N*-acetylated COSs (MW = 229.21–593.12 Da; DP = 1, 2, and 3) produced by the acid hydrolysis of crab chitin, which may contain a small fraction of de-*N*-acetylated residues, was found to inhibit myeloperoxidase activity and decrease free radicals in human myeloid cells, HL-60, or suppress oxidation of DNA and membrane proteins in mouse macrophages Raw 264.7 [29]. In the same work, COSs were shown to also directly scavenge free radical-induced DNA oxidation generated by 2′,7′-dichlorofluorescein, as well as increase intracellular glutathione levels, which confirms the antioxidant properties of COSs in living cells. Enzymatic hydrolysates from housefly larvae powder (HLP) contained mainly hetero-COS composed of GlcNAc and GlcN (DP = 2–6) [30]. Maximum production of COSs with a DP of 5 (51.2 $\mu g \cdot mL^{-1}$) was achieved by hydrolysis of HLP for 72 h and COSs with a DP of 5 exhibited the strongest hydroxyl-scavenging ability and reducing power. COSs obtained from chitosan with a DA of 0.05 have a protective effect on H_2O_2-induced human umbilical vein endothelial cell apoptosis and also from H_2O_2-induced oxidative damage in endothelial cells [31,32]. Five COSs from chitosan with increasing DPs (3–7) were examined and showed effects on the antioxidant activity of soybean (*Glycine max*) seeds during germination [33]. The COSs of different DPs exhibited varying influences on the antioxidant activity in soybean seeds. In particular, COSs with DP = 6 exerted a strong effect and significantly increased the antioxidant activity. Hao et al. [34] produced COS dimers with different sequences, GlcNAc-GlcNAc, GlcN-GlcNAc, GlcNAc-GlcN, and GlcN-GlcN, by the enzymatic method, and the antioxidant activities of all four dimers were studied. When 1 $mg \cdot ml^{-1}$ of the dimers was respectively added, the scavenging effects of the superoxide radicals were about 47 and 66% for GlcN-GlcN and GlcNAc-GlcN, respectively, and 12 and 8% for GlcN-GlcNAc and GlcNAc-GlcNAc. The amino group at the reducing end played a vital role in scavenging superoxide radicals. Furthermore, they found that GlcN-GlcN showed the highest 2,2-diphenyl-1-picrylhydrazyl scavenging activity. However, the scavenging effects of hydroxyl radicals were 45 and 42% for GlcNAc-GlcNAc and GlcN-GlcNAc, and 31 and 32% for GlcN-GlcN and GlcNAc-GlcN upon the addition of 4 $mg \cdot mL^{-1}$ of individual dimers. Although antioxidant activities were dependent on the radical species used in the experiments, GlcN residues in COSs were likely to dominate their antioxidant activity. COSs with a DP of 5–6 appeared to have higher antioxidant activity [30,33]. Antioxidant activities of COSs are summarized in Table 3.

Table 3. A summary of antioxidant bioactivity of COSs.

MW or DP	Source, DA, or Sequence	Activity	Assay	Ref.
COS mixture with DP of 1, 2, and 3	Fully *N*-acetylated crab chitin (DA about 1.0)	• Attenuated peroxidase activity • Attenuated oxidation of DNA/membrane proteins • Exhibited free-radical scavenging effects	Experiments using human myeloid cells, HL-60, and mouse macrophages RAW264.7	[29]
Hetero-COSs with DP of 2–6	From housefly larvae powder	COSs with major DP of 5 exhibited the highest activity	Hydroxyl-scavenging effects	[30]
Chitosan oligomers with DP of 2–6 (weight percentages: 4, 16, 29, 37, and 14%, respectively)	From chitosan with DA < 0.05	Protected from H_2O_2-induced apoptosis or oxidative damage	Experiments using human umbilical vein endothelial cell	[31,32]
COSs with different DPs (3–7)	From chitosan with DA of 0.0	COSs with DP of 6 had the highest activity	Effects on isoflavone contents and antioxidant activity in soyabean seeds during germination	[33]
COS dimers, $(GlcNAc)_2$, GlcN-GlcNAc, GlcNAc-GlcN, and $(GlcN)_2$	Site-specific chitin deacetylase treatments of $(GlcNAc)_2$	• (GlcN)2 > GlcNAc-GlcN >> GlcN-GlcNAc, (GlcNAc)2 toward superoxide radicals • (GlcNAc)2 > GlcN-GlcNAc >> GlcNAc-GlcN, (GlcN)2 toward hydroxyl radicals • (GlcN)2 >> GlcNAc-GlcN, GlcN-GlcNAc, (GlcNAc)2 toward DPPH	All scavenging effects were examined in vitro	[34]

2.4. Antitumor Activity

$(GlcNAc)_6$ was found to display an anti-metastatic effect on tumor cells. Approximately 1 mg.kg^{-1} of $(GlcNAc)_6$ administered intravenously resulted in a significant decrease in the pulmonary nodules formed from the Lewis lung carcinoma in mice [35]. However, the antitumor actions of COSs in vivo included various biological processes, such as absorption of COSs through the intestinal epithelium, distribution and accumulation in the kidney, liver, and spleen, enzymatic transformation, and elimination from the body. The DP, DA, and sequence of COSs may significantly affect the individual processes. Therefore, at present it is impossible to define clearly the structure–activity relationship for antitumor activity in vivo. Salah et al. [36] reported that low-MW chitin (2480 Da) obtained from shrimp shells was more effective against a human leukemia cell line than the higher-MW products (338 kDa chitin and 12 kDa chitosan). COSs with higher DA (0.15) exhibited 50 mg·mL^{-1} of CC_{50}, the concentration of COSs required for 50% cell death of human PC3, A549, and HepG2 cells, while COSs with lower DA (0.01) exhibited 5–25 mg·mL^{-1} of CC_{50} [37]. A decrease in the MW or DA of COSs resulted in enhancement of the in vitro anticancer activity. However, it appeared that a DP of at least 6 is required to exert significant antitumor activity [11,38]. These findings on COS antitumor activities were recently reviewed by Zhai et al. [39]. The antitumor activities of COSs are outlined in Table 4.

Table 4. A summary of antitumor bioactivities of COSs.

MW or DP	Source, DA or Sequence	Activity	Assay	Ref.
$(GlcNAc)_6$	From crab chitin	Decreased the pulmonary nodules	Experiments using Lewis lung carcinoma mice	[35]
Chitin, chitosan 2.5–338 kDa	From shrimp shell chitin	Chitin (2.5 kDa) > chitin (338 kDa), chitosan (12 kDa)	Experiments using human leukemia cells	[36]
Chitosan (DA, 0.015) and its enzymatic digestion products; COSs with DA, 0.0/DP, 3–5 and COSs with DA, 0.15/DP, 6–15	Enzymatic hydrolysis of high molecular weight chitosan with DA of 0.015	COSs with lower MW> COSs with higher MW COSs with lower DA> COSs with higher DA	Experiments using prostate and lung cancer cells, and hepatoma cells	[37]
Chitosan oligomers with DP of 2–6	Chemical and enzymatic hydrolysis of chitosan	DP should be at least 6 for antitumor action	Inhibitory effect on A549 cell proliferation	[38]

2.5. Plant Elicitor Activity

Vander et al. [40] evaluated the capabilities of $(GlcNAc)_n$ (n = 4, 5, 6, 7, 8, 9, and 10), $(GlcN)_n$ (n = 5, 6, and 7) and partially *N*-acetylated chitosan to induce phenylalanine ammonia-lyase (PAL) and peroxidase (POD) in healthy wheat leaves. They found that purified $(GlcN)_n$ were not active as plant elicitors, whereas purified $(GlcNAc)_n$ ($n \geq 7$) strongly elicited POD activities but not PAL activities. Partially *N*-acetylated chitosans elicited both PAL and POD activities, and maximum elicitation was observed with chitosan oligosaccharides with intermediate DAs. Ramakrishna et al. [41] reported the elicitor activities in rice seedlings of $(GlcNAc)_n$ (n = 5–7) produced by a hypertransglycosylating mutant of a chitinase from *Serratia proteamaculans*. $(GlcNAc)_7$ strongly induced an oxidative burst response, as well as peroxidase and PAL activities. These reports confirmed that $(GlcNAc)_n$ (n = 5–7) are most effective in eliciting an immune response in plants. A chitin elicitor receptor kinase 1 (CERK1) was found to be essential for chitin perception by plant cells [42]. CERK1 was composed of an extracellular domain (ectodomain) with three contiguous LysM motifs (LysM1, LysM2, and LysM3), a transmembrane domain, and an intracellular kinase domain. Crystal structure analysis of the COS-bound ectodomain of CERK1 from *Arabidopsis thaliana* (*At*CERK1-ECD) [43] suggested that $(GlcNAc)_8$ acted as a bivalent ligand that bound to the two *At*CERK1-ECD proteins through a continuous groove formed between one LysM2 and another LysM2, inducing the homo-dimerization of two *At*CERK1-ECD proteins. This dimerization formed an active receptor complex and was regarded as crucial for the immune response. In *A. thaliana*, $(GlcNAc)_8$ binding to AtLYK5, another receptor containing three LysM motifs, induced the homo-dimerization, recruiting the two *At*CERK1 proteins and eventually forming an active receptor complex [44]. In rice, a chitin elicitor binding protein (*Os*CEBiP) composed of three LysM motifs [45] may first bind the $(GlcNAc)_8$ elicitor, inducing the dimerization of *Os*CEBiP, and subsequently forming a hetero-tetramer receptor complex composed of two *Os*CEBiP and two *Os*CERK1. Hayafune et al. [46] found that hetero-COS DP-8 containing alternating GlcN and GlcNAc, $(GlcN\text{-}GlcNAc)_4$, did not elicit an immune response in rice plants and hence was likely to inhibit formation of the hetero-tetramer complex. A hetero-COS with a DP of 6, GlcNAc-GlcN-$(GlcNAc)_4$, elicited the immune response in *A. thaliana* but the response was in a lower level caused by $(GlcNAc)_6$. In the case of another hetero-COS with a DP of 6, GlcNAc-GlcN-$(GlcNAc)_2$-GlcN-GlcNAc, the immune response was eliminated almost completely [47]. Although the binding mode of COS elicitor to chitin receptor complex is still controversial [47], the *N*-acetylated residues were confirmed to be essential for the immune response in plants. The biological activities of COSs as plant elicitors are summarized in Table 5.

Table 5. A summary of COSs as plant elicitors.

MW or DP	Source, DA, or Sequence	Activity	Assay	Ref.
$(GlcNAc)_n$ (n = 4, 5, 6, 7, 8, 9, and 10), $(GlcN)_n$ (n = 5, 6, and 7) and partially *N*-acetylated chitosans	Chemical hydrolysis of chitin (fluorolysis)deacetylation of the high-MW chitin	• (GlcN)n were not active as elicitors. (GlcNAc)n with a DP ≥7 elicited POD but not PAL. • Partially N-acetylated chitosans elicited both PAL and POD.	Induction of phenylalanine ammonia-lyase (PAL), peroxidase (POD) in healthy wheat leaves	[40]
$(GlcNAc)_n$ (n = 5, 6, and 7)	Enzymatic transglycosylation	$(GlcNAc)_7$ induced oxidative burst as well as POD and PAL activities.	Induction of phenylalanine ammonia-lyase (PAL), peroxidase (POD), PR protein gene expression in rice seedlings	[41]
$(GlcNAc)_8$ and hetero-COSs with a DP of 8	Enzymatic synthesis	$(GlcNAc)_8$ was active, but $(GlcN\text{-}GlcNAc)_4$ inactive.	Inhibition of CEBiP-dimerization and reactive oxygen generation	[46]
$(GlcNAc)_6$ and two hetero-COSs with DP of 6	Enzymatic deacetylation of $(GlcNAc)_6$	$(GlcNAc)_6$ > GlcNAc-GlcN-$(GlcNAc)_4$ > GlcNAc-GlcN-$(GlcNAc)_2$-GlcN-GlcNAc.	Inhibition of reactive oxygen generation	[47]

3. Synthetic Strategies of COSs with Promising Functions

Although the structure–function relationship of COSs is complex, the available data on COSs of various bioactivities provides information useful for the strategic design of tailor-made COSs with improved biological activities. To obtain COSs with desired sequences different green-chemical strategies have been reported, as described below.

3.1. Use of Transglycosylation (TG) Activity in Chitinolytic Enzymes

Many glycoside hydrolases with anomeric-retaining mechanism have been recognized to catalyze transglycosylation (TG) reactions to some extent [48]. Hen egg-white lysozyme is one of the enzymes that catalyze TG in addition to hydrolysis [49]. TG in the reaction catalyzed by lysozyme was estimated to be efficient from the HPLC-based reaction time-course of COS degradation [50–53] and was utilized for synthesis of COSs with longer chains and a higher degree of crystallinity [54]. Lysozyme TG was also applied to the synthesis of novel inhibitors for hen egg-white lysozyme [55] and a plant elicitor-active oligosaccharides [56]. Like lysozyme, chitinases belonging to the GH18 family were reported to catalyze TG reactions with COSs [57–59]. In these enzymatic TG reactions, native enzymes with full intrinsic hydrolytic activities were used. Therefore, COSs with longer chains produced by TG activity were again hydrolyzed to COSs with shorter chains. It was highly desirable to suppress the hydrolytic activity of these enzymes, although in this case complete loss of the glycosidic bond cleavage should be avoided, because the cleavage of the b-1,4-glycosidic bond activates the anomeric center to the transition state, which is essential for receiving the attack of the acceptor molecule [52,53]. It is necessary to employ some technique that suppresses hydration of the anomeric center in the transition state, while maintaining glycosidic bond cleavage activity. Fukamizo et al. reported enhanced TG activity in a chemically modified lysozyme, in which bulky residues, such as glucosamine and *p*-nitrophenyl-sulfenyl moieties, were introduced into the subsites −4, −3, and −2 of the lysozyme binding cleft [60]. They explained that the enhancement of TG activity was derived from the lower binding ability of the glycone-binding site (negatively numbered subsites), which, in turn, enhanced the acceptor binding to the aglycone-binding site (positively numbered subsites). However, the chemical methods seemed to be applicable only to inexpensive enzymes that are easy to isolate, such as lysozyme.

A GH2 exo-b-glucosaminidase from *Amycolatopsis orientalis* was reported to split off the non-reducing end GlcN unit from the $(GlcN)_n$ (n = 2, 3, 4, . . .) substrates. This enzyme was found to catalyze a TG reaction in addition to hydrolysis; that is, from the $(GlcN)_4$ substrate, $(GlcN)_5$ and $(GlcN)_6$ were formed as TG products [61]. After hydrolysis, the transition state of the GlcN monomer was transferred to the remaining $(GlcN)_4$, which in turn acted as an acceptor, producing $(GlcN)_5$. Similarly, the TG product $(GlcN)_5$ subsequently acted as an acceptor, producing $(GlcN)_6$. A GH5 chitosanase from *Streptomyces griseus* HUT 6037 was also reported to catalyze a TG reaction. When incubated with this enzyme, $(GlcN)_5$ was hydrolyzed to $(GlcN)_2$ + $(GlcN)_3$ or $(GlcNAc)_3$ + $(GlcN)_2$. When the same reaction was conducted in the presence of an excess amount of $(GlcNAc)_3$, TG products, including $(GlcN)_2$-$(GlcNAc)_3$ and $(GlcN)_3$-$(GlcNAc)_3$, were formed [62].

3.2. Mutation Strategies for Enhancing TG Activity

Various mutations have been attempted to enhance the TG activities of GH18 chitinases. Aronson et al. [63] reported for the first time that the mutation of a tryptophan residue in the substrate-binding cleft of *Serratia marcescens* chitinase A strongly enhanced the transglycosylation activity. The mutated tryptophan residue (Trp167) was located at subsite −3; therefore, the mechanism of TG enhancement was similar to that in the chemically modified lysozyme mentioned above [60]. Mutation of the middle Asp residue of the catalytic DxDxE motif of GH18 chitinases was also found to be useful for obtaining an efficiently transglycosylating chitinase [64,65]. Electrostatics in the catalytic cleft may be affected by the Asp mutation, probably resulting in the change in the state of the catalytic water molecule. Madhuprakash et al. [66,67] thoroughly mutated the amino acids localized in the catalytic center and the groove of a bacterial GH18 chitinase, indicating that the triple mutations, which reduce the hydrolytic activity, binding affinity, and stability of intermediate states, provided hypertransglycosylating mutants. Furthermore, they suggested the importance of optimal positioning of the catalytic water molecule and the acceptor molecule surrounding the catalytic center. In addition to these bacterial chitinases, plant chitinases were also employed in producing transglycosylating chitinases. Introducing the Trp side chain into the upper portion of the catalytic center of GH18 chitinases from *Arabidopsis thaliana* (*At*ChiC) and *Cycas revoluta* (*Cr*ChiA) was found to strongly enhance TG activities of the enzymes [68,69]. The enhanced hydrophobicity of this region may change the state of the catalytic water, resulting in the suppression of attack by the water molecule and thus enhancing the TG reaction. In particular, the *Cr*ChiA mutant, in which the Trp side chain was introduced, exhibited hypertransglycosylating activity [69].

3.3. Converting Chitinolytic Enzymes to Glycosynthase

Glycosynthase was first developed from anomer-retaining glycoside hydrolases by mutation of the catalytic nucleophile [70,71]. In 2006, Honda and Kitaoka reported the first glycosynthase from an inverting glycoside hydrolase [72]. Since the retaining GH18 chitinases did not have a corresponding nucleophile in the catalytic center [73], inverting chitinases belonging to GH19 were employed for developing glycosynthase [74,75]. GH19 *Bryum coronatum* chitinase (*Bc*ChiA) mutants, in which the serine residue fixing a catalytic water molecule was mutated to alanine, cysteine, or glycine, successfully catalyzed glycosyl transfer of the activated COS fluoride (COS-F, Figure 3A) to the acceptor COS. Using these *Bc*ChiA mutants, $(GlcNAc)_4$ was synthesized from $(GlcNAc)_2$-fluoride (donor) and $(GlcNAc)_2$ (acceptor). The substrate-binding groove of *Bc*ChiA consists of four subsites, −2, −1, +1, and +2, while that of a GH19 chitinase from rye seeds (RSC-c) consists of eight subsites, −4, −3, −2, −1, +1, +2, +3, and +4. It appeared that glycosynthase derived from RSC-c produces COSs with longer chains. In fact, the double mutants, in which Glu89, acting as a catalytic base and Ser120, fixing a catalytic water molecule, were mutated, produced $(GlcNAc)_7$ from $(GlcNAc)_3$-fluoride (donor) and $(GlcNAc)_4$ (acceptor).

Glycosynthase derived from hen egg-white lysozyme (HEL-D52S) was produced and found to catalyze the glycosyl transfer. Using GlcN-$(GlcNAc)_2$-fluoride obtained from $(GlcNAc)_3$-fluoride by the action of a site-specific chitooligosaccharide *N*-deacetylase, the HEL-D52S glycosynthase allowed the size-controlled synthesis of GlcN-$(GlcNAc)_5$, GlcN-$(GlcNAc)_6$, and GlcN-$(GlcNAc)_7$ from the acceptors $(GlcNAc)_3$, $(GlcNAc)_4$, and $(GlcNAc)_5$, respectively. The use of the site-specific deacetylase avoided a condensation reaction of the donor substrate itself [76].

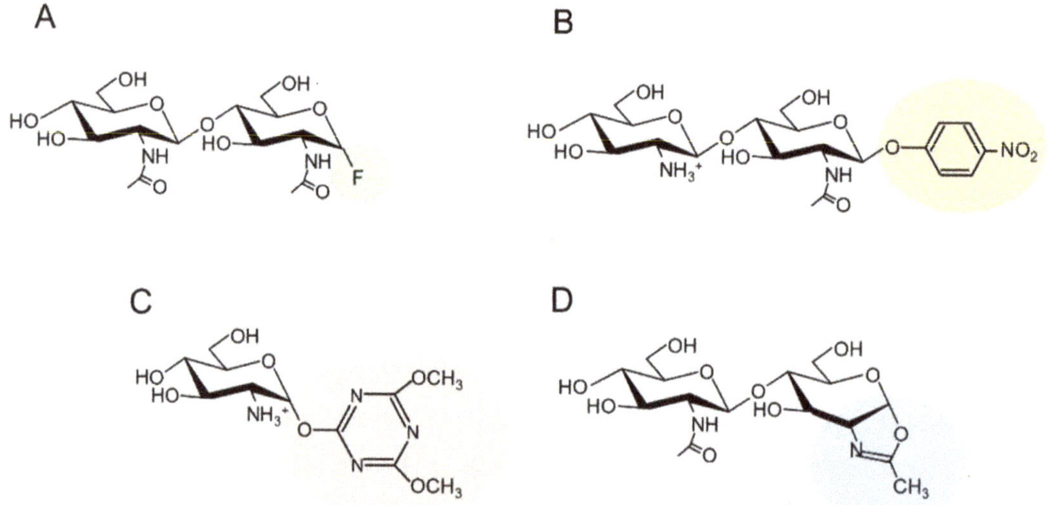

Figure 3. The chemical structures of activated COSs used for different green-chemical synthesis. (**A**) COS-fluoride, (**B**) *p*-nitrophenyl GlcN-GlcNAc dimer, (**C**) GlcN-DMP, (**D**) $GlcNAc)_n$-oxazoline.

3.4. Use of Activated Sugars as Donor Substrates

It is well known that *p*-nitrophenyl glycosides are efficient donor substrates in glycoside synthesis [77]. Harmsen et al. [78] reported the efficient synthesis of COSs with alternating GlcNAc and GlcN using *p*-nitrophenyl GlcN-GlcNAc dimer (GlcN-GlcNAc-*p*NP, Figure 3B), which was obtained by enzymatic deacetylation of GlcNAc-GlcNAc-*p*NP, and *Serratia* GH18 chitinase mutants as the catalysts. The mutants efficiently produced the alternating COSs composed of GlcN and GlcNAc, i.e., $(DA)_2$, $(DA)_3$, $(DA)_4$, and $(DA)_5$, which are possible candidates for inhibitors of chitin-related enzymes and proteins, such as human chitotriosidase, a therapeutic target for systemic sclerosis [78]. $(DA)_4$ was also obtained from oxazoline derivatives of DA and was confirmed to be an inhibitor of chitin elicitor binding protein in plants [46].

Production of reducing end-activated COSs, such as glycosyl fluoride and *p*-nitrophenyl glycoside, used to require protection/deprotection procedures unfamiliar to enzymologists. Nowadays, however, protection-free methods for obtaining activated sugars have been developed and used for reducing end-activation of COSs. Tanaka et al. reported the protection-free synthesis of reducing end-activated sugar derivatives, in which the 4.6-dimethoxy-1,3,5-triazin-2-yl (DMT) group was introduced at the anomeric center of the reducing end [79]. GlcN-DMT (Figure 3C) was synthesized and used as the glycosyl donor for synthesizing $(GlcN)_n$ with longer chains using GH2 exo-b-glucosaminidase as a template [80]. Furthermore, $(GlcNAc)_2$-DMT (Figure 3C) was synthesized and used as a glycosyl donor for synthesizing $(GlcNAc)_4$. A glycosynthase mutant derived from GH19 *Bc*ChiA comprising subsites −2, −1, +1, and +2 successfully synthesized $(GlcNAc)_4$ from $(GlcNAc)_2$-DMT and $(GlcNAc)_2$ [81].

Sugar oxazolines have been frequently used as the glycosyl donors and are synthesized by the protection-free method using the water-soluble dehydrating agents, carbodiimides. Noguchi et al. reported that 2-chloro-1,3-dimethylimidazolinium chloride (DMC) was the most suitable for oxazoline synthesis from COSs. The yields were 70–80% [82]. The oxazoline derivatives from $(GlcNAc)_n$ (Figure 3D; $(GlcNAc)_n$-oxa, n = 2, 3, 4, and 5) were purified by HPLC and used as the donor substrates for synthesizing $(GlcNAc)_n$ with longer chains using an activity-reduced mutant from bacterial GH18 chitinase as a template [83]. Various mutations were introduced into bacterial GH18 chitinases, *Sp*ChiD, and $(GlcNAc)_{10}$ was produced from $(GlcNAc)_5$-oxa in a good yield [84]. Hypertransglycosylating mutants of GH18 chitinases from plant origins, *Arabidopsis thaliana* and *Cycas revoluta*, were also used for $(GlcNAc)_n$ synthesis from the sugar oxazolines [85]. The product distribution was controlled by using different combinations of the substrate size of donor/acceptor.

3.5. Use of Site-Specific Chitin Deacetylases

Hembach et al. [86] reported the enzymatic production of a full set of partially *N*-acetylated chitosan tetramers of all possible sequences consisting of GlcN and GlcNAc, through de-*N*-acetylation of $(GlcNAc)_4$ using various site-specific chitin deacetylases and reverse *N*-acetylation of $(GlcN)_4$ in the presence of acetate (2 M) by the same deacetylases. The structures of the products were successfully identified using ultra high-performance liquid chromatography–electrospray ionization–mass spectrometry (UHPLC/ESI/MS) [15,87], which enabled the full separation of the enzymatic products. In UHPLC/ESI/MS, a hydrophilic interaction chromatography (HILIC) column (Acquity UHPLC BEH Amide) was used to separate COS tetramers with different degrees of acetylation and sequences, and gradient elution was performed with 20–80% acetonitrile in water containing 10 mM NH_4HCO_3 and 0.1% (v/v) formic acid, at a flow rate of 0.8 $mL.min^{-1}$ and 75 °C using an appropriate gradient program controlled by an UHPLC system purchased from Dionex Co. The effluents were detected with an ESI-MS detector in positive mode. The sequencing method of the COS tetramers consists of the following procedures: (1) acetylation of free amino groups in COSs using a deuterated reagent; (2) labeling the reducing end with $H_2{}^{18}O$; (3) quantifying COSs using $[^{13}C_2, {}^2H_3]$-labeled internal standards; (4) sequencing by MS/MS [15]. The enzymatic de-*N*-acetylation/*N*-acetylation of the COSs, combined with the state-of-the-art separation/analysis methods, enabled the production of COSs with desired sequences. This may subsequently lead to defining the structure–function relationships of COSs. This strategy of using site-specific deacetylases was also utilized for producing chitosans with random and non-random acetylation patterns [88].

3.6. Metabolic Engineering Approaches

Studies of the biological activities of COSs usually need large amounts of COSs. However, the enzymatic methods described thus far provided only limited amounts of COSs. As an alternative to the current enzymatic methods, microbial fermentation using recombinant *Escherichia coli* has now attracted researchers' attention, because the method is capable of producing larger amounts of COSs and is environmentally friendly [89]. Recently, a non-pathogenic bacterium, *Bacillus subtilis*, also provided an additional opportunity for the fermentative production of COSs [90]. *B. subtilis* was engineered by introducing a combinatorial pathway for the production of well-defined COSs. Specifically, an exogenous COS synthase was overexpressed in *B. subtilis*, then the GlcNAc synthesis module was also introduced to enhance the intracellular GlcNAc supply. Furthermore, both the de novo pathway and the salvage pathway of UDP-GlcNAc were engineered to further promote the biosynthesis of $(GlcNAc)_n$. The metabolic engineering finally attained the production of $(GlcNAc)_n$ of 4.8 $g{\cdot}L^{-1}$ including $(GlcNAc)_5$ (86%), $(GlcNAc)_4$ (7%), $(GlcNAc)_3$ (5%), and $(GlcNAc)_2$ (2%). Further engineering may provide a cell factory for producing the required amounts of COSs with desired DP, DA, and sequences [91].

4. Conclusions and Future Prospects

Recently, the relationship between the structures and the biological activities of COSs has become more explicit since the biological activities of COSs were satisfactorily characterized with respect to DPs, DAs, and sequences. Based on the data listed in Tables 1–5, it is possible to design COS sequences with desired biological functions. A combination of the strategies summarized in Table 6 enabled the synthesis of tailor-made COSs with specified sequences.

Table 6. Strategies for synthesizing COSs with desired sequences.

Catalysts	Substrates		Products	Ref.
	Donor	Acceptor		
Hen egg-white lysozyme wild type	$(GlcNAc)_3$		$(GlcNAc)_n$ (n = 3–15)	[54]
Hen egg-white lysozyme wild type	$(GlcNAc)_4$	Moranoline (1-deoxynojirimycin)	4-*O*-b-di(tri)-*N*-acetylchitobi(tri)osyl moranoline	[55,56]
Hen egg-white lysozyme Asp101, Trp62-modified	$(GlcNAc)_5$		$(GlcNAc)_9$	[60]
Amycolatopsis orientalis GH2 exo-b-D-glucosaminidase	$(GlcN)_4$		$(GlcN)_5$ and $(GlcN)_6$	[61]
Streptomyces griseus HUT6037 GH5 endo-chitosanase	$(GlcN)_5$	$(GlcNAc)_3$	$(GlcN)_2$-$(GlcNAc)_3$ $(GlcN)_3$-$(GlcNAc)_3$	[62]
Serratia marcescens GH18 chitinase A mutated at Trp at site -3	$(GlcNAc)_4$ or $(GlcNAc)_5$		$(GlcNAc)_6$ or $(GlcNAc)_7$	[63]
Serratia marcescens GH18 chitinases A and B mutated at the middle Asp in the DxDxE motif	$(GlcNAc)_4$		$(GlcNAc)_3$ produced through the transglycosylation product $(GlcNAc)_6$	[64]
Vibrio harveyi GH18 chitinase A mutated at the middle Asp in the DxDxE motif	$(GlcNAc)_4$ $(GlcNAc)_6$		$(GlcNAc)_6$ $(GlcNAc)_8$	[65]
Serratia proteamaculans GH18 chitinase D triple-mutated at the glycon- or aglycon-binding aromatic residues as well as at the middle Asp in the DxDxE motif	$(GlcNAc)_4$		$(GlcNAc)_5$ or $(GlcNAc)_6$	[66]
Serratia proteamaculans GH18 chitinase D single-mutated at the catalytic center and the binding groove	$(GlcNAc)_4$		$(GlcNAc)_5$ or $(GlcNAc)_6$	[67]
Arabidopsis thaliana GH18 chitinase C mutant, in which tryptophan side chain was introduced into the upper portion of the catalytic center	$(GlcNAc)_4$		$(GlcNAc)_3$ produced through the transglycosylation product $(GlcNAc)_6$	[68]
Cycas revoluta GH18 chitinase A mutant, in which tryptophan side chain was introduced into the upper portion of the catalytic center	$(GlcNAc)_4$		$(GlcNAc)_3$ produced through the transglycosylation product $(GlcNAc)_6$	[69]
A glycosynthase derived from *Bryum coronatum* GH19 chitinase A	$(GlcNAc)_2$-fluoride	$(GlcNAc)_2$	$(GlcNAc)_4$	[74]

Table 6. *Cont.*

Catalysts	Substrates		Products	Ref.
	Donor	Acceptor		
A glycosynthase derived from *Secale cereale* GH19 chitinase C	$(GlcNAc)_3$-fluoride	$(GlcNAc)_4$	$(GlcNAc)_7$	[75]
A chitin-oligosaccharide *N*-deacetylase (NodB) and a glycosynthase derived from hen egg-white lysozyme (Asp52 → Ser)	GlcN-$(GlcNAc)_2$-fluoride	$(GlcNAc)_3$ $(GlcNAc)_4$ $(GlcNAc)_5$	GlcN-$(GlcNAc)_5$ GlcN-$(GlcNAc)_6$ GlcN-$(GlcNAc)_7$	[76]
Hypertransglycosylating mutants from *Serratia marcescens* GH18 chitinases A and *Serratia proteamaculans* GH18 chitinase D	GlcN-GlcNAc-*p*NP (*p*-nitorophenylated) obtained by enzymatic de-*N*-acetylation of $(GlcNAc)_2$-*p*NP		$(GlcN\text{-}GlcNAc)_2$ $(GlcN\text{-}GlcNAc)_3$ $(GlcN\text{-}GlcNAc)_4$ $(GlcN\text{-}GlcNAc)_5$	[78]
Amycolatopsis orientalis GH2 exo-b-glucosaminidase	GlcN-DMT (4,6-dimethoxy-1,3,5-triazin-2-yl)	$(GlcNAc)_2$	GlcN-$(GlcNAc)_2$	[80]
A glycosynthase derived from *Bryum coronatum* GH19 chitinase A	$(GlcNAc)_2$-DMT	$(GlcNAc)_2$	$(GlcNAc)_4$	[81]
An activity-reduced mutant from *Bacillus circulans* GH18 chitinase A1	$(GlcNAc)_2$-oxa	$(GlcNAc)_5$	$(GlcNAc)_7$	[83]
Catalytic-site mutants from *Serratia proteamaculans* GH18 chitinase D	$(GlcNAc)_5$-oxa		$(GlcNAc)_{10}$	[84]
Hypertransglycosylating mutants from *Nicotiana tobaccum* GH18 chitinase C and *Cycas revoluta* GH18 chitinase A	$(GlcNAc)_2$-oxa $(GlcNAc)_3$-oxa $(GlcNAc)_4$-oxa $(GlcNAc)_5$-oxa	$(GlcNAc)_5$ $(GlcNAc)_4$ $(GlcNAc)_3$ $(GlcNAc)_2$	$(GlcNAc)_7$	[85]
Site-specific chitin dectylases from fungi	$(GlcNAc)_4$		A full lineup of partially *N*-acetylated chitotetraoses	[86]

It is well known that di-*N*-acetyl chitobiose $(GlcNAc)_2$ is most efficiently produced from enzymatic degradation of chitin [92–97]. We successfully produced a large amount of high-quality $(GlcNAc)_2$ from chitin food wastes using an in-house *Vibrio* chitinase [98], while $(GlcN)_2$ was most efficiently produced by endo-splitting GH46 chitosanases from the chitosan wastes [99,100]. Starting from these chitobioses, various green-chemical strategies are proposed, as shown in Figure 4. COSs of differing chain length can be elongated from a DP of 2 to a DP of 4, 6, 8, or 10 by means of transglycosylation reactions, using chitinase/chitosanase mutants as effective catalysts. Glycosynthase mutants were also useful for chain-length elongation using reducing end-activated COSs as the donor substrates. Subsequent site-specific de-*N*-acetylation/*N*-acetylation may possibly produce tailor-made COSs that possess most plausible sequences with desired biological activities. The concept of tailor-made COSs will further enhance the exploitation of chitin biomass.

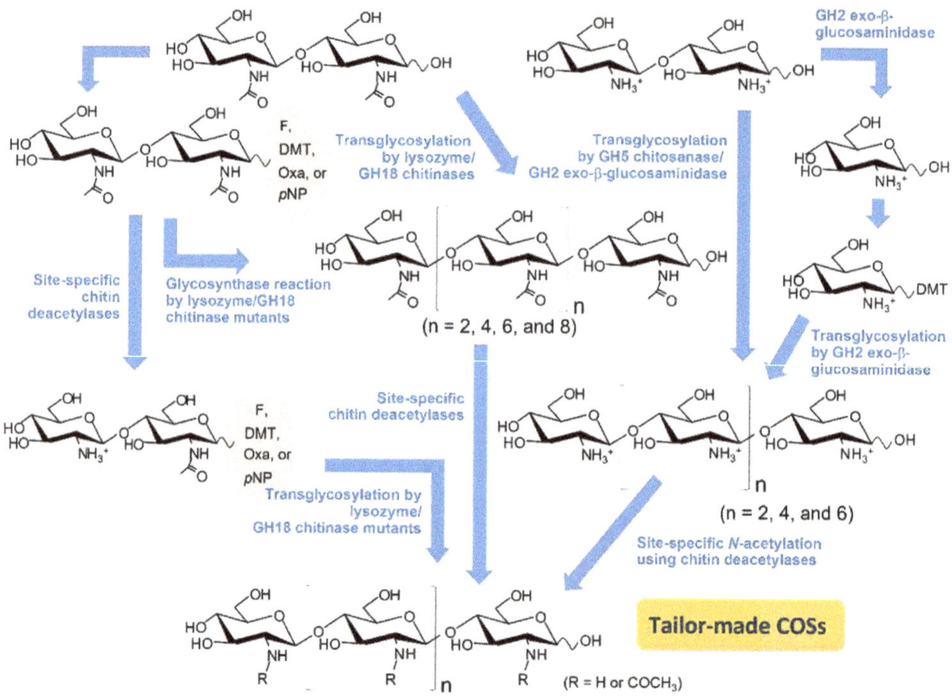

Figure 4. Proposed enzymatic synthesis of tailor-made COSs (DP4—10) through transglycosylation/glycosynthase reactions. Chitobiose and chitosan dimer are used as sugar acceptors and the reducing end-activated COSs are used as donor substrates. Site-specific de-*N*-acetylation/*N*-acetylation are subsequently used in the final step to generate tailor-made COSs.

Author Contributions: Conceptualization, R.T. and W.S.; data curation, R.T. and T.F.; original draft preparation, R.T. and T.F.; review and editing, T.F. and W.S.; project administration, W.S.; funding acquisition, W.S. All authors have read and agreed to the published version of the manuscript.

Funding: This research was supported by the National Research Council of Thailand (grant No. N42A660311).

Informed Consent Statement: Not applicable.

Data Availability Statement: Not applicable.

Acknowledgments: We would like to thank David Apps for valuable and critical proofreading of the manuscript.

Conflicts of Interest: The authors declare no conflict of interest.

Abbreviations

COS, chitooligosaccharide; DP, degree of polymerization; DA, degree of *N*-acetylation; MW, molecular weight; GlcNAc (**A**), 2-acetamido-2-deoxy-D-glucopyranose; GlcN (**D**), 2-amino-2-deoxy-D-glucopyranose; $(GlcNAc)_n$ and $(GlcN)_n$, homo-oligosaccharides composed of GlcNAc and GlcN, respectively, with a DP of n; POD, peroxidase; PAL, phenylalanine ammonia lyase; LysM, lysin motif; CERK1, chitin elicitor receptor kinase 1; CERK1-ECD, CERK1 ectodomain; *At*ChiC, a GH18 chitinase from *Arabidopsis thaliana*; *Nt*ChiV, a GH18 chitinase from *Nicotiana tabacum*; *Cr*ChiA, a GH18 chitinase from *Cycas revoluta*; *Bc*ChiA, a GH19 chitinase from *Bryum coronatum*; RSC-c, a GH19 chitinase from *Secale cereale*; TG, transglycosylation.

References

1. Minke, R.; Blackwell, J. The structure of a-chitin. *J. Mol. Biol.* **1978**, *120*, 167–181. [CrossRef] [PubMed]
2. Sikorski, P.; Hori, R.; Wada, M. Revisit of α-Chitin Crystal Structure Using High Resolution X-ray Diffraction Data. *Biomacromolecules* **2009**, *10*, 1100–1105. [CrossRef] [PubMed]
3. Raabe, D.; Romano, P.; Sachs, C.; Fabritius, H.; Al-Sawalmih, A.; Yi, S.-B.; Servos, G.; Hartwig, H.G. Microstructure and Crystallographic Texture of the Chitin–Protein Network in the Biological Composite Material of the Exoskeleton of the Lobster Homarus Americanus. *Mat. Sci. Eng. A* **2006**, *421*, 143–153. [CrossRef]
4. Jang, M.-K.; Kong, B.-G.; Jeong, Y.-I.; Lee, C.H.; Nah, J.-W. Physicochemical Characterization of a-Chitin, b-Chitin, and g-Chitin Separated from Natural Resources. *J. Polym. Sci. Part A Polym. Chem.* **2004**, *42*, 3423–3432. [CrossRef]
5. Ogawa, K.; Yui, T.; Okuyama, K. Three D Structures of Chitosan. *Int. J. Biol. Macromol.* **2004**, *34*, 1–8. [CrossRef] [PubMed]
6. Okuyama, K.; Noguchi, K.; Kanenari, M.; Egawa, T.; Osawa, K.; Ogawa, K. Structural Diversity of Chitosan and its Complexes. *Carbohydr. Polym.* **2000**, *41*, 237–247. [CrossRef]
7. Kang, X.; Kirui, A.; Muszyński, A.; Widanage, M.C.D.; Chen, A.; Azadi, P.; Wang, P.; Mentink-Vigier, F.; Wang, T. Molecular architecture of fungal cell walls revealed by solid-state NMR. *Nat. Commun.* **2018**, *9*, 2747. [CrossRef]
8. Fernando, L.D.; Dickwella Widanage, M.C.; Penfield, J.; Lipton, A.S.; Washton, N.; Latge, J.-P.; Wang, P.; Zhang, L.; Wang, T. Structural polymorphism of chitin and chitosan in fungal cell walls from solid-state NMR and principal component analysis. *Front. Mol. Biosci.* **2021**, *8*, 727053. [CrossRef]
9. Tharanathan, R.N.; Kittur, F.S. Chitin-the undisputed biomolecule of great potential. *Crit. Rev. Food Sci. Nutr.* **2003**, *43*, 61–87. [CrossRef]
10. Jung, W.J.; Park, R.D. Bioproduction of chitooligosacchrides: Present and perspectives. *Mar. Drugs.* **2014**, *12*, 5328–5356. [CrossRef]
11. Liaqat, F.; Eltem, R. Chitooligosaccharides and their biological activities: A comprehensive review. *Carbohydr. Polym.* **2018**, *184*, 243–259. [CrossRef] [PubMed]
12. Kasaai, M.R. A review of several reported procedures to determine the degree of *N*-acetylation for chitin and chitosan using infrared spectroscopy. *Carbohydr. Polym.* **2008**, *71*, 497–508. [CrossRef]
13. Kasaai, M.R. Determination of the degree of *N*-acetylation for chitin and chitosan by various NMR spectroscopy techniques: A review. *Carbohydr. Polym.* **2010**, *79*, 801–810. [CrossRef]
14. Heux, L.; Brugnerotto, J.; Desbrières, J.; Versali, M.F.; Rinaudo, M. Solid State NMR for Determination of Degree of Acetylation of Chitin and Chitosan. *Biomacromolecules* **2000**, *1*, 746–751. [CrossRef]
15. Cord-Landwehr, S.; Ihmor, P.; Niehues, A.; Luftmann, H.; Moerschbacher, B.M.; Mormann, M. Quantitative mass-spectrometric sequencing of chitosan oligomers revealing cleavage sites of chitosan hydrolases. *Anal. Chem.* **2017**, *89*, 2893–2900. [CrossRef]
16. Mengíbar, M.; Mateos-Aparicio, I.; Miralles, B.; Heras, A. Influence of the physico-chemical characteristics of chitooligosaccharides (COS) on antioxidant activity. *Carbohydr. Polym.* **2013**, *97*, 776–782. [CrossRef]
17. Hao, W.; Li, K.; Ge, X.; Yang, H.; Xu, C.; Liu, S.; Yu, H.; Li, P.; Xing, R. The effect of *N*-acetylation on the anti-inflammatory activity of chitooligosaccharides and its potential for relieving endotoxemia. *Int. J. Mol. Sci.* **2022**, *23*, 8205. [CrossRef]
18. Cord-Landwehr, S.; Moerschbacher, B.M. Deciphering the ChitoCode: Fungal chitins and chitosans as functional biopolymers. *Fungal Biol. Biotechnol.* **2021**, *8*, 19. [CrossRef]
19. Jeon, Y.J.; Park, P.J.; Kim, S.K. Antimicrobial effect of chitooligosaccharides produced by bioreactor. *Carbohydr. Polym.* **2001**, *44*, 71–76. [CrossRef]
20. Li, K.; Xing, R.; Liu, S.; Qin, Y.; Yu, H.; Li, P. Size and pH effects of chitooligomers on antibacterial activity against *Staphylococcus aureus*. *Int. J. Biol. Macromol.* **2014**, *64*, 302–305. [CrossRef]
21. No, H.K.; Park, N.Y.; Lee, S.H.; Meyers, S.P. Antibacterial activity of chitosans and chitosan oligomers with different molecular weights. *Int. J. Food Microbiol.* **2002**, *74*, 65–72. [CrossRef] [PubMed]
22. Aktuganov, G.E.; Safina, V.R.; Galimzianova, N.F.; Gilvanova, E.A.; Kuzmina, L.Y.; Melentiev, A.I.; Baymiev, A.H.; Lopatin, S.A. Constitutive chitosanase from and its potential for preparation of antimicrobial chitooligomers. *World J. Microbiol. Biotechnol.* **2022**, *38*, 167. [CrossRef] [PubMed]
23. Ganan, M.; Lorentzen, S.B.; Agger, J.W.; Heyward, C.A.; Bakke, O.; Knutsen, S.H.; Aam, B.B.; Eijsink, V.G.H.; Gaustad, P.; Sølie, M. Antifungal activity of well-defined chitooligosaccharide preparations against medically relevant yeasts. *PLoS ONE* **2019**, *14*, e0210208. [CrossRef] [PubMed]
24. Kulikov, S.N.; Lisovskaya, S.A.; Zelenikhin, P.V.; Bezrodnykh, E.A.; Shakirova, D.R.; Blagodatskikh, I.V.; Tikhonov, V.E. Antifungal activity of oligochitosans (short-chain chitosans) against some *Candida* species and clinical isolates of *Candida albicans*: Molecular weight-activity relationship. *Eur. J. Med. Chem.* **2014**, *74*, 169–178. [CrossRef]
25. Chung, M.J.; Park, J.K.; Park, Y. Il Anti-inflammatory effects of low-molecular weight chitosan oligosaccharides in IgE–antigen complex-stimulated RBL-2H3 cells and asthma model mice. *Int. Immunopharmacol.* **2012**, *12*, 453–459. [CrossRef]
26. Pangestuti, R.; Bak, S.-S.; Kim, S.-K. Attenuation of pro-inflammatory mediators in LPS-stimulated BV2 microglia by chitooligosaccharides via the MAPK signaling pathway. *Int. J. Biol. Macromol.* **2011**, *49*, 599–606. [CrossRef]
27. Wei, P.; Ma, P.; Xu, Q.-S.; Bai, Q.-H.; Gu, J.-G.; Xi, H.; Du, Y.-G.; Yu, C. Chitosan oligosaccharides suppress production of nitric oxide in lipopolysaccharide-induced N9 murine microglial cells in Vitro. *Glycoconj. J.* **2012**, *29*, 285–295. [CrossRef]

28. Sánchez, Á.; Mengíbar, M.; Fernández, M.; Alemany, S.; Heras, A.; Acosta, N. Influence of preparation methods of chitooligosaccharides on their physicochemical properties and their anti-inflammatory effects in mice and in RAW264.7 Macrophages. *Mar. Drugs* **2018**, *16*, 430. [CrossRef]
29. Ngo, D.-N.; Kim, M.-M.; Kim, S.-K. Chitin oligosaccharides inhibit oxidative stress in live cells. *Carbohydr. Polym.* **2008**, *74*, 228–234. [CrossRef]
30. Zhang, Y.; Zhou, X.; Ji, L.; Du, X.; Sang, Q.; Chen, F. Enzymatic single-step preparation and antioxidant activity of hetero-chitooligosaccharides using non-pretreated housefly larvae powder. *Carbohydr. Polym.* **2017**, *172*, 113–119. [CrossRef]
31. Liu, H.T.; Li, W.M.; Xu, G.; Li, X.Y.; Bai, X.F.; Wei, P.; Yu, C.; Du, Y.G. Chitosan oligosaccharides attenuate hydrogen peroxide-induced stress injury in human umbilical vein endothelial cells. *Pharmacol. Res.* **2009**, *59*, 167–175. [CrossRef] [PubMed]
32. Liu, H.T.; He, J.L.; Li, W.M.; Yang, Z.; Wang, Y.X.; Bai, X.F.; Yu, C.; Du, Y.G. Chitosan oligosaccharides protect human umbilical vein endothelial cells from hydrogen peroxide-induced apoptosis. *Carbohydr. Polym.* **2010**, *80*, 1062–1071. [CrossRef]
33. Jia, Y.; Ma, Y.; Zhou, P.; Cheng, G.; Zhou, J.; Cai, S. Effects of different oligochitosans on isoflavone metabolites, antioxidant activity, and isoflavone biosynthetic genes in soybean (*Glycine max*) seeds during germination. *J. Agric. Food Chem.* **2019**, *67*, 4652–4661. [CrossRef]
34. Hao, W.; Li, K.; Ma, Y.; Li, R.; Xing, R.; Yu, H.; Li, P. Preparation and antioxidant activity of chitosan dimers with different sequences. *Mar. Drugs* **2021**, *19*, 366. [CrossRef]
35. Tsukada, K.; Matsumoto, T.; Aizawa, K.; Tokoro, A.; Naruse, R.; Suzuki, S.; Suzuki, M. Antimetastatic and growth-inhibitory effects of *N*-acetylchitohexaose in mice bearing Lewis lung Carcinoma. *Jap. J. Cancer Res.* **1990**, *81*, 259–265. [CrossRef]
36. Salah, R.; Michaud, P.; Mati, F.; Harrat, Z.; Lounici, H.; Abdi, N.; Drouiche, N.; Mameri, N. Anticancer activity of chemically prepared shrimp low molecular weight chitin evaluation with the human monocyte leukaemia cell line, THP-1. *Int. J. Biol. Macromol.* **2013**, *52*, 333–339. [CrossRef] [PubMed]
37. Park, J.K.; Chung, M.J.; Choi, H.N.; Park, Y.I. Effects of the molecular weight and the degree of deacetylation of chitosan oligosaccharides on antitumor activity. *Int. J. Mol. Sci.* **2011**, *12*, 266–277. [CrossRef]
38. Li, X.; Wang, J.; Chen, X.; Tian, J.; Li, L.; Zhao, M.; Jiao, Y.; Zhou, C. Effect of chitooligosaccharides on cyclin D1, bcl-xl and bcl-2 mRNA expression in A549 cells using quantitative PCR. *Chin. Sci. Bull.* **2011**, *56*, 1629–1632. [CrossRef]
39. Zhai, X.; Li, C.; Ren, D.; Wang, J.; Ma, C.; Abd El-Aty, A.M. The impact of chitooligosaccharides and their derivatives on the in vitro and in vivo antitumor activity: A comprehensive review. *Carbohydr. Polym.* **2021**, *266*, 118132. [CrossRef]
40. Vander, P.; Km, V.R.; Domard, A.; Eddine, E.G.N.; Moerschbacher, B.M. Comparison of the ability of partially *N*-acetylated chitosans and chitooligosaccharides to elicit resistance reactions in wheat leaves. *Plant Physiol.* **1998**, *118*, 1353–1359. [CrossRef]
41. Ramakrishna, B.; Sarma, P.V.S.R.N.; Ankati, S.; Bhuvanachandra, B.; Podile, A.R. Elicitation of defense response by transglycosylated chitooligosaccharides in rice seedlings. *Carbohydr. Res.* **2021**, *510*, 108459. [CrossRef]
42. Miya, A.; Albert, P.; Shinya, T.; Desaki, Y.; Ichimura, K.; Shirasu, K.; Narusaka, Y.; Kawakami, N.; Kaku, H.; Shibuya, N. CERK1, a LysM receptor kinase, is essential for chitin elicitor signaling in *Arabidopsis*. *Proc. Natl. Acad. Sci. USA* **2007**, *104*, 19613–19618. [CrossRef] [PubMed]
43. Liu, T.; Liu, Z.; Song, C.; Hu, Y.; Han, Z.; She, J.; Fan, F.; Wang, J.; Jin, C.; Chang, J.; et al. Chitin-induced dimerization activates a plant immune receptor. *Science* **2012**, *336*, 1160–1164. [CrossRef] [PubMed]
44. Cao, Y.; Liang, Y.; Tanaka, K.; Nguyen, C.T.; Jedrzejczak, R.P.; Joachimiak, A.; Stacey, G. The kinase LYK5 is a major chitin receptor in Arabidopsis and forms a chitin-induced complex with related kinase CERK1. *eLife* **2014**, *3*, e03766. [CrossRef] [PubMed]
45. Liu, S.; Wang, J.; Han, Z.; Gong, X.; Zhang, H.; Chai, J. Molecular mechanism for fungal cell wall recognition by rice chitin receptor OsCEBiP. *Structure* **2016**, *24*, 1192–1200. [CrossRef] [PubMed]
46. Hayafune, M.; Berisio, R.; Marchetti, R.; Silipo, A.; Kayama, M.; Desaki, Y.; Arima, S.; Squeglia, F.; Ruggiero, A.; Tokuyasu, K.; et al. Chitin-induced activation of immune signaling by the rice receptor CEBiP relies on a unique sandwich-type dimerization. *Proc. Natl. Acad. Sci. USA* **2014**, *111*, E404–E413. [CrossRef]
47. Gubaeva, E.; Gubaev, A.; Melcher, R.L.J.; Cord-Landwehr, S.; Singh, R.; El Gueddari, N.E.; Moerschbacher, B.M. 'Slipped Sandwich' model for chitin and chitosan perception in *Arabidopsis*. *Mol. Plant Microbe Interact.* **2018**, *31*, 1145–1153. [CrossRef]
48. Mészáros, Z.; Nekvasilová, P.; Bojarová, P.; Křen, V.; Slámová, K. Advanced glycosidases as ingenious biosynthetic instruments. *Biotechnol. Adv.* **2021**, *49*, 107733. [CrossRef]
49. Sharon, N.; Seifter, S. A transglycosylation reaction catalyzed by lysozyme. *J. Biol. Chem.* **1964**, *239*, PC2398–PC2399. [CrossRef]
50. Kravchenko, N.A. Lysozyme as a transferase. *Proc. R. Soc.* **1967**, *B167*, 429.
51. Chipman, D.M.; Pollock, J.J.; Sharon, N. Lysozyme-catalyzed hydrolysis and transglycosylation reaction of bacterial cell wall oligosaccharides. *J. Biol. Chem.* **1968**, *243*, 487–496. [CrossRef] [PubMed]
52. Masaki, A.; Fukamizo, T.; Otakara, A.; Torikata, T.; Hayashi, K.; Imoto, T. Estimation of rate constants in lysozyme-catalyzed reaction of chitooligosaccharides. *J. Biochem.* **1981**, *90*, 1167–1175. [CrossRef] [PubMed]
53. Fukamizo, T.; Minematsu, T.; Yanase, Y.; Hayashi, K.; Goto, S. Substrate size dependence of lysozyme-catalyzed reaction. *Arch. Biochem. Biophys.* **1986**, *250*, 312–321. [CrossRef] [PubMed]
54. Hattori, T.; Sakabe, Y.; Ogata, M.; Michishita, K.; Dohra, H.; Kawagishi, H.; Totani, K.; Nikaido, M.; Nakamura, T.; Koshino, H.; et al. Enzymatic synthesis of an α-chitin-like substance via lysozyme-mediated transglycosylation. *Carbohydr. Res.* **2012**, *347*, 16–22. [CrossRef] [PubMed]

55. Ogata, M.; Umemoto, N.; Ohnuma, T.; Numata, T.; Suzuki, A.; Usui, T.; Fukamizo, T. A novel transition-state analogue for lysozyme, 4-*O*-β-tri-*N*-acetylchitotriosyl moranoline, provided evidence supporting the covalent glycosyl-enzyme intermediate. *J. Biol. Chem.* **2013**, *288*, 6072–6082. [CrossRef] [PubMed]
56. Akiyama, A.; Kawazu, K.; Kobayashi, A. A novel method for chemo-enzymatic synthesis of elicitor-active chitosan oligomers and partially N-deacetylated chitin oligomers using N-acylated chitotrioses as substrates in a lysozyme-catalyzed transglycosylation reaction system. *Carbohydr. Res.* **1995**, *279*, 151–160. [CrossRef]
57. Madhuprakash, J.; Singh, A.; Kumar, S.; Sinha, M.; Kaur, P.; Sharma, S.; Podile, A.R.; Singh, T.P. Structure of chitinase D from *Serratia proteamaculans* reveals the structural basis of its dual action of hydrolysis and transglycosylation. *Int. J. Biochem. Mol. Biol.* **2013**, *4*, 166–178.
58. Bhuvanachandra, B.; Podile, A.R. A transglycosylating chitinase from *Chitiniphilus shinanonensis* (CsChiL) hydrolyzes chitin in a processive manner. *Int. J. Biol. Macromol.* **2020**, *145*, 1–10. [CrossRef]
59. Wakita, S.; Kobayashi, S.; Kimura, M.; Kashimura, A.; Honda, S.; Sakaguchi, M.; Sugahara, Y.; Kamaya, M.; Matoska, V.; Bauer, P.O.; et al. Mouse acidic mammalian chitinase exhibits transglycosylation activity at somatic tissue pH. *FEBS Lett.* **2017**, *591*, 3310–3318. [CrossRef]
60. Fukamizo, T.; Goto, S.; Torikata, T.; Araki, T. Enhancement of transglycosylation of lysozyme by chemical modification. *Agric. Biol. Chem.* **1989**, *53*, 2641–2651.
61. Fukamizo, T.; Fleury, A.; Côté, N.; Mitsutomi, M.; Brzezinski, R. Exo-b-D-glucosaminidase from *Amycolatopsis orientalis*: Catalytic residues, sugar recognition specificity, kinetics, and synergism. *Glycobiology* **2006**, *16*, 1064–1072. [CrossRef]
62. Tanabe, T.; Morinaga, K.; Fukamizo, T.; Mitsutomi, M. Novel chitosanase from Streptomyces griseus HUT 6037 with transglycosylation activity. *Biosci. Biotechnol. Biochem.* **2003**, *67*, 354–364. [CrossRef]
63. Aronson, N.N., Jr.; Halloran, B.A.; Alexeyev, M.F.; Zhou, X.E.; Wang, Y.; Meehan, E.J.; Chen, L. Mutation of a conserved tryptophan in the chitin-binding cleft of *Serratia marcescens* chitinase A enhances transglycosylation. *Biosci. Biotechnol. Biochem.* **2006**, *70*, 243–251. [CrossRef]
64. Zakariassen, H.; Hansen, M.C.; Jøranli, M.; Eijsink, V.G.H.; Sørlie, M. Mutational effects on transglycosylating activity of family 18 chitinases and construction of a hypertransglycosylating mutant. *Biochemistry* **2011**, *50*, 5693–5703. [CrossRef] [PubMed]
65. Bhuvanachandra, B.; Madhuprakash, J.; Podile, A.R. Active-site mutations improved the transglycosylation activity of Stenotrophomonas maltophilia chitinase A. *Biochim. Biophys. Acta Proteins Proteom.* **2018**, *1866*, 407–414. [CrossRef] [PubMed]
66. Madhuprakash, J.; Dalhus, B.; Rani, T.S.; Podile, A.R.; Eijsink, V.G.H.; Sørlie, M. Key residues affecting transglycosylation activity in family 18 chitinases: Insights into donor and acceptor subsites. *Biochemistry* **2018**, *57*, 4325–4337. [CrossRef]
67. Madhuprakash, J.; Tanneeru, K.; Purushotham, P.; Guruprasad, L.; Podile, A.R. Transglycosylation by chitinase D from *Serratia proteamaculans* improved through altered substrate interactions. *J. Biol. Chem.* **2012**, *287*, 44619–44627. [CrossRef]
68. Umemoto, N.; Ohnuma, T.; Mizuhara, M.; Sato, H.; Skriver, K.; Fukamizo, T. Introduction of a tryptophan side chain into subsite +1 enhances transglycosylation activity of a GH-18 chitinase from *Arabidopsis thaliana, At*ChiC. *Glycobiology* **2013**, *23*, 81–90. [CrossRef] [PubMed]
69. Umemoto, N.; Ohnuma, T.; Osawa, T.; Numata, T.; Fukamizo, T. Modulation of the transglycosylation activity of plant family GH18 chitinase by removing or introducing a tryptophan side chain. *FEBS Lett.* **2015**, *589*, 2327–2333. [CrossRef]
70. Faijes, M.; Planas, A. In vitro synthesis of artificial polysaccharides by glycosidases and glycosynthases. *Carbohydr. Res.* **2007**, *342*, 1581–1594. [CrossRef]
71. Danby, P.M.; Withers, S.G. Advances in enzymatic glycoside synthesis. *ACS Chem. Biol.* **2016**, *11*, 1784–1794. [CrossRef] [PubMed]
72. Honda, Y.; Kitaoka, M. The first glycosynthase derived from an inverting glycoside hydrolase. *J. Biol. Chem.* **2006**, *281*, 1426–1431. [CrossRef] [PubMed]
73. van Aalten, D.M.; Komander, D.; Synstad, B.; Gåseidnes, S.; Peter, M.G.; Eijsink, V.G. Structural insights into the catalytic mechanism of a family 18 exo-chitinase. *Proc. Natl. Acad. Sci. USA* **2001**, *98*, 8979–8984. [CrossRef] [PubMed]
74. Ohnuma, T.; Fukuda, T.; Dozen, S.; Honda, Y.; Kitaoka, M.; Fukamizo, T. A glycosynthase derived from an inverting GH19 chitinase from the moss *Bryum coronatum*. *Biochem. J.* **2012**, *444*, 437–443. [CrossRef]
75. Ohnuma, T.; Dozen, S.; Honda, Y.; Kitaoka, M.; Fukamizo, T. A glycosynthase derived from an inverting chitinase with an extended binding cleft. *J. Biochem.* **2016**, *160*, 93–100. [CrossRef]
76. Rousseau, A.; Armand, S.; Cottaz, S.; Fort, S. Size-controlled synthesis of β(1→4)-GlcNAc oligosaccharides using an endo-glycosynthase. *Chemistry* **2021**, *27*, 17637–17646. [CrossRef]
77. Rastelli, R.A.; Bucke, C. Enzymatic synthesis of oligosaccharides. *Biotechnol. Genet. Eng. Rev.* **1992**, *10*, 253–282.
78. Harmsen, R.A.G.; Aam, B.B.; Madhuprakash, J.; Hamre, A.G.; Goddard-Borger, E.D.; Withers, S.G.; Eijsink, V.G.H.; Sørlie, M. Chemoenzymatic synthesis of chito-oligosaccharides with alternating *N*-D-acetylglucosamine and D-glucosamine. *Biochemistry* **2020**, *59*, 4581–4590. [CrossRef]
79. Tanaka, T.; Noguchi, M.; Kobayashi, A.; Shoda, S. A novel glycosyl donor for chemo-enzymatic oligosaccharide synthesis: 4,6-dimethoxy-1,3,5-triazin-2-yl glycoside. *Chem. Commun.* **2008**, *17*, 2016–2018. [CrossRef]
80. Tanaka, T.; Wada, T.; Noguchi, M.; Ishihara, M.; Kobayashi, A.; Ohnuma, T.; Fukamizo, T.; Brzezinski, R.; Shoda, S.-I. 4,6-Dimethoxy-1,3,5-triazin-2-yl β-D-glycosaminides: Novel substrates for transglycosylation reaction catalyzed by exo-β-D-glucosaminidase from *Amycolatopsis orientalis*. *J. Carbohydr. Chem.* **2012**, *31*, 634–646. [CrossRef]

81. Ohnuma, T.; Tanaka, T.; Urasaki, A.; Dozen, S.; Fukamizo, T. A novel method for chemo-enzymatic synthesis of chitin oligosaccharide catalyzed by the mutant of inverting family GH19 chitinase using 4,6-dimethoxy-1,3,5-triazin-2-yl α-chitobioside as a glycosyl donor. *J. Biochem.* **2019**, *165*, 497–503. [CrossRef] [PubMed]
82. Noguchi, M.; Tanaka, T.; Gyakushi, H.; Kobayashi, A.; Shoda, S. Efficient synthesis of sugar oxazolines from unprotected *N*-acetyl-2-amino sugars by using chloroformamidinium reagent in water. *J. Org. Chem.* **2009**, *74*, 2210–2212. [CrossRef] [PubMed]
83. Yoshida, N.; Tanaka, T.; Noguchi, M.; Kobayashi, A.; Ishikura, K.; Ikenuma, T.; Seno, H.; Watanabe, T.; Kohri, M.; Shoda, S.-I. One-pot chemoenzymatic route to chitoheptaose via specific transglycosylation of chitopentaose-oxazoline on chitinase-template. *Chem. Lett.* **2012**, *41*, 689–690. [CrossRef]
84. Alsina, C.; Sancho-Vaello, E.; Aranda-Martínez, A.; Faijes, M.; Planas, A. Auxiliary active site mutations enhance the glycosynthase activity of a GH18 chitinase for polymerization of chitooligosaccharides. *Carbohydr. Polym.* **2021**, *252*, 117121. [CrossRef] [PubMed]
85. Umemoto, N.; Saito, N.; Noguchi, M.; Shoda, S.-I.; Ohnuma, T.; Watanabe, T.; Sakuda, S.; Fukamizo, T. Plant chitinase mutants as the catalysts for chitooligosaccharide synthesis using the sugar oxazoline derivatives. *J. Agric. Food Chem.* **2022**, *70*, 12897–12906. [CrossRef]
86. Hembach, L.; Cord-Landwehr, S.; Moerschbacher, B.M. Enzymatic production of all fourteen partially acetylated chitosan tetramers using different chitin deacetylases acting in forward or reverse mode. *Sci. Rep.* **2017**, *7*, 17692. [CrossRef]
87. Hamer, S.N.; Cord-Landwehr, S.; Biarnés, X.; Planas, A.; Waegeman, H.; Moerschbacher, B.M.; Kolkenbrock, S. Enzymatic production of defined chitosan oligomers with a specific pattern of acetylation using a combination of chitin oligosaccharide deacetylases. *Sci. Rep.* **2015**, *5*, 8716. [CrossRef]
88. Sreekumar, S.; Wattjes, J.; Niehues, A.; Mengoni, T.; Mendes, A.C.; Morris, E.R.; Goycoolea, F.M.; Moerschbacher, B.M. Biotechnologically produced chitosans with nonrandom acetylation patterns differ from conventional chitosans in properties and activities. *Nat. Commun.* **2022**, *13*, 7125. [CrossRef]
89. Ling, M.; Li, J.; Du, G.; Liu, L. Metabolic engineering for the production of chitooligosaccharides: Advances and perspectives. *Emerg. Top Life Sci.* **2018**, *2*, 377–388.
90. Ling, M.; Wu, Y.; Tian, R.; Liu, Y.; Yu, W.; Tao, G.; Lv, X.; Li, J.; Du, G.; Amaro, R.L.; et al. Combinatorial pathway engineering of *Bacillus subtilis* for production of structurally defined and homogeneous chitooligosaccharides. *Metab. Eng.* **2022**, *70*, 55–66. [CrossRef]
91. Weyer, R.; Hellmann, M.J.; Hamer-Timmermann, S.N.; Singh, R.; Moerschbacher, B.M. Customized chitooligosaccharide production-controlling their length via engineering of rhizobial chitin synthases and the choice of expression system. *Front. Bioeng. Biotechnol.* **2022**, *10*, 1073447. [CrossRef] [PubMed]
92. Lee, S.G.; Koh, H.Y.; Han, S.J.; Park, H.; Na, D.C.; Kim, I.C.; Lee, H.K.; Yim, J.H. Expression of recombinant endochitinase from the Antarctic bacterium, *Sanguibacter antarcticus* KOPRI 21702 in Pichia pastoris by codon optimization. *Protein Expr. Purif.* **2010**, *71*, 108–114. [CrossRef]
93. Yu, P.; Yan, Y.; Gu, Q.; Wang, X. Codon optimisation improves the expression of *Trichoderma viride* sp. endochitinase in *Pichia pastoris*. *Sci. Rep.* **2013**, *3*, 3043. [CrossRef] [PubMed]
94. Itoh, T.; Sugimoto, I.; Hibi, T.; Suzuki, F.; Matsuo, K.; Fujii, Y.; Taketo, A.; Kimoto, H. Overexpression, purification, and characterization of *Paenibacillus* cell surface-expressed chitinase ChiW with two catalytic domains. *Biosci. Biotechnol. Biochem.* **2014**, *78*, 624–634. [CrossRef] [PubMed]
95. Bai, L.; Kim, J.; Son, K.H.; Chung, C.W.; Shin, D.H.; Ku, B.H.; Kim, D.Y.; Park, H.Y. Novel Bi-Modular GH19 Chitinase with Broad pH Stability from a Fibrolytic Intestinal Symbiont of *Eisenia fetida Cellulosimicrobium funkei* HY-13. *Biomolecules* **2021**, *11*, 1735. [CrossRef]
96. Ren, X.B.; Dang, Y.R.; Liu, S.S.; Huang, K.X.; Qin, Q.L.; Chen, X.L.; Zhang, Y.Z.; Wang, Y.J.; Li, P.Y. Identification and Characterization of Three Chitinases with Potential in Direct Conversion of Crystalline Chitin into *N,N′*-diacetylchitobiose. *Mar. Drugs* **2022**, *20*, 165. [CrossRef]
97. Zhao, Q.; Fan, L.; Deng, C.; Ma, C.; Zhang, C.; Zhao, L. Bioconversion of chitin into chitin oligosaccharides using a novel chitinase with high chitin-binding capacity. *Int. J. Biol. Macromol.* **2023**, *244*, 125241. [CrossRef]
98. Thomas, R.; Fukamizo, T.; Suginta, W. Bioeconomic production of high-quality chitobiose from chitin food wastes using an in-house chitinase from *Vibrio campbellii*. *Biores. Bioprocess* **2022**, *9*, 86. [CrossRef]
99. Chen, T.; Cheng, G.; Jiao, S.; Ren, L.; Zhao, C.; Wei, J.; Han, J.; Pei, M.; Du, Y.; Li, J.J. Expression and biochemical characterization of a novel marine chitosanase from *Streptomyces niveus* suitable for preparation of chitobiose. *Mar. Drugs* **2021**, *19*, 300. [CrossRef]
100. Li, Y.; Gou, Y.; Liu, Z.; Xie, T.; Wang, G. Structure-based rational design of chitosanase CsnMY002 for high yields of chitobiose. *Colloids Surf. B Biointerfaces* **2021**, *202*, 111692. [CrossRef]

molecules

Article

Impact of UV Irradiation on the Chitosan Bioactivity for Biopesticide Applications

Solène Meynaud [1], Gaël Huet [1], Daphnée Brulé [2], Christian Gardrat [1], Benoit Poinssot [2] and Véronique Coma [1,*]

1 Laboratoire de Chimie des Polymères Organiques, Université de Bordeaux, CNRS, Bordeaux INP, UMR 5629, 16 Avenue Pey-Berland, F-33600 Pessac, France; solene.meynaud@bordeaux-inp.fr (S.M.); gaelhuet@orange.fr (G.H.); christian.gardrat@u-bordeaux.fr (C.G.)
2 Agroécologie, CNRS, INRAE, Institut Agro, University Bourgogne, F-21000 Dijon, France; daphnee.brule@inrae.fr (D.B.); benoit.poinssot@u-bourgogne.fr (B.P.)
* Correspondence: veronique.coma@u-bordeaux.fr

Abstract: Chitosan is known for its antimicrobial and antifungal properties that make it a promising candidate for plant protection. However, when sprayed in open fields, the bioactivity of chitosan significantly diminishes, suggesting a possible influence of sunlight on chitosan structure. This study aimed to investigate the effects of UV radiation, by using artificial UV sources simulating sunlight, on the stability of chitosan. A powdered chitosan with a low polymerization degree was selected and analyzed using various physicochemical methods, both before and after irradiation. Some minor differences appeared. UV spectra analysis revealed the disappearance of initially present chromophores and the emergence of a new band around 340 nm, potentially indicating the formation of carbonyl compounds. However, elemental analysis, MALDI-TOF spectra, polymerization degree, and infrared spectra did not exhibit any clear structural modifications of chitosan. Interestingly, irradiated powdered chitosan samples maintained their bioactivity, including their eliciting and antifungal properties. In the case of grapevine, irradiated chitosan demonstrated effectiveness in controlling grapevine diseases such as downy mildew, contradicting the assumption that sunlight is responsible for the decreased effectiveness of chitosan in open field conditions.

Keywords: chitosan; UV radiation; sunlight; bioactivity; eliciting properties; grapevine; MAPKs; downy mildew

check for updates

Citation: Meynaud, S.; Huet, G.; Brulé, D.; Gardrat, C.; Poinssot, B.; Coma, V. Impact of UV Irradiation on the Chitosan Bioactivity for Biopesticide Applications. *Molecules* **2023**, *28*, 4954. https://doi.org/10.3390/molecules28134954

Academic Editor: Agnieszka Ewa Wiącek

Received: 16 May 2023
Revised: 14 June 2023
Accepted: 20 June 2023
Published: 23 June 2023

1. Introduction

Chitosan, a natural polymer obtained through the deacetylation of chitin, is mainly extracted from the exoskeletons of crustaceans [1], the cuticles of insects [2] or from the cell walls of fungi [3]. It constitutes ß-(1→4)-D-glucosamine (D-units) linked to N-acetyl-D-glucosamine (A-units) with a variable degree of acetylation (Figure 1).

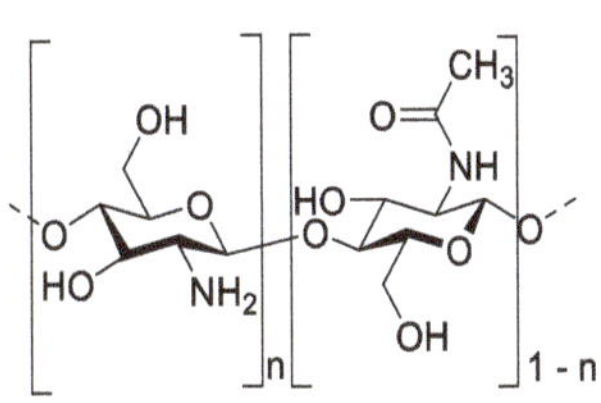

Figure 1. Chemical structure of chitosan with A− and D−units randomly distributed along the chain.

Chitosan has emerged as a potentially applicable polysaccharide in many fields ranging from food packaging [4] to biomedicine [5], pharmaceuticals [6], cosmetics [7], environment [8], and agriculture. In agriculture, chitosan offers a potential solution for reducing the utilization of chemicals and pesticides that contribute to soil pollution and pose health risks to living organisms [9,10]. Chitosan is authorized as a plant protection product according to the European Regulation EC 1107/2009 and Commission Implementing Regulation EU 2022/456 [11]. Chitosan has been widely studied as an efficient tool to trigger immune responses in plants [12,13]. In grapevine, chitosan elicits a variety of defense responses including the phosphorylation of mitogen-activated protein kinases (MAPKs), phytoalexin production, the expression of defense genes, and chitinase or glucanase activities. Together, this leads to resistance against the necrotrophic fungus *Botrytis cinerea*, the biotrophic fungus *Erysiphe necator*, and the obligate biotrophic oomycete *Plasmopara viticola*, the causal agents of grey mold, powdery mildew, and downy mildew, respectively [14–16]. Spraying aqueous chitosan solutions could lead, after the evaporation of water, to the formation of a thin film coat on the plant's surface [17]. However, the effectiveness of plant protection achieved using chitosan is considerably higher in laboratory and greenhouse settings compared to open fields [18]. The reasons for this disparity remain unclear, but two primary hypotheses can be proposed: (i) lixiviation caused by rainfall and morning dew, and/or (ii) the degradation of chitosan due to sunlight exposure in the presence of air, specifically the photo-oxidation of chitosan. This paper focuses on investigating the second hypothesis.

The literature has extensively discussed the impact of photo-irradiation techniques on various biopolymers, including chitosan [19]. Several studies have described the photodegradation of chitosan, both in film form and in solution, in the presence of air. These investigations have utilized UVB irradiation, typically at approximately 254 nm [20–25]. More recently, chitosan films were irradiated using UVA at a wavelength higher than 300 nm [26]. The pairing of UV light and hydrogen peroxide has also been studied [27]. Blends of chitosan with other polymers such as poly(vinylpyrrolidone), poly(ethylene oxide), and pectin [28–30] as well as chitosan modified with keratin, silk fibroin, and tannic acid [31–33] were also irradiated at 254 nm and studied. A recent study focused on examining the photo-oxidation of chitosan in the presence of citric acid, specifically at a wavelength of 340 nm [34]. The main chemical results of UV irradiation indicated a decrease in the molecular weight of chitosan as a result of β-D-(1→4) glycosidic bond scissions and the formation of carbonyl compounds.

In this paper, we aim to investigate the impact of ultraviolet light on a solid chitosan with a low polymerization degree in the presence of air. The chitosan samples used in this study were subjected to irradiation in their as-received state and in their solid acetate form, simulating the chitosan present on leaves after pulverization from an acetic acid formulation. Both irradiated and non-irradiated chitosan samples underwent analysis using various physicochemical methods. Furthermore, the biological properties of these chitosans were investigated, specifically in terms of their elicitation potential and effectiveness in controlling grapevine diseases, such as downy mildew.

2. Results

The solid chitosan was irradiated as received by the supplier and in its solid acetate form. Two irradiation methods at about 340–350 nm were used, which differ through their irradiance value.

2.1. Characterization of Non-Irradiated and Irradiated Chitosans

2.1.1. Physicochemical Characteristics

The following characteristics of the samples (polymerization degree, X-ray photoelectron spectrometry, elemental analysis, deacetylation degree) were studied, and the results are presented in Table 1.

Table 1. Characteristics of the selected chitosan in terms of its degree of polymerization (DP), X-ray photoelectron spectroscopy (XPS), elemental analysis, and deacetylation degree (DD) both before and after 48 h of irradiation in a UV-box.

Chitosan Sample	DP [1]	XPS Atomic %						Elemental Analysis [2]		DD %
		C_{1s}	O_{1s}	N_{1s}	Ca_{2p}	Si_{2p}	Cl_{2p}	%C	%N	
Before irradiation	9	54.32	31.86	7.03	0.24	0.40	6.16	32.46 ± 0.01	5.97 ± 0.01	83 ± 1
After irradiation	8	53.62	32.42	7.29	0.19	0.43	6.06	32.39 ± 0.04	5.95 ± 0.01	82 ± 1

[1]—See the evolution of the DP in Table 2; [2]—Means and standard deviations calculated from three repetitions.

Table 2. Evolution of the polymerization degree vs. the irradiation time.

Irradiation Time (h)	UV-Box DP	Q-Sun XE-1 DP
0	9	9
4	nd *	8
6	9	nd *
24	9	8
48	9	8
96	nd *	8
168	9	nd *

nd *—Not determined.

To determine the deacetylation degree (DD) of chitosan using elemental analysis, it is necessary to ascertain the weight percentage ratio of carbon to nitrogen (C/N) [35]. Additionally, it is crucial to ensure that the samples do not contain nitrogen-containing impurities such as proteins [36]. Furthermore, the presence of water within the chitosan chains does not affect these analyses, as the percentages of oxygen and hydrogen have no influence on the C/N ratio. Therefore, the hydrogen percentage values were not considered in the analysis.

X-ray photoelectron spectroscopy (XPS) was employed to measure the elemental composition of the atoms on the surface of the sample. A typical XPS survey of chitosan revealed the presence of the expected elements, namely, carbon (C), nitrogen (N), and oxygen (O). Additionally, minor traces of calcium (Ca), silicon (Si), and/or chlorine (Cl) may be detected. These components can be attributed to the processing of crustacean exoskeletons, which serve as the source material for chitosan production [37]. Nevertheless, a high proportion of chlorine indicated that the commercial chitosan is in its chlorhydrate form.

The polymerization degree (DP) calculated from ^{1}H NMR measurements remained constant and was not affected by the irradiation time (Table 2).

Finally, no significant differences were observed between non-irradiated and irradiated samples in terms of acetylation degree calculated from elemental analysis, atomic composition of the surface, and polymerization degree of the chitosan.

2.1.2. Spectral Characteristics

- Fourier-Transform Infrared Spectroscopy

FTIR is a valuable tool for investigating structural modifications in chitosan. Infrared spectra are shown in Figure 2. Comparing the spectra before and after irradiation, recovered $\nu_{O\text{-}H}$ and $\nu_{N\text{-}H}$ bands were observed between 3100 and 3500 cm^{-1}, glycosidic $\nu_{C\text{-}O\text{-}C}$ bands were observed between 850 and 1150 cm^{-1}, and the stretching vibration $\nu_{C=O}$ (amide I) appeared at 1616 cm^{-1}, while δ_{NH3+} appeared at 1510 cm^{-1} [38]. As a result, no noticeable impact of irradiation on the infrared spectra was observed.

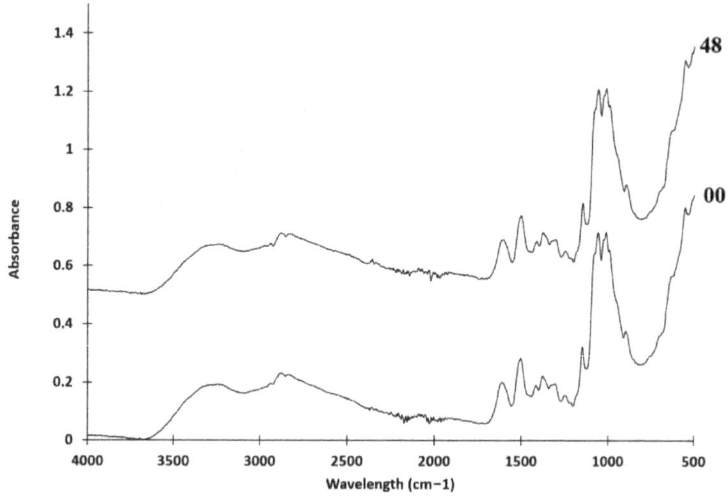

Figure 2. FTIR spectra of powdered non-irradiated (00) and chitosan irradiated for 48 h (48) using a UV-box.

- Mass spectrometry

Chitosan was studied using mass spectrometry, electrospray ionization (ESI), and matrix-assisted laser desorption/ionization time-of-flight mass spectrometry [39,40]. Using ESI, ions with deacetylated D-type structures were identified from D_2 to D_6 accompanied by their dehydrated ions in the spectra of chitosan samples (Figure 3).

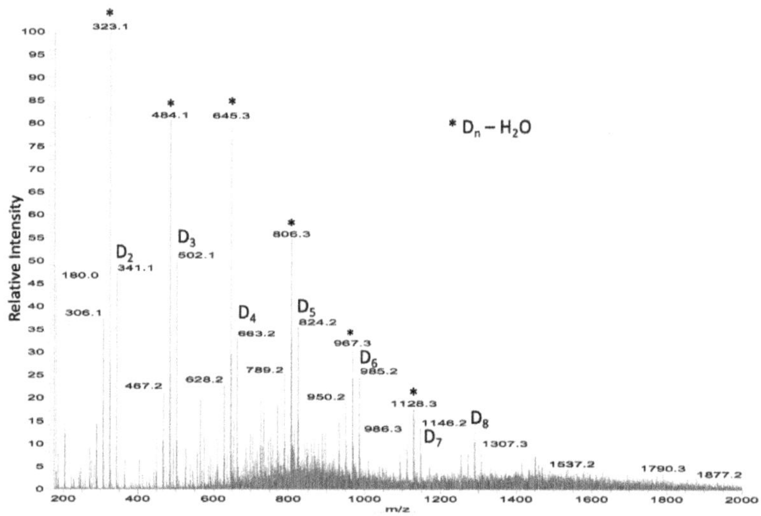

Figure 3. ESI mass spectrum of chitosan. * D_n-H_2O, means dehydrated D_n.

Using collision-induced dissociation, it was demonstrated that the dehydrated ions came from the deacetylated ions through the loss of water. Consequently, it was proposed that the abundance of a global structure, denoted as D_n, was actually the combined result of the abundance of D_n and the (D_n-H_2O) peaks. A comparison was conducted between the electrospray ionization (ESI) spectra of non-irradiated and irradiated chitosan samples exposed to UV-box irradiation for 7 days and Q-Sun XE-1 irradiation for 4 days, respectively (as shown in Table 3). Notably, after chitosan irradiation, there appeared to be an observable increase in the abundances of chitosan oligosaccharides (COS).

Table 3. Comparison of the relative abundances of D_n in non-irradiated and irradiated chitosan. The values presented include uncertainties from three repetitions.

D_n	Non-Irradiated Chitosan	Irradiation Over 7 Days UV-Box	Irradiation Over 4 Days Q-Sun XE-1
D_2	1000	1000	1000
D_3	865 ± 2	1034 ± 2	1065 ± 2
D_4	704 ± 5	909 ± 3	900 ± 2
D_5	617 ± 7	689 ± 6	735 ± 4
D_6	382 ± 10	371 ± 9	422 ± 9

In the MALDI-TOF spectrum of non-irradiated chitosan, some deacetylated ions of the type D_nA_1 were identified when using a reflectron (spectrum not shown). Unfortunately, for unknown reasons, the results were not reproducible in the spectra of the irradiated chitosan. Therefore, it was not possible to determine whether there was a slight deacetylation of the chitosan under UV light.

- Comparison of UV spectra of non-irradiated and irradiated chitosans in UV-box.

Ultraviolet spectra were recorded using chitosan dissolved in acetic acid (Figure 4). The spectra showed similarities, but there were noticeable differences in absorbance between non-irradiated and irradiated chitosan samples. In the case of solid acetate samples, a shoulder appeared around 305 nm, while in solid hydrochloride chitosan, a new band emerged around 340 nm. Additionally, some chromophores near 280 nm disappeared in the irradiated solid hydrochloride chitosan samples.

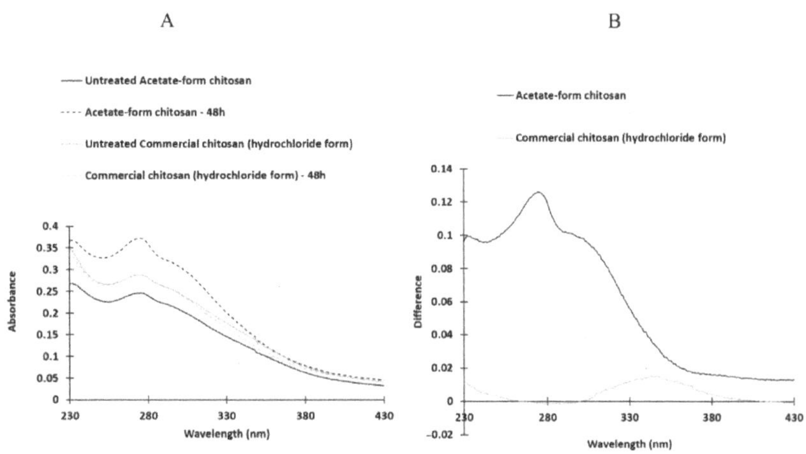

Figure 4. Comparison of UV spectra of non-irradiated and irradiated chitosan samples over a period of 48 h using a UV-box (**A**) and absorbance difference spectra (**B**). The spectra were recorded in acetic acid as the solvent. The chitosan samples were irradiated in their acetate form (represented by the black line) or hydrochloride form (represented by the grey line).

- Comparison of UV spectra between non-irradiated and irradiated chitosan as a function of irradiation time (Figure 5).

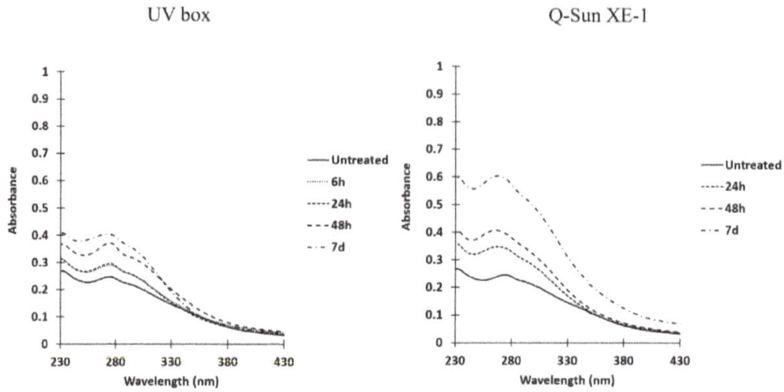

Figure 5. UV spectra of non-irradiated and irradiated chitosan solubilized in acetic acid using a UV-box or a Q-Sun XE-1 system.

The solid chitosan acetate exhibited a slight absorption peak near 310 nm in the ultraviolet spectra, which intensified with longer irradiation times. This absorbance was observed with both of the irradiation devices utilized, but it was more pronounced when the Q-Sun XE-1 system was employed. The comparison of absorbance differences between non-irradiated and irradiated acetate samples revealed the emergence of a shoulder between 280 and 330 nm (spectra not shown).

2.2. Bioactivity of Non-Irradiated and Irradiated Chitosan

To determine the influence of UV on the chitosan's abilities to induce defense responses and resistance in grapevine, we first investigated an early signaling event such as MAPKs phosphorylation. Non-irradiated and irradiated chitosan under its acetate form induced a rapid phosphorylation of two MAPKs with relative molecular masses of 45 and 49 kDa, which was not observed in water-treated control leaf discs. This activation of MAPKs seems to be similar between non-irradiated and irradiated chitosan and does not differ regardless of the method of irradiation (Figure 6).

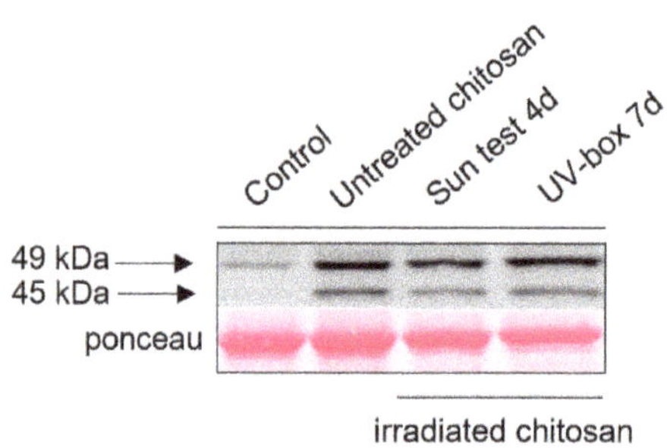

Figure 6. Irradiated-chitosan-induced phosphorylation of MAPKs in grapevine. Activation of two MAPKs detected using immunoblotting with an antibody raised against the human phosphorylated extracellular regulated protein kinase 1/2 (α-pERK1/2) in grapevine leaf discs treated with non-irradiated or Sun test (Q-Sun XE-1 system)- or UV-box-irradiated chitosan (1 mg/mL) or water (negative control). Homogeneous loading was checked using Ponceau red staining.

To further characterize the immune responses triggered by irradiated chitosan, we also investigated its ability to induce resistance against downy mildew in grapevine. Leaves were treated with non-irradiated or irradiated chitosan 48 h prior to inoculation with the biotrophic oomycete *P. viticola*. Non-irradiated and irradiated chitosan treatment significantly reduced *P. viticola* sporulation, and there was no significant difference of protection efficiency between non-irradiated and irradiated chitosan (Figure 7A).

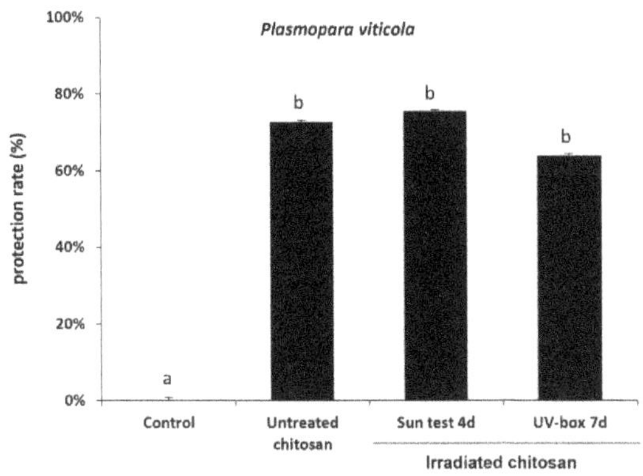

(**A**)

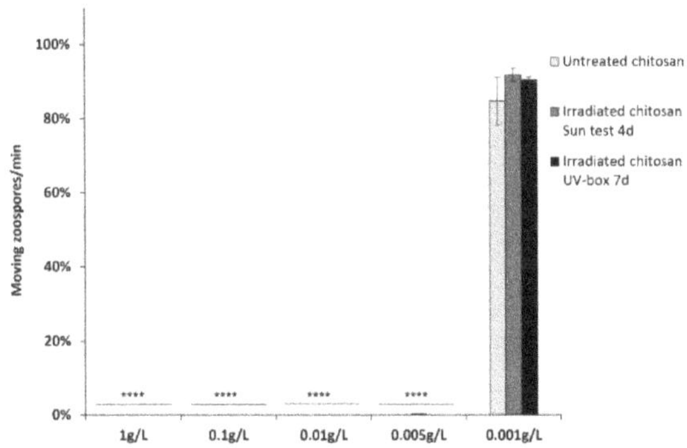

(**B**)

Figure 7. Irradiated-chitosan-induced resistance in grapevine and its toxicity effects on *Plasmopara viticola*. (**A**) Grapevine cuttings were sprayed with non-irradiated and irradiated chitosan (15 mg/L) 48 h before inoculation. Leaf discs were cut 5 dpi and the disease caused by *P. viticola* was assessed at 7 dpi. The sporulating leaf area was evaluated using image analysis Visilog 6.9 software [41]. Values represent the mean of protection rate ± SE (n = 36 discs from three different plants/conditions) from three independent biological experiments. Different letters indicate a statistically significant difference between treatments (Kruskal–Wallis followed by Mann–Whitney post hoc with $p < 0\ 05$). (**B**) Toxicity effects on the motility of *Plasmopara viticola* zoospores. *P. viticola* sporangia were treated with increasing concentrations of non-irradiated and irradiated chitosan, and released zoospores moving on a 1 mm^2 square of a Malassez hemocytometer were counted for one minute. Values represent the mean ± SE (n = 9) of three independent experiments and are expressed as a percentage of the control set as 100%. Asterisks indicate significant differences relative to the control using an unpaired heteroscedastic Student's *t* test; ****, $p < 0.0001$.

The toxicity of irradiated chitosan was then assessed on *P. viticola* by counting the number of moving zoospores after chitosan or water treatment. As with non-irradiated chitosan, irradiated chitosan remains toxic at very low concentrations on *P. viticola* zoospores. Sporangia treated with non-irradiated and irradiated chitosan at concentrations from 1 to 0.005 g/L did not release any moving zoospores (Figure 7B). At 0.001 g/L, there are as many moving zoospores as in the control. Taken together, these results suggest that UV irradiation of chitosan has no impact on its eliciting and antifungal activities and highlights a strong direct biocide effect of non-irradiated and irradiated chitosan.

3. Discussion

As mentioned in the introduction, the use of chitosan as a biopesticide in the field does not yield the expected results, despite the promising bioactivities observed in laboratory or greenhouse settings. Several hypotheses can be proposed, one of which relates to the potential influence of solar light on the structure of chitosan. Conducting experiments in the open field under controlled conditions is challenging due to the rapidly changing weather conditions throughout the day. Therefore, in this work, laboratory experiments were carried out using chitosan with a low degree of polymerization (due to its significant bioactive properties) and dedicated devices for continuous irradiation using ultraviolet lamps to simulate solar radiation. It is important to note that in-depth studies on powdered chitosan have been limited. The chosen chitosan in this study does not form a film and is present only in its powdered form after application via spraying on plants.

Two artificial sources of UV radiation were employed to simulate the ultraviolet contribution of sunlight, primarily at a wavelength range of 340–350 nm. The objective was to assess the potential structural degradation of chitosan and its impact on its biological properties, specifically its eliciting and antifungal activities.

In the UV spectra of solid chitosan in its hydrochloride or acetate form, a band was observed between approximately 280 and 430 nm. This phenomenon has been previously noted by various authors, including Yu et al., in the case of acetate chitosan solutions [42]. To the best of our knowledge, the exact origin of this broad absorption band has not been definitively identified. After irradiating the solid chitosan, the UV spectra exhibited the disappearance of initially present chromophores around 280 nm, and the emergence of a new band around 340 nm for the hydrochloride form and the formation of a band at approximately 300 nm for the acetate form.

The intensity of the new band increased with prolonged irradiation time, although these modifications appeared to be less pronounced in the hydrochloride form. However, according to El-Sawy et al. [43], who conducted irradiation on chitosan under ^{60}Co (γ rays), this absorption band may be attributed to the $n \rightarrow \delta^*$ transition of chitosan's amino groups or the $n \rightarrow \pi^*$ transition of carbonyl or carboxylic groups. Under severe UV conditions (accelerated photo-oxidation at 60 °C), Bussière et al. [26] also suggested that this band could be attributed to the formation of carbonyl groups and the scission of glycosidic bonds. However, in our experimental conditions, it was not possible to reveal any evidence of such degradation in solid chitosan. The FTIR spectra indicated no significant differences between the chemical structures of irradiated and non-irradiated powdered chitosan. No new bands appeared or disappeared. Consequently, it can be concluded that there were no significant chemical transformations in the overall structure of the biomacromolecules. The primary possibility would be the cleavage of ß-D-(1→4) glycosidic bonds. Nevertheless, if such cleavages occurred, other physicochemical modifications should manifest, such as a reduction in the average polymerization degree and a substantial increase in low D_n fragments in ESI mass spectra. In reality, the former was not observed, and the latter was not particularly significant.

After the irradiation of solid chitosans, the UV spectra exhibited the disappearance of chromophores that were originally present in the non-irradiated samples. One possible hypothesis is the photodegradation of amide functions under irradiation. However, if this happened, a change in the carbon content of the sample should be observed, which is not the case.

Consequently, the various physicochemical analyses performed did not reveal any noticeable structural modifications resulting from chitosan irradiation. Additionally, it was found that irradiation had no impact on the elicitation of antifungal activities.

In our specific experimental conditions, it appeared that the UV irradiation did not have sufficient energy to significantly alter the powdered low-molecular-weight chitosan, in accordance with Pandit et al. stating that UV treatment alone is ineffective in degrading chitosan [19]. Additionally, it should be noted that the irradiation procedure employed in this study did not accurately replicate the solar spectrum but rather imposed more rigorous conditions [44].

Consequently, two conclusions can be drawn: Firstly, in the conditions utilized in this study, the UV irradiation alone, which simulates a 5-day exposure in the field, is unlikely to induce substantial modifications in the chitosan structure, as previously observed in the literature [22], and chitosan eliciting and anti-downy mildew properties. Secondly, the lower effectiveness of chitosan in open field applications would not be attributed to sunlight exposure.

4. Materials and Methods

4.1. Materials

The chitosan used in this study was obtained from Elicityl (Crolles, France) through the process of the acidic hydrolysis of chitosan from crustacean shells. The chitosan was stored at room temperature, away from sunlight, to maintain its integrity. Its main characteristics, including its physical and chemical properties, are provided in Table 1 *Before irradiation*. Prior to the bioactivity assays, the chitosan was dissolved in ultrapure water. Glacial acetic acid was supplied by Fisher Chemical (United Kingdom).

Grapevine (*V. vinifera* cv. Marselan) cuttings were grown in a greenhouse until they had developed 6–8 leaves. The first and second youngest adult leaves from each plant were used for experiments. Grapevine downy mildew (*P. viticola*) was routinely maintained on *Vitis vinifera* cv. Marselan plants as previously described [45].

4.2. Methods

4.2.1. Elemental Analyses

Elemental analyses were performed on a ThermoFisher FlashEA-1112 microanalyzer equipped with a sampler changer. Samples were prepared by being weighed precisely (approximatively 1.5 mg) on a tin capsule with a Mettler Toledo XPR2U microbalance. The tin capsules were then compressed into a small cube. In the analyzer, the samples underwent dynamic "flash" combustion at 930 °C. Nitrogen, carbon, and hydrogen were then recombined using a series of reduction and oxidation reactions to give dinitrogen, carbon dioxide, and water. These gases were then separated on a chromatographic column, detected by a katharometer, and then quantified by the integration of a peak resulting from the variation in the thermal conductivity of the gases, leading to the percentages of C, H, and N.

The deacetylation degree was calculated with the C/N weight percentage ratio according to Dos Santos et al. [46].

4.2.2. Degree of Polymerization (DP) by ^{1}H NMR

The measurements of average DPs are based on the integration of the signals corresponding to protons belonging to the non-reducing ends of the polymer chains and those belonging to the reducing ends as described in the literature [47]. ^{1}H-NMR spectra were registered at room temperature using a Liquid-state 400 MHz NMR spectrometer (Bruker ADVANCE I) with 32 scans. Samples were prepared with 20 mg of chitosan and mixed with 1 mL of D_2O and 10 µL of DCl (7.4 M).

4.2.3. Analysis of Chitosan Powder Surface using XPS

A ThermoFisher Scientific K-ALPHA spectrometer was used for XPS surface analysis with a monochromatized Al-Kα source (hν = 1486.6 eV) and a 400 µm X-ray spot size. Powders were pressed onto indium foils. The full spectra (0–1100 eV) were obtained with a constant pass energy of 200 eV, while high-resolution spectra were recorded with a constant pass energy of 40 eV. Charge neutralization was applied during the analysis. High-resolution spectra were quantified using Avantage software provided by ThermoFisher Scientific (Waltham, MA, USA).

4.2.4. Mass Spectrometry

- Electrospray

Electrospray analyses were performed on a linear trap quadrupole (LTQ) mass spectrometer (Thermo Fisher Scientific, Waltham, MA, USA) in positive ion mode using direct infusion of the samples in a mixture of water/methanol (4/1, *v*/*v*) (0.1 mg/mL). Electrospray source parameters were as follows: capillary voltage +20 V, tube lens voltage +90 V, capillary temperature 300 °C, sheath and auxiliary gas flow (N^2) 8 and 5, sweep gas 0, spray voltage 3.6. Spectra were acquired by full range acquisition covering *m/z* 50–2000. Collision-induced dissociations were made with helium.

- MALDI-TOF

MALDI-TOF spectra were registered on a Voyager mass spectrometer (Applied Biosystems, Waltham, MA, USA) equipped with a pulsed nitrogen laser (337 nm) and a time-delayed extracted ion source. Spectra were recorded in positive ion mode using reflectron mode with an accelerating voltage of 20 kV. The chitosan samples were dissolved in a mixture made of H_2O/MeOH (50/50 *v*/*v*) and acetic acid (0.1%, *v*/*v*) at 10 mg/mL. The 2,5-dihydroxybenzoic acid (DHB) matrix solution was prepared by dissolving DHB (10 mg) in MeOH (1 mL). A MeOH solution of cationization agent (NaI, 10 mg/mL) was also prepared. The solutions were combined in a 10:1:1 volume ratio of matrix to sample to cationization agent. One to two microliters of the obtained solution were deposited onto the sample target and vacuum dried.

4.2.5. ATR-FTIR Spectra

Infrared spectra were registered using a FTIR spectrometer with a diamond crystal (PIKE technologies/Gladi Atrvertex 70, Bruker, France) from 4000 to 400 cm^{-1} (32 scans, resolution 4 cm^{-1}).

4.2.6. Ultraviolet Irradiation

Solid commercial chitosan was irradiated as received from the supplier and under its acetate solid form in Petri dishes.

- Preparation of chitosan acetate from commercial chitosan

To prepare the powdered chitosan, 300 mg of it was dissolved in 10 mL of 1% (*v*/*v*) acetic acid and stirred until complete solubilization. The resulting solutions were then transferred into Petri dishes (diameter of 8 cm). Subsequently, the samples were dried at 40 °C for 48 h prior to conducting the UV irradiation tests. The dried powder was evenly spread across the surface of the Petri dishes.

- UV irradiation of solid commercial chitosan and chitosan acetate

Two different methods were used to irradiate the chitosan samples:

(1) The Petri dishes were placed in a laboratory-made irradiation chamber (named "UV-box" in the text) equipped with four black light UV lamps (Mazdafluor TFWN 18) emitting mainly at 350 nm. The distance between the lamp and the sample was ~25 cm, the light intensity was 0.25 mW/cm^2, and the temperature was maintained at 25 °C using a fan.

(2) The Petri dishes were covered with a borosilicate lid to protect the Q-Sun XE-1 test chamber (Q-Lab Corporation, Westlake, OH, USA) from potential gas and dust that could go outside. The device was equipped with a xenon source and a Daylight-Q filter. The irradiance was equal to 0.47 W/m^2 at 340 nm. The chamber was ventilated to maintain the temperature at 50 °C.

The irradiation times were fixed at different values for the UV-box and the Q-Sun XE-1.

- UV–vis analysis of irradiated chitosans

After irradiation, the powdered chitosan samples were dissolved in an aqueous solution of acetic acid (1% *v*/*v*) in order to obtain a concentration of 5 mg/mL. Spectra were recorded on a UV–visible spectrophotometer (Agilent Cary 100) from 250 nm to 430 nm using quartz cells (length 1 cm).

4.2.7. MAPK Activation

Discs of grapevine leaves from greenhouse cuttings were pre-infiltrated with ultrapure water then equilibrated, with the abaxial face on ultrapure water, for 4 h in a 6-well plate. They were then treated by substituting the water with non-irradiated or irradiated chitosan (1 mg/mL) or water (negative control) and harvested 20 min post-treatment. MAPKs activation was detected after immunoblotting of the extracted proteins (20 µg) using an anti-p42/44-phospho-ERK antibody (Cell Signaling). The revealing step was performed on an Amersham™ ImageQuant™800 (Cytiva) using ECL™ Prime as a Western blotting detection reagent. Transfer quality and homogeneous loading were checked using Ponceau red staining. Three independent experiments were performed.

4.2.8. Downy Mildew Assay

For *P. viticola* infection on grapevine cuttings, the lower leaf surface was sprayed with non-irradiated or irradiated chitosan (15 µg/mL) or water (control). Two days post-treatment (dpt), treated leaves were sprayed with a freshly prepared suspension of sporangia ($2 \cdot 10^4$ sp/mL) and plants were maintained in 100% humidity for 4 h. Leaf discs were cut 5 days post-inoculation (dpi) and transferred on moist Whatman paper in a plastic box maintained in 100% humidity under a 10/14 h day/night cycle at 20/18 °C. Infection intensity was assessed as 7 dpi by measuring the sporulating area by using image analysis Visilog 6.9 software [41].

For toxicity assays on *P. viticola* zoospores, a suspension of *P. viticola* sporangia ($1 \cdot 10^5$ sp/mL) prepared in ultrapure water was treated with increasing concentrations of non-irradiated or irradiated chitosan (0.001, 0.005, 0.01, 0.1, and 1 mg/mL). Two hours later, released zoospores moving on a 1 mm^2 square of a Malassez hemocytometer were counted for one minute.

5. Conclusions

The study aimed to assess the influence of sunlight on the stability and bioactivity of powdered chitosan with a low polymerization degree. To simulate sunlight exposure, artificial UV sources were utilized. A range of physico-chemical methods, including elemental analysis, NMR to calculate the polymerization degree, MALDI-TOF, IR, XPS, and UV spectroscopy, were employed to compare the characteristics of chitosan before and after irradiation. Although minor differences were observed in the UV spectra, no significant structural modifications of chitosan were detected under the experimental conditions. Additionally, the anti-downey mildew and eliciting properties of chitosan were unaffected by UV irradiation. These findings suggest that the reduced effectiveness of chitosan in open field conditions cannot be solely attributed to sunlight exposure.

Author Contributions: Conceptualization, C.G., S.M., B.P. and V.C.; methodology, G.H., S.M., D.B., C.G., B.P. and V.C.; investigation, S.M., G.H., D.B. and C.G.; writing—original draft preparation, C.G., G.H. and S.M.; writing—review and editing, C.G., S.M., D.B. and V.C.; supervision, V.C., B.P. and C.G.; funding acquisition, B.P. and V.C. All authors have read and agreed to the published version of the manuscript.

Funding: This work has been financially supported by Agence Nationale de la Recherche (ANR) and Office Français de la Biodiversité (OFB) ("ChitoProtect" project, grant # ANR-19-ECOM-0008).

Institutional Review Board Statement: Not applicable.

Informed Consent Statement: Not applicable.

Data Availability Statement: Not applicable.

Acknowledgments: The authors would like to thank Patrick Rosa and Nathalie Daro (ICMCB, UMR 5026, University of Bordeaux, France) for the microanalysis of chitosans, Elicityl for providing the chitosan samples (Crolles, France), and Guillaume Wantz and Sylvain Chambon (IMS, UMR 5255, University of Bordeaux) for the technical support of the Q-Sun XE-1 test experiments.

Conflicts of Interest: The authors declare no conflict of interest.

Sample Availability: Not applicable.

References

1. Yen, M.-T.; Yang, J.-H.; Mau, J.-L. Physicochemical characterization of chitin and chitosan from crab shells. *Carbohydr. Polym.* **2009**, *75*, 15–21.
2. Guarnieri, A.; Triunfo, M.; Scieuzo, C.; Ianniciello, D.; Tafi, E.; Hahn, T.; Zibek, S.; Salvia, R.; de Bonis, A.; Falabella, P. Antimicrobial properties of chitosan from different developmental stages of the bioconverter insect Hermetia illucens. *Sci. Rep.* **2022**, *12*, 8084. [CrossRef] [PubMed]
3. Sebastian, J.; Rouissi, T.; Brar, S.K.; Hedge, K.; Verma, M. Microwave-assisted extraction of chitosan from *Rhizopus oryzae* NRRL 1526 biomass. *Carbohydr. Polym.* **2019**, *219*, 431–440. [CrossRef] [PubMed]
4. Florez, M.; Guerra-Rodriguez, E.; Cazon, P.; Vazquez, M. Chitosan for food packaging: Recent advances in active and intelligent films. *Food Hydrocoll.* **2022**, *124*, 107328. [CrossRef]
5. Lima, R.; Fernandes, C.; Pinto, M.M.M. Molecular modifications, biological activities, and applications of chitosan and derivatives: A recent update. *Chirality* **2022**, *34*, 1166–1190. [CrossRef]
6. Duceac, I.A.; Coseri, S. Biopolymers and their derivatives: Key components of advanced biomedical technologies. *Biotechnol. Adv.* **2022**, *61*, 108056. [CrossRef]
7. Guzman, E.; Ortega, F.; Rubio, R.G. Chitosan: A promising multifunctional cosmetic ingredient for skin and hair care. *Cosmetics* **2022**, *9*, 99. [CrossRef]
8. Sajid, M. Chitosan-based adsorbents for analytical sample preparation and removal of pollutants from aqueous media: Progress, challenges and outlook. *Trends Environ. Anal. Chem.* **2022**, *36*, e00815. [CrossRef]
9. Damalas, C.A.; Eleftherohorinos, I.G. Pesticide Exposure, Safety Issues, and Risk Assessment Indicators. *Int. J. Environ. Res. Public Health* **2011**, *8*, 1402–1419. [CrossRef]
10. El-Sheikh, E.-S.A.; Ramadan, M.M.; El-Sobki, A.E.; Shalaby, A.A.; McCoy, M.R.; Hamed, I.A.; Ashour, M.-B.; Hammock, B.D. Pesticide Residues in Vegetables and Fruits from Farmer Markets and Associated Dietary Risks. *Molecules* **2022**, *27*, 8072. [CrossRef]
11. European Union. Commission Implementing Regulation (EU) 2022/456 approving the basic substance chitosan in accordance with Regulation (EC) No 1107/2009 of the European Parliament and of the Council concerning the placing of plant protection products on the market, and amending the Annex to Implementing Regulation (EU) No 540/2011. *Off. J. the Eur. Union* **2022**.
12. El Hadrami, A.; Adam, L.R.; El Hadrami, I.; Daayf, F. Chitosan in Plant Protection. *Marine Drugs* **2010**, *8*, 968–987. [CrossRef]
13. Chandra, S.; Chakraborty, N.; Panda, K.; Acharya, K. Chitosan-induced immunity in *Camellia sinensis* (L.) O. Kuntze against blister blight disease is mediated by nitric-oxide. *Plant Physiol. Biochem.* **2017**, *115*, 298–307. [CrossRef]
14. Aziz, A.; Trotel-Aziz, P.; Dhuicq, L.; Jeandet, P.; Couderchet, M.; Vernet, G. Chitosan oligomers and copper sulfate induce grapevine defense reactions and resistance to gray mold and downy midew. *Phytopathology* **2006**, *96*, 1188–1194.
15. Lucini, L.; Baccolo, G.; Rouphael, Y.; Colla, G.; Bavaresco, L.; Trevisan, M. Chitosan treatment elicited defence mechanisms, pentacyclic triterpenoids and stilbene accumulation in grape (*Vitis vinifera* L.) bunches. *Phytochemistry* **2018**, *156*, 1–8. [CrossRef]
16. Brulé, D.; Villano, C.; Davies, L.J.; Trdá, L.; Claverie, J.; Héloir, M.C.; Chiltz, A.; Adrian, M.; Darblade, B.; Tornero, P.; et al. The Grapevine (*Vitis vinifera*) LysM Receptor Kinases VvLYK1-1 and VvLYK1-2 Mediate Chitooligosaccharide-Triggered Immunity. *Plant Biotechnol. J.* **2019**, *17*, 812–825. [CrossRef]
17. Loron, A.; Wang, Y.; Atanasova, V.; Richard-Forget, F.; Gardrat, C.; Coma, V. Chitosan for eco-friendly control of mycotoxinogenic Fusarium graminearum. *Food Hydrocoll.* **2023**, *134*, 108067. [CrossRef]
18. Malausa, T. Le biocontrôle: Introduction, état des lieux, perspectives. In *Actes du Colloque Construisons l'Avenir avec le Biocontrôle*; Rencontres Régionales de la Recherche, du Développement et de la Formation; Chambres d'agriculture Nouvelle Aquitaine: Libourne Montagne, France, 2017; pp. 9–11.
19. Pandit, A.; Indurkar, A.; Deshpande, C.; Jain, R.; Dandeka, P. A systematic review of physical techniques for chitosan degradation. *Carbohydr. Polym. Technol. Appl.* **2021**, *2*, 100033. [CrossRef]
20. Andrady, A.L.; Torikai, A.; Kobatake, T. Spectral sensitivity of chitosan photodegradation. *J. Appl. Polym. Sci.* **1996**, *62*, 1465–1471. [CrossRef]
21. Mucha, M.; Pawlak, A. Complex study on chitosan degradability. *Polimery* **2002**, *47*, 509–516. [CrossRef]
22. Sionkowska, A. Effects of solar radiation on collagen and chitosan films. *J. Photochem. Photobiol. B Biol.* **2006**, *82*, 9–15. [CrossRef]
23. Sionkowska, A.; Kazmarek, H.; Wisniewski, M.; Skopinska, J.; Lazare, S.; Tokarev, V. The influence of UV irradiation on the surface of chitosan films. *Surf. Sci.* **2006**, *600*, 3775–3779. [CrossRef]

24. Sionkowska, A.; Planecka, A.; Kozlowska, J.; Skopinska-Wisniewska, J.; Los, P. Weathering of chitosan films in the presence of low- and high-molecular weight additives. *Carbohydr. Polym.* **2011**, *84*, 900–906. [CrossRef]
25. Sionkowska, A.; Planecka, A.; Lewandowska, K.; Kaczmarek, B.; Szarszewska, P. Influence of UV-irradiation on molecular weight of chitosan. *Prog. Chem. Appl. Chitin Its Deriv.* **2013**, *18*, 21–28. [CrossRef]
26. Bussière, P.O.; Gardette, J.L.; Rapp, G.; Masson, C.; Therias, S. New insights into the mechanism of photodegradation of chitosan. *Carbohydr. Polym.* **2021**, *259*, 117715. [CrossRef]
27. Wang, S.M.; Huang, Q.Z.; Wang, Q.S. Study on the synergetic degradation of chitosan with ultraviolet light and hydrogen peroxide. *Carbohydr. Res.* **2005**, *340*, 1143–1147.
28. Kowalonek, J. Studies of chitosan/pectin complexes exposed to UV radiation. *Int. J. Biol. Macromol.* **2017**, *103*, 515–524. [CrossRef]
29. Kowalonek, J. Surface and thermal properties of UV-irradiated chitosan/poly(ethylene oxide) blends. *J. Photochem. Photobiol. A Chem.* **2017**, *348*, 209–218. [CrossRef]
30. Sionkowska, A.; Wisniewski, M.; Skopinska, J.; Vicini, S.; Marsano, E. The influence of UV irradiation on the mechanical properties of chitosan/poly(vinyl pyrrolidone) blends. *Polym. Degrad. Stab.* **2005**, *88*, 261–267. [CrossRef]
31. Sionkowska, A.; Skopinska-Wisniewska, J.; Planecka, A.; Kozlowska, J. The influence of UV irradiation on the properties of chitosan films containing keratin. *Polym. Degrad. Stab.* **2010**, *95*, 2486–2491. [CrossRef]
32. Sionkowska, A.; Planecka, A.; Lewandowska, K.; Michalska, M. The influence of UV-irradiation on thermal and mechanical properties of chitosan and silk fibroin mixtures. *J. Photochem. Photobiol. B Biol.* **2014**, *140*, 301–305. [CrossRef]
33. Sionkowska, A.; Kaczmarek, B.; Gnatowska, M.; Kowalonek, J. The influence of UV-irradiation on chitosan modified by the tannic acid addition. *J. Photochem. Photobiol. B Biol.* **2015**, *148*, 333–339. [CrossRef]
34. Infurna, G.; Cavallaro, G.; Lazzara, G.; Milioto, S.; Dintcheva, N.T. Effect of different processing techniques and presence of antioxidant on the chitosan film performance. *J. Vinyl Addit. Technol.* **2022**, *28*, 343–351. [CrossRef]
35. Weisspflog, J.; Vehlow, D.; Müller, M.; Kohn, B.; Scheler, U.; Boye, S.; Schwarz, S. Characterization of chitosan with different degree of deacetylation and equal viscosityin dissolved and solid state—Insights by various complimentary methods. *Int. J. Biol. Macromol.* **2021**, *171*, 242–261. [CrossRef]
36. Zimoch-Korzycka, A.; Gardrat, C.; Al Kharboutly, M.; Castellan, A.; Pianet, I.; Jarmoluk, A.; Coma, V. Chemical characterization, antioxidant and anti-listerial activity of non-animal chitosan-glucan complexes. *Food Hydrocoll.* **2016**, *61*, 338.e343. [CrossRef]
37. Matienzo, L.J.; Winnacker, S.K. Dry processes for surface modification of a biopolymer: Chitosan. *Macromol. Mater. Eng.* **2002**, *287*, 871–880. [CrossRef]
38. Kurozumi, S.; Kiyose, M.; Noguchi, T.; Sato, K. A novel hydrochloride free chitosan oligosaccharide production method to improve taste. *Int. J. Biol. Macromol.* **2019**, *140*, 109–118. [CrossRef]
39. Allison, L.; Lutzke, A.; Reynolds, M.M. Identification of low molecular weight degradation products from chitn and chitosan by electrospray ionization time-of-flight mass spectrometry. *Carbohydr. Res.* **2020**, *493*, 108046. [CrossRef]
40. Chen, M.; Zhu, X.; Li, Z.; Guo, X.; Ling, P. Application of matrix-assisted laser desorption/ionization time-of-flight mass spectrometry (MALDI-TOF-MS) in preparation of chitosan oligosaccharides (COS) with degree of polymerization (DP) 5–12 containing well-distributed acetyl groups. *Int. J. Mass Spectrom.* **2010**, *290*, 94–99. [CrossRef]
41. Kim Khiook, I.L.; Schneider, C.; Heloir, M.C.; Bois, B.; Daire, X.; Adrian, M.; Trouvelot, S. Image analysis methods for assessment of H_2O_2 production and Plasmopara viticola development in grapevine leaves: Application to the evaluation of resistance to downy mildew. *J. Microbiol. Methods* **2013**, *95*, 235–244. [CrossRef]
42. Yu, D.; Basumatary, I.B.; Kumar, S.; Ye, F.; Dutta, J. Chitosan modified with bio-extract coating with UV filering feature. *Int. J. Biol. Macromol.* **2023**, *230*, 123145. [CrossRef]
43. El-Sawy, N.M.; Abd El-Rehim, H.A.; Elbarbary, A.M.; Hegazy, E.A. Radiation-induced degradation of chitosan for possible use as a growth promoter in agricultural purposes. *Carbohydr. Polym.* **2010**, *79*, 555–562. [CrossRef]
44. Diffey, B.L. Sources and measurement of ultraviolet radiation. *Methods* **2002**, *28*, 4–13. [CrossRef]
45. Steimetz, E.; Trouvelot, S.; Gindro, K.; Bordier, A.; Poinssot, B.; Adrian, M.; Daire, X. Influence of Leaf Age on Induced Resistance in Grapevine against *Plasmopara viticola*. *Physiol. Mol. Plant Pathol.* **2012**, *79*, 89–96. [CrossRef]
46. Dos Santos, Z.M.; Caroni, A.L.P.F.; Pereira, M.R.; da Silva, D.R.; Fonseca, J.L.C. Determination of deacetylation degree of chitosan: A comparison between conductometric titration and CHN elemental analysis. *Carbohydr. Res.* **2009**, *344*, 2591–2595. [CrossRef]
47. Marzaioli, A.M.; Bedini, E.; Lanzetta, R.; Perino, V.; Parilli, M.; De Castro, C. Preparation and NMR characterization of glucosamine oligomers bearing an azide function using chitosan. *Carbohydr. Polym.* **2012**, *90*, 847–852. [CrossRef]

molecules

MDPI

Article

Impact of Chitosan on the Mechanical Stability of Soils

Agnieszka Adamczuk * and **Grzegorz Jozefaciuk**

Department of Physical Chemistry of Porous Materials, Institute of Agrophysics of Polish Academy of Sciences, Doswiadczalna 4 Str., 20-290 Lublin, Poland; jozefaci@ipan.lublin.pl
* Correspondence: a.adamczuk@ipan.lublin.pl

Abstract: Chitosan is becoming increasingly applied in agriculture, mostly as a powder, however little is known about its effect on soil mechanical properties. Uniaxial compression test was performed for cylindrical soil aggregates prepared from four soils of various properties (very acidic Podzol, acidic Arenosol, neutral Fluvisol and alkaline Umbrisol) containing different proportions of two kinds of chitosan (CS1 of higher molecular mass and lower deacetylation degree, and CS2 of lower molecular mass and higher deacetylation degree), pretreated with 1 and 10 wetting–drying cycles. In most cases increasing chitosan rates successively decreased the mechanical stability of soils that was accompanied by a tendential increase in soil porosity. In one case (Fluvisol treated with CS2) the porosity decreased and mechanical stability increased with increasing chitosan dose. The behavior of acidic soils (Podzol and Arenosol) treated with CS2, differed from the other soils: after an initial decrease, the strength of aggregates increased with increasing chitosan amendment, despite the porosity consequently decreasing. After 10 wetting–drying cycles, the strength of the aggregates of acidic soils appeared to increase while it decreased for neutral and alkaline soils. Possible mechanisms of soil–chitosan interactions affecting mechanical strength are discussed and linked with soil water stability and wettability.

Keywords: aggregate; chitin derivatives; food wastes; stress; strain; Young's modulus

check for updates

Citation: Adamczuk, A.; Jozefaciuk, G. Impact of Chitosan on the Mechanical Stability of Soils. *Molecules* **2022**, *27*, 2273. https://doi.org/10.3390/molecules27072273

Academic Editor: Agnieszka Ewa Wiącek

Received: 1 March 2022
Accepted: 30 March 2022
Published: 31 March 2022

1. Introduction

Due to its non-toxicity, bacteriostatic properties, and polycationic nature, chitosan has found many industrial, medical, pharmaceutical, and environmental applications [1–3]. The availability of industrial quantities of chitosan in the late 1980s enabled it to be tested in agriculture [4]. Up to date, chitosan has been increasingly applied for the improvement of the degradative capacity of contaminated soils by introducing specific consortia of microorganisms (bioaugmentation), providing nutrients to the chemical degrading indigenous microorganisms present at the contaminated sites (biostimulation), biochemical degradation of xenobiotic compounds. Chitosan is used as a carrier for microbial communities that can be encapsulated for further use and for providing resistance to the enzymes from harsh environmental conditions like pH and temperature. It is applied as an antiviral, antibacterial, antifungal and antinematode agent, for seeds coatings, and plant protection by controlling the spreading of diseases [5,6]. Chitosan has been proven to stimulate plant growth and yield, and to induce abiotic and biotic stresses tolerance in various commodities [7,8]. Besides that, chitosan has been employed in soil as a plant nutrient and has shown great efficacy in combination with other industrial fertilizers without affecting the soil's beneficial microbes. Furthermore, it is helpful in reducing fertilizer losses thus reducing the overuse of synthetic fertilizers during crop production, which is important in keeping environmental pollution under check. More recently chitosan is applied as a valuable delivery system for fertilizers, herbicides, pesticides, and micronutrients for crop growth promotion by maintaining balanced and sustained nutrition [1,4,9,10]. Various composite or copolymer systems containing chitosan are used to improve the soil's water-holding capacity [11,12]. Chitosan is also applied for the removal of various types of contaminants from soils [13,14].

Chitosan introduced into a soil in one or more above ways can, as any other biopolymer, interact with soil components in different ways, including adsorption of polymer molecules on surfaces of soil components, covering soil particles with a thin polymer film, formation of polymer ties connecting neighboring particles, adhesion, hydrogen bonding or bridging of soil particles via polyvalent counterions [15,16]. Direct binding of a charged polymer to negatively charged soil components (clay minerals, silica, feldspars, organic matter) with electrostatic forces seems to be a rare feature of chitosan, because, contrary to the majority of natural biopolymers applied for soil treatment (tragacanth, arabic, karaya, gellan, carrageenan, locust bean, xanthan, guar and/or tamarind gums, agar, pectin, alginate, arabinans, amylase, lipids, kasein, cellulose) having anionic or non-ionic character [17–19], it carries a positive charge in a broad range of neutral and acidic pH values [20]. So strong interactions of chitosan with soil components imply that the addition of chitosan should increase soil mechanical resistance leading to soil stabilization, consolidation and hardening. However, among various biopolymers implemented in recent years for soil stabilization, erosion mitigation, and dust control, chitosan has been applied rarely. Huang et al. [17], in their comprehensive review of the application of various polymers for soil stabilization, mentioned only a single application of chitosan. Most probably the application of chitosan for soil stabilization is limited by its very poor solubility in water which changes to some extent depending on molecular weight, deacetylation degree and crystal structure [21–23] and/or by its high biodegradation rate (studies of Mostafa et al. [24] showed the fastest biodegradation of chitosan as compared to the other studied biopolymers). Apart from the successful application of various interpolymer complexes, nanoparticles or composites including chitosan [25–31], a positive effect on soil stabilization by pure chitosan has been achieved only for its highly dispersed jellified forms (chitosan dissolved usually in organic acids). Orts et al. [28] applied chitosan gel for hardening eroded furrows. They observed a strong effect of soil stabilization in laboratory conditions, however, under natural conditions, the effect was observed only in the initial part of the furrow, which was explained by the very strong binding of chitosan to soil components in the area close to the zone of its application. Shariatmadari et al. [32] studied the effects of chitosan dissolved in acetic acid on sandy soil stabilization by using unconfined compression tests. The unconfined compression tests conducted after 7, 14 and 28 days showed the strength of samples treated in dry conditions was higher than for samples cured in saturated water conditions. In contrast, Hataf et al. [27] who studied uniaxial compression of sand–chitosan gel mixtures, observed that chitosan improved soil strength, but its effect decreased with the reduction of water. They concluded that the additional strength that could be achieved with the addition of chitosan to the soil fades as the soil becomes dry. Aguilar et al. [33] investigated the feasibility of using chitosan dissolved in acetic acid as an admixture, or as an external coating, for earthen constructions to improve their resistance in terms of both their key mechanical properties and, more specifically, during the occurrence of water-induced degradation. They showed that in both instances the use of low concentrations of chitosan can significantly improve performance.

However, for agricultural purposes, the application of chitosan gels containing huge amounts of acid may be dangerous for the soil environment because of the potential for inducing aluminum toxicity, mineral destruction, nutrient leaching or heavy metal mobilization [34–36]. In agricultural applications, chitosan is usually introduced into the soil mostly as powders (pure components and admixtures), and also as membranes (chitosan coatings) or particulates (suspensions) [1,2,10,37–41]. Despite this, the literature brings extremely small information on the mechanical properties of soils amended with chitosan, not in a gel but in a powder form. After an extensive search, the authors found only a single paper on this topic by Soldo et al. [42], who studied the mechanical properties of Piedmont well-graded sand with silt, amended with various biopolymers, including chitosan. They mentioned that chitosan was applied in a form as it was supplied by the manufacturer (probably as the powder). After five days of curing, chitosan did not lead to any further increase in the compressive strength even though the specimens with chitosan had the highest compressive strength right after the mixing. Therefore they excluded chitosan from further research of this type.

To fill this knowledge gap, this paper studies the mechanical stability of a few soils amended with different doses of chitosan powder by a uniaxial compression test. Because various chitosan–soil binding mechanisms contribute differently depending on polymer solubilization, concentration, charge, conformation, molecular weight and molecule size, that in turn depend on the ambient solution pH and ionic strength [43–45], four soils of different pH values and two types of chitosan, differing in molecular masses and deacetylation degrees were selected for the testing. Since cyclic wetting–drying affects biopolymer treated soils [46–48] and is used as a measure of the durability of biopolymers [46], the studied soils were subjected also to one and ten wetting–drying cycles.

2. Results

The addition of chitosan of high molecular mass and low deacetylation degree (CS1, see M&M section) in the rates of 0.5, 1, 2, 4 and 8% decreased, in general, the mechanical strength of four studied soils: very acidic Podzol, acidic Arenosol, neutral Fluvisol and alkaline Umbrisol. The intensity of this decrease depended on chitosan dose and the number of wetting–drying cycles with which the soil–chitosan mixtures were pretreated, as is seen from exemplary stress–strain curves of soil cylindrical aggregates shown in Figure 1.

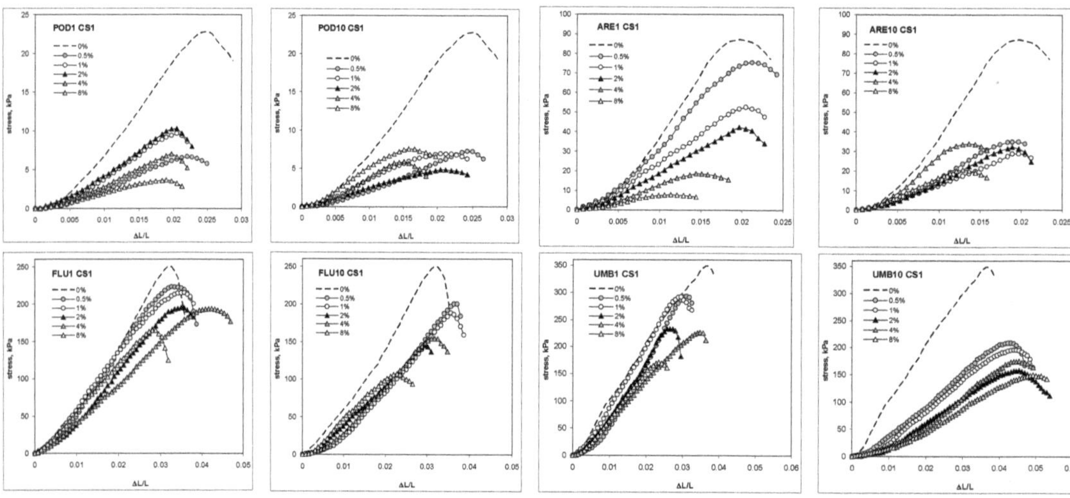

Figure 1. Exemplary compression stress-dilatation dependence for the aggregates of Podzol (POD), Fluvisol (FLU), Umbrisol (UMB) and Arenosol (ARE) amended with different amounts of chitosan CS1, preconditioned with 1 or 10 cycles of wetting–drying. The number of the cycles is written after the abbreviation of each soil.

The addition of chitosan of lower molecular mass and higher deacetylation degree (CS2, see M&M section) in the same rates of 0.5, 1, 2, 4 and 8% decreased the mechanical strength of Podzol, Arenosol and Umbrisol to the lower extent than CS1, and increased the mechanical strength of Fluvisol. Similarly, as for CS1, the intensity of these changes depended on chitosan dose and on the number of wetting–drying cycles with which the soil–chitosan mixtures were pretreated. Exemplary stress–strain curves of soil cylindrical aggregates containing CS2 are shown in Figure 2.

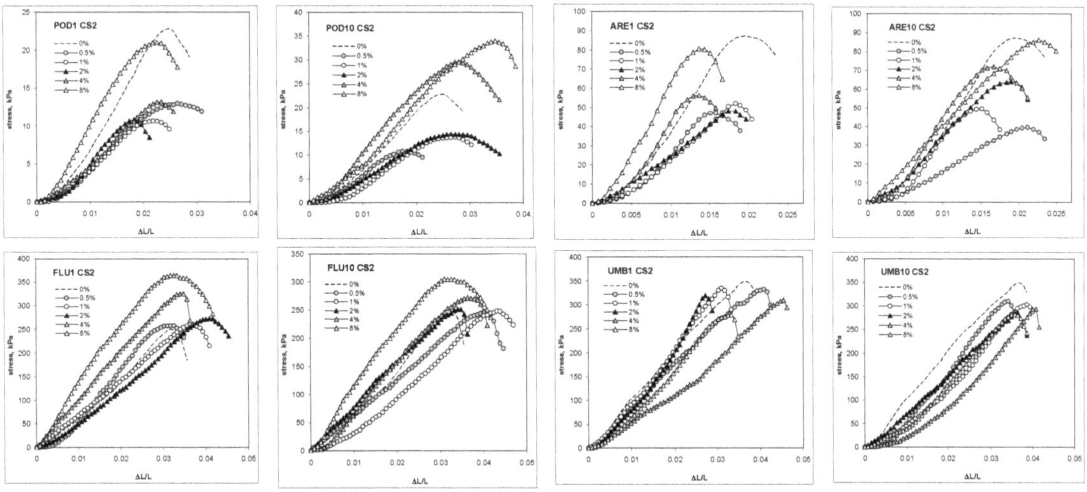

Figure 2. Exemplary compression stress-dilatation dependence for the aggregates of Podzol (POD), Fluvisol (FLU), Umbrisol (UMB) and Arenosol (ARE) amended with different amounts of chitosan CS2, preconditioned with 1 or 10 cycles of wetting–drying. The number of the cycles is written after the abbreviation of each soil.

The maxima of the dependencies shown in Figures 1 and 2, expressing a load per unit average of the cross-sectional area at which the cylindrical specimen of soil fails under compression during the Unconfined Compression Strength, UCS [Pa], are presented in Figure 3.

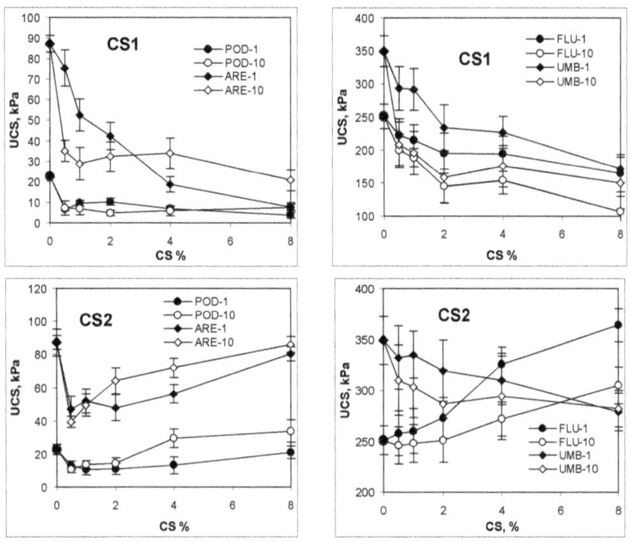

Figure 3. Average values of unconfined compressive strength of the studied aggregates of Podzol (POD), Fluvisol (FLU), Umbrisol (UMB) and Arenosol (ARE) amended with different amounts of chitosan (CS1 and CS2), preconditioned with one or ten cycles of wetting–drying. The number of the cycles is written after the abbreviation of soils. Error bars depict standard deviations.

The UCS of the aggregates containing CS1 generally decreases with the increase in chitosan percentage. A similar effect is observed for CS2 amended Umbrisol. Both acidic soils (Podzol and Arenosol) amended with CS2 behave in a different way: after the initial drop, the UCS increases with increasing chitosan dose, finally approaching the UCS value for the nontreated soil. Quite opposite behavior is noted for CS2 amended Fluvisol, for which the UCS seems to tendentially increase with an increase in the chitosan rate.

The UCS of neutral and alkaline soils (Fluvisol and Umbrisol, respectively) after 10 wetting/drying cycles is lower than the UCS of these soils subjected to a single cycle, whereas the increase in the number of wetting–drying cycles increases the UCS of acidic soils, Podzol and Arenosol, containing higher doses of both kinds of chitosan.

Similar dependencies were observed for Young's modulus and chitosan percentage, since, as commonly observed, UCS and Young's modulus are roughly proportional to each other, as it is presented in Figure 4 for the studied soils.

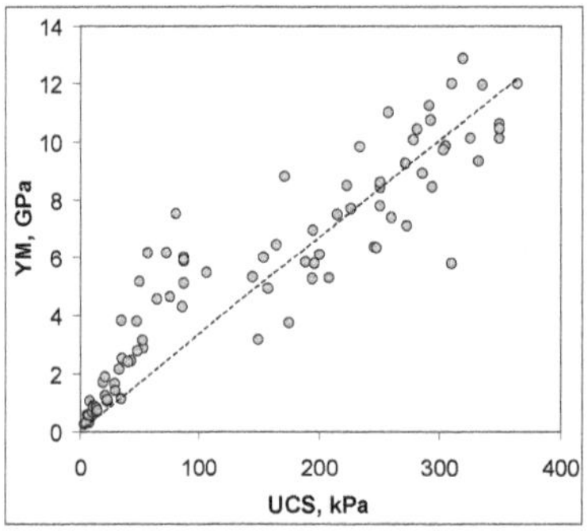

Figure 4. Dependence of Young's (YM) modulus and unconfined compressive strength (UCS) for all studied aggregates.

Structural porosity is considered to be the best measure of the susceptibility of soils and other granular materials to mechanical damage: more porous aggregates would break easier. The porosities of the studied soil/chitosan aggregates, preconditioned with one or ten wetting–drying cycles, are presented in Figure 5.

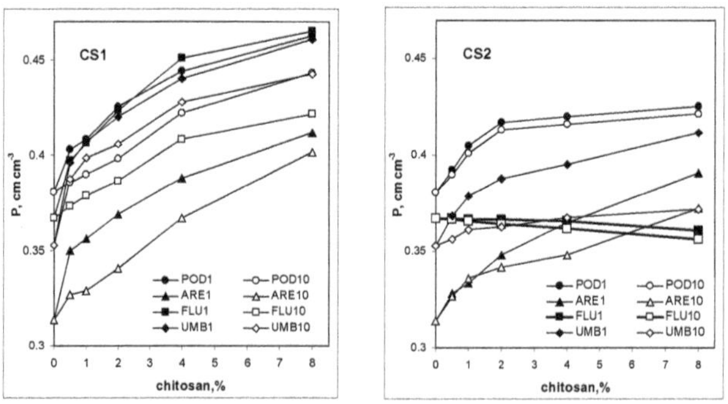

Figure 5. Dependencies of the porosity (P) of soil aggregates on the amendment of the studied chitosan (CS1 and CS2). The soils are abbreviated as: POD—Podzol, ARE—Arenosol, FLU—Fluvisol, UMB—Umbrisol. The number after the soil abbreviation shows the number of wetting–drying cycles applied to soil aggregates. Error bars are covered by the labels of the points.

In general, the addition of both types of chitosan causes a loosening of the soil structure (increasing porosity) with an exception of Fluvisol amended with CS2, for which chitosan addition decreased soil porosity and made the soil more compact. It was observed that CS1 increased soil porosity more than CS2 and the structure became less porous with an increase in the number of wetting–drying cycles.

3. Discussion

Changes in UCS due to the addition of CS1 accompany the opposite changes in porosity for all soils. This trend is valid also for CS2 amended Fluvisol and Umbrisol. However, for very acidic Podzol and acidic Arenosol despite the porosity tendentially increasing with increasing chitosan dose, after the initial drop, the UCS increases also. The increase in porosity observed in the vast majority of the studied soil/chitosan aggregates is most probably caused by the loosening of the soil structure by coarser chitosan particles. However, Kubavat et al. [49] mentioned about soil porosity increase also after the addition of nanoparticles of chitosan copolymerized with methacrylic acid. The higher effect of CS1 on the porosity increase may indicate that CS1 is composed of larger particles than CS2. The exceptional decrease in porosity and a simultaneous increase in UCS for Fluvisol amended with CS2 can be probably connected with the highest content of clay in this soil. In the control soil, the total volume of porous conglomerates of clay particles may exceed the total volume of the soil skeletal pores (free spaces between coarser soil particles) and the coarser particles are "suspended" in the clay phase. In such a case, the addition of nonporous chitosan particles replaces porous clay conglomerates and so the volumetric porosity decreases. In the "suspended" state, rigid skeletal particles rarely contact each other and the mechanical stability of such a system approaches that of the pure clay itself. Additional skeletal porosity created by coarser chitosan particles may be consecutively filled by clay and more contacts between coarser particles occur, and so, the mechanical strength increases. As reported by Jozefaciuk et al. [50] and Horabik and Jozefaciuk [51] there exists a maximum of UCS at a point where all free spaces between coarser particles are totally filled by clay conglomerates and no "free" clay excess is present. Such be the case, increasing the volume of clay makes the UCS smaller. The fact that a similar phenomenon does not occur for CS1 may be because CS1 particles are larger than those of CS2, and, only with the smallest CS1 additions does the volume of newly created free skeletal spaces exceed the volume of the excess of the clay which is available for their total filling.

The other exceptional behavior which needs more detailed examination is that in soils of low pH (Podzol and Arenosol) higher CS2 doses induced the increase in UCS. Electrostatic bonds of positively charged chitosan in a low pH range (PZC of chitosan is usually reported to occur at pH values around 6.5–7.6 [52–54]) with negatively charged soil particles may be responsible for this effect. In soils of higher pH, chitosan around its PZC develops either a small amount of positive charge or a small amount of negative charge on a small surface area, and the electrostatic forces that are generated with negatively charged soil components can range from weakly attractive to weakly repulsive. Additionally, different dissolution/jellification patterns of both studied types of chitosan may exert a significant effect on their interactions with soil components because chitosan gel of highly extended surface can form much more electrostatic bonds than chitosan particles of much smaller surface area. Because CS2 of lower molecular mass dissolves better and faster in soil organic acids (fulvic and humic acids) than CS1 having a higher molecular mass, the effect on UCS increase was observed only in the case of CS2. In a jellified form, chitosan can also glue soil particles together by adhesive forces or by the formation of polymer ties connecting neighboring soil particles which are not in direct contact [55]. Differences between the action of both kinds of chitosan on soils are apparently connected to their molecular properties. As it has been frequently reported, molecular weight and molecule size of polymers play a key role in defining the nature of interactions with various soils and soil minerals. Moen and Richardson [56] found that small-sized polymers distribute more evenly in the microaggregate fraction of soils because of their greater ability to penetrate the fine pores. Richardson et al. [57] also found that high molecular weight polymers may maximize soil–polymer interactions; however, the effectiveness could be affected by limited polymer penetration of the soil surface and failure to attain uniform aggregation adsorption. On the other hand, they found that small-sized polymers could create a more homogeneous soil stabilizing polymer network.

The effect of wetting–drying cycles on soil mechanical resistance depended both on the soil reaction and the kind of chitosan. Ambient solution pH and concentration affect the surface charge and conformation of charged polymers and therefore influence polymer adsorption onto soil particles. A higher concentration of the polymer solutions enables sufficient active functional and structural groups in the polymer to be available for interaction with soil particles and therefore could increase the efficiency of polymer stabilization [45]. Since solubilization and gelling are time-dependent, an increase in the mechanical stability of soil–chitosan mixtures in time occurs in acidic soils. In contrast, for neutral Fluvisol and alkaline Umbrisol, wherein chitosan around its PZC is hardly soluble [58], increasing the number of wetting/drying cycles leads to a decrease in UCS. A similar decrease in the strength of biopolymer treated soils with increasing wetting–drying cycles has been frequently reported [48,59].

As was observed by Adamczuk et al. [60] the water stability of aggregates containing CS1 preconditioned with a single cycle of wetting–drying was generally lower than for aggregates of the control soils. After ten wetting–drying cycles, these aggregates became more water-resistant. In contrast to CS1, the water stability of CS2-containing aggregates reached high water resistance just after the first drying–wetting cycle and it increased only slightly after the next nine wetting–drying cycles. Both types of chitosan increased soil water repellency, which increased further after 10 wetting–drying cycles. The wettability for CS1 amended soils was higher than for CS2. Both water stability and wettability may increase because of the high wetting angles of chitosan, according to Cassie's law [61]. The effect of chitosan on increasing water stability and water repellency might have been caused also by adsorption of the dissolved chitosan molecules on soil components. Even very small amounts of large chitosan molecules can cover the surface of soil particles to a great extent, forming hydrophobic patches. It is possible that this mechanism may decrease the mechanical stability of the aggregates, as well. If, as is likely, the surface covered by chitosan molecules becomes flatter (the surface roughnesses are levelled), the internal friction between smoother soil particles is reduced, and, in consequence, the UCS becomes smaller. It is probable that such an occurrence may be proven by studies of the fractal dimension of chitosan adsorbed soil (a decrease in fractal dimension would indicate surface flattening), which will be a problem in further studies.

Summarizing: coexistence of several antagonistic and synergic mechanisms can explain the observed impact of chitosan on the mechanical and water stability, and wettability of soils. The first is the loosening of soil structure (increase in porosity) due to the addition of coarse chitosan particles that can decrease both mechanical and water stability. However, in clayey soils addition of coarser chitosan particles can make the soil more mechanically resistant. The second mechanism is the formation of electrostatic bonds between positively charged chitosan particles (or jellified/dissolved chitosan molecules) and negatively charged soil components leading to an increase of both mechanical and water stability. This phenomenon would increase with a decrease in soil pH due to better charging and solubilization of chitosan at low pHs. The third mechanism is the formation of adhesive bonds between chitosan and soil components, which, similarly to the previous mechanism, should depend on the amount of the dissolved/jellified chitosan. The fourth mechanism is the adsorption of chitosan molecules on surfaces of soil components causing an increase in soil water stability and hydrophobicity of the soil material along with a possible reduction of the mechanical stability of the soil due to decreasing surface roughness and internal friction. The intensity of all aforementioned mechanisms depending on chitosan jellification, solubilization, surface charge and chain stiffness are governed by the ambient pH [44] and on physicochemical properties of a given kind of chitosan.

4. Materials and Methods

Two kinds of chitosan selected to differ in molecular mass and deacetylation degree were used. The first, abbreviated as CS1, was provided by Sigma Aldrich (St. Louis, MO, USA) and the second (CS2) was provided by Beijing Be-Better Technology Co., Ltd. (Beijing, China). Both forms of chitosan were composed of equal parts of the 0.105–0.053 and 0.177–0.105 mm dry sieved fractions. Soil samples were taken from upper 5–15 cm layers of four soils localized in East Poland, air-dried and screened by a 1 mm sieve. The basic properties of the studied soils and types of chitosan are presented in Table 1. This table recalls data reported by Adamczuk et al. [60] who studied water stability and wettability of the same materials as used in the present paper.

Table 1. Characteristics of the studied soils and chitosan.

Abbreviation	POD	ARE	FLU	UMB	CS1	CS2
Material	Podzol	Arenosol	Fluvisol	Umbrisol	Chitosan1	Chitosan2
Locality E	22°58′41″	22°26′6″	22°59′38″	21°41′59″		
Locality N	51°9′14″	51°2′9″	51°9′43″	50°49′25″		
pH	4.1	5.5	6.5	7.7		
Particle density [g cm^{-3}]	2.54	2.62	2.62	2.68	1.51	1.54
Nitrogen [%]	0.16	0.13	0.46	0.14	7.51	7.79
Total organic carbon [%]	0.65	1.55	3.04	0.9	41.59	41.27
Sand (0.063–2 mm) [%]	72.4	47.1	20.2	10.4		
Silt (0.002–0.063 mm) [%]	25.9	46.2	52.2	72.4		
Clay (<0.002 mm) [%]	1.7	6.7	27.6	17.2		
Grain fraction 0.177–0.105 mm [%]					50	50
Grain fraction 0.105–0.053 mm [%]					50	50
Degree of Deacetylation (DD)					0.77	0.91
Average molecular mass (M), [kDa]					699	280

4.1. Preparation of Soil/Chitosan Aggregates

The soil samples were air-dried and screened by a 1 mm sieve (mesh 18). Carefully homogenized water-saturated pastes were prepared from mixtures of the soils and the chitosan. Distilled water was used for pastes preparation. The content of CS1 and/or CS2 in the mixtures was 0 (control), 0.5, 1, 2, 4, and 8%. Cylindrical aggregates of 20 mm height and 10 mm diameter were formed from the pastes using plastic forms. The first set of aggregates was prepared just after the preparation of the paste and then air-dried (one cycle of wetting–drying) and the second set from the pastes was subjected to 10 wetting–drying cycles (6 days per cycle). All aggregates were then dried until constant mass at laboratory conditions (relative humidity around 60% and temperature around 25 °C). The aggregates are abbreviated further using the abbreviation of a given soil (see Table 1) followed by the number of wetting/drying cycles (e.g., POD1 and POD10 denote aggregates formed from Podzol preconditioned with one and ten wetting/drying cycles, respectively).

4.2. Studies of Soil/Chitosan Aggregates

Unconfined compression tests were performed for ten replicates of each aggregate using the Lloyd LRX material testing machine (Bognor Regis, UK). An aggregate placed vertically on the machine basement was pressed by a piston. The load measured with the accuracy of 0.05 N against displacement of the piston moved with the lowest apparatus speed of 10^{-5} m·s^{-1} was registered. The dependence of the compression stress [Pa], (load divided by the aggregate cross-sectional area) versus strain, $\Delta L/L$ (relative aggregate deformation, equal to piston displacement divided by the aggregate height) was calculated. From these dependencies values of the Unconfined Compression Strength UCS [Pa] (maximum load per unit average cross-sectional area at which the cylindrical specimen of soil falls in compression) and Young's modulus, E [Pa] (slope of the linear parts of the stress–strain dependencies) were estimated.

Volumetric porosity (volumetric fraction of pores on the total volume), P, [cm^3 cm^{-3}] of the aggregates were estimated as follows: at first, the volumes of the solids (soil + chitosan) present in 1 cm^3 of the aggregates, V_s [cm^3], were calculated based on the chitosan percentage in the soil–chitosan mixtures (CS%), the respective particle densities (PD_{soil} and PD_{CS1} or PD_{CS2}, presented in Table 1) and the aggregate bulk densities, BD [g cm^{-3}] reported in [60]:

$$V_s = 1 - [BD \cdot (1 - CS\%/100)/PD_{soil} + BD \cdot (CS\%/100)/PD_{CS}] \quad (1)$$

and next the porosity was calculated as:

$$P = 1 - V_s. \quad (2)$$

4.3. Statistical Analysis

Average values ± standard deviations were calculated.

Author Contributions: Conceptualization, A.A. and G.J.; methodology G.J. and A.A.; experiments: A.A.; software G.J.; data elaboration: G.J.; writing of the original draft: A.A., G.J.; review and editing: G.J. and A.A.; visualization G.J.; supervision G.J.; project administration G.J.; funding acquisition G.J. All authors have read and agreed to the published version of the manuscript.

Funding: Part of this work was funded by the National Science Centre, Poland, within an Opus 15 Project 2018/29/B/ST10/01592: Effect of minerals and different forms of organic carbon on structure, porosity and mechanical and water stability of soil aggregates-model studies.

Institutional Review Board Statement: Not applicable.

Informed Consent Statement: Not applicable.

Data Availability Statement: All data are available from authors after a reasonable request.

Acknowledgments: The authors appreciate much the help of Ewa Paszek for English corrections.

Conflicts of Interest: The authors declare no conflict of interest.

Sample Availability: Samples of the soils and chitosan are available from the authors.

References

1. Pal, P.; Pal, A.; Nakashima, K.; Kumar Yadav, B. Applications of chitosan in environmental remediation: A review. *Chemosphere* **2021**, *266*, 128934. [CrossRef] [PubMed]
2. Majeti, N.V.; Kumar, R. A review of chitin and chitosan applications. *React. Funct. Polym.* **2000**, *46*, 1–27. [CrossRef]
3. Croisier, F.; Jérôme, C. Chitosan-based biomaterials for tissue engineering. *Eur. Polym. J.* **2013**, *49*, 780–792. [CrossRef]
4. Malerba, M.; Cerana, R. Chitosan effects on plant systems. *Int. J. Mol. Sci.* **2016**, *17*, 996. [CrossRef] [PubMed]
5. Jabnoun-Khiareddine, H.; El-Mohamedy, R.S.R.; Abdel-Kareem, F.; Aydi Ben Abdallah, R.; Gueddes-Chahed, M.; Daami-Remadi, M. Variation in chitosan and salicylic acid efficacy towards soil-borne and air-borne fungi and their suppressive effect of tomato wilt severity. *J. Plant Pathol. Microbiol.* **2015**, *6*, 1000325. [CrossRef]
6. Qin, Y.; Li, P. Antimicrobial chitosan conjugates: Current synthetic strategies and potential applications. *Int. J. Mol. Sci.* **2020**, *21*, 499. [CrossRef]
7. Katiyar, D.; Hemantaranjan, A.; Singh, B. Chitosan as a promising natural compound to enhance potential physiological responses in plant: A review. *Ind. J. Plant Physiol.* **2015**, *20*, 1–9. [CrossRef]
8. Rabêlo, V.M.; Magalhães, P.C.; Bressanin, L.A.; Carvalho, D.T.; Oliveira dos Reis, C.; Karam, D.; Doriguetto, A.C.; dos Santos, M.H.; Rodrigues dos Santos Santos Filho, P.; Corrêa de Souza, T. The foliar application of a mixture of semisynthetic chitosan derivatives induces tolerance to water defcit in maize, improving the antioxidant system and increasing photosynthesis and grain yield. *Sci. Rep.* **2019**, *9*, 8164. [CrossRef]
9. Sharif, R.; Mujtaba, M.; Ur Rahman, M.; Shalmani, A.; Ahmad, H.; Anwar, T.; Tianchan, D.; Wang, X. The multifunctional role of chitosan in horticultural crops; a review. *Molecules* **2018**, *23*, 872. [CrossRef]
10. Chouhan, D.; Mandal, P. Applications of chitosan and chitosan based metallic nanoparticles in agrosciences—A review. *Int. J. Biol. Macromol.* **2021**, *166*, 1554–1569. [CrossRef]
11. Basuki, K.T.; Swantomo, D.; Sigit Sanyoto, N.T. Characterization of chitosan-acrylamide hydrogels as soil conditioner. *Adv. Mat. Res.* **2015**, *1112*, 414–417. [CrossRef]
12. Iftime, M.M.; Ailieseia, G.L.; Ungureanub, E.; Marina, L. Designing chitosan based eco-friendly multifunctional soil conditioner systems with urea controlled release and water retention. *Carbohydr. Polym.* **2019**, *223*, 115040. [CrossRef] [PubMed]
13. Guo, J.; Yang, J.; Yang, J.; Zheng, G.; Chen, T.; Huang, J.; Bian, J.; Meng, X. Water-soluble chitosan enhances phytoremediation efficiency of cadmium by *Hylotelephium spectabile* in contaminated soils. *Carbohydr. Polym.* **2020**, *246*, 116559. [CrossRef] [PubMed]

14. Zubair, M.; Adnan Ramzani, P.M.; Rasool, B.; Khan, M.A.; Ur-Rahman, M.; Akhtar, I.; Turan, V.; Tauqeer, H.M.; Farhad, M.; Khan, S.A.; et al. Efficacy of chitosan-coated textile waste biochar applied to Cd-polluted soil for reducing Cd mobility in soil and its distribution in moringa (*Moringa oleifera* L.). *J. Environ. Manag.* **2021**, *284*, 112047. [CrossRef] [PubMed]
15. Chang, I.; Prasidhi, A.K.; Im, J.; Cho, G.C. Soil strengthening using thermo-gelation biopolymers. *Constr. Build. Mater.* **2015**, *77*, 430–438. [CrossRef]
16. Chang, I.; Im, J.; Prasidhi, A.K.; Cho, G.C. Effects of Xanthan gum biopolymer on soil strengthening. *Constr. Build. Mater.* **2015**, *74*, 65–72. [CrossRef]
17. Huang, J.; Kogbara, R.B.; Hariharan, N.; Masad, E.A.; Little, D.N. A state-of-the-art review of polymers used in soil stabilization. *Constr. Build. Mater.* **2021**, *305*, 124685. [CrossRef]
18. Mohammadinejad, R.; Kumar, A.; Ranjbar-Mohammadi, M.; Ashrafizadeh, M.; Han, S.S.; Khang, G.; Roveimiab, Z. Recent advances in natural gum-based biomaterials for tissue engineering and regenerative medicine: A review. *Polymers* **2020**, *12*, 176. [CrossRef]
19. Grzadka, E. Influence of surfactants on the adsorption and elektrokinetic properties of the system: Guar gum/manganese dioxide. *Cellulose* **2013**, *20*, 1313–1328. [CrossRef]
20. Gooneh-Farahani, S.; Naghib, S.M.; Naimi-Jamal, M.R.; Seyfoori, A. A pH-sensitive nanocarrier based on BSA-stabilized graphene-chitosan nanocomposite for sustained and prolonged release of anticancer agents. *Sci Rep.* **2021**, *11*, 17404. [CrossRef]
21. Cho, Y.W.; Jang, J.; Park, C.R.; Ko, S.W. Preparation and solubility in acid and water of partially deacetylated chitins. *Biomacromolecules* **2000**, *1*, 609–614. [CrossRef]
22. Kofuji, K.; Qian, C.J.; Nishimura, M.; Sugiyama, I.; Murata, Y.; Kawashima, S. Relationship between physicochemical characteristics and functional properties of chitosan. *Eur. Polym. J.* **2005**, *41*, 2784–2791. [CrossRef]
23. Park, P.J.; Je, J.Y.; Kim, S.K. Free radical scavenging activities of differently deacetylated chitosans using an ESR spectrometer. *Carbohydr. Polym.* **2004**, *55*, 17–22. [CrossRef]
24. Mostafa, H.M.; Sourell, H.; Bockisch, F.J. Mechanical properties of some bioplastics under different soil types used as biodegradable drip tubes. *Agric. Eng. Int. CIGR J.* **2010**, *12*, 1–16.
25. Kassymova, Z.S.; Orazzhanova, L.K.; Klivenko, A.N.; Mussabayeva, B.K.; Aserzhanov, D.K. Preparation and properties of interpolymer complexes capable of soil structuring. *Russ. J. Appl. Chem.* **2019**, *92*, 208–217. [CrossRef]
26. Chang, I.; Lee, M.; Tran, A.T.P.; Lee, S.; Kwon, Y.M.; Im, J.; Cho, G.C. Review on biopolymer-based soil treatment (BPST) technology in geotechnical engineering practices. *Transp. Geotech.* **2020**, *24*, 100385. [CrossRef]
27. Hataf, N.; Ghadir, P.; Ranjbar, N. Investigation of soil stabilization using chitosan biopolymer. *J. Clean. Prod.* **2018**, *170*, 1493–1500. [CrossRef]
28. Orts, W.J.; Sojka, R.E.; Glenn, G.M. Biopolymer additives to reduce erosion-induced soil losses during irrigation. *Ind. Crops Prod.* **2000**, *11*, 19–29. [CrossRef]
29. Renouard, S.; Hano, C.; Ouagne, P.; Blondeau, J.P.; Lainé, E. Protection of flax fiber-based yarns against natural soil degradation by chitosan. *Mater. Lett.* **2014**, *137*, 269–273. [CrossRef]
30. Ramdas, V.M.; Mandree, P.; Mgangira, M.; Mukaratirwa, S.; Lalloo, R.; Ramchuran, S. Review of current and future bio-based stabilisation products (enzymatic and polymeric) for road construction materials. *Transp. Geotech.* **2021**, *27*, 100458. [CrossRef]
31. Plank, J. Applications of biopolymers and other biotechnological products in building materials. *Appl. Microbiol. Biotechnol.* **2004**, *66*, 1–9. [CrossRef]
32. Shariatmadari, N.; Reza, M.; Tasuji, A.; Ghadir, P.; Javadi, A.A. Experimental study on the effect of chitosan biopolymer on sandy soil stabilization. *E3S Web Conf.* **2020**, *195*, 06007. [CrossRef]
33. Aguilar, R.; Nakamatsu, J.; Ramírez, E.; Elgegren, M.; Ayarza, J.; Kim, S.; Pando, M.A.; Ortega-San-Martin, L. The potential use of chitosan as a biopolymer additive for enhanced mechanical properties and water resistance of earthen construction. *Constr. Build. Mater.* **2016**, *114*, 625–637. [CrossRef]
34. Szabó, S.; Szabó, G.; Bihar, Á. Effects of acid loadings on heavy metal mobilization in Cambisols. *Ann. Geogr.* **2007**, *40*, 72–79.
35. Goulding, K.W. Soil acidification and the importance of liming agricultural soils with particular reference to the United Kingdom. *Soil Use Manag.* **2016**, *32*, 390–399. [CrossRef]
36. Getaneh, S.; Kidanemariam, W. Soil Acidity and Its Managements: A Review. *Int. J. Adv. Res. Biol. Sci.* **2021**, *8*, 70–79.
37. Pandey, P.; Kumar Verma, M.; De, N. Chitosan in agricultural context—A review. *Bull. Env. Pharmacol.* **2018**, *7*, 87–96.
38. Hidangmayum, A.; Dwivedi, P.; Katiyar, D.; Hemantaranjan, A. Application of chitosan on plant responses with special reference to abiotic stress. *Physiol. Mol. Biol. Plants* **2019**, *25*, 313–326. [CrossRef]
39. Kumararaja, P.; Manjaiah, K.M.; Datta, S.C.; Shabeer, A.T.P.; Sarkar, B. Chitosan-g-poly(acrylic acid)-bentonite composite: A potential immobilizing agent of heavy metals in soil. *Cellulose* **2018**, *25*, 3985–3999. [CrossRef]
40. Kanmani, P.; Aravind, J.; Kamaraj, M.; Sureshbabu, P.; Karthikeyan, S. Environmental applications of chitosan and cellulosic biopolymers: A comprehensive outlook. *Bioresour. Technol.* **2017**, *242*, 295–303. [CrossRef]
41. Babla, M.; Katwal, U.; Yong, M.T.; Jahandari, S.; Rahme, M.; Chen, Z.H.; Tao, Z. Value-added products as soil conditioners for sustainable agriculture. *Resour. Conserv. Recycl.* **2022**, *178*, 106079. [CrossRef]
42. Soldo, A.; Miletić, M.; Auad, M.L. Biopolymers as a sustainable solution for the enhancement of soil mechanical properties. *Sci. Rep.* **2020**, *10*, 267. [CrossRef] [PubMed]

43. Nunthanid, J.; Puttipipatkhachorn, S.; Yamamoto, K.; Peck, G.E. Physical properties and molecular behavior of chitosan films. *Drug Dev. Ind. Pharm.* **2001**, *27*, 143–157. [CrossRef] [PubMed]
44. Kasaai, M.R. Calculation of Mark-Houwink-Sakurada (MHS) equation viscometric constants for chitosan in any solvent-temperature system using experimental reported viscometric constants data. *Carbohydr. Polym.* **2007**, *68*, 477–488. [CrossRef]
45. Gu, B.H.; Doner, H.E. The interaction of polysaccharides with silver hill illite. *Clays Clay Miner.* **1992**, *40*, 151–156. [CrossRef]
46. Qureshi, M.U.; Chang, I.; Al-Sadarani, K. Strength and durability characteristics of biopolymer-treated desert sand. *Geomech. Eng.* **2017**, *12*, 785–801. [CrossRef]
47. Fatehi, H.; Ong, D.E.L.; Yu, J.; Chang, I. Biopolymers as green binders for soil improvement in geotechnical applications: A review. *Geosciences* **2021**, *11*, 291. [CrossRef]
48. Chang, I.; Im, J.; Lee, S.-W.; Cho, G.-C. Strength durability of gellan gum biopolymertreated Korean sand with cyclic wetting and drying. *Constr. Build. Mater.* **2017**, *143*, 210–221. [CrossRef]
49. Kubavat, D.; Trivedi, K.; Vaghela, P.; Prasad, K.; Vijay Anand, G.K.; Trivedi, H.; Patidar, R.; Chaudhari, J.; Andhariya, B.; Ghosh, A. Characterization of a chitosan-based sustained release nanofertilizer formulation used as a soil conditioner while simultaneously improving biomass production of *Zea mays* L. *Land Degrad. Dev.* **2020**, *31*, 2734–2746. [CrossRef]
50. Józefaciuk, G.; Skic, K.; Adamczuk, A.; Boguta, P.; Lamorski, K. Structure and strength of artificial soils containing monomineral clay fractions. *Materials* **2021**, *14*, 4688. [CrossRef]
51. Horabik, J.; Jozefaciuk, G. Structure and strength of kaolinite-soil silt aggregates: Measurements and modeling. *Geoderma* **2021**, *382*, 114687. [CrossRef]
52. Neto, J.M.; Bellato, C.R.; Milagres, J.L. Preparation and evaluation of chitosan beads immobilized with Iron(III) for the removal of As(III) and As(V) from water. *J. Braz. Chem. Soc.* **2013**, *24*, 121–132. [CrossRef]
53. Yazdani, M.; Virolainen, E.; Conley, K.; Vahala, R. Chitosan-Zinc(II) Complexes as a Bio-Sorbent for the Adsorptive Abatement of Phosphate: Mechanism of Complexation and Assessment of Adsorption Performance. *Polymers* **2018**, *10*, 25. [CrossRef]
54. Zavareh, S.; Behrouzi, Z.; Avanes, A. Cu (II) binded chitosan/Fe3O4 nanocomomposite as a new biosorbent for efficient and selective removal of phosphate. *Int. J. Biol. Macromol.* **2017**, *101*, 40–50. [CrossRef] [PubMed]
55. Mohamed, S.W.A. *Stabilization of Desert Sand using Water-Born Polymers*; United Arab Emirates University: Abu Dhabi, United Arab Emirates, 2004; p. 437.
56. Moen, D.E.; Richardson, J.L. Ultrasonic dispersion of soil aggregates stabilized by polyvinyl alcohol and T403-glyoxal polymers. *Soil Sci. Soc. Am. J.* **1984**, *48*, 628–631. [CrossRef]
57. Richardson, J.L.; Gunnerson, W.T.; Giles, J.F. Influence of in situ two-phase polymers on aggregate stabilization in various Textured North Dakota Soils. *Can. J. Soil Sci.* **1987**, *67*, 209–213. [CrossRef]
58. Rinaudo, M. Chitin and chitosan: Properties and applications. *Prog. Polym. Sci.* **2006**, *31*, 603–632. [CrossRef]
59. Ayeldeen, M.; Negm, A.; El Sawwaf, M.; Gadda, T. Laboratory study of using biopolymer to reduce wind erosion. *Int. J. Geotech. Eng.* **2016**, *12*, 228–240. [CrossRef]
60. Adamczuk, A.; Kercheva, M.; Hristova, M.; Jozefaciuk, G. Impact of chitosan on water stability and wettability of soils. *Materials* **2021**, *14*, 7724. [CrossRef]
61. Cassie, A.B.D. Contact angles. *Discuss. Faraday Soc.* **1948**, *3*, 11–16. [CrossRef]

molecules

MDPI

Article

Development of Chitosan Green Composites Reinforced with Hemp Fibers: Study of Mechanical and Barrier Properties for Packaging Application

Rim Gheribi [1], Yassine Taleb [1], Louise Perrin [1], Cesar Segovia [2], Nicolas Brosse [3] and Stephane Desobry [1,*]

1 Laboratoire d'Ingénierie des Biomolécules, Université de Lorraine, ENSAIA, 2 Avenue de la Forêt de Haye, BP 20163, 54505 Vandœuvre-lès-Nancy, CEDEX, France; rim.gheribi@univ-lorraine.fr (R.G.); yassine.taleb6@etu.univ-lorraine.fr (Y.T.); louise.perrin@univ-lorraine.fr (L.P.)

2 Centre d'Essais TEchnique LORrain (CETELOR), 27 rue Philippe Seguin, BP 21042, 88051 Épinal, CEDEX 9, France; cesar.segovia@univ-lorraine.fr

3 Laboratoire d'Etudes et de Recherche sur le Matériau Bois, Faculté des Sciences et Technologies, Université de Lorraine, Boulevard des Aiguillettes, BP 70239, 54506 Vandœuvre-lès-Nancy, CEDEX, France; nicolas.brosse@univ-lorraine.fr

* Correspondence: stephane.desobry@univ-lorraine.fr

Abstract: The use of bioresourced packaging materials is an interesting solution for ecological issues. This work aimed to develop novel chitosan-based packaging materials reinforced with hemp fibers (HF). For this purpose, chitosan (CH) films were filled with 15%, 30%, and 50% (*w*/*w*) of two kinds of HF: Untreated fibers cut to 1 mm (UHF) and steam exploded fibers (SEHF). The effect of HF addition and HF treatments on chitosan composites was studied in terms of mechanical properties (tensile strength (TS), elongation at break (EB), and Young's modulus (YM)), barrier properties (water vapor (WVP) and oxygen permeabilities), and thermal properties (glass transition (T_g) and melting temperatures (T_m)). The addition of HF, whether untreated or steam exploded, increased the TS of chitosan composites by 34–65%. WVP was significantly reduced by the addition of HF but no significant change was observed for O_2 barrier property, which was in the range between 0.44 and 0.68 $cm^3 \cdot mm/m^2 \cdot d$. T_m of the composite films increased from 133 °C for CH films to 171 °C for films filled with 15% SEHF. However, no significant modification was observed for T_g (105–107 °C). The present study showed that the developed biocomposites had improved properties, mainly the mechanical resistance. Their use in food packaging will help industrials the move toward a sustainable development and circular economy.

Keywords: mechanical properties; barrier properties; packaging materials; circular economy

Citation: Gheribi, R.; Taleb, Y.; Perrin, L.; Segovia, C.; Brosse, N.; Desobry, S. Development of Chitosan Green Composites Reinforced with Hemp Fibers: Study of Mechanical and Barrier Properties for Packaging Application. *Molecules* **2023**, *28*, 4488. https://doi.org/10.3390/molecules28114488

Academic Editors: Sylvain Caillol and Lin Lin

Received: 10 April 2023
Revised: 27 May 2023
Accepted: 30 May 2023
Published: 1 June 2023

1. Introduction

Nowadays, concerns for environmental safety, ecosystem preservation, and climate change are within the priorities of researchers and industrials. In this context, the use of natural and bioresourced materials, also called biomaterials, is an interesting and promising answer for ecological concerns [1]. These materials are respectful to the environment and human health, they are relatively light in order that their carbon impact could be reduced, and they are biodegradable and/or methanizable by some bacteria species and under controlled conditions [2]. However, in a practical way, this solution remains challenging and difficult to set up. The reasons behind the difficulties of using biomaterials are their cost and technical limits, especially for industrial applications. Biomaterials generally present poor mechanical and barrier properties. Moreover, their low resistance to high temperatures limits industrial application [1].

Several bioresourced polymers were studied and their ability to develop packaging biomaterials was investigated. Among these biopolymers, we found starch, chitosan, alginate, pectin, gelatin, caseinate, and cellulose derivatives [3,4]. Chitosan is one of the most interesting and promising biopolymers for the development of new generation materials. After cellulose, this bioresourced polysaccharide derived by deacetylation from chitin is the second most abundant biopolymer in nature [5]. Chitin is mainly extracted from crustaceans' exoskeletons, insects' exoskeletons, or fungi cell walls. When its degree of deacetylation (DDA) is above 50%, chitin turns into chitosan which is a copolymer made of two structural units: N-acetyl-D-glucosamine and D-glucosamine. Some of its numerous characteristics include biodegradability, biocompatibility, and non-toxicity. Moreover, chitosan is interesting for physicochemical properties (cationic character), antimicrobial and antioxidant capacities, and its ability to be easily modified chemically. Today, chitosan is used in different forms whether as foam, gels, films, or granules and in different fields, namely, cosmetics, medicine, and wastewater treatment [5,6].

For food packaging applications, chitosan was used through different processing methods, such as solvent casting, dip- and spray-coating, layer-by-layer preparation, and extrusion [6]. These chitosan-based materials present good thermal stability and promising physicochemical properties. However, their mechanical and barrier properties are poor and not suitable for food packaging applications. These findings were confirmed by several previous studies, which reported that chitosan films are fragile, highly hygroscopic, and sensitive to environmental variations [7–10].

One of the newly used methods for enhancing the mechanical properties of films is the combination of biopolymers with natural fibers. In fact, natural fibers are lignocellulosic materials with high aspect ratio, and are commonly used for their low cost, durability, good mechanical properties, low weight, and biodegradability [6]. Therefore, natural fibers bring mechanical strength to the polymeric material without altering its ecofriendly attributes. Composites reinforced with natural fibers have a lower environmental impact thanks to their reduced carbon and fuel consumption. Hemp fibers (*Cannabis sativa*) are very abundant plant fibers with a world production of 214×10^3 tons per year and France is one of the main producers of this natural biomass [6,11]. Hemp fibers are basically composed of cellulose (70–74%), hemicellulose (15–20%), and lignin (3.5–5.7%). Even though their mechanical properties depend on various factors, such as plant growing conditions and physical, chemical, and morphological properties, hemp fibers present an interesting tensile strength in the range between 550 and 1110 MPa [11]. Consequently, hemp fibers are considered to be one of the strongest and stiffest natural fibers. Mechanical performance is strongly related to interfacial adhesion between the polymeric matrix and fibers, to the dispersion of fibers in the main matrix, and to the fiber properties, such as dimension, orientation, and surface modification [11]. Hemp fibers have already been used for the reinforcement of several materials, such as polypropylene, polylactic acid, and polybenzoxazine [12–14], in particular, to replace glass fibers; however, they were rarely used in combination with biobased polymers, such as polysaccharides. Several chemical and mechanical treatments have been tested to improve their dispersion and their interfacial adhesion with the polymeric matrix [11]. Steam explosion is one of the mechanical methods and has been described as a method for degumming and producing fine, homogeneous hemp fibers [15,16]. It was shown that this process degraded the middle lamella of the cells by hydrolysis of the pectins and hemicelluloses that it contained.

Chitosan films reinforced with glass, sisal, ramie, or bamboo fibers were studied in the literature [17–19]; however, the incorporation of HF with different mechanical treatments was never investigated. In the present paper, we aimed to study the development of chitosan-based films reinforced with hemp fibers for their potential application as food packaging materials. Untreated and steam exploded hemp fibers were used for this purpose and their effect on mechanical, barrier, and thermal properties was investigated. This work represents a feasibility study in order to have a proof of concept of our hemp fiber-reinforced biomaterials.

2. Results and Discussion

2.1. General Description and Microstructure

Chitosan–hemp fibers composite films (Figure 1A,C) produced in the current study were translucid with a rough touch on the upper side and a smooth touch on the inner side (in contact with the petri dish). Their general appearance was homogeneous and repeatable; however, we could observe that films loaded with UHF were more homogeneous and those filled with SEHF were smoother. No cracks were observed which confirms the cohesiveness and homogeneity of the plasticized composite films. Microstructural properties of pure chitosan films have already been studied in literature [20]. Chitosan films have a compact and homogeneous structure and the surface is smooth and flat. This homogeneous microstructure is generally related to the strong intra- and inter-molecular hydrogen bonds occurring between chitosan functional groups on the one hand, and chitosan, water, and glycerol molecules on the other hand [20].

Once the fibers are added, the composite microstructure was completely different as shown in Figure 1B,D, which represents films with 50% UHF and 50% SEHF, respectively. Micrographs of films filled with 50% of fibers are represented and are representative of the disposition and orientation of fibers in the film structure. The mentioned micrographs showed that, for both treatments, the fibers were randomly dispersed within the film structure, without any precise orientation. Film surface morphology was continuous and homogeneous for both kinds of hemp fibers. SEHF were dispersed more uniformly in the film matrix and covered the entire film surface. In fact, during steam explosion, fibers are blown apart and split into microfibrils due to the process of high pressure and the breakdown of pectin, hemicellulose, and lignin [21,22]. In contrast, the mechanical treatment only modified the length of hemp fibers, but not their chemical or structural properties. The dispersion of UHF in film matrix was less effective and less homogeneous than SEHF.

2.2. Chemical Characterization by Infrared Spectroscopy

To better understand the contribution of hemp fibers to the molecular structure of chitosan composites, fibers were analyzed by FTIR spectroscopy (Figure 2a). Both UHF and SEHF showed similar bands which are specific to plant fibers mostly composed of cellulose, hemicellulose, lignin, and pectin. The broad band observed around 3270–3410 cm^{-1} is assigned to hydroxyl groups of the lignocellulosic material. Peaks observed between 2850 and 2910 cm^{-1} are related to C–H groups of the fibers. C=C stretching bands related to carboxylic or unsaturated acids were detected at 1630 cm^{-1} for UHF and at 1730 and 1639 cm^{-1} for SEHF. This was the main difference shown between UHF and SEHF, which is related to hemicelluloses hydrolysis and delignification during the steam explosion process, as reported by Nader et al. [23]. C–H and C–H_2 bending peaks related to cellulose and lignin were observed at 1200–1430 cm^{-1} for both fibers. Two peaks were detected at 1051 and 1026 cm^{-1} and were attributed respectively to the C–O–C vibration and C–O stretching present in the cellulose structure. A last band was observed at 890 cm^{-1} and was assigned to the C–H vibration of cellulose. All the mentioned peaks specific to lignocellulosic plant fibers were reported in previous works and agreed with our FTIR characterization [24,25].

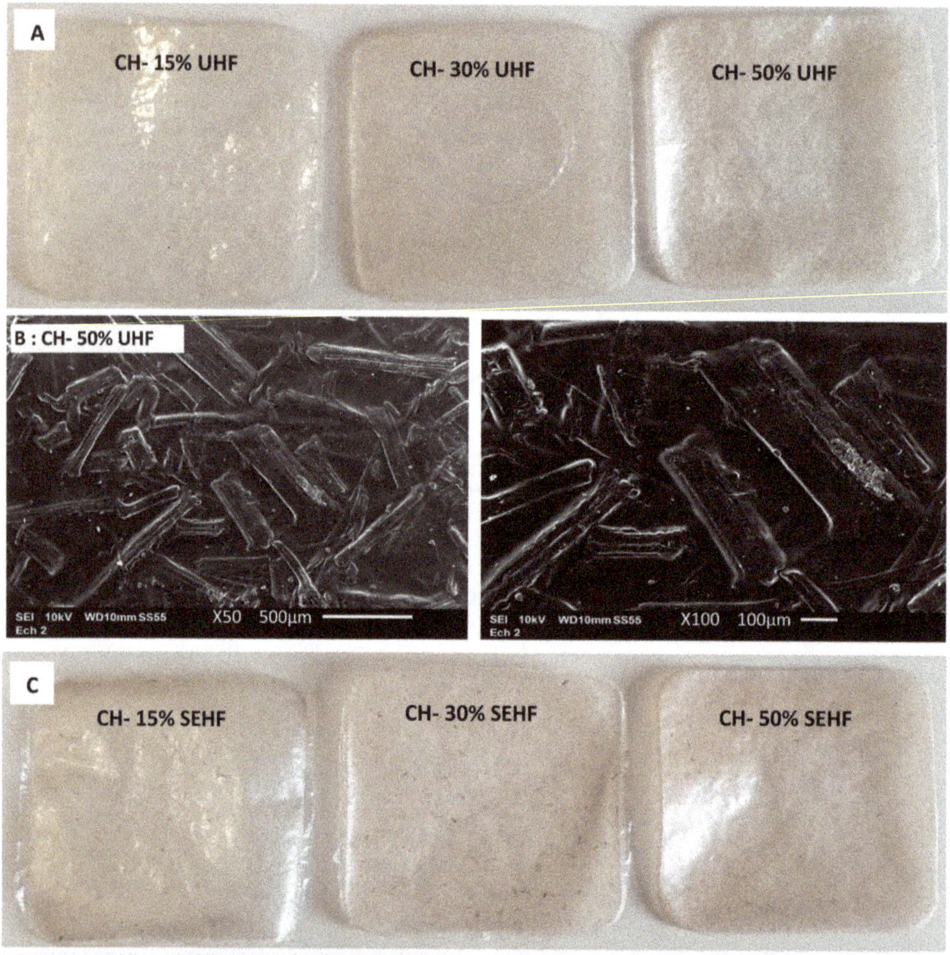

Figure 1. General appearance of chitosan films reinforced with untreated fibers (UHF, (**A**)) and steam exploded fibers (SEHF, (**C**)). Micrographs of films with 50% HF show the fibers at film surface (**B**,**D**) at two different magnifications (×50 on the left and ×100 on the right).

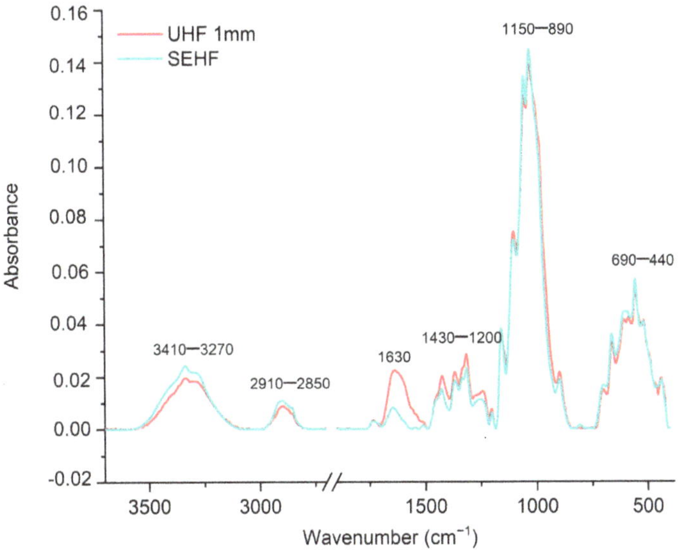

(**a**)

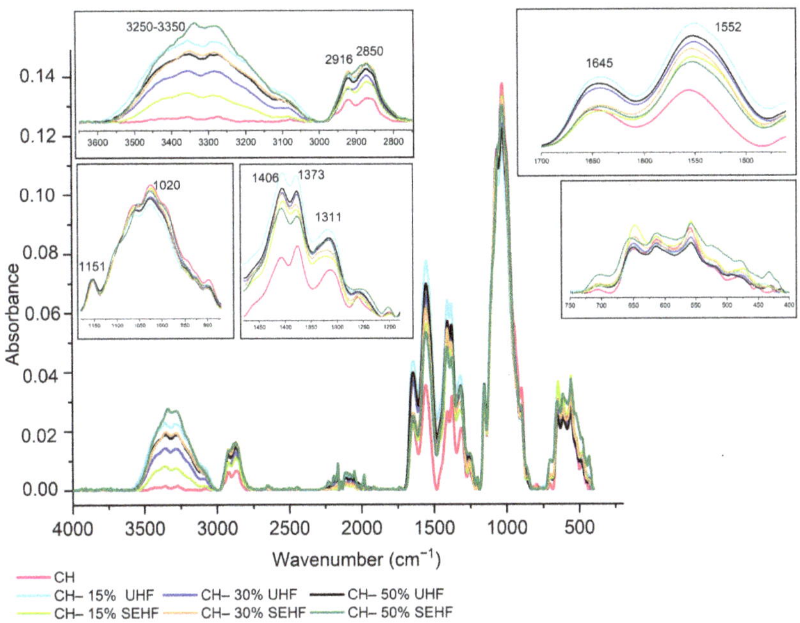

(**b**)

Figure 2. FTIR spectra of untreated and steam exploded hemp fibers (**a**) and the different composites of chitosan–hemp fibers (**b**).

The developed films were analyzed and the related FTIR spectra are presented in Figure 2b. The obtained graphs showed that all composite films had almost the same peaks with some minor variations. CH-based films showed characteristic bands of this branched polysaccharide, mainly amide groups (amide I at 1645 cm^{-1}, amide II at 1552 cm^{-1}, and amide III at 1311 cm^{-1}), C–H stretching vibration at 2916 and 2850 cm^{-1}, C–O symmetric stretching vibration at 1406 cm^{-1}, C–O–C bridge antisymmetric stretching at 1151 cm^{-1}, and C–O stretching at 1020 cm^{-1}. These bands were also identified for chitosan composites by Moalla et al. [20] and Teixeira-Costa et al. [26]. The addition of hemp fibers did not bring any new peaks, which confirms the absence of covalent bonds, but all peaks were shifted to lower wavenumbers which indicates the presence of hydrogen bindings between chemical groups of chitosan polymer and those of hemp fibers (O–H stretching vibration detected mainly at 3250–3350 cm^{-1}). The presence of shifts was detected at 3338, 3356, and 3350 for films with UHF loaded at 15, 30, and 50%, respectively, and observed at 3352, 3355, and 3357 for films with 15, 30, and 50% SEHF, respectively. Moreover, water and glycerol contribute to the creation of these low-energy linkages.

2.3. Mechanical Properties

Figure 3 summarizes the tensile strength (TS), elongation at break (EB), and Young's modulus (YM) of the different composites. The neat chitosan films, developed at 1.5% with a chitosan with a viscosity of 12 mPa·s, present poor tensile strength with values around 32 MPa. These values are comparable to those obtained for films developed with chitosan with different molecular weight, and thus with different viscosities (20–40 mPa·s) [27]. EB did not exceed 3% which was a low value, but is similar to the results obtained by Liu et al. [27]. The addition of 30% UHF did not significantly ($p < 0.05$) affect TS of chitosan-based composites in comparison to the control film (CH). At this ratio, the fibers could not be well-dispersed in the film-forming suspension and surface attraction between the two compounds decreased [9]. However, the addition of hemp fibers, whether untreated or steam exploded, significantly increased the tensile strength for the other tested ratios (15% and 50% UHF, 15%, 30%, and 50% SEHF). The best result was obtained for 50% (w/w) UHF and SEHF, where tensile strength values exceeded 50 MPa (Figure 3). The highest TS values were obtained for 50% of hemp fibers (for both UHF and SEHF). The addition of hemp fibers improved the tensile strength of chitosan films by 34–65%, which means that this method is an efficient, green, and cost-effective way to develop chitosan materials with enhanced mechanical properties. However, it is worth mentioning that the steam explosion treatment improves the compatibility between HF and chitosan, thanks to the high aspect ratio of the steam exploded fibers and their large surface areas, which gives an abundance of hydroxyl groups on the surface [22,28]. Therefore, steam explosion promotes the good and uniform dispersion of individual fibers within the polymeric matrix, which was confirmed by SEM micrographs, and leads to a better mechanical resistance of the composite materials. As UHF gives less homogeneous film structures, the effect of their addition at three different ratios on the mechanical properties of composites could not be relevant.

For EB and YM, no significant increase was observed for all the developed films. However, the obtained results agreed with previous studies reporting that chitosan composites had low EB and YM [26,27]. For the packaging field, the development of CH–HF-based materials with good mechanical resistance could fit well with the requirements of the industrial application, where good mechanical resistance is mandatory.

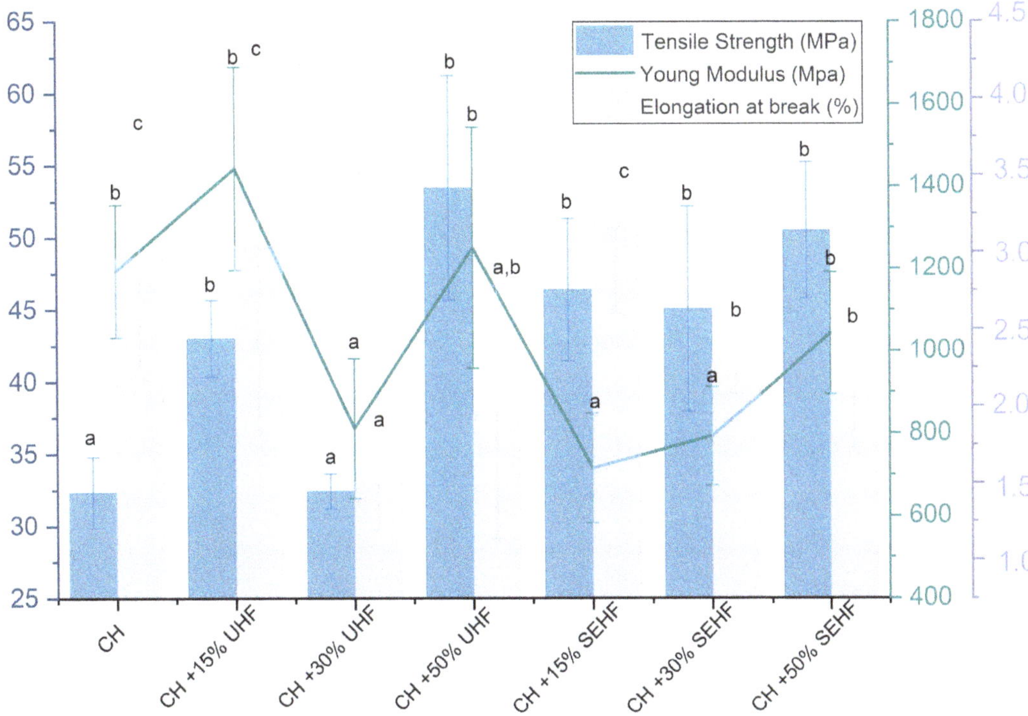

Figure 3. Mechanical properties of the different composites of chitosan–hemp fibers; values with different letters are significantly different ($p < 0.05$).

2.4. Barrier Properties

Barrier properties related to water vapor and oxygen are important parameters for food packaging materials due to the potential diffusion of these two gases through the material from the internal or external environment. Chitosan control films, plasticized with 10% (w/w) glycerol, showed a WVP of 7.5 g·mm/m^2·d·kPa (equal to 86.8 × 10^{-12} g·cm/cm^2·s·Pa), which is comparable to the values obtained by Liu et al. [27] (~2 × 10^{-12} g·cm/cm^2·s·Pa for films based on chitosan with different molecular weights). The addition of hemp fibers reduced significantly the WVP of the studied composites, in comparison with the control film (Table 1). The decrease in WVP was observed for all composite films. However, the lowest WVP values were observed for the films where SEHF were added, especially at 50% (w/w). Steam explosion treatment removes hemicellulose from hemp fibers, which makes it less hydrophilic, and thus decreases its affinity to water molecules. Therefore, the more important the fiber content, the better the water vapor barrier property. It was reported that steam explosion increases the porosity, and thus the tortuosity of fibers, which makes water diffusion more difficult [16]. Therefore, water molecules did not diffuse directly through the film and passed through the different pores of treated HF, which slowed down its crossing and improved its barrier properties.

Table 1. Thickness, water vapor, and oxygen permeabilities of the different chitosan–hemp fibers composites.

	Thickness (μm)	**WVP ($g·mm/m^2·d·kPa$)**	**OP ($cm^3·mm/m^2·d·kPa$)**
CH	167.5 ± 3.5 [b]	7.5 ± 0.2 [a]	0.44 ± 0.17 [a]
CH-15% UHF 1 mm	146.8 ± 6.0 [a]	5.4 ± 0.2 [c]	0.49 ± 0.04 [b]
CH-30% UHF 1 mm	227.2 ± 13.2 [c]	6.5 ± 0.1 [b]	0.79 ± 0.11 [d]
CH-50% UHF 1 mm	271.8 ± 21.0 [d]	6.6 ± 0.1 [b]	0.68 ± 0.18 [c]
CH-15% SEHF	225.0 ± 21.4 [c]	5.6 ± 0.004 [c]	0.67 ± 0.17 [c]
CH-30% SEHF	238.8 ± 23.7 [d]	5.3 ± 0.2 [c]	0.64 ± 0.22 [c]
CH-50% SEHF	235.1 ± 22.8 [c]	3.5 ± 0.1 [d]	0.65 ± 0.21 [c]

CH: Chitosan control (1.5% *w/w*), UHF: Untreated hemp fibers cut to 1 mm, SEHF: Steam exploded hemp fibers, WVP: Water vapor permeability, OP: Oxygen permeability; values are means of n replicates ± standard deviation (n = 10 for thickness; n = 3 for WVP and OP); values with different letters in the same column are significantly different ($p < 0.05$).

For OP, the addition of HF increased film permeability in comparison with control CH films. Incorporation of treated or untreated HF created micropores in the polymeric matrix, which made it more porous, and thus more permeable to gas, such as O_2, CO_2, and N_2. However, all OP results are less than 1 $cm^3·mm/m^2·d·kPa$, which is considered to be a very good barrier property to oxygen, as confirmed by previous studies reporting that chitosan- and polysaccharide-based materials in general, exhibit good gas barrier properties. This could be explained by the well-ordered hydrogen-bonded network in the matrix of films [9].

2.5. X-ray Diffraction

X-ray diffractograms of the CH + HF films were used for the measurement of the degree of crystallinity of the films. As presented in Figure 4, the X-ray diffractogram of neat HF displayed diffractogram peaks at 2θ = 15.5, 22.5, and 34.0, associated with a semi-crystalline structure in the amorphous matrix of fibers. The average crystallinity of the HF was 61.8%. Considering chitosan film, the significant peaks are located at 2θ = 23.5, 34.1. The large peak was due to the wide range of molecular mass of chitosan molecules which was crystallized (75.8% of crystallinity ratio). A major type of crystal appeared (2 theta = 23°).

2.6. Thermal Properties

In Table 2, glass transition temperatures (T_g) ranged from 105 to 110 °C, with no significant differences. The presence of a single T_g for each composite confirms the good miscibility of the different components of film matrix [10]. However, the addition of SEHF significantly increased the melting temperatures (T_m) of the composites, which increased from 134 °C for CH control films to 171, 167, and 142 °C for films filled with 15%, 30%, and 50% of SEHF (*w/w*), respectively. The shift of T_m to higher temperatures by the addition of SEHF could be related to the higher thermal resistance and stability of fibers subjected to steam explosion treatment due to the elimination of hemicellulose (decomposition at 260 °C) and the higher content of cellulose (decomposition at 360 °C) [11]. Moreover, the increased crystallinity of cellulose in the case of steam exploded fibers improves their thermal stability [22]. The high content of free water in film-forming solutions containing 15% of fiber compared to solutions with 30% and 50% SEHF favored crystallization during film formation. Films with 15% of SEHF are more crystallized than those with 30 and 50% SEHF. Crystallization had no effect on T_g, but increased the melting temperature ($T_{m15} > T_{m30} > T_{m15}$). T_m decreased with the increase in SEHF concentration.

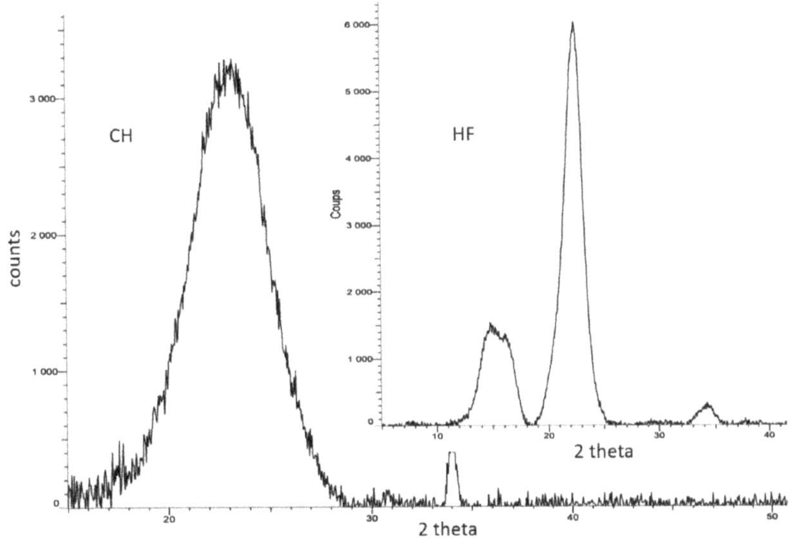

Figure 4. X-ray diffractograms of chitosan (CH) and hemp fibers (HF).

Table 2. Glass transition and melting temperatures of the different composites.

	Glass Transition Temperature T_g (°C)	Melting Temperature T_m (°C)
CH	105 ± 0.4 [a]	133 ± 0.1 [a]
CH-15% UHF 1 mm	105 ± 0.4 [a]	136 ± 0.5 [a]
CH-30% UHF 1 mm	105 ± 0.2 [a]	138 ± 0.1 [a]
CH-50% UHF 1 mm	105 ± 0.4 [a]	133 ± 0.1 [a]
CH-15% SEHF	106 ± 0.7 [a]	171 ± 0.1 [c]
CH-30% SEHF	110 ± 3.1 [b]	167 ± 0.1 [c]
CH-50% SEHF	105 ± 0.9 [a]	141 ± 1.2 [b]

CH: Chitosan control (1.5% *w*/*w*), UHF: Untreated hemp fibers cut to 1 mm, SEHF: Steam exploded hemp fibers; values with different letters in the same column are significantly different ($p < 0.05$).

3. Material and Methods

3.1. Material

Chitosan was purchased from Matexcel (New York, NY, USA) with a degree of deacetylation of 60%, viscosity of 12 mPa·s, and purity of >97%. Glycerol (as a plasticizer) and acetic acid (as a solvent) were purchased from Sigma Aldrich (St. Louis, MO, USA).

Industrial hemp fibers (*Cannabis sativa* L.) were cultivated and harvested by La Chanvrière Company (Troyes, France).

3.2. Methods

3.2.1. Fiber Extraction and Treatments

Hemp fibers (HF) used for the development of chitosan-based composites were subjected to two different extraction methods: Mechanical extraction (industrial process) without any chemical or enzymatic treatment (UHF) and steam explosion (SEHF).

For mechanical treatment, hemp fibers were first opened with the LAROCHE opening machine using the first drum at 800 rpm and low-feeding speed. They were then introduced in the BONINO carding machine at a speed of 15 m/min via a loading system to create a web. The web was fed through the coiler of the carding machine to achieve a sliver of oriented hemp fibers, at a weight of 20 g/m. This sliver was then taken to the Pierret textile cutter, where the fibers were cut to the desired length.

Steam explosion was performed using the method of Chadni et al. [29]. Raw hemp fibers, previously soaked in distilled water for 16 h, were placed into a 2 L reactor with heat jacket and automatic control for steam pressure and sampler residence time. Fibers were steam exploded at 200 °C and 15 bars for 4 min. The steam-saturated biomass was released into a discharge tank. The obtained fibers were later filtered, washed, and dried at 20 °C for 72 h.

3.2.2. Films Development

Films were developed by the casting method according to Liu et al. [27]. Film-forming solutions were prepared by dissolving chitosan (1.5%, *w/w*) in acetic acid solvent (2%, *v/v*) at 70 °C under continuous stirring. Glycerol was then added as plasticizer at 10% (*w/w*, based on chitosan dry weight). For composite films, HF whether cut to 1 mm or steam exploded, were added to the CH film-forming solution at 15, 30, or 50% (*w/w*, according to CH dry weight) and the suspensions were stirred for 1 h. Film-forming solutions were cast into plastic petri dishes (14 cm in diameter), oven-dried for 1 h at 50 °C, and then left to dry for 48 h in a conditioned room at 25 °C and 40% RH. All films were stored at 50% RH and 25 °C prior to mechanical and barrier characterization tests.

3.2.3. Films Characterization

Fourier Transform Infrared Spectroscopy (FTIR)

IR spectroscopy was performed on a Tensor 27 mid-FTIR spectrometer (Bruker, Germany) with a Deuterated Triglycine Sulfate (DTGS) detector and a diamond Attenuated Total Reflectance (ATR) module. FTIR spectra were recorded between 4000 and 400 cm^{-1} with 64 scans and a resolution of 4 cm^{-1}. Dried film samples of 1 × 1 cm were used and the analysis was performed in triplicate.

Scanning Electron Microscopy (SEM)

Surface morphology of films was studied by SEM using a JEOL JSM 6010LA high-resolution scanning electron microscope (Tokyo, Japan). Prior to the analysis, film samples were metallized with gold and the analysis was conducted with an accelerating voltage of 5 or 10 kV.

Thickness

The film thickness was determined using a Mitutoyo micrometer (Tokyo, Japan) and the analysis was performed 10 times on different positions of each film.

Water Vapor Permeability (WVP)

This characterization was performed based on the gravimetric method using ASTM standard method E96/E96 M [30] with modifications (T = 25 °C; ΔRH = 75%). Briefly, film disks (with a surface of 27 cm^2) were placed on hermetic permeation cells filled with 50 g of silica gel. The cells were placed in a chamber containing a saturated solution of sodium chloride to obtain a relative humidity of 75% within the chamber at 25 °C. Weight gain (water adsorption in silica gel) was followed-up for 2 weeks and tests were conducted in triplicate. Water vapor transmission rate (WVPR) was determined by linear regression based on the slope of the cell weight change per square meter of film vs. time ($R^2 > 0.99$) during the steady state.

The WVP (in $g \cdot mm/m^2 \cdot d \cdot kPa$) was calculated as shown in Equation (1):

$$WVP = WVTR \times e/\Delta p \quad (1)$$

where WVTR is the slope ($g/m^2 \cdot d$), e is the film thickness (mm), and Δp is the gradient of partial water pressure ($\Delta p = 2.38$ kPa).

Oxygen Barrier Properties

Oxygen permeability was investigated using an oxygen analyzer (Systech Illinois Instrument, Thame, UK) according to the ASTM D3985-05 method [31] on disk films with a surface of 78.5 cm^2. The oxygen transmission rate (OTR) was measured by the analyzer software. Moreover, the oxygen permeability was calculated at steady state according to Equation (2) and expressed in $cm^3/m^2 \cdot d \cdot kPa$. Tests were performed in triplicate at 23 °C and 50% RH.

$$OP = OTR \times e/\Delta p \quad (2)$$

where OTR is the oxygen transmission rate, e is the thickness of the film, and Δp is the partial pressure difference across the film (kPa).

Mechanical Properties

Mechanical characterization was studied in terms of tensile strength (TS, MPa), elongation at break (E, %), and Young's modulus (YM, MPa). This characterization was performed using an universal tensile machine (Shimadzu AGS-X, Kyoto, Japan), based on the ASTM D882 standard method [32]. Films were cut into rectangular samples of 15 mm × 100 mm and the test was performed with a head speed of 10 mm/min using a double clap separated by 50 mm. Five replicates were tested for each film.

Differential Scanning Calorimetry (DSC)

Thermal properties of chitosan-based composites were studied using a DSC (DSC 2500 Discovery Series from TA Instruments, Hüllhorst, Germany). Film samples (around 40 mg) were sealed in aluminum pans and tests were conducted in triplicate under nitrogen atmosphere. Heating/cooling/heating cycles were run from 0 °C to 180 °C with heating and cooling rates of 10 °C/min.

X-ray Diffraction (XRD)

Crystalline properties of neat chitosan films, untreated hemp, and steam exploded fibers were studied by means of an X-ray diffractometer (BRUKER D8 Advance) using the following conditions: Cu Kα radiation (wavelength of 1.54056 Å) from the anode operating at 40 kV and 40 mA, with a graphite monochromator, an angular range of $2\theta = 5$–$120°$ with a step width of 0.02°/scan. The location and the intensity of the obtained peaks were used for the determination of crystalline ratio (ratio between crystal peaks and amorphous changes in the base line).

Statistical Analysis

All experimental data were subjected to one-way ANOVA analysis with 95% significance level using Origin software (Origin Lab, version 2019, Northampton, MA, USA). Duncan's test was performed to detect the significant differences (<0.05).

4. Conclusions

The effect of different concentrations and treatments of HF on the properties of chitosan films was studied. Mechanical resistance and WVP of composite films were significantly improved by HF addition. TS reached more than 50 MPa for films with 50% of UHF and SEHF. WVP results were particularly interesting for SEHF-charged films. The lowest WVP values were measured for CH films with 50% of SEHF. Similarly, films with SEHF showed higher melting temperatures, which made them suitable for material processing in the packaging industry. This indicated that chitosan and SEHF had good miscibility and compatibility thanks to the fact that steam explosion is a physical method that changes the structural and surface properties of the fiber. The good miscibility of CH and SEHF was supported by SEM, FTIR, XRD, and DSC analyses.

The current study showed that when adding treated or untreated HF, the functional properties of chitosan films were improved, which makes them promising for food packaging applications as films or catering containers. Steam explosion seems to be a good mechanical treatment to improve microstructural surface, to favor crystallinity, and promote interactions between chitosan and SEHF. Incorporating SEHF at 50% based on the chitosan polymer could be particularly interesting to improve functional properties, reduce material processing costs, and respect sustainable development goals.

Author Contributions: Conceptualization, S.D.; Methodology, R.G. and S.D.; Validation, R.G.; Investigation, R.G., Y.T., L.P., C.S., N.B. and S.D.; Resources, R.G., C.S., N.B. and S.D.; Writing—original draft, R.G. and S.D.; Writing—review & editing, R.G., C.S., N.B. and S.D. All authors have read and agreed to the published version of the manuscript.

Funding: This research was funded by Agence Nationale de la Recherche (15-004). This research was also supported by the FEDER and the Grand Est Region (3BR program).

Institutional Review Board Statement: Not applicable.

Informed Consent Statement: Not applicable.

Data Availability Statement: Data sharing is not applicable to this article.

Conflicts of Interest: The authors declare no conflict of interest.

Sample Availability: Not applicable.

References

1. Varghese, S.A.; Pulikkalparambil, H.; Promhuad, K.; Srisa, A.; Laorenza, Y.; Jarupan, L.; Nampitch, T.; Chonhenchob, V.; Harnkarnsujarit, N. Renovation of Agro-Waste for Sustainable Food Packaging: A Review. *Polymers* **2023**, *15*, 648. [CrossRef] [PubMed]
2. Reichert, C.L.; Bugnicourt, E.; Coltelli, M.-B.; Cinelli, P.; Lazzeri, A.; Canesi, I.; Braca, F.; Martínez, B.M.; Alonso, R.; Agostinis, L.; et al. Bio-Based Packaging: Materials, Modifications, Industrial Applications and Sustainability. *Polymers* **2020**, *12*, 1558. [CrossRef]
3. Biswal, T.; BadJena, S.K.; Pradhan, D. Sustainable biomaterials and their applications: A short review. In Proceedings of the National Conference on Trends in Minerals & Materials Technology (MMT), Bhubaneswar, India, 30 October 2019; pp. 274–282.
4. Baranwal, J.; Barse, B.; Fais, A.; Delogu, G.L.; Kumar, A. Biopolymer: A Sustainable Material for Food and Medical Applications. *Polymers* **2022**, *14*, 983. [CrossRef] [PubMed]
5. Kulka, K.; Sionkowska, A. Chitosan Based Materials in Cosmetic Applications: A Review. *Molecules* **2023**, *28*, 1817. [CrossRef]
6. Ilyas, R.A.; Aisyah, H.A.; Nordin, A.H.; Ngadi, N.; Zuhri, M.Y.M.; Asyraf, M.R.M.; Sapuan, S.M.; Zainudin, E.S.; Sharma, S.; Abral, H.; et al. Natural-Fiber- Reinforced Chitosan, Chitosan Blends and Their Nanocomposites for Various Advanced Applications. *Polymers* **2022**, *14*, 874. [CrossRef] [PubMed]
7. Ma, X.; Qiao, C.; Wang, X.; Yao, J.; Xu, J. Structural Characterization and Properties of Polyols Plasticized Chitosan Films. *Int. J. Biol. Macromol.* **2019**, *135*, 240–245. [CrossRef]
8. Kumar, R.; Ul Haq, M.I.; Raina, A.; Anand, A. Industrial applications of natural fibre-reinforced polymer composites–challenges and opportunities. *Int. J. Sustain. Eng.* **2019**, *12*, 212–220. [CrossRef]
9. Pires, J.; Paula, C.D.D.; Souza, V.G.L.; Fernando, A.L.; Coelhoso, I. Understanding the Barrier and Mechanical Behavior of Different Nanofillers in Chitosan Films for Food Packaging. *Polymers* **2021**, *13*, 721. [CrossRef]
10. Bhat, V.G.; Narasagoudr, S.S.; Masti, S.P.; Chougale, R.B.; Vantamuri, A.B.; Kasai, D. Development and evaluation of Moringa extract incorporated Chitosan/Guar gum/Poly (vinyl alcohol) active films for food packaging applications. *Int. J. Biol. Macromol.* **2022**, *200*, 50–60. [CrossRef]
11. Manaia, J.P.; Manaia, A.T.; Rodriges, L. Industrial Hemp Fibers: An Overview. *Fibers* **2019**, *7*, 106. [CrossRef]
12. Hargitai, H.; Rácz, I.; Anandjiwala, R.D. Development of Hemp Fiber Reinforced Polypropylene Composites. *J. Thermoplast. Compos. Mater.* **2008**, *21*, 165–174. [CrossRef]
13. Hu, R.; Lim, J.K. Fabrication and Mechanical Properties of Completely Biodegradable Hemp Fiber Reinforced Polylactic Acid Composites. *J. Compos. Mater.* **2007**, *41*, 1655–1669. [CrossRef]
14. Dayo, A.Q.; Gao, B.C.; Wang, J.; Liu, W.B.; Derradji, M.; Shah, A.H.; Babar, A.A. Natural hemp fiber reinforced polybenzoxazine composites: Curing behavior, mechanical and thermal properties. *Compos. Sci. Technol.* **2017**, *144*, 114–124. [CrossRef]
15. Sauvageon, T.; Lavoie, J.M.; Segovia, C.; Brosse, N. Toward the cottonisation of hemp fibers by steam explosion—Part 1—Defibration and morphological characterization. *Text. Res. J.* **2018**, *88*, 1047–1055. [CrossRef]
16. Ziegler-Devin, I.; Chrusciel, L.; Brosse, N. Steam Explosion Pretreatment of Lignocellulosic Biomass: A Mini-Review of Theorical and Experimental Approaches. *Front. Chem.* **2021**, *9*, 705358. [CrossRef] [PubMed]

17. Arumugam, S.; Kandasamy, J.; Shah, A.U.M.; Sultan, M.T.H.; Safri, S.N.A.; Majid, M.S.A.; Basri, A.A.; Mustapha, F. Investigations on the Mechanical Properties of Glass Fiber/Sisal Fiber/Chitosan Reinforced Hybrid Polymer Sandwich Composite Scaffolds for Bone Fracture Fixation Applications. *Polymers* **2020**, *12*, 1501. [CrossRef] [PubMed]
18. Ji, M.; Li, J.; Li, F.; Wang, X.; Man, J.; Li, J.; Zhang, C.; Peng, S. A Biodegradable Chitosan-Based Composite Film Reinforced by Ramie Fibre and Lignin for Food Packaging. *Carbohydr. Polym.* **2022**, *281*, 119078. [CrossRef]
19. Sun, X.; Jia, X.; Yang, J.; Wang, S.; Li, Y.; Shao, D.; Song, H. Bamboo fiber-reinforced chitosan sponge as a robust photothermal evaporator for efficient solar vapor generation. *J. Mater. Chem. A.* **2021**, *42*, 23891–23901. [CrossRef]
20. Moalla, S.; Ammar, I.; Fauconnier, M.; Danthine, S.; Blecker, C.; Besbes, S.; Attia, H. Development and characterization of chitosan films carrying Artemisia campestris antioxidants for potential use as active food packaging materials. *Int. J. Biol. Macromol.* **2021**, *183*, 254–266. [CrossRef]
21. Reis, R.S.; Tienne, L.G.P.; Souza, D.d.H.S.; Marques, M.d.F.V.; Monteiro, S.N. Characterization of coffee parchment and innovative steam explosion treatment to obtain microfibrillated cellulose as potential composite reinforcement. *J. Mater. Res. Technol.* **2020**, *9*, 9412–9421. [CrossRef]
22. Promhuad, K.; Srisa, A.; San, H.; Laorenza, Y.; Wongphan, P.; Sodsai, J.; Tansin, K.; Phromphen, P.; Chartvivatpornchai, N.; Ngoenchai, P.; et al. Applications of Hemp Polymers and Extracts in Food, Textile and Packaging: A Review. *Polymers* **2022**, *14*, 4274. [CrossRef] [PubMed]
23. Nader, S.; Brosse, N.; Khadraoui, M.; Fuentealba, C.; Ziegler-Devin, I.; Quilès, F.; El-Kirat-Chatel, S.; Mauret, E. A low-cost environmentally friendly approach to isolate lignin containing micro and nanofibrillated cellulose from Eucalyptus globulus bark by steam explosion. *Cellulose* **2022**, *29*, 5593–5607. [CrossRef]
24. Neves, A.C.C.; Rohen, L.A.; Mantovani, D.P.; Carvalho, J.P.R.G.; Vieira, C.M.F.; Lopes, F.P.D.; Tonini Simonassi, N.; Da Luz, F.S.; Monteiro, S.N. Comparative mechanical properties between biocomposites of Epoxy and polyester matrices reinforced by hemp fiber. *J. Mater. Res. Technol.* **2020**, *9*, 1296–1304. [CrossRef]
25. Mongioví, C.; Lacalamita, D.; Morin-Crini, N.; Gabrion, X.; Ivanovska, A.; Sala, F.; Placet, V.; Rizzi, V.; Gubitosa, J.; Mesto, E.; et al. Use of Chènevotte, a Valuable Co-Product of Industrial Hemp Fiber, as Adsorbent for Pollutant Removal. Part I: Chemical, Microscopic, Spectroscopic and Thermogravimetric Characterization of Raw and Modified Samples. *Molecules* **2021**, *26*, 4574. [CrossRef]
26. Teixeira-Costa, B.E.; Ferreira, W.H.; Goycoolea, F.M.; Murray, B.S.; Andrade, C.T. Improved Antioxidant and Mechanical Properties of Food Packaging Films Based on Chitosan/Deep Eutectic Solvent, Containing Açaí-Filled Microcapsules. *Molecules* **2023**, *28*, 1507. [CrossRef] [PubMed]
27. Liu, Y.; Yuan, Y.; Duan, S.; Li, C.; Hu, B.; Liu, A.; Wu, D.; Cui, H.; Lin, L.; He, J.; et al. Preparation and characterization of chitosan films with three kinds of molecular weight for food packaging. *Int. J. Biol. Macromol.* **2020**, *155*, 249–259. [CrossRef]
28. Hongrattanavichit, I.; Aht-Ong, D. Nanofibrillation and characterization of sugarcane bagasse agro-waste using water-based steam explosion and high-pressure homogenization. *J. Clean. Prod.* **2020**, *277*, 123471. [CrossRef]
29. Chadni, M.; Grimi, N.; Bals, O.; Ziegler-Devin, I.; Desobry, S.; Brosse, N. Elaboration of hemicellulose-based films: Impact of the extraction process from spruce wood on the film properties. *Carbohydr. Res.* **2020**, *497*, 108111. [CrossRef]
30. *ASTM E96/E96M-15*; ASTM, Standard Test Methods for Water Vapor Transmission of Materials. Annual Book of ASTM. American Society for Testing and Materials: Philadelphia, PA, USA, 2015.
31. *ASTM D3985-05*; ASTM, Standard Test Method for Oxygen Gas Transmission Rate through Plastic Film and Sheeting Using a Coulometric Sensor. ASTM International: West Conshohocken, PA, USA, 2017.
32. *ASTM D882*; ASTM, Standard Test Methods for Tensile Properties of Thin Plastic Sheeting. Annual book of ASTM. American Society for Testing and Materials: Philadelphia, PA, USA, 2001.

molecules

Review

Biomaterials Based on Chitosan and Its Derivatives and Their Potential in Tissue Engineering and Other Biomedical Applications—A Review

Marta Szulc * and Katarzyna Lewandowska *

Department of Biomaterials and Cosmetic Chemistry, Faculty of Chemistry, Nicolaus Copernicus University in Toruń, Gagarin 7, 87-100 Torun, Poland
* Correspondence: marta.sz@doktorant.umk.pl (M.S.); reol@umk.pl (K.L.); Tel.: +48-56-6114551 (M.S. & K.L.)

Abstract: In the times of dynamically developing regenerative medicine, more and more attention is focused on the use of natural polymers. This is due to their high biocompatibility and biodegradability without the production of toxic compounds, which means that they do not hurt humans and the natural environment. Chitosan and its derivatives are polymers made most often from the shells of crustaceans and are biodegradable and biocompatible. Some of them have antibacterial or metal-chelating properties. This review article presents the development of biomaterials based on chitosan and its derivatives used in regenerative medicine, such as a dressing or graft of soft tissues or bones. Various examples of preparations based on chitosan and its derivatives in the form of gels, films, and 3D structures and crosslinking products with another polymer are discussed herein. This article summarizes the latest advances in medicine with the use of biomaterials based on chitosan and its derivatives and provides perspectives on future research activities.

Keywords: chitosan; chitosan derivatives; cross-linking; biomaterials; polysaccharides

check for updates

Citation: Szulc, M.; Lewandowska, K. Biomaterials Based on Chitosan and Its Derivatives and Their Potential in Tissue Engineering and Other Biomedical Applications—A Review. *Molecules* **2023**, *28*, 247. https://doi.org/10.3390/molecules28010247

Academic Editor: Agnieszka Ewa Wiącek

Received: 29 November 2022
Revised: 20 December 2022
Accepted: 23 December 2022
Published: 28 December 2022

1. Introduction

Tissue engineering is focused on the creation of tissues to repair or replace diseased or damaged organs. Recently, natural polymers have become of increasing interest due to rapidly developing medical applications. This is due to their biodegradability and non-toxicity. They also mimic tissue structure better than synthetic polymers due to their physicochemical similarity. The development of new products based on tissue-mimicking biopolymers that are more robust, non-toxic, and biodegradable is a key issue that will guarantee rapid growth in the development of tissue engineering. Biomimetic natural polymers and hybrid polymer materials have the advantage of combining desired functions with tailored morphology and superior chemical and physical stability. These polymeric materials aim to cover all aspects of the subject, including, for instance, the design of hybrid materials, films, gel, sponge, nanocomposites, and hydrogels, without forgetting studies of structure–property relationships, production of materials with precise structural control and advanced properties, and applications of bioinspired polymers for various fields including tissue engineering, drug delivery systems, or wound dressings [1–4].

The main problems of the resulting materials made from single polymers are insufficient mechanical properties and too rapid biodegradability. Therefore, mixtures of polymers and the use of a cell-free tissue matrix started to be used. Silk fibroin [5–7], collagen [8,9], hyaluronate [10,11], or gelatin [12] were used for this purpose. The materials obtained should be biodegradable and the biodegradation products must be non-toxic and removed from the body without any effect on other tissues. Furthermore, the materials should support cell adhesion, migration, and proliferation through appropriate porosity, pore size, and their appropriate combination. The physicochemical and mechanical properties should be as similar as possible to those of the tissue to be replaced and should be strong enough to allow its implantation during surgery [13,14]. These materials can take the form of thin films, hydrogels, membranes, 3D structures, fibers, and nanofibers.

Herein, we reviewed various examples of chitosan-based biomaterials, mixed with other polymers and cross-linked with chemical agents, in biomedical applications based on previous research.

2. Chitosan and Its Derivatives

Chitosan (CS) (poly(β-(1,4)-2-amino-2-deoxy-D-glucopyranose) is a natural polymer obtained by partial deacetylation of chitin in an alkaline medium (Figure 1). Chitin was produced from the exoskeletons of crustaceans. Chitosan also occurs naturally in the cell walls of some fungi. Chitosan is a polymer with a degree of deacetylation of at least 60%. The polymer's molecular weight and the degree of deacetylation determine its properties such as biodegradability, biocompatibility, viscosity, hydrophilicity, and antibacterial or antifungal properties. The major disadvantage of chitosan is its lack of solubility in water [15–21].

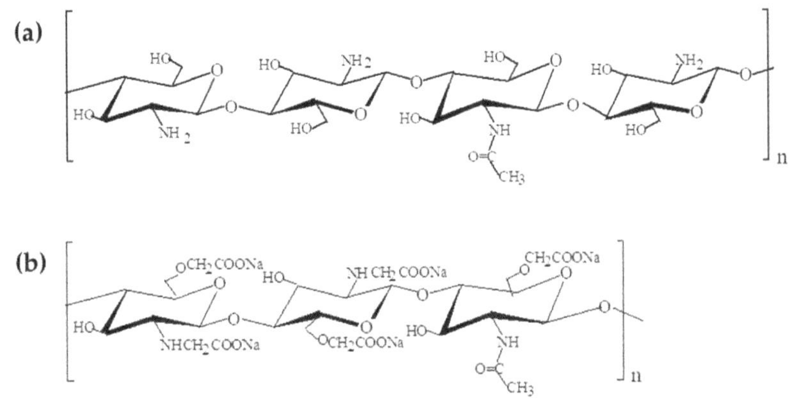

Figure 1. Structures of chitosan (**a**) and carboxymethyl chitosan (**b**).

The most important properties are shown in Figure 2. Chitosan is the only naturally occurring polysaccharide classified as a cationic polyelectrolyte, which allows it to interact with different types of molecules. The polymer's positive charge is responsible for its antibacterial properties, attaching to the negatively charged cell membrane of various microorganisms [14,21].

Carboxymethyl chitosan (CMC) is a chitosan derivative in which the carboxymethyl group is attached to either an amino group or a hydroxyl group (Figure 1).

This chitosan derivative is water soluble and this is one of the main reasons for the increased interest in this polymer by researchers. It can be obtained in many types: N-carboxymethyl chitosan N,N-carboxymethyl chitosan, N,O-carboxymethyl chitosan, and O-carboxymethyl chitosan. During the substitution reaction, the listed types of derivatives or their mixtures can be obtained [22,23]. CMC is characterized by high viscosity, biocompatibility, and biodegradability, and is non-toxic. It also has antimicrobial activity, with O-carboxymethyl chitosan showing greater activity due to the more abundant presence of amino groups. Carboxymethyl chitosan shows improved physicochemical and biological properties relative to chitosan. The properties of CMC are influenced by the average molecular weight, degree of deacetylation, and degree of substitution. In addition, CMC has antioxidant activity, antibacterial or antifungal properties, and the ability to chelate metals [22–26]

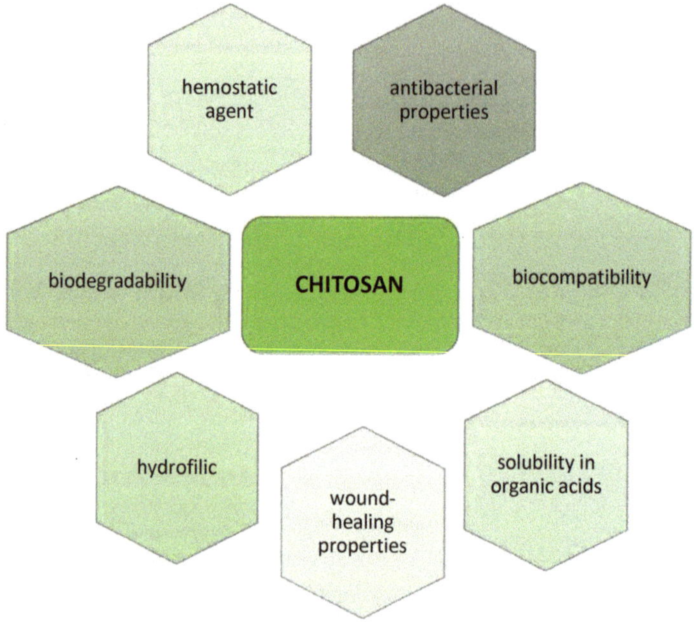

Figure 2. Chitosan properties.

Chitosan acetate is obtained by reaction with acetic acid in an aqueous–ethanol environment. It is water soluble and its solution is more stable than chitosan dissolved in acetic acid, while retaining the physicochemical and biological properties of chitosan. It exhibits stronger antimicrobial activity against Gram-positive bacteria than against Gram-negative bacteria. It is used as a dressing material (Chitopack C®) and a hemostat (Hemcon Bandage®) approved by the FDA [27].

There are other chitosan derivatives such as sulfopropylchitosan, O-quaternary ammonium salt of chitosan, N-succinylchitosan, and others [28,29].

3. Chitosan and Its Derivatives in Medicine

Due to its properties, chitosan and its derivatives can be used in the production of dressing materials, in the manufacture of drugs as a controlled-release active substance carrier, or in tissue engineering involving soft tissues, nerves, cartilage, bones, or arteries. Studies on the use of chitosan are summarized in Table 1 and studies on its derivatives are in Table 2.

Table 1. Application of chitosan-based materials in tissue engineering.

Composition	Method	Application	In Vivo/In Vitro	Advantages	Ref.
Chitosan, genipin	Crosslinking, freeze-drying	Spinal cord tissue engineering	In vivo (rats)	Low cytotoxicity, high histocompatibility, good mechanical properties	[30]
Decellularized extracellular matrix/gelatin/and chitosan, EDC/NHS	Crosslinking, freeze-drying	Skin tissue engineering	In vitro (L929 fibroblasts)	The high modulus of elasticity, biodegradability, non-cytotoxic	[31]

Table 1. *Cont.*

Composition	Method	Application	In Vivo/In Vitro	Advantages	Ref.
Cellulose acetate nanofibers/chitosan/fibroin silk cryogel scaffold, genipin	Electrospinning, crosslinking, freeze-drying	Smooth muscle tissue engineering	In vitro (smooth muscle cell)	Good mechanical properties, good proliferation	[32]
Chitosan/poly (ethylene oxide)	Electrospinning scaffold	Cartilage tissue engineering	In vitro (chondrocyte cells)	Good cell adhesion and proliferation	[33]
Hyaluronic acid/chitosan coacervate-based scaffolds	Centrifuge, incubation	Cartilage tissue engineering	In vitro	Good proliferation and cell viability	[34]
PCL/chitosan-PEO with *A. euchroma* extract	Two-nozzle electrospinning	Skin tissue engineering	In vitro (HDF cells)	Good proliferation and cell viability	[35]
Hydrogels of chitosan/oxidized-modified quince seed gum/curcumin-loaded	Encapsulation	Tissue engineering	In vitro (NIH3T3 fibroblast cells)	Improved thermal stability, swelling ratio, and degradation rate of hydrogels, non-cytotoxicity, good proliferation	[36]
Chitosan scaffolds, sodium hydroxide-crosslinking agent	3D print	Cartilage tissue engineering	In vitro (ATDC5 cells)	Higher elastic modulus, good biocompatibility	[37]
Gelatin/chitosan/polyvinyl alcohol/nano-hydroxyapatite	Freeze-drying	Bone tissue engineering	In vitro (BMSCs cells)	Improved surface bioactivity and biomimetic structure, high osteogenic differentiation ability	[38]
Polycaprolactone–polyurethane/chitosan	Freeze-drying, drying in oven	Bone tissue engineering	In vitro (hBMSCs)	Non-cytotoxicity, good mechanical properties, good promotion of the formation of calcium levels, good gene expression	[39]
Chitosan–hydroxyapatite–carbon	Drying in oven	Bone tissue engineering	In vitro (human osteoblasts)	Good biocompatibility with human osteoblasts, good mechanical properties	[40]
Polycaprolactone/chitosan-g-polycaprolactone/hydroxyapatite	Electrospinning, drying in oven	Bone tissue engineering	In vitro (NIH3T3 fibroblast cells)	High cell viability and proliferation, good mechanical properties	[41]
Chitosan–vanillin–BG (CVB)	Freeze-drying	Bone tissue engineering	In vivo (female mice	Good biocompatibility, bioactivity, strong antibacterial ability, good promotion of osteoblastic differentiation, ectopic bone formation in vivo	[42]
Chitosan-pyrolyzed cork	Freeze-drying	Electrically active biological tissue engineering	In vitro (SH-SY5Y neuroblas-toma cell)	Good biocompatibility, high mechanical strength	[43]

Table 1. *Cont.*

Composition	Method	Application	In Vivo/In Vitro	Advantages	Ref.
Polycaprolactone (PCL)–chitosan/ carboxyl carbon	Electrospinning	Cartilage tissue engineering	In vitro (chondrocytes cells)	High porosity, good mechanical properties, good biocompatibility	[44]
Decellularized Alstroemeria flower stem/chitosan	Freeze-drying	Tissue engineering	In vitro (MC3T3 cells)	Good cell attachment, proliferation and migration, good mechanical properties	[45]
Chitosan/hydroxypropyl methyl cellulose/hydroxyapatite/ lemon grass oil	Freeze gelation method	Bone tissue engineering	In vitro (MC3T3 cells)	Antimicrobial activity (*S. aureus*), non-toxic	[46]
Chitosan/βGP/NaHCO3/ HAp/PECs/gelatin	Gelation in a water batch	Bone tissue engineering	In vitro (MG63 cells)	Good cellular proliferation, osteogenic differentiation	[47]
Chitosan–tripolyphosphate	Exploiting dialysis technique, freeze-drying	Tissue engineering	In vitro (NIH3T3 fibroblast cells)	Good biocompatibility, good mechanical properties	[48]
Chitosan scaffolds with controllable microchannel	Combining a 3D printing microfiber template-leaching method and a freeze-drying method	Tissue engineering	In vitro (NIH3T3 fibroblast cells), in vivo (rats)	Good cell proliferation and distribution, improved cell, tissue growth and vascular formation	[49]
Chitosan/loofah/Poly(3-hydroxybutyric acid-co-3-hydroxyvaleric acid)	Electrospinning, freeze-drying	Tissue engineering	In vitro (human mesenchymal stem cells)	Good cell proliferation and migration, good mechanical and viscoelastic properties, differentiation into adipogenic, osteogenic, and chondrogenic tissues	[50]
Xylan/chitosan/nano-HAp/graphene oxide/reduced graphene oxide	Freeze-drying	Bone tissue engineering	In vitro (MG-63 cell)	Improved mineralization tendency, osteogenic differentiation capability	[51]
Hybrid bionanocomposite of chitosan/poly(vinyl alcohol)/nanobioactive glass/nanocellulose	Drying in oven	Bone tissue engineering	In vitro (red blood cells)	Good porosity, better antibacterial effect (*E. coli, S. aureus*), improved hemocompatibility	[52]
Bacterial cellulose/chitosan/alginate/gelatin	Stirring with heat	Cartilage tissue engineering	In vitro (human mesenchymal stem cells)	Good compressive strength, stability, biocompatibility, good cell proliferation	[53]

Table 1. *Cont.*

Composition	Method	Application	In Vivo/In Vitro	Advantages	Ref.
Chitosan/poly(vinyl alcohol)/nano bioactive glass/nano zinc oxide	Drying in oven	Bone tissue engineering	In vitro (red blood cells)	Better tensile strength, good hemocompatibility, antimicrobial activity (*Enterococcus faecalis, Salmonella typhi*)	[54]
Calcium silicate-coated porous chitosan	Freeze-drying	Dental tissue engineering	In vitro (human dental pulp cells)	Good cell proliferation and mineralization	[55]
Graphene-oxide-containing chitosan	Freeze-drying	Cartilage tissue engineering	In vitro (chondrocytes cells)	Improved physical and mechanical properties, good proliferation	[56]
Injectable chitosan/beta glycerophosphate/ pyrrole oligomers	Stirring	Cartilage tissue engineering	In vitro (fibroblastoid cell CHO-K1)	Good biodegradability, biocompatibility, electro-activity, swelling ratio, and pore size values	[57]
Silk fibroin–chitosan	Freeze-drying	Cartilage tissue engineering	In vitro (human mesenchymal stem cell)	Good porosity, good compressive strength, proliferation, cell viability	[58]
Chitosan/ modified montmorillonite/hydroxyapatite	Microwave irradiation, gas-foaming method, freeze-drying	Bone tissue engineering	In vitro (MG 63 osteoblast cell)	Non-cytotoxic, good biodegradation, swelling properties, and good mechanical properties	[59]
Chitosan-grafted-poly(methyl methacrylate)/ hydroxyapatite scaffold	Freeze-drying	Bone tissue engineering	In vitro (UMR-106 osteoblast-like cells)	Good viability, proliferation, and cells attachment, good mechanical properties, good drug delivery	[60]
Poly-L-lactic acid/chitosan/collagen	Electrospinning	Vascular tissue engineering	In vitro (lymphocyte T cell)	Good cell viability and hemolysis, good mechanical properties, and bust pressure	[61]
Gelatin/chitosan	Electrospinning	Skin tissue engineering	In vitro (human dermal fibroblast cells)	Very good porosity, good mechanical properties, non-cytotoxic, spindle-like shape cells	[62]
l-chitosan/maleic terminated polyethylene glycol	Freeze-drying	Skin tissue engineering	In vitro (HFFF2 cells), in vivo (rats)	Porous structure, high swelling ratio, biocompatibility, fully closed wound with improved vascularization	[63]
Chitosan–vitamin C–lactic acid	Freeze-drying	Skin tissue engineering	In vitro (NIH3T3 fibroblast cells)	Good cell attachment, proliferation and spreading	[64]

Table 2. Applications of chitosan-derivative-based materials in tissue engineering.

Composition	Method	Application	In Vivo/In Vitro	Advantages	Ref.
Carboxymethyl chitosan/genipin	Stirring	Skin tissue engineering	In vitro (HSFs cells) in vivo (rats)	Good cell attachment and proliferation, good wound healing promotion	[65]
Polyvinyl alcohol, carboxymethyl chitosan with silver nanoparticles and borax	Stirring	Skin tissue engineering	In vitro (L929 cells)	Antibacterial properties, good mechanical properties, non-cytotoxic	[66]
Carboxymethyl chitosan/carboxymethyl cellulose hydrogel with heparin and glutaraldehyde	Stirring	Skin tissue engineering	In vivo (rats with diabetes)	Accelerated open wound healing	[67]
Carboxymethyl chitosan/genipin/Zn scaffolds	Freeze-drying	Dental tissue engineering	In vitro (dental pulp stem cells)	Antibacterial properties, good cell proliferation	[68]
Thiolated chitosan and silk fibroin	Incubating at 37 °C	Cartilage tissue engineering	In vitro (chondrocytes cells)	Good mechanical properties, high porosity, good cell proliferation	[69]
Lactoferrin-loaded carboxymethyl cellulose glycol chitosan	Stirring, 3D printing	Tissue engineering applications	In vitro (mouse osteoblastic cells)	Good biocompatibility, good physician properties	[70]
Silk fibroin/carboxymethyl chitosan hydrogel crosslinking by horseradish peroxidase	Stirring	Cartilage tissue engineering	In vitro (chondrocytes cells)	Good biocompatibility, biodegradability, good mechanical and rheological properties	[71]
Carboxymethyl chitosan/oxidized pullulan with methotrexate-loaded mesoporous silica	Stirring	Drug delivery	In vitro (human hepatoma SMMC-7721 and hepatic LO2 cells)	Good biocompatibility, non-cytotoxic, good drug release	[72]
Polymerized CMC-modified adhesive	Mixing the powder with the adhesive	Dental tissue engineering	Antibacterial test	Good antibacterial properties (*S. mutans*)	[73]
Oxidized microcrystalline cellulose/ carboxymethyl chitosan	Stirring	Skin tissue engineering	In vitro blood compatibility test	Good mechanical, self-healing characteristic, good coagulation	[74]
Silk fibroin/carboxymethyl chitosan/strontium substituted hydroxyapatite/cellulose	Freeze-drying	Bone tissue engineering	In vitro (BMSCs cells)	Non-toxic, good hemocompatibility, good gene expression (osteogenic gene markers), high porosity	[75]
Carboxymethyl chitosan-modified glass ionomer cement	Mixing	Dental tissue engineering	In vitro (NIH 3 T3 fibroblast cells)	Good biocompatibility, good attachment, and cell proliferation, better mechanical properties	[76]
Poly(3,4-ethylenedioxythiophene)/ carboxymethyl chitosan	Vibration	Neural tissue engineering		Good biodegradation and electroconductivity, good compressive modulus, better cell adhesion, viability and proliferation	[77]

Table 2. *Cont.*

Composition	Method	Application	In Vivo/In Vitro	Advantages	Ref.
Benzaldehyde-terminated 4-arm PEG/carboxymethyl chitosan/basic fibroblast growth factor	Stirring	Skin tissue engineering	In vitro (blood cells)	Excellent biocompatibility, fast hemostasis capacity, strong wet-tissue adhesion, self-mending, and antibacterial property	[78]
Polycaprolactone /carboxymethyl chitosan/sodium alginate micron-fibrous	Emulsion electrospinning	Periosteal tissue engineering	In vitro (osteoblasts cells)	Excellent tensile strength, no significant cytotoxicity, good cell adhesion	[79]
Carboxymethyl chitosan/sodium alginate hydrogels with polydopamine coatings	Immersion	Skin tissue engineering	In vitro (human umbilical vein endothelial cells), in vivo (rats with MRSA)	Antibacterial, anti-inflammatory properties, good antibacterial properties (*Methicillin-resistant Staphylococcus aureus*), fast wound healing	[80]
Chitosan/carboxymethyl cellulose with silver nanoparticles	Stirring	Skin tissue engineering	In vitro (human skin fibroblasts)	Good mechanical properties, good antibacterial properties (*E.coli*), non-cytotoxic	[81]
Gelatin/carboxymethyl chitosan/ nano-hydroxyapatite	Freeze-drying	Bone tissue engineering	In vitro (human Wharton's jelly MSC microtissue)	High porosity, slow enzymatic degradation, good mechanical properties, good viability, the proliferation of human Wharton's jelly MSC microtissue	[82]
N,O-carboxymethyl chitosan/fucoidan	Freeze-drying	Bone tissue engineering	In vitro (L929 cells)	Good mineralization, good physical properties, good cell proliferation and mineralization	[83]
Diselenide-crosslinked carboxymethyl chitosan nanoparticles with doxorubicin	Stirring, dialysis	Drug delivery	In vitro (tumor cells)	High drug encapsulation efficiency, high drug accumulation, and cytotoxicity in tumor cells	[84]
Thiolated carboxymethyl chitosan-based 3D scaffolds	Freeze-drying	Theragnostic of tissue regeneration	In vitro (human dermo fibroblast cells)	High porosity, good mechanical properties, non-cytotoxic	[85]
Quaternized chitosan/hydroxyapatite curcumin-loaded	Stirring	Bone tissue engineering	In vitro (MG-63 cells)	Good mechanical strength, drug release, good biocompatibility and cell proliferation	[86]
Carboxymethyl chitosan/cellulose nanofiber	Freeze-drying, drying in the oven	Skin tissue engineering	In vivo (rats)	Good blood absorption, and excellent coagulation ability	[87]

Table 2. *Cont.*

Composition	Method	Application	In Vivo/In Vitro	Advantages	Ref.
Carboxymethyl chitosan–plantamajoside	Stirring	Skin tissue engineering	In vitro (L929 cells), in vivo (rats with burn wounds)	Good porosity, good cell viability, proliferation, significantly improved wound healing, granulation tissue proliferation	[88]
Polycaprolactone/galactosylated chitosan	Freeze-drying, electrospinning	Liver tissue engineering	In vitro (HepG2 cells)	Non-cytotoxic, good cell growth, and proliferation	[89]
Cotton fabric/carboxymethyl chitosan/silver nitrate	Pad–dry–cure method, drying in oven	Skin tissue engineering	In vivo (rats with wounds)	Good wound healing properties, antibacterial properties (*E. coli, S. aureus*)	[90]
Chitosan–gelatin–hyaluronic acid	Freeze-drying	Skin tissue engineering	In vitro (fibroblast and keratinocytes cells)	Good mechanical properties, flexible scaffold/cells, artificial skin, good cell proliferation in co-cultures	[91]
Mannose-anchored quaternized chitosan/thiolated carboxymethyl chitosan	Freeze-drying	Drug delivery	In vitro (293T cells)	Non-cytotoxic, high hydrophilicity, good drug release and stability	[92]
Chitosan, carboxymethyl cellulose and silver-nanoparticle-modified cellulose nanowhiskers	Freeze-drying	Bone tissue engineering	In vitro (MG63 cells)	Good mechanical properties, high porosity, excellent antimicrobial activity (*E. coli*), good biomineralization	[93]
N, O-carboxymethyl chitosan/oxidized cellulose containing ε-poly-L-lysine	Freeze-drying	Skin tissue engineering	In vitro (NIH 3T3 cells), in vivo (rabbit)	Good antibacterial properties (*E. coli, S. aureus*), excellent biological security and compatibility in vitro and in vivo	[94]
O-carboxymethyl chitosan/sodium alginate with insulin	Stirring	Drug delivery	In vitro (L929 mouse fibroblast cells), in vivo (rats)	High drug loading capacity and high effectively released drugs as oral drugs, lower glucose level compared with insulin injections	[95]
Polycaprolactone/ carboxymethyl chitosan	Electrospinning	Bone tissue engineering	In vitro (human osteoblast cells MG63)	Good biocompatibility, good cell proliferation	[96]
O-carboxymethyl chitosan nonwoven fabrics	Chitosan needle-punched nonwoven reaction with chloroacetic acid	Skin tissue engineering	In vitro (L929 mouse fibroblast cells), in vivo (rats with a partial-thickness burn)	Good mechanical properties, good cell migration, and proliferation, good healing rate, good angiogenesis	[97]
Recombinant human collagen/carboxylated chitosan	Stirring	Soft tissue engineering	In vitro (NIH 3T3 cells), in vivo (rats with open wounds)	Good biocompatibility, non-cytotoxic, acceleration of the cell infiltration and wound closure	[98]

Table 2. *Cont.*

Composition	Method	Application	In Vivo/In Vitro	Advantages	Ref.
Nano-hydroxyapatite/chitosan/polyethylene glycol	Stirring, filtration, drying in the oven	Bone tissue engineering	In vitro (murine fibroblast L929 cells)	Good thermal stability and swelling ratio, non-cytotoxic	[99]
Norcantharidin-conjugated carboxymethyl chitosan	Vacuum-dried	Drug delivery	In vitro (BEL-7402 cells), in vivo (mice with H22 cells, tumor cells)	Inhibitory effects on the proliferation and migration of cells, changes in cell structure, reduction in the distribution of norcantharidin in heart and kidney tissues, diminished systemic toxicity	[100]
Poly (vinyl alcohol) and fungal mushroom-derived carboxymethyl chitosan	Solution casting technique	Skin tissue engineering	In vitro (skin fibroblasts and keratinocytes)	Good antibacterial properties (*E. coli, S. aureus*), good biocompatibility, good hemolysis	[101]
Carboxymethyl chitosan/oxidized dextran/sodium alginate	Mixing with a double-barreled syringe	Skin tissue engineering	In vitro (L929 cells), in vivo (rat liver injury model and mouse tail amputation model)	Red blood cells could adhere to the surface of hydrogel, good hemostasis, good antibacterial properties (*S. aureus*)	[102]
N,O-carboxymethyl chitosan	Stirring	Drug delivery	In vivo (rabbit)	Good drug delivery, non-cytotoxic to the cornea, good degradability	[103]

The team of Fangsong Zhang et al. [30] used two chemical agents, glutaraldehyde, genipin, and a physical agent, ultraviolet light, to crosslink nerve extracellular matrix/chitosan scaffolds. Scaffolds cross-linked with genipin were characterized by higher porosity and regular structure in contrast to scaffolds cross-linked with glutaraldehyde and UV. The degree of crosslinking for genipin-crosslinked and glutaraldehyde-crosslinked scaffolds were similar to each other. Genipin-crosslinked scaffolds had the lowest degree of cytotoxicity and the highest histocompatibility, with good mechanical properties.

Another team, Jie Xu et al. [31], prepared a scaffold based on decellularized extracellular matrix, gelatin, and chitosan cross-linked EDC/NHS. The resulting scaffolds were characterized by a high modulus of elasticity and biodegradability. The obtained scaffolds are not cytotoxic and provided a good substrate for cell proliferation. The scaffolds were also characterized by antibacterial properties (*E. Coli, S. Aureus*). The scaffolds obtained could be used in skin tissue engineering.

A scaffold for use in muscle tissue engineering is a project by the team of Weiguang Zhao et al. [32]. They used genipin as a crosslinking agent and electrospun cellulose acetate nanofibers that were incorporated into a chitosan/fibroin silk cryogel scaffold. The resulting scaffolds were characterized by larger pores and roughness than the cryogel scaffold itself. They are also a good substrate for smooth muscle cell proliferation, which showed a higher potential for the expression of genes related to muscle contraction. They also exhibit good mechanical properties.

Scaffold for use in cartilage tissue engineering is a study by Christian E. G. Garcia et al. [33]. The properties of chitosan in two forms were compared: thin film and electrospinning scaffold chitosan/poly (ethylene oxide) (PEO). PEO of two different molecular weights was used and different weight ratios of Cs/PEO were applied. Some of the materials obtained were neutralized in order to compare the effect of neutralization on the properties of the scaffolds. The scaffolds after neutralization were characterized by better adhesion of chondrocyte cells and better proliferation; the worst properties were characterized by the chitosan film.

Nihui Zhang and her team [65] used genipin solutions with different concentrations (2.5%, 5%, 10%, and 15%) to produce carboxymethyl chitosan hydrogels. Analysis of HSFs cell proliferation and adhesion showed that the best cell adhesion and proliferation were obtained for the hydrogel with the highest amount of genipin, and the worst for the hydrogel with the lowest proportion of genipin. The best properties for promoting wound healing and reducing the appearance of scars in vivo tests were obtained for the hydrogel carboxymethyl chitosan/genipin 5% (*v/v*). This was confirmed by an in vivo test using female rats. With the additional contribution of aloe vera gel, wound healing results improved even further. In conclusion, genipin-crosslinked chitosan hydrogels are promising candidates for use as a dressing to accelerate wound healing.

Yalei Liu et al. [66] used polyvinyl alcohol, carboxymethyl chitosan, silver nanoparticles, and borax as a crosslinking agent to produce a hydrogel. The resulting hydrogel, due to its dual crosslinking (hydrogen bonds and borate ester bonds), has self-healing properties and is characterized by good mechanical properties. It also exhibits antibacterial properties, as confirmed by a test with *E.coli* and *S. aureus* bacteria. A cytotoxicity test was also performed using L929 cells, which showed that the resulting scaffolds were non-toxic.

Guozhu Chang et al. [67] produced a carboxymethyl chitosan/carboxymethyl cellulose hydrogel using heparin and glutaraldehyde as a crosslinking agent. This allowed the fabrication of a self-healing hydrogel. It is biocompatible with cells and its ability to release drugs has also been studied. An in vivo study was also performed on rats with diabetes, where its effect on accelerating open wound healing was confirmed. It can be concluded that the resulting hydrogel has the potential to be used as a material to accelerate diabetic wound healing.

To form antimicrobial scaffolds, A. Mishra's team [68] used carboxymethyl chitosan, zinc, and genipin. Carboxymethyl chitosan/genipin/Zn scaffolds were obtained. Wet compression analysis showed that the carboxymethyl chitosan/genipin/Zn scaffold was more robust than the non-cross-linked scaffold. Degradation testing was carried out under enzymatic and non-enzymatic conditions. The resulting scaffold also showed good stability. An adhesion and proliferation test was performed using dental pulp stem cells; in addition, a biocompatibility test against red blood cells was performed, which confirmed its good biological properties. An antibacterial test was also performed (*Pseudomonas aeruginosa* ATCC 25619, *S. aureus* ATCC 9144, *S. aureus* ATCC 25923, and *Staphylococcus epidermidis* ATCC 155). No biofilm formed on the surface of the scaffold carboxymethyl chitosan/genipin/Zn. In conclusion, scaffold carboxymethyl chitosan/genipin/Zn can find application in dental tissue engineering due to its antibacterial properties.

Summarizing the data overview on the use of chitosan and its derivatives in tissue engineering (Figure 3a,b), it can be written that the most research on chitosan-based materials concerned bone tissue engineering and the least concerned dental tissue engineering. For materials based on chitosan derivatives, the greatest interest in use was in skin tissue engineering and the least in tissue engineering applications.

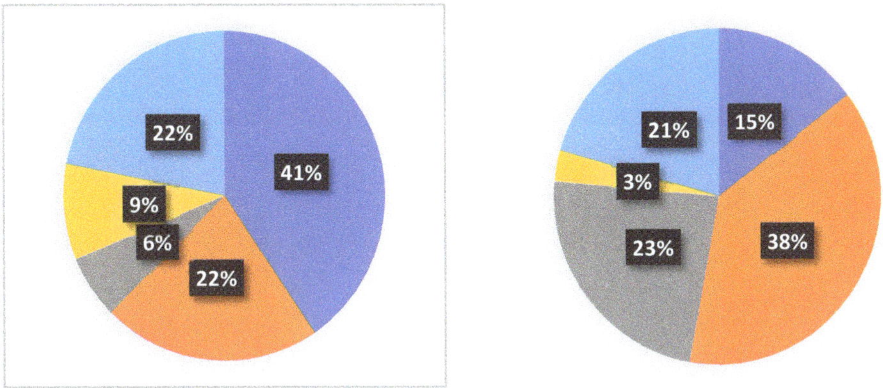

(a) (b)

Figure 3. Applications of chitosan (**a**) and its derivatives (**b**) in tissue engineering. Skin tissue engineering (dark blue), bone tissue engineering (orange), cartilage tissue engineering (grey), dental tissue engineering (yellow), and other tissue engineering and unclassified tissue engineering (light blue).

4. Conclusions

The use of chitosan and its derivatives in medicine offers a huge opportunity in the further development of regenerative medicine. The use of different forms of polymers such as films, hydrogel scaffolds, or the use of strongly developing ways of producing materials such as electrospinning and 3D printing open another door to the medicine of the future. Owing to continuous development, we are able to produce biomaterials that mimic the structure, morphology, and function of various organs such as blood vessels, nerves, soft tissues, or bones. Further research using other solvents, new mixtures, or using a different cross-linking agent may bring us even closer to a perfectly mimicking tissue biomaterial. A constant challenge is to produce in the spirit of green production and ecology in a closed loop using natural polymers where their extraction will not adversely affect the environment.

Author Contributions: Conceptualization, M.S. and K.L.; data curation M.S.; writing—original draft preparation, M.S.; writing—review and editing, M.S. and K.L.; visualization, M.S. and K.L.; supervision, K.L.; project administration, M.S. and K.L.; funding acquisition, K.L. All authors have read and agreed to the published version of the manuscript.

Funding: This research received no external funding.

Institutional Review Board Statement: Not applicable.

Informed Consent Statement: Not applicable.

Data Availability Statement: Not applicable.

Acknowledgments: This work is supported by Grants4NCUStudents, Excellence Initiative – Research University, (Grant IDUB- 4101.00000070), 2022–2023 and Young Scientist Grant, Dean of the Faculty of Chemistry NCU, 2021–2022.

Conflicts of Interest: The authors declare no conflict of interest.

References

1. Zhao, X.; He, X.; Hou, A.; Cheng, C.; Wang, X.; Yue, Y.; Wu, Z.; Wu, H.; Liu, B.; Li, H.; et al. Growth of Cu2O Nanoparticles on Two-Dimensional Zr–Ferrocene–Metal–Organic Framework Nanosheets for Photothermally Enhanced Chemodynamic Antibacterial Therapy. *Inorg. Chem.* **2022**, *61*, 9328–9338. [CrossRef] [PubMed]
2. Zhou, Z.; Wang, Y.; Peng, F.; Meng, F.; Zha, J.; Ma, L.; Du, Y.; Peng, N.; Ma, L.; Zhang, Q.; et al. Intercalation-Activated Layered MoO3 Nanobelts as Biodegradable Nanozymes for Tumor-Specific Photo-Enhanced Catalytic Therapy. *Angew. Chem. Int. Ed.* **2022**, *61*, e202115939. [CrossRef]
3. Qin, L.; Liang, F.; Li, Y.; Wu, J.; Guan, S.; Wu, M.; Xie, S.; Luo, M.; Ma, D. A 2D Porous Zinc-Organic Framework Platform for Loading of 5-Fluorouracil. *Inorganics* **2022**, *10*, 202. [CrossRef]
4. Hu, T.; Gu, Z.; Williams, G.R.; Strimaite, M.; Zha, J.; Zhou, Z.; Zhang, X.; Tan, C.; Liang, R. Layered Double Hydroxide-Based Nanomaterials for Biomedical Applications. *Chem. Soc. Rev.* **2022**, *51*, 6126–6176. [CrossRef]
5. Tu, D.D.; Chung, Y.G.; Gil, E.S.; Seth, A.; Franck, D.; Cristofaro, V.; Sullivan, M.P.; Di Vizio, D.; Gomez, P.; Adam, R.M.; et al. Bladder Tissue Regeneration Using Acellular Bi-Layer Silk Scaffolds in a Large Animal Model of Augmentation Cystoplasty. *Biomaterials* **2013**, *34*, 8681–8689. [CrossRef]
6. Xiao, S.; Wang, P.; Zhao, J.; Ling, Z.; An, Z.; Fu, Z.; Fu, W.; Zhang, X. Bi-Layer Silk Fibroin Skeleton and Bladder Acellular Matrix Hydrogel Encapsulating Adipose-Derived Stem Cells for Bladder Reconstruction. *Biomater. Sci.* **2021**, *9*, 6169–6182. [CrossRef]
7. Cao, N.; Song, L.; Liu, W.; Fan, S.; Jiang, D.; Mu, J.; Gu, B.; Xu, Y.; Zhang, Y.; Huang, J. Prevascularized Bladder Acellular Matrix Hydrogel/Silk Fibroin Composite Scaffolds Promote the Regeneration of Urethra in a Rabbit Model. *Biomed. Mater.* **2018**, *14*, 015002. [CrossRef]
8. Gasanz, C.; Raventós, C.; Temprana-Salvador, J.; Esteves, M.; Fonseca, C.; de Torres, I.; Morote, J. Use of an Acellular Collagen–Elastin Matrix to Support Bladder Regeneration in a Porcine Model of Peritoneocystoplasty. *Cent. Eur. J. Urol.* **2018**, *71*, 353. [CrossRef]
9. Shi, C.; Chen, W.; Chen, B.; Shan, T.; Jia, W.; Hou, X.; Li, L.; Ye, G.; Dai, J. Bladder Regeneration in a Canine Model Using a Bladder Acellular Matrix Loaded with a Collagen-Binding BFGF. *Biomater. Sci.* **2017**, *5*, 2427–2436. [CrossRef]
10. Jalali, S.; Fereidoni, M.; Shahri, N.M.; Lari, R. Effect of Swim Bladder Matrix Treated with Hyaluronic Acid on Wound Healing: An Animal Model Evaluation. *J. Wound Care* **2019**, *28*, 206–213. [CrossRef]
11. Su, Z.; Ma, H.; Wu, Z.; Zeng, H.; Li, Z.; Wang, Y.; Liu, G.; Xu, B.; Lin, Y.; Zhang, P.; et al. Enhancement of Skin Wound Healing with Decellularized Scaffolds Loaded with Hyaluronic Acid and Epidermal Growth Factor. *Mater. Sci. Eng. C* **2014**, *44*, 440–448. [CrossRef] [PubMed]
12. Mathapati, S.; Bishi, D.K.; Venugopal, J.R.; Cherian, K.M.; Guhathakurta, S.; Ramakrishna, S.; Verma, R.S. Nanofibers Coated on Acellular Tissue-Engineered Bovine Pericardium Supports Differentiation of Mesenchymal Stem Cells into Endothelial Cells for Tissue Engineering. *Nanomedicine* **2014**, *9*, 623–634. [CrossRef] [PubMed]
13. Dolcimascolo, A.; Calabrese, G.; Conoci, S.; Parenti, R. Innovative Biomaterials for Tissue Engineering. In *Biomaterial-Supported Tissue Reconstruction or Regeneration*; Barbeck, M., Jung, O., Smeets, R., Koržinskas, T., Eds.; IntechOpen: London, UK, 2019; ISBN 978-1-83880-377-3.
14. Rodríguez-Vázquez, M.; Vega-Ruiz, B.; Ramos-Zúñiga, R.; Saldaña-Koppel, D.A.; Quiñones-Olvera, L.F. Chitosan and Its Potential Use as a Scaffold for Tissue Engineering in Regenerative Medicine. *BioMed Res. Int.* **2015**, *2015*, 821279. [CrossRef] [PubMed]
15. Huang, Y.; Onyeri, S.; Siewe, M.; Moshfeghian, A.; Madihally, S.V. In Vitro Characterization of Chitosan–Gelatin Scaffolds for Tissue Engineering. *Biomaterials* **2005**, *26*, 7616–7627. [CrossRef] [PubMed]
16. Sencadas, V.; Correia, D.M.; Ribeiro, C.; Moreira, S.; Botelho, G.; Gómez Ribelles, J.L.; Lanceros-Mendez, S. Physical-Chemical Properties of Cross-Linked Chitosan Electrospun Fiber Mats. *Polym. Test.* **2012**, *31*, 1062–1069. [CrossRef]
17. Muxika, A.; Etxabide, A.; Uranga, J.; Guerrero, P.; de la Caba, K. Chitosan as a Bioactive Polymer: Processing, Properties and Applications. *Int. J. Biol. Macromol.* **2017**, *105*, 1358–1368. [CrossRef]
18. Aguilar, A.; Zein, N.; Harmouch, E.; Hafdi, B.; Bornert, F.; Offner, D.; Clauss, F.; Fioretti, F.; Huck, O.; Benkirane-Jessel, N.; et al. Application of Chitosan in Bone and Dental Engineering. *Molecules* **2019**, *24*, 3009. [CrossRef]
19. Baranwal, A.; Kumar, A.; Priyadharshini, A.; Oggu, G.S.; Bhatnagar, I.; Srivastava, A.; Chandra, P. Chitosan: An Undisputed Bio-Fabrication Material for Tissue Engineering and Bio-Sensing Applications. *Int. J. Biol. Macromol.* **2018**, *110*, 110–123. [CrossRef]
20. El Knidri, H.; Belaabed, R.; Addaou, A.; Laajeb, A.; Lahsini, A. Extraction, Chemical Modification and Characterization of Chitin and Chitosan. *Int. J. Biol. Macromol.* **2018**, *120*, 1181–1189. [CrossRef]
21. Yang, J.; Tian, F.; Wang, Z.; Wang, Q.; Zeng, Y.-J.; Chen, S.-Q. Effect of Chitosan Molecular Weight and Deacetylation Degree on Hemostasis. *J. Biomed. Mater. Res.* **2008**, *84B*, 131–137. [CrossRef]
22. Jimtaisong, A.; Saewan, N. Utilization of Carboxymethyl Chitosan in Cosmetics. *Int. J. Cosmet. Sci.* **2014**, *36*, 12–21. [CrossRef] [PubMed]
23. Muzzarelli, R.A.A. Carboxymethylated Chitins and Chitosans. *Carbohydr. Polym.* **1988**, *8*, 1–21. [CrossRef]
24. Mourya, V.K.; Inamdara, N.; Ashutosh Tiwari, N. Carboxymethyl Chitosan and Its Applications. *Adv. Mater. Lett.* **2010**, *1*, 11–33. [CrossRef]
25. Sun, T.; Yao, Q.; Zhou, D.; Mao, F. Antioxidant Activity of N-Carboxymethyl Chitosan Oligosaccharides. *Bioorg. Med. Chem. Lett.* **2008**, *18*, 5774–5776. [CrossRef]

26. Shariatinia, Z. Carboxymethyl Chitosan: Properties and Biomedical Applications. *Int. J. Biol. Macromol.* **2018**, *120*, 1406–1419. [CrossRef]
27. Li, Y.; Chen, X.G.; Liu, N.; Liu, C.S.; Liu, C.G.; Meng, X.H.; Yu, L.J.; Kenendy, J.F. Physicochemical Characterization and Antibacterial Property of Chitosan Acetates. *Carbohydr. Polym.* **2007**, *67*, 227–232. [CrossRef]
28. Kahya, N. Water Soluble Chitosan Derivatives and Their Biological Activities: A Review. *Polym. Sci.* **2019**, *5*, 1–16. [CrossRef]
29. Kato, Y. N-Succinyl-Chitosan as a Drug Carrier: Water-Insoluble and Water-Soluble Conjugates. *Biomaterials* **2004**, *25*, 907–915. [CrossRef]
30. Zhang, F.; Zhang, N.; Xu, Q.; Zhang, L.; Zhang, C.; Liu, H.; Yu, Z.; Zhou, S.; Feng, G.; Huang, F. Decellularized Nerve Extracellular Matrix/Chitosan Crosslinked by Genipin to Prepare a Moldable Nerve Repair Material. *Cell Tissue Bank* **2021**, *22*, 419–430. [CrossRef]
31. Xu, J.; Fang, H.; Zheng, S.; Li, L.; Jiao, Z.; Wang, H.; Nie, Y.; Liu, T.; Song, K. A Biological Functional Hybrid Scaffold Based on Decellularized Extracellular Matrix/Gelatin/Chitosan with High Biocompatibility and Antibacterial Activity for Skin Tissue Engineering. *Int. J. Biol. Macromol.* **2021**, *187*, 840–849. [CrossRef]
32. Zhao, W.; Cao, S.; Cai, H.; Wu, Y.; Pan, Q.; Lin, H.; Fang, J.; He, Y.; Deng, H.; Liu, Z. Chitosan/Silk Fibroin Biomimic Scaffolds Reinforced by Cellulose Acetate Nanofibers for Smooth Muscle Tissue Engineering. *Carbohydr. Polym.* **2022**, *298*, 120056. [CrossRef] [PubMed]
33. Garcia, C.E.G.; Lardy, B.; Bossard, F.; Soltero, F.A. Chitosan based biomaterials for cartilage tissue engineering: Chondrocyte adhesion and proliferation. *Food Hydrocolloids for Health* **2021**, *1*, 100018. [CrossRef]
34. Karabıyık Acar, Ö.; Bedir, S.; Kayitmazer, A.B.; Kose, G.T. Chondro-Inductive Hyaluronic Acid/Chitosan Coacervate-Based Scaffolds for Cartilage Tissue Engineering. *Int. J. Biol. Macromol.* **2021**, *188*, 300–312. [CrossRef] [PubMed]
35. Asghari, F.; Rabiei Faradonbeh, D.; Malekshahi, Z.V.; Nekounam, H.; Ghaemi, B.; Yousefpoor, Y.; Ghanbari, H.; Faridi-Majidi, R. Hybrid PCL/Chitosan-PEO Nanofibrous Scaffolds Incorporated with A. Euchroma Extract for Skin Tissue Engineering Application. *Carbohydr. Polym.* **2022**, *278*, 118926. [CrossRef] [PubMed]
36. Yavari Maroufi, L.; Ghorbani, M. Injectable Chitosan-Quince Seed Gum Hydrogels Encapsulated with Curcumin Loaded-Halloysite Nanotubes Designed for Tissue Engineering Application. *Int. J. Biol. Macromol.* **2021**, *177*, 485–494. [CrossRef] [PubMed]
37. Sadeghianmaryan, A.; Naghieh, S.; Alizadeh Sardroud, H.; Yazdanpanah, Z.; Afzal Soltani, Y.; Sernaglia, J.; Chen, X. Extrusion-Based Printing of Chitosan Scaffolds and Their in Vitro Characterization for Cartilage Tissue Engineering. *Int. J. Biol. Macromol.* **2020**, *164*, 3179–3192. [CrossRef]
38. Ma, P.; Wu, W.; Wei, Y.; Ren, L.; Lin, S.; Wu, J. Biomimetic Gelatin/Chitosan/Polyvinyl Alcohol/Nano-Hydroxyapatite Scaffolds for Bone Tissue Engineering. *Mater. Des.* **2021**, *207*, 109865. [CrossRef]
39. Amiryaghoubi, N.; Noroozi Pesyan, N.; Fathi, M.; Omidi, Y. The Design of Polycaprolactone-Polyurethane/Chitosan Composite for Bone Tissue Engineering. *Colloids Surf. A Physicochem. Eng. Asp.* **2022**, *634*, 127895. [CrossRef]
40. Sanchez, A.G.; Prokhorov, E.; Luna-Barcenas, G.; Hernández-Vargas, J.; Román-Doval, R.; Mendoza, S.; Rojas-Chávez, H. Chitosan-Hydroxyapatite-MWCNTs Nanocomposite Patch for Bone Tissue Engineering Applications. *Mater. Today Commun.* **2021**, *28*, 102615. [CrossRef]
41. Shirzaei Sani, I.; Rezaei, M.; Baradar Khoshfetrat, A.; Razzaghi, D. Preparation and Characterization of Polycaprolactone/Chitosan-g-Polycaprolactone/Hydroxyapatite Electrospun Nanocomposite Scaffolds for Bone Tissue Engineering. *Int. J. Biol. Macromol.* **2021**, *182*, 1638–1649. [CrossRef]
42. Hu, J.; Wang, Z.; Miszuk, J.M.; Zhu, M.; Lansakara, T.I.; Tivanski, A.V.; Banas, J.A.; Sun, H. Vanillin-Bioglass Cross-Linked 3D Porous Chitosan Scaffolds with Strong Osteopromotive and Antibacterial Abilities for Bone Tissue Engineering. *Carbohydr. Polym.* **2021**, *271*, 118440. [CrossRef] [PubMed]
43. Scalera, F.; Monteduro, A.G.; Maruccio, G.; Blasi, L.; Gervaso, F.; Mazzotta, E.; Malitesta, C.; Piccirillo, C. Sustainable Chitosan-Based Electrical Responsive Scaffolds for Tissue Engineering Applications. *Sustain. Mater. Technol.* **2021**, *28*, e00260. [CrossRef]
44. Mirmusavi, M.H.; Ahmadian, M.; Karbasi, S. Polycaprolactone-Chitosan/Multi-Walled Carbon Nanotube: A Highly Strengthened Electrospun Nanocomposite Scaffold for Cartilage Tissue Engineering. *Int. J. Biol. Macromol.* **2022**, *209*, 1801–1814. [CrossRef] [PubMed]
45. Esmaeili, J.; Jadbabaee, S.; Far, F.M.; Lukolayeh, M.E.; Kırboğa, K.K.; Rezaei, F.S.; Barati, A. Decellularized Alstroemeria Flower Stem Modified with Chitosan for Tissue Engineering Purposes: A Cellulose/Chitosan Scaffold. *Int. J. Biol. Macromol.* **2022**, *204*, 321–332. [CrossRef] [PubMed]
46. Ali, H.U.; Iqbal, D.N.; Iqbal, M.; Ezzine, S.; Arshad, A.; Zeeshan, R.; Chaudhry, A.A.; Alshawwa, S.Z.; Nazir, A.; Khan, A.F. HPMC Crosslinked Chitosan/Hydroxyapatite Scaffolds Containing Lemongrass Oil for Potential Bone Tissue Engineering Applications. *Arab. J. Chem.* **2022**, *15*, 103850. [CrossRef]
47. Wasupalli, G.K.; Verma, D. Thermosensitive Injectable Hydrogel Based on Chitosan-Polygalacturonic Acid Polyelectrolyte Complexes for Bone Tissue Engineering. *Carbohydr. Polym.* **2022**, *294*, 119769. [CrossRef]
48. Sacco, P.; Borgogna, M.; Travan, A.; Marsich, E.; Paoletti, S.; Asaro, F.; Grassi, M.; Donati, I. Polysaccharide-Based Networks from Homogeneous Chitosan-Tripolyphosphate Hydrogels: Synthesis and Characterization. *Biomacromolecules* **2014**, *15*, 3396–3405. [CrossRef]

49. Jiang, Z.; Zhang, K.; Du, L.; Cheng, Z.; Zhang, T.; Ding, J.; Li, W.; Xu, B.; Zhu, M. Construction of Chitosan Scaffolds with Controllable Microchannel for Tissue Engineering and Regenerative Medicine. *Mater. Sci. Eng. C* **2021**, *126*, 112178. [CrossRef]
50. Baysan, G.; Colpankan Gunes, O.; Akokay, P.; Husemoglu, R.B.; Ertugruloglu, P.; Ziylan Albayrak, A.; Cecen, B.; Havitcioglu, H. Loofah-Chitosan and Poly (−3-Hydroxybutyrate-Co-3-Hydroxyvalerate) (PHBV) Based Hydrogel Scaffolds for Meniscus Tissue Engineering Applications. *Int. J. Biol. Macromol.* **2022**, *221*, 1171–1183. [CrossRef]
51. Ali, A.; Hasan, A.; Negi, Y.S. Effect of Carbon Based Fillers on Xylan/Chitosan/Nano-HAp Composite Matrix for Bone Tissue Engineering Application. *Int. J. Biol. Macromol.* **2022**, *197*, 1–11. [CrossRef]
52. Narmatha, C.P.; Khaleel, B.S.; Sugantha, K.V. Multifunctional Organic and Inorganic Hybrid Bionanocomposite of Chitosan/Poly(Vinyl Alcohol)/Nanobioactive Glass/Nanocellulose for Bone Tissue Engineering. *J. Mech. Behav. Biomed. Mater.* **2022**, *135*, 105427. [CrossRef]
53. Phatchayawat, P.P.; Khamkeaw, A.; Yodmuang, S.; Phisalaphong, M. 3D Bacterial Cellulose-Chitosan-Alginate-Gelatin Hydrogel Scaffold for Cartilage Tissue Engineering. *Biochem. Eng. J.* **2022**, *184*, 108476. [CrossRef]
54. Christy, P.N.; Basha, S.K.; Kumari, V.S. Nano Zinc Oxide and Nano Bioactive Glass Reinforced Chitosan/Poly(Vinyl Alcohol) Scaffolds for Bone Tissue Engineering Application. *Mater. Today Commun.* **2022**, *31*, 103429. [CrossRef]
55. Leite, M.L.; Anselmi, C.; Soares, I.P.M.; Manso, A.P.; Hebling, J.; Carvalho, R.M.; de Souza Costa, C.A. Calcium Silicate-Coated Porous Chitosan Scaffold as a Cell-Free Tissue Engineering System for Direct Pulp Capping. *Dent. Mater.* **2022**, *38*, 1763–1776. [CrossRef]
56. Shamekhi, M.A.; Mirzadeh, H.; Mahdavi, H.; Rabiee, A.; Mohebbi-Kalhori, D.; Baghaban Eslaminejad, M. Graphene Oxide Containing Chitosan Scaffolds for Cartilage Tissue Engineering. *Int. J. Biol. Macromol.* **2019**, *127*, 396–405. [CrossRef]
57. Kashi, M.; Baghbani, F.; Moztarzadeh, F.; Mobasheri, H.; Kowsari, E. Green Synthesis of Degradable Conductive Thermosensitive Oligopyrrole/Chitosan Hydrogel Intended for Cartilage Tissue Engineering. *Int. J. Biol. Macromol.* **2018**, *107*, 1567–1575. [CrossRef]
58. Vishwanath, V.; Pramanik, K.; Biswas, A. Optimization and Evaluation of Silk Fibroin-Chitosan Freeze-Dried Porous Scaffolds for Cartilage Tissue Engineering Application. *J. Biomater. Sci. Polym. Ed.* **2016**, *27*, 657–674. [CrossRef]
59. Kar, S.; Kaur, T.; Thirugnanam, A. Microwave-Assisted Synthesis of Porous Chitosan–Modified Montmorillonite–Hydroxyapatite Composite Scaffolds. *Int. J. Biol. Macromol.* **2016**, *82*, 628–636. [CrossRef]
60. Tithito, T.; Suntornsaratoon, P.; Charoenphandhu, N.; Thongbunchoo, J.; Krishnamra, N.; Tang, I.M.; Pon-On, W. Fabrication of Biocomposite Scaffolds Made with Modified Hydroxyapatite Inclusion of Chitosan-Grafted-Poly(Methyl Methacrylate) for Bone Tissue Engineering. *Biomed. Mater.* **2019**, *14*, 025013. [CrossRef]
61. Fiqrianti, I.; Widiyanti, P.; Manaf, M.; Savira, C.; Cahyani, N.; Bella, F. Poly-L-Lactic Acid (PLLA)-Chitosan-Collagen Electrospun Tube for Vascular Graft Application. *JFB* **2018**, *9*, 32. [CrossRef]
62. Pezeshki-Modaress, M.; Zandi, M.; Rajabi, S. Tailoring the Gelatin/Chitosan Electrospun Scaffold for Application in Skin Tissue Engineering: An in Vitro Study. *Prog. Biomater.* **2018**, *7*, 207–218. [CrossRef] [PubMed]
63. Jafari, A.; Hassanajili, S.; Azarpira, N.; Bagher Karimi, M.; Geramizadeh, B. Development of Thermal-Crosslinkable Chitosan/Maleic Terminated Polyethylene Glycol Hydrogels for Full Thickness Wound Healing: In Vitro and in Vivo Evaluation. *Eur. Polym. J.* **2019**, *118*, 113–127. [CrossRef]
64. Madni, A.; Khan, R.; Ikram, M.; Naz, S.S.; Khan, T.; Wahid, F. Fabrication and Characterization of Chitosan–Vitamin C–Lactic Acid Composite Membrane for Potential Skin Tissue Engineering. *Int. J. Polym. Sci.* **2019**, *2019*, 4362395. [CrossRef]
65. Zhang, N.; Gao, T.; Wang, Y.; Liu, J.; Zhang, J.; Yao, R.; Wu, F. Modulating Cationicity of Chitosan Hydrogel to Prevent Hypertrophic Scar Formation during Wound Healing. *Int. J. Biol. Macromol.* **2020**, *154*, 835–843. [CrossRef] [PubMed]
66. Liu, Y.; Mao, J.; Guo, Z.; Hu, Y.; Wang, S. Polyvinyl Alcohol/Carboxymethyl Chitosan Hydrogel Loaded with Silver Nanoparticles Exhibited Antibacterial and Self-Healing Properties. *Int. J. Biol. Macromol.* **2022**, *220*, 211–222. [CrossRef]
67. Chang, G.; Dang, Q.; Liu, C.; Wang, X.; Song, H.; Gao, H.; Sun, H.; Zhang, B.; Cha, D. Carboxymethyl Chitosan and Carboxymethyl Cellulose Based Self-Healing Hydrogel for Accelerating Diabetic Wound Healing. *Carbohydr. Polym.* **2022**, *292*, 119687. [CrossRef]
68. Mishra, A.H.; Mishra, D. Evidences of Biomimetic and Nonantibiotic Characteristics of the Zinc–Carboxymethyl Chitosan–Genipin Organometallic Complex and Its Biocompatibility Aspects. *Biomacromolecules* **2020**, *21*, 688–700. [CrossRef]
69. Liu, J.; Yang, B.; Li, M.; Li, J.; Wan, Y. Enhanced Dual Network Hydrogels Consisting of Thiolated Chitosan and Silk Fibroin for Cartilage Tissue Engineering. *Carbohydr. Polym.* **2020**, *227*, 115335. [CrossRef]
70. Janarthanan, G.; Tran, H.N.; Cha, E.; Lee, C.; Das, D.; Noh, I. 3D Printable and Injectable Lactoferrin-Loaded Carboxymethyl Cellulose-Glycol Chitosan Hydrogels for Tissue Engineering Applications. *Mater. Sci. Eng. C* **2020**, *113*, 111008. [CrossRef]
71. Li, T.; Song, X.; Weng, C.; Wang, X.; Gu, L.; Gong, X.; Wei, Q.; Duan, X.; Yang, L.; Chen, C. Silk Fibroin/Carboxymethyl Chitosan Hydrogel with Tunable Biomechanical Properties Has Application Potential as Cartilage Scaffold. *Int. J. Biol. Macromol.* **2019**, *137*, 382–391. [CrossRef]
72. Rui, Q.; Gao, J.; Yin, Z.-Z.; Li, J.; Cai, W.; Wu, D.; Kong, Y. A Biodegradable PH and Glutathione Dual-Triggered Drug Delivery System Based on Mesoporous Silica, Carboxymethyl Chitosan and Oxidized Pullulan. *Int. J. Biol. Macromol.* **2022**, *224*, 1294–1302. [CrossRef] [PubMed]
73. Yao, S.; Chen, S.; Wang, R.; Zhang, K.; Lin, X.; Mai, S. Antibacterial Activity and Bonding Performance of Carboxymethyl Chitosan–Containing Dental Adhesive System. *Int. J. Adhes. Adhes.* **2022**, *119*, 103269. [CrossRef]

74. Yin, H.; Song, P.; Chen, X.; Huang, Q.; Huang, H. A Self-Healing Hydrogel Based on Oxidized Microcrystalline Cellulose and Carboxymethyl Chitosan as Wound Dressing Material. *Int. J. Biol. Macromol.* **2022**, *221*, 1606–1617. [CrossRef] [PubMed]
75. Zhang, X.; Chen, Y.; Han, J.; Mo, J.; Dong, P.; Zhuo, Y.; Feng, Y. Biocompatiable Silk Fibroin/Carboxymethyl Chitosan/Strontium Substituted Hydroxyapatite/Cellulose Nanocrystal Composite Scaffolds for Bone Tissue Engineering. *Int. J. Biol. Macromol.* **2019**, *136*, 1247–1257. [CrossRef]
76. Kashyap, P.K.; Chauhan, S.; Negi, Y.S.; Goel, N.K.; Rattan, S. Biocompatible Carboxymethyl Chitosan-Modified Glass Ionomer Cement with Enhanced Mechanical and Anti-Bacterial Properties. *Int. J. Biol. Macromol.* **2022**, *223*, 1506–1520. [CrossRef]
77. Xu, C.; Guan, S.; Wang, S.; Gong, W.; Liu, T.; Ma, X.; Sun, C. Biodegradable and Electroconductive Poly(3,4-Ethylenedioxythiophene)/ Carboxymethyl Chitosan Hydrogels for Neural Tissue Engineering. *Mater. Sci. Eng. C* **2018**, *84*, 32–43. [CrossRef]
78. Hao, Y.; Zhao, W.; Zhang, H.; Zheng, W.; Zhou, Q. Carboxymethyl Chitosan-Based Hydrogels Containing Fibroblast Growth Factors for Triggering Diabetic Wound Healing. *Carbohydr. Polym.* **2022**, *287*, 119336. [CrossRef]
79. Tao, F.; Cheng, Y.; Tao, H.; Jin, L.; Wan, Z.; Dai, F.; Xiang, W.; Deng, H. Carboxymethyl Chitosan/Sodium Alginate-Based Micron-Fibers Fabricated by Emulsion Electrospinning for Periosteal Tissue Engineering. *Mater. Des.* **2020**, *194*, 108849. [CrossRef]
80. Liu, T.; Feng, Z.; Li, Z.; Lin, Z.; Chen, L.; Li, B.; Chen, Z.; Wu, Z.; Zeng, J.; Zhang, J.; et al. Carboxymethyl Chitosan/Sodium Alginate Hydrogels with Polydopamine Coatings as Promising Dressings for Eliminating Biofilm and Multidrug-Resistant Bacteria Induced Wound Healing. *Int. J. Biol. Macromol.* **2022**, *in press*. [CrossRef]
81. Osorio Echavarría, J.; Gómez Vanegas, N.A.; Orozco, C.P.O. Chitosan/Carboxymethyl Cellulose Wound Dressings Supplemented with Biologically Synthesized Silver Nanoparticles from the Ligninolytic Fungus Anamorphous Bjerkandera Sp. R1. *Heliyon* **2022**, *8*, e10258. [CrossRef]
82. Maji, S.; Agarwal, T.; Das, J.; Maiti, T.K. Development of Gelatin/Carboxymethyl Chitosan/Nano-Hydroxyapatite Composite 3D Macroporous Scaffold for Bone Tissue Engineering Applications. *Carbohydr. Polym.* **2018**, *189*, 115–125. [CrossRef] [PubMed]
83. Lu, H.-T.; Lu, T.-W.; Chen, C.-H.; Lu, K.-Y.; Mi, F.-L. Development of Nanocomposite Scaffolds Based on Biomineralization of N,O-Carboxymethyl Chitosan/Fucoidan Conjugates for Bone Tissue Engineering. *Int. J. Biol. Macromol.* **2018**, *120*, 2335–2345. [CrossRef] [PubMed]
84. Wang, J.; Liu, J.; Lu, D.-Q.; Chen, L.; Yang, R.; Liu, D.; Zhang, B. Diselenide-Crosslinked Carboxymethyl Chitosan Nanoparticles for Doxorubicin Delivery: Preparation and in Vivo Evaluation. *Carbohydr. Polym.* **2022**, *292*, 119699. [CrossRef] [PubMed]
85. Medeiros Borsagli, F.G.L.; de Souza, A.J.M.; Paiva, A.E. Ecofriendly Multifunctional Thiolated Carboxymethyl Chitosan-Based 3D Scaffolds with Luminescent Properties for Skin Repair and Theragnostic of Tissue Regeneration. *Int. J. Biol. Macromol.* **2020**, *165*, 3051–3064. [CrossRef] [PubMed]
86. Murugan, E.; Akshata, C.R.; Ilangovan, R.; Mohan, M. Evaluation of Quaternization Effect on Chitosan-HAP Composite for Bone Tissue Engineering Application. *Colloids Surf. B Biointerfaces* **2022**, *218*, 112767. [CrossRef]
87. Zhou, M.; Liao, J.; Li, G.; Yu, Z.; Xie, D.; Zhou, H.; Wang, F.; Ren, Y.; Xu, R.; Dai, Y.; et al. Expandable Carboxymethyl Chitosan/Cellulose Nanofiber Composite Sponge for Traumatic Hemostasis. *Carbohydr. Polym.* **2022**, *294*, 119805. [CrossRef]
88. Yu, N.; Li, Y.; Wang, Y.; Xu, H.; Ye, F.; Fu, Q. Healing Effect of Carboxymethyl Chitosan-Plantamajoside Hydrogel on Burn Wound Skin. *Burns* **2022**, *48*, 902–914. [CrossRef]
89. Ghahremanzadeh, F.; Alihosseini, F.; Semnani, D. Investigation and Comparison of New Galactosylation Methods on PCL/Chitosan Scaffolds for Enhanced Liver Tissue Engineering. *Int. J. Biol. Macromol.* **2021**, *174*, 278–288. [CrossRef]
90. Shaheen, T.I.; Abdelhameed, M.F.; Zaghloul, S.; Montaser, A.S. In Vivo Assessment of the Durable, Green and in Situ Bio-Functional Cotton Fabrics Based Carboxymethyl Chitosan Nanohybrid for Wound Healing Application. *Int. J. Biol. Macromol.* **2022**, *209*, 485–497. [CrossRef]
91. Liu, H.; Mao, J.; Yao, K.; Yang, G.; Cui, L.; Cao, Y. A Study on a Chitosan-Gelatin-Hyaluronic Acid Scaffold as Artificial Skin in Vitro and Its Tissue Engineering Applications. *J. Biomater. Sci. Polym. Ed.* **2004**, *15*, 25–40. [CrossRef]
92. Jin, Z.; Hu, G.; Zhao, K. Mannose-Anchored Quaternized Chitosan/Thiolated Carboxymethyl Chitosan Composite NPs as Mucoadhesive Carrier for Drug Delivery. *Carbohydr. Polym.* **2022**, *283*, 119174. [CrossRef] [PubMed]
93. Hasan, A.; Waibhaw, G.; Saxena, V.; Pandey, L.M. Nano-Biocomposite Scaffolds of Chitosan, Carboxymethyl Cellulose and Silver Nanoparticle Modified Cellulose Nanowhiskers for Bone Tissue Engineering Applications. *Int. J. Biol. Macromol.* **2018**, *111*, 923–934. [CrossRef] [PubMed]
94. Cheng, F.; Xu, L.; Dai, J.; Yi, X.; He, J.; Li, H. N, O-Carboxymethyl Chitosan/Oxidized Cellulose Composite Sponge Containing ε-Poly-l-Lysine as a Potential Wound Dressing for the Prevention and Treatment of Postoperative Adhesion. *Int. J. Biol. Macromol.* **2022**, *209*, 2151–2164. [CrossRef] [PubMed]
95. Zhang, H.; Gu, Z.; Li, W.; Guo, L.; Wang, L.; Guo, L.; Ma, S.; Han, B.; Chang, J. PH-Sensitive O-Carboxymethyl Chitosan/Sodium Alginate Nanohydrogel for Enhanced Oral Delivery of Insulin. *Int. J. Biol. Macromol.* **2022**, *223*, 433–445. [CrossRef]
96. Sharifi, F.; Atyabi, S.M.; Norouzian, D.; Zandi, M.; Irani, S.; Bakhshi, H. Polycaprolactone/Carboxymethyl Chitosan Nanofibrous Scaffolds for Bone Tissue Engineering Application. *Int. J. Biol. Macromol.* **2018**, *115*, 243–248. [CrossRef]
97. Jiang, Z.; Li, L.; Li, H.; Xia, L.; Hu, H.; Wang, S.; Liu, C.; Chi, J.; Yang, Y.; Song, F.; et al. Preparation, Biocompatibility, and Wound Healing Effects of O-Carboxymethyl Chitosan Nonwoven Fabrics in Partial-Thickness Burn Model. *Carbohydr. Polym.* **2022**, *280*, 119032. [CrossRef]
98. Yang, Y.; Campbell Ritchie, A.; Everitt, N.M. Recombinant Human Collagen/Chitosan-Based Soft Hydrogels as Biomaterials for Soft Tissue Engineering. *Mater. Sci. Eng. C* **2021**, *121*, 111846. [CrossRef]

99. Shakir, M.; Jolly, R.; Khan, M.S.; Iram, N.; Sharma, T.K.; Al-Resayes, S.I. Synthesis and Characterization of a Nano-Hydroxyapatite/Chitosan/Polyethylene Glycol Nanocomposite for Bone Tissue Engineering: Nano-hydroxyapatite/chitosan/polyethylene glycol nanocomposite. *Polym. Adv. Technol.* **2015**, *26*, 41–48. [CrossRef]
100. Chi, J.; Jiang, Z.; Chen, X.; Peng, Y.; Liu, W.; Han, B.; Han, B. Studies on Anti-Hepatocarcinoma Effect, Pharmacokinetics and Tissue Distribution of Carboxymethyl Chitosan Based Norcantharidin Conjugates. *Carbohydr. Polym.* **2019**, *226*, 115297. [CrossRef]
101. Rao, K.M.; Sudhakar, K.; Suneetha, M.; Won, S.Y.; Han, S.S. Fungal-Derived Carboxymethyl Chitosan Blended with Polyvinyl Alcohol as Membranes for Wound Dressings. *Int. J. Biol. Macromol.* **2021**, *190*, 792–800. [CrossRef]
102. Xie, M.; Zeng, Y.; Wu, H.; Wang, S.; Zhao, J. Multifunctional Carboxymethyl Chitosan/Oxidized Dextran/Sodium Alginate Hydrogels as Dressing for Hemostasis and Closure of Infected Wounds. *Int. J. Biol. Macromol.* **2022**, *219*, 1337–1350. [CrossRef] [PubMed]
103. Yang, L.; Lan, Y.; Guo, H.; Cheng, L.; Fan, J.; Cai, X.; Zhang, L.; Chen, R.; Zhou, H. Ophthalmic Drug-Loaded N,O-Carboxymethyl Chitosan Hydrogels: Synthesis, in Vitro and in Vivo Evaluation. *Acta Pharm. Sin.* **2010**, *31*, 1625–1634. [CrossRef] [PubMed]

molecules

MDPI

Article

Chitosan Modulates Volatile Organic Compound Emission from the Biocontrol Fungus *Pochonia chlamydosporia*

Jorge Mestre-Tomás [1,2,*], David Esgueva-Vilà [1], Alba Fuster-Alonso [3], Federico Lopez-Moya [1,*] and Luis V. Lopez-Llorca [1]

1 Laboratory of Plant Pathology, Department of Marine Sciences and Applied Biology, University of Alicante, 03690 Alicante, Spain; lv.lopez@ua.es (L.V.L.-L.)
2 Institute for Integrative Systems Biology (CSIC-UV), Spanish National Research Council, 46980 Paterna, Spain
3 Institut de Ciències del Mar (ICM-CSIC), Renewable Marine Resources Department, 08003 Barcelona, Spain; afuster@icm.csic.es
* Correspondence: jorge.mestre.tomas@csic.es (J.M.-T.); federico.lopez@ua.es (F.L.-M.)

Abstract: Fungal volatile organic compounds (VOCs) are responsible for fungal odor and play a key role in biological processes and ecological interactions. VOCs represent a promising area of research to find natural metabolites for human exploitation. *Pochonia chlamydosporia* is a chitosan-resistant nematophagous fungus used in agriculture to control plant pathogens and widely studied in combination with chitosan. The effect of chitosan on the production of VOCs from *P. chlamydosporia* was analyzed using gas chromatography–mass spectrometry (GC-MS). Several growth stages in rice culture medium and different times of exposure to chitosan in modified Czapek–Dox broth cultures were analyzed. GC-MS analysis resulted in the tentative identification of 25 VOCs in the rice experiment and 19 VOCs in the Czapek–Dox broth cultures. The presence of chitosan in at least one of the experimental conditions resulted in the *de novo* production of 3-methylbutanoic acid and methyl 2,4-dimethylhexanoate, and oct-1-en-3-ol and tetradec-1-ene in the rice and Czapek–Dox experiments, respectively. Other VOCs changed their abundance because of the effect of chitosan and fungal age. Our findings suggest that chitosan can be used as a modulator of the production of VOCs in *P. chlamydosporia* and that there is also an effect of fungal age and exposure time.

Keywords: volatile organic compounds; chitosan; *Pochonia chalmydosporia*; gas chromatography; mass spectrometry

check for updates

Citation: Mestre-Tomás, J.; Esgueva-Vilà, D.; Fuster-Alonso, A.; Lopez-Moya, F.; Lopez-Llorca, L.V. Chitosan Modulates Volatile Organic Compound Emission from the Biocontrol Fungus *Pochonia chlamydosporia*. *Molecules* **2023**, *28*, 4053. https://doi.org/10.3390/molecules28104053

Academic Editor: Agnieszka Ewa Wiącek

Received: 5 April 2023
Revised: 9 May 2023
Accepted: 10 May 2023
Published: 12 May 2023

1. Introduction

There is an increasing incidence of pests in crops. This is aggravated by globalization and climate change [1]. The extensive use of phytosanitaries selects resistant pests, posing a threat to global food security [2]. Therefore, there is an increasing need for new tools that can be useful in different aspects of human life, such as volatile organic compounds (VOCs) [3–5].

The denomination of volatile organic compounds refers to organic compounds whose vapor pressure is at least 0.01 kPa at 20 °C [6]. Fungal VOCs are receiving increasing attention due to their potential in a wide variety of applications in the biotechnological sector, mainly in the field of agriculture, industry, and medicine [7,8]. Fungal VOCs can be used as biopesticides or as activators of plant defense [9,10]. They are also used in the perfume and food industries [8,11] and are a potential alternative to fossil fuels [12,13].

The origin of VOCs may determine the presence and amount of particular fungal volatiles available. This may, for instance, restrict their mass production. Therefore, finding new sources of VOC production could greatly facilitate their application in new fields. The production of fungal VOCs may be affected by growth medium, moisture, temperature, fungal growth, and growth modulators [8,14,15]. Here, we report the use of chitosan to modify VOC production by *Pochonia chlamydosporia* in composition and abundance.

Pochonia chlamydosporia (=*Metacordyceps chlamydosporia*) (Goddard) Zare and Gams 2001 (*Pc*) is a nematophagous fungus widely used as a biocontrol agent due to its ability to parasitize nematode eggs [16,17] and to induce systemic resistance and plant growth [18,19]. *Pc* is distributed worldwide and has a tritrophic (saprotrophic, endophytic, and nematophagous) lifestyle [20].

Chitosan is a linear biopolymer derived from the partial deacetylation of chitin consisting of β-(1→4)-*N*-acetyl-D-glucosamine and β-(1→4)-D-glucosamine [21]. Chitosan is a natural fungicide with antimicrobial activity [22] and an elicitor of plant defenses [23,24]. However, chitosan can promote the growth and sporulation of resistant fungi, such as *Pc*, or other fungal parasites of invertebrates, such as entomopathogenic fungi [25]. *Pc* can tolerate high chitosan doses and can degrade it and use it as a nutrient source [26]. *Pc* has enzymatic machinery, including a large number of chitosanases, which are induced during nematode egg parasitism [27,28]. Chitosan induces the chitosanolytic activity of *Pc* by modifying its gene expression [29].

Several studies have analyzed the ability of chitosan to induce VOCs in plants [30–32]. The production of these VOCs can be modulated with chitosan concentration and time of exposure [33,34]. The effect of chitosan on *Pc* at the molecular level has been widely studied [35]. However, to the best of our knowledge, no studies have been conducted on the effect of chitosan on VOC production by fungi.

The goal of this work was to determine the effect of chitosan on the production of VOCs by *Pochonia chlamydosporia* isolate 123. We investigated the influence of fungal growth and time of exposure to chitosan on VOC production. We focused on identifying which compounds were produced *de novo*, inhibited, or with a change in production with chitosan. Our hypothesis is that chitosan could be used as a modulator of the production of VOCs by *Pc*, leading to the discovery of new molecules that could have potential applications across multiple fields.

2. Results

We investigated whether the addition of chitosan to the culture medium, the age of the fungus culture, the time of exposure to chitosan, and the type of culture medium may influence the patterns of emission of volatile organic compounds by *P. chlamydosporia*.

2.1. Fungal VOC Profiles—Culture in Solid Medium (Rice)

P. chlamydosporia grown in Erlenmeyer flasks with rice yielded a total of 25 different VOCs in the three growth stages studied (15, 25, and 35 days after inoculation (DAI)) (Figures 1 and S2). Based on tentative identification, the most abundant compounds detected for *Pc* grown on rice were 1,3-dimethoxybenzene, (6*Z*)-7,11-dimethyl-3-methylene-1,6,10-dodecatriene, and oct-1-en-3-ol (Tables 1 and 2). Among the VOCs detected, 3-methylbutanoic acid and methyl 2,4-dimethylhexanoate were only found in the samples that were treated with chitosan, and 3-hydroxybutan-2-one, octa-1,3-diene, and five other VOCs were found to have increased in abundance at at least one of the different culture ages. In addition, (6*Z*)-7,11-Dimethyl-3-methylene-1,6,10-dodecatriene, (4*S*)-1-Methyl-4-(6-methylhepta-1,5-dien-2-yl)cyclohexene, and seven further VOCs showed a decrease at different fungal ages (Figure S1 and Table S1).

The PCA showed that samples were mainly grouped by fungal age, while for the 15-day old culture subgroup, there were also two associations, one for samples with chitosan and one for controls (Figure 1B). After 25 and 35 days of culture, controls and treatments were not well differentiated. Thus, the main differences in the grouping were mainly due to the culture age, although chitosan treatment seemed to also have an impact at each culture age.

Table 1. Mean peak heights and standard deviations of VOCs detected in rice control samples of *P. chlamydosporia* after different culture times (mean (SD)). Abbreviations: VOC = volatile organic compound; DAI = days after inoculation.

	Control Buffer Solution		
Major VOCs	**15 DAI**	**25 DAI**	**35 DAI**
3-Hydroxybutan-2-one	66,402.67 (52,092.53)	125,554.33 (27,721.47)	0 (0)
Octa-1,3-diene	56,919.33 (41,951.78)	55,423.67 (53,923.95)	16,699 (5750.02)
Butane-2,3-diol	6820.33 (11,813.16)	0 (0)	0 (0)
Methoxybenzene	69,604 (29,288.09)	91,969.33 (50,952.5)	39,812.67 (50,272.28)
Oct-1-en-3-ol	649,414 (228,962.08)	0 (0)	0 (0)
Octan-3-one	0 (0)	399,307.33 (46,213.61)	0 (0)
Octan-3-ol	119,415 (42,769.72)	114,748 (41,654.88)	0 (0)
1,3-Dimethoxybenzene	30,228,196.67 (4,503,694.87)	23,852,155.67 (3,776,948.08)	24,519,265 (853,358.59)
Methyl 2-phenylacetate	21,108.33 (36,560.71)	26,653 (8728.81)	0 (0)
3-Methoxyphenol	57,144.67 (13,644.68)	82,675.67 (46,591.62)	176,919.67 (282,216.43)
1,2,3-Trimethoxybenzene	0 (0)	0 (0)	79,005 (70,432.6)
3,4-Dimethoxyphenol	20,412 (6363.42)	16,339.67 (4189.58)	47,431 (28,622)
1,2,4-Trimethoxybenzene	0 (0)	0 (0)	69,126.33 (59,929.47)
(1*S*,5*S*,6*R*)-2,6-Dimethyl-6-(4-methylpent-3-enyl)bicyclo[3.1.1]hept-2-ene	91,357 (9722.48)	95,971 (51,607.35)	0 (0)
(6*Z*)-7,11-Dimethyl-3-methylene-1,6,10-dodecatriene	1,102,849.33 (104,001.08)	1,402,859 (620,066.31)	442,461.67 (217,469.06)
	Control Buffer Solution		
Minor VOCs	**15 DAI**	**25 DAI**	**35 DAI**
3-Methylbutanoic acid	0 (0)	0 (0)	0 (0)
Hept-2-enal	0 (0)	7376.33 (12,776.18)	0 (0)
1,2,7,7-Tetramethylbicyclo[2.2.1]hept-2-ene	16,335.33 (28,293.63)	20,317 (8170.16)	0 (0)
Methyl 2,4-dimethylhexanoate	0 (0)	0 (0)	0 (0)
1,3-Dimethoxy-2-methylbenzene	14,198.33 (7099.4)	0 (0)	0 (0)
(4*S*)-1-Methyl-4-(6-methylhepta-1,5-dien-2-yl)cyclohexene	20,789 (2941.92)	21,290 (11,113.46)	14,563 (2446.21)
1,4-Dichloro-2,5-dimethoxybenzene	0 (0)	0 (0)	22,518.67 (22,153.03)
2,4-Dimethylquinoline	10,420.33 (3401.84)	0 (0)	0 (0)
1,2,3,4-Tetramethoxybenzene	20,149.33 (6534.38)	0 (0)	0 (0)
2,7,7,10-Tetramethyl-3-oxatetracyclo[7.3.0.0^{2,4}.0^{6,8}]dodecane	0 (0)	21,819.67 (8513.99)	0 (0)

When comparing the VOCs occurring in samples with chitosan and controls at different ages of the fungus, 19 VOCs were found in the youngest growth stage of *P. chlamydosporia* (15 days). Two minor VOCs were uniquely expressed when chitosan was present in the sample (3-methylbutanoic acid and methyl 2,4-dimethylhexanoate). In addition, four compounds showed moderate evidence of being emitted in higher abundance in the chitosan-treated samples than in the controls: 3-hydroxybutan-2-one (BF 10.08), octa-1,3-diene (BF 4.08), butane-2,3-diol (BF 3.79), and 1,2,3,4-tetramethoxybenzene (BF 6.32). On the other hand, 1,3-dimethoxybenzene (BF 0.31), 3,4-dimethoxyphenol (BF 0.13), and 1,3-dimethoxy-2-methylbenzene (BF 0.18) decreased their emissions, thus having higher abundance in the controls.

Table 2. Mean peak heights and standard deviations of VOCs detected in chitosan-treated rice samples of *P. chlamydosporia* after different culture times (mean (SD)). Abbreviations: VOC = volatile organic compound; DAI = days after inoculation.

	Chitosan Solution		
Major VOCs	**15 DAI**	**25 DAI**	**35 DAI**
3-Hydroxybutan-2-one	140,203.67 (53,930.99)	388,898.33 (323,491.84)	0 (0)
Octa-1,3-diene	89,883.33 (40,449.43)	69,747 (19,223.05)	40,255.33 (22,192.96)
Butane-2,3-diol	38,923.67 (47,948.61)	0 (0)	0 (0)
Methoxybenzene	51,741.33 (13,805.67)	152,280.33 (17,822.66)	3329 (5766)
Oct-1-en-3-ol	399,266 (405,852.5)	0 (0)	0 (0)
Octan-3-one	0 (0)	277,246.67 (54,831.37)	0 (0)
Octan-3-ol	130,339.33 (39,439.22)	46,439.33 (15,604.4)	0 (0)
1,3-Dimethoxybenzene	23,396,956.33 (2,794,659)	27,827,414 (1,515,361.21)	8,751,248.67 (11,680,587.39)
Methyl 2-phenylacetate	48,876.67 (3305.18)	31,430.67 (5845.67)	0 (0)
3-Methoxyphenol	73,917 (28,675.39)	191,472.33 (20,750.74)	0 (0)
1,2,3-Trimethoxybenzene	0 (0)	0 (0)	31,746.67 (54,986.84)
3,4-Dimethoxyphenol	12,502 (3616.14)	25,793 (1228)	5007.67 (8673.53)
1,2,4-Trimethoxybenzene	0 (0)	0 (0)	25,044 (43,377.48)
(1*S*,5*S*,6*R*)-2,6-Dimethyl-6-(4-methylpent-3-enyl)bicyclo[3.1.1]hept-2-ene	84,528.33 (16,562.62)	44,375 (13,065.53)	0 (0)
(6*Z*)-7,11-Dimethyl-3-methylene-1,6,10-dodecatriene	922,376.67 (40,678.69)	1,068,108.67 (492,573.01)	176,422.67 (118,053.64)
	Chitosan Solution		
Minor VOCs	**15 DAI**	**25 DAI**	**35 DAI**
3-Methylbutanoic acid	14,420 (2770.17)	0 (0)	0 (0)
Hept-2-enal	0 (0)	16,133 (5651.12)	0 (0)
1,2,7,7-Tetramethylbicyclo[2.2.1]hept-2-ene	28,434.67 (10,528.13)	19,826.33 (3413.38)	0 (0)
Methyl 2,4-dimethylhexanoate	23,886.67 (1965.31)	0 (0)	0 (0)
1,3-Dimethoxy-2-methylbenzene	8232.33 (4957.03)	0 (0)	0 (0)
(4*S*)-1-Methyl-4-(6-methylhepta-1,5-dien-2-yl)cyclohexene	20,508 (3187.7)	14,716.33 (6302.52)	3672.67 (6361.25)
1,4-Dichloro-2,5-dimethoxybenzene	0 (0)	0 (0)	7982.67 (13,826.38)
2,4-Dimethylquinoline	11,458 (4963.33)	0 (0)	0 (0)
1,2,3,4-Tetramethoxybenzene	32,050 (11,831.09)	0 (0)	0 (0)
2,7,7,10-Tetramethyl-3-oxatetracyclo[7.3.0.0^{2,4}.0^{6,8}]dodecane	0 (0)	18,851 (22,189.55)	0 (0)

In the 25-day culture samples (second growth stage), 15 VOCs were found in total and shared in both conditions. However, four of them (3-hydroxybutan-2-one, methoxybenzene, 3-methoxyphenol, and 3,4-dimethoxyphenol) were more expressed in the samples with chitosan, and four other VOCs less so (octan-3-one, octan-3-ol, (1*S*,5*S*,6*R*)-2,6-Dimethyl-6-(4-methylpent-3-enyl)bicyclo[3.1.1]hept-2-ene, and (4*S*)-1-Methyl-4-(6-methylhepta-1,5-dien-2-yl)cyclohexene). 3-Hydroxybutan-2-one increased in abundance from 15 to 25 days of culture, but the difference between controls and chitosan-treated samples was greater, from 10.08 BF after 15 days to 41.33 BF after 25 days. On the other hand, 3,4-dimethoxyphenol went from having higher values in controls after 15 days of culture to having higher values in chitosan in a higher growth stage (25 days of culture).

Finally, in the oldest growth stage (35 days), the diversity of compounds was reduced to 10 VOCs, with 1 of which being only emitted in the absence of chitosan. The presence of chitosan resulted in an increase in only one compound, octa-1,3-diene, with a Bayes Factor of 14.22. On the other hand, five compounds showed a decrease in abundance in the presence of chitosan.

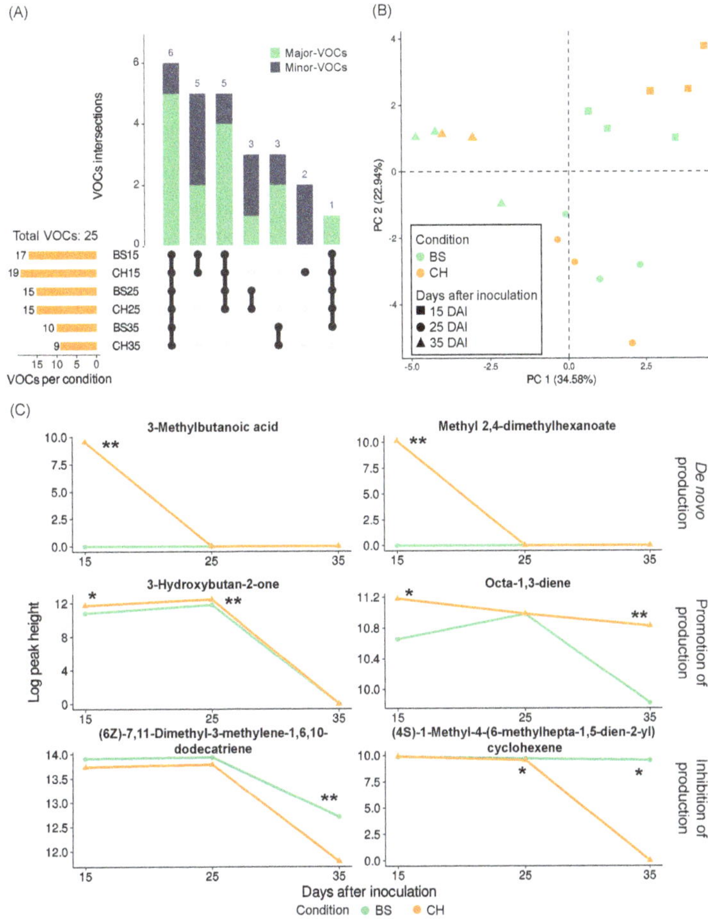

Figure 1. Effect of chitosan on VOC production by *P. chlamydosporia* on solid medium (rice grains). (**A**) UpSet plot of the intersection of VOCs found under culture conditions of *P. chlamydosporia* in rice. (**B**) PCA of the VOCs of *P. chlamydosporia* in rice. (**C**) Line plots with the median values of the log-transformed data (peak heights of VOCs + 1). Abbreviations: VOC = volatile organic compound; CH = chitosan treatment; BS = control buffer solution; * = Bayes Factor 3–10 or $\frac{1}{10} - \frac{1}{3}$; ** = Bayes Factor > 10 or $< \frac{1}{10}$.

VOC diversity rapidly decreased as *P. chlamydosporia* aged, going from 19 VOCs after 15 days of growth to 10 VOCs after 35 days. Seven compounds were only emitted in the youngest stage of the fungus, two of which only when treated with chitosan. After 25 days, 15 different compounds were emitted, 2 of which were only emitted in this growth stage. Finally, after 35 days, six compounds from the previous stages and four new ones were released.

Of these compounds, octa-1,3-diene, methyl 2-phenylacetate, (4*S*)-1-Methyl-4-(6-methylhepta-1,5-dien-2-yl)cyclohexene, and 1,2,7,7-Tetramethylbicyclo[2.2.1]hept-2-ene showed a tendency to decrease with culture age. Furthermore, depending on whether the sample had been treated with chitosan or not, we found that the differences in abundance among the times markedly varied, as was the case of (4*S*)-1-Methyl-4-(6-methylhepta-1,5-dien-2-yl)cyclohexene, which had a less drastic decrease in controls than when treated with chitosan.

2.2. Fungal VOC Profiles—Culture in Liquid Medium (Modified Czapek–Dox)

We also evaluated the effect of chitosan on VOC production by *Pc* in liquid medium (Czapek–Dox) (Figure S4). We detected a total of 19 VOCs, 2 of which (oct-1-en-3-ol and tetradec-1-ene) were only detected in the samples with chitosan (Figure 2). In addition, five VOCs were exclusively detected in the samples treated with chitosan at a specific exposure time (Tables 3 and 4).

In the PCA, samples seemed to be more spread out than samples grown on rice (Figure 2B). However, we continued to see a tendency to cluster primarily by exposure time or age of culture, and within those big clusters, we found subgroups depending on whether the sample had been treated with chitosan or not.

Table 3. Mean peak heights and standard deviations of VOCs detected in control *P. chlamydosporia* samples grown in modified Czapek–Dox for different exposure times (mean (SD)). Abbreviations: VOC = volatile organic compound.

	Control Buffer Solution		
Major VOCs	**24 h**	**48 h**	**72 h**
Oct-1-en-3-ol	0 (0)	0 (0)	0 (0)
8-Methylheptadecane	64,212.67 (27,072.87)	66,092.33 (31,020.32)	33,097 (8613.77)
2,6,11-Trimethyldodecane	38,103.67 (12,330.32)	34,817.33 (5539.97)	18,099.33 (4608.43)
Hexacosane	23,811.67 (7685.3)	64,745.33 (14,575.82)	24,891.33 (3341.71)
2,4-ditert-butylphenol	116,881 (39,599.25)	78,442.67 (730.06)	59,575.33 (3437.02)
2,6,10,15-Tetramethylheptadecane	0 (0)	42372.67 (16256.22)	0 (0)
Tetradec-1-ene	0 (0)	0 (0)	0 (0)
Hexadecan-1-ol	361,050.67 (110,953.32)	224,019.67 (135,154.82)	6539.67 (11,327.03)
Octadec-9-en-1-ol	171,950 (42,297.08)	90,861.33 (40,895.74)	4027.67 (6976.12)
	Control Buffer Solution		
Minor VOCs	**24 h**	**48 h**	**72 h**
2-Methylpyrazine	0 (0)	0 (0)	5531 (9579.97)
3,7-Dimethyldecane	15,702.67 (13,631.84)	0 (0)	0 (0)
5-Methylundecane	13,636 (11,954.55)	0 (0)	0 (0)
Naphthalene	0 (0)	10,350 (17,926.73)	0 (0)
1,3-ditert-butylbenzene	16,117.67 (14,093.38)	0 (0)	0 (0)
3,4,6-Trimethylundecane	12,117.67 (10,601.65)	5167 (8949.51)	0 (0)
2,6,10-Trimethyldodecane	4661 (8073.09)	9740 (8511.64)	0 (0)
3-Ethyl-5-(2-ethylbutyl)octadecane	9472.67 (8294.68)	0 (0)	0 (0)
Nonadecane	9776.67 (8620.81)	30,276.33 (6070.58)	0 (0)
2,6,11,15-Tetramethylhexadecane	0 (0)	13,774 (12,290.88)	0 (0)

Table 4. Mean peak heights and standard deviations of VOCs detected in *P. chlamydosporia* samples grown in modified Czapek–Dox for different times of exposure to chitosan (hours) (mean (SD)). Abbreviations: VOC = volatile organic compound.

	Chitosan Solution		
Major VOCs	**24 h**	**48 h**	**72 h**
Oct-1-en-3-ol	218,516.67 (130,830.14)	132,448.33 (113,897.66)	55,408.67 (24,658.18)
8-Methylheptadecane	87,610.33 (33,754.89)	69,603.33 (37,892.55)	74,636.33 (48,102.79)
2,6,11-Trimethyldodecane	59,529 (24,339.84)	35,308.67 (9317.39)	47,015.67 (30,486.18)
Hexacosane	51,634.33 (21,668.02)	58,443.33 (20,214.85)	59,734.67 (28,128.13)
2,4-ditert-butylphenol	131,087.33 (33,937.52)	69,674.33 (8951)	75,155.67 (33,691.57)
2,6,10,15-Tetramethylheptadecane	27,279.33 (26,850.31)	40,188.67 (5461)	0 (0)
Tetradec-1-ene	34,551.33 (30,503.11)	0 (0)	0 (0)
Hexadecan-1-ol	1,036,705 (126,868.82)	191,852.33 (29,836.23)	17,328.67 (15,497.53)
Octadec-9-en-1-ol	626,179.33 (96,866.49)	118,469 (33,443.61)	21,290.33 (6700.12)

Table 4. *Cont.*

	Chitosan Solution		
Minor VOCs	24 h	48 h	72 h
2-Methylpyrazine	0 (0)	0 (0)	5023 (8700.09)
3,7-Dimethyldecane	11179.67 (9686.23)	0 (0)	0 (0)
5-Methylundecane	10,221 (8929.62)	0 (0)	0 (0)
Naphthalene	0 (0)	6897.67 (5980.28)	0 (0)
1,3-ditert-butylbenzene	25,663 (9942.73)	0 (0)	0 (0)
3,4,6-Trimethylundecane	16,235.67 (14,948.86)	10,693.33 (9270.68)	13,517.67 (12,817.11)
2,6,10-Trimethyldodecane	11,949 (11,228.94)	0 (0)	17,214.67 (16,102.73)
3-Ethyl-5-(2-ethylbutyl)octadecane	14,332.67 (13,385.78)	0 (0)	11,371.33 (10,643.96)
Nonadecane	27,725.33 (11,623.31)	27,063 (4893.47)	24,760 (21,735.48)
2,6,11,15-Tetramethylhexadecane	0 (0)	16,144 (949.76)	0 (0)

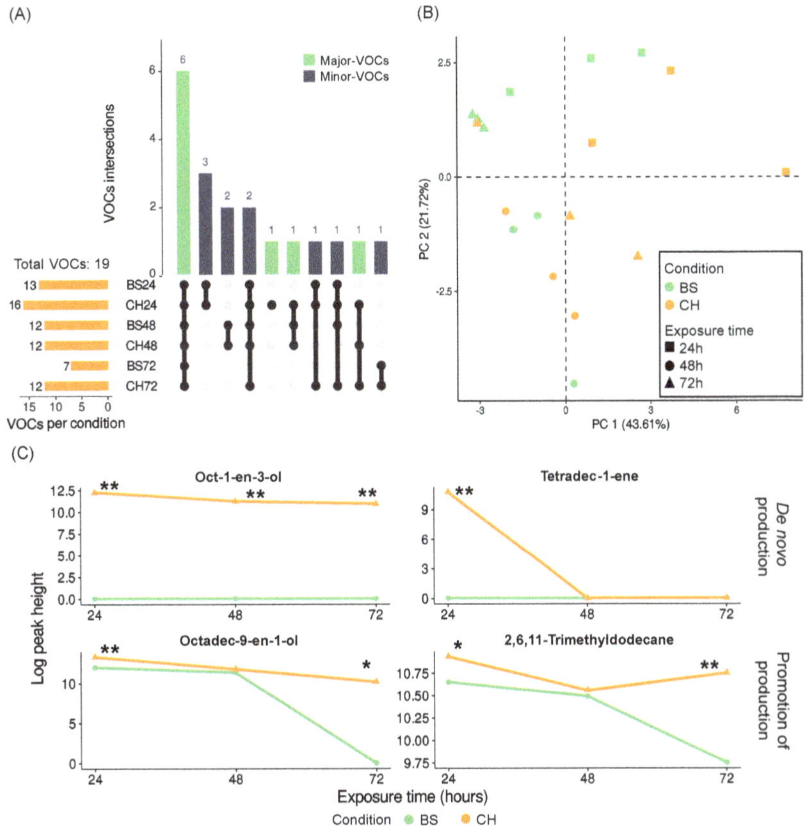

Figure 2. Effect of chitosan on VOC production by *P. chlamydosporia* in liquid medium (Czapek–Dox). (**A**) UpSet plot of the intersection of VOCs found under culture conditions of *P. chlamydosporia* in modified Czapek–Dox. (**B**). PCA of the VOCs from *P. chlamydosporia* in modified Czapek–Dox. (**C**) Line plots with the median values of the log-transformed data (peak heights of VOCs + 1). Abbreviations: VOC = volatile organic compound; CH = chitosan treatment; BS = control buffer solution; * = Bayes Factor 3–10 or $\frac{1}{10} - \frac{1}{3}$; ** = Bayes Factor > 10 or $< \frac{1}{10}$.

When independently evaluating the three exposure times (24, 48, and 72 h in contact with chitosan), we saw that oct-1-en-3-ol was found all three times only when treated with chitosan. After five days of culture and 24 h with chitosan, we detected 16 different compounds (Figure S3 and Table S2), 3 of which were only present when treated with chitosan (oct-1-en-3-ol, 2,6,10,15-tetramethylheptadecane, and tetradec-1-ene) and 5 of which had higher abundance. Hexacosane (BF 21.89), hexadecan-1-ol (BF 150.67), and octadec-9-en-1-ol (BF 683.09) showed very high values of Bayes Factor. After two days with chitosan, controls and treatments showed fewer differences. We found 13 VOCs, of which oct-1-en-3-ol was only found in the chitosan samples, and 2,6,10-trimethyldodecane, in the controls; there were no differences in abundance for the rest of the VOCs. Finally, three days of the samples being in contact with chitosan was when the largest differences were found. Twelve VOCs were detected, five of which came only from chitosan-treated samples and four of which were produced in higher abundance. Five of the new or induced VOCs found after 72 h of exposure to chitosan were also detected at higher levels after 24 h of chitosan exposure (oct-1-en-3-ol, 8-methylheptadecane, 2,6,11-trimethyldodecane, hexacosane, and octadec-9-en-1-ol).

The comparison of VOCs among different exposure times revealed an increase in VOC diversity with longer exposure time or more culture days. The greatest differences were found between the treatment and the control after 3 days of exposure to chitosan. The abundance of compounds such as octadec-9-en-1-ol and hexadecan-1-ol tended to decrease when the time of exposure to chitosan was longer; however, the controls also decreased, though not in a such drastic way. There were other cases, such as 2,6,11-trimethyldodecane, nonadecane, 8-methylheptadecane, or 3,4,6-trimethylundecane, whose abundance decreased with time but remained almost constant when treated with chitosan.

3. Discussion

Volatile organic compounds (VOCs) are essential to many biological processes and have a wide range of uses from an anthropogenic perspective. Given the variety of industrial applications, many studies actually focus on VOCs from an economic perspective [8]. VOCs have good information-carrying properties, because they can disperse through the air and soil [36–38]. Fungi are capable of producing numerous VOCs, making their volatilomes a rich source of secondary metabolites [39]. In addition to the fungal species, other elements, such as the substrate, temperature, the age of the fungus, and the volatile extraction method itself, affect the composition of the volatilome [8]. Multiple studies have emphasized how chitosan acts as a defense inhibitor in plants and modifies secondary metabolism by triggering the synthesis of new VOCs [31,32,40]. The fungus *P. chlamydosporia* has exceptional enzymatic machinery for degrading chitosan and using it as a nutrient source [26,27,29]. To determine the changes in *Pc* volatilome when subjected to chitosan stress, the VOC profile was analyzed using HS-SPME/GC-MS in solid medium in different ages of the fungus and in liquid medium over different times of exposure to chitosan.

The findings indicate that chitosan could be a modulator of *Pc* secondary metabolism not only by triggering the production of new VOCs but also by increasing or decreasing the abundance of others. This experiment provides new insights into the chitosan modulatory effect that has already been observed in other organisms [32,33]. The compounds detected after applying chitosan may be due not only to the fungus-synthesized VOCs but also to substrate degradation products (either from rice, Czapek–Dox, or chitosan itself) [8]. *Pc* is a chitosan-resistant fungus [26] whose gene expression is modified when in contact with chitosan [29], inducing chitinases and chitosanases during nematode egg parasitism [27]. Together with other fungi that are resistant to chitosan, it is capable of producing valuable bioproducts by degrading chitosan [41]. This gene expression modification could be related to volatile production, suggesting the modulatory effect of chitosan on the volatilome of *Pc*.

In addition, the experiment with rice as a culture medium suggests that the volatilome evolves with the age of the fungus, which is in accordance with previous studies on other fungal species [10,15,42]. Pan et al. [42] analyzed the VOC profile of *Alternaria brassicae* and reported a reduction in the diversity of VOCs as the fungus aged, which is the same trend we found in *Pc*, where the 35-day-old culture had less VOC diversity than the 15-day-old culture. However, as Zhang et al. [15] noted, the early growth phases of *Neurospora dictyophora* (3–10 days of culture) reported fewer distinct VOCs than 15-day cultures, indicating that it may also be related to the fungal growth rate. The liquid-medium experiment suggests that the time of exposure to chitosan also has an effect on VOC production. After 72 h of exposure to chitosan, the volatilome was more similar to that after 48 h with or without chitosan than to that after 72 h without chitosan. Therefore, chitosan could slow down the age effect on the volatilome.

Pc produces compounds derived from benzene, hydrocarbons, ketones, and alcohol groups. VOCs found in *Pc* by Lozano-Soria et al. [10] were similar to but did not exactly match those determined in this study, possibly due to the different experimental conditions that affect VOC profile composition. In modified Czapek–Dox broth culture, oct-1-en-3-ol, which is a typical compound found in the odor of many fungi [43], was only detected in the presence of chitosan. Oct-1-en-3-ol has been reported to be a repellent of *Cosmopolites sordidus* [10] and has a role in fungal spore germination and dispersal [44]. Tetradec-1-ene was also only found in chitosan-treated cultures after 24 h of exposure, and a weak effect of attraction to the flour beetle *Tribolium confusum* has been reported [45]. 3-Methylbutanoic acid was registered with chitosan treatment after 10 days of growth in rice and has a strong association with the inhibition of germination and mycelium growth in several fungi [46,47].

In conclusion, our findings suggest that the effect of chitosan, and the relationship with culture age and chitosan exposure time alter the VOCs produced by *Pochonia chlamydosporia*. Chitosan modified the production of VOCs in *P. chlamydosporia* and triggered the synthesis of others. Future research should be performed to understand how other chitosan-resistant fungi respond to chitosan as well as to investigate and exploit the bioproducts produced.

4. Materials and Methods

4.1. Fungi and Source of Chitosan

Pochonia chlamydosporia var. *chlamydosporia* (=*Metacordyceps chlamydosporia* var. *chlamydosporium*) isolate 123 (*Pc*) (ATCC No. MYA-4875; CECT No. 20929) from the Phytopathology Laboratory (University of Alicante) fungal collection was used for experiments. *Pc* was isolated from nematode (*Heterodera avenae*) infected eggs.

Chitosan T8 with a molecular weight of 70 kDa and a deacetylation degree of 90.1% was provided by Marine BioProducts GmbH (Bremerhaven, Germany) and produced using crustacean exoskeletons. Chitosan solution was prepared as described by Palma-Guerrero et al. [25]. Briefly, chitosan was dissolved in 0.25 mol $\cdot$ L^{-1} HCl, and the pH was adjusted to 5.6 with NaOH under constant stirring. The resulting solution was dialyzed for 48 h for salt removal and autoclaved (120 °C for 20 min).

Chitosan solutions (CH) for the experiments (1 and 5 $mg \cdot mL^{-1}$) were prepared using autoclaved distilled water (120 °C, 20 min). A control buffer solution (BS) was also prepared in the same manner, without addition of chitosan.

4.2. Experimental Design

To obtain a wide view of the effect of chitosan on the synthesis of volatile organic compounds (VOCs) from *Pc* and to analyze the whole volatilome, two experiments were conducted.

In the first experiment (solid medium), *Pc* conidia (a suspension of 10^6 conidia in 5 mL of autoclaved distilled water) were inoculated into 100 mL Erlenmeyer flasks, each containing 25 g of hydrated rice as solid culture medium, previously autoclaved (120 °C, 20 min). Flasks were covered with cotton and aluminum foil and incubated at 23–25 °C with natural lightning for (i) 10, (ii) 20, or (iii) 30 days. After these periods, 5 mL of chitosan (1 $mg \cdot mL^{-1}$ final concentration) was added to each flask and mixed well. Flasks were incubated for 5 more days. Controls were prepared by adding 5 mL of the buffer solution instead of chitosan. Moreover, flasks with rice only, rice plus chitosan solution, and rice plus buffer were also prepared. The experiment was replicated 3 times.

In the second experiment (liquid cultures), *Pc* was cultured in 20 mL screw-capped vials with 5 mL of modified Czapek–Dox broth medium prepared as described by Olivares-Bernabeu and López-Llorca [48]. *Pc* conidia were inoculated at a final concentration of 10^6 conidia·mL^{-1}. Vials were covered with cotton and aluminum foil and incubated at 25 °C under shaking at 120 rpm. After 5 days, modified Czapek–Dox liquid medium was removed to only leave *Pc* mycelia in the vials. A volume of 5 mL of chitosan (5 mg·mL^{-1}) was then added to vials with *Pc* mycelia. Vials were further incubated sealed (to avoid VOC losses) for 24, 48, or 72 h at 25 °C. Afterwards, vials were taken for GC-MS analysis. Controls had 5 mL of buffer solution instead of chitosan. The experiment was replicated 3 times.

4.3. SPME and GC-MS Analysis

Pc VOCs were analyzed using headspace solid-phase microextraction (HS-SPME) and gas chromatography–mass spectrometry (GC MS). The GC-MS analysis was conducted using Agilent 7890B GC System equipped with a GERSTEL multipurpose autosampler (MPS).

In the first experiment, which involved solid medium, 5 g of homogenized *Pc* culture was added to 20 mL GERSTEL vials for VOC analysis. The vials were then sealed and left to equilibrate for 15 min. In the second experiment, *Pc* was already grown in the vials containing the liquid medium with different times of exposure to chitosan.

The VOCs were absorbed for 15 min using SPME divinylbenzene/carbon wide range/polydimethylsiloxane (DVB/C-WR/PDMS) fiber (10 mm in length, 80 µm in phase thickness, and 23 in needle gauge) at 60 °C. The desorption of the VOCs from the fiber took 240 s (4 min) in the GC injector, which was operated in splitless mode at a temperature of 250 °C. After the removal of the SPME fiber, chromatography continued for 53 min. Separation was performed with an Agilent J&W DB-624 capillary GC column (30 m × 0.25 mm × 1.4 µm). The column oven was initially set at 40 °C and ramped up at a rate of 5 °C·min^{-1} to reach the maximum temperature of 230 °C, which was held for 10 min. The carrier gas, helium, was maintained at a constant flow rate of 1 mL·min^{-1}. Mass spectra were obtained using electron impact ionization with a mass range of 25 to 400 atomic mass units (a.m.u.), a scan speed of 1.562 u·s^{-1}, and a frequency of 3.8 scans·s^{-1}. The ionization source temperature was 230 °C and worked at 70 eV. The quadrupole temperature was maintained at 150 °C.

4.4. Tentative Identification of VOCs

The tentative identification of VOCs was performed using the NIST11.L Mass Spectral database. VOCs whose match with the mass spectra library was less than 50% were discarded, while the rest were evaluated using Agilent MassHunter Qualitative Analysis Navigator B.08.00 software to overlay chromatograms. Those compounds under each condition of the experiment that were only found in one of the three replicates were eliminated. VOCs detected two or more times under the same condition were selected.

Moreover, to identify VOCs exclusively produced by *Pc*, we removed artifacts such as silicones and column bleeding from our sample analyses. Additionally, volatiles from samples with rice only with either chitosan or buffer were discarded from the samples, assuming that they had not been produced by *Pc*.

Data were directly calculated from the peak heights of total ion current (TIC) profiles. The remaining VOCs were classified into two groups according to their abundance: major VOCs (peak height ≥ 50,000 ppm) and minor VOCs (peak height < 50,000 ppm).

4.5. Statistical Analysis

As part of the exploratory analysis of the data, Principal Component Analysis (PCA) was performed using the "prcomp" function from the R stats package (version 4.2.3). The data were standardized before PCA.

Regarding statistical modeling, we propose a Generalized Linear Model (GLM) [49] for each volatile and time of study (DAI for the rice experiment and exposure time for the

Czapek experiment). The variable of interest or dependent variable is the concentration, and the explanatory or independent variable is the control/treatment factor.

$$Y_{ij} \sim Gamma(\mu_{ij}, \phi), i = 1, \ldots, n; j = 1, 2;$$
$$log(\mu_{ij}) = \beta_0 + \alpha_{ij} d,$$

where i is the number of observations; j is the factor's level; Y_{ij} is the concentration and follows a Gamma distribution; μ_{ij} is the mean of the distribution, which is linked to the linear predictor by the log-link function; and the linear predictor consists of an intercept β_0 and a fixed factor $\alpha_{ij} d$ (control/treatment).

Parameter and hyperparameter estimation was performed in the context of Bayesian inference [50,51]. For that, computational approaches and numerical approximations are needed to estimate the posterior distribution of parameters. In our study, we applied Integrated Nested Laplace Approximation (INLA) using R-INLA (version 23.03.19) to approximate the posterior marginal distributions of the model's parameters and hyperparameters [52]. Finally, vaguely informative prior distributions were used for all the parameters.

Once we sampled the posterior distributions of μ_{ij}, we computed the differences among them in order to analyze the effect of chitosan. Then, Bayesian hypothesis contrast was performed on the difference in these posterior distributions (θ). For that, we calculated the Bayes Factor (BF) using the coefficients of the probability of being different from 0 ($\alpha_0 = P(\theta < 0|data)$ and $\alpha_1 = P(\theta > 0|data)$). The BF quantifies the evidence of H_1 with respect to H_0, where large values of BF indicate greater support for H_1, and *vice versa* [53,54]. Jeffreys [53] established one of the best-known criteria for the interpretation of the Bayes Factor [55]. A Bayes Factor greater than 1 supports the alternative hypothesis, while a value less than 1 favors the null hypothesis. Bayes Factors between 1/3 and 3 are generally considered to provide barely worth mentioning evidence. Values between 3 and 10 (or between 1/10 and 1/3) indicate substantial evidence, those between 10 and 30 (or between 1/30 and 1/10) indicate strong evidence, and those above 30 (or below 1/30) indicate very strong evidence for one hypothesis over the other.

5. Conclusions

Our findings suggest that chitosan can modulate the production of volatile organic compounds (VOCs) by *P. chlamydosporia* and perhaps similar biocontrol fungi. This could be a novel approach to generate VOCs with various applications in agrobiotechnology and medical fields.

6. Patents

The results of this paper have been filed for a Spanish Patent (P202230103), with Jorge Mestre-Tomás, David Esgueva-Vilà, Federico Lopez-Moya, and Luis V. Lopez-Llorca as inventors.

Supplementary Materials: The following supporting information can be downloaded at: https://www.mdpi.com/article/10.3390/molecules28104053/s1, Table S1: Bayes Factor between treatment and control of VOCs from *P. chlamydosporia* in rice cultures on different DAI, Table S2: Bayes Factor between treatment and control of VOCs from *P. chlamydosporia* in Czapek–Dox broth cultures after different times of exposure to chitosan, Figure S1: Line plots with the median values and standard deviations of the peak heights of VOCs produced by *P. chlamydosporia* in rice cultures, Figure S2: GC-MS chromatogram of headspace volatile compounds produced by *P. chlamydosporia* 15 days after inoculation with control buffer solution (BS) and with chitosan solution (CH), Figure S3: Line plots with the median values and standard deviations of the peak heights of VOCs produced by *P. chlamydosporia* in modified Czapek–Dox broth cultures, Figure S4: GC-MS chromatogram of headspace volatile compounds produced by *P. chlamydosporia* cultured in modified Czapek–Dox broth for 5 days after 24 h with control buffer solution (BS) and 24 h of exposure to chitosan (CH).

Author Contributions: Conceptualization, J.M.-T., F.L.-M. and L.V.L.-L.; methodology, J.M.-T., D.E.-V., F.L.-M. and L.V.L.-L.; investigation and validation; J.M.-T. and D.E.-V.; data curation; J.M.-T. and D.E.-V.; formal analysis, J.M.-T. and A.F.-A.; resources, F.L.-M. and L.V.L.-L.; writing—original draft preparation, J.M.-T., D.E.-V. and A.F.-A.; writing—review and editing, J.M.-T., D.E.-V., A.F.-A., F.L.-M. and L.V.L.-L.; visualization, J.M.-T.; supervision, F.L.-M. and L.V.L.-L.; project administration, L.V.L.-L.; funding acquisition, F.L.-M. and L.V.L.-L. All authors have read and agreed to the published version of the manuscript.

Funding: This research was funded by PID2020-119734RB-I00, Project from the Spanish Ministry of Science and Innovation, and by European Project H2020 MUSA, grant number 727624.

Institutional Review Board Statement: Not applicable.

Informed Consent Statement: Not applicable.

Data Availability Statement: The data presented in this study are available in Supplementary Materials.

Acknowledgments: The authors would like to thank the technical support of P. Blasco from Research Technical Services (SSTTI) of University of Alicante.

Conflicts of Interest: The authors declare no conflict of interest. The funders had no role in the design of the study; in the collection, analyses, or interpretation of data; in the writing of the manuscript; or in the decision to publish the results.

Sample Availability: Not applicable.

Abbreviations

The following abbreviations are used in this manuscript:

Pc	*Pochonia chlamydosporia*
DAI	Days after inoculation
GC	Gas chromatography
HS-SPME	Headspace solid-phase microextraction
MS	Mass spectrometry
VOC	Volatile organic compound

References

1. Schneider, L.; Rebetez, M.; Rasmann, S. The effect of climate change on invasive crop pests across biomes. *Curr. Opin. Insect Sci.* **2022**, *50*, 100895. [CrossRef]
2. Hawkins, N.J.; Bass, C.; Dixon, A.; Neve, P. The evolutionary origins of pesticide resistance. *Biol. Rev.* **2019**, *94*, 135–155. [CrossRef] [PubMed]
3. Kanchiswamy, C.N.; Malnoy, M.; Maffei, M.E. Bioprospecting bacterial and fungal volatiles for sustainable agriculture. *Trends Plant Sci.* **2015**, *20*, 206–211. [CrossRef] [PubMed]
4. Brilli, F.; Loreto, F.; Baccelli, I. Exploiting plant volatile organic compounds (VOCs) in agriculture to improve sustainable defense strategies and productivity of crops. *Front. Plant Sci.* **2019**, *10*, 264. [CrossRef]
5. Maurya, A.K. Application of plant volatile organic compounds (VOCs) in agriculture. In *New Frontiers in Stress Management for Durable Agriculture*; Springer: Singapore, 2020; pp. 369–388.
6. Pagans, E.; Font, X.; Sánchez, A. Emission of volatile organic compounds from composting of different solid wastes: Abatement by biofiltration. *J. Hazard. Mater.* **2006**, *131*, 179–186. [CrossRef]
7. Kramer, R.; Abraham, W.R. Volatile sesquiterpenes from fungi: What are they good for? *Phytochem. Rev.* **2012**, *11*, 15–37. [CrossRef]
8. Inamdar, A.A.; Morath, S.; Bennett, J.W. Fungal volatile organic compounds: More than just a funky smell? *Annu. Rev. Microbiol.* **2020**, *74*, 101–116. [CrossRef] [PubMed]
9. Schalchli, H.; Tortella, G.; Rubilar, O.; Parra, L.; Hormazabal, E.; Quiroz, A. Fungal volatiles: An environmentally friendly tool to control pathogenic microorganisms in plants. *Crit. Rev. Biotechnol.* **2016**, *36*, 144–152. [CrossRef]
10. Lozano-Soria, A.; Picciotti, U.; Lopez-Moya, F.; Lopez-Cepero, J.; Porcelli, F.; Lopez-Llorca, L.V. Volatile organic compounds from entomopathogenic and nematophagous fungi, repel banana black weevil (*Cosmopolites sordidus*). *Insects* **2020**, *11*, 509. [CrossRef]
11. Splivallo, R.; Ottonello, S.; Mello, A.; Karlovsky, P. Truffle volatiles: From chemical ecology to aroma biosynthesis. *New Phytol.* **2011**, *189*, 688–699. [CrossRef]
12. Naik, B.S. Volatile hydrocarbons from endophytic fungi and their efficacy in fuel production and disease control. *Egypt. J. Biol. Pest Control* **2018**, *28*, 1–9. [CrossRef]

13. Achimón, F.; Brito, V.D.; Pizzolitto, R.P.; Zygadlo, J.A. Effect of carbon sources on the production of volatile organic compounds by *Fusarium verticillioides*. *J. Fungi* **2022**, *8*, 158. [CrossRef]
14. Fiedler, K.; Schütz, E.; Geh, S. Detection of microbial volatile organic compounds (MVOCs) produced by moulds on various materials. *Int. J. Hyg. Environ. Health* **2001**, *204*, 111–121. [CrossRef]
15. Zhang, H.; Xu, L.; Kong, X.; Ji, X.; Yang, S. Changes and bioactivities on volatile organic compounds of endophytic fungi *Neurospora dictyophora* 3ZF-02 in different ages. *J. Basic Microbiol.* **2023**, *63*, 104–115. [CrossRef] [PubMed]
16. Manzanilla-Lopez, R.H.; Esteves, I.; Finetti-Sialer, M.M.; Hirsch, P.R.; Ward, E.; Devonshire, J.; Hidalgo-Díaz, L. *Pochonia chlamydosporia*: Advances and challenges to improve its performance as a biological control agent of sedentary endo-parasitic nematodes. *J. Nematol.* **2013**, *45*, 1. [PubMed]
17. Forghani, F.; Hajihassani, A. Recent advances in the development of environmentally benign treatments to control root-knot nematodes. *Front. Plant Sci.* **2020**, *11*, 1125. [CrossRef] [PubMed]
18. Coutinho, R.R.; Pacheco, P.V.M.; Monteiro, T.S.A.; Balbino, H.M.; Moreira, B.C.; de Freitas, L.G. Root colonization and growth promotion of cover crops by *Pochonia chlamydosporia*. *Rhizosphere* **2021**, *20*, 100432. [CrossRef]
19. Escudero, N.; Lopez-Llorca, L.V. Effects on plant growth and root-knot nematode infection of an endophytic GFP transformant of the nematophagous fungus *Pochonia chlamydosporia*. *Symbiosis* **2012**, *57*, 33–42. [CrossRef]
20. Suarez-Fernandez, M.; Aragon-Perez, A.; Lopez-Llorca, L.V.; Lopez-Moya, F. Putative LysM effectors contribute to fungal lifestyle. *Int. J. Mol. Sci.* **2021**, *22*, 3147. [CrossRef]
21. Younes, I.; Rinaudo, M. Chitin and chitosan preparation from marine sources. Structure, properties and applications. *Mar. Drugs* **2015**, *13*, 1133–1174. [CrossRef]
22. Abd El-Hack, M.E.; El-Saadony, M.T.; Shafi, M.E.; Zabermawi, N.M.; Arif, M.; Batiha, G.E.; Khafaga, A.F.; Abd El-Hakim, Y.M.; Al-Sagheer, A.A. Antimicrobial and antioxidant properties of chitosan and its derivatives and their applications: A review. *Int. J. Biol. Macromol.* **2020**, *164*, 2726–2744. [CrossRef] [PubMed]
23. El-Tantawy, E. Behavior of tomato plants as affected by spraying with chitosan and aminofort as natural stimulator substances under application of soil organic amendments. *Pak. J. Biol. Sci. PJBS* **2009**, *12*, 1164–1173. [CrossRef] [PubMed]
24. Iriti, M.; Faoro, F. Abscisic acid is involved in chitosan-induced resistance to tobacco necrosis virus (TNV). *Plant Physiol. Biochem.* **2008**, *46*, 1106–1111. [CrossRef]
25. Palma-Guerrero, J.; Jansson, H.B.; Salinas, J.; Lopez-Llorca, L. Effect of chitosan on hyphal growth and spore germination of plant pathogenic and biocontrol fungi. *J. Appl. Microbiol.* **2008**, *104*, 541–553. [CrossRef] [PubMed]
26. Palma-Guerrero, J.; Gómez-Vidal, S.; Tikhonov, V.E.; Salinas, J.; Jansson, H.B.; Lopez-Llorca, L.V. Comparative analysis of extracellular proteins from *Pochonia chlamydosporia* grown with chitosan or chitin as main carbon and nitrogen sources. *Enzym. Microb. Technol.* **2010**, *46*, 568–574. [CrossRef]
27. Aranda-Martinez, A.; Lenfant, N.; Escudero, N.; Zavala-Gonzalez, E.A.; Henrissat, B.; Lopez-Llorca, L.V. CAZyme content of *Pochonia chlamydosporia* reflects that chitin and chitosan modification are involved in nematode parasitism. *Environ. Microbiol.* **2016**, *18*, 4200–4215. [CrossRef]
28. Escudero, N.; Lopez-Moya, F.; Ghahremani, Z.; Zavala-Gonzalez, E.A.; Alaguero-Cordovilla, A.; Ros-Ibañez, C.; Lacasa, A.; Sorribas, F.J.; Lopez-Llorca, L.V. Chitosan increases tomato root colonization by *Pochonia chlamydosporia* and their combination reduces root-knot nematode damage. *Front. Plant Sci.* **2017**, *8*, 1415. [CrossRef]
29. Suarez-Fernandez, M.; Sambles, C.; Lopez-Moya, F.; Nueda, M.J.; Studholme, D.J.; Lopez-Llorca, L.V. Chitosan modulates *Pochonia chlamydosporia* gene expression during nematode egg parasitism. *Environ. Microbiol.* **2021**, *23*, 4980–4997. [CrossRef]
30. He, P.Q.; Tian, L.; Chen, K.S.; Hao, L.H.; Li, G.Y. Induction of volatile organic compounds of *Lycopersicon esculentum* Mill. and its resistance to *Botrytis cinerea* Pers. by burdock oligosaccharide. *J. Integr. Plant Biol.* **2006**, *48*, 550–557. [CrossRef]
31. Yin, H.; Fretté, X.C.; Christensen, L.P.; Grevsen, K. Chitosan oligosaccharides promote the content of polyphenols in Greek oregano (*Origanum vulgare* ssp. *hirtum*). *J. Agric. Food Chem.* **2012**, *60*, 136–143. [CrossRef]
32. Badiali, C.; De Angelis, G.; Simonetti, G.; Brasili, E.; Tobaruela, E.d.C.; Purgatto, E.; Yin, H.; Valletta, A.; Pasqua, G. Chitosan oligosaccharides affect xanthone and VOC biosynthesis in *Hypericum perforatum* root cultures and enhance the antifungal activity of root extracts. *Plant Cell Rep.* **2018**, *37*, 1471–1484. [CrossRef] [PubMed]
33. Peiqing, H.; Xuezheng, L.; Jihong, S.; Xiaohang, H.; Kaoshan, C.; Guangyou, L. Induction of volatile organic compound in the leaves of *Lycopersicon esculentum* by chitosan oligomer. *High Technol. Lett.* **2005**, *11*, 95–100.
34. Zhang, P.; Chen, K. Age-dependent variations of volatile emissions and inhibitory activity toward *Botrytis cinerea* and *Fusarium oxysporum* in tomato leaves treated with chitosan oligosaccharide. *J. Plant Biol.* **2009**, *52*, 332–339. [CrossRef]
35. Lopez-Moya, F.; Suarez-Fernandez, M.; Lopez-Llorca, L.V. Molecular mechanisms of chitosan interactions with fungi and plants. *Int. J. Mol. Sci.* **2019**, *20*, 332. [CrossRef] [PubMed]
36. Asensio, D.; Peñuelas, J.; Filella, I.; Llusià, J. On-line screening of soil VOCs exchange responses to moisture, temperature and root presence. *Plant Soil* **2007**, *291*, 249–261. [CrossRef]
37. Morath, S.U.; Hung, R.; Bennett, J.W. Fungal volatile organic compounds: A review with emphasis on their biotechnological potential. *Fungal Biol. Rev.* **2012**, *26*, 73–83. [CrossRef]
38. McBride, S.G.; Osburn, E.D.; Lucas, J.M.; Simpson, J.S.; Brown, T.; Barrett, J.E.; Strickland, M.S. Volatile and Dissolved Organic Carbon Sources Have Distinct Effects on Microbial Activity, Nitrogen Content, and Bacterial Communities in Soil. *Microb. Ecol.* **2022**, *85*, 659–668. [CrossRef]

39. Hung, R.; Lee, S.; Bennett, J.W. Fungal volatile organic compounds and their role in ecosystems. *Appl. Microbiol. Biotechnol.* **2015**, *99*, 3395–3405. [CrossRef]
40. Karuppusamy, S. A review on trends in production of secondary metabolites from higher plants by in vitro tissue, organ and cell cultures. *J. Med. Plants Res.* **2009**, *3*, 1222–1239.
41. Kaczmarek, M.B.; Struszczyk-Swita, K.; Li, X.; Szczęsna-Antczak, M.; Daroch, M. Enzymatic modifications of chitin, chitosan, and chitooligosaccharides. *Front. Bioeng. Biotechnol.* **2019**, *7*, 243. [CrossRef]
42. Pan, Y.; Jin, H.; Yang, S.; Liu, H. Changes of volatile organic compounds and bioactivity of *Alternaria brassicae* GL07 in different ages. *J. Basic Microbiol.* **2019**, *59*, 713–722. [CrossRef] [PubMed]
43. Zawirska-Wojtasiak, R. Optical purity of (R)-(-)-1-octen-3-ol in the aroma of various species of edible mushrooms. *Food Chem.* **2004**, *86*, 113–118. [CrossRef]
44. Chitarra, G.S.; Abee, T.; Rombouts, F.M.; Posthumus, M.A.; Dijksterhuis, J. Germination of *Penicillium paneum* conidia is regulated by 1-octen-3-ol, a volatile self-inhibitor. *Appl. Environ. Microbiol.* **2004**, *70*, 2823–2829. [CrossRef] [PubMed]
45. Kheloul, L.; Kellouche, A.; Bréard, D.; Gay, M.; Gadenne, C.; Anton, S. Trade-off between attraction to aggregation pheromones and repellent effects of spike lavender essential oil and its main constituent linalool in the flour beetle *Tribolium confusum*. *Entomol. Exp. Appl.* **2019**, *167*, 826–834. [CrossRef]
46. Chaves-López, C.; Serio, A.; Gianotti, A.; Sacchetti, G.; Ndagijimana, M.; Ciccarone, C.; Stellarini, A.; Corsetti, A.; Paparella, A. Diversity of food-borne *Bacillus* volatile compounds and influence on fungal growth. *J. Appl. Microbiol.* **2015**, *119*, 487–499. [CrossRef]
47. Huang, C.; Zhang, L.; Johansen, P.G.; Petersen, M.A.; Arneborg, N.; Jespersen, L. *Debaryomyces hansenii* strains isolated from Danish cheese brines act as biocontrol agents to inhibit germination and growth of contaminating molds. *Front. Microbiol.* **2021**, *12*, 662785. [CrossRef]
48. Olivares-Bernabeu, C.M.; López-Llorca, L.V. Fungal egg-parasites of plant-parasitic nematodes from Spanish soils. *Rev. Iberoam. Micol.* **2002**, *19*, 104–110.
49. Hardin, J.W.; Hilbe, J.M. *Generalized Linear Models and Extensions*; Stata Press: College Station, TX, USA, 2007.
50. Van de Schoot, R.; Kaplan, D.; Denissen, J.; Asendorpf, J.B.; Neyer, F.J.; Van Aken, M.A. A gentle introduction to Bayesian analysis: Applications to developmental research. *Child Dev.* **2014**, *85*, 842–860. [CrossRef]
51. Muth, C.; Oravecz, Z.; Gabry, J. User-friendly Bayesian regression modeling: A tutorial with rstanarm and shinystan. *Quant. Methods Psychol.* **2018**, *14*, 99–119. [CrossRef]
52. Rue, H.; Martino, S.; Chopin, N. Approximate Bayesian inference for latent Gaussian models by using integrated nested Laplace approximations. *J. R. Stat. Soc. Ser. B* **2009**, *71*, 319–392. [CrossRef]
53. Jeffreys, H. *Theory of Probability*, 3rd ed.; Clarendon Press: Oxford, UK, 1961.
54. Kass, R.E.; Raftery, A.E. Bayes factors. *J. Am. Stat. Assoc.* **1995**, *90*, 773–795. [CrossRef]
55. Dittrich, D.; Leenders, R.T.A.; Mulder, J. Network autocorrelation modeling: A Bayes factor approach for testing (multiple) precise and interval hypotheses. *Sociol. Methods Res.* **2019**, *48*, 642–676. [CrossRef]

molecules MDPI

Article

The Influence of Polysaccharides/TiO_2 on the Model Membranes of Dipalmitoylphosphatidylglycerol and Bacterial Lipids

Agata Ładniak [1,2,*], Małgorzata Jurak [1], Marta Palusińska-Szysz [3] and Agnieszka Ewa Wiącek [1]

1 Department of Interfacial Phenomena, Institute of Chemical Sciences, Faculty of Chemistry, Maria Curie-Skłodowska University, M. Curie-Skłodowska Sq. 3, 20-031 Lublin, Poland; malgorzata.jurak@poczta.umcs.lublin.pl (M.J.); a.wiacek@poczta.umcs.lublin.pl (A.E.W.)
2 Laboratory of X-ray Optics, Centre for Interdisciplinary Research, Faculty of Science and Health, The John Paul II Catholic University of Lublin, Konstantynów 1J, 20-708 Lublin, Poland
3 Department of Genetics and Microbiology, Institute of Biological Sciences, Faculty of Biology and Biotechnology, Maria Curie-Skłodowska University, Akademicka 19, 20-033 Lublin, Poland; marta.palusinska-szysz@mail.umcs.pl
* Correspondence: agata.gozdecka@poczta.umcs.lublin.pl

Abstract: The aim of the study was to determine the bactericidal properties of popular medical, pharmaceutical, and cosmetic ingredients, namely chitosan (Ch), hyaluronic acid (HA), and titanium dioxide (TiO_2). The characteristics presented in this paper are based on the Langmuir monolayer studies of the model biological membranes formed on subphases with these compounds or their mixtures. To prepare the Langmuir film, 1,2-dipalmitoyl-*sn*-glycero-3-phospho-rac-(1-glycerol) (DPPG) phospholipid, which is the component of most bacterial membranes, as well as biological material-lipids isolated from bacteria *Escherichia coli* and *Staphylococcus aureus* were used. The analysis of the surface pressure-mean molecular area (π-A) isotherms, compression modulus as a function of surface pressure, $C_S^{-1} = f(\pi)$, relative surface pressure as a function of time, $\pi/\pi_0 = f(t)$, hysteresis loops, as well as structure visualized using a Brewster angle microscope (BAM) shows clearly that Ch, HA, and TiO_2 have antibacterial properties. Ch and TiO_2 mostly affect *S. aureus* monolayer structure during compression. They can enhance the permeability of biological membranes leading to the bacteria cell death. In turn, HA has a greater impact on the thickness of *E. coli* film.

Keywords: chitosan; hyaluronic acid; titanium dioxide; DPPG; bacterial lipids; Langmuir film; π-A isotherm; Brewster angle microscopy

check for updates

Citation: Ładniak, A.; Jurak, M.; Palusińska-Szysz, M.; Wiącek, A.E. The Influence of Polysaccharides/TiO_2 on the Model Membranes of Dipalmitoylphosphatidylglycerol and Bacterial Lipids. *Molecules* **2022**, 27, 343. https://doi.org/10.3390/molecules27020343

Academic Editor: Gianfranco Romanazzi

Received: 14 November 2021
Accepted: 2 January 2022
Published: 6 January 2022

1. Introduction

The cell envelope of Gram-negative bacteria consists of two distinct layers, the outer (OM) and the inner (IM) membranes, separated by the periplasm with a thin layer of peptidoglycan. Both membrane leaflets of the IM are mostly composed of phospholipids. The OM is an asymmetric lipid bilayer containing mainly phospholipids in the inner leaflet, whereas the outer leaflet is mainly composed of lipopolysaccharide (LPS). This strict OM asymmetry is important for the proper functioning of the bacterial cell, as it constitutes a permeability barrier, preventing the penetration of toxic compounds inside of the cell [1]. Due to difficulties in studying the biological processes that take place on or within the natural membranes, resulting from the high complexity of such structures, research in this area is performed with the use of biomimetic model membranes [2]. It has been proven that their application allows the assessment of changes in the physicochemical properties of lipid layers caused by foreign substances [3].

Among the available models of biological membranes, the greatest interest has focused on biomimetic monolayer films formed using the Langmuir technique [4,5]. Although this method enables the characterization of only monomolecular films at the interface, there is a strong correlation between properties (pressure, surface area per lipid molecule, phase transition, compressibility) of mono- and bilayers prepared from cell membrane components, at surface pressures of 30–35 mN/m [6–10]. Owing to this, it is possible to predict, with high probability, interactions at the level of living organisms. This makes the Langmuir technique an ideal tool for obtaining information about the properties and behavior of membranes (or their components), as well as for assessing the influence of various chemical substances, e.g., biologically active, biomolecules, on the behavior of biologically important systems [8–10]. However, extrapolating the results obtained from such studies to actual studies of biophysical situations may require some additional considerations. To visualize the interface organization of the lipid components of a monolayer or changes in interfacial behavior resulting from the introduction of an interesting compound into the monolayer and/or subphase, and thus for a better understanding of the structure–property relation of organized films, the Langmuir technique can be easily combined with optical Brewster angle microscopy (BAM). BAM is useful for determining the structure of the tested films, detecting phase changes, emerging domains, or multi-molecular structures as well as changes in monolayer thickness and other morphological effects [11,12]. Owing to this coupling (Langmuir trough with BAM), it is possible to observe the changes occurring during compression in real-time. The analysis of these changes in relation to the π-A isotherms can provide a more precise characterization of the interactions. Knowledge about the organization of monolayers and their interactions with the components of the support and the possibility of their control at the molecular level determine the potential usefulness of the received bioproducts and enable the development of a wide range of science and technology.

Chitosan (Ch) is a linear copolymer of β-(1→4)-D-glucosamine and β-(1→4)-*N*-acetyl-D-glucosamine and has a polycationic nature in acidic solutions. The positively charged backbone of chitosan (controlled by the pH) plays a crucial role in its interaction with other macromolecules or with negatively charged or neutral phospholipids of biomembranes. Clearly, the protonated amino groups NH_3^+ of chitosan are involved in the mechanism of adsorption (forced by electrostatic interactions) of this positively charged polyelectrolyte on the lipid membrane of vesicles [13,14]. Contrarily, hyaluronic acid is a polyanion. It consists of glucuronic acid and *N*-acetyl-glucosamine. HA is hydrophilic, biocompatible, biodegradable, and negatively charged in a wide pH range (pKa 3–4) [15]. Its presence in thermo-reactive hydrogels leads to the improvement of viscoelastic and mucoadhesive properties. In addition, it proves that Ch and HA also can create strong interactions with inorganic substances, e.g., titanium dioxide particles (TiO_2). Moreover, the presence of free carboxylic, amine, and hydroxyl groups at the surface of the HA and Ch films allows to bind stably the titania nanoparticles to the polymeric substrate, without the need to functionalize the nanoparticles. Moreover, these interactions lead to different physical and chemical surface adsorption, including the simple adsorption by electrostatic attraction, hydrogen bonding, and chemical binding through the formation of ester like linkage, bridging, and chelating [16,17]. TiO_2 nanoparticles have found application in tissue engineering. They can be a component of biocomposites. Nanocomposites with bactericidal properties would be excellent materials for biomedical applications. TiO_2 nanoparticles are being tested for use in cancer therapies. In addition, some studies confirm that both TiO_2 and chitosan have antimicrobial (antibacterial, antifungal) properties and are also non-toxic. One of the proposed mechanisms of the antimicrobial action of Ch and TiO_2 is based on biological disorders of the membranes of microorganisms. Therefore, it is also important to develop a mechanism of action at the molecular level, e.g., using the Langmuir monolayer technique by studying the effect of these substances on bacterial membrane components.

Knowledge about the possibility of creating or not the chemical bonds/interactions between a biomaterial and biological membrane is very important when making decisions about the specific use of a given preparation. Such an analysis is necessary regardless of the type of product prepared, and at the same time permits to direct research in order to obtain further characteristics. The effective combination of these ingredients will allow receiving a multifunctional material for applications in the field of medicine.

Therefore, we made an effort to determine whether chitosan, TiO_2, and hyaluronic acid and/or their mixtures cause disturbances in the structure of model biological membranes, formed of the synthetic material 1,2-dipalmitoyl-*sn*-glycero-3-phospho-rac-(1-glycerol) sodium salt (DPPG) phospholipid, which is the component of most bacterial membranes, as well as biological material (lipids isolated from bacteria *Escherichia coli*, a Gram-negative organism model and *Staphylococcus aureus*, a Gram-positive organism model). This allowed us to confirm the assumption that the mechanism of the antibacterial action of these compounds and their mixtures is based on disorders of the membrane of microorganisms. That aspect was also checked by microbiological methods, using the same species of bacteria. This provided the verification of the proposed method. We assumed that the study of material isolated from microorganisms would make it possible to predict the results of microbiological tests. On the other hand, the knowledge of the mechanism of these substances interactions with membranes/monolayer is an indispensable element in the production of carriers of active substances (e.g., drugs), as it allows controlling the amount of the pharmaceuticals incorporated into the carrier (and then released from it).

2. Materials and Methods

2.1. Samples and Subphases Preparation

Chitosan (Ch; 100,000–300,000 g/mol; DD = 82 ± 2% [18] ACRōS Organics, Göteborg, Sweden) was dissolved (using homogenizer (SilentCrusher M, Heidolph, Schwabach, Germany)) in acetic acid (0.1% solution of glacial AA; 99.5–99.9%; Avantor Performance Materials, Gliwice, Poland) at a concentration of 0.1 mg/mL. Titanium dioxide (TiO_2 particle size 10–30 nm [18]; Sigma Aldrich, Poznań, Poland) was mixed with AA in amount of 1.2 mg/mL after purification, as described previously [18,19]. Hyaluronic acid (HA; high-molecular; 1.60–1.80 MDa; 1% solution (Poznań, Poland)) was stirred with AA in proportion 0.5 mL/L. The solutions/dispersions of various combinations of these ingredients were prepared, namely: AA, AA/Ch, AA/HA, AA/TiO_2, AA/Ch/TiO_2, AA/HA/TiO_2, AA/Ch/HA, AA/Ch/HA/TiO_2, and their concentrations were simultaneously maintained. In addition, the TiO_2-containing suspensions were placed in ultrasonic cleaner for 15 min. All samples were prepared before measurements in room temperature. As a reference, subphase deionized water from the Milli-Q system (Millipore, Burlington, MA, USA, conductivity 18.2 MΩ cm) was used.

2.2. Isolation of Lipids from Bacteria

Lipids from the bacterial mass of *E. coli* and *S. aureus* were isolated using the Bligh and Dyer method [20] with a chloroform-methanol-water mixture (1:2:1.8 $v/v/v$).

2.3. Model Bacterial Membrane Preparation (Langmuir Films)

For the preparation of model membranes, 1,2-dipalmitoyl-*sn*-glycero-3-phospho-rac-(1-glycerol) sodium salt (phospholipid DPPG) (≥99%, Sigma Aldrich, Poznań, Poland), and bacterial lipid of *Escherichia coli* and *Staphylococcus aureus* were used. DPPG was dissolved in chloroform (98.5%):methanol (99.8%) solution, in 4:1 ratio. *E. coli* lipids were dissolved in chloroform:methanol mixture (9:1) while *S. aureus* lipids in pure chloroform. All samples were prepared in concentration of 1 mg/mL. All solvents were purchased from Avantor Performance Materials (Gliwice, Poland).

2.4. π-A Isotherms and Stability of the Lipid Monolayers Measurements

The π-A isotherms were recorded using a double-barrier Langmuir trough (KSV NIMA, total area 783 cm^2) placed on an anti-vibration table in a dust-reduced environment. The surface pressure was measured by the Wilhelmy plate method. Temperature of the aqueous subphase was kept constant at 20 ± 0.1 °C by a circulating water system from the thermostat (Lauda). The lipid solution was applied dropwise on subphase surface.

After 10 min (necessary for solvent evaporation), compression of the lipid monolayer was followed at the constant pre-set speed of 10 mm/min (29 cm^2/min). As a result of monolayer compression, the π-A isotherms were registered. Based on these plots, the following parameters were taking into account during analysis: A_0, the lift-off area, defines the area per molecule at which the transition from the gas phase (G) to the expanded liquid (LE) occurs, A_{lim}, the limit area, is a parameter that defines the area per molecule in a closely packed monolayer (determined by extrapolation of the linear part of the isotherm to zero surface pressure), π_{coll} is a surface pressure at the monolayer collapse. Based on data received from the π-A isotherms, the compression modulus was calculated using the formula:

$$C_S^{-1} = -A\left(\frac{d\pi}{dA}\right)_{n,\,T}$$

This parameter is useful to identify the physical state of a monolayer [21–23] to be classified as liquid-expanded: Cs^{-1} in the range of 12.5−50 mN/m, liquid: 50−100 mN/m, liquid-condensed 100−250 mN/m, or as condensed >250 mN/m. The minima in the plots of C_S^{-1} vs. π, on the other hand, correspond to phase transitions.

Additionally, the one-hour stability of the compressed to 35 mN/m (for DPPG) or 30 mN/m (for both bacterial extract) monolayers and compression-decompression measurements (to π = 30 or 35 mN/m; with relaxation times after compression = 1000 s) were recorded. The monolayer stability was expressed as a π/π_0 ratio, where π-surface pressure value at given time, π_0-surface pressure value to which the monolayer was compressed, i.e., equal 30 mN/m for *E. coli* and *S. aureus* monolayers, equal 35 mN/m for DPPG film.

2.5. Measurements of the Monolayer's Morphology

The morphology of monolayers was visualized with a Brewster angle microscope (BAM; Accurion, Goettingen, Germany) equipped with a 50 mW laser emitting *p*-polarized light at a wavelength of 658 nm, a polarizer, an analyzer, and a CCD camera. The BAM apparatus was coupled with the Langmuir trough. The procedure of monolayer formation was identical to that described above. BAM images presented in this paper show monolayer fragments of 360 × 200 μm^2. Due to the application of appropriate parameters of the polarizer and analyzer, the aqueous solution (without the monolayer) at Brewster's angle takes a shade of black, and the lipid domains appearing in this field take a shade of gray, which makes it possible to determine the thickness of the monolayer. Moreover, owing to the single-layer optical model [24,25], it is also possible to estimate the thickness of the phospholipid layer formed during the compression.

2.6. Microbiological Tests

Antimicrobial activity of chitosan Ch (AA/Ch), AA/TiO_2, AA/HA, AA/Ch/HA, AA/Ch/TiO_2, AA/HA/TiO_2, and AA/Ch/HA/TiO_2 against *Escherichia coli* and *Staphylococcus aureus* was determined using a colony forming unit (CFU)-counting assay and Live/Dead staining combined with fluorescent intensity measurements. *E. coli* (ATCC 25922) and *S. aureus* (ATCC 25923) were grown in 5 mL Luria-Bertani (LB) medium at 37 °C with agitation at 200 r.p.m.

A bacterial suspension with OD_{600} of 0.1 was diluted 10^{-4}-fold in LB medium. Next, 5 µL of the last dilution of the bacterial culture was transferred into a sterile Eppendorf tube, mixed with 5 µL of LB medium (control) or suspension of the appropriate compound (final concentration was the same as for subphase, e.g., AA: 0.1%, Ch: 0.1 mg/mL, TiO_2: 1.2 mg/mL, HA: 0.5 mL/L (*v*/*v*)), and incubated for 1 h at 37 °C. The mixtures were transferred onto agar plates and the colonies were counted after 1 day of incubation at 37 °C (BTL). The controls defined the total (100%) survival of *E. coli* and *S. aureus* cells incubated in identical conditions but without compound addition. Results from the antimicrobial assay represent the mean of three independent experiments, with three replicates per experiment.

Hence, 100 µL of the test compounds (final concentrations AA: 0.1%, Ch: 0.1 mg/mL, TiO_2: 1.2 mg/mL, HA: 0.5 mL/L (*v*/*v*)) were added to 100 µL of bacterial suspension (*E. coli*,

S. *aureus*) with OD = 0.1. After one hour of incubation of all samples at 37 °C, the samples containing TiO_2 were centrifuged at 500× *g*, 1 min, TR (Labnet Prism R). The supernatants were transferred into a sterile Eppendorf tube and stained using the Live/Dead BacLight bacteria viability assay kit (ThermoFisher, Waltham, Massachusetts, United States). The Live/Dead BacLight bacteria viability kit allows to discriminate viable bacterial cells from membrane compromised dead cells. The BacLight is composed of the nucleic acid-binding stains SYTO 9 and propidium iodide (PI) which differ in their spectral characteristics, as well as in their ability to penetrate bacterial cells. SYTO 9 penetrates bacterial membranes resulting in a green fluorescing signal, whereas PI is membrane impermeant and penetrates cells with damaged membranes only. Live/Dead staining of the bacteria was performed by incubation for 15 min in Live/Dead staining solution (5 µM Syto9 and 30 µM PI in 3% DMSO) in RT.

Images of the bacteria were collected using an Axiovert 200 M confocal microscope with an LSM 5 PASCAL scanning head (Carl Zeiss, Jena, Germany). The Live/Dead test was performed on the basis of 20 images of each sample. Images were captured with AxioVision 4.8 software (Carl Zeiss) in the multichannel fluorescence technique, with the AxioCam HR3 camera, using 470 nm and 546 nm filters for the green and red channels, respectively, with the same exposure time for each pair of images. Analyses were performed using ImageJ 1.50i (Wayne Rasband, National Institutes of Health, Kensington, MD, USA). The sum of the lighting values was analyzed separately for each channel in a pair of images, which corresponds to the percentage of live and dead bacteria.

3. Results

3.1. Effect of Ch, HA, TiO_2 on the DPPG Model Membranes

3.1.1. Behavior of the DPPG Monolayer during Compression

In general, the basis for the analysis of the interactions between the components within the monolayer, as well as the components constituting the film and those present in the subphase, is the dependence of changes in surface pressure (π) as a function of the area per one molecule in the monolayer (A) (called π-A isotherms), recorded during compression measurements. Surface pressure, defined as the difference between the water surface tension without and with the presence of the Langmuir film, is measured by the Wilhelmy plate method. During this process, the physical state of the monolayer changes (gaseous (G), expanded liquid (LE), condensed liquid (LC), solid (S) are distinguished), and is influenced by the intermolecular interactions occurring at the interface, whose strength and range change due to approaching the molecules to each other. Particular orientations and packing of molecules on the subphase surface correspond to the specific states. On the other hand, under high surface pressure, the monolayer may bend or collapse, which is manifested by a sudden drop/change in the surface pressure. Therefore, the analysis of the shape, course, and position of the π-A curves provides information on the phase transitions, monolayer phase state, and its organization and stability (packing density and molecular orientation or conformation).

The π-A isotherm formed during compression of the DPPG monolayer on the water subphase (H_2O) was called a reference (Figure 1). Additionally, for better visualization of results, the values of A_0 and A_{lim} parameters are presented in Table 1.

Table 1. Lift-off area (A_0) and limit area (A_{lim}) determined from the π-A isotherms for the DPPG monolayer on individual subphases.

Subphase	DPPG Monolayer on Different Subphases	
	A_0 [Å^2]	A_{lim} [Å^2]
H_2O	52.5	48.3
AA	51.9	46.5
AA/Ch	107.2	63.5

Table 1. *Cont.*

Subphase	DPPG Monolayer on Different Subphases	
	A_0 [Å^2]	A_{lim} [Å^2]
AA/TiO_2	72.8	54.3
AA/HA	51.0	47.9
AA/Ch/TiO_2	117.0	67.1
AA/HA/TiO_2	68.2	50.9
AA/Ch/HA	110.5	65.3
AA/Ch/HA/TiO_2	116.4	64.5

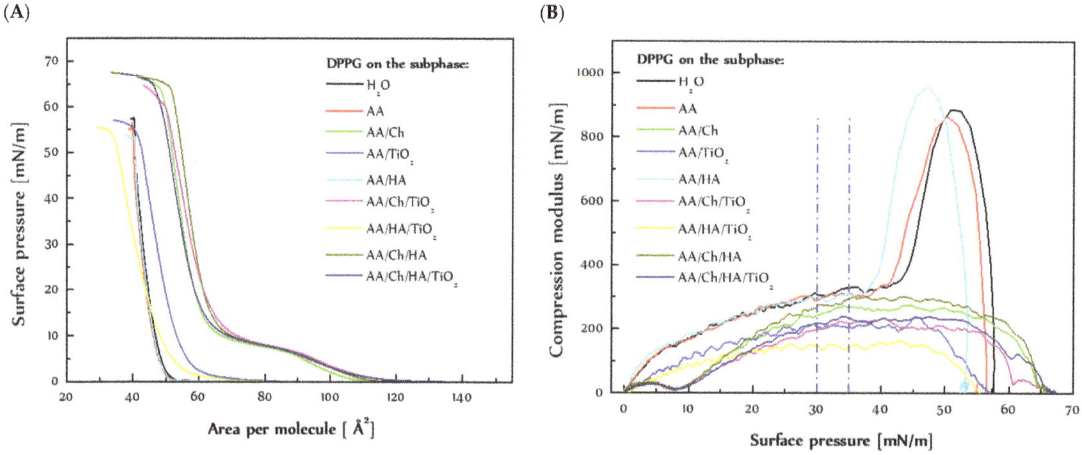

Figure 1. (**A**) π = f(A) isotherms obtained by means of the Langmuir technique and determined based on them (**B**) C_S^{-1} = f(π) relationships for the DPPG monolayer on individual subphases.

The π-A isotherms of DPPG monolayers registered on subphases containing chitosan, TiO_2, and/or hyaluronic acid had a completely different course in relation to the reference (Figure 1A). Interestingly, the use of acetic acid (AA) and acetic acid with the addition of hyaluronic acid (AA/HA) as a subphase for the lipid films did not cause the visible changes during their compression. The course/shape of isotherms for the DPPG monolayers on AA and AA/HA was the same as for the reference isotherm. Only the curves were shifted towards slightly smaller areas per molecule (in relation to the isotherm of DPPG on H_2O) and collapsed at slightly lower surface pressure values.

The A_0 parameter (Table 1) increases with the addition of further components to the subphase, which indicates that the chitosan and TiO_2 get between the DPPG molecules weakening the attraction forces between the phospholipid molecules. Only the use of acetic acid and acetic acid with the addition of hyaluronic acid has the opposite effect, that is, it slightly enhances the attraction forces between the DPPG molecules. This conclusion also arrived from analysis of limit area values (A_{lim}). In addition, the limit area per molecule in the monolayer grows with the increasing amounts of components in the subphase. This means that even at high compression, subphase components (excluding AA, and AA/HA) mostly affect the monolayer behavior.

The DPPG phospholipid (pKa 3.5–4) bears the negative charge due to the presence of the negatively charged PO_4^- group and non-deprotonated glycerol [22]. However, in the DPPG monolayers on AA and AA/HA, despite the presence of acetic acid and hyaluronic acid in the subphase (pH 3.4, 3.7 respectively), the arrangement of the DPPG molecules during compression did not change. The G/LE phase transition (A_0 parameter) was not shifted significantly, and the LC/S transition was not lost (remain visible; see graph C_S^{-1} = f(π), Figure 1)). Moreover, the AA and HA presence did not affect the A_{lim}

values which means that those substances did not substantially affect the organization of the tightly packed DPPG monolayer. Under applied conditions, a strong intermolecular hydrogen bonding between the glycerol hydroxyl and the phosphate of the neighboring lipid molecules can prevail, resulting in dense packing and rigidity of the monolayer [22].

The most significant changes in the course of the π-A isotherms were noted after additional introduction of TiO_2 to those (AA, AA/HA) subphases. In the case of TiO_2-containing dispersions (AA/TiO_2, AA/HA/TiO_2), the curves moved towards larger values of area per molecule, suggesting a weakening of the attraction forces between the DPPG molecules. In addition, TiO_2 caused a milder increase in surface pressure when the condensed DPPG domains formed during transition from the gas to expanded liquid phase and then to the condensed liquid phase. Moreover, the monolayers collapsed at a surface pressure similar to that of the monolayer spread on water but probably according to a different mechanism induced by TiO_2.

The negatively charged heads of the phospholipid create favorable conditions for interaction with positively charged TiO_2 particles. TiO_2 penetrates between DPPG molecules at the beginning of compression, causing a reduction in the available area at the interface, thus accelerating the G-LE phase transition. However, the slight difference of π_{coll} values to the reference suggests that the TiO_2 particles could be pushed out of the monolayer at high surface pressures. On the other hand, the change in the shape of the isotherms for DPPG monolayers on the TiO_2-containing subphases is indicative of DPPG-TiO_2 interactions, and thus TiO_2 particles may be located just below the surface of the lipid film. The close presence of TiO_2 imposes a different arrangement of both the charged hydrophilic heads and the hydrophobic chains of DPPG molecules at the interface. The DPPG molecule consists of two saturated chains which, under high surface pressure, due to the lack of spatial restrictions, position themselves vertically in relation to the liquid surface. In turn, the head of the phospholipid has both a negatively charged phosphate group and glycerol, which is a very weak acid (pKa = 14.4), occurring in the protonated form. The appearance of a positive charge (TiO_2) in the vicinity may determine the formation of strong interaction with PO_4^-, and thus change the arrangement of the entire fragment of the molecule.

Moreover, the analysis of the $C_s^{-1} = f(\pi)$ relationship (Figure 1B) suggests that the addition of chitosan and TiO_2 to the subphase caused deviations from the typical course of the curve. However, acetic acid and hyaluronic acid caused only shifts, but allowed maintaining the correct shape.

On the typical π-A isotherms for DPPG, the LE-LC phase transition region is not visible because in the DPPG monolayers the condensed domains are formed at lower values of surface pressure [26]. In addition, the compression modulus value at approximately 45 mN/m proves that the state of the DPPG monolayer changes from the condensed liquid phase to a solid. The lack of LE-LC phase transition if the acetic acid or acetic acid with the addition of hyaluronic acid subphases are used confirms that these components do not affect the DPPG monolayer properties. They only affect the value at which the LC-S phase transition occurs.

The presence of TiO_2 causes a complete disintegration of the DPPG monolayer, which manifests itself as the disappearance of the LC-S phase transition and difficulty in the formation of a compact film. The compression modulus value has been significantly reduced, which means that in the presence of TiO_2 the DPPG monolayers exist in the less packed and ordered state. This drastic reduction in packing suggests great weakness of the attraction forces between DPPG molecules, causing the monolayer to be highly elastic independently of the surface pressure values.

Drastic changes in the course of the π-A isotherms occurred in the case of chitosan-containing subphases. The isotherms shift towards larger areas as compared to the reference which indicates the penetration of chitosan into the DPPG monolayer, thereby causing increasing repulsive Coulombic forces between negatively charged DPPG headgroups. This gives rise to looser molecular packing in DPPG monolayers, where the LE-LC phase transition is observed [22]. After compression of the monolayers, the π-A isotherms registered for chitosan-containing subphases were collapsed at much higher values of surface pressure than that of the monolayer on water. This can suggest its presence in a closely packed monolayer.

As mentioned above, the introduction of chitosan into the subphase causes the appearance of the LE-LC phase transition region, which does not occur for the DPPG monolayer on water. This denotes that the Ch presence slows the creation of condensed domains in the DPPG monolayers. As an effect, the LE-LC phase transition is revealed. Similar results were obtained by Piosik et al. [27], who examined the effect of nanoparticles with a Fe_3O_4 core and a chitosan crown on the behavior and structure of the DPPG monolayer. The presence of the Fe_3O_4-Ch nanoparticles in the water also causes the appearance of the liquid expanded phase between the gaseous state and liquid condensed phase. Additionally, the shift of the π-A isotherms of DPPG in the presence Fe_3O_4-Ch and thus the expansion increases with the rising concentration of the Fe_3O_4-Ch nanoparticles, indicating the adsorption of the investigated nanoparticles into the phospholipid monolayers, which were probably located preferably into loosen monolayer areas in the liquid expanded phase.

Referring again to our chitosan-containing subphases, the graph shows a plateau which is indicative of LE-LC phase transition (Figure 1A). The precise value of the LE-LC phase transition pressure can be read based on the dependency of the compression modulus as a function of the surface pressure. The graph (Figure 1B) shows a minimum in the course of the function, dividing the graph into two maxima that correspond to the expanded liquid (LE) and the condensed liquid (LC). The presence of chitosan contributed to the fluidity of the DPPG monolayers, keeping them in the state of LC at high values of surface pressure. On the other hand, DPPG monolayer on the chitosan-containing subphases showed slightly more stability than the reference one. They were collapsed at much higher surface pressure values. The same observations were obtained by [26], where the presence of Fe_3O_4-Ch also caused a decrease of the maximal ${C_S}^{-1}$ value associated with the disappearance of the solid state at the DPPG final compression stage. In addition, it is noticeable to reduce the monolayer packing after an additional introduction of TiO_2 into the Ch-containing subphases (lower values of the compression modulus were received). Ultimately, the shape/course both the π-A isotherms and the ${C_S}^{-1} = f(\pi)$ relationships was very similar to those obtained for DPPC monolayers [28]. This means that as a result of the interactions of Ch with DPPG molecules, similar conditions occurred during compression. DPPC has the same hydrophobic part as DPPG but differs in the hydrophilic part. DPPC is zwitterionic, i.e., it has two differentially charged groups, ${PO_4}^-$ and -$N(CH_3)_3{}^+$. On the one hand, this may suggest that (I) Ch penetrates between DPPG molecules, interacting with the phosphate groups, thus reducing the charge of the hydrophilic heads, (II) Ch causes the vertical orientation of the DPPG molecules at the interface, (III) Ch remains in the monolayer even during its high compression (A_{lim} value was significantly higher than the reference), (IV) Ch determines some kind of stabilization of the lipid film (π_{coll} was much higher than the reference), but at the same time (V) Ch increases the fluidization of the film (keeping it in the LC state). This observation may be a result of separation of the phosphate-groups by the positively charged Ch in the DPPG headgroup region which prevents a formation of rigid and tightly packed monolayer. As a consequence, the LE-LC phase transition is revealed.

In order to more exactly characterize the interactions between the subphase components (TiO_2, HA, and Ch) and DPPG molecules, the stability of DPPG monolayers (expressed as $\pi/\pi_0 = f(t)$) previously compressed to $\pi = 35$ mN/m, was tested.

3.1.2. Stability of the DPPG Monolayer

The lateral pressure on the monolayer during its compression forces the appropriate behaviour and orientation of DPPG molecules. When compression is stopped (and thus surface pressure does not increase), phospholipid molecules reorganize and any changes in surface pressure during relaxation indicate their different responses to the mechanical compression. Among other things, the changes taking place may also suggest the loss of lipid material from the interface. This in turn could be confirmed by the compression-decompression measurements. The observation of the stability of the monolayers over time, combined with the analysis of hysteresis loops, provides detailed information on the interactions between the components of the subphase and the model membrane.

Figure 2A shows changes of a relative surface pressure over one hour relaxation of DPPG monolayers, previously compressed to 35 mN/m on different subphases. As can be seen, a rapid decrease of π/π_0 ratio took place within 15 min for all cases. After that time, the DPPG monolayer showed comparatively constant values of relative surface pressure. The trend of π/π_0 ratio variations with time for DPPG monolayer on different subphases was close to the reference one. Only more intense changes occurred for DPPG film on the AA/TiO_2 and AA/HA/TiO_2 subphases, where a greater pressure drop was noted at first 10 min. However, the quite stable level of π/π_0 ratio was achieved in a comparable time as for DPPG layers on the remaining subphases. For these two cases, specific/characteristic hysteresis loops were also noted during the compression-decompression measurements with downtime 1000 s (Figure 2B). Although a slight downtime plateau was observed, there was a drastic drop in surface pressure as soon as the decompression cycle was initiated (the descending curve is almost perpendicular to the *x*-axis). This may suggest the intercalation of TiO_2 particles between DPPG molecules that are able to form strong bonds with PO_4^- not governed by electrostatic forces alone, but also of specific adsorption [29–32]. TiO_2 interacts with adjacent DPPG molecules, during the entire compression cycle, on the one hand, to some extent stabilizing the monolayer. On the other hand, they make it difficult for DPPG monolayers to take the place of individual phase transitions and create a compact structure. TiO_2 increases the monolayer fluidity, as observed in Figure 1B.

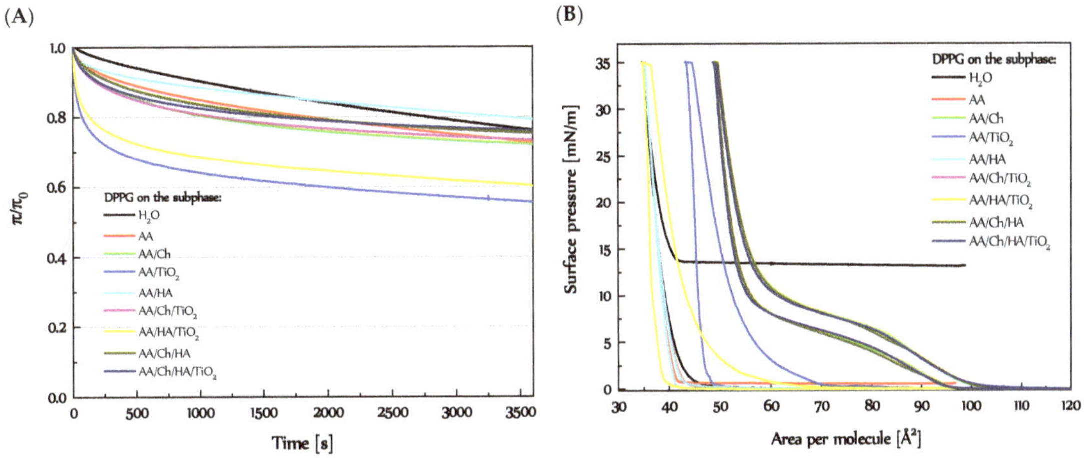

Figure 2. (**A**) π/π_0 = f(t) (more details Figure S1) and (**B**) hysteresis loops obtained by compression and decompression of the DPPG monolayers on the individual subphases.

Lower stability of the DPPG monolayers (expressed as π/π_0 = f(t), Figure 2A) in the presence of AA/TiO_2 and AA/HA/TiO_2 in relation to the others, resulting from the fact that DPPG molecules adsorbed on TiO_2 surfaces, even during high compression, provide greater possibilities of the spatial arrangement of hydrophobic chains (because they were separated by TiO_2 particles). In turn, a dramatic decrease of pressure reflected in decompression curves (Figure 2B) suggests that these interactions are so strong that DPPG molecules do not loosen during the expansion of monolayers, but the rupture of the lipid film probably takes place.

The use as a subphase of AA and AA/HA for DPPG monolayers did not reveal any specific interactions. The relative surface pressure of the DPPG films on the AA and

AA/HA during 1 h did not differ from the typical changes recorded for the reference (DPPG on the water), and their thicknesses over the entire compression cycles were very similar. The compression-decompression curves were also identical, with almost no hysteresis loops. This suggests that AA and HA did not disturb the structure, packing, or stability of the DPPG monolayers.

In contrast, a special type of interaction occurred between DPPG molecules and chitosan. All DPPG films recorded on subphases containing this biopolymer (AA/Ch, AA/Ch/TiO_2, AA/Ch/HA, AA/Ch/HA/TiO_2), despite demonstrating deviations from the typical course of the π-A curve (Figure 1A), showed stability at an almost identical level (Figures 2A and S1). Moreover, the same hysteresis loops during compression-decompression measurements were recorded for them. This indicates that the contact of DPPG with Ch creates strong interactions which at the same time enable the production of a slightly looser packing (Figure 1B) but stable monolayer. These interactions take place during the initial contact and are stable under all conditions (compression, decompression, for 1 h in a closely packed monolayer). More precisely, these interactions are so strong and dominant that it is impossible to see the influence of other substances present in the subphase (HA, TiO_2), regardless of their combination.

3.1.3. Thickness and Structure of the DPPG Monolayers

An important aspect of lipid film stability in highly compressed states is the packing of lipid molecules and the presence of ordered and disordered regions. Real-time images from the Brewster angle microscope can provide information in this respect. In addition, based on measurements of the reflectivity, the relative thickness of DPPG monolayers (Table 2) deposited at the subphase/air interface was investigated.

Table 2. Thickness of the DPPG monolayers during compression.

Surface Pressure [mN/m]	Thickness of DPPG Monolayer Obtained on Different Subphases [nm]				
	H_2O	AA	AA/Ch	AA/HA	AA/Ch/HA
0.5	0.3	2.1	0.9	-	1.1
5	2.5	2.7	1.5	2.6	1.6
10	2.6	2.7	1.4	2.7	1.9
15	2.7	2.8	2.7	2.8	0.9
20	2.7	2.8	3.5	2.8	1.8
25	2.8	2.8	3.7	2.8	3.6
30	2.8	2.8	3.8	2.8	3.9
35	2.8	2.8	3.8	2.9	4.0
40	2.8	2.9	3.9	2.8	4.0
45	2.8	2.9	3.9	2.9	4.1
50	2.8	2.9	3.9	2.9	4.1
55	2.8	2.9	3.9	2.9	4.1
60			3.8		4.0
65			3.5		3.6

The π-A isotherm (Figure 1A) and BAM images recorded for the DPPG monolayer compressed on the pure water subphase (Figure 3) are consistent with those presented in the literature [26,27,33–35]. Although the presence of AA and HA in the subphase did not cause significant differences in the $\pi = f(A)$ and $C_S^{-1} = f(\pi)$ plots (Figure 1) for DPPG monolayers, observation with BAM (Figure 3) showed that these substances influenced the structure of the film during compression. They contributed to the creation of a more compact and homogeneous structure but did not change its thickness relative to the reference (DPPG on the water). That means the HA presence did not disturb the DPPG monolayer, but only caused an "improvement" of the compression process due to accelerating the appropriated phase transitions (from gas through expanded and condensed liquid to solid state). This may be an effect of repulsion interaction between negatively charged PO_4^- headgroups of DPPG and carboxylic groups of HA.

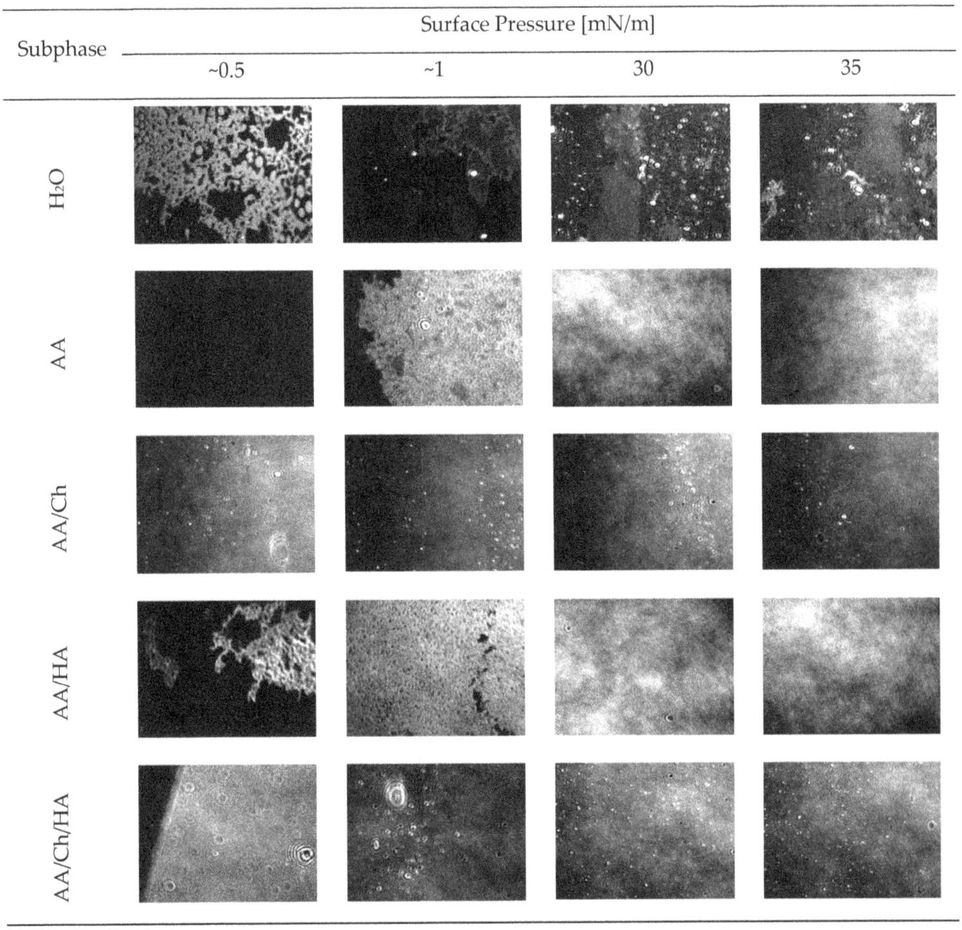

Figure 3. Structure of the DPPG monolayers during their compression (at surface pressures of 0.5, 1, 30 and 35 mN/m) on the different subphases (AA—acetic acid, AA/Ch—chitosan on acetic acid, AA/HA—hyaluronic acid in acetic acid, AA/Ch/HA—chitosan in acetic acid with the addition of hyaluronic acid).

On the other hand, the DPPG film on the AA/Ch subphase below π = 15 mN/m showed a thickness of about 1 nm smaller and above this pressure-about 1 nm greater compared to the film on H_2O (Table 2). Moreover, as mentioned before, Ch caused an increase of DPPG monolayer fluidity. Therefore, in the BAM images, the characteristic condensed domains of DPPG that occurred for the water subphase were not observed. In addition, during compression, the grain-shaped areas were visible, whose amount accreted with increasing the surface pressure. These small domains comprise the condensed phase of DPPG monolayer. These LC phase islands appear porous, as the dark gaseous phase coexists with the bright domain structures. This suggests that Ch is probably located preferably in loose monolayer areas in the liquid expanded phase, creating Ch-rich and lipid-poor regions. Chitosan can not only interact electrostatically with charged groups, but can also produce hydrophobic interactions with the lipid tails [36]. Consequently, Ch may insert between DPPG molecules, causing the larger expansion of the examined monolayers, which was also reflected as the decrease in the monolayer packing (Figure 1B). In effect, these physical interactions between Ch molecules and phospholipids possibly account for the enhanced permeability of biological membranes induced by Ch molecules. Similar changes for the DPPG monolayer were also observed by [33] using DMSO solution as the subphase. This effect was assigned the condensing and caging effect of DMSO on the DPPG monolayers, although it is a negatively charged lipid. The observed condensing effect of DMSO molecules on phospholipids physically changes the morphology of the film, creating DMSO-rich and lipid-poor regions. Moreover, the adsorption kinetics recorded during penetration experiments carried out by [27] revealed the strongest adsorption of the Fe_3O_4-Ch nanoparticles in the DPPG monolayer.

Such a structure of DPPG film was also observed on the AA/Ch/HA subphase. Therefore, in the most compressed state, the expansion of the DPPG films occurs and the Ch molecules remain adsorbed into them. Here, however, the appearance of a lipid-poor region at 15 mN/m was additionally observed as a drastic decrease in monolayer thickness, which confirms the above suggestions.

3.2. *The Influence of Ch, HA and TiO_2 on the S. aureus Lipids Monolayer*

3.2.1. Behavior and Packing of *S. aureus* Lipid Monolayers during Compression

The π-A isotherm for the film composed of lipids isolated from *S. aureus* species registered on the H_2O subphase was called the reference (Figure 4). The π-A isotherm plotted has a mild course without any visible inflections indicative of phase transitions. The surface pressure began to increase at an area of A_0 = 93.9 Å^2. The monolayer collapse occurred at π = 47.3 mN/m (Figure 4A). Used as subphases the AA and AA/HA solutions caused a slight shift of the *S. aureus* lipid isotherms to smaller areas, A_0 = 89.6 Å^2 and A_0 = 88.1 Å^2, respectively. The monolayers collapsed at almost the same surface pressure π_{coll} = 46.7 mN/m and 46.8 mN/m, respectively. The shapes of the π-A plots were maintained, although a little inflection at π~27.5 mN/m took place. This was reflected in the graph of compression modulus ($C_S{}^{-1}$) versus surface pressure. Clearly, for the *S. aureus* lipid film on the H_2O subphase, the $C_S{}^{-1}$ values rose from the compression beginning, reaching a maximum at 30 mN/m ($C_S{}^{-1}$ = 95.2 mN/m). In turn, in the case of AA and AA/HA subphases (Figure 4B), two maxima were visible at π = 25 mN/m ($C_S{}^{-1}$ ca. 85 mN/m) and 40 mN/m ($C_S{}^{-1}$ ca. 75 mN/m), and minimum at 30 mN/m ($C_S{}^{-1}$ ca. 70 mN/m).

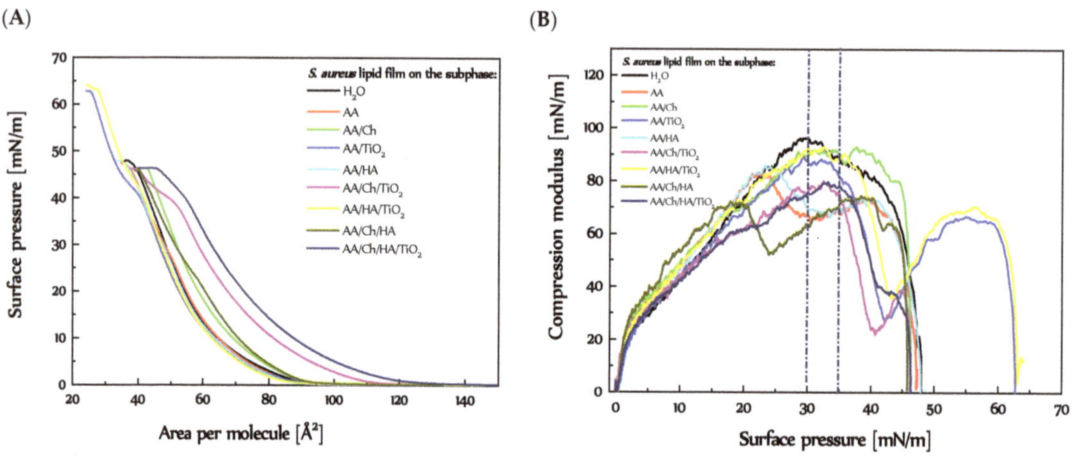

Figure 4. (**A**) π-A isotherms obtained by means of the Langmuir technique and determined based on them (**B**) $C_S{}^{-1}$ = f(π) relationships for *S. aureus* lipid monolayers on the individual subphases (more details Figure S2).

A different course of the π-A plot occurred for the *S. aureus* lipid film on the AA/TiO_2 subphase. Despite a similar progression at the beginning of compression, i.e., the G-LE transition took place at A = 94.7 Å^2 (near the reference), followed by an increase in π with tendency similar to that on the H_2O subphase. However, two inflections were observed, the first one at $\pi \sim 40$ mN/m and the second one at $\pi = 62.5$ mN/m. The same shape of a graph was received for *S. aureus* lipid film on the AA/HA/TiO_2. Only slight shift of the plot to smaller molecular areas took place, $A_0 = 88.7$ Å^2. In Figure 4B, two maxima of C_S^{-1} values were observed, at $\pi = 32.5$ mN/m and 55 mN/m. The inflections at π~40 mN/m were visible as a minimum of $C_S^{-1} = f(\pi)$ in Figure 4B. That drastic decrease in packing degree may suggest the expulsion of at least one of the ingredients. For the sake of clarity, the *S. aureus* cells consist of cardiolipin (CL) in 42%, and phosphatidylglycerol (PG) in 58% [37,38]. Based on the literature data concerning the behavior of the pure CL [23,39] and PG monolayers (previous section) on different kinds of subphases, it can be concluded that the CL monolayer collapsed earlier than that of PG. The two inflections were also observed for the *S. aureus* lipid films on the AA/Ch/TiO_2 and AA/Ch/HA/TiO_2 subphases. Thus, in Figure 4B, for these cases, a minimum of the $C_S^{-1} = f(\pi)$ function was also visible (at $\pi = 40$ mN/m), dividing the graph into two maxima, at π ca. 32.5 mN/m and ca. 44 mN/m. A difference was the surface pressure value of collapse, π~46 mN/m, smaller even than a reference one. In addition, these isotherms (Figure 4A) were shifted to larger molecular areas, $A_0 = 113.4$ Å^2, and $A_0 = 128.5$ Å^2, respectively.

For the *S. aureus* lipid monolayer on the AA/Ch subphase, the shape of the π-A plot was maintained (relative to a reference), only shifted to slightly higher areas, $A_0 = 96.2$ Å^2, and collapsed at some lower pressure value, $\pi_{coll} = 46.2$ mN/m (Figures 4A and S2). This was also reflected in the $C_S^{-1} = f(\pi)$ relationship. For lipid film on that subphase, only one maximum of the compression modulus values was noted (at π ca. 37.5 mN/m). However, in the case of the AA/Ch/HA subphase, despite the fact that the take-off area of the π-A isotherm for *S. aureus* lipid film was similar ($A_0 = 95.5$ Å^2) to that on AA/Ch, and the collapse pressure was the same, a change in the course of plot was observed. More precisely, the small inflection at π ca. 22.5 mN/m appeared (Figures 4A and S2). That was visible in Figure 4B as a minimum of the graph between two maxima, at π ca. 20 mN/m and π ca. 38 mN/m.

Generally, despite all changes in the course of the plot that occurred in Figure 4B in relation to the reference, the *S. aureus* lipid films on all subphases possessed compression modulus values in the range characteristic for the liquid phase.

Interestingly, for the *S. aureus* lipid films on the H_2O, AA/TiO_2, and AA/HA/TiO_2 subphases, similar values of A_{lim} parameter were noted, 64.4 Å^2, 65.0 Å^2, and 65.5 Å^2, respectively. This suggests that these components of the subphase (mainly TiO_2) did not penetrate into the *S. aureus* monolayer, but lipids molecules could adsorb on the TiO_2 particles causing an expulsion of the CL or PG. Slightly larger A_{lim} values were received for the *S. aureus* lipid film on the AA (68.5 Å^2) and AA/HA (66.7 Å^2) subphases. This indicates that the acidic environment leads to the weakness of the attraction forces between lipid molecules in the Langmuir layer.

A substantial increase in the A_{lim} values occurred for the *S. aureus* lipid film on the Ch-containing subphases, namely AA/Ch (71.2 Å^2), AA/Ch/HA (77.9 Å^2), AA/Ch/TiO_2 (85.6 Å^2), and AA/Ch/HA/TiO_2 (91.8 Å^2). This implies that Ch can penetrate the monolayer made of *S. aureus* lipids, weakening the attraction forces between them. Moreover, a Ch-induced effect has been reinforced by further components.

3.2.2. Stability of the *S. aureus* Lipid Monolayers

To analyze the changes in the structure of *S. aureus* lipid film taking place at the pressure corresponding to the pressure of biological membranes in more detail, the stability of the *S. aureus* lipid film at $\pi = 30$ mN/m over one hour and compression-decompression measurements were done (Figure 5).

The stability of *S. aureus* lipid monolayer compressed to 30 mN/m was reflected as π/π_0 ratio versus time (Figure 5A). Reference monolayer had very stable relative pressure over one hour. A minimal decrease of π/π_0 value was observed. Interestingly, almost the same level was reached by *S. aureus* lipid film on the AA, AA/Ch, and AA/Ch/HA subphases. In turn, for membrane on AA/HA subphases from the beginning the decrease of surface pressure was occurred, more intense after 29 min. This suggests that the interactions between Ch and lipid molecules during compression were so strong that they made it impossible of relaxation lipids molecules at the interface. On the other hand, HA did not influence a membrane during compression, but only had the ability to interact on their packed form. A slightly greater decrease of π/π_0, at the beginning, appeared for *S. aureus* lipid film on the AA/TiO_2 and AA/Ch/TiO_2 subphases, in relation to the reference, but a relatively stable level was reached after just 10 min. A different situation occurred in the case of AA/HA/TiO_2 and AA/Ch/HA/TiO_2 subphases. The lipid films on both subphases were not stable in time. Only for *S. aureus* lipid film on AA/Ch/HA/TiO_2 subphase was a constant π/π_0 value reached after 45 min.

The obtained results are discussed more accurately in the next section.

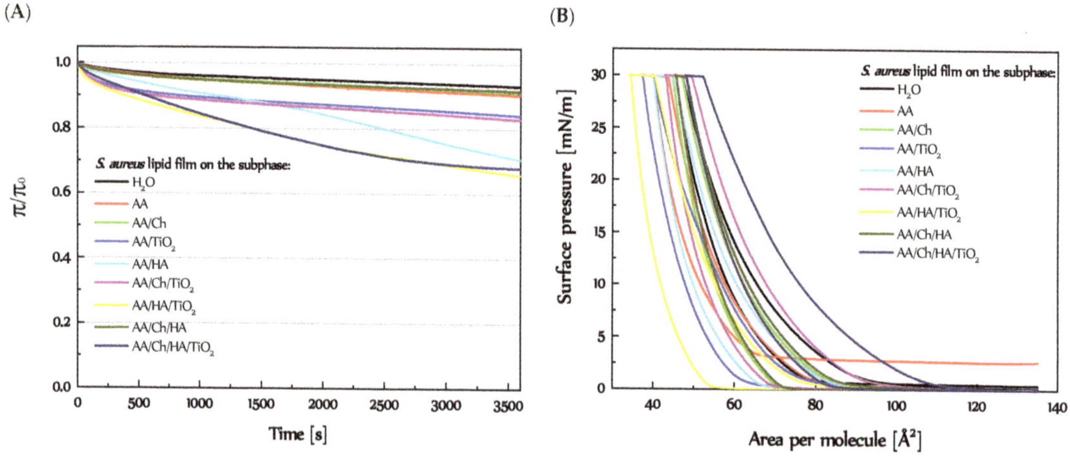

Figure 5. (**A**) π/π_0 = f(t) and (**B**) hysteresis loops (more details Figures S3 and S4) obtained by compression and decompression of the *S. aureus* lipid monolayers on the individual subphases.

The magnitude of the hysteresis loops obtained during compression-decompression measurements provides information about the existence of interactions between lipids molecules (in a closely packed layer) and subphase components, as a result of which the loss of material from interface could have occurred. Both compression and decompression curves for *S. aureus* lipid film had similar courses on each of the subphases (except AA) (Figure 5). In all cases, the descending curves were shifted to smaller molecular areas, besides reference. In effect, hysteresis loops of considerable sizes were obtained (Figures S3 and S4).

Interestingly, the decompression curve of the *S. aureus* lipid film on H_2O was shifted to the larger areas per molecule. At the beginning of the decompression process, it was about 2 Å^2, but total relaxation was achieved at area larger by about 12.3 Å^2. This suggests that during compression the highly packed aggregates, separated by empty areas, could be created, which at constant surface pressure value (at downtime) relax. In turn, using the AA solution as a subphase caused a shift descending curve (at the start) about 4 Å^2 to the left side of the graph, but total decompression occurred at π ca. 2.5 mN/m (did not achieve zero surface pressure) at area ca. 18.4 Å^2 smaller. A slightly bigger plateau at downtime in relation to the reference on the AA/Ch, AA/TiO_2, and AA/Ch/HA was noted (2.6 Å^2, 2.8 Å^2, and 2.9 Å^2, respectively). Ultimately, in these cases, the decompression monolayer occurred at 18.3 Å^2, 19.7 Å^2, and 18.3 Å^2 in the smaller areas, in reference to the compression curves. The greater hysteresis loops were obtained for *S. aureus* lipid films on the AA/Ch/TiO_2, AA/HA/TiO_2, and AA/Ch/HA/TiO_2, and the differences in the G-LE phase transition area between ascending and descending curves were 28.0 Å^2, 26.8 Å^2, and 27.6 Å^2, respectively (Figure S4). Meanwhile, the length of the plateau at downtime was in sequence 6.7 Å^2, 6.3 Å^2, and 4.6 Å^2. In turn, a decompression curve of *S. aureus* lipid monolayer on an AA/HA subphase was shifted to the left side of the graph about 5.4 Å^2 at 30 mN/m and 19.8 Å^2 at π~0 mN/m (Figure S3). Taking into consideration effects involving by AA/Ch, AA/HA, and AA/TiO_2 subphases on model *S. aureus* lipid monolayer, it could be concluded that Ch and TiO_2 affect the monolayer structure during compression, downtime, and decompression in a constant manner. In turn, HA has a greater impact on a packed monolayer (during downtime). This is related to the charge of the individual components of the subphases. HA with a negative charge causes repulsion of the lipid molecules (also negatively charged), which is more intense at downtime. In effect, the attraction forces between lipids become stronger, creating more condense and packed regions occupying smaller areas. More considerations have been made as regards the structure and thickness of *S. aureus* lipid film (in the next section).

3.2.3. Thickness and Structure of the *S. aureus* Lipid Monolayers

Taking into account that the Langmuir film is not a pure component but rather a mixture of lipids (in this case CL and PG), in order to draw more precise conclusions, it is necessary to analyze the membrane behavior under other conditions in more detail.

Therefore, the structure of *S. aureus* lipid films was analyzed during compression using the Brewster angle microscope, obtaining information in real-time on the shape of domains formed (Figure 6) and the thickness of the membranes (Table 3). The morphology of the monolayers obtained on individual subphases is presented in Figure 6.

Table 3. Thickness of the *S. aureus* lipid monolayers during compression.

Surface Pressure [mN/m]	Thickness of *S. aureus* Lipid Monolayer Obtained on Different Subphases [nm]				
	H_2O	AA	AA/Ch	AA/HA	AA/Ch/HA
0.5	1.4	1.7	1.7	1.5	1.6
5	1.8	1.9	2.0	1.9	1.8
10	2.0	2.2	2.0	2.1	2.0
15	2.1	2.3	2.5	2.3	2.2
20	2.2	2.4	2.6	2.4	2.3
25	2.3	2.6	3.1	2.5	2.5
30	2.4	2.7	2.8	2.6	2.5
35	2.5	2.8	3.0	2.8	2.7
40	2.7	2.8	3.0	2.8	2.8
45	2.7	2.7	3.1	2.7	2.8

The *S. aureus* lipid film on the H_2O subphase formed condensed, bright domains even at low values of the surface pressure $\pi > 5$ mN/m. The structure of the film resembled a grid with round, symmetrical holes of various diameters. As the compression progressed, these holes decreased to form a more compact and uniform layer. However, up to $\pi < 5$ mN/m, slight defects in the form of black-colored holes were observed, indicating the existence of small regions of non-condensed state. The number of these defects decreased with the increase of π value. On the other hand, at $\pi > 20$ mN/m, domains in the form of bright points began to appear, and their amount increased with the progress of compression. The highest quantity was observed at $\pi = 35$–40 mN/m, which could indicate partial miscibility of the lipid components in the monolayer and possible ejection of those having a lower collapse surface pressure value [40]. Both $\pi = f(A)$ and $C_S{}^{-1} = f(\pi)$ relationships plotted for this lipid film suggest that its collapse took place above surface pressure value, $\pi > 45$ mN/m. Moreover, the thickness (Table 3) of the obtained film increases, until $\pi = 40$–45 mN/m. Overall, *S. aureus* lipid film thickness during compression slowly increases with surface pressure π, from 1.4 nm for $\pi = 0.5$ mN/m to 2.7 nm for $\pi = 40$–45 mN/m.

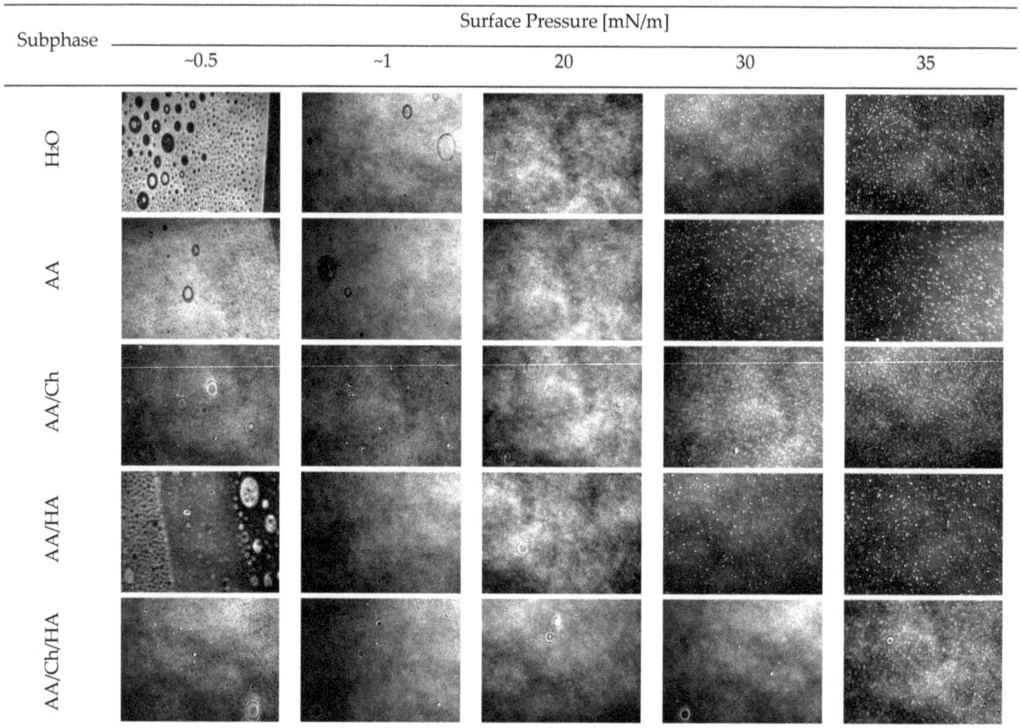

Figure 6. Structure of the *S. aureus* lipid monolayer during their compression (at surface pressures of 0.5, 1, 20, 30 and 35 mN/m) on the different subphases (AA—acetic acid, AA/Ch—chitosan on acetic acid, AA/HA—hyaluronic acid in acetic acid, AA/Ch/HA—chitosan in acetic acid with the addition of hyaluronic acid).

The *S. aureus* lipid film compressed on the acetic acid subphase showed an upper level of condensation at the same pressure values compared to the *S. aureus* lipid film on the H_2O, while the entire structure was retained. This is also confirmed by the values of the thickness of *S. aureus* lipid layer (Table 3) at the respective surface pressures, which were from 0.1 to 0.3 nm higher compared to the values obtained on H_2O. At low surface pressure values, less condensed areas in the form of black craters were also observed. Almost no bright highly condensed fragments were present at the start of compression. On the other hand, at $\pi > 25$ mN/m, aggregates appeared in the form of white, bright dots, which remained to $\pi = 35$ mN/m, then their amount significantly decreased.

As expected, the use of a mixture of acetic and hyaluronic acid as a subphase caused a visible change in the structure of the *S. aureus* lipid film (Figure 6). The coexistence of a reticular structure with small, round holes and elliptical, bright (more condensed) areas was observed. A homogeneous structure at $\pi < 1$ mN/m was formed, but with numerous black holes/craters. As the pressure increases, at the π~4 mN/m (data not shown), thread-like bright packed structures were also observed, and bright "glowing" dots/aggregates resembling the structure of the DPPG monolayer on H_2O subphase (Figure 3). However, above surface pressure value $\pi > 5$ mN/m, a relatively homogeneous lipid layer of *S. aureus* was obtained, without any defects. On the other hand, above $\pi > 25$ mN/m, white points began to appear, as they appeared in *S. aureus* lipid films prepared on other subphases. In this case (on the AA/HA subphase), their greatest amount occurred at $\pi = 40$ mN/m. As a result of these changes, the thickness (Table 3) of the *S. aureus* lipid monolayer was also slightly higher than the reference one (differences at the level of 0.1–0.3 nm).

In the presence of chitosan in the subphase (AA/Ch) the complete disappearance of the reticular structure of *S. aureus* lipid film was obtained (Figure 6). At the beginning of compression, a condensed layer with numerous aggregates was observed. The monolayer did not create a fully dense structure throughout the compression cycle. Two types of structure were visible at all times, the first in the form of black holes/craters, evidencing the co-existence of domains in more and less condensed phases, and the second in the form of white spots (revealed at the pressure of 10 mN/m < π < 20 mN/m), indicating the existence of highly condensed regions. In addition, specific elliptical "blinking" structures appeared, which had already been observed previously for monolayers of other phospholipids registered on chitosan-containing subphases [Ładniak et al., personal communication]. All defects occurred regardless of the prevailing π, and their amount dropped sharply above π > 5 mN/m. On the other hand, at π > 30 mN/m, bright points were revealed (also described for other subphases). The appearance of a compact (but with numerous domains) structure of the *S. aureus* lipid film was also reflected in its thickness (Table 3), which was greater than the reference one from 0.2 to 0.8 nm (taking into account the entire compression cycle).

A situation similar to that of the above-discussed monolayer on the AA/Ch subphase occurred for the lipid film on the subphase containing both biopolymers, i.e., AA/Ch/HA. In this case also the monolayer was not homogeneous, it had the same domains (Figure 6). The difference is that the structures with greater condensation were visible in significant amounts at value π > 35 mN/m. However, the thickness of the *S. aureus* lipid layer on AA/Ch/HA subphase was not as significantly different as that on the AA/Ch subphase. It was only slightly less than the reference one by 0.1–0.2 nm (Table 3).

In summary, the presence of Ch and Ch/HA in the subphase had the greatest effect on the changes of the structure of *S. aureus* lipid layer during the compression process. Simultaneously, BAM images allowed observation of the growing lipid domains (Figure 6). Their presence was evidenced by bright points appearing on the dark subphase background, which combine with increasing pressure to form a uniform film. On the other hand, bright white aggregates appearing at π~30 mN/m may also denote the existence of partial miscibility of lipid components in the monolayer. As a consequence, strong hydrogen bonds can form between the same group of film components, more specifically between neighboring lipid molecules. The uprising of these permanent interactions means that during expansion it was not possible to return them to the state before compression. Consequently, the decompress curve was shifted towards much smaller molecular surfaces, creating large hysteresis loops and a significant plateau at maximum compression pressure (during downtime) (Figure 5B), which can suggest the loss of lipid material from the interface.

The situation of the monolayer in the region at surface pressure π = 30 mN/m, and as a result of strong interactions between the components of the lipid film, causes "partial" stabilization of this state. Therefore, when recording the stability over time, no significant changes in the relative surface pressure were observed. The lack of visible relaxation of molecules in the lipid film after ending the compression process proves that the molecules have already adapted their spatial orientation to the prevailing surface conditions.

Only in the case of the *S. aureus* lipid film created on the AA/HA subphase was a sharp decrease of relative surface pressure after 0.5 h observed (Figure 5A). In this case, however, at π = 30 mN/m, a smaller amount (than in the monolayer for AA/Ch subphase) of monolayer inhomogeneities was observed (Figure 6). This can mean that under this pressure, the monolayer is more homogeneous. In turn, that suggests that HA affects the *S. aureus* lipid monolayer, but not as significantly as Ch. This is probably due to differences in the charge of both polysaccharides, HA is a polyanion with a strong negative charge and Ch is a polycation which can strongly interact with CL (one of components of the *S. aureus* monolayer with negative charge). This conclusion may be supported by the fact that CL, because of the restricted movement resulting from its size, is not able to form intra- or intermolecular hydrogen bonds that can stabilize the negative charge of the head.

Therefore, the unshielded negative charge can easily interact with protons and cations in solution, e.g., chitosan molecules. The interaction reduces the effective size of the polar head, promoting the formation of nonlamellar structures [41]. In addition, CL induces the folding structure locally on the membrane, before reaching the collapsed state upon further compression [39]. That effect can be also strengthened by the Ch-presence. Consequently, we can observe the increase of the thickness and packing degree of the *S. aureus* lipid monolayer. The model for a folding structure assumes that the increased interfacial area forming a curvature helps the CL molecules to minimize the repulsive force caused by the unsaturated hydrocarbon chains and the negatively charged headgroups. However, the collapsed structure is an even more thermodynamically favorable process. In the collapsed structure, the internal repulsion between hydrocarbon chains is drastically reduced, and the negatively charged headgroups repulsion is completely gone [39].

Due to technical reasons, i.e., too much scattering of the laser light emitted by the BAM by the large TiO_2 particles, it was impossible to record the structure and thickness of *S. aureus* lipid films on the subphases containing this component. However, taking into account the above dependencies, we can conclude that both the π/π_0 ratio over time for *S. aureus* lipid films on AA/TiO_2 and AA/Ch/TiO_2 subphases and the significant size of hysteresis loops (mainly on AA/Ch/TiO_2) also may indicate the partial miscibility of lipid components in the membrane at π = 30 mN/m or its collapse. As a result, both subphases have a destructive effect on the monolayer made of lipids isolated from *S. aureus*.

In turn, the π/π_0 = f(t) ratio determined for the *S. aureus* lipid films on the AA/HA/TiO_2 and AA/Ch/HA/TiO_2 subphases decreased with time. On the one hand, this would suggest no monolayer collapse at π = 30 mN/m, so the decrease in π/π_0 would result from the relaxation of the lipid mixture at the interface. On the other hand, taking into account the fact that after 2500 s the *S. aureus* lipid film on AA/Ch/HA/TiO_2 reached a stable level of π/π_0 ratio, while on AA/HA/TiO_2 it continued to decline after analogous time, we can conclude that their effect on the membrane was not the same and each of the cases discussed must be considered separately.

Based on the results obtained for AA/Ch and AA/Ch/TiO_2 subphases, the following changes can be visible as an effect of TiO_2 addition to Ch during *S. aureus* lipid film formation: (I) shift of the *S. aureus* lipid film π-A isotherm towards higher molecular areas, (II) decrease in the C_S^{-1} values, (III) lowering π/π_0, but maintaining stability for 1 h, (IV) increasing the hysteresis loop; (in relation to data received for *S. aureus* lipid film on the AA/Ch subphase). The similar modifications of the *S. aureus* lipid film were observed (except for point II) after the introduction of TiO_2 particles into the mixed AA/Ch/HA subphase. It is also worth noting that the *S. aureus* lipid film on the AA/Ch/HA subphase showed higher inhomogeneity, suggesting partial miscibility of lipid components, than existed on the AA/HA subphase. Considering, in the same way, the influence of TiO_2 addition to the AA/HA subphase on the behavior of the *S. aureus* lipid film, we do not observe the above-mentioned relationships (I–IV). Thus, we can conclude that the use of the AA/Ch/HA/TiO_2 subphase could also have had a destructive effect on *S. aureus* lipid monolayer. Moreover, the decrease in π/π_0 could be dictated not only by lipid relaxation, but also by additional interference of such a complex subphase on the lipid components present in the membrane. On the other hand, the combination of TiO_2 with HA could have contributed to the further reduction of the disturbance effect on the *S. aureus* lipid monolayer. As a result of the existence of competition for interactions between the negatively charged lipid components of the *S. aureus* lipid layer and HA molecules with TiO_2 particles.

3.3. The Influence of the Ch, HA and TiO_2 on the E. coli Lipids Monolayers

3.3.1. Behavior and Packing of *E. coli* Lipid Monolayers during Compression

The π-A isotherm for the film composed of lipids isolated from *E. coli* registered on the H_2O subphase was called the reference. For this case, the G-LE phase transition occurred at A = 85.5 Å^2. As the compression progressed, the surface pressure increased, finally, the monolayer collapsed at surface pressure value π = 47.4 mN/m. The shape of the isotherm did not reveal any inflections. On the other hand, the dependence of the compression modulus (C_S^{-1}) as a function of surface pressure indicated that the monolayer was mainly present in the expanded liquid state. Only in the pressure range of 20–30 mN/m, the C_S^{-1} had the values characteristic for condensed liquid.

The use of AA solution as a subphase resulted in a faster pressure increase in the range of π 0–25 mN/m (A_0 = 86.2 Å^2). On the other hand, at higher pressure values, the isotherm course for *E. coli* lipids was similar to the reference. After all, layer collapse occured at slightly lower value π = 46.3 mN/m. Moreover, the film in the entire pressure range did not exceed the C_S^{-1} values characteristic for the state of expanded liquid. A similar relationship was observed for the *E. coli* lipid film on the subphase with hyaluronic acid (AA/HA). There was also a shift in the G-LE transition towards larger areas (A_0 = 87.8 Å^2). However, above the pressure of 25 mN/m, the isotherm course was the same as that on the AA subphase (Figure S5). Moreover, the plot of the C_S^{-1} = f (π) relationship was similar. In both cases (AA and AA/HA subphases), the maximum C_S^{-1} values for *E. coli* lipid film were observed at 15, 25, and 40 mN/m surface pressures.

The use of TiO_2 in the subphase (AA/TiO_2) resulted in visible changes in the course of the isotherm for the *E. coli* lipids layer. The G-LE transition took place at A = 92.2 Å^2. Thereafter, there was a much milder increase in pressure as the compression progressed. Additionally, an inflection was observed at a pressure of approximately 40 mN/m, which can indicate a collapse of one or more constituents of the monolayer, or their partial miscibility. In turn, the graph of the compression modulus as a function of surface pressure plotted two maxima of values at π value 25 mN/m and 53 mN/m. On the other hand, the observed minimum at 40 mN/m indicates partial miscibility and/or expulsion of one kind of the components (phosphatidylethanolamine (PE), CL, or PG) from the monolayer, rather than phase transition, because the C_S^{-1} value does not drop to zero [40]. When such a situation takes place, in the monolayer the coexisting areas with packed and not-condensed domains can be observed. Higher surface pressure values may come to reveal that situation, which is reflected in a decrease in the value of the compression modulus. After combining HA with TiO_2 in the subphase, the same course of the isotherm for the *E. coli* lipid film was recorded like that on the AA/TiO_2 subphase. The difference is that it started at 79.7 Å^2 and finally collapsed at π value ca. 61 mN/m. This was also reflected in the C_S^{-1} = f(π) plot. Two peaks at π = 27 mN/m and 54 mN/m were also observed. However, in the π range 0–15 mN/m, the C_S^{-1} values were similar to those obtained for *E. coli* lipid layer on the AA/HA subphase. The similar course of isotherm was received also by [42] for the POPE/POPG (3:1 ratio) on the 15 mM NaCl solution subphase. Additionally, after adding the CL to this Langmuir film, disappearance of the plateau at ~40 mN/m was noted [43]. Considering the fact that in our case *E. coli* lipid film consists of PE, PG, and CL, it can be concluded that the presence of positive charged TiO_2 particles and further HA (containing Na^+ ions) provides an expulsion of CL from the monolayer, simultaneously stabilizing the created membrane.

However, the use of chitosan with acetic acid as a subphase did not affect the shape of the *E. coli* lipid film isotherm compared to the reference. There was only a shift in its course towards larger areas (A_0 = 90.5 Å^2). A plateau suggesting a collapse took place at surface pressure value π = 46.2 mN/m. A similar relationship occurred on the subphase containing both biopolymers (AA/Ch/HA). The *E. coli* lipid film during compression plotted the similar isotherm as on AA/Ch, with its position approaching the reference isotherm (A_0 = 88.6 Å^2). The values of the compression modulus for lipid films on these subphases (AA/Ch and AA/Ch/HA) were also practically comparable in the entire pressure range (Figure S5). The only difference was that in the AA/Ch subphase three maxima of C_S^{-1} values were observed (at π ca. 20, 30, and 37 mN/m), but on the AA/Ch/HA subphase only two maxima (at π = 24 and 37 mN/m).

After applying a mixed subphase of TiO_2 with chitosan, despite the fact that the isotherm for the lipid film showed a similar inflection in the region of π = 40 mN/m (as for AA/HA and AA/HA/TiO_2), a shift of its course towards larger areas occurred (A_0 = 98.1 Å^2). Additionally, a much smoother increase of surface pressure π was noted when condensed domains were formed in the monolayer. Above π = 25 mN/m, the course of the curve was almost identical to that obtained on H_2O subphase. Finally, at π = 46.7 mN/m, a plateau was visible, suggesting a collapse of the membrane. However, significant differences were noticed for the value of the compression modulus at individual stages of compression process. At the beginning of the compression, an increase in the C_S^{-1} value was observed with a tendency comparable to that obtained for *E. coli* lipid film on AA/TiO_2. In the range of π = 10–20 mN/m, a relatively constant C_S^{-1} value was obtained, followed by a sharp increase, reaching a maximum at π = 35 mN/m. Then, at π = 40 mN/m, the minimum C_S^{-1} values suggested partial miscibility and/or an ejection of some components (CL, PE, or PG) from the monolayer, as there was another increase in the modulus value, reaching a maximum at π ca. 43 mN/m. Finally, the monolayer was collapsed at π = 46.2 mN/m.

In turn, the HA-addition to the AA/Ch/TiO_2 subphase (i.e., AA/Ch/HA/TiO_2) contributed to the softening of the inflection at π = 40 mN/m. Moreover, the increase of π occurred at A_0 = 93.3 Å^2 (G-LE phase transition). Next, π values in the monolayer kept rising steadily as the compression progressed until it reached a plateau at π = 46.4 mN/m. Similarly, C_S^{-1} initially increased with the trend also plotted by the *E. coli* lipid film on the AA/Ch/TiO_2 subphase, but above π > 15 mN/m, the increase was slightly slower, until to a maximum of about π = 35 mN/m.

The different course of the plots for *E. coli* lipid monolayers on the AA/TiO_2 and AA/HA/TiO_2 subphases was caused by the enhancement of the attracting interactions between the Langmuir film components (in the case of *E. coli*, these are: CL, PG, PE), which may be result of repulsion between lipid molecules and subphase components, or the penetration of TiO_2 particles and/or HA molecules into the monolayer (A_{lim} = 65.5 Å^2, A_{lim} = 63.4 Å^2, respectively). In the remaining subphases, the lipid molecules in the films occupied a larger area in a tightly packed monolayer (for AA A_{lim} = 69.8 Å^2, for AA/Ch/HA A_{lim} = 71.3 Å^2, for AA/HA A_{lim} = 71.7 Å^2, for AA/Ch A_{lim} = 72.9 Å^2, for AA/Ch/HA/TiO_2 A_{lim} = 73.1 Å^2, for AA/Ch/TiO_2 A_{lim} = 76.9 Å^2). The obtained packing parameters can prove the weakening of the attractive interactions/forces between the membrane components caused by the presence of the subphase ingredients. Moreover, it can be seen that the presence of HA in the subphase, compared to the corresponding subphase without this biopolymer, shifts the isotherm for *E. coli* lipid film towards smaller areas per molecule. Such a relationship was also observed previously when examining the effect of HA, Ch and TiO_2 on the DPPC membrane (except AA/Ch/HA/TiO_2) [28]. On the other hand, during an investigation of the effect of the same substances on the DOPC monolayer behaviour, a completely opposite relationship was obtained [Ładniak et al., personal communication], i.e., an increase in the A_{lim} parameter on the subphases containing HA compared to those without it.

Moreover, the observed changes resulting from the mixing in the subphase, i.e., Ch with TiO_2 or HA with TiO_2, in relation to differences caused by individual components, confirm our previous reports [18,28,44–47]. Clearly, in the case of the AA/HA/TiO_2 subphase, TiO_2 dominates in the interactions with Langmuir film components. This is completely different behaviour for AA/Ch/TiO_2, where Ch limits/eliminates changes caused by TiO_2. This can suggest that TiO_2 particles are dispersed in the HA matrix in a way that did not limit their contact with the external environment. On the other hand, the Ch molecules adsorb on the surface of TiO_2 particles, preventing their interaction with the environment. As was mentioned earlier, the differences for these polysaccharides are due to the presence of opposite charges in their molecules.

Larger discrepancies in the position of isotherms for *S. aureus* lipid films (Figure 4) than *E. coli* (Figure 7) were an effect of the higher sensibility of PG affecting the ion presence in the subphases (caused an expansion of the isotherm) than PE. Both phospholipids (PG and PE) have the capability to form hydrogen bonds. However, in the case of PG molecules, these interactions are weaker due to electrostatic repulsions existing between negatively charged phosphatidylglycerol moieties [42].

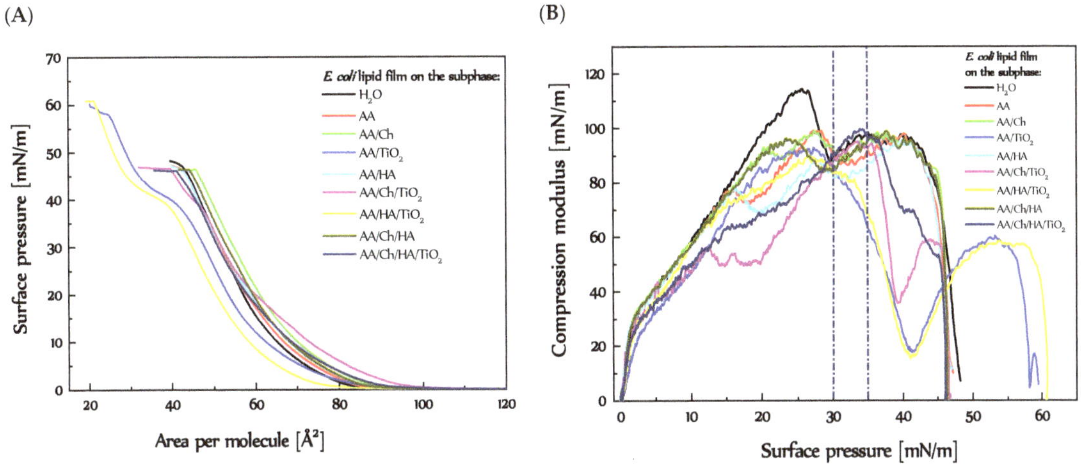

Figure 7. (**A**) π-A isotherms obtained by means of the Langmuir technique and determined based on them (**B**) $C_S{}^{-1} = f(\pi)$ relationship for *E. coli* lipid monolayers on the individual subphases (more details Figure S5).

3.3.2. Stability of the *E. coli* Lipid Monolayers

Figure 8A presents the one-hour relative stability of the *E. coli* lipid films compressed to π = 30 mN/m on the different subphases. The most interesting fact is that the *E. coli* lipid films on almost all subphases (except AA/TiO_2 and AA/HA/TiO_2) achieved constant values of surface pressure the faster, than reference one. Contrariwise, for *E. coli* lipid monolayers on AA/TiO_2 and AA/HA/TiO_2, a greater decrease in surface pressure values occurred, although after 20 min a milder drop was observed. On the other hand, in the case of H_2O, AA, and AA/HA subphases, a uniform decrease of π value took place all the time. In turn, *E. coli* lipid film on the other subphases got a permanent value of fewer than 20 min.

The data from compression-decompression measurements are presented in Figure 8B. For all samples, the descending curves were shifted to the left side of the graph (except *E. coli* lipid films on the H_2O and AA subphases). The lipids film on H_2O and AA subphases did not return to the starting state, i.e., stable surface pressure suggesting transition monolayer to the gas phase occurred at ca.3 and 5 mN/m, respectively. In addition, a downtime plateau was 1.1 Å^2 and 5.0 Å^2, accordingly. The *E. coli* lipid films on the AA/HA, AA/TiO_2, and AA/HA/TiO_2 subphases revert to the state before compression, but the descending curves were shifted at 30 mN/m about 0.7 Å^2, 3.4 Å^2, and 3.2 Å^2, at near-zero surface pressure about 6.0 Å^2, 13.2 Å^2, and 6.0 Å^2, respectively (Figures S6 and S7). The biggest hysteresis loop was observed for *E. coli* lipid film on the AA/TiO_2 subphase, where the start of decompression occurred at 3.4 Å^2 smaller area and finished at 13.2 Å^2 smaller area, in relation to the compression curve as an effect of TiO_2 particles.

None of the *E. coli* lipid films on the Ch-containing subphases achieved zero surface pressure after decompression, although the inflection informing about the transitions of the monolayers to the gas phase was visible, but at slightly higher π values (below 1 mN/m). Additionally, the differences in the areas of the compression and decompression curves at π = 30 mN/m and 0 mN/m were as follows: AA/Ch: 2.1 Å^2 and 3.0 Å^2, AA/Ch/TiO_2: 3.4 Å^2 and 2.8 Å^2, AA/Ch/HA: 0.9 Å^2 and 8.2 Å^2, AA/Ch/HA/TiO_2: 2.1 Å^2 and 8.5 Å^2 (Figures S6 and S7).

Besides, analyzing the changes of structure and stability for both lipid films (*E. coli* and *S. aureus*) on different subphases in relation to the composition of lipid layers (*E. coli* 5% CL, 15% PG and 80% PE or *S. aureus* 42% CL and 58% PG [37,38]), it can be suggested that the monolayer plateau occurring at ~40 mN/m refers mainly to CL. It is a component of both lipid films, but in various proportions. However, it is clearly visible regarding compression-decompression measurements that the size of downtime plateau closely correlates with CL amount in the appropriate bacterial lipid film.

(A) (B)

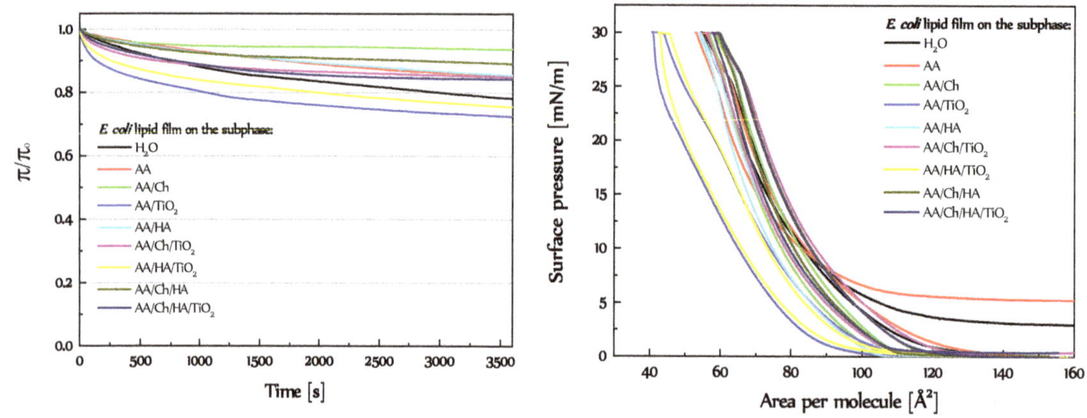

Figure 8. (**A**) π/π_0 = f(t) and (**B**) hysteresis loops (more details Figures S6 and S7) obtained by compression and decompression of the *E. coli* lipid monolayers on the individual subphases.

3.3.3. Thickness and Structure of the *E. coli* Lipid Monolayers

Visualization of the *E. coli* lipid film (Figure 9) on the H_2O subphase showed that at low π values < 1 mN/m, condensed domains were formed, and the entire structure was in the form of a network with holes of various diameters. Additionally, the high-packing areas were separated by empty bands/spaces. In the π range of 1–10 mN/m, a compact, homogeneous structure with no visible defects was formed. Above π = 10 mN/m, aggregates appeared in the form of brighter points, and their amount increased with the progress of compression process. Their greatest quantity occurred at surface pressure values π > 45 mN/m. This is also reflected in the thickness of the film (Table 4), which at π = 0.5 mN/m was 1.5 nm and gradually increased to the value of 2.5 nm ($\pi \geq 40$ mN/m).

Table 4. Thickness of the *E. coli* lipid monolayers during compression.

Surface Pressure [mN/m]	Thickness of *E. coli* Monolayer Obtained on Different Subphases [nm]				
	H_2O	AA	AA/Ch	AA/HA	AA/Ch/HA
0.5	1.5	1.4	1.4	0.7	1.6
5	1.7	1.7	1.7	1.2	1.7
10	1.9	1.8	1.9	1.4	1.7
15	2.0	2.0	2.0	1.6	2.4
20	2.1	2.1	2.2	1.7	2.5
25	2.2	2.2	2.3	1.8	2.5
30	2.3	2.3	2.3	1.8	2.7
35	2.4	2.3	2.4	1.9	2.8
40	2.5	2.4	2.5	2.1	2.9
45	2.5	2.5	2.6	2.1	2.9

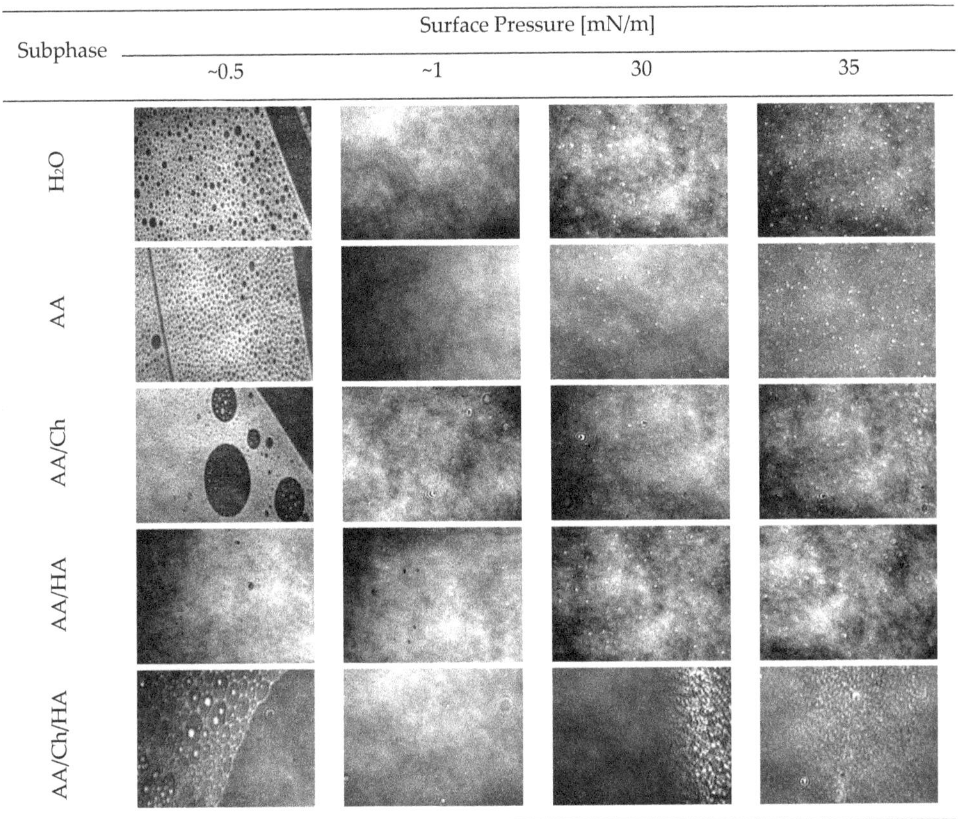

Figure 9. Structure of the *E. coli* lipid monolayer during their compression (at surface pressures of 0.5, 1, 30 and 35 mN/m) on the different subphases (AA—acetic acid, AA/Ch—chitosan on acetic acid, AA/HA—hyaluronic acid in acetic acid, AA/Ch/HA—chitosan in acetic acid with the addition of hyaluronic acid).

The use of acetic acid as a subphase for the *E. coli* lipid film had no significant effect on its structure and thickness. Both features were comparable to that obtained with H_2O subphase. Moreover, at a pressure near 1 mN/m, a compact, uniform layer was formed with visible domains above $\pi > 10$ mN/m. However, at value $\pi > 40$ mN/m a characteristic band of focused bright points was observed.

The presence of chitosan in the subphase (AA/Ch) slightly changed the structure of the obtained *E. coli* lipid film. It caused additional less condensed regions within the packed mesh structure. Moreover, it was observed that the regions with a tight-knit (reticulated) structure were more densely packed than those obtained on H_2O and AA subphases. Ultimately, however, a compact and homogeneous layer was obtained with no visible damage/disturbance. Defects in the form of small dark craters/holes began to appear at $\pi > 15$ mN/m. At larger surface pressure value $\pi > 25$ mN/m, bright points were also observed (similar to that occurred in other cases). The greatest amount of them was found at $\pi > 35$ mN/m. However, the observed changes did not significantly alter the thickness of the *E. coli* lipid film compared to that obtained with H_2O subphase.

Also, the mixing of acetic acid with hyaluronic acid in subphase (AA/HA) contributed to the acceleration of the formation of a compact structure of the *E. coli* lipid film, but with twice smaller thickness. At first, a meshwork structure with small holes was observed which quickly formed a uniform layer. However, at $\pi > 20$ mN/m, domains in the form of bright points gradually appeared, and their amounts increased significantly above $\pi > 40$ mN/m. The changes caused by the presence of HA in the subphase also influenced the formation of an *E. coli* lipid film of much lower thickness during the entire compression cycle, compared to all tested samples (0.4–0.5 nm in the range of $5 \leq \pi \leq 45$ mN/m, relative to the reference).

The mixing of biopolymers in the subphase (AA/Ch/HA) also influenced the structure of the *E. coli* lipid film. At low-pressure values, foam-like structures were visible, which confirm the coexistence of the gas and liquid phases. However, a uniform, dense layer was quickly formed, with small structures in both forms, e.g., black holes and bright spots (seen in other cases). Additionally, above surface pressure of 25 mN/m, clusters of bright points in the form of band-zones were observed (which also occurred in the *E. coli* lipid film on the AA subphase). Despite the occurrence of these above structures, the *E. coli* lipid film on AA/Ch/HA had the greatest thickness among all the tested films (above $\pi = 10$ mN/m it had a value greater by about 0.4 nm in relation to the *E. coli* lipid film on H_2O subphase).

Compared to BAM images obtained on the AA, AA/Ch, AA/HA, and AA/Ch/HA subphases for *S. aureus* lipid films (Figure 6), the faster appearance of domains was observed for *E. coli* films on the same subphases. As already mentioned, this was an effect of higher sensibility of PG affecting the ion presence in the subphases than PE [42]. On the other hand, Sennato et al. [48] suggest that the increase of DPPE concentration promotes the phase separation of the film into components with a different degree of stability, in addition the less stable one collapses very close to the CL collapse pressure. Thus, in this case, i.e., *E. coli* lipid monolayers, because they consist in 80% of PE, more discrepancies in the course of the π-A and $C_S^{-1} = f(\pi)$ relationships were observed.

Moreover, the use of the HA subphase (AA/HA) had a greater impact on the thickness of the resulting *E. coli* (Figure 9) than *S. aureus* lipid film (Figure 6). The structure of lipid monolayer of *S. aureus* (Figure 6) was most influenced by Ch. A consequence of adding the cationic antimicrobial agent (e.g., Ch) interacting with anionic lipids (PG, CL) could result in the emerging complex membrane domains becoming unstable and/or having altered properties. If these natural membrane domains are biologically relevant, modifying their properties may also lead to cell death [38]. In turn, the carboxylic ($-COO^-$), hydroxyl (-OH), and amine groups ($-NH_3^+$) of HA can strongly interact with positively charged ethanolamine and negatively charged phosphorus groups of PE.

3.4. *Microbiological Tests*

E. coli, a Gram-negative organism model, and *S. aureus*, a Gram-positive one, were used to study the antimicrobial activity of AA/Ch, AA/TiO_2, AA/HA, AA/Ch/TiO_2, AA/HA/TiO_2, AA/Ch/HA, and AA/Ch/HA/TiO_2 mixtures. All tested compounds showed antimicrobial activity against *E. coli* and *S. aureus* to a greater or lesser extent. However, there were considerable differences in the sensitivity to the compounds between these bacteria and methods used. The sensitivity of *E. coli* and *S. aureus* to the compounds is presented in Table 5 and Figure 10. The highest anti *E. coli* effect was caused by AA/Ch and AA/Ch/HA/TiO_2 dispersions. The AA/Ch dispersion was responsible for the 96% mortality rate in *E. coli* in the CFU counting test and 74% in Live/Dead staining. The AA/Ch/HA/TiO_2 dispersion decreased the viability of *E. coli* by 93% in the CFU counting test and 91% in Live/Dead staining. AA/HA/TiO_2, AA/Ch/HA, and AA/Ch/TiO_2 were also highly active against *E. coli*. Among the tested mixtures, AA/TiO_2 and AA/HA showed the lowest activity against *E. coli* in both tests used.

Table 5. Mortality (%) of *S. aureus* and *E. coli* in the presence of the compounds measured using a colony-counting assay (A) and live/dead staining with fluorescent intensity measurements (B).

Sample	Mortality (%)			
	E. coli		*S. aureus*	
	A	B	A	B
AA/Ch	96 ± 2	74 ± 6	96 ± 1	72 ± 7
AA/TiO_2	78 ± 4	58 ± 7	70 ± 5	65 ± 3
AA/HA	73 ± 5	49 ± 4	25± 4	35±5
AA/Ch/TiO_2	92 ± 2	73 ± 8	43 ± 1	41 ± 2
AA/HA/TiO_2	94 ± 2	74 ± 6	0	52 ± 2
AA/Ch/HA	92 ± 1	82 ± 5	62 ± 1	60 ± 4
AA/Ch/HA/TiO_2	93 ± 2	91 ± 1	42 ± 4	41 ± 1

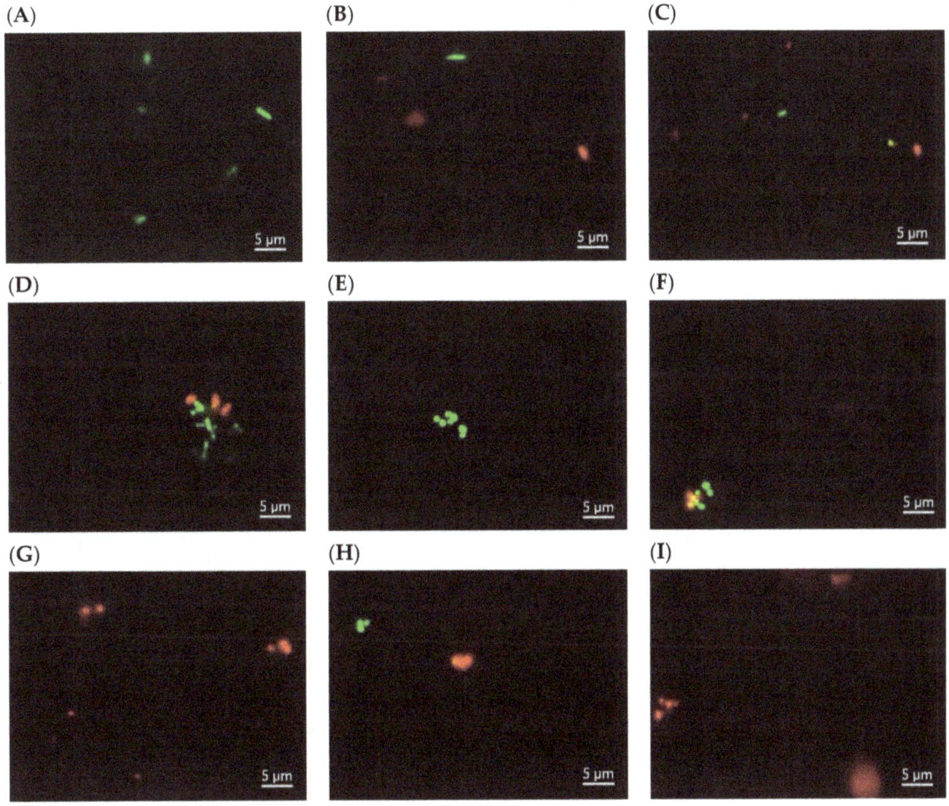

Figure 10. *E. coli* with Live/Dead staining and imaged with LSCM. The green signal is due to the dye SYTO9, indicating alive cells while the red signal is due to PI which marks the dead cells. (**A**) *E. coli* control, (**B**) *E. coli* treated with AA/TiO_2, (**C**) *E. coli* treated with AA/Ch, (**D**) *E. coli* treated with AA/Ch/HA/TiO_2. *S. aureus* with Live/Dead staining and imaged with LSCM. (**E**) *S. aureus* control, (**F**) *S. aureus* treated with AA/TiO_2, (**G**) *S. aureus* treated with AA/Ch, (**H**) *S. aureus* treated with AA/HA, (**I**) *S. aureus* treated with AA/Ch/TiO_2. Magnification 1000×.

Similar to *E. coli*, chitosan was also very active against *S. aureus*, as the killing effect was 96% in the CFU counting test and 72% in Live/Dead staining. AA/TiO_2 was the second most lethal suspension against *S. aureus*. Although chitosan and TiO_2 alone were the most active against *S. aureus*, 96% and 70% (72% and 65% in Live/Dead staining), respectively, no synergistic effect of their antibacterial activity in mixed sample AA/Ch/TiO_2 was observed.

On the other hand, AA/HA was characterized by the lowest killing activity against *S. aureus*, amounting to 25% in the CFU counting assay and 35% in Live/Dead staining. Despite its high activity against *E. coli*, AA/HA/TiO_2 showed no bactericidal activity against *S. aureus* (0%) in the CFU counting assay. However, Live/Dead staining showed the lethal effect of AA/HA/TiO_2 on *S. aureus* (52% mortality). Probably, these differences are due to the fact that AA/HA/TiO_2 causes *S. aureus* cells to enter a dormancy state termed viable but non-culturable (VBNC).

Comparing the effect of AA/Ch, AA/HA, and AA/TiO_2 on the mortality of both bacteria, it is clearly visible (and simultaneously confirmed) that AA/Ch has the biggest antibacterial activity. Nonetheless, the synergism/strengthening of that property was observed for the mixture of AA/HA/TiO_2 and AA/Ch/HA/TiO_2 against *E. coli* in Live/Dead staining measurements. Taking into the consideration that Ch and HA possess a similar structure to the LPS in OM of Gram-negative bacteria (consisting of repeating oligosaccharide units of various sugar residues connecting by glycosidic linkages), it probably facilitates the interactions of biopolymers with the bacterial membranes. Then, the TiO_2 particles surrounded by polysaccharides being in contact with bacteria can be successively released. In turn, the combination of Ch with TiO_2, does not affect the *E. coli* mortality (for both AA/Ch and AA/Ch/TiO_2 mixtures the same level of antibacterial activity is noted).

The rate of bacterial mortality under the influence of the tested compounds correlates with the values of physicochemical parameters obtained for model lipid bacterial membranes using the Langmuir technique and Brewster angle microscopy. As we expect, given that Ch has the most antibacterial activity, the greatest impact on the change of A_0, and A_{lim} parameters, hysteresis loops, thickness, and a conformation of molecules in monolayers (Figures 6 and 9) is observed for this biopolymer. In the case of Ch-containing subphases, the largest packing disturbance is manifested in the larger area occupied by molecule in the compact layer (A_{lim} was in the range of 6.8–27.4 Å^2 for *S. aureus* lipid films, and 3.5–9.1 Å^2 for *E. coli* lipid films) in relation to the reference lipid monolayer on water. Additionally, a strong interaction between subphase components (mainly Ch) and lipid molecules in the monolayer during compression makes the relaxation of lipid molecules at the interface difficult (Figures 5A and 8A). Besides, Ch (alone or in combination with HA) has the most influence on the monolayer structures, leading to the appearance of a large amount of condensed domains (Figures 6 and 9), and causing the significant increase (average of about 0.4nm) of their thickness (Tables 3 and 4) in comparison to the reference. Despite the fact that microbiological tests did not reveal the synergism of antibacterial properties of Ch in combination with TiO_2 and/or HA against *S. aureus*, in the Langmuir measurements an enhancement of this effect was observed. For *S. aureus* monolayers on the AA/Ch/TiO_2, AA/Ch/HA, and AA/Ch/HA/TiO_2, bigger A_{lim} values and hysteresis loops in relation to the bacterial layer on the AA/Ch subphase were noted. This may be because the sensitivity of the two methods is different. As far as the application aspect is concerned, it is important, because even subtle changes in the typical bacterial monolayer may, in the long run, suggest the intensification of the antibacterial properties of the tested phase.

However, considering the mortality rate of the bacteria in the microbiological tests, we have to take into account that the bacterial membranes consist of not only lipids, but also of other components, such as proteins and polysaccharides. This fact and knowledge about the structure of cell wall Gram(−) (containing IM and OM [1]) and Gram(+) (having a thick murein layer) bacteria allow to explain the different levels of bacterial mortality. Between Ch, HA, and TiO_2 compounds and bacterial components, other specific interactions could also occur. For example, researchers [49] suggest that HA, as a component of the mammalian extracellular matrix, is used by pathogens as a carbon source during tissue invasion and replication, which may preferentially attract bacteria to the nanogel surface. Therefore, it can be an effective coating for antibacterial substances, which, as a result of "digesting" HA, can successively be released and interact with phospholipids of the bacterial membrane, increasing its permeability and causing subsequent disruption of the cell wall. Antibacterial ability was also observed by Makvandi et al. [50], who did not show the antibacterial properties of HA against Gram(+) *S. aureus*, *Bacillus subtilis*, and Gram(−) *Pseudomonas aeruginosa*, *E. coli*, but introduced Ag particles into the coatings. In turn, Abdelrahman et al. [51] observed the antibacterial effect only after the combination of HA with Ch. Similarly, Chen et al. [52] made a similar report for HA adduct obtained by N-halogenation reaction of HA.

4. Conclusions

The presence of Ch, TiO_2, and HA in the subphases influenced the DPPG membrane structure, mainly by the modifications of packing degree. The greatest changes of compression modulus and average surface area per molecule occurred in the presence of Ch and/or TiO_2 at pressures values close to those corresponding to the biological membranes. At this pressure, the DPPG monolayers formed a less condensed film on all tested subphases, compared to standard layer occurring on water. The Ch and TiO_2 caused significant deviations from the typical behaviour of DPPG membrane in all conducted experiments. The Ch penetration caused the change in orientation of the DPPG molecules and slowed the creation of condensed domains in the monolayers. The Ch was located preferably into loose monolayers creating the Ch-rich and lipid-poor regions, producing hydrophobic interactions with the lipid tails. As an effect, increasing repulsive Coulombic forces between negatively charged DPPG headgroups occurred. In contrast, TiO_2 particles could be located just below the surface of the lipid film and form strong bonds with PO_4^- governed not only by electrostatic forces alone, but also of specific adsorption. This indirectly showed that one of the mechanisms of antibacterial action of mixtures containing chitosan, TiO_2, and/or hyaluronic acid is probably based on bacterial membrane disturbance.

The monolayer morphology of all bacterial membranes showed inhomogeneities strictly dependent on the specific subphases compositions and in the case of *E. coli* and *S. aureus* lipid films suggested the expulsion of one of the components (CL, PE, or PG) from the monolayer or partial miscibility.

All components of the subphases influenced the structure of *E. coli* and *S. aureus* lipids monolayer but through different mechanisms. Analyzing the hysteresis loops on model *S. aureus* monolayer, it could be concluded that Ch and TiO_2 affected the monolayer structure during compression. In turn, HA had a greater impact on a packed monolayer. This correlated to the different compositions of lipid membranes. The carboxylic, hydroxyl, and amine groups of HA could strongly interact with a positive charge of ethanolamine and an anionic charge of phosphorus groups of PE. Therefore, the use of the HA had a stronger influence on the thickness of the *E. coli* (80% PE) than on *S. aureus* lipid film (0% PE). The structure of *S. aureus* monolayer was most influenced by Ch molecules interacting with anionic lipids (PG, CL), which could result in the emerging complex membrane domains becoming unstable and/or having altered properties. In effect, these interactions between Ch molecules and phospholipids enhanced the permeability of biological membranes and may be one of the mechanisms leading to cell death of the bacteria as shown in microbiological tests.

The obtained materials with in-depth physicochemical characteristics brought us closer to functional systems useful in the production of biocompatible antibacterial films and/or skin substitutes.

Supplementary Materials: The following supporting information can be downloaded online, Figure S1: $\pi/\pi0$ = f(t) relationships for the DPPG monolayer on biopolymers-containing subphases (inset: for TiO_2-containing subphases); Figure S2: π-A isotherms obtained by means of the Langmuir technique and determined based on them C_S^{-1} = f(π) relationships (inset) for the *S. aureus* lipid film on biopolymers-containing subphases; Figure S3: Hysteresis loops obtained by compression and decompression of the *S. aureus* lipid film on biopolymers-containing subphases; Figure S4: Hysteresis loops obtained by compression and decompression of the *S. aureus* lipid film on TiO_2-containing subphases; Figure S5: π-A isotherms obtained by means of the Langmuir technique and determined based on them $C_S^{-1} = f(\pi)$ relationships (inset) for the *E. coli* lipid film on biopolymers-containing subphases; Figure S6. Hysteresis loops obtained by compression and decompression of the *E. coli* lipid film on biopolymers-containing subphases; Figure S7. Hysteresis loops obtained by compression and decompression of the *E. coli* lipid film on TiO_2-containing subphases.

Author Contributions: Conceptualization, A.Ł., M.J. and A.E.W.; methodology, A.Ł., M.J., A.E.W. and M.P.-S.; investigation, A.Ł. and M.P.-S.; writing—original draft preparation, A.Ł. and M.P.-S.; writing—review & editing, A.Ł., M.J. and A.E.W.; visualization, A.Ł., M.J. and A.E.W.; supervision, M.J. and A.E.W. All authors have read and agreed to the published version of the manuscript.

Funding: This research received no external funding.

Institutional Review Board Statement: Not applicable.

Informed Consent Statement: Not applicable.

Conflicts of Interest: The authors declare no conflict of interest.

Sample Availability: *E. coli* and *S. aureus* bacteria are available from the authors.

References

1. Sohlenkamp, C.; Geiger, O. Bacterial membrane lipids: Diversity in structures and pathways. *FEMS Microbiol. Rev.* **2016**, *40*, 133–159. [CrossRef] [PubMed]
2. Eeman, M.; Deleu, M. From Biological membranes to biomimetic model membranes. *Biotechnol. Agron. Soc.* **2010**, *14*, 719–736.
3. Guzmán, E.; Santini, E.; Ferrari, M.; Liggieri, L.; Ravera, F. Interaction of particles with Langmuir monolayers of 1,2-dipalmitoyl-*sn*-glycero-3-phosphocholine: A matter of chemistry? *Coatings* **2020**, *10*, 469. [CrossRef]
4. Brockman, H. Lipid monolayers: Why use half a membrane to characterize protein-membrane interactions? *Curr. Opin. Struct. Biol.* **1999**, *9*, 438–443. [CrossRef]
5. Maget-Dana, R. The monolayer technique: A potent tool for studying the interfacial properties of antimicrobial and membrane-lytic peptides and their interactions with lipid membranes. *Biochim. Biophys. Acta* **1999**, *1462*, 109–140. [CrossRef]
6. Marsh, D. Lateral pressure in membranes. *Biochim. Biophys. Acta* **1996**, *1286*, 183–223. [CrossRef]
7. Chapman, M.N.; Micelles, D. *Monolayers and Biomembranes*; Wiley-Liss: New York, NY, USA, 1995; Volume XII, ISBN 0-471-56139-8.
8. Stefaniu, C.; Brezesinski, G.; Möhwald, H. Langmuir monolayers as models to study processes at membrane surfaces. *Adv. Colloid Interface Sci.* **2014**, *208*, 197–213. [CrossRef] [PubMed]
9. Guzmán, E.; Santini, E. Lung surfactant-particles at fluid interfaces for toxicity assessments. *Curr. Opin. Colloid Interface Sci.* **2019**, *39*, 24–39. [CrossRef]
10. Garcia-Mouton, C.; Hidalgo, A.; Cruz, A.; Pérez-Gil, J. The lord of the lungs: The essential role of pulmonary surfactant upon inhalation of nanoparticles. *Eur. J. Pharm. Biopharm.* **2019**, *144*, 230–243. [CrossRef]
11. Hoenig, D.; Moebius, D. Direct Visualization of monolayers at the air-water interface by Brewster angle microscopy. *J. Phys. Chem.* **1991**, *95*, 4590–4592. [CrossRef]
12. Vollhardt, D. Brewster angle microscopy: A preferential method for mesoscopic characterization of monolayers at the air/water interface. *Curr. Opin. Colloid Interface. Sci.* **2014**, *19*, 183–197. [CrossRef]
13. Quemeneur, F.; Rammal, A.; Rinaudo, M.; Pépin-Donat, B. Large and giant vesicles "decorated" with chitosan: Effects of pH, salt or glucose stress, and surface adhesion. *Biomacromolecules* **2007**, *8*, 2512–2519. [CrossRef] [PubMed]
14. Quemeneur, F.; Rinaudo, M.; Pépin-Donat, B. Influence of molecular weight and pH on adsorption of chitosan at the surface of large and giant vesicles. *Biomacromolecules* **2008**, *9*, 396–402. [CrossRef] [PubMed]
15. Mero, A.; Campisi, M. Hyaluronic Acid bioconjugates for the delivery of bioactive molecules. *Polymers* **2014**, *6*, 346–369. [CrossRef]
16. Zhang, Q.-L.; Du, L.-C.; Weng, Y.-X.; Wang, L.; Chen, H.-Y.; Li, J.-Q. Particle-size-dependent distribution of carboxylate adsorption sites on TiO_2 nanoparticle surfaces: Insights into the surface modification of nanostructured TiO_2 electrodes. *J. Phys. Chem. B* **2004**, *108*, 15077–15083. [CrossRef]
17. Wang, M.; Guo, Y.; Wang, Q.; Zhang, X.; Huang, J.; Lu, X.; Wang, K.; Zhang, H.; Leng, Y. Density functional theory study of interactions between glycine and TiO_2/graphene nanocomposites. *Chem. Phys. Lett.* **2014**, *599*, 86–91. [CrossRef]
18. Wiącek, A.E.; Gozdecka, A.; Jurak, M. Physicochemical characteristics of chitosan–TiO_2 biomaterial. 1. Stability and swelling properties. *Ind. Eng. Chem. Res.* **2018**, *57*, 1859–1870. [CrossRef]
19. Gozdecka, A.; Wiącek, A.E. Effect of UV radiation and chitosan coating on the adsorption-photocatalytic activity of TiO_2 particles. *Mater. Sci. Eng. C* **2018**, *93*, 582–594. [CrossRef]
20. Bligh, E.G.; Dyer, J.W. A rapid method of total lipid extraction and purification. *Can. J. Biochem. Physiol.* **1959**, *37*, 911–917. [CrossRef] [PubMed]
21. Davies, J.T. *Interfacial Phenomena*, 2nd ed.; Academic Press: London, UK, 1963; ISBN 978-0-323-16166-4.

22. Krajewska, B.; Wydro, P.; Jańczyk, A. Probing the modes of antibacterial activity of chitosan. Effects of pH and molecular weight on chitosan interactions with membrane lipids in Langmuir films. *Biomacromolecules* **2011**, *12*, 4144–4152. [CrossRef]
23. Mildner, J.; Wnętrzak, A.; Dynarowicz-Latka, P. Cholesterol and cardiolipin importance in local anesthetics–membrane interactions: The Langmuir monolayer study. *J. Membr. Biol.* **2019**, *252*, 31–39. [CrossRef] [PubMed]
24. Rodríguez Patino, J.M.; Sánchez, C.C.; Rodríguez Niño, M.R. Morphological and structural characteristics of monoglyceride monolayers at the air–water interface observed by Brewster angle microscopy. *Langmuir* **1999**, *15*, 2484–2492. [CrossRef]
25. Winsel, K.; Hönig, D.; Lunkenheimer, K.; Geggel, K.; Witt, C. Quantitative Brewster angle microscopy of the surface film of human broncho-alveolar lavage fluid. *Eur. Biophys. J.* **2003**, *32*, 544–552. [CrossRef] [PubMed]
26. Vollhardt, D.; Fainerman, V.B.; Siegel, S. Thermodynamic and textural characterization of DPPG phospholipid monolayers. *J. Phys. Chem. B* **2000**, *104*, 4115–4121. [CrossRef]
27. Piosik, E.; Ziegler-Borowska, M.; Chełminiak-Dudkiewicz, D.; Martyński, T. Effect of aminated chitosan-coated Fe_3O_4 nanoparticles with applicational potential in nanomedicine on DPPG, DSPC, and POPC Langmuir monolayers as cell membrane models. *Int. J. Mol. Sci.* **2021**, *22*, 2467. [CrossRef]
28. Ładniak, A.; Jurak, M.; Wiącek, A.E. Langmuir monolayer study of phospholipid DPPC on the titanium dioxide–chitosan–hyaluronic acid subphases. *Adsorption* **2019**, *25*, 469–476. [CrossRef]
29. Pasqui, D.; Golini, L.; Giovampaola, C.D.; Atrei, A.; Barbucci, R. Chemical and biological properties of polysaccharide-coated titania nanoparticles: The key role of proteins. *Biomacromolecules* **2011**, *12*, 1243–1249. [CrossRef] [PubMed]
30. Neouze, M.-A.; Schubert, U. Surface modification and functionalization of metal and metal oxide nanoparticles by organic ligands. *Monatsh. Chem.* **2008**, *139*, 183–195. [CrossRef]
31. Guerrero, G.; Mutin, P.H.; Vioux, A. Anchoring of phosphonate and phosphinate coupling molecules on titania particles. *Chem. Mater.* **2001**, *13*, 4367–4373. [CrossRef]
32. Michelmore, A.; Gong, W.; Jenkins, P.; Ralston, J. The interaction of linear polyphosphates with titanium dioxide surfaces. *Phys. Chem. Chem. Phys.* **2000**, *2*, 2985–2992. [CrossRef]
33. Chen, X.; Huang, Z.; Hua, W.; Castada, H.; Allen, H.C. Reorganization and caging of DPPC, DPPE, DPPG, and DPPS monolayers caused by dimethylsulfoxide observed using Brewster angle microscopy. *Langmuir* **2010**, *26*, 18902–18908. [CrossRef]
34. Wydro, P.; Flasiński, M.; Broniatowski, M. Molecular organization of bacterial membrane lipids in mixed systems-A comprehensive monolayer study combined with grazing incidence X-Ray diffraction and Brewster angle microscopy experiments. *Biochim. Biophys. Acta* **2012**, *1818*, 1745–1754. [CrossRef]
35. Torrano, A.A.; Pereira, Â.S.; Oliveira, O.N.; Barros-Timmons, A. Probing the interaction of oppositely charged gold nanoparticles with DPPG and DPPC Langmuir monolayers as cell membrane models. *Colloids Surf. B Biointerfaces* **2013**, *108*, 120–126. [CrossRef]
36. Wydro, P.; Krajewska, B.; Hąc-Wydro, K. Chitosan as a lipid binder: A Langmuir monolayer study of chitosan–lipid interactions. *Biomacromolecules* **2007**, *8*, 2611–2617. [CrossRef]
37. Epand, R.F.; Savage, P.B.; Epand, R.M. Bacterial lipid composition and the antimicrobial efficacy of cationic steroid compounds (ceragenins). *Biochim. Biophys. Acta* **2007**, *1768*, 2500–2509. [CrossRef]
38. Epand, R.M.; Epand, R.F. Domains in bacterial membranes and the action of antimicrobial agents. *Mol. BioSyst.* **2009**, *5*, 580–587. [CrossRef] [PubMed]
39. Phan, M.D.; Shin, K. Effects of cardiolipin on membrane morphology: A Langmuir monolayer study. *Biophys. J.* **2015**, *108*, 1977–1986. [CrossRef]
40. Jurak, M.; Szafran, K.; Cea, P.; Martín, S. Analysis of molecular interactions between components in phospholipid-immunosuppressant-antioxidant mixed Langmuir films. *Langmuir* **2021**, *37*, 5601–5616. [CrossRef] [PubMed]
41. Unsay, J.D.; Cosentino, K.; Subburaj, Y.; García-Sáez, A.J. Cardiolipin effects on membrane structure and dynamics. *Langmuir* **2013**, *29*, 15878–15887. [CrossRef]
42. Wydro, P. The Influence of cardiolipin on phosphatidylglycerol/phosphatidylethanolamine monolayers—Studies on ternary films imitating bacterial membranes. *Colloids Surf. B Biointerfaces* **2013**, *106*, 217–223. [CrossRef] [PubMed]
43. Domènech, Ò.; Sanz, F.; Montero, M.T.; Hernández-Borrell, J. Thermodynamic and structural study of the main phospholipid components comprising the mitochondrial inner membrane. *Biochim. Biophys. Acta* **2006**, *1758*, 213–221. [CrossRef] [PubMed]
44. Ładniak, A.; Jurak, M.; Wiącek, A. Surface characteristics of DPPC monolayers deposited from titanium dioxide-chitosan-hyaluronic acid subphases on a glass support. *Prog. Chem. Appl. Chitin Deriv.* **2019**, *24*, 106–118. [CrossRef]
45. Ładniak, A.; Jurak, M.; Wiącek, A.E. Effect of chitosan, hyaluronic acid and/or titanium dioxide on the physicochemical characteristic of phospholipid film/glass surface. *Physicochem. Probl. Miner. Process.* **2019**, *55*, 1535–1548. [CrossRef]
46. Ładniak, A.; Jurak, M.; Wiącek, A.E. Wettability of DPPC monolayers deposited from the titanium dioxide–chitosan–hyaluronic acid subphases on glass. *Colloids Interfaces* **2019**, *3*, 15. [CrossRef]
47. Ładniak, A.; Jurak, M.; Wiącek, A.E. Physicochemical characteristics of chitosan-TiO_2 biomaterial. 2. Wettability and biocompatibility. *Colloid Surf. A-Physicochem. Eng. Asp.* **2021**, *630*, 127546. [CrossRef]
48. Sennato, S.; Bordi, F.; Cametti, C.; Coluzza, C.; Desideri, A.; Rufini, S. Evidence of domain formation in cardiolipin—Glycerophospholipid mixed monolayers. A thermodynamic and AFM Study. *J. Phys. Chem. B* **2005**, *109*, 15950–15957. [CrossRef]
49. Simonson, A.W.; Lawanprasert, A.; Goralski, T.D.P.; Keiler, K.C.; Medina, S.H. Bioresponsive peptide-polysaccharide nanogels—A versatile delivery system to augment the utility of bioactive cargo. *Nanomedicine* **2019**, *17*, 391–400. [CrossRef] [PubMed]

50. Makvandi, P.; Ali, G.W.; Della Sala, F.; Abdel-Fattah, W.F.; Borzacchiello, A. Biosynthesis and characterization of antibacterial thermosensitive hydrogels based on corn silk extract, hyaluronic acid and nanosilver for potential wound healing. *Carbohydr. Polym.* **2019**, *223*, 115023. [CrossRef] [PubMed]
51. Abdelrahman, R.M.; Abdel-Mohsen, A.M.; Zboncak, M.; Frankova, J.; Lepcio, P.; Kobera, L.; Steinhart, M.; Pavlinak, D.; Spotaz, Z.; Sklenářévád, R.; et al. Hyaluronan biofilms reinforced with partially deacetylated chitin nanowhiskers: Extraction, fabrication, *In-Vitro* and antibacterial properties of advanced nanocomposites. *Carbohydr. Polym.* **2020**, *235*, 115951. [CrossRef]
52. Chen, W.; Zhu, Y.; Zhang, Z.; Gao, Y.; Liu, W.; Borjihan, Q.; Qu, H.; Zhang, Y.; Zhang, Y.; Wang, Y.-J.; et al. Engineering a multifunctional *N*-halamine-based antibacterial hydrogel using a super-convenient strategy for infected skin defect therapy. *Chem. Eng. J.* **2020**, *379*, 122238. [CrossRef]

molecules

Article

Multi-Functional Chitosan Nanovesicles Loaded with Bioactive Manganese for Potential Wound Healing Applications

Edwin Davidson [1,2,*], Jorge Pereira [1,2], Giuliana Gan Giannelli [2,3], Zachary Murphy [1], Vasileios Anagnostopoulos [1] and Swadeshmukul Santra [1,2,3,*]

1 Department of Chemistry, University of Central Florida, Orlando, FL 32826, USA; jorge.pereira@ucf.edu (J.P.); zachary.murphy@ucf.edu (Z.M.); vasileios.anagnos@ucf.edu (V.A.)
2 NanoScience Technology Center, University of Central Florida, Orlando, FL 32826, USA; giuliana.giannelli@ucf.edu
3 Burnett School of Biomedical Sciences, University of Central Florida, Orlando, FL 32826, USA
* Correspondence: edwin.davidsonbarahona@ucf.edu (E.D.); swadeshmukul.santra@ucf.edu (S.S.)

Abstract: Chronic skin wound is a chronic illness that possesses a risk of infection and sepsis. In particular, infections associated with antibiotic-resistant bacterial strains are challenging to treat. To combat this challenge, a suitable alternative that is complementary to antibiotics is desired for wound healing. In this work, we report multi-functional nanoscale chitosan vesicles loaded with manganese (Chi-Mn) that has potential to serve as a new tool to augment traditional antibiotic treatment for skin wound healing. Chi-Mn showed antioxidant activity increase over time as well as antimicrobial activity against *E. coli* and *P. aeruginosa* PA01. The modified motility assay that mimicked a skin wound before bacterial colonization showed inhibition of bacterial growth with Chi-Mn treatment at a low area density of 0.04 μg of Mn per cm^2. Furthermore, this study demonstrated the compatibility of Chi-Mn with a commercial antibiotic showing no loss of antimicrobial potency. In vitro cytotoxicity of Chi-Mn was assessed with macrophages and dermal cell lines (J774A.1 and HDF) elucidating biocompatibility at a wide range (2 ppm–256 ppm). A scratch wound assay involving human dermal fibroblast (HDF) cells was performed to assess any negative effect of Chi-Mn on cell migration. Confocal microscopy study confirmed that Chi-Mn tested at the MIC (16 ppm Mn) has no effect on cell migration with respect to control. Overall, this study demonstrated the potential of Chi-Mn nanovesicles for wound healing applications.

Keywords: chitosan; nanovesicles; wound healing

check for updates

Citation: Davidson, E.; Pereira, J.; Gan Giannelli, G.; Murphy, Z.; Anagnostopoulos, V.; Santra, S. Multi-Functional Chitosan Nanovesicles Loaded with Bioactive Manganese for Potential Wound Healing Applications. *Molecules* **2023**, *28*, 6098. https://doi.org/10.3390/molecules28166098

Academic Editor: Agnieszka Ewa Wiącek

Received: 13 July 2023
Revised: 8 August 2023
Accepted: 11 August 2023
Published: 17 August 2023

1. Introduction

For the past decades, wound care has become a global medical challenge [1,2]. Wounds or skin incisions can happen in our daily life due to a wide variety of reasons, such as medical procedures, cuts, excessive pressure, burns, or underlying medical conditions. Wounds can become a serious and chronic illness in many patients. The impact is significant in healthcare system with an estimated cost of more than USD 28 billion on wound care in 2014 [3]. In recent years, a vast amount of effort has been dedicated to understanding the wound healing processes and developing new therapeutic strategies to provide effective care and reduce health costs [4,5].

Wound healing is comprised of overlapping multi-phased processes (e.g., inflammation, new tissue/proliferation, and maturation/remodeling) [6] involving a myriad of complex biochemical signals and cellular interaction. When these processes are orchestrated in a timely manner, they can achieve optimal restoration of tissue integrity and functionality [7,8]. Nevertheless, the restoration process takes time, which makes wounds vulnerable to bacterial infections and highlights the need to effectively prevent infection to avoid complications, such as sepsis and septic shock [9]. Antibiotic-resistant bacteria are an emerging threat to the wound care of patients. To combat against antibiotic-resistant pathogens, there is a pressing need to develop suitable multi-functional therapeutic tools. Ideally, this tool will possess multiple antimicrobial mechanisms of action. An alternative would be to augment antibiotic therapy by supplementing with biocompatible antimicrobial nanocomposite. These alternatives are expected to provide enhanced protection against infections and mitigation of bacterial resistance incidence [10,11].

There is an immense body of literature outlining the negative impact of excessive reactive oxygen species (ROS) during the healing process [12–14]. In particular, chronic wounds have prolonged inflammatory responses creating an environment that fosters an increased accumulation of ROS, surpassing the inherent antioxidant capacity of cells [15]. This high level ROS accumulation hinders the normal physiological healing phase. As a result, the progression to the cell proliferation phase is affected negatively, which is needed to complete the healing process of a wound [15–18]. Therefore, development of an antimicrobial formulation that incorporates antioxidant properties is highly desirable for wound healing applications.

Currently, wound healing strategies involve the use of wound dressings (e.g., hydrogels, scaffolds) [19,20], ointments [21], growth factors [22], RNA [23], and/or antibiotic treatments [24], among others [25]. Clinically, the most frequently prescribed treatments for wound care include wound dressings and antibiotics due to application simplicity, good efficacy profile, and cost-effectiveness [26,27]. Any advanced multi-functional wound care strategy must demonstrate good compatibility with antibiotics. This research addresses the above by introducing a multi-functional wound healing formulation which has potential to be administrated as a wound dressing, a topical therapeutic, and/or in combination with antibiotics.

In this work, we assessed for the first time the potential of nanoscale chitosan loaded with manganese for wound healing applications. Chitosan is a promising material due to its degradability, biocompatibility, and antimicrobial activity [28–30]. Furthermore, chitosan is known to provide hemostatic and healing stimulation [31]. Chitosan KA01 is a US FDA approved wound dressing that is made of 100% non-woven chitosan fiber material [32].

Our attention was drawn to manganese since it is economical, abundant, and is an essential micronutrient for many metabolic and enzymatic processes in humans [33]. In particular, the presence of manganese ions has been reported to provide stimulation on the proliferation of keratinocytes and fibroblasts cell lines, by inducing the expression of integrins during the proliferation phase on in vivo studies of cell monolayers [34,35]. Additionally, compared to other transition metals, manganese ions provide higher antioxidant effects in the inhibition of microsomal lipid peroxidation and peroxyl radical quenching [36]. Therefore, in this study, manganese (Mn) was incorporated as a bioactive agent within the chitosan (Chi) vesicles (Chi-Mn) to enhance the wound healing process. It is expected that the Chi-Mn when combined with standard antibiotic treatment would exhibit multifunctional attributes (antimicrobial, antioxidant, and cell proliferation capabilities).

2. Results and Discussion

2.1. Characterization of Nanoscale Chitosan Vesicles Loaded with Bioactive Manganese Ions

The Chi-Mn was designed to serve as biocompatible nanovesicles carrying antioxidant Mn ions for wound healing applications. One-step facile synthesis of Chi-Mn resulted in the formation of stable colloidal suspension. The SEM images (Figure 1A) revealed the formation of spherical vesicles with a sharp contrast demarcating the outer and inner regions of the vesicles. The average hydrodynamic diameter as measured by the DLS was 150 nm (Figure 1B). The SEM particle size and size distribution corroborated well with the DLS results. The zeta potential value of +24 mV confirmed the positive surface charge of the Chi-Mn nanovesicles. This result is expected as the primary amines groups of chitosan are exposed to the outer surface of the nanovesicles, possibly interacting with the Mn ions. This positive surface charge of chitosan is advantageous for wound healing due to their affinity to negatively charged cells/tissue surface as well as their ability to recruit red blood cells, facilitating hemostasis [37,38].

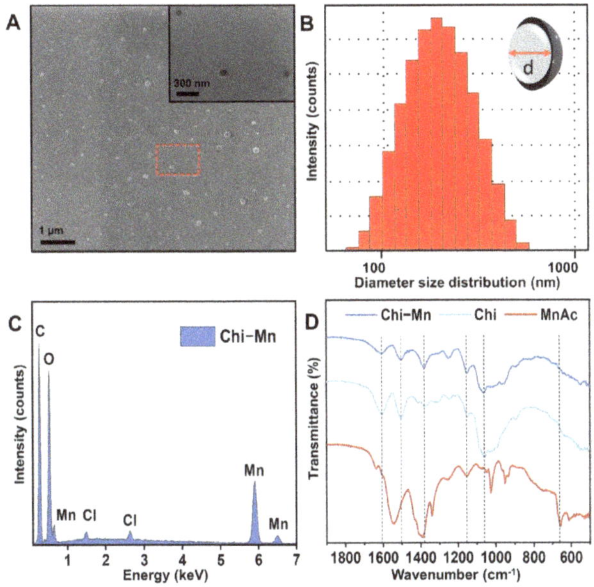

Figure 1. (**A**) SEM images of Chi-Mn. Inset corresponds to a magnified SEM image taken from the region within the red box in (**A**) showing the nanosized vesicles. (**B**) DLS histogram showing the hydrodynamic diameter size distribution of Chi-Mn. (**C**) EDS elemental analysis of Chi-Mn showing the presence of elemental Mn. (**D**) FTIR spectra of Chi, MnAc, and Chi-Mn.

The UV-Vis spectra of Chi-Mn nanovesicles (Figure S1 of Supplementary Materials) showed the characteristic absorbance peak at 380 nm, which is slightly red shifted when compared to the spectra previously reported by our group [39], suggesting the interaction of chitosan nanovesicle with Mn ions. The formation of the nanovesicles under our modified synthesis protocol is in accordance with Gartner et al. and other groups that evaluated acidic depolymerization of low molecular weight chitosan and the effect of acetate ion induced crosslinking of depolymerized chitosan [40,41].

To demonstrate the presence of Mn within the chitosan vesicles, we conducted EDS mapping of the lyophilized dry powder showing a high signal of Mn (Figure 1C). Furthermore, we recorded and compared the FTIR spectra of the lyophilized powder of Chi, Chi-Mn vesicles, and the manganese (II) acetate hexahydrate control. Table S1 of Supplementary Materials summarizes the major peaks of Chi-Mn, Chi, and MnAc, showing no significant difference between Chi-Mn and Chi spectral signature except for the C=O stretch. The observed Chi-Mn major bands are 1603 cm^{-1} for C=O stretch, 1502 cm^{-1} for N-H bend, 1379 cm^{-1} for C-H bend, 1150 cm^{-1} for C-O stretch, and 1061 cm^{-1} for C-O-H stretch [42]. It is worth noting that Chi-Mn has minor band shifts when compared to chitosan unloaded vesicles in the peaks corresponding to the N-H bend, C-O-H stretch, and C=O stretch. The MnAc characteristic band at 613 cm^{-1} for Mn-O vibration band is not present in the Chi-Mn spectrum [43]. Altogether, these results suggest that the detected Mn in the chitosan vesicles is interacting with the carbonyl (residual acetyl group) and amine groups of chitosan. This finding is similar to what has been previously reported in the literature that Mn ions chelate with chitosan, forming complexes [44–47]. Guan et al. reported that metal ions (e.g., Mn) could be present partly in solution (free form) as well as remain associated with chitosan surface (metal-ligand complex) without causing flocculation [48,49].

Finally, the ICP-MS quantified the amount of Mn in the purified Chi-Mn sample. The metallic Mn content of Chi-Mn was estimated to be 2138 $\pm$ 4 ppm.

2.2. Determination of Antioxidant Activity

We conducted a DPPH assay to determine the percentage of free radical scavenging activity, which correlates the higher value of scavenging activity to antioxidant activity. It was found that Chi-Mn radical scavenging activity was approximately 60% over the course of 2 h (Figure S2 of Supplementary Materials). However, to further understand the antioxidant potential of the Chi-Mn, we included appropriate controls of MnAc, Chi, and ascorbic acid (AA). The AA is a well-known antioxidant agent that has strong radical scavenging activities. AA exhibited about 93% scavenging activity (Figure S2) in the first 30 min. This level of activity sustained over 4 h of recorded duration. Alternatively, Chi achieved less than 20% antioxidant activity (Figure S2). The above antioxidant activity results of Chi and Chi-Mn suggest that Mn ion is primarily responsible for exhibiting the antioxidant activity, and therefore it is a bioactive species. MnAc exhibited similar behavior to the Chi-Mn, suggesting no loss of Mn bioactivity despite its association with Chi polymeric matrix. Even though AA exhibited higher antioxidant activity than our Chi-Mn, the Mn loaded nanovesicles demonstrated the advantage of sustained antioxidant activity. This is a desirable property of a wound healing treatment. According to previously reported studies, treatments with slower antioxidant kinetics are beneficial for wound healing applications due to sustained ROS protection [15,50–53]. The antioxidant activity shown by Chi-Mn was higher than the chitosan nanoparticles reported by Divya et al. [54], but comparable to the activity of previously reported manganese nanoparticles by Zhang et al. [55].

2.3. Antimicrobial Studies

Recently, Puca et al. reported a microbial isolate analysis of infected wounds and antimicrobial resistance species. In this work, Gram-negative bacteria were detected as the most prevalent species in most patients and *P. aeruginosa* and *E. coli* had the highest occurrence [56]. Additionally, *P. aeruginosa* is designated as a serious threat by the US Center for Disease Control, accounting for about 13% of infections attributed to multi-drug resistant strains [57]. On the other hand, Peralta et al. reported *E. coli* as the most frequent Gram-negative bacteria causing bloodstream infection associated with antibiotic resistance [58]. Overall, both bacteria have a high risk of developing drug resistance by biofilm formation, leading to a severe challenge to surmount in infected wounds [59,60]. For this reason, the antimicrobial activity of Chi-Mn was assessed through MIC and MBC against these common bacterial strains, *E. coli* and *P. aeruginosa*. Similarly, other studies related to wound infection have focused on these two bacterial strains [61–63].

In Table 1, we summarized the MIC and MBC for both bacterial strains. The Chi and Chi-Mn treatments exhibited significantly improved (32×) antibacterial efficacy with respect to MIC and MBC values when compared to MnAc (Figure S3 of Supplementary Materials), confirming the interaction of cationic Chi with the negatively charged bacterial cell surface, a contact killing mechanism. In contrast, both Chi and Chi-Mn exhibited reduced antimicrobial activity against *P. aeruginosa* PA01, requiring minimum treatment concentration of 256 μg/mL to prevent bacterial growth. Notably, both Chi and MnAc exhibited distinct MIC and MBC, which suggests that each material might have its own mode of action to inhibit and kill bacteria.

Table 1. MIC and MBC of MnAc, Chi, and Chi-Mn.

	MnAc		Chi		Chi-Mn	
Bacteria	**MIC**	**MBC**	**MIC**	**MIC**	**MIC**	**MBC**
E. coli K-12	256	>256	8	8	8	8
P. aeruginosa PA01	256	>256	256	256	256	>256

It is worth noting that our nanoformualtion exhibited significantly higher antimicrobial properties of approximately 10 times more inhibition compared to a previously reported chitosan nanoparticle with manganese ions by Du et al. [64]. In their work, they used an ionic gelation process with Mn ions reporting an MIC and MBC for *E. coli* of 73 and 97 ppm, respectively. Similarly, other reports with chitosan nanoparticles also reported the need of higher concentration of chitosan to achieve the inhibition of *E. coli* [65,66]. Although, some of these chitosan nanoparticles reports [67] have lower MIC against *P. aeruginosa* in the broth dilution method, our nanoformulation exhibited higher potency in a more relevant antimicrobial assay (e.g., motility assay) pertinent to wound healing.

To combat against the emergence of antibiotic-resistant bacterial strains, a wound dressing platform involving more than one mechanism of action is highly desirable [68]. Wound dressing in combination with oral or topical antibiotic treatments is a standard practice to expedite the wound healing processes and prevent infection [69–71]. It is therefore important to evaluate the compatibility of the Chi-Mn with a clinically used antibiotic that is prescribed to mitigate sepsis or severe wound infection. The compatibility of colistin (CO) with Chi-Mn was assessed using a checkerboard assay against *P. aeruginosa* PA01. CO is a re-emerging antibiotic used to combat multi-drug resistant bacterial infection, making it a suitable treatment against resistant *P. aeruginosa* strains [72].

The results from the checkerboard assay demonstrate that CO inhibits bacterial growth at a MIC concentration as low as 1 mg/mL, while Chi-Mn exhibited an MIC of 256 mg/mL of Mn (Figure 2A). None of the treatment combinations showed antagonistic interaction. These findings demonstrate that Chi-Mn is compatible with CO and could be used against *P. aeruginosa* PA01. Furthermore, we assessed the bactericidal activity (Figure 2B) of the Chi-Mn + CO combination at different concentrations. The results show that Chi-Mn at a concentration of 256 mg/mL Mn is able to lower the minimum bactericidal concentration (MBC) of CO by four times (<1 ppm). This reduction in CO bactericidal concentration is significant, demonstrating the potential benefit of Chi-Mn in improving the antimicrobial potency of CO.

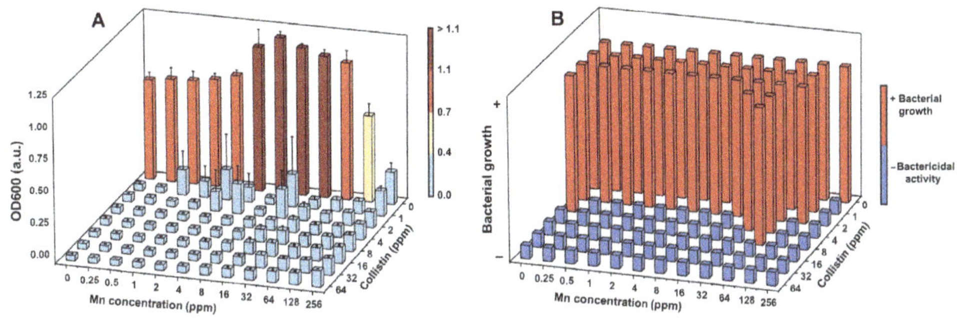

Figure 2. (**A**) Quantification of bacterial growth for *P. aeruginosa* PA01 in response to different concentrations and combinations of Chi-Mn and colistin. The bacterial growth was quantified by measuring OD600 after 24 h of exposure and represented using a heat scale to facilitate visualization. The heat scale values represent the threshold for each level. (**B**) Evaluation of minimum bactericidal concentration based on the in vitro antimicrobial activity shown in (**A**) against *P. aeruginosa* PA01. The error bars represent the standard deviation (n = 3).

To assess the performance of the nanoformulation as a wound dressing, we performed a modified bacterial motility assay by depositing a thin layer of material over agar plates. This assay was designed to mimic a non-infected skin wound before exposure to bacteria (Figure 3A) and has a platform to evaluate the protection of Chi-Mn film barrier against *P. aeruginosa* PA01 inoculation. Bacterial growth was evaluated up to day 5 to mimic common wound dressing practices. The results showed that bacterial motility was inhibited at a concentration of 0.04 μg of Mn per cm^2, while complete bacterial growth inhibition was achieved above 0.09 μg of Mn per cm^2 (Figure 3B,C). Furthermore, bacterial growth was significantly delayed at a concentration as low as 0.04 μg Mn per cm^2, showing only 0.82 mm of bacterial growth at day 5. The statistical analysis showed a significant difference in inhibition between the control and the treatment at a concentration as low as 0.04 μg of Mn per cm^2. It is important to mention that although Chi-Mn exhibited limited in vitro antimicrobial potency against *P. aeruginosa* PA01 in the microbroth assay, it can be very effective in preventing bacterial growth at a very low concentration when used as a film barrier. The bacterial motility assay results further confirmed the contact killing potency of Chi-Mn film in comparison to Chi-Mn suspension. Overall, the above results demonstrate the versatility of the Chi-Mn for wound protection as a stand-alone topical therapeutic or as a wound dressing substrate in combination with other antibiotic treatments.

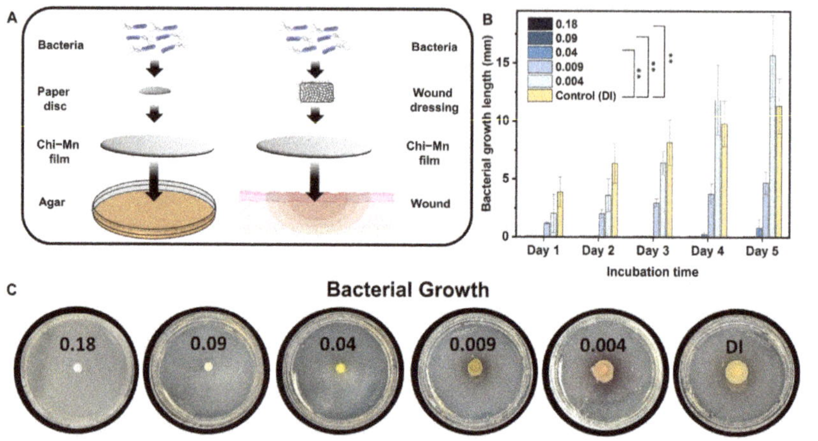

Figure 3. (**A**) Schematic representation of the motility assay mimicking a fresh wound before bacterial exposure. (**B**) Length of *P. aeruginosa* PA01 growth after exposure to different area densities of Chi-Mn in μg of Mn per cm^2. The error bars represent the standard deviation ($n = 3$). Statistical analysis was performed using one-way ANOVA (** $p = 0.0021$) with a p value of 0.05 for Tukey's post-hoc analysis. (**C**) Optical images captured at day 5 after exposure to different area densities of Chi-Mn in μg of Mn per cm^2.

2.4. Biocompatibility Studies

Biocompatibility studies involved HDF as model cell line due to their relevance to wound healing. HDF are primary skin cells with critical roles in the complex process of healing, from interactions with keratinocytes cells to synthesis of growth factors [73]. Additionally, we used J774A.1 mouse macrophage to evaluate the effects of Chi-Mn on the immunology with a systemic non-differentiated immune. It should be noted that J774A.1 is widely used in immunology research as a model cell line. Considering that macrophages are also a key cell line during the early stages of healing process providing immune protection and cells organization to accelerate the recovery [74].

In vitro biocompatibility was evaluated through an assessment of cell viability of two model cells (HDF and J774A.1) using AlamarBlue cell viability assay that measures the metabolic rates of cells upon exposure to Chi-Mn and corresponding controls (Figure 4). In accordance with ISO 10993-5:2009, percentages of cell viability above 80% are considered non-cytotoxic and lower than 40% are strong cytotoxic [75]. In this study, it was observed that J774A.1 cells experienced severe cytotoxicity at concentration above 8.0 ppm of Mn with Chi, MnAc, and Chi-Mn. This suggests the susceptibility of this cell line to the higher concentration of chitosan and manganese ions, whereas HDF cells do not exhibit cytotoxicity at any of the tested concentrations. As expected, Chi had the highest viability across different concentrations, highlighting why there are many studies and commercial products with chitosan as treatment for wound injuries [37,38]. Comparable HDF viability results were previously reported in other chitosan nanoparticle studies, showing the great biocompatibility of chitosan with this specific cell line [76,77].

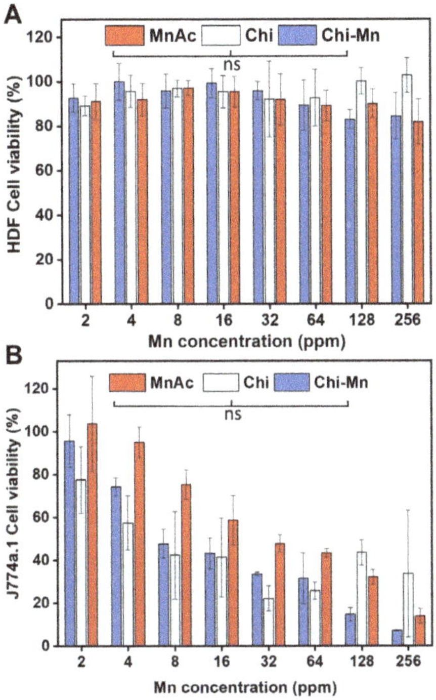

Figure 4. (**A**) Cell viability of human dermal fibroblasts (HDFs) and (**B**) macrophages (J774A.1) upon treatment with Chi-Mn, Chi, and MnAc. Cells were exposed to the respective treatments for 24 h, followed by AlamarBlue reagent for 2 h. Fluorescence was measured at Ex/Em 560/590 nm. Error bars represent standard deviation ($n = 9$). Statistical analysis was performed using Brown–Forsythe and Welch ANOVA test with a p value of 0.05 for Dunnett T3 post-hoc analysis.

These results support the duality of Mn ions in a biological system. The excess of Mn can lead to a detrimental impact in cells viability, but under appropriate concentrations it can render beneficial effects. Thus, we demonstrated the biocompatibility of our Chi-Mn when used within the therapeutic window. The statistical analysis by concentration showed no significant difference between Chi-Mn and its individual components, which corroborates the safe combination of MnAc and Chi with no antagonistic effect toward cell viability. More importantly, although the therapeutic window for macrophages is below 8 ppm Mn, our formulation still has bactericidal properties in broth and solid state against *E. coli* and *P. aeruginosa* PA01, respectively. On the other hand, HDF cell line presented a much broader therapeutic window which is a desirable characteristic for our wound healing platform, as we envision Chi-Mn to be used topically in a wound dressing in direct contact with dermal and epidermal cell lines.

Furthermore, to obtain a better understanding of the applicability of Chi-Mn for wound healing, we performed a scratch wound assay to assess HDF cell migration in the closure of an artificially generated wound. Regarding wounds, cell's migration is a pivotal step in the process of wound healing [78]. For these experiments, we utilized open source software ImageJ to analyze the wound area by changing contrast to effectively discriminate the boundary between the cell monolayer and the open wound area. The metallic concentration of Mn was kept constant at 16 ppm on all the experiments involving Mn. The wound closure was assessed at 16 h to evaluate the effects of the treatments on HDF cell migration prior to cell proliferation doubling time. It was observed that the Chi-Mn provided an average wound closure of 57.84%, which was higher than the control

(49.30%) and MnAc (32.32%) (Figure 5). Confocal microscopic images showed that the Chi-Mn treatment facilitated cell migrating to the open space of the wound, bridging the gap between both extremities of the cell's monolayer. Similarly, Chi provided a comparable wound closure of 55.23%, which reiterates the importance of chitosan materials for wound healing.

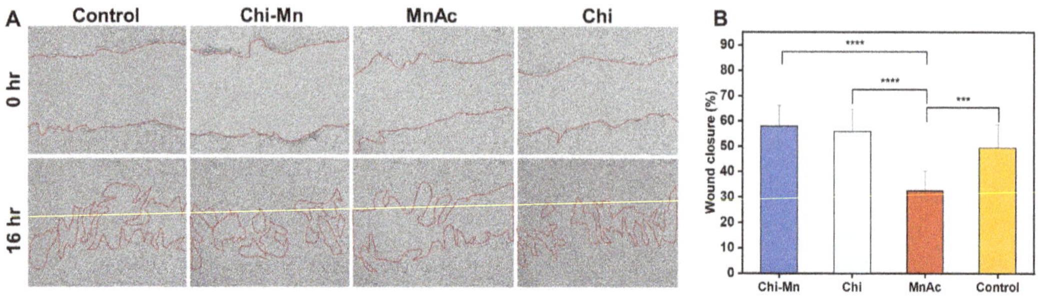

Figure 5. (**A**) Confocal microscopy images of wound scratch assay in HDF cells at time 0 and 16 h after exposure to Chi-Mn, Chi, MnAc, and control (media). Scale bar 100 μm. (**B**) Wound closure percentage represents the mean and error bars represent standard deviation (n = 12). Statistical analysis was performed using one-way ANOVA (**** $p = 0.0001$ and *** $p = 0.0002$) with a p value of 0.05 for Tukey's post-hoc analysis.

Altogether, these results corroborate the biocompatibility of Chi-Mn and its safe potential usage, as a wound dressing or topical therapeutic. This study indeed demonstrated the potential of Chi-Mn nanoformulation as an effective wound dressing providing cell migration and proliferation. Both the combination of proliferation and migration are imperative processes involved in wound closure. At certain concentration range (as demonstrated in the checkerboard assay), the Chi-Mn, in combination with antibiotic treatment could effectively prevent wound infection to facilitate healing.

3. Materials and Methods

3.1. Synthesis and Characterization

3.1.1. Preparation of Nanoscale Chitosan Vesicles Loaded with Manganese Ions

Water-soluble chitosan vesicles (Chi) were prepared following the protocol of Basumallick et al. with some modifications [39]. Briefly, 150 mg of chitosan low molecular weight (≥75% deacetylated; Sigma Aldrich, St Louis, MO, USA) and 300 mg of manganese (II) acetate tetrahydrate (MnAc) (Acros Organics, Morris Plains, NJ, USA) are dissolved in 30 mL of 1 M hydrochloric acid (Thermo Fisher Scientific, Waltham, MA, USA) to form a semi-fluid solution. Next, the mixture was transferred to a Teflon container (30 mL capacity) and carefully placed within a stainless-steel hydrothermal chamber. Next, the chamber was placed inside an oven (Thermo Scientific, Waltham, MA, USA) pre-heated at 150 °C. After 1.5 h of hydrothermal treatment, the chamber was removed from the oven and cooled down to room temperature. The Chi-Mn product mixture was filtered through a 0.45 μm syringe filter (Thermo Fisher Scientific, Waltham, MA, USA) to remove larger size vesicles. Finally, the product mixture was dialyzed for 24 h in a solution containing equal concentration of manganese acetate to prevent losses of the bioactive, and then the solution pH was raised to approximately 6.0. Hereafter, the resulting purified nanovesicle formulation will be referred to as Chi-Mn.

3.1.2. Fourier-Transform Infrared Spectroscopy (FTIR)

Prior to FTIR studies, the samples were frozen and lyophilized (FreeZone 4.5 L Freeze Dry System, Labconco, Kansas City, MO, USA) to obtain a dry powder for further analysis. To demonstrate the presence of Mn ions within the chitosan vesicles (Chi), the FTIR spectra of powder Chi-Mn and corresponding controls (MnAc and Chi) were recorded using a Perkin Elmer Spectrum 100 ATR FTIR Spectrometer (PerkinElmer, Waltham, MA, USA).

3.1.3. Scanning Electron Microscopy (SEM)

To evaluate the morphology of Chi-Mn, the sample solution was drop-casted onto a silicon wafer and left to dry overnight in a desiccator with silica gel. The sample was imaged using a Zeiss Nvision 40 (Zeiss, Oberkochen, Germany) with a 5 kV acceleration voltage.

3.1.4. Energy-Dispersive X-ray Spectroscopy (EDS)

The EDS mapping was carried out on samples, which were already prepared for the SEM imaging study. Briefly, SEM images were collected at acceleration voltage of 15 kV with a TM3000 tabletop microscope (Hitachi, Ibaraki, Japan) equipped with an EDS detector followed by the elemental mapping with the Bruker Quantax 70 software version 1.3.

3.1.5. Dynamic Light Scattering (DLS) and Zeta Potential (ζ)

Prior to DLS measurements, the Chi-Mn suspension was filtered through a 0.45 μm syringe filter, dialyzed, and then analyzed with a Zetasizer ZS90 (Malvern Panalytical, Malvern, United Kingdom) to determine the average hydrodynamic size, size distribution, and zeta potential. Particle suspension was prepared using DI water for the DLS and zeta potential measurements.

3.1.6. Ultraviolet-Visible (UV-Vis) Spectroscopy

All the UV-Visible spectra were recorded on an Evolution 220 UV-Vis Spectrophotometer (Thermo Scientific, Waltham, MA, USA) using a 1 cm quartz cuvette.

3.1.7. Inductive Coupled Plasma Mass Spectrometry (ICP-MS)

To determine total Mn content in the nanoformulation, ICP-MS (iCAP RQ, Thermo Scientific, Waltham, MA, USA) was used. Internal standards of scandium (Inorganic Ventures, Christiansburg, VA, USA) and germanium (SPEX CertiPrep, Metuchen, NJ, USA) were used to correct any instrumental drift. A calibration curve of 10–500 ppb Mn was prepared, and samples were diluted to approximately 100 ppb Mn with 2% nitric acid for analysis. Measurements were conducted using kinetic energy discrimination (KED) to account for diatomic interferences.

3.2. Determination of Antioxidant Activity

The antioxidant activity of a formulation is commonly assessed through free radical scavenging assays. To determine the free radical scavenging capacity, the 1,1-diphenyl-2-picryl hydrazyl (DPPH, Fisher scientific, Waltham, MA, USA) assay was performed following the reported methods with some modifications [79,80]. Briefly, 190 μL of a 0.25 mg/mL DPPH solution was added to a 96-well plate along with 10 μL of Chi-Mn and corresponding controls (Ascorbic acid, MnAc and Chi) at a concentration of 5 mg/mL. Then, the mixture was left for incubation in the dark from 0.5 to 4 h. The absorbance of the suspension was recorded at 517 nm in different time intervals up to 4 h using a plate reader (Infinite 200Pro Tecan absorbance, Männedorf, Switzerland). The radical scavenging activity percentage was then calculated using the following equation:

$$RSA\ \% = \frac{Abs_{control} - (Abs_{treatment} - Abs_{blank})}{Abs_{control}} * 100$$

Abs treatment is the absorbance of the treatment, Abs blank is the absorbance of the treatment with no DPPH, and Abs control is the absorbance of DPPH with no treatment. The scavenging activity experiment was conducted in triplicates.

3.3. Antimicrobial and Biocompatibility Studies

3.3.1. Determination of Minimum Inhibitory Concentration (MIC) and Minimum Bactericidal Concentration (MBC)

The antimicrobial properties of Chi-Mn were investigated by calculating MIC and MBC values, elucidating the lowest concentration needed to inhibit bacterial growth and to completely prevent bacterial colony formation, respectively. MIC experiments were performed using the broth microdilution method following the guidelines from Clinical and Laboratory Standards Institute [81]. We tested a concentration range from 2 to 256 μg of Mn/mL of the Chi-Mn formulation and the controls (MnAc and Chi). Both bacterial strains, *E. coli* K-12 (ATCC 29181, Manassas, VA, USA) and *P. aeuroginosa* PA01 (ATCC 15692, Manassas, VA, USA), were grown with nutrient broth and LB Miller broth (Fisher scientific, Waltham, MA, USA), respectively. In all antimicrobial studies, the final bacterial concentration in each well was 5×10^5 CFU/mL. After the addition of the treatments, the plates were covered, sealed, and incubated for 24 h at 37 °C under 150 rpm agitation. A plate reader (Infinite 200Pro Tecan absorbance, Männedorf, Switzerland) was used to measure the absorbance at 600 nm (OD600). To determine the MBC, 100 μL of the bacterial suspensions treated with two concentrations above and two below the MIC value were plated on agar petri dishes. Then, the petri dishes were incubated for 24 h at 37 °C under 150 rpm agitation. All experiments were conducted in triplicates.

3.3.2. Checkerboard Assay

To assess the effect of Chi-Mn on the antimicrobial activity of colistin (CO), a checkerboard assay was carried out using a published protocol [82]. Initially, 50 μL of Difco LB Miller broth (Fisher scientific, Waltham, MA, USA) was dispensed in all wells of a sterile 96-well plate. Subsequently, 50 μL of Chi-Mn was pipetted into the first column, except for the first row in which 50 μL of 2× Chi-Mn was dispensed. This treatment was then serially diluted until the penultimate column to achieve a final concentration gradient of 256–0.25 mg/mL of Mn. Subsequently, 50 μL of CO solution was added to the first row and serially diluted until the second to last row of the well plate to achieve a final concentration gradient of 64–1 mg/mL of CO. Finally, 50 μL of LB broth and then 100 μL of bacterial suspension at 10^6 CFU/mL were added to each well. The plates were then sealed with parafilm and incubated for 24 h at 37 °C under 150 rpm agitation. After incubation, the absorbance at 600 nm (OD_{600}) was measured with the Infinite 200Pro Tecan plate reader.

To assess the bactericidal activity of the drug combination, 25 μL of the treated bacterial solutions were inoculated into sterile 96-well plates containing 50 μL of LB Miller agar. These plates were inspected for colony growth after 24 h of incubation at 37 °C. All experiments were conducted in triplicates.

3.3.3. Bacterial Motility Assay

The motility of *P. aeruginosa* PA01 was evaluated following the previously reported protocol of Bernal-Mercado et al. with some modifications [83]. First, we deposited Chi-Mn at different concentrations on the surface of Difco LB Miller agar (Fisher scientific, Waltham, MA, USA) petri dishes to form a film. Separately, bacterial suspensions were grown in LB Miller broth for 24 h at 37 °C and 150 rpm. Afterwards, the bacterial suspensions were adjusted to ~10^8 CFU/mL and then diluted to ~10^6 CFU/mL before inoculating a paper disc in the center of the petri dishes with 20 μL of the bacteria suspension. The plates were incubated at 37 °C and the length of bacterial growth around the disc was measured daily for 5 days. All tests were conducted in triplicates. The statistical analysis was performed with GraphPad Prism 9.4.1 (GraphPad Software Inc., Boston, MA, USA) using a one-way ANOVA (p value 0.05) with Tukey's post-hoc analysis to determine statistical differences between the treatments.

3.3.4. Determination of Cell Viability

To evaluate the effect of Chi-Mn on cell viability in vitro, we conducted a cytotoxicity assay following the guidelines from ISO 10993-5:2009 [75]. Human dermal fibroblasts (HDF, ATCC PCS-201-012, Manassas, VA, USA) and murine macrophages (J774A.1, ATCC TIB-67, Manassas, VA, USA) were obtained from the American Type Culture Collection (ATCC). Dulbecco's modified Eagle medium (DMEM, Thermo Scientific, Waltham, MA, USA) with 1% penicillin/streptomycin and 10% heat-inactivated fetal bovine serum (Thermo Scientific, Waltham, MA, USA) was used to culture cells in a T-75 flask under incubation at 37 °C at 5% CO_2. Both cell types were cultured until 80–90% confluency and then seeded into a 96-well tissue culture microplate at a density of 1×10^4 cells per well. After 24 h of incubation, the old media containing treatments were replaced with a new medium containing 10% AlamarBlue cell viability reagent (Thermo Scientific, Waltham, MA, USA). The cells were then incubated for 2 h at 37 °C at 5% CO_2, allowing viable cells to metabolize the dye to highly fluorescent product. Using a plate reader (Infinite 200Pro Tecan absorbance, Männedorf, Switzerland), fluorescence intensity was measured at 590 nm upon excitation at 560 nm. The cell viability percentage (CV%) was calculated using the following equation:

$$CV\% = \frac{F_{treatment}}{F_{untreated}} * 100$$

F treatment corresponds to the fluorescence intensity of the treatment and F untreated is the fluorescence intensity of the untreated control. All experiments were conducted in triplicates. The statistical analysis was performed with GraphPad Prism 9.4.1 (GraphPad Software Inc., Boston, MA, USA) using a Brown–Forsythe and Welch ANOVA (*p* value 0.05) with Dunnett T3 post-hoc analysis to determine statistical differences between the treatments.

3.3.5. Determination of In Vitro Wound Healing Activity

In order to evaluate the wound healing activity of Chi-Mn, we performed a scratch wound assay following the protocol of Martinotti et al. with some modifications [84]. The human dermal fibroblast (HDF) was utilized as the model cell line to mimic a skin wound incision as well as for the determination of the cell migration, an important step in wound healing. HDF was cultured following the same conditions and protocol as described in the cell viability assay (Section 3.3.4). Once 100% confluency was achieved, the cells were seeded to a 24-well tissue culture microplate at a density of 1.3×10^5 cells per well and left for overnight incubation. Then, we proceeded to scrape the cell monolayer in a straight line using a 10 μL pipette tip from west to east and north to south to create a cross. Subsequently, the medium was aspirated from the well and the monolayer was washed with PBS to remove detached cells. Next, Chi-Mn and the controls were added at a concentration of 16 μg of Mn/mL to the cell monolayer. Immediately after addition of the treatment, the first set of images were acquired using a laser confocal microscope (Keyence BZ-X800, Osaka, Japan) in phase contrast mode with a 10× objective. Finally, we followed the method of Suarez-Arnedo to manually measure the wound area coverage using open access ImageJ software version 1.53t [85]. The quantitative data were obtained by measuring the wound area coverage at time point 0 and after 16 h. Then, the wound closure percentage was calculated using the following equation:

$$Wound\ Closure\ \% = \frac{A_{t=0} - A_{t=16}}{A_{t=0}} * 100$$

where $A_{t=0}$ is the initial wound opening area (in μm^2) at time 0 and $A_{t=16}$ is the wound area after 16 h from the initial scratch. All experiments were conducted in triplicates. The statistical analysis was performed with GraphPad Prism 9.4.1 (GraphPad Software Inc., Boston, MA, USA) using one-way ANOVA (*p* value 0.05) with Tukey's post-hoc analysis to determine statistical differences between the treatments.

4. Conclusions

In the present proof-of-concept study, we successfully demonstrated the usefulness of hydrothermal-based synthesis process (one-step and facile), leading to the formation of multifunctional nanoscale chitosan vesicles incorporated with bioactive Mn ions. This study demonstrated the antimicrobial activity and compatibility of Chi-Mn as well as its potential use in combination with an antibiotic treatment to prevent wound infections. Additionally, the Chi-Mn nanoformulation was biocompatible fostering cell proliferation and cell migration within the safe therapeutic window. This demonstrated the promising functionality of Chi-Mn nanoformulation to protect against infection and promote the wound healing process. The Chi-Mn demonstrated slow antioxidant kinetics providing a temporal increase in the scavenging of free radicals' activity, which in theory, are expected to provide a prolonged protection against ROS species.

Future studies are required to explore the mode of action of Chi-Mn against clinically relevant bacterial strains and the Mn release kinetics in the tissue level. To the best of our knowledge, this is the first report of hydrothermally synthesized nanoscale chitosan vesicles with manganese acetate, demonstrating multi-functional activities (antioxidant, antimicrobial, and cell proliferation booster), which are important for effective wound healing. We believe that the Chi-Mn has the potential to serve as a wound dressing substrate in a hydrogel or as a topical therapeutic for cutaneous wound injuries, offering a novel treatment alternative.

Supplementary Materials: The following supporting information can be downloaded at: https://www.mdpi.com/article/10.3390/molecules28166098/s1. Table S1: FTIR spectral data comparison of MnAc, Chi, and Chi-Mn; Figure S1: UV-Visible spectra of Chi-Mn, Chi, and MnAc; Figure S2: Radical scavenging activity percentage of Mn-Chi in comparison with Ascorbic Acid, MnAc, Chi, and Chi-Mn. Error bars represent standard deviations ($n = 3$); Figure S3: In vitro antibacterial activity of Chi-Mn, Chi, and MnAc against E. coli K-12 and P. aeruginosa PA01. Error bars represent standard deviations ($n = 9$).

Author Contributions: Conceptualization, E.D., J.P. and S.S.; methodology, E.D., J.P. and G.G.G.; investigation, E.D., J.P., Z.M. and G.G.G.; data curation, E.D. and J.P.; visualization, E.D. and J.P.; formal analysis, E.D., J.P., G.G.G., Z.M. and V.A.; writing—original draft preparation, E.D.; writing—review and editing, E.D., J.P., G.G.G., Z.M. and V.A.; resources, V.A. and S.S.; supervision, S.S.; project administration, S.S. All authors have read and agreed to the published version of the manuscript.

Funding: This research received no external funding.

Data Availability Statement: The data presented in this study are available on request from the corresponding author.

Acknowledgments: Authors acknowledge the UCF Department of Chemistry and the NanoScience Technology Center for providing appropriate resources and facilities to support material characterization.

Conflicts of Interest: The authors declare no conflict of interest.

References

1. Xu, Z.; Liu, G.; Zheng, L.; Wu, J. A polyphenol-modified chitosan hybrid hydrogel with enhanced antimicrobial and antioxidant activities for rapid healing of diabetic wounds. *Nano Res.* **2023**, *16*, 905–916. [CrossRef]
2. Matoori, S.; Veves, A.; Mooney, D.J. Advanced bandages for diabetic wound healing. *Sci. Transl. Med.* **2021**, *13*, eabe4839. [CrossRef] [PubMed]
3. Nussbaum, S.R. An Economic Evaluation of the Impact, Cost, and Medicare Policy Implications of Chronic Nonhealing Wounds. *Value Health* **2018**, *21*, 27–32. [CrossRef]
4. Takeo, M.; Lee, W.; Ito, M. Wound healing and skin regeneration. *Cold Spring Harb. Perspect. Med.* **2015**, *5*, a023267. [CrossRef]
5. Velnar, T.; Bailey, T.; Smrkolj, V. The wound healing process: An overview of the cellular and molecular mechanisms. *J. Int. Med. Res.* **2009**, *37*, 1528–1542. [CrossRef] [PubMed]
6. Gurtner, G.C. Wound repair and regeneration. *Nature* **2008**, *453*, 314–321. [CrossRef]
7. Broughton, G.I.; Janis, J.E.; Attinger, C.E. The Basic Science of Wound Healing. *Plast. Reconstr. Surg.* **2006**, *117*, 12S–34S. [CrossRef]

8. Tottoli, E.M. Skin Wound Healing Process and New Emerging Technologies for Skin Wound Care and Regeneration. *Pharmaceutics* **2020**, *12*, 735. [CrossRef]
9. Järbrink, K. The humanistic and economic burden of chronic wounds: A protocol for a systematic review. *Syst. Rev.* **2017**, *6*, 15. [CrossRef]
10. Khameneh, B. Breakthroughs in bacterial resistance mechanisms and the potential ways to combat them. *Microb. Pathog.* **2016**, *95*, 32–42. [CrossRef]
11. Mouton, J. Combination therapy as a tool to prevent emergence of bacterial resistance. *Infection* **1999**, *27* (Suppl. 2), S24–S28. [CrossRef]
12. Polaka, S. Emerging ROS-Modulating Technologies for Augmentation of the Wound Healing Process. *ACS Omega* **2022**, *7*, 30657–30672. [CrossRef]
13. Bilgen, F. The effect of oxidative stress and Raftlin levels on wound healing. *Int. Wound J.* **2019**, *16*, 1178–1184. [CrossRef] [PubMed]
14. Dunnill, C. Reactive oxygen species (ROS) and wound healing: The functional role of ROS and emerging ROS-modulating technologies for augmentation of the healing process. *Int. Wound J.* **2017**, *14*, 89–96. [CrossRef] [PubMed]
15. Xu, Z. Advances and impact of antioxidant hydrogel in chronic wound healing. *Adv. Healthc. Mater.* **2020**, *9*, 1901502. [CrossRef] [PubMed]
16. Deng, Z. M1 macrophage mediated increased reactive oxygen species (ROS) influence wound healing via the MAPK signaling in vitro and in vivo. *Toxicol. Appl. Pharmacol.* **2019**, *366*, 83–95. [CrossRef] [PubMed]
17. Koo, M.-A. Effective stacking and transplantation of stem cell sheets using exogenous ROS-producing film for accelerated wound healing. *Acta Biomater.* **2019**, *95*, 418–426. [CrossRef]
18. Saoudi, M. The Role of *Allium subhirsutum* L. in the attenuation of dermal wounds by modulating oxidative stress and inflammation in Wistar albino rats. *Molecules* **2021**, *26*, 4875. [CrossRef]
19. Hanna, J.R.; Giacopelli, J.A. A review of wound healing and wound dressing products. *J. Foot Ankle Surg.* **1997**, *36*, 2–14. [CrossRef]
20. Negut, I.; Dorcioman, G.; Grumezescu, V. Scaffolds for Wound Healing Applications. *Polymers* **2020**, *12*, 2010. [CrossRef]
21. Valizadeh, A. Preparation and Comparison of Effects of Different Herbal Oil Ointments as Wound-Healing Agents. *Cells Tissues Organs* **2019**, *207*, 177–186. [CrossRef]
22. Koria, P. Delivery of Growth Factors for Tissue Regeneration and Wound Healing. *BioDrugs* **2012**, *26*, 163–175. [CrossRef] [PubMed]
23. Cheema, S.K. Regulation and guidance of cell behavior for tissue regeneration via the siRNA mechanism. *Wound Repair Regen.* **2007**, *15*, 286–295. [CrossRef] [PubMed]
24. O'Meara, S. Systematic reviews of wound care management: (3) antimicrobial agents for chronic wounds; (4) diabetic foot ulceration. *Health Technol. Assess.* **2000**, *4*, 1–237. [CrossRef] [PubMed]
25. Chouhan, D. Emerging and innovative approaches for wound healing and skin regeneration: Current status and advances. *Biomaterials* **2019**, *216*, 119267. [CrossRef]
26. Ho, J.; Walsh, C.; Yue, D.; Dardik, A.; Cheema, U. Current Advancements and Strategies in Tissue Engineering for Wound Healing: A Comprehensive Review. *Adv. Wound Care* **2017**, *6*, 191–209. [CrossRef]
27. Mirhaj, M. Emerging treatment strategies in wound care. *Int. Wound J.* **2022**, *19*, 1934–1954. [CrossRef]
28. Dai, T. Chitosan preparations for wounds and burns: Antimicrobial and wound-healing effects. *Expert Rev. Anti-Infect. Ther.* **2011**, *9*, 857–879. [CrossRef]
29. Patrulea, V. Chitosan as a starting material for wound healing applications. *Eur. J. Pharm. Biopharm.* **2015**, *97*, 417–426. [CrossRef]
30. Yilmaz, E. Chitosan: A versatile biomaterial. *Biomater. Mol. Eng. Tissue* **2004**, *553*, 59–68.
31. Muzzarelli, R. Biochemistry, histology and clinical uses of chitins and chitosans in wound healing. *EXS* **1999**, *87*, 251–264.
32. Whang, H.S. Hemostatic Agents Derived from Chitin and Chitosan. *J. Macromol. Sci. Part C* **2005**, *45*, 309–323. [CrossRef]
33. Haque, S.; Tripathy, S.; Patra, C.R. Manganese-based advanced nanoparticles for biomedical applications: Future opportunity and challenges. *Nanoscale* **2021**, *13*, 16405–16426. [CrossRef] [PubMed]
34. Tenaud, I. In vitro modulation of keratinocyte wound healing integrins by zinc, copper and manganese. *Br. J. Dermatol.* **1999**, *140*, 26–34. [CrossRef] [PubMed]
35. Barrat, J. Le gluconate de cuivre et de manganèse dans le traitement des crevasses du sein lors de l'allaitement maternel. *Rev. Fr. Gynécol. Obstét.* **1992**, *87*, 49–51.
36. Coassin, M.; Ursini, F.; Bindoli, A. Antioxidant effect of manganese. *Arch. Biochem. Biophys.* **1992**, *299*, 330–333. [CrossRef]
37. Ahmadi, F. Chitosan based hydrogels: Characteristics and pharmaceutical applications. *Res. Pharm. Sci.* **2015**, *10*, 1–16.
38. Clay, J.G.; Grayson, J.K.; Zierold, D. Comparative Testing of New Hemostatic Agents in a Swine Model of Extremity Arterial and Venous Hemorrhage. *Mil. Med.* **2010**, *175*, 280–284. [CrossRef]
39. Basumallick, S. Hydrothermally treated chitosan spontaneously forms water-soluble spherical particles stable at a wide pH range. *Int. J. Polym. Mater. Polym. Biomater.* **2016**, *65*, 751–758. [CrossRef]
40. Gartner, C.; López, B.L. Stiffness of chitosan in solution through intrinsic viscosity measurem. *Rev. Fac. Ing. Univ. Antioq.* **2010**, *2010*, 20–29.

41. Lee, S.-H.; Park, S.-M.; Kim, Y. Effect of the concentration of sodium acetate (SA) on crosslinking of chitosan fiber by epichlorohydrin (ECH) in a wet spinning system. *Carbohydr. Polym.* **2007**, *70*, 53–60. [CrossRef]
42. Larkin, P.J. Chapter 6—IR and Raman Spectra–Structure Correlations: Characteristic Group Frequencies. In *Infrared and Raman Spectroscopy*, 2nd ed.; Larkin, P.J., Ed.; Elsevier: Amsterdam, The Netherlands, 2018; pp. 85–134.
43. Cakic, S. Spectroscopic characteristics of highly selective manganese catalysis in acqueous polyurethane systems. *Sensors* **2006**, *6*, 1708–1720. [CrossRef]
44. Kamiński, W.; Modrzejewska, Z. Application of chitosan membranes in separation of heavy metal ions. *Sep. Sci. Technol.* **1997**, *32*, 2659–2668. [CrossRef]
45. Ali, M.E.A. Chitosan nanoparticles extracted from shrimp shells, application for removal of Fe(II) and Mn(II) from aqueous phases. *Sep. Sci. Technol.* **2018**, *53*, 2870–2881. [CrossRef]
46. Huang, Y. Evaluation of diethylenetriaminepentaacetic acid–manganese(II) complexes modified by narrow molecular weight distribution of chitosan oligosaccharides as potential magnetic resonance imaging contrast agents. *Magn. Reson. Imaging* **2011**, *29*, 554–560. [CrossRef]
47. Muzzarelli, R.A.; Rocchetti, R.; Muzzarelli, M.G. The isolation of cobalt, nickel, and copper from manganese nodules by chelation chromatography on chitosan. *Sep. Sci. Technol.* **1978**, *13*, 153–163. [CrossRef]
48. Guan, B. Removal of Mn(II) and Zn(II) ions from flue gas desulfurization wastewater with water-soluble chitosan. *Sep. Purif. Technol.* **2009**, *65*, 269–274. [CrossRef]
49. Guibal, E. Interactions of metal ions with chitosan-based sorbents: A review. *Sep. Purif. Technol.* **2004**, *38*, 43–74. [CrossRef]
50. Zhu, Y. Sustained release of stromal cell derived factor-1 from an antioxidant thermoresponsive hydrogel enhances dermal wound healing in diabetes. *J. Control. Release* **2016**, *238*, 114–122. [CrossRef]
51. Liu, M. Antioxidant thermogelling formulation for burn wound healing. *Chem. Asian J.* **2022**, *17*, e202200396. [CrossRef]
52. Xiao, J. A cooperative copper metal–organic framework-hydrogel system improves wound healing in diabetes. *Adv. Funct. Mater.* **2017**, *27*, 1604872. [CrossRef] [PubMed]
53. Guan, Y. Sustained oxygenation accelerates diabetic wound healing by promoting epithelialization and angiogenesis and decreasing inflammation. *Sci. Adv.* **2021**, *7*, eabj0153. [CrossRef] [PubMed]
54. Divya, K.; Smitha, V.; Jisha, M.S. Antifungal, antioxidant and cytotoxic activities of chitosan nanoparticles and its use as an edible coating on vegetables. *Int. J. Biol. Macromol.* **2018**, *114*, 572–577. [CrossRef] [PubMed]
55. Zhang, Y. Multienzymatic Antioxidant Activity of Manganese-Based Nanoparticles for Protection against Oxidative Cell Damage. *ACS Biomater. Sci. Eng.* **2022**, *8*, 638–648. [CrossRef]
56. Puca, V. Microbial species isolated from infected wounds and antimicrobial resistance analysis: Data emerging from a three-years retrospective study. *Antibiotics* **2021**, *10*, 1162. [CrossRef]
57. Centre for Disease, Control and Prevention. *Antibiotic Resistance Threats in the United States, 2013*; Centers for Disease Control and Prevention. US Department of Health and Human Services: Atlanta, GA, USA, 2013.
58. Peralta, G. Impact of antibiotic resistance and of adequate empirical antibiotic treatment in the prognosis of patients with *Escherichia coli* bacteraemia. *J. Antimicrob. Chemother.* **2007**, *60*, 855–863. [CrossRef]
59. Ito, A. Increased antibiotic resistance of *Escherichia coli* in mature biofilms. *Appl. Environ. Microbiol.* **2009**, *75*, 4093–4100. [CrossRef]
60. Thi, M.T.T.; Wibowo, D.; Rehm, B.H. Pseudomonas aeruginosa biofilms. *Int. J. Mol. Sci.* **2020**, *21*, 8671. [CrossRef]
61. Nagoba, B.S. A simple and effective approach for the treatment of chronic wound infections caused by multiple antibiotic resistant *Escherichia coli*. *J. Hosp. Infect.* **2008**, *69*, 177–180. [CrossRef]
62. Zhao, G. Delayed wound healing in diabetic (db/db) mice with *Pseudomonas aeruginosa* biofilm challenge: A model for the study of chronic wounds. *Wound Repair Regen.* **2010**, *18*, 467–477. [CrossRef]
63. Gupta, P. Bacteriophage Therapy of Chronic Nonhealing Wound: Clinical Study. *Int. J. Low. Extrem. Wounds* **2019**, *18*, 171–175. [CrossRef] [PubMed]
64. Du, W.-L. Antibacterial activity of chitosan tripolyphosphate nanoparticles loaded with various metal ions. *Carbohydr. Polym.* **2009**, *75*, 385–389. [CrossRef]
65. Sarwar, A.; Katas, H.; Zin, N.M. Antibacterial effects of chitosan–tripolyphosphate nanoparticles: Impact of particle size molecular weight. *J. Nanoparticle Res.* **2014**, *16*, 2517. [CrossRef]
66. Wu, T. Integration of lysozyme into chitosan nanoparticles for improving antibacterial activity. *Carbohydr. Polym.* **2017**, *155*, 192–200. [CrossRef] [PubMed]
67. Chandrasekaran, M.; Kim, K.D.; Chun, S.C. Antibacterial Activity of Chitosan Nanoparticles: A Review. *Processes* **2020**, *8*, 1173. [CrossRef]
68. Owen, L.; Laird, K. Synchronous application of antibiotics and essential oils: Dual mechanisms of action as a potential solution to antibiotic resistance. *Crit. Rev. Microbiol.* **2018**, *44*, 414–435. [CrossRef]
69. Georgescu, M. Bioactive wound dressings for the management of chronic wounds. *Curr. Org. Chem.* **2017**, *21*, 53–63. [CrossRef]
70. Ramalingam, R. Core–shell structured antimicrobial nanofiber dressings containing herbal extract and antibiotics combination for the prevention of biofilms and promotion of cutaneous wound healing. *ACS Appl. Mater. Interfaces* **2021**, *13*, 24356–24369. [CrossRef]
71. Mi, F.L. Control of wound infections using a bilayer chitosan wound dressing with sustainable antibiotic delivery. *J. Biomed. Mater. Res. Off. J. Soc. Biomater. Jpn. Soc. Biomater. Aust. Soc. Biomater. Korean Soc. Biomater.* **2002**, *59*, 438–449. [CrossRef]

72. Nation, R.L.; Li, J. Colistin in the 21st century. *Curr. Opin. Infect. Dis.* **2009**, *22*, 535–543. [CrossRef]
73. Werner, S.; Krieg, T.; Smola, H. Keratinocyte–fibroblast interactions in wound healing. *J. Investig. Dermatol.* **2007**, *127*, 998–1008. [CrossRef] [PubMed]
74. Rodero, M.P.; Khosrotehrani, K. Skin wound healing modulation by macrophages. *Int. J. Clin. Exp. Pathol.* **2010**, *3*, 643–653.
75. Standardization, I. *Biological Evaluation of Medical Devices—Part 5: Tests for In Vitro Cytotoxicity*; ISO: Geneva, Switzerland, 2009.
76. Biranje, S.S. Hemostasis and anti-necrotic activity of wound-healing dressing containing chitosan nanoparticles. *Int. J. Biol. Macromol.* **2019**, *121*, 936–946. [CrossRef]
77. Fahimirad, S. Antimicrobial Activity, Stability and Wound Healing Performances of Chitosan Nanoparticles Loaded Recombinant LL37 Antimicrobial Peptide. *Int. J. Pept. Res. Ther.* **2021**, *27*, 2505–2515. [CrossRef]
78. Shaw, T.J.; Martin, P. Wound repair at a glance. *J. Cell Sci.* **2009**, *122*, 3209–3213. [CrossRef]
79. Blois, M.S. Antioxidant determinations by the use of a stable free radical. *Nature* **1958**, *181*, 1199–1200. [CrossRef]
80. Kedare, S.B.; Singh, R.P. Genesis and development of DPPH method of antioxidant assay. *J. Food Sci. Technol.* **2011**, *48*, 412–422. [CrossRef]
81. Wikler, M.A. *Methods for Dilution Antimicrobial Susceptibility Tests for Bacteria that Grow Aerobically: APPROVED Standard*; CLSI: Wayne, PA, USA, 2006; Volume 26, pp. M7–A7.
82. Bellio, P. New and simplified method for drug combination studies by checkerboard assay. *MethodsX* **2021**, *8*, 101543. [CrossRef]
83. Bernal-Mercado, A.T. Hydrophobic Chitosan Nanoparticles Loaded with Carvacrol against Pseudomonas aeruginosa Biofilms. *Molecules* **2022**, *27*, 699. [CrossRef]
84. Martinotti, S.; Ranzato, E. Scratch Wound Healing Assay. In *Epidermal Cells*; Springer: Berlin/Heidelberg, Germany, 2019; pp. 225–229.
85. Suarez-Arnedo, A. An image J plugin for the high throughput image analysis of in vitro scratch wound healing assays. *PLoS ONE* **2020**, *15*, e0232565. [CrossRef]

molecules

Article

Preparation and Surface Characterization of Chitosan-Based Coatings for PET Materials

Klaudia Szafran [1,*], Małgorzata Jurak [1], Robert Mroczka [2] and Agnieszka Ewa Wiącek [1,*]

1 Department of Interfacial Phenomena, Institute of Chemical Sciences, Faculty of Chemistry, Maria Curie-Skłodowska University, 20031 Lublin, Poland

2 Laboratory of X-ray Optics, Department of Chemistry, Institute of Biological Sciences, Faculty of Medicine, The John Paul II Catholic University of Lublin, 20708 Lublin, Poland

* Correspondence: klaudia.wozniak@poczta.umcs.lublin.pl (K.S.); agnieszka.wiacek@mail.umcs.pl (A.E.W.)

Abstract: Poly(ethylene terephthalate)—PET—is one of the most frequently used polymers in biomedical applications. Due to chemical inertness, PET surface modification is necessary to gain specific properties, making the polymer biocompatible. The aim of this paper is to characterize the multicomponent films containing chitosan (Ch), phospholipid 1,2-dioleoyl-*sn*-glycero-3-phosphocholine (DOPC), immunosuppressant cyclosporine A (CsA) and/or antioxidant lauryl gallate (LG) which can be utilized as a very attractive material for developing the PET coatings. Chitosan was employed owing to its antibacterial activity and also its ability to promote cell adhesion and proliferation favorable for tissue engineering and regeneration purposes. Moreover, the Ch film can be additionally modified with other substances of biological importance (DOPC, CsA and LG). The layers of varying compositions were prepared using the Langmuir—Blodgett (LB) technique on the air plasma-activated PET support. Then their nanostructure, molecular distribution, surface chemistry and wettability were determined by atomic force microscopy (AFM), time-of-flight secondary ion mass spectrometry (TOF-SIMS), X-ray photoelectron spectroscopy (XPS), contact angle (CA) measurements and the surface free energy and its components' determination, respectively. The obtained results show clearly the dependence of the surface properties of the films on the molar ratio of components and allow for a better understanding of the coating organization and mechanisms of interactions at the molecular level both inside the films and between the films and the polar/apolar liquids imitating the environment of different properties. The organized layers of this type can be helpful in gaining control over the surface properties of the biomaterial, thus getting rid of the limitations in favor of increased biocompatibility. This is a good basis for further investigations on the correlation of the immune system response to the presence of biomaterial and its physicochemical properties.

Keywords: chitosan; cyclosporine A; lauryl gallate; XPS; AFM; TOF-SIMS; contact angle; surface free energy

Citation: Szafran, K.; Jurak, M.; Mroczka, R.; Wiącek, A.E. Preparation and Surface Characterization of Chitosan-Based Coatings for PET Materials. *Molecules* **2023**, *28*, 2375. https://doi.org/10.3390/molecules28052375

Academic Editor: Már Másson

Received: 30 November 2022
Revised: 18 February 2023
Accepted: 20 February 2023
Published: 4 March 2023

1. Introduction

PET is one of the most commonly used polyesters in biomedical applications. Due to flexibility and strength PET fibers are very interesting materials in surgery and orthopedic devices applied mainly as surgical suture membranes, surgical meshes, heart valves, endovascular stent grafts, urinary and vascular catheters, scaffolds for ligament and tendon repair [1,2].

The biostability of this polymer is associated with the presence of hydrophobic aromatic groups of terephthalic acid with long crystallinity which reduces hydrolytic degradation. In spite of these beneficial features, PET fibers have some limitations in terms of the poor blood compatibility and the poor cell-adhesion properties, which are relevant for biomedical applications [3,4]. Accordingly, synthetic scaffolds can induce high incidences of postoperative infection, chronic immune response [1], compliance mismatch aside, and thrombogenicity being the major reason for artificial body substitute failure. Consequently, there is an urgent need to reduce the thrombosis process, and in parallel to strengthen the PET polymer-cell interactions and endothelial cell adhesion.

For the improvement of general and specific applications of chemically inert PET, intensive research attempts have been made for surface modification. In order to introduce different functional groups on the PET surface without changing the original bulk properties, various useful methods have been employed in many laboratories. Most of these studies are based on hydrolysis, aminolysis, graft co-polymerization, enzymatic modification, radiation, or plasma treatment [5,6].

Hydrolysis generates a mixture of $-OH$ and $-COO^-$ functional groups while aminolysis introduces primary $-NH_2$ or secondary $> NH$ amino groups on the PET surfaces. Both types of reactions can result in significant sample degradation due to the surface chain cleavage. Grafting is the other way, wherein the monomers can be joined onto the polymer chains through covalent bonds. This process can be activated by the chemical reaction, photoirradiation, and plasma treatment [5,6]. Plasma processing can cause the formation of the polar groups including $-OH$, $-COOH$, $-NH_2$, and $-SO_4^-$ on the polymer surfaces using different gases such as air, NH_3, SO_2, CO_2, and other organic compounds [6]. Plasma has been gaining a lot of popularity as a basic treatment, [7] while grafting and coating are utilized for further functionalization [6,8]. The functional groups are a key parameter in the multi-anchoring processes. The optimization strategy of biomaterials' fabrication is focused on the multifunctional ways of immobilizing biomolecules which employ rather a mixture of biomolecules than single ones ([6] and references herein). One of the prospective approaches comprises the coverage of PET using biologically friendly substances for enhanced biocompatibility for the surrounding tissues.

To address this issue specifically, biocompatible and biodegradable natural polymers like chitosan can be used as the PET surface coatings and/or matrices for the controlled drug release [9]. Indeed, the type of the polymer is a main factor responsible for the drug release profile as well as the method applied to cover the body implant with the polymer/drug [10].

Chitosan (Figure 1a), a natural polysaccharide, exhibits a unique combination of favorable biological properties such as biocompatibility, biodegradability, bioadhesiveness, non-immunogenicity, non-toxicity, wound-healing acceleration, and anti-microbial properties [10,11]. In addition, owing to its cationic character, it attracts peptides, proteins, and drugs through biological tissues and performs key functions in the endothelial cell attachment and growth [12]. Matrices or biomaterials with the chitosan coating tender many advantages in drug delivery and implant devices, including slowing the rate of degradation, the modulation of cell responses and the local release of drugs and/or growth factors from the surface of implants and tissue engineering scaffolds [10]. The polymer coating itself should be resistant to fracture and degradation, as well as firmly biocompatible and anti-thrombogenic. Moreover, a number of potential postoperative complications at the site of implantation such as recurrent narrowing, inflammatory disease or thrombosis can be minimized by the local release of drugs from the biodegradable polymers [13]. However, in spite of the great progress in artificial body implant development and interventional surgery, a lot of aspects like thrombosis and restenosis still remain to be resolved.

Undoubtedly, chitosan plays a relevant role in all processes of hemostasis. It induces the aggregation of morphotic blood components to promote the blood clotting and vasoconstriction at the injured site. The coagulant properties are strongly dependent on the interactions of positively charged chitosan with negatively charged erythrocytes, platelets, and plasma proteins. These interactions are determined mainly by the degree of deacetylation (DD) and relative molecular weight (MW) of polysaccharide [14,15]. Therefore, the appropriate selection of chitosan properties (DD, MW) in combination with the other charge-modifying components can ensure the proper balance of positive charges of chitosan-based coatings with the negative charges of blood cells and certain proteins. In consequence, it can be applied as a biocompatible structural material in the tissue reconstruction and as a matrix for the drug-released systems. This can provide the intimate contact with the tissues surrounding the biomaterial surface without causing negative side effects such as thrombosis.

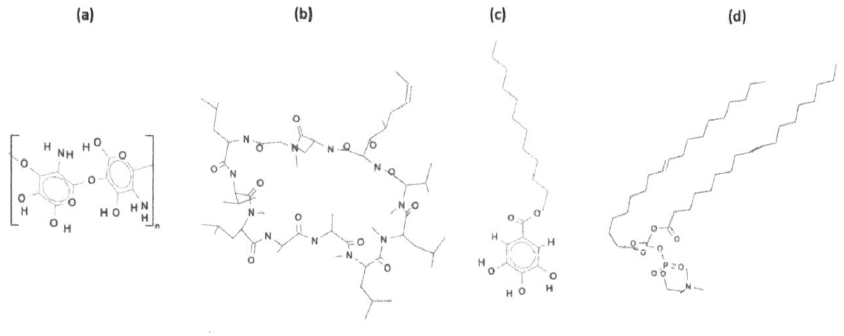

Figure 1. Chemical structure of chitosan monomer (Ch) (**a**), cyclosporine A (CsA) (**b**), lauryl gallate (LG) (**c**) and 1,2-dioleoyl-*sn*-glycero-3-phosphocholine (DOPC) (**d**).

Lately chitosan and its derivatives have shown perspective in the restenosis treatment. For instance, sirolimus containing chitosan-coated liposomes were found to be a novel local anti-restenosis drug carrier of sustained release behavior over conventional liposomes [16]. The stent covered by electrospinning with the chitosan/poly-cyclodextrin based nanofibers loaded with simvastatin was reported as promising for restenosis prevention [17]. Moreover, the core-shell drug encapsulated nanofibers fabricated by the co-assembly of chitosan and paclitaxel for the metal stent coating, inhibited platelet adhesion significantly, and exhibited relatively good hemocompatibility [18]. In addition, the metal stents coated with the chitosan/hyaluronic acid loading antibodies can repair vascular damages by capturing the own hematopoietic stem cells, thus contributing to their rapid natural repair, avoiding inflammation and rejection, thrombosis, and restenosis [19]. Finally, the in vitro tests with sirolimus-eluting from the chitosan-silica hybrid coating for the coronary stents evidenced great cytocompatibility and hemocompatibility, with a slight risk of in-stent restenosis and thrombogenicity [12]. As follows from the above studies, the application of chitosan-coated stents has become a new therapeutic approach for the cardiovascular disease.

Accordingly, our attention was focused on the chitosan-based coatings for PET materials which can act as efficient matrices for the localized release of the therapeutic agent-cyclosporine A (CsA, Figure 1b). This is an immunosuppressive drug used commonly for the prevention of organ transplant rejection, in part owing to its unique ability to inhibit the activation of T cells without impairing the myeloid cell activity [20]. CsA is a cyclic undecapeptide of high lipophilicity and low permeability through the biological barriers (gastrointestinal tract, skin, and cornea) which entails inefficient administration of this poorly water–soluble drug via the oral route. Moreover, the oral formulations of CsA show high pharmacokinetic variability and low bioavailability confirmed by the unstable and incomplete absorption which is disadvantageous for efficient therapy [21,22]. In principle, the insertion of a drug on the covered implant can provide a better outcome. On the other hand, the major clinical concern is CsA-induced adverse side effects, including hepatotoxicity, nephrotoxicity, cardiotoxicity, and neural toxicity [20,23]. As the elucidation, some studies demonstrated that the CsA treatment provokes the excessive oxidative stress caused by the reactive oxygen species (ROS) and free radicals which injure chemically the biological molecules like lipids, proteins, nucleic acids, thus leading to inflammation, cell aging and death [23,24]. Although the molecular mechanism of CsA-induced toxicity is still unclear, it is obvious that antioxidants can play a beneficial role in mitigating these side effects.

As safety is a crucial issue to consider during the novel delivery system's development, it is essential to design new chitosan-based biocoatings for PET materials that could be used for drug incorporation, its localized release and absorption in blood. Additionally, to overcome the shortcomings faced by the CsA drug, an antioxidant can be used simultaneously. Studies on the effect of the combination of gallic acid as a strong antioxidant and cyclosporine A revealed the improvement in cardiac function [25]. The application of the gallic acid derivative, lauryl gallate (LG, Figure 1c) with the C12 hydrophobic chains, would be more interesting due to its greater antioxidant [26], membrane [27] and surface activities. The LG is required to be adsorbed at the solution/air interface and creates the prospect for employment of the Langmuir and Langmuir-Blodgett technology in the biocoating production. Both techniques provide the easy control of molecular composition, density, elasticity, and arrangement suitable for the preparation of highly ordered thin films with the molecular-level precision to be exploited in the tissue engineering [28].

Obviously, the fundamental aspect of biomaterials' development intended for contact with living biological cells or fluids is the precise characteristics of their surface. Generally, the physicochemical properties such as topography, chemical composition, and wettability of the biomaterial used as blood-contacting devices play an important role in the activation of platelets and in the initial cell attachment determining the overall proliferation and differentiation yield of the biomaterials [29,30]. Some studies indicate that surfaces with a balanced distribution of hydrophilic and hydrophobic microdomains exhibit an optimized hemocompatibility [29,31]. Thus, the surface structure and chemistry govern a reduction in the protein adsorption and coagulation resulting in enhanced biocompatibility.

In this paper, the Langmuir monolayer technique was utilized to fabricate the multicomponent monolayers at the chitosan solution/air interface, and to determine optimized conditions for the chitosan-based films' transfer onto the PET substrate by means of the Langmuir-Blodgett methodology. Since chitosan by itself exhibits a poor surface activity [32], it was a component of the subphase. On the other hand, based on the amphiphilic properties of biomolecules, which are capable of self-assembling at the interface, LB technology made it possible to prepare the phospholipid 1,2-dioleoyl-*sn*-glycero-3-phosphocholine (DOPC, Figure 1d) monolayer mimicking a cell membrane and thereby facilitating the incorporation of biologically relevant molecules, i.e., immunosuppressant CsA, and antioxidant LG. Since these three compounds (DOPC, CsA, and LG) are poorly soluble in the subphase they formed stable mixed Langmuir monolayers of varying compositions at the chitosan solution/air interface which ensured an accurate control of the molecular arrangement and a homogeneous deposition onto the PET plates. Moreover, to guarantee the attachment of the molecules to the polymer plate, its surface was pre-treated with low temperature air plasma. Finally, the LB films prepared on the PET substrate were characterized in terms of wettability, topography, molecular orientation (spatial arrangement) and chemical composition by means of the combination of contact angle measurements, atomic force microscopy (AFM), time-of-flight secondary ion mass spectrometry (TOF-SIMS), and X-ray photoelectron spectroscopy (XPS).

Biomaterial surface characteristics at the molecular level is without doubt a very promising way for the development of compatible coatings with the capability of efficient drug delivery and minute investigations of biological processes at the foreign material/tissue interface.

2. Results and Discussion

To design biocompatible coatings for the implants used in tissue engineering there is a need to examine the physicochemical properties of the films before (Langmuir monolayers technique) and after their transfer on the solid support by means of the Langmuir-Blodgett method. Previously the studies of the interactions occurring between the molecules forming the monolayers at the air/liquid interface were carried out [33]. The thermodynamic parameters (excess area per molecule, excess and total Gibbs energy of mixing), the changes in the surface potential, and the apparent dipole moment as well as morphology at the air/AA and air/Ch interfaces were determined. All examined monolayers at the air/liquid interface were found to be in the liquid state (L) which was proven by the compression modulus values. Additionally, it was evidenced that the chitosan presence in the subphase affects the organization of the molecules forming a monomolecular film as well as the type and magnitude of the interactions between them [33]. In this research the thickness of the monolayers at the air/liquid interface (surface pressure of 10 mN m^{-1}) was determined by means of the Brewster angle microscopy (BAM). Then thin films were deposited onto the air plasma-activated PET substrate by means of the Langmuir-Blodgett technique. The modified PET surfaces were characterized by the determination of the surface chemistry (XPS), the root mean square roughness (S_q) and the height profiles (AFM), the molecular arrangement (TOF-SIMS), and the surface wettability (CA).

2.1. Langmuir and Langmuir—Blodgett Monolayer Analysis

In the previous paper the morphology of all monolayers was examined by means of the Brewster angle microscopy (Supplementary materials in [33]). The BAM images confirmed that the obtained monolayers are homogeneous. In the whole range of the surface pressure no domains were visible, which is in good agreement with the monolayer miscibility expressed by the negative values of excess Gibbs energy and total Gibbs energy of mixing [33]. In this paper the relative thickness (d), changes in the area per molecule as a function of time at the air/AA and air/Ch interfaces as well as the transfer ratio (TR) values are estimated and listed in Table 1.

Table 1. Thickness (d) and transfer ratio (TR) values for the indicated monolayers obtained at the air/AA and air/Ch interfaces as well as the S_q roughness parameter gained by means of the AFM for the modified PET substrates.

Subphase	0.1% AA			0.1% mg mL^{-1} Ch		
Monolayer	***d* (nm)**	***TR***	**S_q (nm)**	***d* (nm)**	***TR***	**S_q (nm)**
DOPC	2.0 ± 0.1	1.2 ± 0.1	1.31 ± 0.05	2.0 ± 0.1	1.3 ± 0.1	2.42 ± 0.06
CsA	1.9 ± 0.1	1.0 ± 0.1	2.70 ± 0.04	1.7 ± 0.1	0.9 ± 0.1	1.22 ± 0.08
DC 0.50	1.7 ± 0.1	1.1 ± 0.3	1.79 ± 0.28	1.6 ± 0.1	1.0 ± 0.2	1.15 ± 0.13
DCL 0.25	2.0 ± 0.1	1.0 ± 0.1	0.81 ± 0.27	1.8 ± 0.1	0.9 ± 0.1	1.75 ± 0.20
DCL 0.50	2.0 ± 0.1	1.2 ± 0.1	1.07 ± 0.08	1.9 ± 0.1	0.9 ± 0.2	1.21 ± 0.17
DCL 0.75	2.1 ± 0.1	1.2 ± 0.2	1.89 ± 0.07	2.3 ± 0.1	0.8 ± 0.1	1.03 ± 0.06
LG	2.1 ± 0.1	1.0 ± 0.1	2.21 ± 0.08	2.5 ± 0.1	0.9 ± 0.2	1.26 ± 0.21

Firstly, according to Equation (1) the relative thickness was estimated for the monolayers studied on both AA and Ch subphases [34].

$$R = \frac{I_r}{I_0} = \left(\pi\frac{d}{\lambda}\right)^2 \frac{\left(n_1^2 - n_2^2 - 1 + \frac{n_2^2}{n_1^2}\right)^2}{1 + n_2^2} \quad (1)$$

where I_0 means the incident intensity, I_r is the reflected intensity, n_1 and n_2 indicate the refractive indices of the film and pure subphase, respectively; λ is the wavelength of the incident light.

For 0.1% acetic acid the monolayer thickness values are smaller than or equal to 2.1 nm. These are strictly related to the chemical structures of the examined compounds. The smallest d value is gained for the single DOPC-CsA (Table 1). This is a result of the flatness of the CsA molecule (stiff aminoacid ring, Figure 1b) which can change its conformation in relation to the polarity of the environment [35–37]. On the contrary, the single LG monolayer exhibits the highest d value on the Ch subphase. The molecules forming this layer can be more vertically localized by the fact that saturated hydrocarbon chains occur in their structure (Figure 1c). The presence of the unsaturated hydrocarbon chains in the DOPC structure (Figure 1d) promotes a slightly smaller relative thickness value. Owing to the *cis* double bonds, the chains can be more inclined than in the single LG monolayer. After adding LG to DOPC-CsA to form ternary (DOPC-CsA-LG) layers, the thickness values were found to be intermediated between them (Table 1). Among the mixed monolayers similar *d* values are gained. The addition of chitosan molecules to the acidic liquid phase slightly affects the monolayer thickness values (Table 1). Meanwhile, an increase in the thickness of the phospholipid monolayers in the presence of chitosan was shown by Cámara et al. [38] and Wydro et al. [39]. The changes in the *d* value are related to the Ch capability of movement from the bulk phase to the interface. Thus, the Ch can locate in the subsurface and is able to interact with the polar heads of the phospholipid [40].

As a result of the interactions between the chitosan and DOPC molecules, the monolayer formed at the air/liquid interface is more densely packed [33]. Similarly, greater compression modulus values were obtained for the single LG and ternary DOPC-CsA-LG monolayers [33]. Philippova et al. found that both acetylated and deacetylated units of chitosan are capable of forming hydrogen bonds [41]. Hydrogen interactions between the $-NH_2$ and/or $-NH_3$ groups and numerous amide groups of cyclosporine A may contribute to a decrease in the thickness of the CsA monolayer. Moreover, in the aqueous medium the CsA molecules can form the so-called open conformation and the intramolecular bonds cannot be formed [35]. Thus, the interactions between the Ch and CsA molecules can be promoted.

In the next stage of the experiments, the changes in the area per molecule in time were determined to examine the stability of the monolayer and capability of its penetration by the subphase molecules. Any loss of molecules due to desorption from the interface should be considered in the transfer ratio calculation. Figure 2 shows the relative changes in the area per molecule after the monolayer compression to the surface pressure of 10 mN m^{-1} and keeping it stable for 1 h with the constant speed of the barriers equal 5 mm min^{-1}. The single DOPC monolayer at the air/AA interface is characterized by 23% reduction of the area per molecule (Figure 2a). This is an evidence that the added AA molecules can interact strongly with the phospholipid molecules promoting their desorption. Such interactions result in the monolayer molecules' removal from the interface into the bulk liquid phase. On the other hand, the single CsA (Figure 2b) and LG (Figure 2g) films are more stable (only 2% and 6% decline, respectively). The mixed monolayers which contain the DOPC molecules exhibit the A/A_0 decrease of 8–16% (Figure 2c–f).

The presence of the Ch molecules in the subphase results in better stability of the examined monolayers (Figure 2) except for the single LG layer (Figure 2g). Surprisingly, the smallest desorption is observed for the single DOPC monolayer (1%, Figure 2a). This proves the existence of strong interactions between the phospholipid and chitosan molecules through the Lifshitz-van der Waals forces and Coulombic interactions. Therefore, it can be stated that the chitosan molecules form electrostatic complexes with the lipid and/or are accommodated in the lipid monolayer through the hydrophobic interactions [39].

After the chitosan addition to the subphase for the single CsA layer, the 2% decrease in the relative molecular area in time is registered (Figure 2b). This can prove that the interactions between the CsA and chitosan molecules are weaker than between the DOPC and chitosan ones. This can be a consequence of the hydrophobic nature of the CsA and Ch. However, the interactions between the molecules through the hydrogen bonds (chitosan $-NH_2/-NH_3^+$ groups and cyclosporine amide groups, Figure 1) can take place. In the case of the single LG, the interactions with chitosan occur between the hydrophobic chain (LG, Figure 1c) and the polysaccharide skeleton (Ch, Figure 1a) through the Lifshitz-van der Waals forces. Moreover, numerous hydroxyl groups of the LG can interact with the polar groups of the chitosan. For the mixed (binary and ternary) monolayers the smaller desorption occurs in the chitosan molecules' presence (5–9% decline, Figure 2c–f), whereby the smallest one is observed for the DOPC-CsA-LG 0.50 and 0.75 (Figure 2e,f, respectively). This is related to the strong attractive interactions between the molecules forming monolayers (negative Gibbs energy of mixing, the minimum for χ_{LG} = 0.50 [33]).

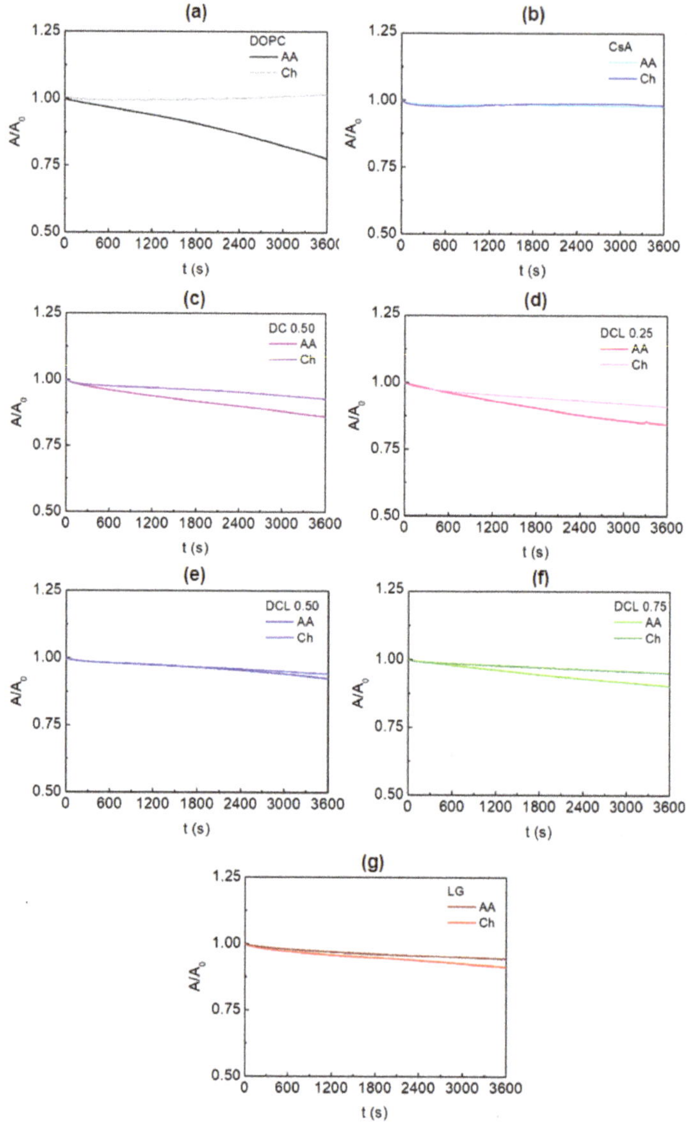

Figure 2. Relative molecular area (A/A_0) changes during the one-hour stabilization at 10 mN m^{-1} with the barrier speed of 5 mm min^{-1} for DOPC (**a**), CsA (**b**), DOPC-CsA 0.50 (**c**), DOPC-CsA-LG 0.25 (**d**) DOPC-CsA-LG 0.50 (**e**), DOPC-CsA-LG 0.75 (**f**), and LG (**g**) monolayers.

In order to confirm the effectiveness of the monolayer deposition the transfer ratio was accomplished. Deposition of the monolayers was carried out directly after reaching the surface pressure of 10 mN m^{-1} by withdrawal from the liquid subphase. This is due to the fact that in the first minutes the molecules forming the monolayers' removal to the liquid phase is the smallest (Figure 2). Based on the negative changes of the mean molecular area in time, there is a need to take into account the possible loss of the molecules in the liquid subphase in the transfer ratio determination. The corrected TR values calculated according to Equation (2) are listed in Table 1.

$$TR = \frac{\Delta A_m}{A_s} \quad (2)$$

where ΔA_m denotes the decrease in the monolayer surface area on the subphase and A_s indicates the substrate coated area.

For almost all examined monolayers TR values are close to unity for both AA and Ch subphases. This indicates that the molecules are transferred successfully onto the air plasma-activated PET substrate and the obtained LB film can be assumed to be a monolayer.

Among the single monolayers the greatest TR values are obtained for the single DOPC layer (TR = 1.2 and 1.3, for AA and Ch subphases, respectively). The transfer ratios are smaller than those obtained for the water subphase which is related to the lower desorption of the molecules into the liquid phase when the acetic acid and chitosan are added to the subphase [42,43]. For the single CsA layer similar TR values are obtained for all examined compositions of the subphases. After mixing the CsA and DOPC molecules forming the binary DOPC-CsA layer, smaller values of TR are obtained. This is due to the repulsion between the molecules forming the monolayer [44]. The addition of the third component-LG, changes the character of the interactions to the attraction ones. The DOPC and CsA are not matched in shape (DOPC-inverted truncated cone, CsA-ring), thus LG can be a linker between them [33,44]. Whereby, the TR values obtained during the monolayer transfer from the air/AA interface are higher than in the presence of the Ch. This is due to the larger desorption process in the absence of the chitosan molecules (Table 1). Thus, the excess molecules transfer is promoted.

2.2. XPS Analysis

The XPS spectra gained for PET, PET_{air}, and PET_{air}/AA/DOPC are presented in Figure 3. The obtained concentration of particular elements is shown in Table 2. Both spectra and position of peaks originating from the functional groups are in agreement with those found in the literature [45–48].

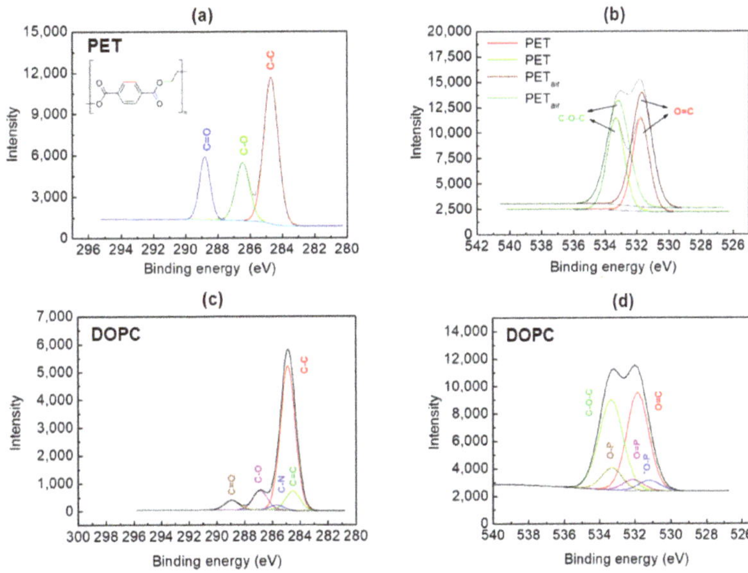

Figure 3. XSP spectra C 1s for unmodified PET (**a**), O 1s for unmodified and air plasma-activated PET (**b**), C 1s for PET_{air}/AA/DOPC (**c**), and O 1s for PET_{air}/AA/DOPC (**d**).

Table 2. XPS data for unmodified and modified PET surfaces (conc. in at.%).

Element	PET	PET_{air}	PET_{air}/AA/DOPC	PET_{air}/Ch	PET_{air}/Ch/DOPC
C 1s	77.7	72.2	75.8	74.2	72.4
N 1s	1.0	2.7	-	0.9	3.0
O 1s	21.3	25.1	23.3	24.9	23.8
P 2p	-	-	0.8	-	0.7

Plasma treatment of the PET surface causes the increase in the nitrogen (from 1% to 2.7%) and oxygen (from 21.3% to 25.1%) concentration. The increased amount of oxygen is visible in greater peak intensities (Figure 3). Additionally, the $-N =$ group is identified. This confirms that plasma action adds new functional groups containing these atoms onto the PET surface. The presence of the functional groups improves the PET surface hydrophilicity as well as enables the attachment of the transferred molecules by chemical bonds. In the case of O 1s XPS spectra obtained for the activated PET_{air} surface, the peaks are slightly broadened in relation to PET which can indicate that the polymer structure is changed after the plasma treatment.

The analysis of phospholipid DOPC film on PET_{air} detects the signals of C, O, N, and P. The main peak corresponds to the aliphatic tails of the lipid [49]. The concentration of the oxygen decreases to 23.3% while that of carbon is slightly larger in comparison to PET_{air}. The obtained peak areas are proportional to the number of the individual atoms within the DOPC molecule which proves the monolayer formation. The position of the combination of phosphorous with oxygen was stated according to Wagstaffe et al. [50]. The O 1s band contains contribution of several bonds such as $C-O$, $C=O$, $P-O$ and $P=O$ thus the deconvolution is difficult and can be encumbered with an error. Moreover, the obtained peaks' positions are similar to those described by Panajotović et al. [49] for very similar phospholipid DPPC that possesses all saturated hydrocarbon chains. The introduction of the Ch molecules is related to the presence of the nitrogen atoms, which is not observed for the $PET_{air}/AA/DOPC$ surface. Furthermore, the Ch film resulted in a decrease in oxygen and nitrogen concentration with respect to PET_{air}. This is indicative of the formation of new bonds between the activated PET surface and the Ch molecules. The analysis of mixed monolayers is very difficult due to overlapping of many peaks. A more thorough analysis was performed using mass spectrometry and is described in Section 2.4. TOF-SIMS demonstrates a significantly better detection limit than XPS and provides certain identification of monolayer species.

2.3. *AFM Analysis*

The AFM technique was used to characterize the topography of the low temperature air plasma-modified PET surfaces with deposited chitosan film and/or Langmuir monolayers. For this purpose, the roughness parameter Root Mean Square, S_q, was determined according to Equation (3).

$$S_q = \sqrt{\frac{1}{MN}\sum_{k=0}^{M-1}\sum_{l=0}^{N-1}[z(x_k, y_l)]^2} \tag{3}$$

where x and y are the coordinates, z is the perpendicular deviation from the ideally smooth surface, M is the number of points in the x direction and N is the number of points in the y direction.

Figures 4–7 show the AFM micrographs. Figures 4 and 6 present the micrographs with the scanned area of 20 × 20 μm^2 and with the marked zoomed area, while Figures 5 and 7 the determined profiles.

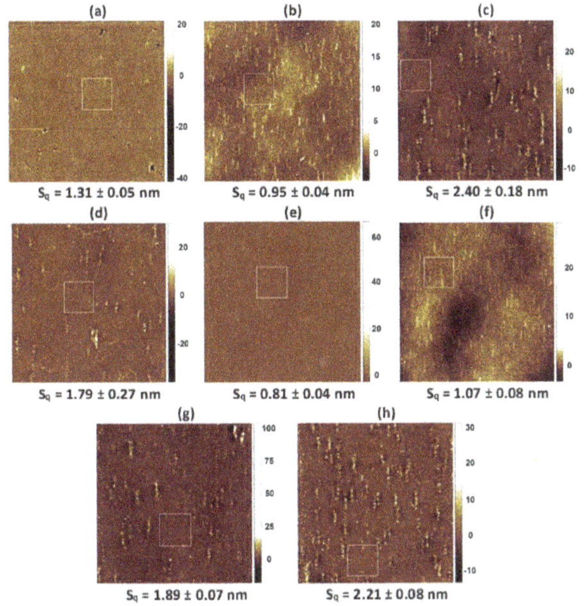

Figure 4. AFM micrographs for the 20 × 20 μm^2 scanned areas: PET$_{air}$/AA (**a**), PET$_{air}$/AA/DOPC (**b**), PET$_{air}$/AA/CsA (**c**), PET$_{air}$/AA/DOPC-CsA 0.50 (**d**), PET$_{air}$/AA/DOPC-CsA-LG 0.25 (**e**), PET$_{air}$/AA/DOPC-CsA-LG 0.50 (**f**), PET$_{air}$/AA/DOPC-CsA-LG 0.75 (**g**), PET$_{air}$/AA/LG (**h**). Z scale: nm.

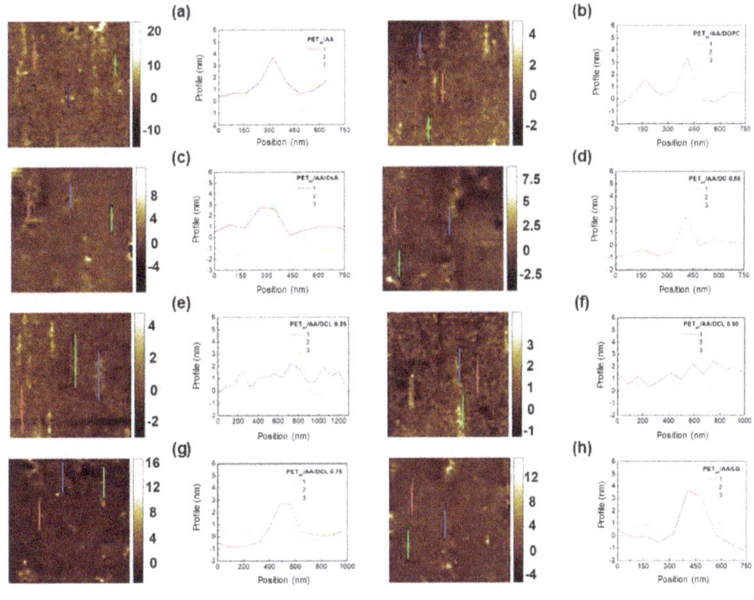

Figure 5. AFM zoomed micrographs and profiles obtained for: PET$_{air}$/AA (**a**), PET$_{air}$/AA/DOPC (**b**), PET$_{air}$/AA/CsA (**c**), PET$_{air}$/AA/DOPC-CsA 0.50 (**d**), PET$_{air}$/AA/DOPC-CsA-LG 0.25 (**e**), PET$_{air}$/AA/DOPC-CsA-LG 0.50 (**f**), PET$_{air}$/AA/DOPC-CsA-LG 0.75 (**g**), PET$_{air}$/AA/LG (**h**). Z scale: nm.

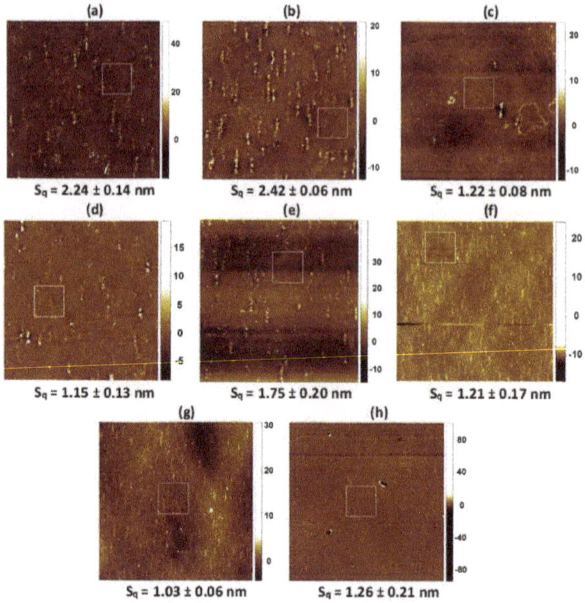

Figure 6. AFM micrographs for the 20 × 20 μm^2 scanned areas: PET_{air}/Ch (**a**), PET_{air}/Ch/DOPC (**b**), PET_{air}/Ch/CsA (**c**), PET_{air}/Ch/DOPC-CsA 0.50 (**d**), PET_{air}/Ch/DOPC-CsA-LG 0.25 (**e**), PET_{air}/Ch/DOPC-CsA-LG 0.50 (**f**), PET_{air}/Ch/DOPC-CsA-LG 0.75 (**g**), PET_{air}/Ch/LG (**h**). Z scale: nm.

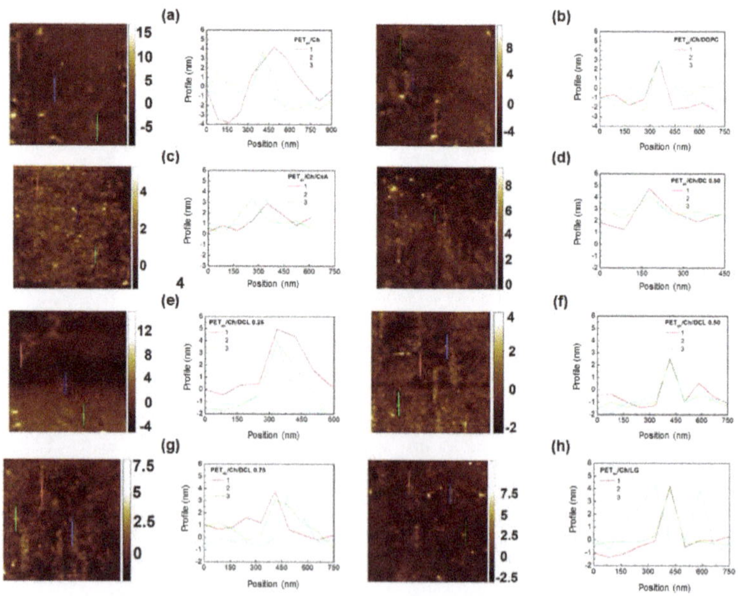

Figure 7. AFM zoomed micrographs and profiles obtained for: PET_{air}/Ch (**a**), PET_{air}/Ch/DOPC (**b**), PET_{air}/Ch/CsA (**c**), PET_{air}/Ch/DOPC-CsA 0.50 (**d**), PET_{air}/Ch/DOPC-CsA-LG 0.25 (**e**), PET_{air}/Ch/DOPC-CsA-LG 0.50 (**f**), PET_{air}/Ch/DOPC-CsA-LG 0.75 (**g**), PET_{air}/Ch/LG (**h**). Z scale: nm.

As it was previously reported, the unmodified PET surface is smooth ($S_q \approx 2$ nm), while after the plasma modification the roughness increases to a great extent ($S_q \approx 4$ nm) [42]. The greater roughness results from the addition of the new function groups containing oxygen, nitrogen, and carbon ($-OH, C-O, O=C-O, C=O, N-CO-N$). Owing to them, the process of the monolayers' transference is more effective which was confirmed by the XPS measurements (Section 2.2) as well as by the other authors [45,51,52]. Deposition of the Langmuir monolayers and/or the chitosan film results in a smoother surface (smaller S_q parameter, Table 1, Figures 4 and 6). Nevertheless, some domains are visible in the AFM micrographs, although the Brewster angle microscopy did not expose them on the microscale [33]. It proves that the molecules forming monolayers undergo reorganization during the deposition process from the liquid phase (AA, Ch) to the plasma-modified PET surface (solid support).

Of the single layers, the DOPC monolayer transfer from the AA subphase results in the biggest PET surface smoothing (Figure 2b). This finding is in good agreement with the surface roughness obtained after the deposition of the single DOPC layer from the air/water interface [42] indicating that acetic acid has no effect on the evolution of PET_{air}/DOPC surface asperity. However, this time vertical domains parallel to the PET_{air} withdrawal direction with the height ca. 3 nm and the width 247 nm are visible (Figure 5b). This can be due to the change in the transfer ratio value greater than 1 which suggests that more than one monolayer is deposited onto the PET_{air} substrate (TR = 1.2, Table 1). This is probably related to the changes in the interactions between the molecules at the air/AA interface and after the film transfer onto the PET_{air} substrate. These interactions are a key point in the packing and arrangement of the thin layer onto the solid support.

More inhomogeneous and rough surfaces are gained after the deposition of the single CsA and LG layers (Figure 4c,h, respectively). The height of the protrusions for the single CsA is in the range from ca. 2.5 nm to even 5 nm (Figure 5c) while the width is similar to that of the protrusions on the unmodified PET (380 nm) [42]. This suggests that the CsA molecules are densely packed around the protrusions similarly to those observed previously for the CsA film transferred from the water subphase [42]. This is a result of the largest stiffness of CsA among the single monolayers at the surface pressure of 10 mN m^{-1} (compression modulus equal 51.1 mN m^{-1}, [33]). The TR value is equal to 1 (Table 1) which confirms the single layer deposition onto the PET_{air} substrate.

After deposition of the binary DOPC-CsA film onto the polymer substrate, a smoother surface is obtained with respect to the PET_{air}/AA/CsA (S_q = 1.79 nm, Figure 4d). The intermediate roughness between the DOPC and CsA films arises from the smaller packing of the mixed monolayer in comparison to the single CsA layer [33]. The height of the protrusions reaches 2.6 nm (Figure 5d) which is very close to those of the single DOPC and CsA layers. Similarly, the TR value is approximated to 1 (TR = 1.1).

The addition of the third component-LG, causes the smoothing of the examined surfaces (Table 1). In the group of the ternary DOPC-CsA-LG monolayers, decreasing roughness is obtained after the deposition of the DOPC-CsA-LG 0.25 and 0.50 (Figure 4e,f, respectively). It is in line with the greater packing of the monolayers obtained at the air/AA interface and better miscibility expressed by the negative values of the excess and total Gibbs energy of mixing [33]. The visible protrusions for the ternary monolayer at the LG molar fraction of 0.25 are with the height 1.7 nm and the width 226 nm (Figure 4e) while at χ_{LG} = 0.50 with almost the same height but smaller width (Figure 4f). This observation can be a reason for the changes of the surface roughness parameter (Figure 2e,f). Surprisingly, the higher LG molar fraction (χ_{LG} = 0.75) does not cause the largest smoothing of the PET_{air} surface. After the deposition of the DOPC-CsA-LG 0.75, the surface roughness is similar to that of PET_{air}/AA/DOPC-CsA 0.50 (Figure 2g,e, respectively). The reason for that can be found in the smaller packing of the monolayer at the air/AA interface (C_s^{-1} = 41.7 mN m^{-1}) than for the binary DOPC-CsA layer (C_s^{-1}= 44 mN m^{-1}) [33]. Moreover, the attraction interactions between the DOPC-CsA-LG molecules are weaker for the LG molar fraction of 0.75 than for 0.25 and 0.50.

After the single LG monolayer deposition, the biggest roughness of the coated PET surface is obtained if the single layers are taken into account (Figure 4h). The PET_{air}/AA/LG surface is characterized by the protrusions with the height of 3.3 nm and width of 195 nm (Figure 4h). This result is contrary to the surface roughness gained after the LG film transfer from the air/water interface [42] despite the greater packing of the monolayer [33,44]. On the other hand, the thickness of the single LG layer at the air/AA interface is smaller than at the air/water interface [42]. Thus, it can be said that the acetic acid molecules influence the tilt and/or organization of the molecules forming the monolayer. Moreover, due the mismatch in the LG structure (large polar head and hydrocarbon tail, Figure 1c) the LG layer cannot form a dense LB film and as a result the molecular ion is difficult for detection in the TOF-SIMS spectra (see Section 2.4).

The adsorption of the chitosan film onto the air plasma-activated PET (PET_{air}) surface is confirmed by the smaller S_q parameter value (S_q = 2.24 nm, Figure 6a). The effective adsorption is possible due to the interactions between the PET_{air} and chitosan hydroxyls as well as amino functional groups [53] and/or ester bonds [54]. In the AFM micrographs there are visible local aggregates which is a characteristic feature for many polysaccharides.

Additionally, the deacetylation degree (DD) is a property of a key importance in the surface roughness characterization [55]. Chitosan with the DD equal to 82.3% used in this research has in its structure both acetylated and deacetylated groups whereby the latter are more numerous. As it was reported by the Thepsak et al. the strong binding of the chitosan film and plasma activated PE can take place [54]. The polar functional groups containing oxygen can interact with the hydroxyl groups of the chitosan to form ester linkages. This process is assisted in the acidic environment. Additionally, there are possible intermolecular hydrogen bonds between the functional groups on the PET_{air} and $-OH$ and $-NH_2$ groups of chitosan [54]. For this reason, strong binding of the chitosan film to the PET_{air} surface can be obtained.

The surface roughness of the DOPC film deposited from the chitosan subphase onto the PET_{air} substrate (PET_{air}/Ch/DOPC) is larger (Figure 6b) in comparison to PET_{air}/AA/DOPC (Figure 4b). It can be stated that the presence of the Ch film impacts the DOPC molecules' organization due to the interactions between the positively charged amino groups in the acidic environment ($pK_a \approx 6.5$ [39]) and the polar choline head of the DOPC. A similar observation was reported by Cámara et al. [38]. The researchers examined the interactions between the distearoylphosphatidylglycerol (DSPG) and chitosan at the air/water interface. They correlated the increase in the packing with the fact that the chitosan molecules can locate at the interface. Additionally, on the basis of the electrostatic interactions Silva et al. proved that chitosan could form complexes with mucin and dimyristoylphosphatidic acid DMPA [56].

The largest smoothing is obtained for the PET_{air} substrate with the Ch film and the deposited single CsA layer (Figure 6c). The protrusions are with the height 2.3 nm and the width ca. 300 nm. These results are very similar to those obtained for the unmodified PET [42]. This can be a result of the greatest packing of the monolayer molecules obtained at the air/Ch interface for all the examined Langmuir layers [33]. The TR value is close to 1 which confirms the monomolecular layer deposition (Table 1). Additionally, the CsA monolayer shows the greatest stability over time on the chitosan subphase (Figure 2b). The presence of the Ch film results in the twice lower S_q value for the PET_{air} substrate with the deposited Ch and binary DOPC-CsA 0.50 layers (Figure 6d) and is characterized by a smaller protrusion (height ~2.1 nm, Figure 7d) than for PET_{air}/AA/DOPC. This confirms strongly that chitosan influences the interactions between the DOPC and CsA molecules. Moreover, Ch can fill free gaps developed from the structural mismatch of molecules.

The deposition of the ternary DOPC-CsA-LG monolayers alters the surface roughness of the PET_{air}/Ch substrates (Figure 6e–g). For the three-component DOPC-CsA-LG χ_{LG} = 0.25 surface roughness increases with respect to the binary DOPC-CsA despite the greater packing of the molecules [33]. This is visible in the size of the protrusions (height 4.4 nm, width 255 nm). The reason for that is a specific reorganization of the film-forming molecules after their transfer to the PET_{air} substrate from the air/Ch interface. The smoothing of the surface is gained after the DOPC-CsA-LG 0.50 and 0.75 layers' deposition from the chitosan liquid phase (Figure 6f,g). This is related to the strong attractive interactions between the molecules forming the thin films and their greater packing at the air/Ch interface in relation to the binary DOPC-CsA 0.50 layer [33]. The size of the protrusions for both ternary monolayers (DOPC-CsA-LG 0.50 and 0.75) is similar. Protrusions height and width are equal to 3.4 nm and 166 nm and 2.9 nm and 220 nm for χ_{LG} = 0.50 an 0.75, respectively.

After the single LG layer deposition from the chitosan subphase the smooth surface of PET_{air}/Ch/LG is obtained (S_q = 1.26 nm, Figure 6h). Surprisingly, there are visible height protrusions (Figure 7h, 4.1 nm). This can be related to the changes in the LG orientation induced by the interactions with chitosan through the hydrogen bonding and dispersion forces as well as smaller packing of the molecules forming monolayers [33].

Generally, the obtained TR values for all examined monolayers deposited onto the PET_{air} substrates from the chitosan subphase are close to 1 which confirms the single layer deposition. Additionally, the desorption of the molecules forming monolayers is smaller when the Ch molecules are present in the liquid subphase which indicates that monolayers are more stable. Moreover, the determined excess Gibbs energy and the total Gibbs energy of mixing proved that the obtained monolayers are more stable (stronger attraction between DOPC, CsA, LG molecules [33]) which guarantees the effective transfer of the layers.

2.4. TOF-SIMS Analysis

2.4.1. PET and PET_{air}

The TOF-SIMS measurements were made for the PET and plasma-treated PET and the characteristic peaks in the positive TOF-SIMS spectra were identified. To the best of our knowledge the peaks determined for the plasma-activated PET (Table 3) were not reported previously.

Table 3. Peaks identified for the PET and PET_{air} surfaces in the positive TOF-SIMS mass spectra.

Assignment	m/z	Identification
$(NH_2)^+$	16	PET_{air}
$(NH_3)^+$	17	PET_{air}
$(CNH_2)^+$	28	PET_{air}
$(CH_2NN_2)^+$	30	PET_{air}
$(CH_6N)^+$	32	PET_{air}
$(NOH_4)^+$	34	PET_{air}
$(C_2H_4N)^+$	42	PET_{air}
$(C_4H_{10}N_2)^+$	86	PET_{air}
$(C_5H_{14}N_2)^+$	102	PET_{air}
$(C_5H_{11}N_3)^+$	113	PET_{air}
$(C_7H_{18}N_2)^+$	130	PET_{air}
$(C_7H_{15}N_4)^+$	155	PET_{air}
$(C_{10}H_9N_4)^+$	193	PET, PET_{air}

As it is shown in Table 3 there is only one common peak: $(C_{10}H_9N_4)^+$ (m/z = 193) identified in the TOF-SIMS mass spectra for PET and PET_{air}. The remaining peaks were identified only for PET_{air}. It can be clearly seen that during the plasma treatment the PET surface is functionalized by the NH_2 and NH_3^+ groups. The distribution of fragments for the PETair surface is presented in Figure 8. The intensity of the $(NH_2)^+$ fragment is about ten times smaller than NH_3^+. It means that the nitrogen atom is bound to the carbon atom via the single bond. The NH_3^+ fragment is yielded as follows (Equation (4)):

$$NH_2^{\bullet} + H^{\bullet} \rightarrow NH_3^+ \tag{4}$$

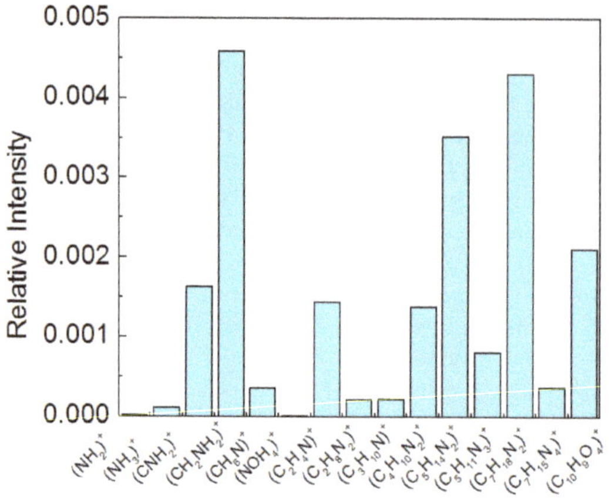

Figure 8. Intensity distribution of the fragments for PET_{air}.

The existence of NH_2 rather than NH on the PET substrate is supported also by the large intensity fragment $(CNH_2)^+$ (Table 3). The fragment CHNH is not identified in the spectra. Moreover, it is important to distinguish that the NH_2 group is bound to the phenyl ring or hydrocarbon chain in the PET monomer. The $(C_6H_3NH_2)^+$ fragment is not identified while the $(CHNH_2)^+$ and $(CH_2NH_2)^+$ fragments demonstrate a great intensity (Figure 8). Moreover, the NH_2 group is bound to only one carbon in the hydrocarbon chain (Figure 9) since only the $(C_2H_4N)^+$ and $(CH_2NH_2)^+$ fragments are detected while the $(C_2H_8N_2)^+$ fragment is not observed. Furthermore, the existence of numerous peaks obtained due to the recombination of $(CNH_2)^+$ fragments with other free radical ions, suggests unambiguously that the NH_2 group resides within the hydrocarbon chain of the PET monomer unit. For example, the second prominent fragment $(C_7H_{18}N_2)^+$ (Figure 8) is yielded by cleaving the $C-O$ bonds (Figure 9), and recombines with the other free radical fragments as follows (Equation (5)):

$$CH_2CH_2NH_2^\bullet + CH_2CH_2NH_2^\bullet + CH_2CH_2^\bullet + CH_2^\bullet \rightarrow (C_7H_{18}N_2)^+ \quad (5)$$

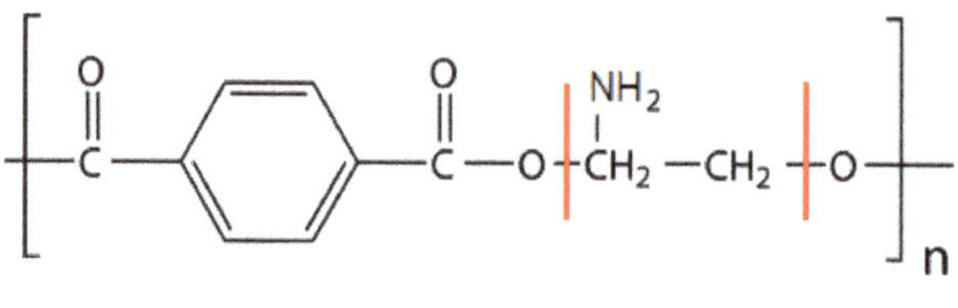

Figure 9. Proposed molecular structure of PET after the plasma treatment.

The XPS measurements were made for the selected samples. For the plasma-activated PET, the $-N=$ functional group was identified. It plays the role of a binding agent for the deposited LB films.

In the TOF-SIMS spectra the NH_2 group is identified. However, one should keep in mind that during the Bi^+ bombardment plenty of highly active free hydrogen radical ions are yielded which can react easily with the nitrogen atom containing the free electron pairs. For this reason, the intensity of the $(CNH_2)^+$ fragment is significantly smaller than $(CH_2NH_2)^+$ since the latter fragment does not contain the free electron pairs.

However, the identification of the $-N=$ group by XPS is uncertain since the peak position (399.67 eV) for the PET_{air} sample corresponds to the amide and amino groups [57].

On the other hand, according to [58] after the deconvolution this energy corresponds to the $-N =$ group.

Taking into account the above, one can conclude that existence of the $(NH_2)^+$ group on the PET_{air} surface is proved by TOF-SIMS while we cannot rule out that nitrogen can exist in the highly reactive $-N =$ form that plays a role of the binding agent for the DOPC, LG and CsA species.

The existence of DOPC is determined by the XPS spectra (Figure 3). The areas of the peaks are proportional to the number of carbon atoms in DOPC. Chitosan species are not identified due to the insufficient detection limit and the energy resolution of the XPS instrument used. It is possible that XPS equipped with a synchrotron beam will be useful for this purpose. However, we do not expect any additional information. Moreover, TOF-SIMS demonstrates a significantly better detection limit than XPS and provides certain identification of the monolayer species that in the case of XPS is very often problematic due to overlapping of many peaks. For this reason, the deconvolution can lead to improper conclusions.

2.4.2. Films Deposited onto PET_{air}

The assignment m/z and identification of the most characteristic fragments of DOPC, CsA and LG for the deposited LB films on the PET_{air} substrate, prepared from acetic acid (AA) or chitosan (Ch) subphase are listed in Table 4. Furthermore, the fragments' distribution is shown in Figure 10. There is a lack of the molecular ions of the single DOPC and LG if the LB monolayers transferred from the AA subphase are considered. The molecular ion with the mass equal to the molecular mass of the examined compound can be used as an indicator for the ordering (arrangement) of the monolayer forming molecules. The possible reason for the absence of this ion is related to the roughness and heterogeneous structure of the plasma activated PET substrate surface which was discussed in detail in our earlier paper [42]. The unsaturated bonds in the hydrocarbon chains of DOPC (Figure 1d) give rise to the steric hindrance that does not allow to form a densely packed and well-ordered layer. In the case of the LG layer, the lack of the molecular ion is in relation with the disproportion of the size of the polar head and apolar chain (Figure 1c) which provokes chain tilting to compensate for the head-tail mismatch. For this reason, monolayers occur in the liquid state with loosely packed molecules on the acetic acid subphase [33]. Assuming that the physical phase of the monolayers does not change significantly during their transfer to the solid support, the LB film molecules are prone to fragmentation due to the bismuth ion bombardment. For the same reason, the molecular ions were not determined for the single DOPC and LG layers transferred onto the PET_{air} substrate from the water subphase [42].

Table 4. The most characteristic positive fragments in the TOF-SIMS mass spectra obtained for all examined monolayers.

Assignment	m/z	Identification	References
$(C_5H_{14}NO)^+$	104	DOPC	[59]
$(C_5H_{13}PO_3N)^+$	166	DOPC	[59]
$(C_5H_{15}NPO_4)^+$	184	DOPC	[59]
$(C_7H_5O_4)^+$	153	LG	[59]
$(C_7H_6O_5)^+$	170	LG	[59]
$(C_7H_7O_5)^+$	171	LG	[59]
$(C_6H_{14}N)^+$	100	CsA	[60]
$(C_{61}H_{107}N_{10}O_{12})^+$	1172	CsA pseudo-molecular ion	[60]
$(C_{62}H_{112}N_{11}O_{12})^+$ $(CsA + H)^+$	1202	CsA molecular ion	[60]
$(C_2H_4NO)^+$	58	Chitosan	[61]
$(C_2H_5NO)^+$	59	Chitosan	
$(C_2H_6NO)^+$	60	Chitosan	[53]

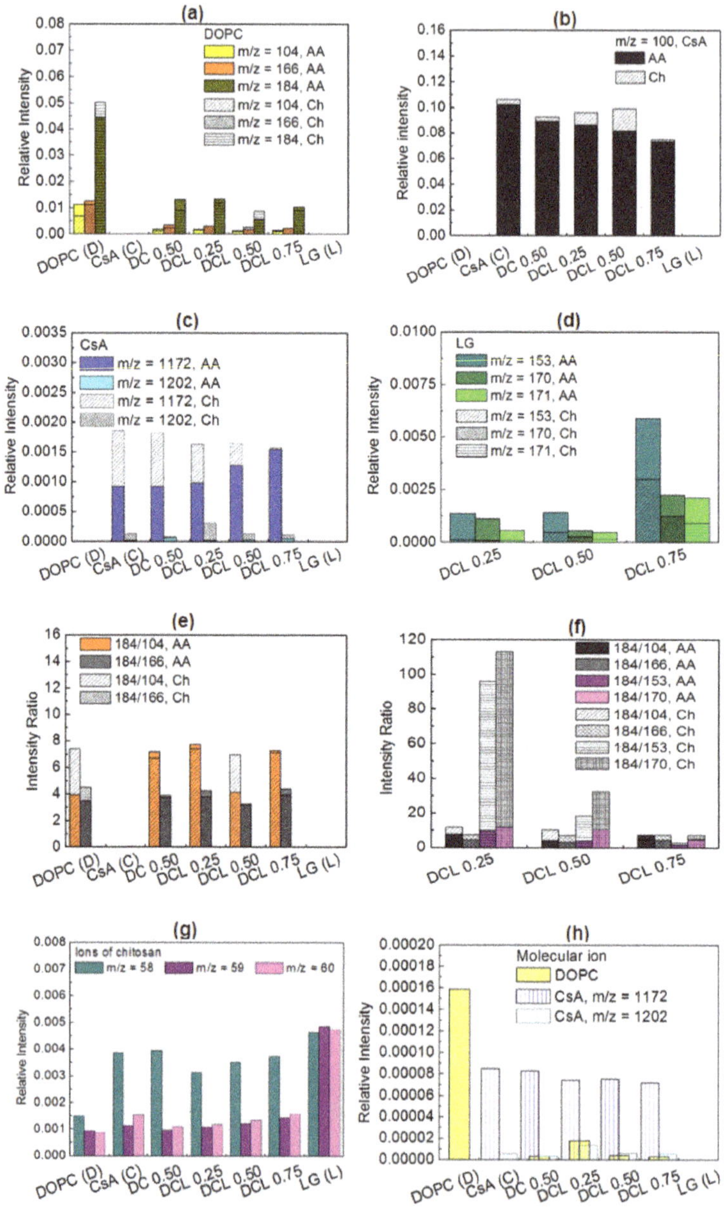

Figure 10. Distribution of 1,2-dioleoyl-*sn*-glycero-3-phosphocholine m/z = 104, 166, and 184 (phosphocholine) fragments (**a**), cyclosporine m/z = 100 fragment (**b**), cyclosporine m/z = 1172, 1202 fragments (**c**), lauryl gallate m/z = 153, 170, 171 fragments (**d**), intensity ratio of DOPC to DOPC fragments (**e**) intensity ratio of DOPC to DOPC and DOPC to LG fragments (**f**), distribution of the chitosan m/z = 58, 59, 60 fragments (**g**), molecular ions of the DOPC and CsA single layers (**h**) in the single, binary and ternary monolayers deposited onto PET_{air} at the air/AA and air/Ch interfaces.

The intensity of the most intensive fragments of the DOPC phosphocholine head with m/z = 184 $(C_5H_{15}NPO_4)^+$ dominates over the intensity of the fragments m/z = 166 $(C_5H_{13}PO_3N)^+$ and m/z = 104 $(C_5H_{14}NO)^+$ (Figure 10a). This is in good agreement with the literature data [62–64]. A comparable distribution of the choline head fragments was obtained for the DOPC layer without acetic acid deposited onto the PET_{air} [42] and on mica [59]. As it was previously reported, the most prominent fragments for the CsA molecule are m/z = 100 $(C_6H_{14}N)^+$, 1172 $(C_{61}H_{107}N_{10}O_{12})^+$ and 1202 $(C_{62}H_{112}N_{11}O_{12})^+$. The positive spectra fragments are consistent with the data published by the other researchers [60,65]. After mixing the DOPC and CsA molecules to gain the binary DOPC-CsA 0.50 film, the intensity of the above-mentioned DOPC fragments (m/z = 184, 166, 104) decreases c.a. 3-times (Figure 10a). On the other hand, the intensity of the most abundant fragment of CsA (m/z = 100) is reduced by about 10% for the binary film (Figure 10b). This indicates a smaller CsA amount (surface coverage) which is in line with the reduced CsA content in the DOPC-CsA monolayer at the interface (before deposition) in relation to the single CsA film. Analogous intensities of the CsA fragments were observed without the presence of acetic acid in the subphase (pure water) while the intensities of the DOPC choline head were twice smaller for the single DOPC layer and five-times smaller for the binary DOPC-CsA film [42].

This can suggest that a smaller amount of the DOPC molecules was transferred to the PET_{air} substrate. Due to higher ions relative intensities in the presence of the acetic acid in the liquid subphase it can be said that it stands for the more effective deposition process of the DOPC onto the PET_{air} (higher PET_{air} surface coverage with the DOPC molecules) while it does not exhibit any impact on the transfer of the CsA molecules. On the other hand, under the smaller PET_{air} surface coverage with CsA (smaller intensity of (m/z = 100, Figure 10b) intensity of the pseudomolecular ion of CsA (m/z = 1172, $(C_{61}H_{107}N_{10}O_{12})^+$) is maintained while that of the molecular ion $(C_{62}H_{112}N_{11}O_{12})^+$ increases in comparison to the single CsA layer (Figure 10c). This is determined by the reorientation of the CsA molecules to be more perpendicular towards the PET_{air} substrate. That latter process is more intensive than for the DOPC-CsA monolayer without acetic acid as it was presented in our earlier paper [42]. A more upright position of the CsA molecules is a natural consequence of the greater surface coverage with DOPC in the DOPC-CsA layer that reduces the sites accessible to CsA in comparison to the single CsA layer.

The presence of the third component, lauryl gallate (LG) in the mixed monolayers transferred on the PET_{air} substrate is revealed by the occurrence of m/z = 153 $(C_7H_5O_4)^+$, 170 $(C_7H_6O_5)^+$ and 171 $(C_7H_7O_5)^+$ fragments (Figure 10d). The LG amount at the molar fraction of 0.25 causes a slight increase in the intensity of the CsA pseudomolecular ion and significant reduction that of the molecular ion (Figure 10c). This inconsistency can be determined by the unknown mechanism of the yield of the pseudo-molecular ion that is formed by the subtraction of two oxygen atoms from the molecular ion. Despite that fact, with the smaller amount of CsA for the ternary DOPC-CsA-LG 0.25 due to the smaller intensity of the fragment with m/z = 100 (Figure 10b), the CsA molecules have to be less tilted towards the substrate than in the case of the single CsA monolayer. This results from a slightly larger intensity of the fragment m/z = 1172 and a similar intensity of the molecular ion (Figure 10c) in comparison to the single CsA layer but under the smaller surface coverage. The changes of the geometrical orientation of CsA in the three-component DOPC-CsA-LG 0.25 film in comparison to the binary DOPC-CsA 0.50 one is difficult to estimate due to the mentioned inconsistency in changes of the pseudomolecular/molecular ion intensities. For the ternary DOPC-CsA-LG 0.50 significant reduction of the surface coverage with CsA is observed (Figure 10b,c). The intensity of the m/z = 100 $(C_6H_{14}N)^+$ decreases (Figure 10b) while the intensity of the pseudomolecular and molecular ions increases (Figure 10c). The latter behavior exhibits unambiguously a strong reorientation of the CsA molecules that take the most perpendicular orientation towards the PET_{air} substrate for the DOPC-CsA-LG 0.75 layer (the greatest intensity of the molecular ions, Figure 10c).

Based on the relative intensity the surface coverage with DOPC demonstrates a minimum for the ternary DOPC-CsA-LG 0.50 monolayer while for DOPC-CsA-LG 0.75 it is slightly smaller than for DOPC-CsA-LG 0.25 (Figure 10a). Similarly, an unexpected distribution of intensity of the most characteristic fragments of LG (Figure 10d) is observed. For the DOPC-CsA-LG 0.50 film the intensity of the most abundant fragment (m/z = 153) remains unchanged while that of the fragments with m/z = 170 and 171 is even reduced (with respect to the DOPC-CsA-LG 0.25). However, since the fragment with m/z = 153 is produced by a simple cleavage of the bond $C - O$ [59] it illustrates more appropriately the surface coverage with LG than the fragments m/z = 170 and 171 which need the additional hydrogen ions for yielding. For the greatest molar fraction of LG (χ_{LG}= 0.75) in the three-component layer, a significant increase in all LG fragments' intensity can be clearly seen (Figure 10d).

The unexpectedly smaller amount of the DOPC and LG for the ternary DOPC-CsA-LG 0.50 layer deposited on the PET_{air} substrate can be determined only by the reduced transfer of the DOPC and LG from the AA subphase onto the solid support in relation to pure water. The reason for that behavior for the DOPC-CsA-LG 0.50 film is unknown. During the deposition of the mixed DOPC-CsA-LG monolayer without acetic acid a different behavior was observed in our previous studies [42].

On the other hand, the intensity of all DOPC fragments for DOPC-CsA-LG 0.75 is smaller than for DOPC-CsA-LG 0.25 that can correspond to the content of DOPC in the mixed monolayers at the air/liquid interface. This fact suggests that there are no direct interactions between the hydroxyl groups of LG and the phosphate group of DOPC that affects greatly the increase in the intensity of all DOPC fragments through the transfer of hydrogen ions as it was proved in the previous paper [59]. This conclusion is additionally supported by the distribution of the 184/104 and 184/166 intensity ratio (Figure 10e).

In the mixed DOPC-CsA-LG monolayers only a slight increase in the 184/104 and 184/166 ratios is observed suggesting the rather weak direct interactions between the DOPC and LG molecules shown above. Similarly, for ternary DOPC-CsA-LG 0.75 there are no significant changes of the 184/104 and 184/166 ratios. Moreover, the DOPC-CsA-LG 0.50 monolayer is not a representative one due to the small surface coverage (small intensity of the DOPC choline head, Figure 10a) onto the PET_{air} surface. Therefore, the sample cannot be considered in this part of the discussion. On the other hand, a significant increase in the 184/104 intensity ratio in the DOPC-CsA 0.50 layer in comparison to the single DOPC layer is observed (Figure 10e). This indicates evidently that the interactions between the CsA and DOPC molecules determine the transfer of hydrogen ions from the CsA molecules to oxygen atom bound with the phosphorus atom next to the hydrocarbon chains in the DOPC phosphate group. In this way, it gives a rise to the yield of the $(C_5H_{15}NPO_4)^+$ fragment (m/z = 184, Figure 10a).

The latter observation is governed by the presence of acetic acid since without this acid the direct interactions between the hydroxyl groups of LG and the phosphate group of DOPC [59] rather than between DOPC and CsA were observed. In order to improve the biocompatibility of the surface and attractive interactions between DOPC, CsA, and LG in the deposited thin film, the Ch molecules were added to the acidic liquid phase.

The distribution of the intensity of the most characteristic fragments of chitosan is presented in Figure 10g. Similarly, the other authors reported the same mass to ratio fragment of this polysaccharide [66]. The smallest intensity of the most intensive fragment of chitosan m/z = 58 $(C_2H_4NO)^+$ corresponds to the largest surface coverage with the monolayer deposited onto chitosan and vice versa, the largest intensity means the smallest surface coverage with that layer. The intensities of the identified molecular ions of the single DOPC and the molecular and pseudomolecular ions of the CsA are also presented in Figure 10a,h and c,h, respectively. The DOPC molecular ion (m/z = 786) was identified for the single DOPC monolayer. The existence of the molecular ion is determined by a densely packed, uniform DOPC monolayer and indicates a great affinity of the DOPC molecules for chitosan (Figure 10h). Such a kind of monolayer was previously identified on the mica substrate [59]. Moreover, the presence of the molecular ion of different phospholipid 1,2-dipalmitoyl-*sn*-glycero-3-phosphocholine (DPPC) on the chitosan layer deposited on the PET_{air} substrate was reported in the previous paper [53].

Moreover, the TOF-SIMS experiments confirmed that the obtained DOPC film is the monolayer. As it was reported in the literature [67], the molecular escape depth from the organic films (tetraglyme) was equal to 1.8 nm. We can assume that the molecular escape depth for DOPC can be maintained on the similar level. It means that if the films can organize into a bilayer or a multilayer with the thickness greater than 2–3 nm, the underlying chitosan film should not be detected. As it was shown in Figure 10g, for all chitosan-containing layers there were detected the polysaccharide fragments with a varying intensity. The most prominent chitosan fragment (m/z = 58) demonstrates the smallest intensity for the single DOPC layer. This indicates that in that case the monolayer is the most ordered, dense and molecules are oriented perpendicularly to the surface. Even for this layer, chitosan is detected due to the layer thickness smaller than the molecular escape depth. If the molecular escape depth is maintained in the range 2–3 nm from DOPC/chitosan interface as it was calculated for tetraglyme, the chitosan layer can be detected. If the DOPC molecules form a bilayer, one can expect the film thickness to be in the range of c.a. 4 nm. At these circumstances, the chitosan ions should not be detected. Similarly, due to a smaller concentration of DOPC in the remaining monolayers the chitosan intensity increases due to the free gaps around the molecular species (CsA and LG).

The other argument for the monolayer formation is the occurrence of molecular ions of DOPC and CsA in the single films. The molecular ion can be identified when an appropriately dense and ordered monolayer is developed. When the intermolecular distance between the same species (DOPC or CsA) is too long, the molecular ion is not yielded due to larger fragmentation of molecular species tilted towards the substrate. This phenomenon is well known and reported in a number of papers [42,53,59].

After mixing the DOPC with the CsA molecules (DOPC-CsA 0.50) the DOPC molecular ion disappears. A small intensity of the molecular ion is observed for the ternary DOPC-CsA-LG monolayer with a small molar fraction of LG (χ_{LG} = 0.25). The intensities' distribution of the smaller mass fragments of the single DOPC depicted in Figure 10a corresponds to the PET_{air}/Ch surface coverage with DOPC. The significantly greater surface coverage with DOPC is related to the large intensity of the molecular ion while the second largest intensity (DOPC-CsA-LG 0.25) determines a densely packed DOPC monolayer confirmed by identification of the molecular ion (Figure 10h). The latter fact points out to the positive role of the LG in developing a more uniform and ordered DOPC monolayer. It is also possible that in the presence of LG, DOPC can be transferred more preferentially than from the binary DOPC-CsA monolayer. The latter hypothesis is strongly supported by the largest intensity of the $(C_5H_{15}NPO_4)^+$ (m/z = 184) fragment among two and three component layers (Figure 10a). As a result, greater amounts of the DOPC are transferred from the DOPC-CsA-LG 0.25 layer than from the binary DOPC-CsA 0.50 despite a larger molar fraction of DOPC in the binary layer than in the ternary DOPC-CsA-LG 0.25. At a higher LG molar fraction, the amount of DOPC is reduced but it is similar for DOPC-CsA-LG 0.50 and 0.75 and it is comparable to the amount transferred for the binary DOPC-CsA 0.50. This points out that a supporting role of LG in transfer of DOPC is maintained. This phenomenon is determined by the direct interactions between the DOPC and LG molecules, and it was discussed elsewhere [53,59].

The mutual position of the functional groups of DOPC and LG can be estimated from the distribution of the intensity ratio 184/104 and 184/166 (Figure 10f). However, the results depicted in Figure 10f do not show significant changes in the distribution of the 184/104 and 184/166 ratios. This indicates that the LG molecules on the chitosan layer can surround and penetrate the DOPC molecules at a different location of the DOPC polar choline head. As a result, any characteristic fragment: of m/z = 184, 166 and 104 (Figure 10a) is not preferentially yielded [53].

The surface coverage with cyclosporine A is estimated from the intensity distribution of $(C_6H_{14}N)^+$ fragment (Figure 10b). The largest amount of CsA is observed for the single CsA layer. After mixing DOPC with the CsA molecules the surface coverage with CsA is reduced by about 15%. Since the corresponding molecular ion intensity (Figure 10c,h) is simultaneously reduced by about 50%, it can be assumed that the orientation of CsA is changed to more horizontal towards the PET_{air} substrate. On the other hand, the intensity of the pseudomolecular ion is only slightly smaller that can suggest maintenance of the CsA orientation. However, the exact pathway of yield of the pseudomolecular ion is unclear. It is safer to use a molecular ion for estimation of geometrical orientation of the CsA molecules. Additionally, the intensities of the molecular and pseudomolecular ions are greater in the presence of the Ch film than the AA molecules.

Introducing the LG as the third component of the monolayer (DOPC-CsA-LG) leads to a slight increase in the surface coverage with CsA and a strong increase in the molecular ion intensity (Figure 10b,c). As a result, the CsA molecules are the most packed and demonstrate the most perpendicular orientation towards the PET_{air} for the LG molar fraction of 0.25. Similar behavior is observed for the DOPC (see earlier discussion for DOPC). The latter observation shows that the LG at the small molar fraction of 0.25 plays a role of an efficient agent for the DOPC and CsA molecules binding to the chitosan layer. On the other hand, the intensity of the characteristic LG fragments is very small (Figure 10d) which can be related to the competitive adsorption of the monolayer molecules from the liquid phase to the PET_{air} substrate and/or to the change of the molecules' orientation. Furthermore, the LG molecules can be removed from the PET substrate due to the high vacuum action. For the DOPC-CsA-LG 0.50 monolayer the amount of the CsA increases further (Figure 10b) while the CsA molecules are more tilted as the intensity of molecular ion decreases (Figure 10c). That phenomenon is possible due to the smaller amount of the DOPC molecules and in consequence, larger sites accessible to the chitosan sublayer. For the largest molar fraction of LG (DOPC-CsA-LG 0.75) the amount of CsA on the chitosan layer is significantly reduced while the orientation of molecules is changed to be a more perpendicular. The intensity of the molecular ion is similar for the ternary DOPC-CsA-LG 0.50 and 0.75 (Figure 10c). The packing of the monolayer molecules and in consequence, their reorientation is determined by a significant amount of the LG molecules identified in the deposited monolayer (Figure 10c,d).

In order to examine the influence of the monolayer composition and/or the chitosan film presence on the wettability of the modified PET substrates contact angle measurements were made.

2.5. Contact Angles (CA) and Surface Free Energy

In the next stage of the experiments, the water (θ^W), formamide (θ^F) and diiodomethane (θ^D) advancing contact angles were measured onto the unmodified and modified PET substrates. The exemplary images of the contact angles for water and formamide are shown in Figure S1 and Figure S2, respectively, (Supplementary Materials). The selection of such three test liquids was made on the basis of the previously conducted experiments by Jańczuk et al. [68]. Then based on the contact angle values surface free energy and its components were calculated using the Lifshitz-van der Waals acid-base (LWAB, [69,70]) model. Based on this approach the work of adhesion (W_A^a) of liquid (l) to solid (s) takes the form:

$$W_A^a = \gamma_l(1 + \cos\theta_A) = 2\sqrt{\gamma_s^{LW}\gamma_l^{LW}} + 2\sqrt{\gamma_s^{+}\gamma_l^{-}} + 2\sqrt{\gamma_s^{-}\gamma_l^{+}} \quad (6)$$

The authors of the LWAB approach defined the total surface free energy as the sum of apolar γ_s^{LW} and polar γ_s^{AB} interactions (Equations (7) and (8)).

$$\gamma_s^{tot} = \gamma_s^{LW} + \gamma_s^{AB} \quad (7)$$

$$\gamma_s^{AB} = 2\sqrt{\gamma_s^{-}\gamma_s^{+}} \quad (8)$$

where θ_A is the average advancing contact angle, γ_s^{tot} is the total surface free energy, γ_s^{LW} is the Lifshitz-van der Waals component, γ_s^{AB} is the Lewis acid-base component, $\gamma_{s/l}^{-}$ is the electron-donor parameter of solid or liquid, $\gamma_{s/l}^{+}$ is the electron-acceptor parameter of solid or liquid.

To calculate the γ_s^{tot} value, it is essential to solve three equations of Equation (7) with the three unknowns.

Figure 11 presents the measured contact angle values while Figure 12 the calculated surface free energy and its components with the standard deviations as numerical errors.

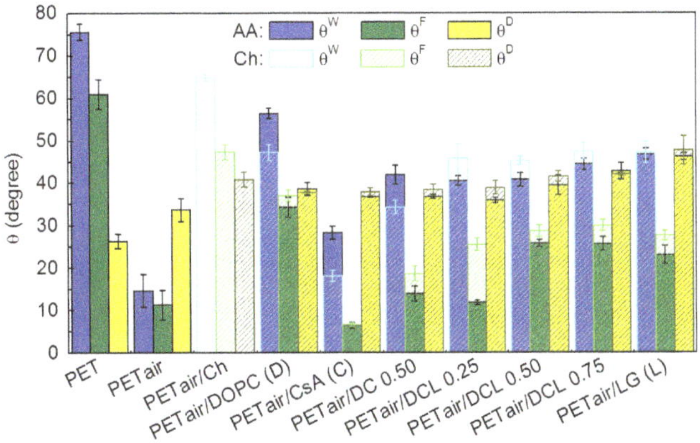

Figure 11. Contact angle values measured for the unmodified and modified by the LB films deposited from the air/AA and air/Ch interfaces, PET surfaces.

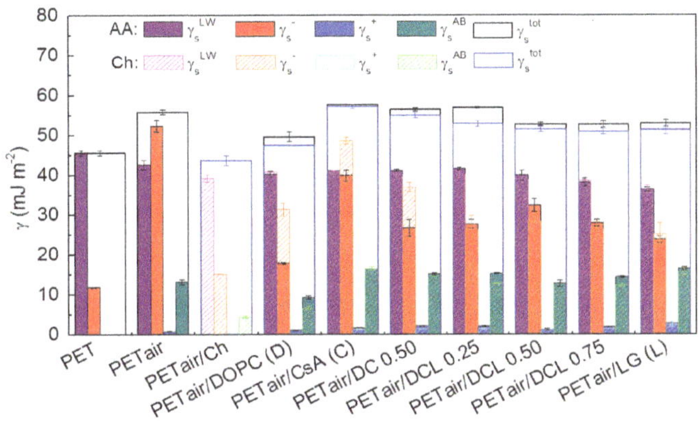

Figure 12. Surface free energy and its components calculated based on the LWAB approach for the unmodified and modified by the LB layers deposited from the air/AA and air/Ch interfaces onto the PET surfaces.

As it was previously mentioned the unmodified PET substrate is of the hydrophobic character (θ^W = 75.6° ± 1.9°, θ^F = 61.0° ± 3.5°, θ^D = 26.4° ± 1.7°) [42,71–73], which is in good agreement with the data reported by Yang et al. [7] while the plasma action changes it to be more hydrophilic (Figure 11). Such behavior confirms the addition of the new functional groups containing nitrogen, oxygen, and carbon ($-OH$, $C-O$, $O=C-O$, $C=O$, $N-CO-N$ [45,52,54,74,75]). The reduction of the contact angle value entails larger surface free energy values (Figure 12). The unmodified PET surface is characterized by the γ_s^{tot} equal to the Lifshitz-van der Waals component γ_s^{LW} (45.6 mJ m^{-2}, Figure 12). After the plasma activation the acid-base interactions (γ_s^{AB}) are found. Thus, the activated surface and the liquid molecules can interact mostly by the hydrogen bonds. The greater wettability is accompanied by the increased surface roughness of the plasma modified PET substrate [42,45].

The measured contact angle values for PET_{air} with the Langmuir monolayers deposited from the air/AA interface are similar to those transferred from the air/water interface [42]. This is a result of a small amount of AA in the solution whereby the dielectric constant of the AA subphase is practically equal to that of water [76]. Therefore, as for the water subphase, deposition of the monomolecular film from it onto the activated PET substrate changes the surface character to be more hydrophobic in relation to PET_{air} and more hydrophilic with respect to the unmodified PET substrate. Changes in the surface hydrophilic-hydrophobic character are correlated with the packing of the monolayer (compression modulus), morphological structure (BAM analysis) as well as interactions between the molecules forming a monolayer (excess and total Gibbs energy of mixing) [33]. The greater attractive interactions between the molecules forming the mixed monolayers, the larger packing of the layers, and thereby the modified PET surface exhibits the smaller roughness. Thus, better wettable properties are gained.

The presence of the single (DOPC, CsA, or LG) monolayers results in the greater values of contact angles for all test liquids in comparison to the PET_{air} substrate. Among them the biggest θ^W is obtained for the one-component DOPC layer (56.4° ± 1.2°, Figure 11). Thus, the smaller total surface free energy is obtained where the electron-donor parameter (γ_s^-) has a significantly smaller contribution (Figure 12). Such changes can be caused by the exposition of the hydrocarbon chains of the DOPC molecules to the air whereas the polar parts are shielded from the polar liquids' molecules. However, due to the *cis* double bonds present in the hydrocarbon chains (Figure 1d), the LB layer is loosely packed, and the chains can be inclined with respect to the PET_{air} substrate, making head groups more accessible to the polar liquid molecules. This is revealed in the γ_s^- parameter value (17.9 mJ m^{-2}). The inclination of the molecules is also confirmed by the absence of the molecular ion in the TOF-SIMS spectra (see Section 2.4).

The most wettable surface is found for the single CsA layer, γ_s^{tot} = 57.2 mJ m^{-2}. In that case the greatest γ_s^- and γ_s^{AB} interactions with water and formamide occur among the LB films. Despite the largest packing which was confirmed by the compression modulus (C_s^{-1} = 51.1 mN m^{-1}) [33] and TOF-SIMS (molecular ion, see Section 2.4), the contact angles are characterized by the smallest value. This can be associated with the CsA structure (Figure 1b). The presence of the aminoacid ring causes the formation of free spaces being penetrated by the test liquids molecules making contact with the plasma modified PET substrate.

Mixing the DOPC and CsA molecules to form the binary DOPC-CsA layer entails the decrease in the contact angle value with respect to the single DOPC layer (Figure 11), thus the increase in γ_s^{tot} (56.4 mJ m^{-2}). The biggest changes are observed for θ^F, which both for the single CsA and binary DOPC-CsA acquire the smallest values (θ^F = 6.5° and 14.0°, respectively, Figure 11). Thus, it can be stated that the presence of the CsA molecules is mainly responsible for the acid-base interactions (γ_s^{AB}). Additionally, the TOF-SIMS analysis confirmed that the CsA molecules in the binary DOPC-CsA monolayer are orientated more perpendicularly towards PET_{air} than in the single CsA layer, thereby the amide groups are more accessible to the molecules of the test liquids. Moreover, the formamide molecule has in its structure the amino and carbonyl groups capable of interacting by the hydrogen bonds with the CsA molecules.

The addition of the third component-LG to the monolayer results in a decrease in the γ_s^{LW}, γ_s^+, γ_s^{AB} values while an increase in the γ_s^- value (with respect to DOPC-CsA, Figure 12). This suggests that occurrence of the LG in the ternary DOPC-CsA-LG films weakens the dispersive interactions in relation to the DOPC-CsA monolayer. All the abovementioned changes are associated with the LG structure. The hydroxyl groups bound with the aromatic ring (Figure 1c) can interact with the liquid molecules by the H-bonds. Thus, increased γ_s^- proved that the interactions through the electron-donor groups are stronger and result in a more profound contact of the monolayers with the polar (water, formamide) test liquids. Additionally, the aromatic ring in which the π-π interactions can occur [77] also impacts the acid-base interactions. Among the ternary DOPC-CsA-LG monolayers the largest contact angles of the polar test liquids are obtained for the DOPC-CsA-LG 0.50 (θ^W = 40.8° ± 1.6°, θ^F = 25.8° ± 0.8°) and 0.75 (θ^W = 44.4° ± 1.3°, θ^F = 25.7° ± 1.6°, Figure 11). This is associated with the strong attractive interactions between the molecules forming the layer, whereby the minimum Gibbs energy for the DOPC-CsA-LG 0.50 as well as the largest packing (C_s^{-1} = 50.1 mN m^{-1}) were found [33]. As it was proved previously [42], the total surface free energy for the three-component DOPC-CsA-LG monolayers with the LG molar fractions of 0.50 and 0.75 is very similar and changes in the way of the interactions are evinced by the polar interactions (γ_s^-, γ_s^+, γ_s^{AB}). These differences can be ascribed to the packing of the monolayers which impacts the surface roughness (see Section 2.3).

The single LG layer yields the largest water (θ^W = 46.8° ± 1.4°) and diiodomethane (θ^D = 46.2° ± 0.9°) contact angles among the single monolayers. Thus, the increase in the electron-acceptor (γ_s^+) and almost the double decrease in the electron-donor (γ_s^-) parameters occur which results in a slight γ_s^{AB} increase, in relation to PET$_{air}$ (Figure 12). This is correlated with the tilting of the molecules with respect to PET$_{air}$ which makes the polar heads of LG more accessible for the interactions with the polar test liquids. Additionally, the inclination of the lauryl gallate molecules is confirmed by the mass spectrometry, where the LG molecular ion is not observed (see Section 2.4).

The air plasma-activated PET still has in the surface the aromatic rings which are a source of the π electrons. These electrons can interact readily with the positively charged $-NH_3^+$ chitosan groups. Thus, the adhesion of the chitosan film to the PET$_{air}$ surface is strong enough to form a stable film due to the acid-base interactions between the PET functional groups and the hydroxyl and amino groups of chitosan [53]. However, the formation of the chemical bonds cannot be excluded as the XPS analysis shown. For instance, Theapsak et al. proposed the mechanism of the chitosan coating onto the polyethylene via the ester linkage formation [54]. The chitosan coating of the PET$_{air}$ surface causes the increase in the contact angle value (θ^W = 64.9° ± 0.8°, θ^F = 47.4° ± 1.8° and θ^D = 40.9° ± 1.8°) in relation to PET$_{air}$ (Figure 11). This indicates that the Ch layer leads to the weakening of the strength of the PET$_{air}$ interactions with the test liquids. The larger contact angle values correlate with the decreased γ_s^{tot} which is similar to the unmodified PET substrate (Figure 12). The basic difference is that the Ch film reveals the ability of interactions with the liquids not only by dispersive forces but also through hydrogen bonding (γ_s^- = 15.2 mJ m^{-2} and γ_s^+ = 0.30 mJ m^{-2}).

The deposition of the examined monolayers onto the PET$_{air}$ from the chitosan subphase substrate results in an increase in the surface polarity in relation to PET$_{air}$/Ch (Figure 11). Among the single monolayers, the smallest contact angle of the polar test liquids and, thus the largest surface free energy (γ_s^{tot} = 57.2 mJ m^{-2}) are obtained for the CsA monolayer (PET$_{air}$/Ch/CsA). Additionally, θ^W and θ^F are lower than those obtained for PET$_{air}$/AA/CsA (Figure 11). The reason for that phenomenon are the changes in the orientation and/or conformation of the polypeptide molecules in the presence of Ch. This is proved by the mass spectrometry; the pseudomolecular and molecular ions have larger relative intensities (Figure 10c). Moreover, the intensity of the most prominent fragment (m/z = 100) points out that the surface coverage with the CsA layer is greater. Surprisingly, this behavior cannot be justified by the compression modulus value which is almost the same in the presence and absence of the chitosan layer [33].

The presence of the single DOPC and LG monolayers also contributes to the decrease in the water and formamide contact angle values in relation to PET$_{air}$/Ch (Figure 11). The smaller θ^D is also obtained for the PET$_{air}$/Ch/DOPC surface (θ^D = 37.7 ± 0.3°). Different dependence of diiodomethane contact angles is observed for the single LG monolayer. Its increase can be seen (θ^D = 47.7 ± 3.4°). Thus, the transfer of the single LG film changes significantly the contribution of γ_s^-, γ_s^+, and γ_s^{AB}, while after the single DOPC layer deposition the biggest changes in the γ_s^{tot} values are related to the increase of the electron-donor parameter (γ_s^-) in comparison to the PET$_{air}$/Ch (Figure 12). This indicates that the polar liquids molecules can have contact with the polar groups of the monolayers ($-OH$ of LG and $-OPO_3^-$ and $-N^+(CH_3)_3-$ of DOPC).

The binary DOPC-CsA 0.50 monolayer transfer onto the PET_{air} surface from the air/Ch interface increases the polarity of the surface in relation to the PET_{air}/Ch/DOPC (θ^W smaller by almost 13°, and θ^F decreased by more than 18°, Figure 11). The smaller values of the contact angle of polar liquids result in the electron-donor parameter increase (37 mJ m^{-2}, Figure 12). The addition of the LG results in the increase in the contact angles of all three test liquids (Figure 11). The higher is the LG molar fraction the more hydrophobic character of the surface is obtained (in comparison to the binary DOPC-CsA layer). On the other hand, the changes in the hydrophilic-hydrophobic character are not significant. This can be related to the similar packing of the monolayer at the air/Ch interface [33].

The wettability of the biomaterial surface is very important for assessment of the interactions of the solid with blood and surrounding tissues [29]. It was proved that the surface with the precisely designed hydrophilic-hydrophobic character can prevent from protein adsorption and coagulation [78,79]. Additionally, it was reported that the hydrophobic surfaces enhance cell affinity, while highly hydrophilic ones can prevent from cell-cell interactions [80] which are of key importance in the tissue engineering. In this experiment the deposition of the thin layers caused the increase in the contact angle value in relation to the PET_{air}. Of the single monolayers the greatest wettable properties are exhibited by cyclosporine A. Due to the largest packing of this monolayer [33], the CsA molecules can be much more vertically orientated toward the PET_{air} substrate, thus its polar groups can interact readily with the liquids. In the mixed monolayers the modified PET surface has more hydrophobic character which is related to the structure of the molecules forming films. The smaller hydrophilicity is related to the strong attraction forces between the film molecules and thus their larger packing is obtained. The chitosan film presence influences the molecules orientation and wetting properties. The measured contact angles of all test liquids on the PET_{air}/Ch are larger in comparison to PET_{air}/AA. All ternary DOPC-CsA-LG monolayers deposited onto PET_{air}/Ch seem to have desired wettable properties in the field of the tissue engineering.

The mixed films can be used successfully to cover other substrates of medical importance, such as metals and polymers. Our previous studies showed that the DOPC-CsA-LG monolayers can be transferred effectively onto gold [43]. The obtained LB films on the gold surfaces were characterized by integrity of coating for ensuring reliability and safety of the medical devices; and uniformity and stability of the coating for avoiding localized corrosion due to local defects [43]. Additionally, the application of air plasma to the other polymers such as polyetheretherketone (PEEK) yields analogous functional groups which can ensure a similar bonding between the deposited film and the substrate surface as it was obtained in the case of PET. In this aspect, the examined films seem to be universal. The other studies presented a possibility of covering the PEEK polymer surface with CsA, phospholipid DPPC, chitosan, and very recently with bioglass [60,80,81]. The bioglass-chitosan films on the air plasma-activated PEEK were found to be stable during the 21-day incubation in the simulated body fluid.

However, in this stage of research it is difficult to determine unequivocally the LB films' stability after their placement in the human body. The studies presented in this paper are of fundamental character and aimed at the LB layers' characterization by means of available techniques. This research still needs to be extended before such systems can be tested in vitro and in vivo. Nevertheless, these investigations are necessary to provide unique information about the molecular organization of thin films which give the opportunity to obtain a biocompatible coating for medical devices.

3. Materials and Methods

3.1. Solution Preparation

The 1 mg mL^{-1} solutions of the single 1,2-dioleoyl-*sn*-glycero-3-phosphocholine (DOPC, ≥99%, Sigma, St. Louis, MI, USA), cyclosporine A (CsA, ≥99%, Alfa Aesar,

Kandel, Germany), lauryl gallate (LG, ≥99%, Aldrich, St. Louis, MI, USA), binary DOPC-CsA 0.50 (at the 1:1 molar ratio), and ternary DOPC-CsA-LG (the LG molar fraction equal to 0.25, 0.50, 0.75) were obtained. The binary and ternary solutions were prepared by mixing proper amounts of the single ones dissolved in 4:1, v:v chloroform:methanol (chloroform 99.8%, Macron Fine Chemicals, methanol ≥99.9%, Fluka^{TM}) solvents.

3.2. Subphase Solution Preparations

The 0.1 mg mL^{-1} chitosan (Ch) and 0.1% acetic acid (AA) solutions were used as liquid subphases. Chitosan (MW 100,000-300,000 Da, degree of deacetylation 82.3% ± 0.9, Acrōs Organics, Göteborg, Sweden) was dissolved in 0.1% acetic acid diluted from the concentrate (99.5–99.9%, Avantor Performance Materials Poland S.A. Gliwice, Poland) with water from the Milli-Q Plus 185 system (Millipore, Burlington, MA, USA) with the specific resistivity of 18.2 MΩ cm.

3.3. PET Substrates Modification

PET plates (20 × 20 × 3 mm^3, Bayer Material Science, Leverkusen, Germany) were cleaned with methanol (99.8%, Avantor Performance Materials S.A. Gliwice, Poland) and deionized Milli-Q water in the ultrasonic bath (UM4, Unima, Olsztyn, Poland). Then the plates were dried in the exsiccator for 24 h at room temperature (20 ± 1°) and placed in the plasma generator chamber (Plasma type system, Diener Electronic, Ebhausen, Germany). The process of the PET surface activation with low temperature and low pressure (0.20 mbar) air plasma lasted 1 min as in our previous studies [41,69]. The power of air plasma was 460 W, and the plasma process ran with a continuous gas flow of 22 sccm. After the one-minute action, the plates were removed for further modifications. Next, the PET plates were immersed immediately in the liquid (chitosan or acetic acid) subphase.

3.4. Langmuir Monolayer Formation

Firstly, the Langmuir-Blodgett KSV 2000 Standard trough (KSV, Helsinki, Finland) was wiped with the acetone (99.5%, Avantor Performance Materials S.A. Gliwice, Poland) and methanol and then rinsed three times with the Milli-Q water. Then the proper volume (30–70 μL) of the solutions was placed onto the liquid subphase using the microsyringe (Hamilton-Bonaduz, Bonaduz, Switzerland). The surface tension was measured by means of the Wilhelmy plate with the 0.1 mN m^{-1} accuracy. After placing the solution on the subphase and solvent evaporation (duration of 10 min), the symmetrical compression was performed using mobile barriers with the speed of 20 mm min^{-1}. At the same time, the surface pressure-area per molecule ($\pi - A$) isotherm was registered. Additionally, the possible loss of the monolayer molecules into the liquid subphase before their transfer onto the PET_{air} was examined. For this purpose, changes in the area per molecule after gaining the given surface pressure of 10 mN m^{-1} were registered with barriers moving at the speed of 5 mm min^{-1}.

3.5. Brewster Angle Microscopy (BAM) Analysis

The Langmuir trough coupled with the Brewster angle microscope (BAM, Accurion GmbH, Goettingen, Germany) with the lateral resolution of 2 μm was used to estimate the monolayer relative thickness at the air/liquid interface (d). At first, before placing the solution on the liquid subphase, the camera calibration was made. Thereby the plot of the grey level as a function of incidence angle was gained. The incidence angle with the smallest reflectivity constituted the minimum of the parabolic fit. After the solution spreading and the solvent evaporation, the monolayer was compressed. Simultaneously, the grey scale data were converted into the reflectivity (R) and then into the relative film thickness (d) according to Equation (1) in compliance with the single-layer optical model [34].

3.6. *Langmuir—Blodgett (LB) Monolayer Formation*

In order to deposit efficiently the monolayers onto the low temperature air plasma-activated PET substrate, the thin film at the air/liquid interface was compressed to the surface pressure of 10 mN m^{-1}. After the given surface pressure had gained the PET_{air}, substrate was withdrawn from the subphase through the compressed film with the barrier speed of 5 mm min^{-1}. Then the PET_{air} substrate with the deposited LB layers was dried to remove water molecules and placed in the dark glass exsiccator before the next stages of the experiments (XPS, AFM, TOF-SIMS, CA).

The transfer ratio (TR) as the indicator of the efficiency of Langmuir monolayer transfer from the liquid phase to the polymer substrate and of the LB film quality was calculated according to Equation (2).

Determination of the transfer ratio value required providing the correction for the loss of the molecules during the transfer of the monolayer due to the desorption of the film material kept at the constant surface pressure.

3.7. *X-ray Photoelectron Spectroscopy (XPS) Analysis*

In order to analyze the chemical composition of the examined surfaces the X-ray photo-electron spectroscopy (XPS) studies were carried out using the multi-chamber UHV system (PREVAC). Spectra were collected by means of a hemispherical Scienta R4000 electron analyzer. The Scienta SAX-100 X-ray source (Al Kα, 1486.6 eV, 0.8 eV band) equipped with the XM 650 X-ray Monochromator (0.2 eV band) was utilized as a complementary piece of equipment. The pass energy of the analyzer was set to 200 eV for the survey spectra (with 750 meV step), and 50 eV for the regions (high resolution spectra): C 1s, O 1s and N 1s with a 50 meV step. The base pressure in the analysis chamber was 5×10^{-9} mbar. During the spectra collection it was not larger than 3×10^{-8} mbar.

3.8. *Atomic Force Microscopy (AFM) Analysis*

To examine the LB layers in the nanoscale the atomic force microscope (5600LS AFM, Aglient Technologies) was used. The tests were carried out in the non-contact mode (tip radius <7 nm, resonance frequency 330 kHz) with the resolution of 256 $\times$ 256 and the scanning area of 20 $\times$ 20 μm^2. The measurements were made in at least three randomly selected locations on the sample. The post processing data analysis was performed using the Probe Image Processor (SPIP) v. 5.1.4 software (Image Metrology, Hørsholm, Denmark).

3.9. *Time of Flight Secondary Ion Mass Spectrometry (TOF-SIMS) Analysis*

The modified PET substrates were placed in the ultra-high vacuum ($p < 10^{-9}$ mbar) chamber of the TOF-SIMS.5 instrument (ION-TOF GmbH, Münster, Germany) prior to the measurements. Subsequently, the TOF-SIMS spectra were obtained. The primary ion source of Bi^+ was used at 30 keV and corresponded to the 1.0 pA primary beam current in the spectrometry mode where the scanning area of the secondary ions was 200 $\times$ 200 μm^2 with 256 $\times$ 256 pixels (1 shot/pixel). All measurements were performed under the static positive conditions (dose < 1×10^{12} ions cm^{-2}). To neutralize the charge left on the surface an electron flood gun (20 eV) and the surface potential (U = −360 V) were applied. The post-processing data analysis was conducted using SurfaceLab 6.7 software (ION-TOF) and Origin 2019 (OriginLab, Northampton, MA, USA). The spectra calibration was applied by means of the positions of CH_3^+, $C_2H_3^+$, $C_2H_5^+$ fragments. All fragments' intensities were normalized to the total intensity.

3.10. *Contact Angle Measurements (CA)*

To determine the hydrophilic-hydrophobic character of the modified PET substrates the contact angle measuring was carried out (DGD ADR, GBX S.A.R.L, Romans-sur-Isére, France). For this purpose three test liquids with well-known surface tension: Milli-Q water, formamide (99.5%, Acrõs Organics, Geel, Belgium), and diiodomethane (99%, Sigma-Aldrich, St. Louis, MI, USA) were applied. For every modified PET substrate, and every test liquid, the same volume of liquid (6 µL) was placed and then using computer software Windrop++ the advancing contact angle was measured from the shape of settled droplet.

The values of water, formamide, and diiodomethane contact angles were used for further calculation of surface free energy and its components.

4. Conclusions

In this paper the single (DOPC, CsA, LG), binary (DOPC-CsA) and ternary (DOPC-CsA-LG) monolayers were deposited onto the air plasma-activated PET surface from the AA and Ch subphases by means of the Langmuir-Blodgett technique. Firstly, the stability of the monolayers and their relative thickness at the air/liquid interface were determined. After transferring the monolayers onto the PET_{air} the XPS, AFM, TOF-SIMS, CA measurements were made for the surface characteristics. Furthermore, the influence of the Ch film on the physicochemical properties of the modified PET substrate was examined.

The obtained results indicated that the biggest stability of monolayers occurred in the presence of chitosan in the subphase. However, it decreased significantly with time which imposed the necessity of deposition immediately after reaching the predetermined surface pressure. The transfer ratio values estimated under these conditions were close to unity, confirming the deposition of the monomolecular films. The XPS analysis evidenced that the air plasma action added new oxygen and nitrogen containing functional groups to the PET surface. They allowed for chemical bond formation with molecules of the tested monolayers. The micrographs revealed the protrusions which can be ascribed to the molecular reorganization during the transfer process of monolayers from the subphase (AA, Ch) to the solid PET surface. The Ch film presence modulated the roughness of the examined surfaces which is promising for the hemocompatibility of the modified PET. The molecular ions identified by the TOF-SIMS analysis proved that chitosan molecules improved transfer of more packed and well-organized DOPC-CsA-LG monolayers. The water contact angle values measured for the mixed films were in the range 30–50° which can ensure the optimal hydrophilic-hydrophobic character of the biomaterials used in the tissue engineering.

The studies allowed for the comprehensive characterization of physicochemical properties of the coatings such as composition, topography, molecular arrangement, and wettability on the solid support, useful for their potential applications in biomedicine. Based on the above findings it can be concluded that regardless of composition (DOPC-CsA-LG 0.25, 0.50, 0.75) all chitosan-containing films on PET_{air} have appropriate characteristics to be employed successfully. We hope that such organized coatings after the in vitro and in vivo tests will be used to increase the biocompatibility of medical devices.

Supplementary Materials: The following supporting information can be downloaded at: https://www.mdpi.com/article/10.3390/molecules28052375/s1. Figure S1: Water contact angle images for the indicated LB films on PET_{air}, Figure S2: Formamide contact angle images for the indicated LB films deposited on PET_{air}.

Author Contributions: Conceptualization, K.S., M.J. and A.E.W.; methodology, K.S., M.J. and R.M., formal analysis, K.S., M.J., R.M. and A.E.W.; investigation K.S. and R.M.; writing—original draft preparation, K.S., M.J. and R.M.; writing—review and editing, K.S., M.J., R.M. and A.E.W.; visualization, K.S., M.J., R.M. and A.E.W.; supervision, M.J. and A.E.W. All authors have read and agreed to the published version of the manuscript.

Funding: This research received no external funding.

Institutional Review Board Statement: Not applicable.

Informed Consent Statement: Not applicable.

Data Availability Statement: Not applicable.

Conflicts of Interest: The authors declare no conflict of interest.

References

1. Longo, U.G.; Lamberti, A.; Maffulli, N.; Denaro, V. Tendon augmentation grafts: A systematic review. *Br. Med. Bull.* **2010**, *94*, 165–188. [CrossRef] [PubMed]
2. Maitz, M.F. Applications of synthetic polymers in clinical medicine. *Biosurf. Biotribol.* **2015**, *1*, 161–176. [CrossRef]
3. Vesel, A.; Recek, N.; Motlan, H.; Mozetic, M. Endothelialization of polyethylene terephthalate treated in SO_2 plasma determined by the degree of material cytotoxicity. *Plasma* **2018**, *1*, 12–22. [CrossRef]
4. Rodrigues, A.; Figueiredo, L.; Diogo, H.; Bordado, J. Mechanical behavior of PET fibers and textiles for stent-grafts using video extensometry and image analysis. *Ciência Tecnol. Mater.* **2018**, *30*, 23–33. [CrossRef]
5. Swar, S.; Zajícová, V.; Rysová, M.; Lovětinská-Šlaborová, I.; Voleský, L.; Stibor, I. Biocompatible surface modification of poly(ethylene terephthalate) focused on pathogenic bacteria: Promising prospects in biomedical applications. *J. Appl. Polym. Sci.* **2017**, *134*, 1–11. [CrossRef]
6. Drobota, M.; Ursache, S.; Aflori, M. Surface functionalities of polymers for biomaterial applications. *Polymers* **2022**, *14*, 2307. [CrossRef]
7. Yang, L.; Chen, J.; Guo, Y.; Zhang, Z. Surface modification of a biomedical polyethylene terephthalate (PET) by air plasma. *Appl. Surf. Sci.* **2009**, *255*, 4446–4451. [CrossRef]
8. Mao, Y.; Li, Q.; Wu, C. Surface modification of PET fiber with hybrid coating and its effect on the properties of PP composites. *Polymers* **2019**, *11*, 1726. [CrossRef]
9. Safdar, R.; Omar, A.A.; Arunagiri, A.; Regupathi, I.; Thanabalan, M. Potential of chitosan and its derivatives for controlled drug release applications—A review. *J. Drug Deliv. Sci. Technol.* **2019**, *49*, 642–659. [CrossRef]
10. Jennings, J.A.; Wells, C.M.; McGraw, G.S.; Velasquez Pulgarin, D.A.; Whitaker, M.D.; Pruitt, R.L.; Bumgardner, J.D. Chitosan coatings to control release and target tissues for therapeutic delivery. *Ther. Deliv.* **2015**, *6*, 855–871. [CrossRef]
11. Merzendorfer, H.; Cohen, E. Chitin/chitosan: Versatile ecological, industrial, and biomedical applications. In *Extracellular Sugar-Based Biopolymers Matrices*; Biologically-inspired systems; Cohen, E., Merzendorfer, H., Eds.; Springer International Publishing: Cham, Switzerland, 2019; pp. 541–624.
12. Jang, T.-S.; Cheon, K.-H.; Ahn, J.-H.; Song, E.-H.; Kim, H.-E.; Jung, H.-D. In-vitro blood and vascular compatibility of sirolimus-eluting organic/inorganic hybrid stent coatings. *Colloids Surf. B Biointerfaces* **2019**, *179*, 405–413. [CrossRef]
13. Sevostyanov, M.A.; Baikin, A.S.; Sergienko, K.V.; Shatova, L.A.; Kirsankin, A.A.; Baymler, I.V.; Shkirin, A.V.; Gudkov, S.V. Biodegradable stent coatings on the basis of PLGA polymers of different molecular mass, sustaining a steady release of the thrombolityc enzyme streptokinase. *React. Funct. Polym.* **2020**, *150*, 104550. [CrossRef]
14. Hu, Z.; Lu, S.; Cheng, Y.; Kong, S.K.; Li, S.; Li, C.; Yang, L. Investigation of the effects of molecular parameters on the hemostatic properties of chitosan. *Molecules* **2018**, *23*, 3147. [CrossRef]
15. Fan, P.; Zeng, Y.; Zaldivar-Silva, D.; Agüero, L.; Wang, S. Chitosan-based hemostatic hydrogels: The concept mechanism, application, and prospects. *Molecules* **2023**, *28*, 1743. [CrossRef]
16. Haeri, A.; Sadeghian, S.; Rabbani, S.; Anvari, M.S.; Ghassemi, S.; Radfar, F.; Dadashzadeh, S. Effective attenuation of vascular restenosis following local delivery of chitosan decorated sirolimus liposomes. *Carbohydr. Polym.* **2017**, *157*, 1461–1469. [CrossRef]
17. Kersani, D.; Mougin, J.; Lopez, M.; Degoutin, S.; Tabary, N.; Cazaux, F.; Janus, L.; Maton, M.; Chai, F.; Sobocinski, J.; et al. Stent coating by electrospinning with chitosan/poly-cyclodextrin based nanofibers loaded with simvastatin for restenosis prevention. *Eur. J. Pharm. Biopharm.* **2020**, *150*, 156–167. [CrossRef]
18. Tang, J.; Liu, Y.; Zhu, B.; Su, Y.; Zhu, X. Preparation of paclitaxel/chitosan co-assembled core-shell nanofibers for drug-eluting stent. *Appl. Surf. Sci.* **2017**, *393*, 299–308. [CrossRef]
19. Zhang, S.; Zhang, F.; Feng, B.; Fan, Q.; Yang, F.; Shang, D.; Sui, J.; Zhao, H. Hematopoietic stem cell capture and directional differentiation into vascular endothelial cells for metal stent-coated chitosan/hyaluronic acid loading CD133 antibody. *Tissue Eng. Part.* **2015**, *21*, 1173–1183. [CrossRef]
20. Tedesco, D.; Haragsim, L. Cyclosporine: A review. *J. Transplant.* **2012**, *2012*, 230386. [CrossRef]
21. Wiącek, A.E.; Jurak, M.; Ładniak, A.; Przykaza, K.; Szafran, K. Cyclosporine CsA—The physicochemical characterization of liposomal and colloidal systems. *Colloids Interfaces* **2020**, *4*, 46. [CrossRef]
22. Hao, X.; Zhang, H.; Liu, R.; Che, J.; Zhang, D.; Liang, J.; Sun, L. Red blood cell membrane functionalized biomimetic nanoparticles for systemic lupus erythematosus treatment. *Mater. Today Adv.* **2022**, *16*, 100294. [CrossRef]
23. Akbulut, S.; Elbe, H.; Eris, C.; Dogan, Z.; Toprak, G.; Yalcin, E.; Otan, E.; Turkoz, Y. Effects of antioxidant agents against cyclosporine-induced hepatotoxicity. *J. Surg. Res.* **2015**, *193*, 658–666. [CrossRef] [PubMed]
24. Chin, Y.-T.; Tu, H.-P.; Lin, C.-Y.; Kuo, P.-J.; Chiu, H.-C.; Liu, S.-H.; Lee, S.-Y.; Fu, E. Antioxidants protect against gingival overgrowth induced by cyclosporine, A. *J. Periodontal. Res.* **2021**, *56*, 397–407. [CrossRef] [PubMed]
25. Badavi, M.; Sadeghi, N.; Dianat, M.; Samarbafzadeh, A. Gallic acid and cyclosporine mixture and their effects on cardiac dysfunction induced by ischemia/reperfusion and ENOS/INOS expression. *Int. J. Cardiovasc. Sci.* **2017**, *30*, 207–218.
26. Kikuzaki, H.; Hisamoto, M.; Hirose, K.; Akiyama, K.; Taniguchi, H. Antioxidant properties of ferulic acid and its related compounds. *J. Agric. Food Chem.* **2002**, *50*, 2161–2168. [CrossRef]
27. Walters, K.A.; Bialik, W.; Brain, K.R. The effects of surfactants on penetration across the skin*. *Int. J. Cosmet. Sci.* **1993**, *15*, 260–271. [CrossRef]

28. Oliveira, O.N.; Caseli, L.; Ariga, K. The past and future of Langmuir and Langmuir-Blodgett films. *Chem. Rev.* **2022**, *122*, 6459–6513.
29. Menzies, K.L.; Jones, L. The impact of contact angle on the biocompatibility of biomaterials. *Optom. Vis. Sci.* **2010**, *87*, 387–399. [CrossRef]
30. Jurak, M.; Wiącek, A.E.; Ładniak, A.; Przykaza, K.; Szafran, K. What affects the biocompatibility of polymers? *Adv. Colloid Interface Sci.* **2021**, *294*, 102451.
31. Deppisch, R.; Storr, M.; Buck, R.; Göhl, H. Blood material interactions at the surfaces of membranes in medical applications. *Sep. Purif. Technol.* **1998**, *14*, 241–254. [CrossRef]
32. Elsabee, M.Z.; Morsi, R.E.; Al-Sabagh, A.M. Surface active properties of chitosan and its derivatives. *Colloids Surf. B Biointerfaces* **2009**, *74*, 1–16. [CrossRef]
33. Szafran, K.; Jurak, M.; Wiącek, A.E. Effect of Chitosan on the interactions between phospholipid DOPC, cyclosporine A and lauryl gallate in the Langmuir monolayers. *Colloids Surf. A Physicochem. Eng. Asp.* **2022**, *652*, 129843. [CrossRef]
34. Winsel, K.; Hönig, D.; Lunkeneimer, K.; Geggel, K.; Witt, C. Quantitative Brewster angle microscopy of the surface film of human broncho-alveolar lavage fluid. *Eur. Biophys. J.* **2003**, *32*, 544–552.
35. Witek, J.; Keller, B.G.; Blatter, M.; Meissner, A.; Wagner, T.; Riniker, S. Kinetic models of cyclosporine A in polar and apolar environments reveal multiple congruent conformational states. *J. Chem. Inf. Model.* **2016**, *56*, 1547–1562. [CrossRef]
36. Dougherty, P.G.; Sahni, A.; Pei, D. Understanding cell penetration of cyclic peptides. *Chem. Rev.* **2019**, *119*, 10241–10287. [CrossRef]
37. Lomize, A.L.; Pogozheva, I.D. Physics–based method for modeling passive membrane permeability and translocation pathways of bioactive molecules. *J. Chem. Inf. Model.* **2019**, *59*, 3198–3213. [CrossRef]
38. Cámara, C.I.; Quiroga, M.V.C.; Wilke, N.; Jimenez-Kairuz, A.; Yudi, L.M. Effect of chitosan on distearoylphosphatidylglycerol films at air/water and liquid/liquid interfaces. *Electrochim. Acta* **2013**, *94*, 124–133.
39. Wydro, P.; Krajewska, B.; Hąc-Wydro, K. Chitosan as a lipid binder: A Langmuir monolayer study of chitosan-lipid interactions. *Biomacromolecules* **2007**, *8*, 2611–2617.
40. Ahmed, I.; Dildar, L.; Haque, A.; Patra, P.; Mukhopadhyay, M.; Hazra, S.; Kulkarni, M.; Thomas, S.; Plaisier, J.R.; Dutta, S.B.; et al. Chitosan-fatty acid interaction mediated growth of Langmuir monolayer and Langmuir-Blodgett films. *J. Colloid Interface Sci.* **2018**, *514*, 433–442. [CrossRef]
41. Philipova, O.E.; Volkov, E.V.; Sitnikova, N.L.; Khokhlov, A.R. Two types of hydrophobic aggregates in aqueous solutions of chitosan and its derivative. *Biomacromolecules* **2001**, *2*, 483–490.
42. Szafran, K.; Jurak, M.; Mroczka, R.; Wiącek, A.E. Surface properties of the polyethylene terephthalate (PET) substrate modified with the phospholipid-polypeptide-antioxidant films. Design of functional biocoatings. *Pharmaceutics* **2022**, *14*, 2815. [CrossRef] [PubMed]
43. Jurak, M.; Szafran, K.; Cea, P.; Martín, S. Characteristics of phospholipid-immunosuppressant-antioxidant mixed Langmuir-Blodgett films. *J. Phys. Chem. B* **2022**, *126*, 6936–6947. [CrossRef]
44. Jurak, M.; Szafran, K.; Cea, P.; Martín, S. Analysis of molecular interactions between components in phospholipid-immunosuppressant-antioxidant mixed Langmuir films. *Langmuir* **2021**, *37*, 5601–5616. [CrossRef] [PubMed]
45. Bui, V.T.; Liu, X.; Ko, S.H.; Choi, H.S. Super-amphiphilic surface of nano silica/polyurethane hybrid coated PET film via a plasma treatment. *J. Colloid Interface Sci.* **2015**, *453*, 209–215. [CrossRef] [PubMed]
46. Beamson, G. *The XPS of Polymer Database, Surface Spectra 2000 from High Resolution XPS of Organic Polymers-the Scienta ESCA300 Database*; Wiley: Hoboken, NJ, USA, 1992.
47. Koimura, M.; Tateshi, H.; Hatakeyama, K.; Miyamoto, S.; Ogata, C.; Funatsu, A.; Taniguchi, T.; Matsumoto, Y. Analysis of reduced graphene oxides by X-ray photoelectron spectroscopy and electrochemical capacitance. *Chem. Lett.* **2013**, *42*, 924–926.
48. Rabchinskii, M.K.; Ryzhkov, S.A.; Kirilenko, D.A.; Ulin, N.V.; Baidakova, M.V.; Shnitov, V.V.; Pavlov, S.I.; Chumakov, R.G.; Stolyarova, D.Y.; Besedina, N.A.; et al. From graphene oxide towards aminated graphene: Facile synthesis, its structure and electronic properties. *Sci. Rep.* **2020**, *10*, 6902. [CrossRef]
49. Panajotović, R.; Ptasinska, S.; Lyamayev, V.; Pronce, K. Low-energy electron damage of DPPC molecules—A NEXAFS study. *Rad. Appl.* **2016**, *1*, 46–50.
50. Wagstaffe, M.; Thomas, A.G.; Jackman, M.J.; Torres-Molina, M.; Syres, K.L.; Handrup, K. An experimental investigation of the adsorption of a phosphonic acid on the anatase TiO_2(101) surface. *J. Phys. Chem. C* **2016**, *120*, 1693–1700. [CrossRef]
51. Pandiyaraj, K.N.; Selvarajan, V.; Rhee, Y.H.; Kim, H.W.; Shah, S.I. Glow discharge plasma-induced immobilization of heparin and insulin on polyethylene terephthalate film surfaces enhances anti-thrombogenic properties. *Mater. Sci. Eng. C* **2009**, *29*, 796–805. [CrossRef]
52. Abdel-Fattah, E. Surface modifications of PET in argon atmospheric pressure plasma: Gas flow rate effect. *Surf. Interface Anal.* **2022**, *54*, 794–802. [CrossRef]
53. Jurak, M.; Wiącek, A.E.; Mroczka, R.; Łopucki, R. Chitosan/phospholipid coated polyethylene terephthalate (PET) polymer surfaces activated by air plasma. *Colloids Surf. A Physicochem. Eng. Asp.* **2017**, *532*, 155–164. [CrossRef]
54. Theapsak, S.; Watthanaphanit, A.; Rujiravanit, R. Preparation of Chitosan-coated polyethylene packaging films by DBD plasma treatment. *ACS Appl. Mater. Interfaces* **2012**, *4*, 2474–2482. [CrossRef]
55. Chou, S.-F.; Lee, C.-H.; Cho, C.-H.; Lai, J.-Y. Relationships between surface roughness/stiffness of chitosan coatings and fabrication of corneal keratocyte spheroids: Effect of degree of deacetylation. *Colloids Surf. B Biointerfaces* **2016**, *142*, 105–113. [CrossRef]

56. Silva, C.A.; Nobre, T.M.; Pavinatto, F.J.; Oliveira, O.N. Interaction of chitosan and mucin in a biomembrane model environment. *J. Colloid Interface Sci.* **2012**, *376*, 289–295. [CrossRef]
57. Arrigo, R.; Hävecker, M.; Wrabetz, S.; Blume, R.; Lerch, M.; McGregor, J.; Parrott, E.P.J.; Zeitler, J.A.; Gladden, L.F.; Knop-Gericke, A.; et al. Tuning the acid/base properties of nanocarbons by functionalization via amination. *J. Am. Chem. Soc.* **2010**, *132*, 9616–9630. [CrossRef]
58. Raicopol, N.; Andronescu, C.; Atasiei, R.; Hanganu, a.; Pilan, L. Post-polymerization electrochemical functionalization of a conducting polymer: Diazonium salt electroreduction at polypyrrole electrodes. *J. Electrochem. Soc.* **2014**, *12*, G103–G113. [CrossRef]
59. Jurak, M.; Mroczka, R.; Łopucki, R. Properties of artificial phospholipid membranes containing lauryl gallate or cholesterol. *J. Membr. Biol.* **2018**, *251*, 277–294. [CrossRef]
60. Przykaza, K.; Jurak, M.; Wiącek, A.E.; Mroczka, R. Characteristics of hybrid chitosan/phospholipid-sterol, peptide coatings on plasma activated PEEK polymer. *Mater. Sci. Eng. C Mater. Biol. Appl.* **2021**, *120*, 111658. [CrossRef]
61. Finšgar, M.; Ristić, T.; Fardim, P.; Zemljič, L.F. Time-of-flight secondary ion mass spectrometry analysis of chitosan-treated viscose fibres. *Anal. Biochem.* **2018**, *557*, 131–141. [CrossRef]
62. Vaezian, B.; Anderton, C.R.; Kraft, M.L. Discriminating and imaging different phosphatidylcholine species within phase-separated model membranes by principal component analysis of TOF-secondary ion mass spectrometry images. *Anal. Chem.* **2011**, *83*, 5337–5343. [CrossRef]
63. Lankeoff, I.; Sjövall, P.; Ewing, A.G. Relative quantification of phospholipid accumulation in the PC12 cell plasma membrane following phospholipid incubation using TOF-SIMS imaging. *Anal. Chem.* **2010**, *82*, 10006–10014. [CrossRef] [PubMed]
64. Gulin, A.; Mochalova, M.; Denisov, N.; Nadtochenko, V. Secondary ion mass spectrometric signal enhancement of phosphatidylcholine dioleoyl on enlarged nanoparticles surface. *Appl. Surf. Sci.* **2014**, *316*, 36–41. [CrossRef]
65. Bąchor, R.; Kluczyk, A.; Stefanowicz, P.; Szewczuk, Z. Preparation of novel deuterated cyclosporin A standards for quantitative LC-MS analysis. *J. Mass Spectrom.* **2017**, *52*, 817–822. [CrossRef] [PubMed]
66. D'Almeida, M.; Attik, N.; Amalric, J.; Brunon, C.; Renaud, F.; Abouelleil, H.; Toury, B.; Grosgogeat, B. Chitosan coating as an antibacterial surface for biomedical applications. *PLoS ONE* **2017**, *12*, e0189537. [CrossRef] [PubMed]
67. Muramoto, S.; Brison, J.; Castner, D.G. Exploring the surface sensitivity of TOF-secondary ion spectrometry by measuring the implantation and sampling depths of Bi_n and C_{60} ions in organic films. *Anal Chem.* **2012**, *84*, 365–372. [CrossRef]
68. Jańczuk, B.; Chibowski, E.; Bruque, J.M.; Kerkeb, M.L.; González Caballero, F. On the consistency of surface free energy components as calculated from contact angles of different liquids: An application to the cholesterol surface. *J. Colloid Interface Sci.* **1993**, *159*, 421–428. [CrossRef]
69. Van Oss, C.J.; Chaudhury, M.K.; Good, R.J. Interfacial Lifshitz-van der Waals and polar interactions in macroscopic systems. *Chem. Rev.* **1988**, *88*, 927–941. [CrossRef]
70. Van Oss, C.J. *Interfacial Forces in Aqueous Media*; Marcel Dekker: New York, NY, USA, 1994.
71. Szafran, K.; Jurak, M.; Wiącek, A.E. Effect of chitosan and lipid layers deposited onto polyethylene terephthalate (PET) on its wetting properties. *Prog. Chem. Appl. Chitin Deriv.* **2021**, *26*, 210–221.
72. Woźniak, K.; Jurak, M.; Wiącek, A.E. Wetting properties of phospholipid-polypeptide monolayers deposited onto polyethylene terephthalate. *Ann. Sect. AA* **2019**, *74*, 69–88.
73. Wiącek, A.E.; Jurak, M.; Gozdecka, A.; Worzakowska, M. Interfacial properties of PET and PET/starch polymers developed by air plasma processing. *Colloids Surf. A Physicochem. Eng. Asp.* **2017**, *532*, 323–331. [CrossRef]
74. Vesel, A.; Mozetic, M.; Zalar, A. XPS study of oxygen plasma activated PET. *Vacuum* **2007**, *2*, 248–251. [CrossRef]
75. Vesel, A.; Junkar, I.; Cvelbar, U.; Kovac, J.; Mozetic, M. Surface modification of polyester by oxygen- and nitrogen-plasma treatment. *Surf. Interface Anal.* **2008**, *40*, 1444–1453. [CrossRef]
76. Mohsen-Nia, M.; Amiri, H. Measurement and modelling of static dielectric constants of aqueous solutions of methanol, ethanol and acetic acid at T = 293.15 K and 91.3 kPa. *J. Chem. Termodyn.* **2013**, *57*, 67–70. [CrossRef]
77. Lamarche, R.M.; DeWolf, C. Strong headgroup interactions drive highly directional growth and unusual phase co-existence in self-assembled phenolic films. *ACS Appl. Mater. Interfaces* **2019**, *11*, 45354–45363. [CrossRef]
78. Okano, T.; Aoyagi, T.; Kataoka, K.; Abe, K.; Sakurai, Y. Hydrophilic-hydrophobic microdomain surfaces having an ability to suppress platelet aggregation and their in vitro antithrombogenicity. *J. Biomed. Mater. Res.* **1986**, *20*, 919–927. [CrossRef]
79. Fauchex, N.; Schweiss, R.; Lützow, K.; Werner, C.; Growth, T. Self-assembled monolayers with different terminating groups as model substrates for cell adhesion studies. *Biomaterials* **2004**, *25*, 2721–2730. [CrossRef]
80. Wiącek, A.E.; Terpiłowski, K.; Jurak, M.; Worzakowska, M. Low-temperature air plasma modification of chitosan-coated PEEK biomaterials. *Polym. Test.* **2016**, *50*, 325–334. [CrossRef]
81. Przykaza, K.; Jurak, M.; Kalisz, G.; Mroczka, R.; Wiącek, A.E. Characteristics of hybrid bioglass-chitosan coatings on the plasma activated PEEK polymer. *Molecules* **2023**, *28*, 1729. [CrossRef]

molecules

Article

Chitosan–Glycolic Acid Gel Modification of Chloride Ion Transport in Mammalian Skin: An In Vitro Study

Olga Zavyalova [1], Dominika Dąbrowska-Wisłocka [1], Konrad Misiura [1] and Iga Hołyńska-Iwan [2,*]

[1] Department of Chemical Technology and Pharmaceuticals, Faculty of Pharmacy, Ludwik Rydygier Collegium Medicum in Bydgoszcz, Nicolaus Copernicus University in Torun, 87-100 Torun, Poland; zavolg@cm.umk.pl (O.Z.); dominika.dabrowska@cm.umk.pl (D.D.-W.)

[2] Department of Pathobiochemistry and Clinical Chemistry, Faculty of Pharmacy, Ludwik Rydygier Collegium Medicum in Bydgoszcz, Nicolaus Copernicus University in Torun, 87-100 Torun, Poland

* Correspondence: igaholynska@cm.umk.pl

Abstract: Chitosan, a polyaminosaccharide with high medical and cosmetic potential, can be combined with the beneficial properties of glycolic acid to form a gel that not only moisturizes the skin, but also has a regenerative effect. Its involvement in the activation of biochemical processes may be associated with the activity of skin ion channels. Therefore, the aim of the research was to evaluate the immediate (15 s) and long-term (24 h) effect of chitosan–glycolic acid gel (CGG) on the transepithelial electric potential and the transepithelial electric resistance (R) of skin specimens tested in vitro. Stimulation during immediate and prolonged application of CGG to skin specimens resulted in a significant decrease in the measured minimal transepithelial electric potential (PDmin). The absence of any change in the R after the CGG application indicates that it does not affect the skin transmission, or cause distortion, microdamage or changes in ion permeability. However, the reduction in potential may be due to the increased transport of chloride ions, and thus water, from outside the cell into the cell interior. Increased secretion of chloride ions is achieved by stimulating the action of the CFTR (cystic fibrosis transmembrane conductance). It can be assumed that chitosan gently stimulates the secretion of chlorides, while maintaining a tendency to reduce the transport of sodium ions, without causing deformation or tissue damage.

Keywords: chitosan; epithelial sodium channels; chloride ion transport; cystic fibrosis transmembrane regulator; electric potential; electric resistance; skin

check for updates

Citation: Zavyalova, O.; Dąbrowska-Wisłocka, D.; Misiura, K.; Hołyńska-Iwan, I. Chitosan–Glycolic Acid Gel Modification of Chloride Ion Transport in Mammalian Skin: An In Vitro Study. *Molecules* **2023**, *28*, 6581. https://doi.org/10.3390/molecules28186581

Academic Editor: Agnieszka Ewa Wiącek

Received: 15 August 2023
Revised: 6 September 2023
Accepted: 7 September 2023
Published: 12 September 2023

1. Introduction

Chitosan is a linear polyaminosaccharide obtained in the process of chemical or enzymatic deacetylation of chitin. It is a biologically derived raw material and the second most common polymer after cellulose. Chitosan is well-known for its applicability as a biocompatible, biodegradable and non-toxic material for biomedical, food-related, supplemental and cosmetic use [1–4]. The unique properties of chitosan result from the presence of amine and hydroxyl functional groups in its structure [3–5]. Its special features and biological activity, including antimicrobial, antioxidant, anti-cancer, anti-inflammatory, hemocompatible and hemostatic properties, also result from its chemical structure and make it an excellent raw material for biomedical applications [6,7] (Figure 1).

The antimicrobial effect is closely related to the presence of positively charged amino groups in the structure of the molecule, which interact with the negatively charged surface of the cell walls of microorganisms. This leads to damage to the membranes and destruction of the internal structures of the pathogens [8]. Chitosan also has a high chelating ability, supporting antimicrobial activity [5,6]. This makes chitosan a potential replacement for some antibiotics that are effective against drug-resistant bacteria [9]. Moreover, chitosan acts as an antioxidant that neutralizes free radicals by creating stable radicals with its functional groups [5].

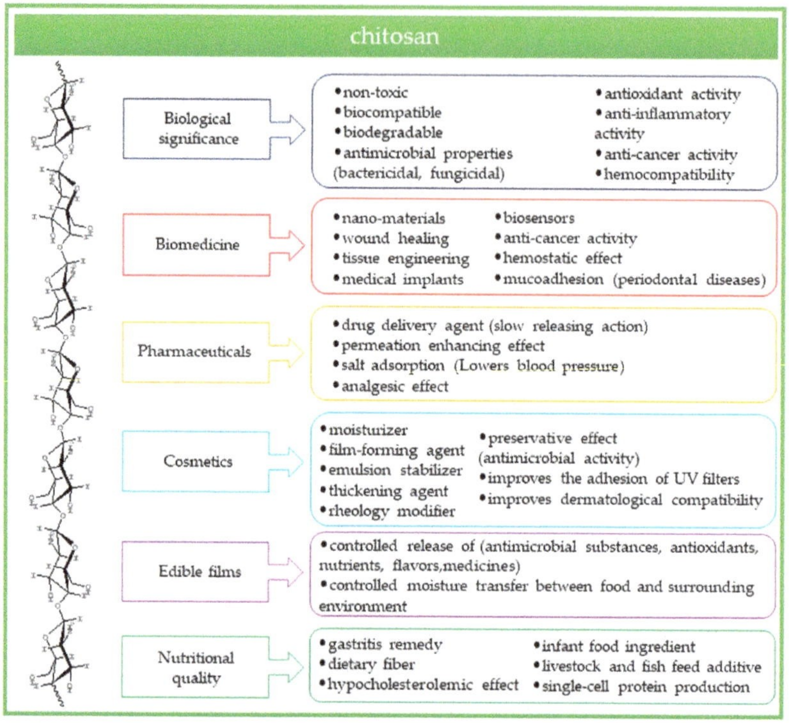

Figure 1. Properties and applications of chitosan.

The polymer, due to targeting pro-inflammatory cytokines and inhibiting their expression, exerts an anti-inflammatory effect [5]. In addition, it stimulates humoral and cellular immunity. It has been proven that chitosan inhibits the growth of cancer cells and is helpful in gene therapy aimed towards the direct delivery of genetic material [10]. The bioactivity of the polymer is conditioned by the degree of deacetylation and molecular weight [8].

The presence of amino and hydroxyl groups in the chitosan molecule make it easy to modify it with other bioactive compounds. Chemical modifications include cross-linking, graft copolymerization, carboxymethylation, etherification, esterification, O-acetylation, hydroxyalkylation, sulfonation, acetylation, quaternization, Schiff base reaction, grafting and several others. These modifications not only improve solubility and affect rheological properties, thermal stability and oxidation resistance, but they also impart new biological activity [4]. Chitosan modification products with desired properties are, in various forms, components of biomaterials—for example, hydrogels [11], nanofibers [12], microparticles [13], nanoparticles [14] or scaffolds [14–16].

Chitosan is of particular importance in skin regeneration processes. It has the ability to increase the influx of phagocytic cells to the site of infection and affects the proliferation of fibroblasts. It also stimulates the production of cytokines and activates macrophages and neutrophils, which cleanse wounds. This polymer is an inhibitor of metalloproteinase 2 (MMP-2), which is present in skin fibroblasts and a hydrolyzing type IV collagen. Inhibiting the action of MMP-2 enables the proper reconstruction of damaged tissues in the case of chronic wounds [17,18]. These properties make chitosan a great potential tool in skin tissue engineering [5,19,20].

The study of chitosan properties requires its prior dispersion in an aqueous environment, as is the case with most polysaccharides. Chitosan dissolves only in acidic solutions, at a pH value lower than pH6. This is due to the presence of intermolecular hydrogen bonds that prevent the dissolution of chitosan in water or organic solvents. As a result of protonation of amino groups in an acidic environment (Figure 2), chitosan forms a water-soluble cationic polyelectrolyte [3,21].

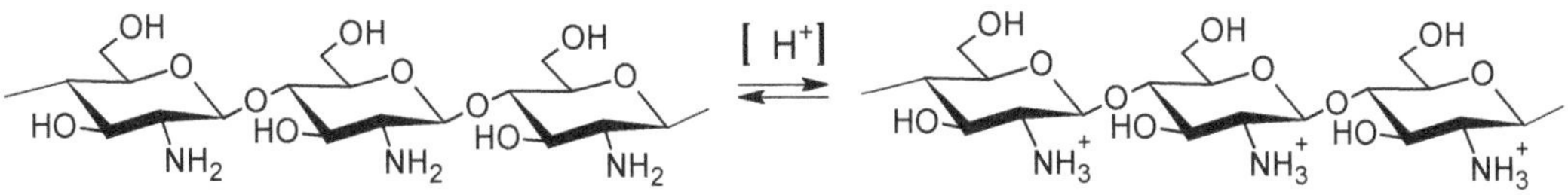

Figure 2. Scheme of chitosan protonation under acidic conditions.

In this study, glycolic acid was selected to lower the pH of the solution and form a water-soluble chitosan polycation. Glycolic acid (hydroxyacetic acid, Figure 3) is the simplest hydroxycarboxylic acid. It belongs to the group of alpha hydroxy acids, among which it has the smallest molecule [22].

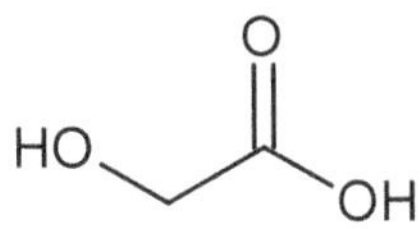

Figure 3. Chemical structure of glycolic acid.

Glycolic acid is present in many sugar-rich plants, including sugar cane, sugar beet, apples and grapes. And for years it has been well known among specialists in the cosmetics and dermatology community as a multifunctional skin care product [22,23]. The small size of the acid molecule determines its excellent penetration into the skin layers, intensifying the degradation of corneodesmosomes responsible for corneocyte adhesion. Due to this, glycolic acid is a popular exfoliant. The effectiveness of glycolic acid as a peeling exfoliant depends primarily on the amount that is in the most biologically active free acid form, its pH value and its concentration. As with all alpha hydroxy acids, glycolic acid may cause irritation and erythema of the skin, which is intensified at high acid concentration and lower pH [22].

Depending on the concentration used, glycolic acid can have an anti-inflammatory effect. At a concentration of 0.1 mM, it has a significant photoprotective effect on human keratinocytes by regulating the secretion of cytokines induced by UVB radiation and the secretion of chemokines in keratinocytes [24].

CGG is formed as a result of interactions between the positively charged protonated amine groups of chitosan and the negative charges present in the glycolic acid molecule, with the formation of various types of bonds connecting different chitosan chains (Figure 4) [21]. The obtained gel can be used not only in various skin anti-aging agents, but also in wound dressing materials [4] and improved drug delivery systems [25].

Due to the high molecular weight of chitosan, it seems that only the superficial effect of water retention on the skin can be expected. However, changes in the skin under the influence of chitosan indicate the activation of more advanced physiological processes. Studies of changes in electrophysiological parameters of the skin, such as skin electrical potential or tissue resistance, can shed light on the nature of skin changes.

Mammals' skin, due to the close intercellular connections and its multilayer, bipolar structure, is characterized by an electrophysiological parameter of electrical resistance and electric potential [26–29]. The multilayered skin structure, tight intercellular connections and the high content of keratin and lipids in cells determine high resistance values [26,29,30]. The value of electrical resistance also depends on the activity of sweat glands, the degree of skin hydration and the integrity of the epidermis, because damage may cause a decrease in electrical resistance [26,29–31]. Rabbit skin seems to be an appropriate experimental model to study the effect of various substances on the properties of tissue in relation to human skin. There are even reports that rabbit skin is more sensitive than that of humans [27,31–33]. The properties of the skin can be assessed by checking its integrity and the transport of ions through the membranes of its building cells [30,34]. This can be done by examining the changes in the electric potential of tissues exposed to various factors [26,27,31,32,35]. The transport of skin ions is essential for the proper functioning of the entire organism [36,37]. The functioning of epithelial sodium channels (ENaCs) in keratinocytes is related to the transport of water through the skin layer, the action of immunocompetent cells and wound healing. The transport of chloride ions is via the transmembrane regulator of cystic fibrosis (CFTR) and the chloride channel (CLCA) [38,39]. The activity of chloride channels seems to be important for water flow and for dehydration and overhydration of the skin microenvironment [37,40]. CFTR channels act on cellular regulators that influence, for example, the functioning of ENaCs [37,38]. Changes in the functioning of sodium and/or chloride channels may underlie problems with hypersensitivity to pain [41,42], the onset of hypersensitivity and/or allergy [41,43], issues with regeneration and healing [40] and atopic dermatitis [39].

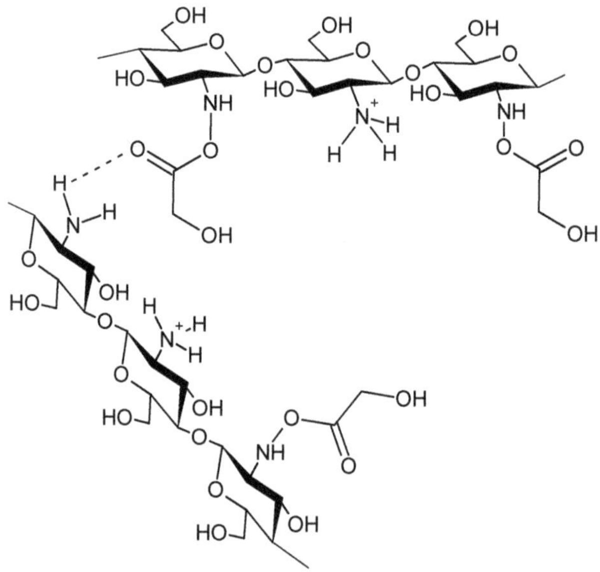

Figure 4. Scheme of chitosan–glycolic acid interaction.

The aim of the study was to evaluate the immediate (15 s) and long-term (24 h) effects of chitosan gel with glycolic acid on both the transepithelial electric potential and the transepithelial electric resistance of the skin specimens tested in vitro. The transepithelial electric potential reflects changes in transepithelial ion transport pathways following the use of chitosan gel with glycolic acid. Currently, there are no direct scientific reports on the effects of chitosan on electrophysiological parameters of the skin, which indicate changes in the functioning of ion channels.

2. Results

Transepithelial electrical resistance was measured initially before and after mechanical and mechanical–chemical stimulations for skin fragments treated with both iso-osmotic Ringer's solution and chitosan. Results ranged from 2474 Ω*cm^2 (median, R final chitosan incubation) to 3450 Ω*cm^2 (median, R final RS incubation). There were no statistically significant differences in any of the study groups (Table 1, Wilcoxon test) or between the groups (Table 1, Mann–Whitney test). The resistance of the skin specimens did not change under the influence of chitosan.

Table 1. Transepithelial resistance (R) and transepithelial electric potential (PD) measured under steady-state conditions with immediate and prolonged effects of chitosan–glycolic acid gel solution on skin specimens.

		PD Initial (mV)	PD Final (mV)	Wilcoxon Test (p)	R Initial ($\Omega*cm^2$)	R Final ($\Omega*cm^2$)	Wilcoxon Test (p)
Incubation: RS (n = 40)	median	−0.53	−0.86	0.000103	2721	3450	0.667697
	lower quartile	−0.24	−1.31		1422	1467	
	upper quartile	0.23	−0.15		4855	5024	
Incubation: CGG (n = 32)	median	−1.11	−0.53	0.099541	2817	2474	0.379375
	lower quartile	−1.48	−0.86		1050	1013	
	upper quartile	−0.09	0		5916	5203	
Mann–Whitney test (p)	RS vs. Chitosan	0.002873	0.219698		0.66735	0.62078	

Abbreviations: RS—iso-osmotic Ringer solution, CGG—chitosan–glycolic acid gel in RS, PD—transepithelial potential difference of epithelial skin surface measured in stationary conditions (mV), R—resistance measured in stationary conditions ($\Omega*cm^2$), $p < 0.05$.

The initial PD, measured without stimulation for the skin fragments held in the iso-osmotic RS, was −0.53 mV (median). However, the final one was statistically significantly reduced (Table 1, Wilcoxon test) to the level of −0.86 mV (median). The initial PD measured for the chitosan treated skin specimens was −1.11 mV (median). The final PD measured after a series of mechanical and mechanical–chemical stimulations tended to increase to −0.53 mV (median), but this was not a statistically significant increase. Comparing the initial PD value measured for the control tissues and those treated with chitosan, a statistically significant reduction was demonstrated (Table 1, Mann–Whitney test). Meanwhile, the final PD was similar in both tissue groups.

The immediate effects of chitosan included a statistically significant decrease in measured PDmin to a value of −0.57 mV (median), compared to −0.4 mV (median) for the samples that underwent RS stimulation (Table 2, Wilcoxon test, RS incubation). Long-term retention of tissues in the chitosan solution and stimulation with the chitosan and Ringer's solution resulted in a significant decrease of measured PDmin to the value of −1.31 mV (median) for chitosan stimulation and −1.06 mV (median) for RS stimulation. The PDmin values measured for these stimulations were similar (Table 2, Wilcoxon test, CGG incubation). The comparison of the two incubation conditions showed a significant decrease in the PDmin measured after stimulation with RS and with chitosan solution, performed under long-term stationary conditions (Table 3, Mann–Whitney test).

The PDmax measured under the conditions of incubation in Ringer's solution and during stimulation with this solution, before chitosan administration, was 0.37 mV (median) positive, while during chitosan stimulation it decreased to 0.23 mV (median). These values did not differ significantly (Table 2, Wilcoxon test, RS incubation). During incubation in chitosan solution and stimulation with its solution and with RS, the PDmax was similar and amounted to 0 mV (median). Comparison of the two incubation conditions showed no significant change in PDmax for RS stimulation (Table 3, Mann–Whitney test). The analysis of PDmax values measured for stimulation with chitosan solution showed a statistically significant reduction of this parameter (Table 3, Mann–Whitney test).

The administration of sodium (A) and chloride (B) ion transport blockers caused changes in PDmax and PDmin in a similar manner, regardless of whether chitosan was administered for 15 s during the stimulation or whether incubation for 24 h was used (Table 4).

Table 2. Minimal (PDmin) and maximal (PDmax) transepithelial electric potential, measured during 15 s mechanical (RS) and mechanical–chemical (chitosan–glycolic acid gel) stimulations of skin specimens treated with immediate and prolonged exposure to chitosan–glycolic acid gel solution.

Incubation: RS (n = 40)				Incubation: CGG (n = 32)			
Stimulation		**PDmin (mV)**	**PDmax (mV)**	**Stimulation**		**PDmin (mV)**	**PDmax (mV)**
RS	median	−0.4	0.37	CGG	Median	−1.31	0
	lower quartile	−0.98	−0.23		lower quartile	−2.06	−0.97
	upper quartile	0	0.97		upper quartile	−0.36	0.58
CGG	median	−0.57	0.23	RS	Median	−1.06	0
	lower quartile	−1.16	−0.34		lower quartile	−2.04	−0.26
	upper quartile	−0.15	0.69		upper quartile	−0.27	0.45
	Wilcoxon test (*p*)	0.014024	0.112656		Wilcoxon test (*p*)	0.378920	0.243286

Abbreviations: RS—iso-osmotic Ringer solution, CGG—chitosan–glycolic acid gel in RS, PDmin—minimal transepithelial potential difference during 15 s stimulation of epithelial skin surface (mV), PDmax—maximal transepithelial potential difference during 15 s stimulation of epithelial skin surface (mV), $p < 0.05$.

Table 3. Mann–Whitney test results for minimal (PDmin) and maximal (PDmax) transepithelial electric potential, measured during 15 s mechanical (RS) and mechanical–chemical (chitosan–glycolic acid gel) stimulations of skin specimens exposed to chitosan–glycolic acid gel solution immediately and for a prolonged period.

PDmin	*p*
RS stimulation in RS incubation vs. RS stimulation in CGG incubation	0.000502
CGG stimulation in RS incubation vs. CGG stimulation in CGG incubation	0.002292
PDmax	*p*
RS stimulation in RS incubation vs. RS stimulation in CGG incubation	0.090533
CGG stimulation in RS incubation vs. CGG stimulation in CGG incubation	0.043357

Abbreviations: RS—iso-osmotic Ringer solution, CGG—chitosan–glycolic acid gel in RS, PDmin—minimal transepithelial potential difference during 15 s stimulation of epithelial skin surface (mV), PDmax—maximal transepithelial potential difference during 15 s stimulation of epithelial skin surface (mV), $p < 0.05$.

The applied series of stimulations caused reproducible changes in the measured PDmin and PDmax potential, regardless of the applied incubation conditions and the type of stimulation (Tables 2 and 4). PDmin and PDmax values measured during stimulation were statistically significantly different from PD measured under stationary conditions, i.e., without stimulation (Table 5, Wilcoxon test).

Table 4. Minimal (PDmin) and maximal (PDmax) transepithelial electric potential measured during 15 s mechanical–chemical stimulations of skin specimens treated with immediate and prolonged exposure to chitosan–glycolic acid gel solution.

Incubation		**RS (n = 40)**		**Chitosan (n = 32)**	
Stimulation		**PDmin (mV)**	**PDmax (mV)**	**PDmin (mV)**	**PDmax (mV)**
B	Median	−0.64	0.08	−0.63	0.36
	lower quartile	−1.45	−0.43	−1.28	−0.06
	upper quartile	−0.2	0.67	−0.19	1.94

Table 4. *Cont.*

Incubation		RS (n = 40)		Chitosan (n = 32)	
Stimulation		PDmin (mV)	PDmax (mV)	PDmin (mV)	PDmax (mV)
A	Median	−0.86	−0.08	−0.96	0.06
	lower quartile	−1.28	−0.72	−3.03	0
	upper quartile	−0.21	0.69	−0.26	0.98
AB	Median	−0.54	0.23	−0.54	0.12
	lower quartile	−0.74	−0.57	−1.16	−0.59
	upper quartile	0.07	0.92	−0.12	2.12

Abbreviations: RS—iso-osmotic Ringer solution, CGG—chitosan–glycolic acid gel in RS, A—amiloride (0.1 mM) solution; B—bumetanide (0.1 mM) solution; AB—mixture A (0.1 mM) and B (0.1 mM) solutions; PDmin—minimal transepithelial potential difference during 15 s stimulation of epithelial skin surface (mV); PDmax—maximal transepithelial potential difference during 15 s stimulation of epithelial skin surface (mV).

Table 5. Results of Wilcoxon test for minimal (PDmin) and maximal (PDmax) transepithelial electric potential measured during 15 s mechanical (Ringer solution) and mechanical–chemical (chitosan solution) stimulation of skin specimens treated with immediate and prolonged exposure to chitosan glycolic acid gel solution.

Incubation: RS (n = 40)	*p*
PD vs. PDmin for RS stimulation	<0.001
PD vs. PDmax for RS stimulation	<0.001
PD vs. PDmin for CGG stimulation	<0.001
PD vs. PDmax for CGG stimulation	<0.001
Incubation: CGG (n = 32)	*p*
PD vs. PDmin for CGG stimulation	<0.001
PD vs. PDmax for CGG stimulation	<0.001
PD vs. PDmin for RS stimulation	<0.001
PD vs. PDmax for RS stimulation	<0.001

Abbreviations: RS—iso-osmotic Ringer solution, CGG—chitosan–glycolic acid gel in RS, PDmin—minimal transepithelial potential difference during 15 s stimulation of epithelial skin surface (mV), PDmax—maximal transepithelial potential difference during 15 s stimulation of epithelial skin surface (mV), $p < 0.05$.

3. Discussion

In this study, the influence of chitosan–glycolic acid gel on the electrophysiological parameters, such as electric potential and resistance of tissues of the skin, has been investigated.

The positive effects of chitosan application have been confirmed in many scientific studies. Chitosan is beneficial for the skin, hair and nails [11–16,44,45]. Despite its positive effects on the skin, chitosan is not absorbed by it. It creates a hydrophilic film on the skin surface, which effectively reduces transepidermal water loss from the epidermis and supports the renewal of the skin's natural hydro-lipid coat [44,45]. Chitosan has antimicrobial and antioxidant properties, supports skin regeneration processes, and has anti-inflammatory and anti-cancer effects.

The advantageous effect of glycolic acid has been confirmed in relation to various skin problems connected to keratinization disorders. Glycolic acid peeling treatments affect the lightening of melasma and post-inflammatory hyperpigmentation, are an element of anti-acne therapy in adolescents and adults and reduce actinic or seborrheic keratosis [46–48]. Glycolic acid also improves the penetration of other exfoliants. The anti-aging effect of glycolic acid is related to its ability to stimulate the production of hyaluronic acid and collagen [49].

The use of glycolic acid creates an acidic environment that allows chitosan to be introduced into the solution. Glycolic acid, having the smallest molecule among hydroxy acids, is a weak acid (pKa = 3.83). The chitosan chain will interact not only with the hydrogen ions formed during acid dissociation, but also with counter-anions (glycolate), which contributes to better dispersal of the chitosan [50–52].

It is difficult to explain the beneficial influence of chitosan on the skin only on the basis of its superficial effects. It can be assumed that chitosan is involved in the activation of biochemical processes responsible for the functioning of the skin, including the activity of ion channels. However, there are no direct studies regarding the effect of chitosan on such processes.

Changes in ion transport in the skin can be measured using a modified Ussing chamber [31,32]. The modification consists of placing the tissue in a horizontal position and applying a stimulus. Stimulation is based on the free release of fluid onto the surface of the examined tissue. Full-thickness skin fragments with preserved vitality, layering and physiological intercellular spaces, as well as active nerve endings and functioning channels and ion pumps, are examined. The assessment of ion transport is important for the evaluation of processes such as tissue hydration; the ability to receive stimuli; and the function of immunocompetent cells and melanocytes [31,32]. So far, the influence of chitosan and glycolic acid on the electrophysiological parameters of the skin has not been studied. The effect of CGG on the transport of sodium ions, chlorides and water in a multilayer structure seems to be extremely important for inferring the moisturizing, immunogenic effect or the maintenance of a uniform skin tone.

Our studies were carried out in stationary conditions (RS, CGG), during which mechanical and/or mechanical–chemical (A, B, AB) stimulation caused changes in the ion transport.

The lack of changes in the electrical resistance after the use of chitosan proves that the incubation of the skin specimens in the applied concentration of chitosan solution does not change the skin permeability (Table 1, Mann–Whitney). It also does not cause deformations, microdamage or changes in the ion permeability. The resistance values are stable and the tissue was alive and reactive throughout the experiment (Table 2, Wilcoxon, control and CGG). All skin specimens used were alive and retained full structure and compactness as well as active nerve fibers. Chitosan gel with glycolic acid applied to the outer skin layer did not change the vitality and compactness of the tissues. It also did not increase the ion and water permeability of the analyzed fragments. The ability to respond to the applied mechanical and/or mechanical–chemical stimulus was not affected by CGG, as tissues treated with CGG reacted analogously to control fragments (Table 5, Wilcoxon test).

Maintaining a negative charge on the surface of skin cells depends on the proper transport of chloride ions to the surface of the skin and the penetration of sodium ions into the cells [27,31,32]. The use of ion transport inhibitors amiloride and bumetanide allowed the inhibition of the entire pathway for individual ions and obtaining layers, cells that cannot secrete chloride or absorb sodium [1,2,27,31,32]. A reduction of the potential measured in stationary conditions (PDinitial) was demonstrated in comparison with preparations treated with Ringer's solution and with CGG (Table 1, Mann–Whitney, PDinitial). The reduction of the potential may result from the increased transport of chloride ions from the cell, and thus cause an influx of water in the intracellular direction [1,2]. After a series of stimulations, the potential value (PDfinal) does not differ from the control. The cells transported available chloride ions and water, so there is no room for excessive ion/water influx and cell swelling (Table 1, Mann–Whitney, PDfinal). Chitosan and glycolic acid gel does not increase the activity of the sodium–potassium pump maintaining the difference in potential under stationary conditions. The physiological activity is preserved (Table 5, Wilcoxon, CGG). The constantly occurring, ion transport enabling reaction to external and internal stimuli was not affected by the CGG solution. The generation of a more electronegative PD after incubation in CGG is most likely related to the intensification of chloride secretion to the surface. It is not possible to absorb such a large amount of sodium ions without interfering with the cell volume, which was ruled out after examining R, which was not changed after the application of CGG.

The absence of changes in the transepithelial transport of sodium ions after the use of CGG, both during stimulation and in case of prolonged action, is important for maintaining skin tissue homeostasis [31,33,36,37,43,53]. Each time, the intensification of the transport of sodium ions is associated with the movement of water towards the outer layers and its loss [1,31,37]. Additionally, the displacement of sodium ions is associated with the response to sensory stimuli [42] and the response to the triggers of local inflammation [43,53]. Substances changing the sodium transport in the skin may cause hypersensitivity reactions, pain and hyperreactivity [32]. The applied CGG solution did not cause such reactions under the proposed experimental conditions (Tables 1, 2 and 4).

Increased secretion of chloride ions is achieved thanks to the activation of the CFTR transporter (cystic fibrosis transmembrane conductance). This channel also acts out the role of a cell volume regulator, by inhibiting the ENaC channel (epithelial Na channel) and stimulating other chloride ion transport channels present in the apical membrane of skin cells [1,27,31,32]. The stimulation of the CFTR channel is also associated with the inhibition of the 2Cl-K-Na cotransporter [54]. In addition, CFTR stimulation is associated with the maintenance of the physiological pH of cells in the cell cytoplasm due to the interaction with the sodium–potassium pump and potassium channels [55]. Access to magnesium ions is essential for the activation and regulatory action of CFTR [27]. The CFTR channel present in the sweat channels is inhibited by bumetanide [1,2,26]. It can be assumed that the applied chitosan solution gently stimulates CFTR to chloride secretion, maintaining a tendency to decrease sodium ion transport (changes in potential—PD, Tables 3 and 4) and to maintain a constant cell volume. At the same time, it causes a slight increase in the amount of water in the intracellular spaces, without causing deformations or tissue damage (no change in R, Table 1). This confirms the data [21,23,24,45] that chitosan and glycolic acid have a moisturizing effect on the skin tissue, allowing the preservation and/or replenishment of water in the cells. The thesis about the effect of chitosan on ion channels is confirmed by reports on the effectiveness of the use of chitosan-based nanosystems in order to modify the operation of ion channels, e.g., reducing ENaC activity [51] and stimulating CFTR activity [52].

4. Materials and Methods

4.1. Animals

Specimens were excised from adult albino New Zealand rabbits of both sexes, weighing 3.5 to 4.0 kg and ranging in age from three to four months. Animals were subjected to asphyxiation with CO_2. The gas causes respiratory depression, reduces the contractility of the heart muscle and has a beneficial effect on the neuromuscular system. The death of each animal was confirmed by two methods, by a qualified person. The obtained preparations, isolated from the skin of ears, were stripped of hair and cleaned. The experiment consisted in examining tissues taken from sacrificed animals. To maximize the use of sacrificed animals, each rabbit's trachea, intestines, skin, liver, kidneys and muscles were also collected for other experimental procedures. The isolated tissues were submerged and incubated in the appropriate solution according to the experimental protocol. Rabbit skin prepared this way contains corneocytes, keratinocytes (95%), fibroblasts, immunocompetent cells, hair follicles, and nerve fiber endings.

4.2. Experimental Procedure

A modified Ussing chamber was used in the experiments. The tissue was mounted in a horizontal position, which allowed the stimulus to be applied to the examined surface [12]. The nozzle outlet was located at a safe distance from the tissue structure (approx. 5–7 mm). The surface of the external side of the tissue, which was gently rinsed with the solution, was 1 cm^2. The fluid flowing through the nozzle moved perpendicularly to it. A single stimulation lasted 15 s and the fluid was administered in a volume of 1 mL (0.06 mL/s), during which the analyzer recorded noticeable changes in the transepithelial electric potential. The modified Ussing chamber used consisted of two parts, and the tissue placed between the half-chambers acted as a partition. In order to equalize the pressure after administration of the stimulation fluid, went-holes had been placed in the upper half-chamber to allow the removal of excess solution.

Mechanical stimulation was performed using a Ringer's solution (RS), a stimulating fluid without a chemical component that only mechanically affected the isolated tissue. The combined mechanical–chemical stimulation was achieved through the use of a stimulus, CGG and/or amiloride and/or bumetanide. The immediate effect was tested with a 15 s mechanical–chemical stimulation by chitosan solution of the skin specimens. The prolonged effect was tested after applying the chitosan solution for 24 h to the external surface of the skin taken from the ear of a rabbit.

Tissues were kept in the dark at constant temperature (25 °C) and humidity (55%). After that time, measurements of electrophysiological parameters were performed:

(1) In stationary conditions: transepithelial electric potential (PD, mV) continuously measured, transepithelial electric resistance (R, Ω^*cm^2) measured after stimulation and counted according to Ohm's law.
(2) During 15 s stimulation: minimal transepithelial electric potential (PDmin), maximal transepithelial electric potential (PDmax).

Measurement of electrophysiological parameters lasted 30 min for each tissue specimen. The following parameters were measured during the experiment:

- R—transepithelial electrical resistance recorded while the tissue sample was exposed to a current with a stimulus intensity of ± 10 µA; then, after measuring the voltage change, calculations were made according to the Ohm's law (Ω^*cm^2).
- PD—changes in transepithelial electric potential difference measured in stationary conditions, i.e., without stimulation, recorded continuously (mV).
- PDmax and PDmin—minimal and maximal transepithelial electric potential difference measured during a 15 s stimulation (mV).

4.3. Chemicals

For the experiment the following solutions were used:

- Ringer's solution (RS)—(K^+ 4.0 mM; Na^+ 147.2 mM; Ca^{2+} 2.2 mM; Mg^{2+} 2.6 mM; Cl^- 160.8 mM; Hepes 10.0 mM), solution with iso-osmotic properties.
- Mineral compounds (NaCl, $CaCl_2$, KCl, $MgCl_2$) were purchased in Avantor Performance Materials Poland S.A., Poland.
- Chitosan–glycolic acid gel in RS (CGG) was prepared by dissolving chitosan (2.6% *w/v*) in 30 mL of aqueous glycolic acid solutions and put away for 24 h in a dark place. After this time, it was diluted with 500 mL of the Ringer's solution. Chitosan and glycolic acid were obtained from ACROS Organics, Poland, and used without further purification.
- A—amiloride, 3,5-diamino-6-chloro-2-carboxylic acid, 0.1 mM, (Sigma-Aldrich, St. Louis, MO, USA), inhibitor of ENaCs, used as an inhibitor of transepithelial sodium transport pathways.
- B—bumetanide, 3-butylamino-4-phenoxy-5-sulfamoylbenzoic acid, 0.1 mM (Sigma-Aldrich, St. Louis, MO, USA), inhibitor of Na-K-Cl cotransporter, used as an inhibitor of transepithelial chloride transport pathways.
- AB—a solution of amiloride (A, 0.1 mM) and bumetanide (B, 0.1 mM).

4.4. Data Analysis

The Ussing chamber was connected to the EVC 4000 measuring instrument (WPI, Worcester, MA, USA) coupled with the MP150 (Biopac, Goleta, CA, USA), a computer system which enables the recording of experimental results. The non-parametricity of the data distribution was confirmed by the Kolmogorov–Smirnov test, with the Lilliefors correction.

The results were presented as median and lower and upper quartile. Due to the fact that incorrect distribution of data was demonstrated, non-parametric tests, Wilcoxon test and U Mann–Whitney test with significance level $p < 0.005$ were used for the analysis. The obtained values were statistically analyzed using the Statistica 13.1. Programs.

5. Conclusions

The influence of chitosan–glycolic acid gel on the electrophysiological parameters, such as electric potential and resistance of skin tissues, was investigated. The studies were carried out in stationary conditions (RS, CGG) during which mechanical and/or mechanical–chemical (A, B, AB) stimulations caused changes in ion transport. A modified Ussing chamber was used in the experiments.

Studies have shown that stable values of electrical resistance are observed after the use of chitosan. This means that incubation in the applied concentration of chitosan solution does not increase the skin's permeability, the tissue remains alive and reactive. The observed decrease in stationary potential (PDinitial) probably indicates an increased transport of chloride ions, and thus water, from outside the cell into the cell interior. After the start of the stimulations, the potential value (PDfinal) did not differ from the control, the cells transported the available chloride ions and water and there was no excess ion/water flux and no cell swelling.

Chitosan maintains the potential difference under stationary conditions, but does not increase the activity of the sodium–potassium pump. Physiological activity is maintained, which is important for maintaining skin tissue homeostasis. Increased secretion of chloride ions is initiated by stimulating the action of CFTR. This causes a slight increase in the amount of water in the intercellular spaces without causing distortions and damage to the tissue.

Author Contributions: O.Z.—conceptualization, data curation, writing—original draft preparation; D.D.-W.—data curation, investigation, writing—review and editing; K.M.—writing—review and editing; I.H.-I.—conceptualization, investigation, methodology, writing—original draft preparation. All authors have read and agreed to the published version of the manuscript.

Funding: This research received no external funding.

Institutional Review Board Statement: The present experiment did not include living animals and according to the Polish and European Union law, the bioethical committee agreement was not required. Animal care was in accordance with the guidelines and regulations as stipulated by the Polish Animal Protection Act and the European Directive on the Protection of Animals Used for Scientific Purposes (2010/63/EU). All applicable institutional and national guidelines for the care and use of animals were followed.

Informed Consent Statement: Not applicable.

Data Availability Statement: Data will be able on request, email: igaholynska@cm.umk.pl.

Conflicts of Interest: The authors declare no conflict of interest.

Sample Availability: Samples of the compounds are available from the authors.

References

1. Dobrzeniecka, W.; Daca, M.; Nowakowska, B.; Sobiesiak, M.; Szewczyk-Golec, K.; Woźniak, A.; Hołyńska-Iwan, I. The Impact of Diclofenac Gel on Ion Transport in the Rabbit (*Oryctolagus cuniculus*) Skin: An In Vitro Study. *Molecules* **2023**, *28*, 1332. [CrossRef] [PubMed]
2. Hołyńska-Iwan, I.; Sobiesiak, M.; Kowalczyk, W.; Wróblewski, M.; Cwynar, A.; Szewczyk-Golec, K. Nickel ions influence the transepithelial sodium transport in the trachea, intestine and skin. *Sci. Rep.* **2023**, *13*, 6931. [CrossRef] [PubMed]
3. Guarnieri, A.; Triunfo, M.; Scieuzo, C.; Ianniciello, D.; Tafi, E.; Hahn, T.; Zibek, S.; Salvia, R.; De Bonis, A.; Falabella, P. Antimicrobial properties of chitosan from different developmental stages of the bioconverter insect Hermetia illucens. *Sci. Rep.* **2022**, *12*, 8084. [CrossRef] [PubMed]
4. Fatullayeva, S.; Tagiyev, D.; Zeynalov, N.; Mammadova, S.; Aliyeva, E. Recent advances of chitosan-based polymers in biomedical applications and environmental protection. *J. Polym. Res.* **2022**, *29*, 259. [CrossRef]
5. Madni, A.; Kousar, R.; Naeem, N.; Wahid, F. Recent Advancements in Applications of Chitosan-based Biomaterials for Skin Tissue Engineering. *J. Biores Bioprod.* **2021**, *6*, 11–25. [CrossRef]
6. Jiménez-Gómez, C.P.; Cecilia, J.A. Chitosan: A Natural Biopolymer with a Wide and Varied Range of Applications. *Molecules* **2020**, *25*, 3981. [CrossRef] [PubMed]

7. Foster, L.J.; Ho, S.; Hook, J.; Basuki, M.; Marçal, H. Chitosan as a Biomaterial: Influence of Degree of Deacetylation on Its Physiochemical, Material and Biological Properties. *PLoS ONE* **2015**, *10*, e0135153. [CrossRef]
8. Chandrasekaran, M.; Kim, K.; Chun, S. Antibacterial Activity of Chitosan Nanoparticles: A Review. *Processes* **2020**, *8*, 1173. [CrossRef]
9. Li, J.; Zhuang, S. Antibacterial Activity of Chitosan and Its Derivatives and Their Interaction Mechanism with Bacteria: Current State and Perspectives. *Eur. Polym. J.* **2020**, *138*, 109984. [CrossRef]
10. Dhillon, G.S.; Kaur, S.; Sarma, S.J.; Brar, S.K.; Verma, M.; Surampalli, R.Y. Recent Development in Applications of Important Biopolymer Chitosan in Biomedicine, Pharmaceuticals and Personal Care Products. *Curr. Tissue Engineering.* **2013**, *2*, 20–40. [CrossRef]
11. Rodríguez-Rodríguez, R.; Espinosa-Andrews, H.; Velasquillo-Martínez, C. Composite hydrogels based on gelatin, chitosan and polyvinyl alcohol to biomedical applications: A review. *Int. J. Polym. Mat. Pol. Biomat.* **2018**, *69*, 1–20. [CrossRef]
12. Pouranvari, S.; Ebrahimi, F.; Javadi, G.; Maddah, B. Chemical cross-linking of chitosan/polivinyl Alcohol electrospun nanofibers. *Mater. Technol.* **2016**, *50*, 663–666. [CrossRef]
13. Udrea, L.E.; Hritcu, D.; Popa, M.I.; Rotariu, O. Preparation and characterization of polyvinyl alcohol-chitosan biocompatible magnetic microparticles. *J. Magn. Magn. Mater.* **2011**, *323*, 7–13. [CrossRef]
14. Singh, A.; Mittal, A.; Benjakul, S. Chitosan nanoparticles: Preparation, food applications and health benefits. *SciAsia* **2021**, *47*, 1–10. [CrossRef]
15. Beleño Acosta, B.; Advincula, R.C.; Grande-Tovar, C.D. Chitosan-Based Scaffolds for the Treatment of Myocardial Infarction: A Systematic Review. *Molecules* **2023**, *28*, 1920. [CrossRef] [PubMed]
16. Li, Z.; Chen, X.; Bao, C.; Liu, C.; Liu, C.; Li, D.; Yan, H.; Lin, Q. Fabrication and Evaluation of Alginate/Bacterial Cellulose Nanocrystals-Chitosan-Gelatin Composite Scaffolds. *Molecules.* **2021**, *26*, 5003. [CrossRef] [PubMed]
17. Wang, C.-H.; Cherng, J.-H.; Liu, C.-C.; Fang, T.-J.; Hong, Z.-J.; Chang, S.-J.; Fan, G.-J.; Hsu, S.-D. Procoagulant and antimicrobial effects of chitosan in wound healing. *Int. J. Mol. Sci.* **2023**, *22*, 7067. [CrossRef]
18. Sakthiguru, N.; Sithique, M.A. Fabrication of bioinspired chitosan/gelatin/allantoin biocomposite film for wound dressing application. *Int. J. Biol. Macromol.* **2020**, *152*, 873–883. [CrossRef]
19. Moeini, A.; Pedram, P.; Makvandi, P.; Malinconico, M.; Gomez d'Ayala, G. Wound healing and antimicrobial effect of active secondary metabolites in chitosan-based wound dressings: A review. *Carbohydr. Polym.* **2020**, *233*, 115839. [CrossRef]
20. Bano, I.; Arshad, M.; Yasin, T.; Ghauri, M.A.; Younus, M. Chitosan: A potential biopolymer for wound management. *Int. J. Biol. Macromol.* **2017**, *102*, 380–383. [CrossRef]
21. Zavyalova, O.; Gajewska, S.; Dąbrowska-Wisłocka, D.; Sionkowska, A. Characteristics of physicochemical and rheological properties of chitosan hydrogels based on selected hydroxyl acids. *Eng. Biomat.* **2021**, *161*, 2–7. [CrossRef]
22. Lukic, M.; Filipovic, M.; Pajic, N.; Lunter, D.; Bozic, D.; Savic, S. Formulation of topical acidic products and acidification of the skin—Contribution of glycolic acid. *Int. J. Cosmet. Sci.* **2021**, *43*, 419–431. [CrossRef] [PubMed]
23. Fartasch, M.; Teal, J.; Menon, G.K. Mode of action of glycolic acid on human stratum corneum: Ultrastructural and functional evaluation of the epidermal barrier. *Arch. Dermatol. Res.* **1997**, *289*, 404–409. [CrossRef] [PubMed]
24. Tang, S.C.; Yang, J.H. Dual Effects of Alpha-Hydroxy Acids on the Skin. *Molecules* **2018**, *23*, 863. [CrossRef] [PubMed]
25. Kählig, H.; Hasanovic, A.; Biruss, B.; Höller, S.; Grim, J.; Valenta, C. Chitosan-glycolic acid: A possible matrix for progesterone delivery into skin. *Drug Dev. Ind. Pharm.* **2009**, *35*, 997–1002. [CrossRef] [PubMed]
26. Ferreira, D.; Silva, C.; Souza, M. Electrical impedance model for evaluation of skin irritation in rabbits and human. *Skin. Res. Technol.* **2007**, *13*, 259–267. [CrossRef]
27. Hołyńska-Iwan, I.; Szewczyk-Golec, K. Analysis of changes in sodium and chloride ion transport in the skin. *Sci. Rep.* **2020**, *10*, 18094. [CrossRef]
28. Aioi, A.; Okuda, M.; Matsui, M.; Tonogaito, H.; Hamada, K. Effect of high population density environment on skin barrier function in mice. *J. Dermatol. Sci.* **2001**, *25*, 189–197. [CrossRef]
29. Abdayem, R.; Callejon, S.; Portes, P.; Kirilov, P.; Demame, F.; Pirot, F.; Jannin, V.; Haftek, M. Modulation of transepithelial electric resistance (TEER) in reconstructed human epidermis by excipients known to permeate intestinal tight junctions. *Exp. Dermatol.* **2015**, *24*, 686–691. [CrossRef]
30. Boer, M.; Duchnik, E.; Maleszka, R.; Marchlewicz, M. Structural and biophysical characteristics of human skin in maintaining proper epidermal barrier function. *Adv. Dermatol. Allergol.* **2016**, *33*, 1–5. [CrossRef]
31. Hołyńska-Iwan, I.; Smyk, P.; Chrustek, A.; Olszewska-Słonina, D.; Szewczyk-Golec, K. The influence of hydration status on ion transport in the rabbit (Oryctolagus cuniculus) skin-An in vitro study. *PLoS ONE* **2021**, *16*, e0255825. [CrossRef] [PubMed]
32. Hołyńska-Iwan, I.; Sobiesiak, M. Cisplatin influences the skin ion transport: An in vitro study. *Biomed. Pharmacol.* **2020**, *129*, 1–11. [CrossRef] [PubMed]
33. Jung, E.; Maibach, H. Animal models for percutaneous absorption. *J. Appl. Toxicol.* **2015**, *35*, 1–10. [CrossRef] [PubMed]
34. Raj, N.; Voegeli, R.; Rawlings, A.V.; Doppler, S.; Imfeld, D.; Munday, M.R.; Lane, M.E. A fundamental investigation into aspects of the physiology and biochemistry of the stratum corneum in subjects with sensitive skin. *Int. J. Cosmet. Sci.* **2017**, *39*, 2–10. [CrossRef] [PubMed]
35. Denda, M.; Ashida, Y.; Inoue, K.; Kumazawa, N. Skin surface electric potential induced by ion-flux through epidermal cell layers. *Biochem. Biophys. Res. Commun.* **2001**, *284*, 112–117. [CrossRef] [PubMed]

36. Tucker, M.; Six, A.; Moyen, N.; Satterfield, A.; Ganio, M. Effect of hypohydration on postsynaptic cutaneous vasodilation and sweating in healthy men. *Am. J. Physiol. Regul. Integr. Comp. Physiol.* **2017**, *312*, R637–R642. [CrossRef]
37. Hanukoglu, I.; Boggula, V.R.; Vaknine, H.; Sharma, S.; Klevman, T.; Hanukoglu, A. Expression of epithelial sodium channel (ENaC) and CFTR in the human epidermis and epidermal appendages. *Histochem. Cell Biol.* **2017**, *147*, 733–748. [CrossRef]
38. Chiu, W.-T.; Vi Tran, T.T.; Pan, S.-C.; Huang, H.-K.; Chen, Y.-C.; Wong, W.-T. Cystic Fibrosis Transmembrane Conductance Regulator: A possible new target for photodynamic therapy enhances wound healing. *Adv. Wound Care* **2019**, *8*, 476–486. [CrossRef]
39. Seltmann, K.; Meyer, M.; Sulcova, J.; Kockmann, T.; Wehkamp, U.; Weidinger SAuf Dem Keller, U.; Werner, S. Humidity-regulated, CLCA2 protects the epidermis from hyperosmotic stress. *Sci. Transl. Med.* **2018**, *10*, eaao4650. [CrossRef]
40. Chen, J.; Chen, Y.; Chen, Y.; Yang, Z.; You, B.; Ruan, Y.C.; Peng, Y. Epidermal CFTR Suppresses MAPK/NF-κB to Promote Cutaneous Wound Healing. *Cell Physiol. Biochem.* **2016**, *39*, 2262–2274. [CrossRef]
41. Yang, H.-Y.; Charles, R.-P.; Hummler, E.; Baines, D.L.; Isseroff, R.R. The epithelial sodium channel mediates the directionality of galvanotaxis in human keratinocytes. *J. Cell Sci.* **2013**, *126*, 1942–1951. [CrossRef] [PubMed]
42. Baumauer, K.M.; DeBarry, J.J.; Adelman, P.C.; Miller, R.H.; Hashisuka, J.; Lee, K.H.; Ross, S.E.; Koerber, H.R.; Davis, B.M.; Albers, K.M. Keratinocytes can modulate and directly initiate nociceptive responses. *eLife* **2015**, *4*, e09674. [CrossRef] [PubMed]
43. Xu, W.; Hong, S.J.; Zeitchek, M.; Cooper, G.; Jia, S.; Xie, P.; Quereshi, H.A.; Zhong, A.; Portetfield, M.D.; Galiano, R.D.; et al. Hydration status regulates sodium flux and inflammatory pathways through epithelial sodium channel (ENaC) in the skin. *J. Investig. Dermatol.* **2015**, *135*, 796–806. [CrossRef] [PubMed]
44. Ferreira, P.; Ferreira, V.F.; de Carvalho da Silva, F.; Fretias, C.S.; Pereira, P.R.; Flosi Paschoalin, V.M. Chitosans and Nanochitosans: Recent Advances in Skin Protection, Regeneration, and Repair. *Pharmaceutics* **2022**, *14*, 1307. [CrossRef] [PubMed]
45. Tangkijngamvong, N.; Phaiyarin, P.; Wanichwecharungruang, S.; Kumtornrut, C. The anti-sebum property of chitosan particles. *J. Cosmet. Dermatol.* **2020**, *19*, 2135–2140. [CrossRef] [PubMed]
46. Kubiak, M.; Mucha, P.; Rotsztejn, H. Comparative study of 15% trichloroacetic acid peel combined with 70% glycolic acid and 35% trichloroacetic acid peel for the treatment of photodamaged facial skin in aging women. *J. Cosmet. Dermatol.* **2020**, *19*, 137–146. [CrossRef] [PubMed]
47. Pavithra, S.; Gopalakrishnan, K.; Shanmugam, J. Efficacy of 70% Glycolic Acid Peel versus 30% Salicylic Acid Peel in the Treatment of Mild to Moderate Acne Vulgaris: A Retrospective Study. *J. Clin. Diagn. Res.* **2022**, *16*, 5–8. [CrossRef]
48. Nautiyal, A.; Wairkar, S. Management of hyperpigmentation: Current treatments and emerging therapies. *Pigment. Cell Melanoma Res.* **2021**, *34*, 1000–1014. [CrossRef]
49. Narda, M.; Trullas, C.; Brown, A.; Piquero-Casals, J.; Granger, C.; Fabbrocini, G. Glycolic acid adjusted to pH 4 stimulates collagen production and epidermal renewal without affecting levels of proinflammatory TNF-alpha in human skin explants. *J. Cosmet. Dermatol.* **2021**, *20*, 513–521. [CrossRef]
50. Soares, L.; Perim, R.; Alvarenga, E.; Guimarães, L.; Carvalho Teixeira, A.; Coimbra, J.; Oliveira, E. Insights on physicochemical aspects of chitosan dispersion in aqueous solutions of acetic, glycolic, propionic or lactic acid. *Int. J. Biol. Macromol.* **2019**, *128*, 140. [CrossRef]
51. Kolonko, A.; Bangel-Ruland, N.; Goycoolea, F.; Weber, W. Chitosan Nanocomplexes for the Delivery of ENaC Antisense Oligonucleotides to Airway Epithelial Cells. *Biomoleculs.* **2020**, *10*, 553. [CrossRef] [PubMed]
52. Kolonko, A.; Fernández Fernández, E.; Santos-Carballal, B.; Goycoolea, F.M.; Weber, W.-M. Functional restoring of defect CFTR by transfection of CFTR- mRNA using chitosan. *JSM Genet. Genomics.* **2016**, *3*, 1016.
53. Xu, W.; Hong, S.J.; Zhong, A.; Xie, P.; Jia, S.; Sie, Z.; Zeitchek, M.; Niknam-Bienia, S.; Zhao, J.; Porterfield, D.M.; et al. Sodium channel Nax is a regulator in epithelial sodium homeostasis. *Sci. Transl. Med.* **2015**, *7*, 312ra177. [CrossRef] [PubMed]
54. Reddy, M.; Quinton, P. Bumetanide blocks CFTR GCl in the native sweat duct. *Am. J. Physiol.* **1999**, *276*, C231–C237. [CrossRef] [PubMed]
55. Reddy, M.; Quinton, P. Rapid regulation of electrolyte absorption in sweat duct. *J. Membr. Biol.* **1994**, *140*, 57–67. [CrossRef]

molecules

MDPI

Article

Cytotoxicity Enhancement of α-Mangostin with Folate-Conjugated Chitosan Nanoparticles in MCF-7 Breast Cancer Cells

Yedi Herdiana [1,*], Nasrul Wathoni [1], Shaharum Shamsuddin [2,3,4] and Muchtaridi Muchtaridi [5]

1 Department of Pharmaceutics and Pharmaceutical Technology, Faculty of Pharmacy, Universitas Padjadjaran, Sumedang 45363, Indonesia; nasrul@unpad.ac.id
2 School of Health Sciences, Universiti Sains Malaysia, Kubang Kerian 16150, Malaysia; shaharum1@usm.my
3 Nanobiotech Research Initiative, Institute for Research in Molecular Medicine (INFORMM), Universiti Sains Malaysia, Penang 11800, Malaysia
4 USM-RIKEN Interdisciplinary Collaboration on Advanced Sciences (URICAS), Universiti Sains Malaysia, Penang 11800, Malaysia
5 Department of Pharmaceutical Analysis and Medicinal Chemistry, Faculty of Pharmacy, Universitas Padjadjaran, Sumedang 45363, Indonesia; muchtaridi@unpad.ac.id
* Correspondence: y.herdiana@unpad.ac.id

Abstract: α-mangostin (AM) is a promising natural anticancer agent that can be used in cancer research. However, its effectiveness can be limited by poor solubility and bioavailability. To address this issue, chitosan-based nanoparticles (CSNPs) have been investigated as a potential delivery system to enhance the cytotoxicity to cancer cells and improve selectivity against normal cells. In this study, we developed folate-conjugated chitosan nanoparticles (F-CS-NPs) using a carbodiimide-based conjugation method to attach folate to chitosan (CS), which have different molecular weights. The NPs were crosslinked using tripolyphosphate (TPP) via ionic gelation. To characterize the F-CS-NPs, we utilized various analytical techniques, including transmission electron microscopy (TEM) to evaluate the particle size and morphology, Fourier-transform infrared spectroscopy (FTIR) to confirm the presence of functional groups, and ultraviolet-visible spectroscopy (UV-Vis) to measure the absorption spectrum and confirm the presence of folate. The particle size of AM-F-CS-NPs ranged from 180 nm to 250 nm, with many having favorable charges ranging from +40.33 ± 3.4 to 10.69 ± 1.3 mV. All NPs exhibited the same spherical morphology. The use of F-CS-NPs increased drug release, followed by a sustained release pattern. We evaluated the cytotoxicity of AM, AM-F-CS-HMW, and AM-F-CS-LMW NPs against MCF-7 cells and found IC_{50} values of 8.47 ± 0.49, 5.3 ± 0.01, and 4.70 ± 0.11 μg/mL, respectively. These results confirm the improved cytotoxicity of AM in MCF-7 cells when delivered via F-CS-NPs. Overall, our in vitro study demonstrated that the properties of F-CS-NPs greatly influence the cytotoxicity of AM in MCF-7 breast cancer cells (significantly different ($p < 0.05$)). The use of F-CS-NPs as a drug-delivery system for AM may have the potential to develop novel therapies for breast cancer.

Keywords: chitosan; conjugates; drug release; active targeting; nanocarrier

check for updates

Citation: Herdiana, Y.; Wathoni, N.; Shamsuddin, S.; Muchtaridi, M. Cytotoxicity Enhancement of α-Mangostin with Folate-Conjugated Chitosan Nanoparticles in MCF-7 Breast Cancer Cells. *Molecules* **2023**, *28*, 7585. https://doi.org/10.3390/molecules28227585

Academic Editor: Agnieszka Ewa Wiącek

Received: 28 September 2023
Revised: 2 November 2023
Accepted: 9 November 2023
Published: 14 November 2023

1. Introduction

Breast cancer is a significant cause of mortality among women worldwide [1,2], and current treatment methods, such as chemotherapy, surgery, radiation, and hormonal therapy, have limitations, including lack of effectiveness, damage to healthy organs, and development of drug resistance in cancer cells [3,4]. The tumor microenvironment is a critical factor in cancer therapy. It allows for more significant drug accumulation at the tumor site and targeted and regulated drug release, while maintaining selectivity to normal cells [5]. Mangosteen (*Garcinia mangostana Linn.*) is well known as the queen of fruits, and it grows in the Southeast Asian region. α-mangostin (AM), the main xanthone derivative contained in mangosteen pericarp, has many pharmacological activities, such as antioxidant, antiproliferation, anti-inflammatory, and anticancer [6,7]. AM has demonstrated potential therapeutic benefits for many types of cancer cells, with antitumor effects inhibiting key phases of cancer development [8–11], increasing cell cycle arrest and death in breast cancer cells [10,11], and growing selectivity [12]. However, AM's hydrophobic nature, poor water solubility, limited bioavailability [6,13–15], and accumulation in target organs pose significant challenges [16,17]. Folate-conjugated chitosan nanoparticles (F-CS-NPs) have been developed to enhance AM cytotoxicity and to selectively target MCF-7 breast cancer cells. Using F-CS-NPs for targeted drug delivery may help to overcome the limitations of conventional treatment methods and increase the effectiveness of AM in breast cancer therapy.

Nanomicelle formation and polymeric nanoparticle encapsulation have been proven to enhance the solubility of AM and facilitate its targeted delivery to specific organs [18]. Various studies, conducted both in vitro and in vivo, have demonstrated the enhanced cytotoxicity, antiproliferative activity, and induction of apoptosis when AM is delivered through these nanoparticle systems, such as transferrin-conjugated lipid–polymer hybrid NPs [19], AM-encapsulated poly(lactic-co-glycolic) acid (PLGA) NPs [20], AM-CS/sodium alginate NPs [21], AM-CS/carrageenan NPs [22] and hyaluronic acid-coated CS-NPs [23]. Additionally, NPs constructed with crosslinked CS and glyoxal have demonstrated controlled-release properties, making them a promising drug-delivery system for hydrophobic compounds such as AM [24].

Modified chitosan nanoparticles (CSNPs) have emerged as a promising drug-delivery system for cancer therapy, addressing critical challenges in bioavailability, stability, cellular uptake, protein adsorption, and drug distribution [25–27]. Recent studies have demonstrated that depolymerized CS can increase the cytotoxicity of AM in MCF-7 cells. The properties of CSNPs are influenced by the inherent properties of the CS polymer [28,29]. CS's unique chemical and physical characteristics determine the degree of crystallinity, surface smoothness, porosity, surface charge, size, and shape. The increased deacetylation degree of CS enhances its hydrophilicity, making it more suitable for formulating drug carriers. As a result, these properties directly affect the drug-loading capacity, release behavior, biocompatibility, targeted delivery, cellular uptake, and overall therapeutic efficacy of CSNPs [30,31].

Folate conjugation is vital in actively targeting cancer cells, as many tumors overexpress folate receptors [32]. This targeted approach enhances drug uptake by cancer cells and reduces damage to healthy tissues, minimizing the side effects associated with conventional chemotherapy [2,4,33]. F-CS-NPs, obtained by attaching folate conjugates to CS using a carbodiimide reaction, have demonstrated increased cytotoxicity against breast cancer cells, highlighting their potential for breast cancer therapy [34]. CS-based drug carriers show great potential in cancer treatment when modified and targeted with folate.

This research paper aims to further this study by functionalizing CSNPs with folate for targeted delivery of AM using CS, which have different molecular weights. Adding folate will provide a new vector for cancer cells, serving as an active target for drug delivery and enhancing the cytotoxicity of AM in a multimodal manner (Scheme 1). However, further studies are needed to investigate the potential enhancement of the cytotoxicity of AM using F-CS-NPs, based on release studies. This work represents an essential step toward developing nanosystems for treating breast cancer by combining CS, derivatized CS, and natural anticancer drugs.

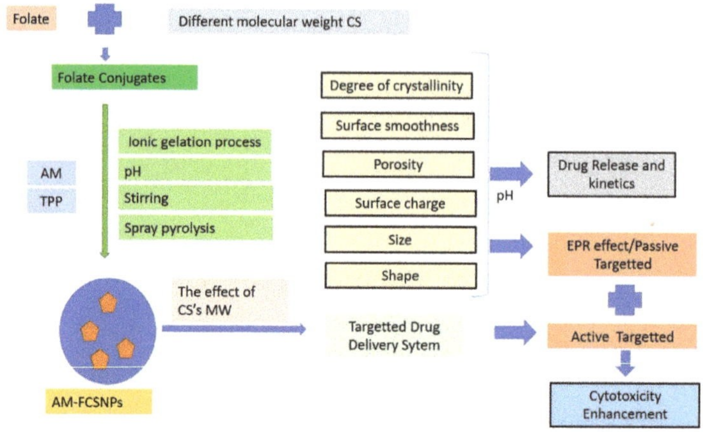

Scheme 1. The idea of AM-F-CS NPs as an active, targeted drug.

2. Results and Discussion

2.1. Characterization of F-CS

The interaction between CS and folate was investigated using FTIR spectroscopy. The measurements were obtained at 4000–400 cm^{-1} under vacuum pressure of 60 kN within 15 min.

F-CS, as depicted in Figure 1, demonstrates comparable IR spectra to CS due to their structural similarities. CS and F-CS have similar chemical compositions and functional groups, contributing to the overlapping peaks observed in their infrared spectra. Specifically, they exhibit peaks within the 3300- to 3500-cm^{-1} range, corresponding to the stretching vibrations of hydroxyl (OH) groups in both CS and F-CS. Additionally, both materials show peaks at approximately 1642 cm^{-1}, representing the stretching vibrations of N-H bonds, further highlighting their shared characteristics. These similarities in the IR spectra suggest that F-CS retains the essential features of CS while undergoing specific modifications or treatments [35]. The presence of an amide bond in F-CS can be identified by the C=O stretching band, which typically appears in the 1650 and 750 cm^{-1} ranges. This band is usually strong and sharp, characteristic of all amide compounds. Therefore, the appearance of the characteristic C=O stretching band in the IR spectra of F-CS confirms the presence of an amide bond between CS and folate. While CS and F-CS should have similar IR spectra, modifying CS with folate may result in a slight shift or change in the intensity of specific peaks, particularly in the regions where the amide bond is present (1650 and 750 cm^{-1}). The observed shift in peak intensity can confirm the successful conjugation of folate in CS, ultimately leading to the formation of F-CS.

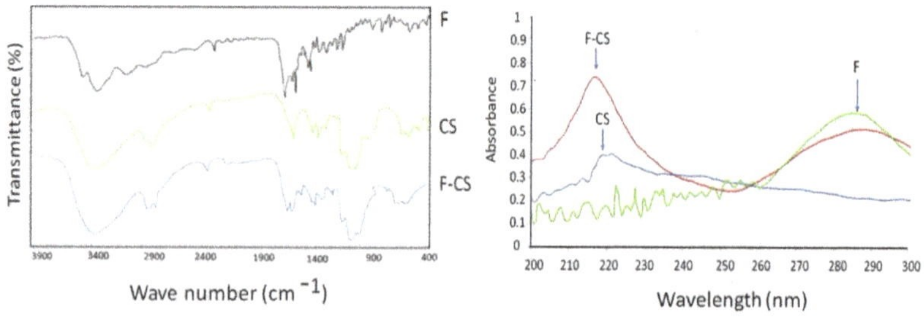

(**A**) (**B**)

Figure 1. (**A**) IR spectral of F-CS; (**B**) UV spectral absorption of F-CS.

Peak absorption spectroscopy is a valuable technique for identifying specific substances in a sample and quantifying their concentrations. The peak of the absorption spectrum corresponds to the wavelength at which the substance absorbs light most strongly. One can determine the presence of specific compounds by comparing the absorption spectrum of an unknown sample to known absorption spectra of various substances. In this case, peak absorption at 282 nm could indicate the presence of substances such as amino acids, which typically exhibit a peak absorption at around 280 nm due to aromatic amino acids that absorb light in that region. In the case of F-CS, two absorption bands appeared in the spectral ranges of 200–300 and 300–400 nm, which were assigned to the π-π^* and n-π^* transitions, respectively (Table 1) [36]. Further analysis using UV-Vis and FTIR spectroscopy confirmed the attachment of folate to CS, as indicated by the characteristic absorption bands of the amide bond in the IR spectra [23]. However, F-CS yield may be reduced due to the sequential processes required in its manufacture, such as dialysis purification. During dialysis, unbound folate and soluble EDC hydrochloride can diffuse outside the dialysis membrane and be removed from the F-CS solution.

Table 1. UV absorption spectroscopy of CS, F, and F-CS.

Compound	Max Absorbance (λ)	Chromophore Groups
CS	218 nm	N-acetyl glucosamine and glucosamine
F	287 nm	$\pi \rightarrow \pi^*$ of the aromatic ring
F-CS	218 nm and 282 nm	The n$\rightarrow\pi^*$ and $\pi\rightarrow\pi^*$ transitions are related to the formation of amide bonds and the transition of C=C bonds, respectively.

2.2. *Preparation of AM-F-CS-NPs*

Folate-functionalized AM-F-CS-NPs with uniform shapes were successfully synthesized using the ionic gelation process. The SEM and TEM data showed that the NPs had a homogenous sub-200 nm spherical shape. Na TPP ($Na_5P_3O_{10}$) was dissolved in water to carry hydroxyl and phosphoric ions in the ionic gelation process. This process formed an ionic complex via electrostatic interactions between the amine groups of CS and the trivalent TPP anion [37,38] due to the polycationic nature of CS in acidic solutions (pH 4.0–6.5). Cross-linking enhanced several essential characteristics of the CSNPs, including their stability, mechanical strength, swelling capacity, solubility, and drug release capabilities [39–46].

2.3. *Characterization of AM-F-CS-NPs.*

2.3.1. Physical and Morphology Nanoparticles

Nanoparticle design plays a crucial role in the efficient delivery of drugs due to various biophysical limitations [47]. The AM-F-CS-NP demonstrated the desired diameter, encapsulation efficiency, and spherical shape. Using 20-kDa CS reduced the NP size from 250.9 ± 23 to 180.5 ± 12 nm (Table 2). This smaller size is advantageous, as it favors renal filtration, which has a margin of 3.5 nm, thereby protecting the drug from rapid elimination [48]. Additionally, the smaller size allows for greater penetration into solid tumors with larger vascular cavities (200 to 780 nm) through which the drug-containing NPs can enter [49].

Table 2. Loading efficiency and loading capacity of the NPs: (A) AM-F-CS-HMW-NPs; (B) AM-F-CS-LMW-NPs.

Formula	Size (nm)	ZP (mV)	Shape	Loading Efficiency (%)	Loading Capacity (%)	PDI
F1	250.9 ± 23	+40.33 ± 3.4	Spherical	88.34 ± 1.30	5.93 ± 0.43	0.513 ± 0.05
F2	180.5 ± 12	+10.69 ± 1.3	Spherical	80.35 ± 1.40	5.43 ± 0.33	0.429 ± 0.03

The difference in nanoparticle sizes can be attributed to the type of molecular weight of CS used in the formulation. High molecular weight CS usually contains more amino groups along its long chain, which can result in larger particle sizes when cross-linked to form CSNPs. As a result, more available amino groups allow for more cross-link sites, resulting in larger aggregates or nanoparticles. Second, the molecular weight also represents the length of the CS chain, which can affect the particle-formation process and the resulting nanoparticle structure. Longer CS chains may have a more significant steric effect, with the size and spatial arrangement of the polymer chains affecting the accessibility of the crosslinking sites. This process can lead to more collapsed or denser particles in larger locations due to the limited availability of reactive groups for crosslinking.

Moreover, the ratios of CS, folate, and Na TPP concentrations can influence the surface charge of NPs. The zeta potential of the NPs increased from 10.69 ± 1.3 mV to 40.33 ± 3.4 mV, indicating an increase in their stability. In general, boosting the zeta potential of NPs increases their surface charge, leading to strong repulsive interactions and more excellent stability and uniform size [50]. This improved stability is crucial for effective drug delivery and targeted therapies, making CSNPs a promising strategy in pharmaceutical applications.

In this study, we observed that decreasing the molecular weight of CS significantly reduced the zeta potential of the NPs from 40.33 ± 3.4 mV to 10.69 ± 1.3 mV. CSNPs typically have a positive surface charge due to the protonation of amine groups in acidic pH conditions. When high-molecular weight CS is employed, it can increase the number of positively charged NH_3^+ groups on the surface, thereby increasing the surface charge. However, for pH levels other than 4 and 6, neither the type of CS nor its molecular weight significantly impacted the surface charge of the NPs [51,52].

Most studies have found that CSNPs exhibit a spherical morphology (Figure 2) [53,54]. The molecular weight of CS does not appear to affect the spherical shape of the particles [55]. Instead, the complexation between oppositely charged species shapes CS into spherical particles [56]. Cancer cells use NPs based on their size, shape, and surface functionalization [57,58].

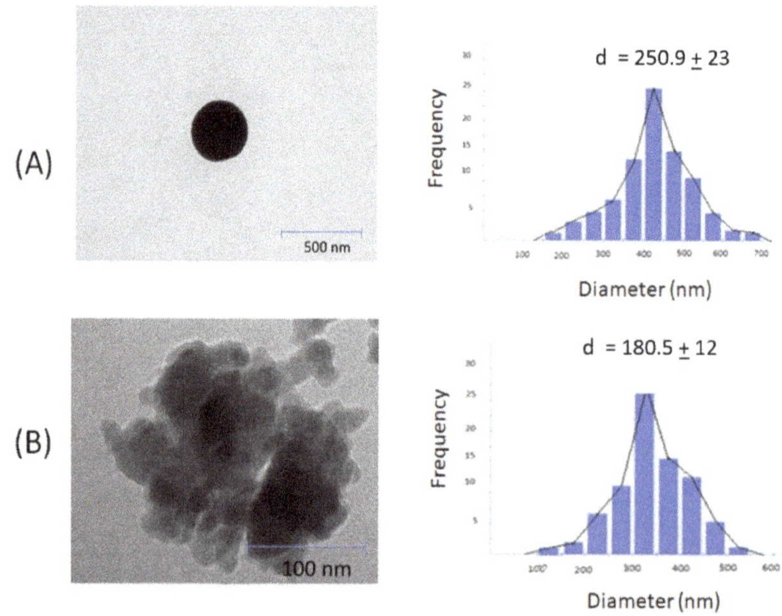

Figure 2. TEM images and size distribution: (**A**) F1 (AM-F-CS-HMW); (**B**) F2 (AM-F-CS-LMW).

The smoothness of nanoparticle surfaces (Figure 3) significantly impacts drug characteristics, particularly in terms of their stability, bioavailability, and targeting ability [59,60]. A smoother surface can enhance the stability of nanoparticles by reducing defects or irregularities that might lead to particle aggregation or degradation. Moreover, a smooth surface facilitates improved drug bioavailability, as it enhances absorption rates and lowers the likelihood of clearance by the immune system. Additionally, smooth nanoparticles are more effective in targeting specific cells or tissues by minimizing non-specific interactions with other cells or molecules [61]. Achieving a smooth nanoparticle surface is crucial for optimizing drug delivery and enhancing therapeutic outcomes [62].

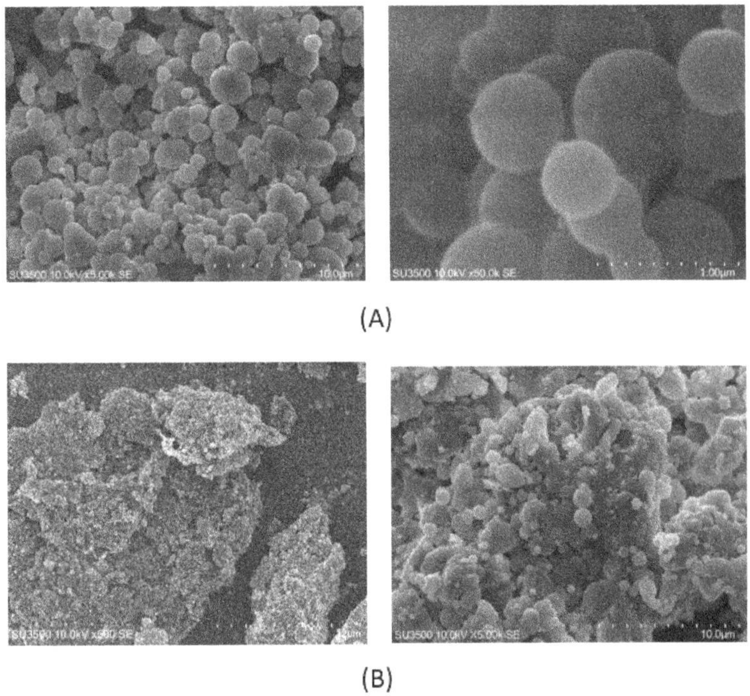

Figure 3. SEM of CSNP surfaces: (**A**) F1 (AM-F-CS-HMW); (**B**) F2 (AM-F-CS-LMW).

Several factors can influence the smoothness of nanoparticle surfaces, including the synthesis method, surface modification, size and shape, agglomeration, environmental conditions, quality control, surface energy, and composition of the NPs. For instance, in the case of CSNPs, hydrophobicity can affect surface smoothness. Hydrophobic drugs may interact with the hydrophobic regions of CS molecules, leading to aggregate formation and rough surfaces. Conversely, hydrophilic drugs may interact with the hydrophilic regions of CS molecules, resulting in more homogeneous drug distribution and a smoother surface [63]. Furthermore, the synthesis method and surface modification used in preparing CSNPs can also influence the smoothness of the particles. For instance, electrostatic interaction-based methods may result in a more irregular surface due to collision and aggregation. At the same time, ionic gelation may lead to smoother surfaces through cross-linking of the CS [64,65].

Recent studies have shown that the size reduction effect of folate conjugation to CSNPs may depend on various factors, such as the degree of deacetylation of CS and the folate-to-CS ratio [66,67]. A higher degree of deacetylation of CS may result in a more significant reduction in nanoparticle size upon folate conjugation. Additionally, the folate-to-CS ratio can affect the size reduction effect, with a higher folate-to-CS ratio leading to a more substantial decrease in size.

Moreover, the smoothness of CSNPs can affect their drug release, stability, and targeting ability and impact cellular uptake and toxicity. It has been reported that rough surfaces of NPs can induce more cellular uptake and cause higher toxicity due to increased interaction with cell membranes. Therefore, maintaining a smooth surface can reduce toxicity and improve biocompatibility.

In addition, the surface charge of CSNPs can also affect their biological properties. A recent study showed that positively charged CSNPs can enhance cellular uptake and improve therapeutic efficacy compared to negatively charged NPs. However, positively charged NPs may also lead to increased toxicity, which should be carefully evaluated.

Overall, the conjugation of folate to CSNPs can significantly impact their size and biological properties, and these factors should be carefully considered in the design and development of targeted drug-delivery systems [66,68,69].

2.3.2. FTIR Analysis

Using FTIR to assess AM–CS interactions, the ability of the ionic gelation mechanism to form AM–CS–NPs was evaluated (Figure 4).

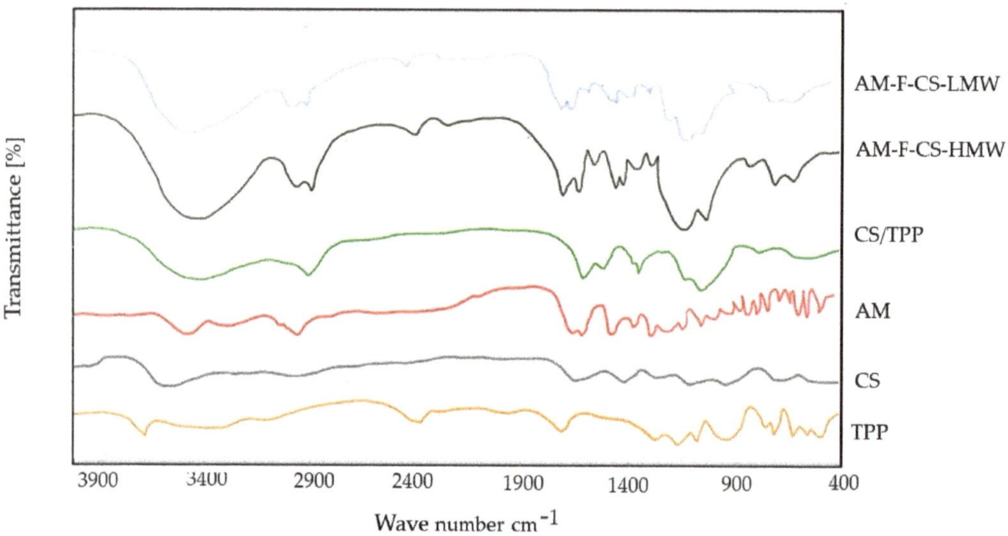

Figure 4. FTIR spectra of AM–F-CS NPs.

Figure 4 displays the FTIR spectrum of CSNPs, where the broad peak in the 3500- to 3300-cm^{-1} region corresponds to hydrogen-bonded O-H stretching vibrations. The primary amine and amide form II N-H peaks overlap, the C-O-C asymmetric stretch peak appears at around 1150 cm^{-1}, and the C-N stretch peak of amine type I is observed at 1317 cm^{-1}. Notably, in the NPs, the N-H bending vibration of amine I at 1600 cm^{-1} and the carbonyl stretch of amide II at 1650 cm^{-1} shifted to 1540 cm^{-1} and 1630 cm^{-1}, respectively, indicating successful nanoparticle formation and functionalization [70,71].

Additionally, the FTIR spectrum of low molecular weight CS (CS-LMW) matched that of the original CS. No band was detected between 1650 and 1900 cm^{-1}, indicating the absence of oxidative groups in CS-LMW [72,73].

2.4. *In Vitro Drug Release*

The impact of F-modification on AM release from CSNPs was investigated in vitro, and it was found that there was no significant effect. As shown in Figure 3, the release of the drug is dependent on both pH and time. pH significantly impacts AM-F-CSNP release, as improved protonation of CS amino groups can result in a loose nanoparticle structure and increased AM solubility, which may be responsible for the accelerated drug release at a lower pH. This pH-dependent hydrolysis is advantageous for delivering anticancer medications in the tumor microenvironment, which has a higher acidic pH (6.0–7.0) than normal plasma (7.4) [49].

The in vitro release profile of AM-F-CS-NPs was conducted over 8 h, and the drugs exhibited sustained release behavior, with a gradual increase in total drug release up to 20 h. The formulation displayed continuous release activity. The initial burst release observed can be attributed to the small NPs. As the particle diameter decreased, so did the distance to the drug's surface, resulting in sustained drug release [34].

The release data from this study showed that the pH and molecular weight of CS significantly influenced the release of AM from AM-F-NPs (Figure 5). At a low pH, the surface charge of the CSNP becomes positive, reducing electrostatic interactions and facilitating drug release. This pH responsiveness is beneficial under physiological conditions (pH 7.4), in which most of the drug remains within the nanoparticles, causing prolonged circulation and reducing side effects on healthy tissue. However, once drug-containing CSNPs are endocytosed by tumor cells, the lower pH in the tumor microenvironment promotes rapid drug release, potentially increasing the effectiveness of cancer therapy. This pH-dependent drug release mechanism in AM-F-NPs offers a promising approach for targeted and controlled drug delivery in cancer treatment. High molecular CS has a larger number of amino groups, and the steric effect can result in the formation of CSNPs, which are larger and potentially more compact than those formed using CS with a lower molecular weight. Additionally, the particle size of the NPs can impact drug dissolution following the Noyes–Whitney equation. The Noyes–Whitney equation describes the dissolution rate of a solid drug particle and states that the dissolution rate is directly proportional to the surface area of the drug particles [74,75].

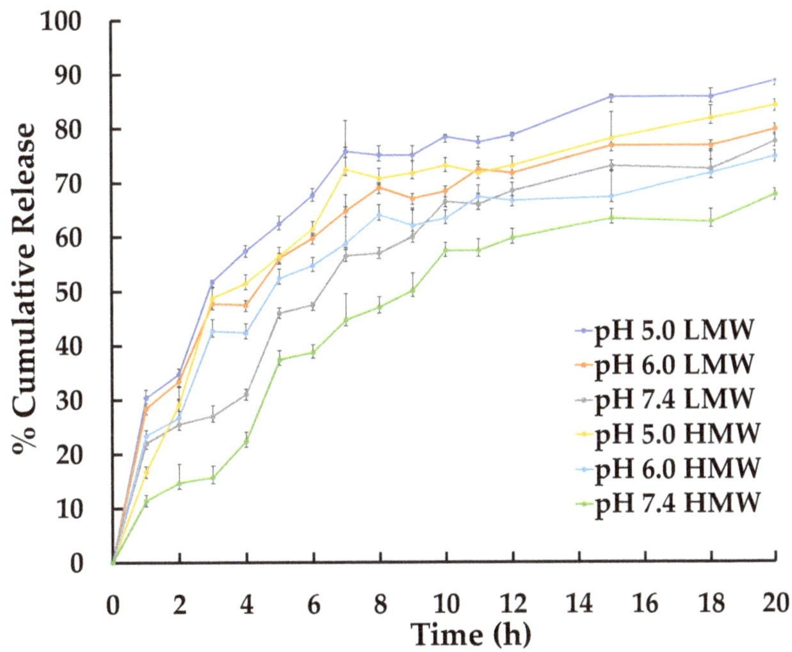

Figure 5. In vitro AM release profiles from AM-F-CS-NPs at 37 °C.

It is important to note that drug release and cytotoxicity often have a close relationship. Drug release refers to the process by which drugs are released from the delivery system and become available to the body for therapeutic effects. The rate and duration of drug release can significantly impact the efficacy and safety of a drug. For example, a slow release rate may result in a lower peak drug concentration, reducing toxicity and leading to a lower overall therapeutic effect. Conversely, a drug-delivery system that releases the drug quickly may lead to a higher peak drug concentration and, therefore, higher cytotoxicity. However, a drug-delivery system that releases the drug slowly may lead to a lower peak drug concentration and, hence, lower cytotoxicity and overall therapeutic effect.

Several variables can affect drug release from CSNPs, including the polymer molecular weight, deacetylation, nanoparticle size, porosity, and shape. The ability of drug molecules to interact electrostatically and be incorporated into CS structures depends on their length and conformation [76]. Thus, optimizing these variables can significantly impact the efficacy and safety of drug-delivery systems based on CSNPs.

To understand the kinetic release of AMG from CSNPs within the first 24 h, we conducted an analysis using various mathematical models: zero-order, first-order, Higuchi, and Korsmeyer-Peppas [77]. The correlation coefficients (r) obtained from the CSNP data (shown in Tables 3 and 4) indicate that there are two types of AMG release from F-CS-NPs:

- At a pH of 5, the release model follows the Higuchi model, suggesting a matrix-type release mechanism based on Fickian diffusion;
- However, at pH values of 6 and 7, the release follows the Korsmeyer–Peppas model, indicating a non-Fickian diffusion process.

Table 3. Higuchi regression parameter of AM from CSNPs in PBS in 24 h.

Parameter	pH 5.0		pH 6.0		pH 7.4	
	CS–LMW	CS–HMW	CS–LMW	CS–HMW	CS–LMW	CS–HMW
Slope (%h$^{-0.5}$)	28.23 ± 1.48	22.43 ± 2.21	26.77 ± 1.31	19.51 ± 0.42	20.02 ± 2.08	19.10 ± 0.75
Correlation coefficient (r)	0.95 ± 0.02	0.93 ± 0.02	0.92 ± 0.03	0.91 ± 0.01	0.79 ± 0.03	±0.03

Table 4. Korsmeyer–Peppas regression parameter of AM from CSNPs in PBS in 24 h.

Parameter	pH 5.0		pH 6.0		pH 7.4	
	CS–LMW	CS–HMW	CS–LMW	CS–HMW	CS–LMW	CS–HMW
Slope (%h$^{-0.5}$)	0.37 ± 0.01	0.361 ± 0.02	0.58 ± 0.01	0.46 ± 0.03	0.51 ± 0.03	0.74 ± 0.03
Correlation coefficient (r)	0.94 ± 0.01	0.92 ± 0.01	0.94 ± 0.03	0.93 ± 0.02	0.92 ± 0.09	0.92 ± 0.01

The Korsmeyer–Peppas release model determines the value of "n" from a regression equation, with obtained values typically ranging from 0.45 to 1. A value of "n" between 0.45 and 1.0 indicates non-Fickian release, suggesting contributions from mechanisms other than Fickian diffusion. Non-Fickian conditions can lead to faster or slower drug release rates than expected from Fickian diffusion alone. The drug-release mechanism can be further confirmed using Korsmeyer–Peppas plots, with slope values less than 0.5 indicating diffusion-controlled release [78], contributing to more sustained drug release [77]. Analyzing drug release under different pH conditions helps in the design of drug-delivery systems with optimized and controlled release profiles, ultimately enhancing therapeutic outcomes.

2.5. *In Vitro Cytotoxicity*

AM offers anticancer characteristics and is used to treat breast cancer. This research customized CSNPs with folate and AM to target FR-positive cells. No cytotoxic effect on cells was observed for CS-TPP (Figure 6). There was a significant difference between the cytotoxicity of AM and AM-F-CS-NPs. The IC_{50} of AM, AM-F-CS-HMW-NPs, and AM-F-CS-LMW-NPs was 8.47 ± 0.49, 5.3 ± 0.01, and 4.70 ± 0.11 μg/mL, respectively.

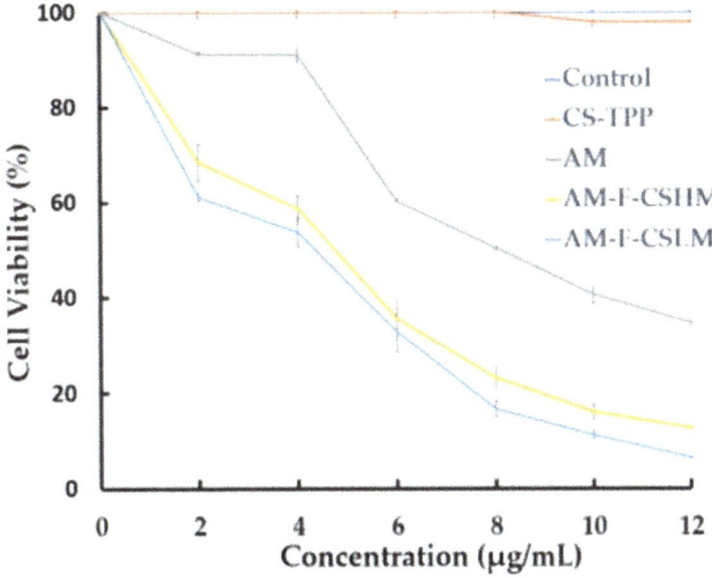

Figure 6. In vitro cytotoxicity of AM-F-CS-NPs.

The MTT assay is commonly used to assess cytotoxicity. In the current study, the presence of free amine groups and reduced steric hindrance resulting from folate conjugation may have facilitated the growth of CS-NPs. Folate has a strong affinity for folate receptors (FRs) and is stable, cost-effective, and not very immunogenic. F-CS-NPs have been used to encapsulate various anticancer drugs, including doxorubicin, 5-fluorouracil, ursolic acid [79], and cytarabine [46], and as DNA-delivery vehicles [35]. Conjugating polymers with targeting ligands, such as folate, has become a popular approach for targeted drug delivery. The increased internalization of NPs through endocytosis is mostly due to the specific interaction of folate receptors overexpressed in breast cancer cells [69]. In previous studies, passive targeting based on physicochemical properties, such as size and surface charge, has been easily adjustable by adjusting the component molecules or fabrication method [57]. F-CS-AM-NPs may increase AM effectiveness in two ways (Figure 7).

Active targeting through folate-directed AM-F-CS-NPs enhances cytotoxicity, as the MTT results indicate, with more significant inhibition of MCF-7 cancer cell growth and survival compared to AM alone. This capability makes AM-F-CS-NPs a potential solution for overcoming drug resistance in tumors by specifically targeting cells expressing folate receptors, which are often overexpressed in cancer cells [35]. Additionally, AM-F-CS-NPs exhibit high specificity in cellular uptake by folate receptor-expressing cells, offering a suitable vehicle for addressing tumor drug resistance. Regarding passive targeting mechanisms, the shape and size of nanoparticles play crucial roles [80,81]. Spherical polymer NPs of a specific size facilitate membrane penetration and absorption, while the size range of 100–200 nm displayed by AM-F-CS-NPs takes advantage of unique tumor-associated phenomena, such as holes or gaps between endothelial cells in the tumor vasculature and tissue, enabling targeted delivery to cancer cells [82].

Moreover, CS polycationic characteristics allow positively charged AM-F-CS-NPs to be readily absorbed by cancer cells with negatively charged membranes through electrostatic interactions. The zeta potential exceeding 25 mV ensures stable nanoparticle dispersion and absorption by tumor cells. Furthermore, protonation of the CS amino group at a lower pH promotes drug release from the porous matrix of CSNPs, displaying pH adaptive swelling and decreased solubility. These combined active and passive targeting features make AM-F-CS-NPs a promising strategy for overcoming drug resistance in tumors and improving cancer treatment efficacy [83]. Folate conjugation with CS in cancer drug-delivery systems offers significant advantages, such as targeted drug delivery, biocompatibility, and sustained release [32]. However, there are challenges related to synthesis complexity, variable targeting efficiency, and potential immune responses [32,84,85]. Despite these disadvantages, researchers continue to explore and optimize CS-based drug-delivery systems for improved cancer therapy outcomes [86,87].

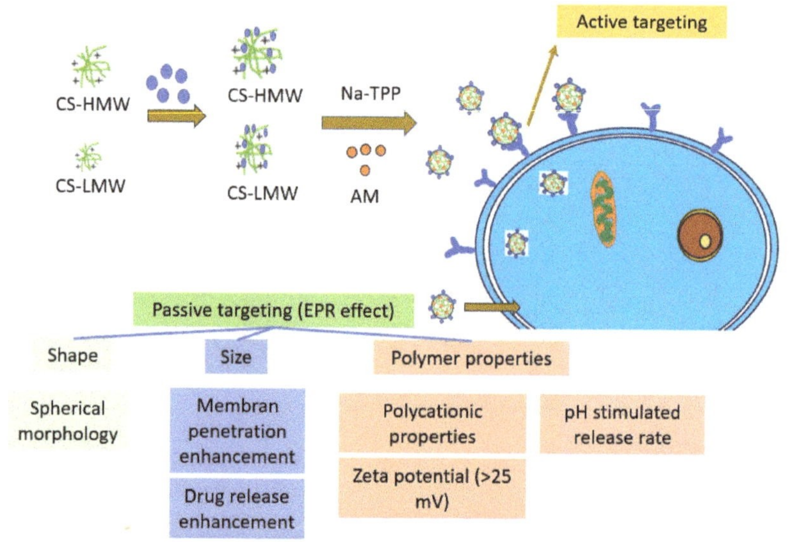

Figure 7. The main aspect of the cytotoxicity enhancement of AM-F-CS-NPs.

Molecular modelling and structural aspects are crucial in drug development and delivery [88,89]. They use computer methods to study how drugs interact with target molecules, predict drug behavior, and design delivery systems for better results (targeting, controlled release, stability, biocompatibility, and pharmacokinetics) [90,91]. Using these computational tools, we can discover and optimize new drugs faster, leading to more effective and targeted treatments for different diseases.

3. Materials and Methods

3.1. Materials

CS (300 kDa) and sodium tripolyphosphate (Na TPP) were purchased from Interlab, Ltd. (Jakarta, Indonesia). Folate (F) and 1-(3-dimethylaminoproply)-3-ethylcarbodiimide hydrochloride (EDC) were purchased from Sigma-Aldrich (St. Louis, MO, USA). MCF-7 cancer cells were provided by the American Type Culture Collection (Manassas, VA, USA).

3.2. Preparation of F-CS

Low-molecular weight CS (20 kDa) was prepared as in previous studies [57]. F-CS was synthesized using an aminoacylation process. For the attachment of FA to CS, a 0.2 mol ratio of FA to 1 mol CS was used. Initially, 200 mg of CS (1.24 mmol) was dissolved in 50 mL of 1 M acetic acid. A solution of 100 mg (0.23 mmol) of FA and 50 mg (0.26 mmol) of EDC hydrochloride in 20 mL of DMSO was then added to the CS solution, and the mixture was continuously stirred using a magnetic stirrer for 18 h in the dark. After 18 h, the pH was adjusted to 9 using 1 M sodium hydroxide. CS-modified folic acid was precipitated by centrifuging the mixture at 2500 rpm. F-CS was dissolved in 50 mL of water and dialyzed against PBS for three days, followed by four days against water. The F-CS-containing solution was then freeze-dried and kept at 4 °C.

3.3. Characterization F-CS

Fourier-transform infrared spectrophotometer (Model IR Prestige-21, Kyoto, Japan) and a UV-spectrophotometer (Shimadzu UV-1601, Kyoto, Japan) at 363 nm were used to study the CS-chemical F's structure [35].

3.4. Preparation AM-F-CS-NPs

The CSNPs were produced using a modified ionic gelation method, as described in Table 5. Initially, 20 mg of AM was diluted with 20 mL of ethanol and mixed with a 0.1% *w*/*v* (200 mg/200 mL) solution of F-CS in 1% acetic acid. This combination was stirred overnight at a pH of 4.7–4.8. In a separate container, 40 mg of TPP was dissolved in 10 mL of cold (25 °C), filtered (0.22 μm) distilled water [92]. Next, 10 mL of the TPP solution was added to 200 mL of the AM-F-CS solution, using this process to prepare Formula (2) [93].

Table 5. AM polymeric nanoparticle formulation.

Formulation	F1	F2
AM (mg)	20	20
F-CS-HMW (mg)	200	
F-CS-LMW (mg)		200
Na TPP (mg)	40	40

3.5. Characterization of AM-F-CS NPs

3.5.1. Physical Properties and Morphology Nanoparticles

Zetasizer SZ 100 Horiba was used to analyze samples for particle size and zeta potential (Kyoto, Japan) [72]. SEM (Thermo Scientific, Braunschweig, Germany) was used to analyze the NP surface morphology (SEM). Nanoparticle powder was attached on a stub. The powder was conductive with a narrow platinum beam for 30 s at 10 mA. The 10-kV picture is magnified. TEM (Thermo Scientific, Braunschweig, Germany) was utilized to analyze the morphologies of all CSNPs. Before analysis, carbon-coated samples were inspected under a microscope.

3.5.2. Fourier-Transform Infrared Analysis

F-CS was characterized by a Fourier-transform infrared (FTIR) spectrophotometer (Model IR Prestige-21, Kyoto, Japan) at 4000–400 cm^{-1} [70].

3.6. Loading Efficiency (LE) and Loading Capacity (LC)

UV-VIS spectroscopy was used to measure the LE of AM and LC of CSNPs. Twenty milligrams of sample NPs were diluted in ethyl acetate before centrifugation (3000 rpm, 10 min). The absorbance of filtrate at 245 nm was determined using UV-visible spectrophotometry to assess the amount of free AM [94]. The total quantity of AM was estimated, and the silt was reconstituted in ethanol to ascertain the LC. The various concentrations (2–20 g/mL) recorded at 245 nm were used to generate a standard curve.

The LE and LC of AM contained in NPs were determined using Equations (1) and (2), respectively [95,96]:

$$\text{Loading Efficiency } (\%) = \frac{\text{mass of AM present in nanoparticle (mg)}}{\text{mass of AM used (mg)}} \times 100\% \quad (1)$$

$$\text{Loading Capacity } (\%) = \frac{\text{mass of AM present in nanoparticle (mg)}}{\text{the total mass of nanoparticle (mg)}} \times 100\% \quad (2)$$

3.7. In Vitro Drug Release

To dialyze the AM-F-CS-NPs, 100 mg of the particles were suspended in 50 mL of deionized water for injection and then placed inside a membrane dialysis bag (Ward Science, West Henrietta, NY, USA, MW cut-off 14,000 Da) in PBS at pH values of 5.0, 6.0, and 7.4 with constant stirring at 37 °C. Five milliliters of dialysis buffer was withdrawn from the dialysis bag at regular intervals. Samples were taken periodically for 2, 4, 6, and 8 h, and the AM content was determined by UV-spectrophotometry analysis (Shimadzu UV-1601, Kyoto, Japan) [97,98]. pH 5.0, pH 6.0, and pH 7.4 were used to test AM release from NPs [99–101]. We obtained the release profile by plotting the total amount of AM released from the matrix against time in PBS.

The in-vitro drug release data were applied to several kinetic models (zero order, first order, Higuchi's kinetics, and Korsmeyer's equation) for formulation in various release media (Table 6). The in-vitro drug release data were fitted to multiple kinetic models to comprehend the drug release and rate-controlling mechanisms for the varied drug release. Every release experiment was conducted in triplicate. For each kinetic model, the drug release mechanism and linearization were calculated by finding the fit quality (R2) and the number of residuals squared (SSR).

Table 6. Kinetic models and the equations of kinetic models.

No	Kinetic Models	The Equations of Kinetic Models
1	Zero Order	$C = K_0t$
2	First Order	$\text{Log } C = \text{Log } C_o - kt/2.303$
3	Higuchi's kinetics	$Q = Kt^{1/2}$
4	Korsmeyer's equation	$Mt/M\infty = Kt^n$

3.8. In Vitro Cytotoxicity

In a 96-well culture plate, 1×10^4 MCF-7 cells were seeded in 10,000 culture medium and incubated for 24 h. Various quantities of F-CS-AM-NPs (1, 0.5, 0.25, 0.05, or 0.01 mg/mL) were added to the cells and cultured for 48 h at 37 °C. Ten microliters of 0.5% MTT solution was added to each well's cells, which were then cultured for an additional 4 h before 100 µL of DMSO was added. After dissolving formazan, the plate was agitated. Calculating the metabolism of the tetrazolium substrate, MTT, allowed for the determination of cell growth. The absorbance was at 570 nm. Cell viability declined as absorption dropped. IC_{50} was computed using linear regression's best-fit line [102]. Similarly, we measured pure AM, CS-TPP, and F-CS-TPP as controls.

3.9. Statistical Analysis

The standard deviation of the mean was given, and all measurements were performed in triplicate. The results were analyzed using the *t*-test or Student's *t*-test. Results were considered statistically significant if their *p*-value was less than or equal to 0.05.

4. Conclusions

CSNPs have been the subject of extensive research and are ideal delivery systems for cytotoxic anticancer agents. Compared to conventional NPs, which rely on passivity for tumor targeting and pure AM, the cytotoxicity of CNSPs is increased. F-CS-AM-NP is a potential tool for targeting MCF-7 cells and offers a promising strategy for optimizing therapeutic effectiveness. Further research is needed to clarify how various molecular weights of CS in CSNPs act on other breast cancer cell lines and normal cell lines and their feasibility for clinical application. This study still requires confirmation in vivo and selectivity tests on normal cells.

Author Contributions: Conceptualization, Y.H.; methodology, N.W.; validation, M.M.; formal analysis, S.S.; investigation, Y.H.; resources, N.W.; data curation, M.M.; writing—original draft preparation, Y.H.; writing—review and editing, Y.H., S.S., N.W. and M.M.; visualization, Y.H.; supervision, S.S., N.W. and M.M.; project administration, M.M.; funding acquisition, M.M. All authors have read and agreed to the published version of the manuscript.

Funding: This research was funded by the Rector of Universitas Padjadjaran for Hibah Percepatan Lektor Kepala 2023 (No: 1549/UN6.3.1/PT.00/2023).

Institutional Review Board Statement: Not applicable.

Informed Consent Statement: Not applicable.

Data Availability Statement: The data samples are provided by the corresponding author on request.

Acknowledgments: We would like to thank the Rector of Universitas Padjadjaran for funding this research.

Conflicts of Interest: The authors declare no conflict of interest.

References

1. Iqbal, J.; Abbasi, B.A.; Mahmood, T.; Kanwal, S.; Ali, B.; Shah, S.A. Plant-derived anticancer agents: A green anticancer approach. *Asian Pac. J. Trop. Biomed.* **2017**, *7*, 1129–1150. [CrossRef]
2. Shetty, V.; Jakhade, A.; Shinde, K.; Chikate, R.; Kaul-Ghanekar, R. Folate mediated targeted delivery of cinnamaldehyde loaded and FITC functionalized magnetic nanoparticles in breast cancer: In vitro, in vivo and pharmacokinetic studies. *New J. Chem.* **2021**, *45*, 1500–1515. [CrossRef]
3. Ruman, U.; Buskaran, K.; Pastorin, G.; Masarudin, M.J.; Fakurazi, S.; Hussein, M.Z. Synthesis and characterization of chitosan-based nanodelivery systems to enhance the anticancer effect of sorafenib drug in hepatocellular carcinoma and colorectal adenocarcinoma cells. *Nanomaterials* **2021**, *11*, 497. [CrossRef] [PubMed]
4. Sartaj, A.; Qamar, Z.; Qizilbash, F.F.; Alhakamy, N.A.; Baboota, S.; Ali, J. Polymeric Nanoparticles: Exploring the Current Drug Development and Therapeutic Insight of Breast Cancer Treatment and Recommendations. *Polymers* **2021**, *13*, 4400. [CrossRef] [PubMed]
5. Angelopoulou, A.; Kolokithas-Ntoukas, A.; Fytas, C.; Avgoustakis, K. Folic Acid-Functionalized, Condensed Magnetic Nanoparticles for Targeted Delivery of Doxorubicin to Tumor Cancer Cells Overexpressing the Folate Receptor. *ACS Omega* **2019**, *4*, 22214–22227. [CrossRef]
6. Ibrahim, M.Y.; Hashim, N.M.; Mariod, A.A.; Mohan, S.; Abdulla, M.A.; Abdelwahab, S.I.; Arbab, I.A. α-Mangostin from *Garcinia mangostana* Linn: An updated review of its pharmacological properties. *Arab. J. Chem.* **2016**, *9*, 317–329. [CrossRef]
7. Setyawati, L.U.; Nurhidayah, W.; Khairul Ikram, N.K.; Mohd Fuad, W.E.; Muchtaridi, M. General toxicity studies of alpha mangostin from *Garcinia mangostana*: A systematic review. *Heliyon* **2023**, *9*, e16045. [CrossRef]
8. Rizeq, B.; Gupta, I.; Ilesanmi, J.; Alsafran, M.; Rahman, M. The Power of Phytochemicals Combination in Cancer Chemoprevention. *J. Cancer* **2020**, *11*, 4521–4533. [CrossRef]
9. Lee, H.N.; Jang, H.Y.; Kim, H.J.; Shin, S.A.; Choo, G.S.; Park, Y.S.; Kim, S.K.; Jung, J.Y. Antitumor and apoptosis-inducing effects of α-mangostin extracted from the pericarp of the mangosteen fruit (*Garcinia mangostana* L.) in YD-15 tongue mucoepidermoid carcinoma cells. *Int. J. Mol. Med.* **2016**, *37*, 939–948. [CrossRef]
10. Salehi, B.; Fokou, P.V.T.; Yamthe, L.R.T.; Tali, B.T.; Adetunji, C.O.; Rahavian, A.; Mudau, F.N.; Martorell, M.; Setzer, W.N.; Rodrigues, C.F.; et al. Phytochemicals in prostate cancer: From bioactive molecules to upcoming therapeutic agents. *Nutrients* **2019**, *11*, 1483. [CrossRef]
11. Vemu, B.; Nauman, M.C.; Veenstra, J.P.; Johnson, J.J. Structure activity relationship of xanthones for inhibition of Cyclin Dependent Kinase 4 from mangosteen (*Garcinia mangostana* L.). *Int. J. Nutr.* **2019**, *4*, 38–45. [CrossRef] [PubMed]
12. Novilla, A.; Djamhuri, D.S.; Fauziah, N.; Maesaroh, M.; Balqis, B.; Widowati, W. Cytotoxic Activity of Mangosteen (*Garcinia mangostana* L.) Peel Extract and α-Mangostin toward Leukemia Cell Lines (HL-60 and K-562). *J. Nat. Remedies* **2016**, *16*, 52. [CrossRef]
13. Li, I.; Han, A.; Hamburger, M.; Kinghorn, D.; Butterweck, V.L.B. Determination of alpha-mangostin in rat plasma by HPLC-MS and its application to pharmacokinetic studies. *Planta Med.* **2010**, *76*, P338. [CrossRef]
14. Zhao, Y.; Tang, G.; Tang, Q.; Zhang, J.; Hou, Y.; Cai, E.; Liu, S.; Lei, D.; Zhang, L.; Wang, S. A Method of Effectively Improved α-Mangostin Bioavailability. *Eur. J. Drug Metab. Pharmacokinet.* **2016**, *41*, 605–613. [CrossRef] [PubMed]
15. Kondo, M.; Zhang, L.; Ji, H.; Kou, Y.; Ou, B. Bioavailability and antioxidant effects of a xanthone-rich mangosteen (*Garcinia mangostana*) product in humans. *J. Agric. Food Chem.* **2009**, *57*, 8788–8792. [CrossRef]
16. Yao, L.; Gu, X.; Song, Q.; Wang, X.; Huang, M.; Hu, M.; Hou, L.; Kang, T.; Chen, J.; Chen, H.; et al. Nanoformulated alpha-mangostin ameliorates Alzheimer's disease neuropathology by elevating {LDLR} expression and accelerating amyloid-beta clearance. *J. Control. Release* **2016**, *226*, 1–14. [CrossRef]

17. Herdiana, Y.; Wathoni, N.; Shamsuddin, S.; Muchtaridi, M. α-Mangostin nanoparticles cytotoxicity and cell death modalities in breast cancer cell lines. *Molecules* **2021**, *26*, 5119. [CrossRef]
18. Wathoni, N.; Rusdin, A.; Motoyama, K.; Joni, I.M.; Lesmana, R.; Muchtaridi, M. Nanoparticle drug delivery systems for α-mangostin. *Nanotechnol. Sci. Appl.* **2020**, *13*, 23–36. [CrossRef]
19. Sakpakdeejaroen, I.; Muanrit, P.; Panthong, S.; Ruangnoo, S. Alpha-Mangostin-Loaded Transferrin-Conjugated Lipid-Polymer Hybrid Nanoparticles: Development and Characterization for Tumor-Targeted Delivery. *Sci. World J.* **2022**, *2022*, 9217268. [CrossRef]
20. Verma, R.K.; Yu, W.; Shrivastava, A.; Shankar, S.; Srivastava, R.K. α-Mangostin-encapsulated PLGA nanoparticles inhibit pancreatic carcinogenesis by targeting cancer stem cells in human, and transgenic (KrasG12D, and KrasG12D/tp53R270H) mice. *Sci. Rep.* **2016**, *6*, 32743. [CrossRef]
21. Wathoni, N.; Rusdin, A.; Febriani, E.; Purnama, D.; Daulay, W.; Azhary, S.Y.; Panatarani, C.; Joni, I.M.; Lesmana, R.; Motoyama, K.; et al. Formulation and Characterization of α-Mangostin in Chitosan Nanoparticles Coated by Sodium Alginate, Sodium Silicate, and Polyethylene Glycol. *J. Pharm. Bioallied Sci.* **2019**, *11*, 619–627. [CrossRef] [PubMed]
22. Wathoni, N.; Meylina, L.; Rusdin, A.; Mohammed, A.F.A.; Tirtamie, D.; Herdiana, Y.; Motoyama, K.; Panatarani, C.; Joni, I.M.; Lesmana, R.; et al. The Potential Cytotoxic Activity Enhancement of α-Mangostin in Chitosan-Kappa Carrageenan-Loaded Nanoparticle against MCF-7 Cell Line. *Polymers* **2021**, *13*, 1681. [CrossRef]
23. Meylina, L.; Muchtaridi, M.; Joni, I.M.; Elamin, K.M.; Wathoni, N. Hyaluronic Acid-Coated Chitosan Nanoparticles as an Active Targeted Carrier of Alpha Mangostin for Breast Cancer Cells. *Polymers* **2023**, *15*, 1025. [CrossRef] [PubMed]
24. Samprasit, W.; Opanasopit, P. Chitosan-Based Nanoparticles for Controlled-Release Delivery of α–Mangostin. *Int. J. Pharma Med. Biol. Sci.* **2020**, *9*, 1–5. [CrossRef]
25. Gatoo, M.A.; Naseem, S.; Arfat, M.Y.; Dar, A.M.; Qasim, K.; Zubair, S. Physicochemical Properties of Nanomaterials: Implication in Associated Toxic Manifestations. *Biomed Res. Int.* **2014**, *2014*, 498420. [CrossRef] [PubMed]
26. Khan, I.; Saeed, K.; Khan, I. Nanoparticles: Properties, applications and toxicities. *Arab. J. Chem.* **2019**, *12*, 908–931. [CrossRef]
27. Hoosain, F.G.; Choonara, Y.E.; Tomar, L.K.; Kumar, P.; Tyagi, C.; du Toit, L.C.; Pillay, V. Bypassing P-Glycoprotein Drug Efflux Mechanisms: Possible Applications in Pharmacoresistant Schizophrenia Therapy. *Biomed Res. Int.* **2015**, *2015*, 484963. [CrossRef]
28. Yanat, M.; Schroën, K. Preparation methods and applications of chitosan nanoparticles; with an outlook toward reinforcement of biodegradable packaging. *React. Funct. Polym.* **2021**, *161*, 104849. [CrossRef]
29. Moraru, C.; Mincea, M.; Menghiu, G.; Ostafe, V. Understanding the Factors Influencing Chitosan-Based Nanoparticles-Protein Corona Interaction and Drug Delivery Applications. *Molecules* **2020**, *25*, 4758. [CrossRef]
30. Herdiana, Y.; Wathoni, N.; Shamsuddin, S.; Joni, I.M.; Muchtaridi, M. Chitosan-Based Nanoparticles of Targeted Drug Delivery System in Breast Cancer Treatment. *Polymers* **2021**, *15*, 879. [CrossRef]
31. Jafernik, K.; Ładniak, A.; Blicharska, E.; Czarnek, K.; Ekiert, H.; Wiącek, A.E.; Szopa, A. Chitosan-Based Nanoparticles as Effective Drug Delivery Systems—A review. *Molecules* **2023**, *28*, 1963. [CrossRef]
32. Farran, B.; Montenegro, R.C.; Kasa, P.; Pavitra, E.; Huh, Y.S.; Han, Y.-K.; Kamal, M.A.; Nagaraju, G.P.; Rama Raju, G.S. Folate-conjugated nanovehicles: Strategies for cancer therapy. *Mater. Sci. Eng. C* **2020**, *107*, 110341. [CrossRef] [PubMed]
33. Pawar, A.; Singh, S.; Rajalakshmi, S.; Shaikh, K.; Bothiraja, C. Development of fisetin-loaded folate functionalized pluronic micelles for breast cancer targeting. *Artif. Cells, Nanomed. Biotechnol.* **2018**, *46*, 347–361. [CrossRef] [PubMed]
34. Ortiz-Islas, E.; Sosa-Arróniz, A.; Manríquez-Ramírez, M.E.; Rodríguez-Pérez, C.E.; Tzompantzi, F.; Padilla, J.M. Mesoporous silica nanoparticles functionalized with folic acid for targeted release Cis-Pt to glioblastoma cells. *Rev. Adv. Mater. Sci.* **2021**, *60*, 25–37. [CrossRef]
35. Cheng, L.; Ma, H.; Shao, M.; Fan, Q.; Lv, H.; Peng, J.; Hao, T.; Li, D.; Zhao, C.; Zong, X. Synthesis of folate-chitosan nanoparticles loaded with ligustrazine to target folate receptor positive cancer cells. *Mol. Med. Rep.* **2017**, *16*, 1101–1108. [CrossRef]
36. Baibarac, M.; Smaranda, I.; Nila, A.; Serbschi, C. Optical properties of folic acid in phosphate buffer solutions: The influence of pH and UV irradiation on the UV-VIS absorption spectra and photoluminescence. *Sci. Rep.* **2019**, *9*, 14278. [CrossRef]
37. Popova, E.V.; Zorin, I.M.; Domnina, N.S.; Novikova, I.I.; Krasnobaeva, I.L. Chitosan—Tripolyphosphate Nanoparticles: Synthesis by the Ionic Gelation Method, Properties, and Biological Activity. *Russ. J. Gen. Chem.* **2020**, *90*, 1304–1311. [CrossRef]
38. Bhumkar, R.D.; Pokharkar, V.B. Studies on effect of pH on cross-linking of Chitosan with sodium tripolyphosphate: A technical note. *AAPS PharmSciTech* **2006**, *7*, 50. [CrossRef]
39. Jóźwiak, T.; Filipkowska, U.; Szymczyk, P.; Rodziewicz, J.; Mielcarek, A. Effect of ionic and covalent crosslinking agents on properties of chitosan beads and sorption effectiveness of Reactive Black 5 dye. *React. Funct. Polym.* **2017**, *114*, 58–74. [CrossRef]
40. Westlake, J.R.; Laabei, M.; Jiang, Y.; Yew, W.C.; Smith, D.L.; Burrows, A.D.; Xie, M. Vanillin Cross-Linked Chitosan Film with Controlled Release of Green Tea Polyphenols for Active Food Packaging. *ACS Food Sci. Technol.* **2023**, *3*, 1680–1693. [CrossRef]
41. Sharmin, N.; Rosnes, J.T.; Prabhu, L.; Böcker, U.; Sivertsvik, M. Effect of Citric Acid Cross Linking on the Mechanical, Rheological and Barrier Properties of Chitosan. *Molecules* **2022**, *27*, 5118. [CrossRef]
42. Han, X.; Wen, H.; Luo, Y.; Yang, J.; Xiao, W.; Xie, J. Effects of chitosan modification, cross-linking, and oxidation on the structure, thermal stability, and adsorption properties of porous maize starch. *Food Hydrocoll.* **2022**, *124*, 107288. [CrossRef]
43. Mathew, S.A.; Arumainathan, S. Crosslinked Chitosan-Gelatin Biocompatible Nanocomposite as a Neuro Drug Carrier. *ACS Omega* **2022**, *7*, 18732–18744. [CrossRef] [PubMed]

44. Dey, S.; Majumdar, S.; Hasnain, M.S.; Nayak, A.K. Chapter 11—Cross-linking of chitosan in drug delivery. In *Chitosan in Drug Delivery*; Hasnain, M.S., Beg, S., Nayak, A.K.B.T.-C., Eds.; Academic Press: Cambridge, MA, USA, 2022; pp. 277–299. ISBN 978-0-12-819336-5.
45. Tariq, H.; Rehman, A.; Kishwar, F.; Raza, Z.A. Citric Acid Cross-Linking of Chitosan Encapsulated Spearmint Oil for Antibacterial Cellulosic Fabric. *Polym. Sci. Ser. A* **2022**, *64*, 456–466. [CrossRef]
46. Geethakumari, D.; Bhaskaran Sathyabhama, A.; Raji Sathyan, K.; Mohandas, D.; Somasekharan, J.V.; Thavarool Puthiyedathu, S. Folate functionalized chitosan nanoparticles as targeted delivery systems for improved anticancer efficiency of cytarabine in MCF-7 human breast cancer cell lines. *Int. J. Biol. Macromol.* **2022**, *199*, 150–161. [CrossRef] [PubMed]
47. Ye, H.; Shen, Z.; Yu, L.; Wei, M.; Li, Y. Manipulating nanoparticle transport within blood flow through external forces: An exemplar of mechanics in nanomedicine. *Proc. R. Soc. A Math. Phys. Eng. Sci.* **2018**, *474*, 1–25. [CrossRef]
48. Claveau, S.; Nehlig, É.; Garcia-Argote, S.; Feuillastre, S.; Pieters, G.; Girard, H.A.; Arnault, J.-C.; Treussart, F.; Bertrand, J.-R. Delivery of siRNA to Ewing Sarcoma Tumor Xenografted on Mice, Using Hydrogenated Detonation Nanodiamonds: Treatment Efficacy and Tissue Distribution. *Nanomaterials* **2020**, *10*, 553. [CrossRef]
49. Wang, Z.; Deng, X.; Ding, J.; Zhou, W.; Zheng, X.; Tang, G. Mechanisms of drug release in pH-sensitive micelles for tumour targeted drug delivery system: A review. *Int. J. Pharm.* **2018**, *535*, 253–260. [CrossRef]
50. Nokhodi, F.; Nekoei, M.; Goodarzi, M.T. Hyaluronic acid-coated chitosan nanoparticles as targeted-carrier of tamoxifen against MCF7 and TMX-resistant MCF7 cells. *J. Mater. Sci. Mater. Med.* **2022**, *33*, 1–24. [CrossRef]
51. Katas, H.; Hussain, Z.; Ling, T.C. Chitosan Nanoparticles as a Percutaneous Drug Delivery System for Hydrocortisone. *J. Nanomater.* **2012**, *2012*, 372725. [CrossRef]
52. Bruinsmann, F.A.; Pigana, S.; Aguirre, T.; Souto, G.D.; Pereira, G.G.; Bianchera, A.; Fasiolo, L.T.; Colombo, G. Chitosan-Coated Nanoparticles: Effect of Chitosan Molecular Weight on Nasal Transmucosal Delivery. *Pharmaceutics* **2019**, *11*, 86. [CrossRef] [PubMed]
53. Agarwal, M.; Agarwal, M.K.; Shrivastav, N.; Pandey, S.; Das, R.; Gaur, P. Preparation of Chitosan Nanoparticles and their In-vitro Characterization. *Int. J. Life. Sci. Sci. Res.* **2018**, *4*, 1713–1720. [CrossRef]
54. Hembram, K.C.; Prabha, S.; Chandra, R.; Nimesh, S.; Hembram, K.C.; Prabha, S.; Chandra, R.; Ahmed, B.; Nimesh, S. Advances in preparation and characterization of chitosan nanoparticles for therapeutics. *Artif. Cells Nanomed. Biotechnol.* **2016**, *44*, 305–314. [CrossRef] [PubMed]
55. Sreekumar, S.; Goycoolea, F.M.; Moerschbacher, B.M.; Rivera-Rodriguez, G.R. Parameters influencing the size of chitosan-TPP nano- and microparticles. *Sci. Rep.* **2018**, *8*, 4695. [CrossRef]
56. Giri, T.K. *Nanoarchitectured Polysaccharide-Based Drug Carrier for Ocular Therapeutics*; Elsevier Inc.: Amsterdam, The Netherlands, 2016; ISBN 9780323477222.
57. Herdiana, Y.; Wathoni, N.; Shamsuddin, S.; Muchtaridi, M. Cytotoxicity Enhancement in MCF-7 Breast Cancer Cells with Depolymerized Chitosan Delivery of α-Mangostin. *Polymers* **2022**, *14*, 3139. [CrossRef]
58. Dasgupta, S.; Auth, T.; Gompper, G. Shape and Orientation Matter for the Cellular Uptake of Nonspherical Particles. *Nano Lett.* **2014**, *14*, 687–693. [CrossRef]
59. Ghorbani, F.; Kokhaei, P.; Ghorbani, M.; Eslami, M. Application of different nanoparticles in the diagnosis of colorectal cancer. *Gene Rep.* **2020**, *21*, 100896. [CrossRef]
60. Yaqoob, A.A.; Sekeri, S.H.; Othman, M.B.H.; Ibrahim, M.N.M.; Feizi, Z.H. Thermal degradation and kinetics stability studies of oil palm (*Elaeis Guineensis*) biomass-derived lignin nanoparticle and its application as an emulsifying agent. *Arab. J. Chem.* **2021**, *14*, 103182. [CrossRef]
61. Bierhalz, A.C.K.; Westin, C.B.; Moraes, Â.M. Comparison of the properties of membranes produced with alginate and chitosan from mushroom and from shrimp. *Int. J. Biol. Macromol.* **2016**, *91*, 496–504. [CrossRef]
62. Liu, F.; Xu, J.; Wu, L.; Zheng, T.; Han, Q.; Liang, Y.; Zhang, L.; Li, G.; Yang, Y. The Influence of the Surface Topographical Cues of Biomaterials on Nerve Cells in Peripheral Nerve Regeneration: A Review. *Stem Cells Int.* **2021**, *2021*, 8124444. [CrossRef]
63. George, D.; Maheswari, P.U.; Begum, K.M.M.S. Chitosan-cellulose hydrogel conjugated with L-histidine and zinc oxide nanoparticles for sustained drug delivery: Kinetics and in-vitro biological studies. *Carbohydr. Polym.* **2020**, *236*, 116101. [CrossRef] [PubMed]
64. Richardson, J.J.; Björnmalm, M.; Caruso, F. Technology-driven layer-by-layer assembly of nanofilms. *Science* **2015**, *348*, 1–11. [CrossRef] [PubMed]
65. Ahmed, M.E.; Saber, D.; Abd Elaziz, K.; Alghtani, A.H.; Felemban, B.F.; Ali, H.T.; Megahed, M. Chitosan-based nanocomposites: Preparation and characterization for food packing industry. *Mater. Res. Express* **2021**, *8*, 025017. [CrossRef]
66. Ullah, S.; Azad, A.K.; Nawaz, A.; Shah, K.U.; Iqbal, M.; Albadrani, G.M.; Al-Joufi, F.A.; Sayed, A.A.; Abdel-Daim, M.M. 5-Fluorouracil-Loaded Folic-Acid-Fabricated Chitosan Nanoparticles for Site-Targeted Drug Delivery Cargo. *Polymers* **2022**, *14*, 2010. [CrossRef] [PubMed]
67. Radnia, F.; Mohajeri, N.; Hashemi, F.; Imani, M.; Zarghami, N. Design and development of folate-chitosan/CD nanogel: An efficient fluorescent platform for Cancer-specific delivery of AntimiR-21. *React. Funct. Polym.* **2021**, *160*, 104814. [CrossRef]
68. San, H.H.M.; Alcantara, K.P.; Bulatao, B.P.I.; Sorasitthiyanukarn, F.N.; Nalinratana, N.; Suksamrarn, A.; Vajragupta, O.; Rojsitthisak, P.; Rojsitthisak, P. Folic Acid-Grafted Chitosan-Alginate Nanocapsules as Effective Targeted Nanocarriers for Delivery of Turmeric Oil for Breast Cancer Therapy. *Pharmaceutics* **2023**, *15*, 110. [CrossRef]

69. Ruman, U.; Buskaran, K.; Bullo, S.; Pastorin, G.; Masarudin, M.J.; Fakurazi, S.; Hussein, M.Z. Sorafenib and 5-Fluorouracil Loaded Dual Drug Nanodelivery Systems for Hepatocellular Carcinoma and Colorectal Adenocarcinoma. *Res. Sq.* **2021**. [CrossRef]
70. Herdiana, Y.; Handaresta, D.F.; Joni, I.M.; Wathoni, N.; Muchtaridi, M. Synthesis of nano-α mangostin based on chitosan and Eudragit S 100. *J. Adv. Pharm. Technol. Res.* **2020**, *11*, 95–100. [CrossRef]
71. Saraf, N.S. Formulation and evaluation of antifungal agent in a hydrogel containing nanoparticle of low molecular weight chitosan. *Int. J. Res. Pharm. Sci.* **2020**, *11*, 247–259.
72. Zheng, X.; Yin, Y.; Jiang, W.; Xing, L.; Pu, J. Synthesis and Characterization of Low Molecular Weight Chitosan. *BioResources* **2015**, *10*, 2338–2349. [CrossRef]
73. Ma, Z.; Wang, W.; Wu, Y.; He, Y.; Wu, T. Oxidative Degradation of Chitosan to the Low Molecular Water-Soluble Chitosan over Peroxotungstate as Chemical Scissors. *PLoS ONE* **2014**, *9*, 3–9. [CrossRef] [PubMed]
74. Yang, H.C.; Hon, M.H. The effect of the degree of deacetylation of chitosan nanoparticles and its characterization and encapsulation efficiency on drug delivery. *Polym. Plast. Technol. Eng.* **2010**, *49*, 1292–1296. [CrossRef]
75. do Nascimento, E.G.; de Caland, L.B.; de Medeiros, A.S.A.; Fernandes-Pedrosa, M.F.; Soares-Sobrinho, J.L.; dos Santos, K.S.C.R.; da Silva-Júnior, A.A. Tailoring drug release properties by gradual changes in the particle engineering of polysaccharide chitosan based powders. *Polymers* **2017**, *9*, 253. [CrossRef] [PubMed]
76. Lucio, D.; Martínez-Ohárriz, M.C. Chitosan: Strategies to Increase and Modulate Drug Release Rate. *Biol. Act. Appl. Mar. Polysacch.* **2017**. [CrossRef]
77. Fernández-Romero, A.M.; Maestrelli, F.; Mura, P.A.; Rabasco, A.M.; González-Rodríguez, M.L. Novel findings about double-loaded curcumin-in-HPβcyclodextrin-in liposomes: Effects on the lipid bilayer and drug release. *Pharmaceutics* **2018**, *10*, 256. [CrossRef]
78. Basak, S.; Kumar, K.; Ramlingam, M. Design and release characteristics of sustained release tablet containing metformin HCl. *Brazilian J. Pharm. Sci.* **2008**, *44*, 477–482. [CrossRef]
79. Jin, H.; Pi, J.; Yang, F.; Jiang, J.; Wang, X.; Bai, H.; Shao, M.; Huang, L.; Zhu, H.; Yang, P.; et al. Folate-Chitosan Nanoparticles Loaded with Ursolic Acid Confer Anti-Breast Cancer Activities in vitro and in vivo. *Sci. Rep.* **2016**, *6*, 30782. [CrossRef]
80. Augustine, R.; Hasan, A.; Primavera, R.; Wilson, R.J.; Thakor, A.S.; Kevadiya, B.D. Cellular uptake and retention of nanoparticles: Insights on particle properties and interaction with cellular components. *Mater. Today Commun.* **2020**, *25*, 101692. [CrossRef]
81. Hatem, S.; Elkheshen, S.A.; Kamel, A.O.; Nasr, M.; Moftah, N.H.; Ragai, M.H.; Elezaby, R.S.; El Hoffy, N.M. Functionalized chitosan nanoparticles for cutaneous delivery of a skin whitening agent: An approach to clinically augment the therapeutic efficacy for melasma treatment. *Drug Deliv.* **2022**, *29*, 1212–1231. [CrossRef]
82. Herdiana, Y.; Wathoni, N.; Gozali, D.; Shamsuddin, S.; Muchtaridi, M. Chitosan-Based Nano-Smart Drug Delivery System in Breast Cancer Therapy. *Pharmaceutics* **2023**, *15*, 879. [CrossRef]
83. Almualla, M.A.; Mousa, M.N.; Sattar, M. Chemical modification and characterization of chitosan for pharmaceutical applications. *Egypt. J. Chem.* **2021**, *64*, 3635–3649. [CrossRef]
84. Yusuf, A.; Almotairy, A.R.Z.; Henidi, H.; Alshehri, O.Y.; Aldughaim, M.S. Nanoparticles as Drug Delivery Systems: A Review of the Implication of Nanoparticles' Physicochemical Properties on Responses in Biological Systems. *Polymers* **2023**, *15*, 1596. [CrossRef] [PubMed]
85. Mitchell, M.J.; Billingsley, M.M.; Haley, R.M.; Wechsler, M.E.; Peppas, N.A.; Langer, R. Engineering precision nanoparticles for drug delivery. *Nat. Rev. Drug Discov.* **2021**, *20*, 101–124. [CrossRef] [PubMed]
86. Li, X.; Gao, Y.; Li, H.; Majoral, J.-P.; Shi, X.; Pich, A. Smart and bioinspired systems for overcoming biological barriers and enhancing disease theranostics. *Prog. Mater. Sci.* **2023**, *140*, 101170. [CrossRef]
87. Lu, Y.; Luo, Q.; Jia, X.; Tam, J.P.; Yang, H.; Shen, Y.; Li, X. Multidisciplinary strategies to enhance therapeutic effects of flavonoids from Epimedii Folium: Integration of herbal medicine, enzyme engineering, and nanotechnology. *J. Pharm. Anal.* **2023**, *13*, 239–254. [CrossRef]
88. Bunker, A.; Róg, T. Mechanistic Understanding from Molecular Dynamics Simulation in Pharmaceutical Research 1: Drug Delivery. *Front. Mol. Biosci.* **2020**, *7*, 1–36. [CrossRef]
89. Adelusi, T.I.; Oyedele, A.-Q.K.; Boyenle, I.D.; Ogunlana, A.T.; Adeyemi, R.O.; Ukachi, C.D.; Idris, M.O.; Olaoba, O.T.; Adedotun, I.O.; Kolawole, O.E.; et al. Molecular modeling in drug discovery. *Inform. Med. Unlocked* **2022**, *29*, 100880. [CrossRef]
90. Katiyar, R.S.; Jha, P.K. Molecular simulations in drug delivery: Opportunities and challenges. *WIREs Comput. Mol. Sci.* **2018**, *8*, e1358. [CrossRef]
91. Ferreira, L.G.; Dos Santos, R.N.; Oliva, G.; Andricopulo, A.D. *Molecular Docking and Structure-Based Drug Design Strategies*; Schrödinger: New York, NY, USA, 2015; Volume 20, ISBN 5516337380.
92. Srikulkit, K. Preparation of Depolymerized Chitosan and Its Effect on Dyeability of Preparation of Depolymerized Chitosan and Its Effect on Dyeability of Mangosteen Dye. *Chiang Mai J. Sci.* **2011**, *38*, 473–484.
93. Al-Nerawi, N.K.; Alsharif, S.S.M.; Dave, R.H. Preparation of Chitosan-TPP Nanoparticles: The Influence of Chitosan Polymeric Properties and Formulation Variables. *Int. J. Appl. Pharm.* **2018**, *10*, 60–65. [CrossRef]
94. Asasutjarit, R.; Meesomboon, T.; Adulheem, P. Physicochemical properties of alpha-mangostin loaded nanomeulsions prepared by ultrasonication technique. *Heliyon* **2019**, *5*, e02465. [CrossRef] [PubMed]

95. Garms, B.C.; Poli, H.; Baggley, D.; Han, F.Y.; Whittaker, A.K.; Anitha, A.; Grøndahl, L. Evaluating the effect of synthesis, isolation, and characterisation variables on reported particle size and dispersity of drug loaded PLGA nanoparticles. *Mater. Adv.* **2021**, *2*, 5657–5671. [CrossRef]
96. Lee, K.H.; Khan, F.N.; Cosby, L.; Yang, G.; Winter, J.O. Polymer Concentration Maximizes Encapsulation Efficiency in Electrohydrodynamic Mixing Nanoprecipitation. *Front. Nanotechnol.* **2021**, *3*, 1–14. [CrossRef]
97. Mulia, K.; Rachman, D.; Krisanti, E.A. Preparation, characterization and release profile of chitosan alginate freeze dried matrices loaded with mangostins Preparation, characterization and release profile of chitosan alginate freeze dried matrices loaded with mangostins. *J. Phys.* **2019**, *1*, 012009. [CrossRef]
98. Aisha, A.F.A.; Abu-Salah, K.M.; Ismail, Z.; Shah, A.M.; Majid, A. Determination of total xanthones in *Garcinia mangostana* fruit rind extracts by ultraviolet (UV) spectrophotometry. *J. Med. Plants Res.* **2013**, *7*, 29–35.
99. Cao, M.; Luo, X.; Wu, K.; He, X. Targeting lysosomes in human disease: From basic research to clinical applications. *Signal Transduct. Target. Ther.* **2021**, *6*, 379. [CrossRef]
100. Lee, S.; Shanti, A. Effect of exogenous ph on cell growth of breast cancer cells. *Int. J. Mol. Sci.* **2021**, *22*, 9910. [CrossRef]
101. Pérez-Herrero, E.; Fernández-Medarde, A. The reversed intra- and extracellular pH in tumors as a unified strategy to chemotherapeutic delivery using targeted nanocarriers. *Acta Pharm. Sin. B* **2021**, *11*, 2243–2264. [CrossRef]
102. Abolhasani, M.H.; Safavi, M.; Goodarzi, M.T.; Kassaee, S.M.; Azin, M. Identification and anti-cancer activity in 2D and 3D cell culture evaluation of an Iranian isolated marine microalgae. *DARU J. Pharm. Sci.* **2018**, *26*, 105–116. [CrossRef]

molecules

MDPI

Article

Chitosan Soft Matter Vesicles Loaded with Acetaminophen as Promising Systems for Modified Drug Release

Loredana Nicoleta Hilițanu [1], Liliana Mititelu-Tarțău [1,*], Eliza Grațiela Popa [2,*], Beatrice Rozalina Bucă [1,†], Irina Luciana Gurzu [3,†], Paula Alina Fotache [1], Ana-Maria Pelin [4], Daniela Angelica Pricop [5] and Liliana Lăcrămioara Pavel [6]

1 Department of Pharmacology, Faculty of Medicine, 'Grigore T. Popa' University of Medicine and Pharmacy, 700115 Iasi, Romania; ln.rusu@yahoo.com (L.N.H.); beatrice-rozalina.buca@umfiasi.ro (B.R.B.); fotachepaula@yahoo.com (P.A.F.)
2 Department of Pharmaceutical Technology, Faculty of Pharmacy, 'Grigore T. Popa' University of Medicine and Pharmacy, 700115 Iasi, Romania
3 Department of Preventive Medicine and Interdisciplinarity, Faculty of Medicine, 'Grigore T. Popa' University of Medicine and Pharmacy, 700115 Iasi, Romania; irina-luciana.gurzu@umfiasi.ro
4 Department of Pharmaceutical Sciences, Faculty of Medicine and Pharmacy, 'Dunarea de Jos' University, 800010 Galati, Romania; anapelin@gmail.com
5 Research Center with Integrated Techniques for Atmospheric Aerosol Investigation in Romania, RECENT AIR, Laboratory of Astronomy and Astrophysics, Astronomical Observatory, Physics, 'Al. I. Cuza' University, 700506 Iasi, Romania; daniela.a.pricop@gmail.com
6 Department of Morphological and Functional Sciences, Faculty of Medicine and Pharmacy, 'Dunarea de Jos' University, 800010 Galati, Romania; doctorpavel2012@yahoo.com
* Correspondence: lylytartau@yahoo.com (L.M.-T.); eliza.popa@umfiasi.ro (E.G.P.)
† These authors contributed equally to this work.

Abstract: Our study was designed to acquire, characterize and evaluate the biocompatibility of novel lipid vesicles loaded with acetaminophen (APAP) and coated with chitosan (CS). We investigated the in vitro and in vivo drug release kinetics from these systems, and we conducted assessments for both in vitro hemocompatibility and in vivo biocompatibility. For the in vivo biocompatibility evaluation, the mice were randomly divided into four groups of six animals and were treated orally as follows: control group: 0.1 mL/10 g body weight of double-distilled water; CS group: 0.1 mL/10 g body weight 1% CS solution; APAP group: 150 mg/kg body weight APAP; APAP-v group: 150 mg/kg body weight APAP-loaded lipid vesicles. The impact of APAP-v on various hematological, biochemical, and immune parameters in mice were assessed, and the harvested tissues were subjected to histopathological examination. The innovative formulations effectively encapsulating APAP within soft vesicles exhibited reasonable stability in solution and prolonged drug release in both in vitro and in vivo studies. The in vitro hemolysis test involving APAP-loaded vesicles revealed no signs of damage to red blood cells. The mice treated with APAP-v showed neither significant variances in hematological, biochemical, and immune parameters, nor structural changes in the examined organ samples, compared to the control group. APAP-v administration led to prolonged drug release. We can conclude that the APAP-v are innovative carrier systems for modifying drug release, making them promising candidates for biomedical applications.

Keywords: acetaminophen; lipid; vesicles; chitosan; biocompatibility; mice

check for updates

Citation: Hilițanu, L.N.; Mititelu-Tarțău, L.; Popa, E.G.; Bucă, B.R.; Gurzu, I.L.; Fotache, P.A.; Pelin, A.-M.; Pricop, D.A.; Pavel, L.L. Chitosan Soft Matter Vesicles Loaded with Acetaminophen as Promising Systems for Modified Drug Release. *Molecules* **2024**, *29*, 57. https://doi.org/10.3390/molecules29010057

Academic Editors: Dimitrios Bikiaris and Agnieszka Ewa Wiącek

Received: 8 November 2023
Revised: 10 December 2023
Accepted: 19 December 2023
Published: 21 December 2023

1. Introduction

APAP is a derivative of para-aminophenol (N-(4-Hydroxyphenyl)-acetamide). It possesses analgesic and antipyretic properties, but does not exhibit significant anti-inflammatory effects. It acts by indirectly inhibiting the COX-1, COX-2, and COX-3, thus raising the pain threshold and reducing painful sensations [1–3].

After oral administration, APAP exhibits favorable absorption characteristics, with a peak concentration in the bloodstream achieved approximately 90 min following ingestion. Its maximum effect is typically observed between 30 min and 2 h after oral intake, while rectal administration results in a maximum effect within 3–4 h [4]. When taken orally, the duration of its action ranges from 3 to 7 h and results in a bioavailability of 88% [5,6]. The lack of total absorption resides in its property of low permeability, being a class 3 BCS drug (high solubility/low permeability). APAP binds to plasma proteins to a varying extent, typically around 10–25% [7,8]. APAP has a half-life ranging from 1 to 4 h, and its primary route of excretion is through the urine, with approximately 5% being eliminated in its free, unconjugated form [6].

In the existing literature, there are numerous considerations concerning the development of lipid vesicles to integrate drugs, constituting a broad area of research interest. Our methodology for crafting these systems stands out as an original approach that, as of now, has not been undertaken by any research team. In our chitosan-based lipid vesicles, APAP serves as the model drug, representing an innovative strategy. The uniqueness of chitosan-based vesicles encapsulating active substances stems from the combination of chitosan's distinctive properties with the advantages inherent in vesicular drug-delivery systems [9,10]. A chitosan coating can improve the stability of lipid vesicles, protecting their cargo from degradation by enzymes or harsh environments [11]. This fusion presents opportunities for enhancing drug effectiveness, minimizing adverse effects, achieving targeted delivery, and improving patient adherence [12,13]. Researchers consistently delve into refining and optimizing these formulations to unlock their complete potential in pharmaceutical applications [14]. Hence, the studies highlighted in the paper align with the current landscape of modern research endeavors within this field.

The clinical use of APAP has demonstrated potential gastrointestinal irritation and related side effects. Encapsulating it within lipid vesicles offers a promising strategy to mitigate these adverse effects through controlled release and targeted delivery, potentially reducing direct interaction with the gastrointestinal tract. Tackling these challenges through the development of APAP-v represents a significant stride in pharmaceutical research.

Currently, there is a limited availability of modified-release acetaminophen tablets on the market, in particular, those containing 650 mg per tablet. Through the development of these vesicles, we are broadening the scope for novel extended-release acetaminophen formulations. These formulations may encompass a combined total of two therapeutic doses, aiming to address chronic moderate pain. They can be used either independently or in synergistic combinations with other analgesics.

Various methods and different active ingredients have been used to design nanosystems encapsulating acetaminophen. Using surfactant and cholesterol as ingredients, a group of researchers prepared multilamellar niosomes containing acetaminophen through the thin film hydration method [15]. Acetaminophen was also loaded into PEGylated nano graphene oxide particles via the sonication technique [16] or in core-shell biodegradable microspheres [17]. Other types of nanoparticles with acetaminophen based on poly(lactide-co-glycolide acid) were acquired using the emulsion-solvent evaporation procedure [18].

Since their discovery in the 1960s, liposomes have been studied intensively for the formulation of many dosage forms, in order to overcome some of the inconveniences posed by drugs for various administration routes. Liposomes [19] can trap hydrophobic and hydrophilic drugs, thus improving drug delivery and bioavailability, or modifying the drug-release kinetics.

In recent decades, a novel approach has surfaced involving the development of increasingly sophisticated liposomal vesicles within the realm of nanoformulation. These new drug-delivery systems are designed to target drug delivery to specific tissues and limit some of their side effects, especially liver toxicity [20,21].

Liposomes are lipid vesicles with a spherical morphology, containing a central hydrophilic core, surrounded by a phospholipidic membrane. The scientific community has shown increasing interest in liposomal systems, due to their biocompatibility, biodegradability, non-immunogenicity and lack of toxicity [22,23]; thus, multiple drug-delivery systems based on liposomes for various administration routes have been designed. However, liposomal systems have their limitations: rapid blood clearance when administered via intravenous injection, and the tendency to combine/fuse, causing drug leakage during storage [24–26]. One of the solutions to these problems is to induce surface changes (i.e., using a different agent to coat the liposomes), in order to enhance their stability, extend their life in the bloodstream and ensure modified release of the entrapped drug [27,28].

Many synthetic, semisynthetic or natural polymers (collagen, gelatin, quercetin) used for coating the liposomal surface have been investigated for their potential as drug carriers with possible use in the biomedical field. The main problem of the liposomal in vivo fate is their uptake by the reticuloendothelial system cells, which considerably lowers their bioavailability. The most implemented method to reduce this phenomenon is by coating the unilamellar lipid vesicles with hydrophilic substances (i.e., CS), which results in reduced phagocytic uptake, decreases the interaction with other proteins, and improves vesicle stability and transmucosal oral absorption [29]. It has been shown that the most effective process for the development of lipid-based drug carriers is sonication homogenization, which provides a narrow particle-size distribution, higher particle content in dispersions and avoids organic solvents. The systemic utilization of coating materials primarily offers crucial advantages such as maintaining physical stability, promoting dispersion, and ensuring the presence of colloidal particles in the bloodstream [30].

CS is a naturally occurring polysaccharide compound derived from marine crustaceans, mollusks, insects, and fungi; it has been studied intensively as a biomedical material [31–33], due to its anti-tumor, anti-ulcer, immune-stimulatory, antidiabetic, antioxidant, and antimicrobial properties [34]. It can be formulated in various dosage forms such as solutions, gels, mixtures, medicated sponges, tablets, membranes, pastes for different applications [35].

CS nanoparticles fulfill the requirements that are sought after in such systems, including effectiveness, affordability, compatibility with the body, degradability, lack of toxicity, and absence of immunogenic properties [35,36]. Also, the use of CS-coated nanoparticles has demonstrated some intrinsic properties, such as high antioxidant action, anti-hyperlipidemic and liver-protective action against fat consumption or alcohol-induced steatosis [37].

Swelling reduces liposomal stability and can cause the quick release of the enclosed agents (such as natural products, antibacterial substances, anti-inflammatory medicines, cardiovascular drugs, hormonal compounds); several studies have shown that the CS layer acts like a wall that prevents swelling, improving the stability of soft lipid vesicles [38]. As a linear polysaccharide with a positive charge, CS has the ability to create stable complexes with negatively charged compounds like phospholipids, making it a potential candidate for the entrapment and controlled delivery of active substances [33,39,40]. It is a cationic polyelectrolyte with a high proportion of primary amines, which can interact easily with various targeted surfaces, such as cell membranes or the lipid bilayer of vesicles [41]. CS is soluble in acidic aqueous solutions, in correlation with the pH value of the medium and the percentage of de-acetylation, due to the protonation of amino groups [42]. Differences in molecular mass, the degree of de-acetylation and in the structural position of residue acetyl groups on the main polymer chain can lead to different properties of CS grades [39,43].

The purpose of our study was the design, characterization and the biocompatibility evaluation of original CS-coated vesicles entrapping APAP.

2. Results and Discussion

From a physical perspective, lipid vesicles prove versatile in encapsulating both hydrophilic and lipophilic drugs, safeguarding the active substance from external factors. These concentric vesicles encase an aqueous compartment entirely enclosed by a lipid membrane, primarily composed of phosphatidylcholine and cholesterol, the latter serving as a fluidity buffer. However, one drawback observed in lipid vesicles against their usefulness is their susceptibility to instability in colloidal solutions, leading to agglomeration. To counteract this issue, we have successfully stabilized them through chitosan coating. Additionally, our findings indicate that these lipid vesicles containing a drug are leaky in nature, resulting in the rapid release of the active substance, emphasizing the necessity for the supplementary use of chitosan.

Coating lipid vesicles with hydrophilic substances like chitosan has resulted in decreased cellular uptake by the reticuloendothelial system and minimized interaction with other proteins. This coating of colloidal carriers has demonstrated enhancements in particle stability and the facilitation of oral transmucosal transport. It has been established that the most effective method for producing lipid-based drug carriers involves homogenization through sonication. This process ensures a precise particle-size distribution, increased particle content in dispersions, and eliminates the use of organic solvents.

2.1. pH Value of Solutions with APAP

It was noted that the addition of CS to the suspension with APAP-loaded vesicles led to a decrease in pH from 6.00 (for APAP solution) to 4.65 (for non-dialyzed APAP-v solution) (Table 1). Dialysis enabled both purification of the dispersion by removing the non-encapsulated drug residues and modification of the pH to values close to physiological values (pH of APAP-v solution = 6.63) (Table 1).

Table 1. pH values of the solutions containing APAP.

Solution	pH
APAP solution (APAP)	6.00
APAP vesicles with CS (prior to the dialysis process) (APAP-v nondialyzed)	4.65
Dialyzed APAP vesicles with CS (APAP-v)	6.63

2.2. Size Distribution of APAP-v

The image obtained with the Malvern Zetasizer Nano ZS ZEN-3500 device showed that for APAP-v without CS, the mean hydrodynamic distribution was situated at values of 901.2 nm (Figure 1a) and for APAP-v, the mean hydrodynamic distribution was at 603 nm (Figure 1b). The polydispersity index of 0.268 indicated a high degree of monodispersity, suggesting the presence of vesicles with fairly close dimensions, as seen in the size histogram (Figure 1a).

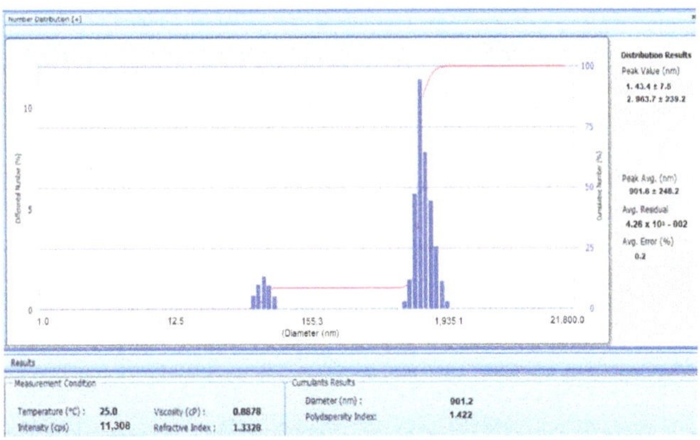

(a)

Figure 1. *Cont.*

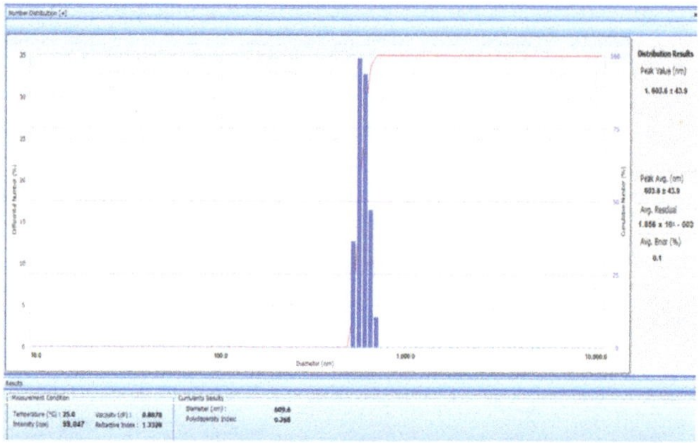

(**b**)

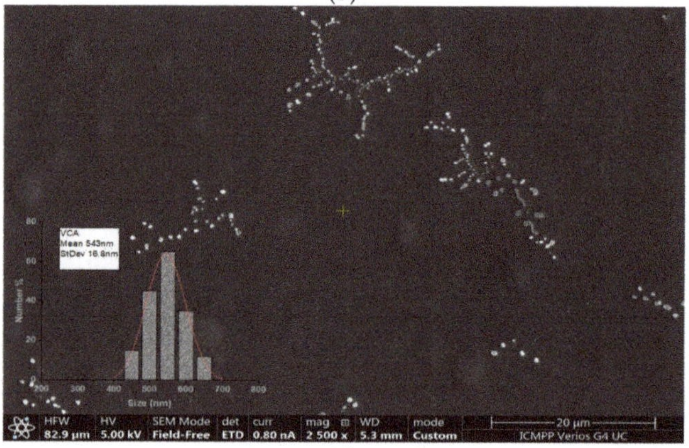

(**c**)

Figure 1. Size histogram (**a**) of APAP-v without CS and (**b**) APAP-v (**c**) SEM micrograph and size histogram of APAP-v obtained by direct measurement using Image J software, variant 1.8.0.

SEM micrographs of dialyzed APAP-v suspensions showed clusters of vesicles with relative spherical shapes and sizes ranging between 500 and 900 nm (Figure 1).

The SEM image highlights the arrangement of vesicles in clusters with branching structures of different lengths. The size histogram obtained from SEM micrograph analysis suggests the formation of vesicles larger than 543 nm (Figure 1b). It can be seen that the vesicles containing the drug were roughly round in shape and well dispersed. On the other hand, the presence of blisters that did not contain medicine, with irregular and elongated shapes could also be observed. The hydrodynamic size revealed by the diameter distribution led us to conclude that there was only one type of vesicles, with an average size of 609 nm.

2.3. Zeta Potential of APAP-v

The stability of colloidal solutions is defined according to the average zeta potential value. Zeta potential determination is a significant technique for the characterization of nanocrystals to estimate surface charge, which can be used to understand the physical stability of nanodispersions [44].

The stability of the APAP-v was assessed using a zeta potential stability model designed for positively charged spherical particles suspended within a uniform medium. According to this model, the affinity between particle surface and water generates a zone that is organized in layers of water, and this new structure highly modifies the particle–water–polymer properties. The electrostatic field on the particle surface, in combination with the thermal motion of the surrounding ions, creates opposite charges and therefore shields the electric charge on the particle surface. The actual electric charge of the diffuse shielding layer is equal in magnitude to the actual surface charge, but has the opposite charge [45].

Based on the zeta potential distribution, our conclusion is that all the vesicle systems exhibited a predominant positive charge. In the case of the APAP solution, the measured zeta potential was +18.26 ± 1.43 mV, thus corresponding to an average level of dispersion stability. The average zeta potential value of +5.61 ± 0.36 mV of APAP-v without CS indicating that the colloidal solution had a low dispersion, being at the threshold of agglomeration (Figure 2). It was observed that, by coating the lipid vesicles with CS, the zeta potential was +28.45 ± 2.75 mV (Figure 2), suggesting a moderate stability level of the colloidal dispersion. We believe that coating the APAP vesicles with CS results in an augmentation of the positive charge on the vesicle surfaces and an escalation of the repulsive forces acting between them.

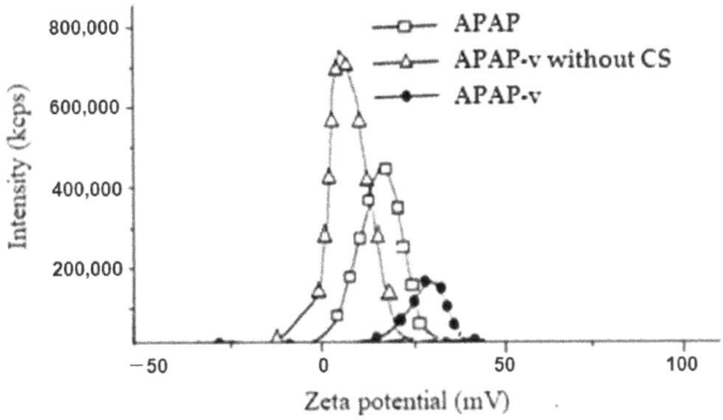

Figure 2. Zeta potential of APAP-v.

The physicochemical analysis of the obtained systems revealed that coating the APAP-loaded vesicles with a 1% CS layer enabled large vesicles of over 500 nm with very low polydispersity to be obtained, thus being quite uniform and having a spherical morphology. Following the stabilization of the lipid vesicles with CS, the zeta potential of 28.45 mV indicated a medium level of stability in the dispersion. The significant vesicle size could potentially result in reduced drug release into the tissue but the zeta potential can counterbalance this effect, ensuring effective permeability of the liposomes through the mucosal tissue. By using a neutral lipid, such as phosphatidylcholine, to obtain the vesicles, further coating them with CS led to an increase in the positive charge on the vesicle surface. Increasing the zeta potential led to higher electrostatic repulsion forces between APAP vesicles, with maintenance of dispersion stability.

The dispersions remained transparent for over six months after being prepared and stored at room temperature. This prolonged clarity might be attributable to the effective coating with 1% CS, which appears to safeguard the phospholipid membranes against oxidation when stored under various temperature conditions.

It was apparent that the new APAP-v demonstrated enhanced properties compared to APAP-v without CS, despite similar drug encapsulation efficiency. These improvements were characterized by notably smaller sizes (603.6 ± 43.9 versus 901.6 ± 248.2), a more uniform size distribution (polydispersity index 0.268 versus 1.422), and enhanced stability in the colloidal solution (zeta potential +28.45 ± 2.75 mV versus +5.61 ± 0.36 mV). An elevated polydispersity index of 1.422 suggested an inhomogeneous morphology of the APAP-v without CS. The efficiency of the release of the drug in the tissues is conditioned by a polydispersity index as low as possible, so as to avoid the agglomeration of the vesicles. The reduced size of APAP-v and its remarkably low polydispersity index supported their outstanding dispersion within the biological release medium.

2.4. The Efficacy of APAP Entrapment in Vesicles

The calibration curve obtained of APAP in aqueous solution had the mean square deviation (R^2) value of 0.999 (Figure 3). The UV spectra revealed that the APAP solution had maximum absorbance at 243 nm, while the APAP-v exhibited a maximum absorbance at 243.5 nm. Since the initial mass of APAP in solution was 16.67 mg/mL, and 3.48 g/mL of APAP was removed from the APAP-v, we assessed a 79.12% efficacy of drug entrapment in these microsystems.

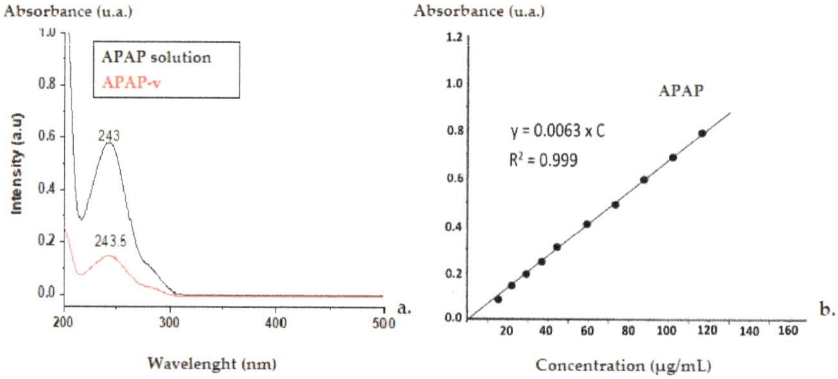

Figure 3. (**a**) UV spectra for APAP-v and APAP-solution (a.u.—absorption units) (**b**) calibration curve.

2.5. In Vitro Release Kinetics of APAP from APAP-v

Studying the release kinetics in vitro, we found a slower release of APAP from the nanovesicles stabilized with CS, in comparison to the drug release pattern from the solution.

In the dissolution testing, we noticed that 91.4% of APAP was released from the solution after 15 min, while only 1.6% was released from APAP-v (Figure 4). At 30 min, it was revealed that from the simple solution, the drug was released in a proportion of 98.5%, and after 45 min the drug was completely released (Figure 4). This is in accordance with the standards set in the compendial guidelines for the release of APAP from solid dosage forms (e.g., tablets). Almost half of the amount of APAP (49.3%) was released after two hours from the APAP-v, 97.4% after 6 h, and the total amount after 8 h (Figure 4).

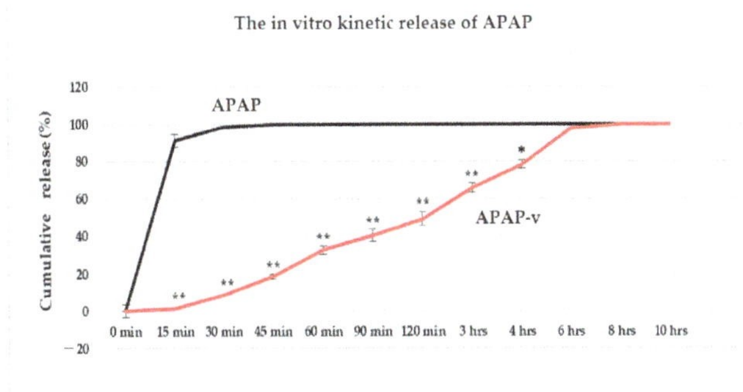

Figure 4. The in vitro kinetics profile (cumulative release percentage versus time) of APAP from APAP solution and from APAP-v, obtained using the permeation method. * $p < 0.05$, ** $p < 0.01$.

The finding that APAP was released in a slightly lower percentage from the vesicles may be ascribed to the enhanced dispersion of the drug as individual molecules tightly enclosed within the nanovesicles stabilized with CS. In our experimental setup, laboratory animals were not exposed to isolated lipid vesicles but rather to a colloidal solution in which these nanovesicles were dispersed.

2.6. In Vivo Release Kinetics of APAP from APAP-v

The highest plasma concentration of APAP following oral administration of APAP-v occurred at the two-hour mark (25.3 ± 1.17 μg/mL), in contrast to APAP-treated animals, where the maximum blood concentration (27.4 ± 1.21 μg/mL) was achieved at 30 min (Figure 5). Peak APAP levels were observed within the 15–60 min time frame, while notably elevated plasma-APAP concentrations resulting from drug release from APAP-v were seen between the two- and four-hour mark (Figure 5). This observation highlights a sustained release of the drug from the microsystems.

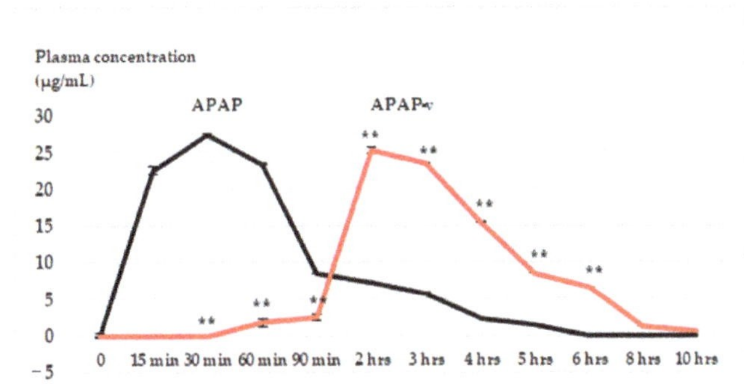

Figure 5. The in vivo kinetics profile (plasma concentration versus time) of APAP from APAP solution and from APAP-v, obtained using the HPLC method. The data are displayed as arithmetic mean ± S.D. of the average values for 6 animals per group. ** $p < 0.01$.

In the in vitro dissolution tests for both APAP and APAP-v, the outcomes are presented as the percentage concentration over a specified duration. Conversely, the in vivo release graph depicts results in plasma concentrations measured in μg/mL. To make comparisons uniform, converting these values to a percentage of the drug released allows for a direct assessment. In the in vivo assessment, at 1.5 h, the release of APAP from the vesicles measures 2.5 μg/mL in plasma, approximately equivalent to an 8% release of the drug from the vesicles. This rate of release appears significantly slower compared to the in vitro test, which demonstrates an approximate 40% release at 1.5 h.

In comparison to the in vitro release pattern of the vesicles, the in vivo kinetics exhibited a significant release after 2 h, reaching nearly 80%. This pattern resembles a gastro-resistant release model, albeit incomplete. The release of some APAP prior to 2 h can be attributed to the free fraction of APAP surrounding the vesicles. Observations at the 6 h mark on the in vivo graph indicated that approximately 20% of the drug remained, indicating a slow clearance of APAP from the body, with complete clearance occurring after 10 h. This extended duration contrasts with the in vitro test, indicating a prolonged presence of APAP in the system. Overall, even if the in vitro/in vivo graphs do not overlap, there is clearly an in vivo prolonged release of APAP-v.

A recent study revealed that utilizing a high molecular weight chitosan with elevated viscosity and a high deacetylation index, as employed in our lipid-vesicle production, leads to reduced solubility in the acidic conditions of the stomach [46]. Previous research demonstrated that a higher deacetylation index in chitosan results in increased steric hindrance, promoting the formation of a robust electrostatic complex with bile salts, thereby diminishing their efficacy [47]. The notable contrast between the brief drug-release duration of APAP and the extended latency period of APAP-v may be elucidated by the swift disintegration of lipid membranes in the gastric setting in comparison to the resistance created by chitosan with a high deacetylation index.

Conventional lipid vesicles are sensitive to damage caused by harsh chemical and enzymatic gastro-intestinal environments, resulting in reduced oral bioavailability [48]. The chitosan layer improved liposome stability in simulated gastric fluid [49], explained by the enhanced interactions between the chitosan and liposome surface under low pH in simulated gastric fluid (pH 1.2) due to the amino groups protonation in chitosan (pKa 6.5). The molecular configuration of chitosan also became more expanded, leading to a stronger affinity for the liposome surface [39]. The sustained release of encapsulated drug results from drug diffusion from the lipid bilayer and the adhesive chitosan layer [50]. Liposomes coated with chitosan released the drug in a slower manner, with this effect being attributed to the existence of the chitosan layer, which delayed the drug diffusion into the medium.

2.7. *In Vitro Hemocompatibility*

When the erythrocyte suspension was exposed to Triton X-100, a significant breakdown of the cell membrane occurred, leading to a notable level of hemolysis (88.45 ± 4.17%) (Table 2).

Table 2. In vitro hemocompatibility of CS, APAP, APAP-v. Data are displayed as arithmetic mean ± S.D. of the mean values for six animals in a group. ** $p < 0.01$ statistically significant when compared to the negative control group.

Group	Triton X-100	Negative Control	CS	APAP	APAP-v
Hemolysis %	88.45 ± 4.17 **	0.07 ± 0.01	1.64 ± 0.15	2.38 ± 0.25	2.52 ± 0.21

This difference was statistically significant (** $p < 0.01$) when compared to the negative control group (0.07 ± 0.01%). Following the incubation of the erythrocyte suspension with the CS solution, only a slight degree of hemolysis was observed (1.64 ± 0.15%) (Table 2). This reduction was not substantially different from the saline solution group, indicating favorable in vitro hemocompatibility. Upon exposure to APAP, the red blood cells exhibited minor hemolysis (2.38 ± 0.25%), nonsignificant compared to the negative control group, indicating no harmful effect of APAP on the erythrocytes (Table 2). Incubating the erythrocyte suspension with APAP-v resulted in negligible hemolysis (2.52 ± 0.21%), without notable differences from the saline group (Table 2). This finding indicates that APAP entrapped in APAP-v did not induce the destruction of these blood elements.

Considering the literature-reported data, a hemolysis rate below 5% is regarded as acceptable, while values below 2% are deemed biologically insignificant. This implies that the administered substance can be safely used in vivo [51,52]. Scientific investigations have indicated that using CS with a high degree of deacetylation (exceeding 80%) is associated with a reduced risk of erythrocyte hemolysis [53]. Furthermore, when CS possesses a high molecular mass, its chains have low mobility, which greatly reduces binding to red blood cells [54,55]. Since CS is administered in a colloidal solution form, its impact on red blood cells is minimal. The finding that APAP-v exhibited low hemolysis suggests that these nanovesicles are compatible with blood and do not present a systemic toxicity concern.

2.8. *In Vivo Biocompatibility*

Following the preparation of APAP-v, we assessed their biocompatibility by examining specific serum parameters. In vivo studies serve as initial investigations that can illustrate the impact of administering these novel systems on plasma parameters. These indicators are pertinent for estimating the potential presence of liver inflammation or kidney dysfunction, oxidative processes, and any eventual structural changes in organs.

2.8.1. Hematological Tests

No statistically significant variation in RBC counts were detected among the groups of mice receiving CS, APAP, or APAP-v, when compared to the reference batch with distilled water at 24 h and at one week during the laboratory tests (Figure 6). Similarly, no momentous fluctuations in blood Hb were found among the batches studied in comparison to those receiving the control substance during the evaluation (Figure 6). Additionally, no noteworthy changes in Ht values were observed between the groups with entrapped and non-entrapped APAP; respectively, the group with polymer solution and the control lot, at the two time points designated for measurement (Figure 6).

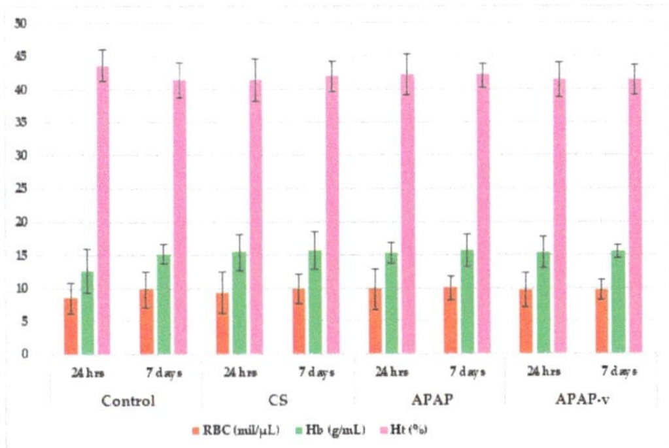

Figure 6. Hematological parameters: red blood cells (RBC), the hemoglobin (Hb) and hematocrit (Ht) values in CS, APAP, APAP-v groups. The data are displayed as arithmetic mean ± S.D. of the average values for 6 animals per group.

The use of the polymer solution did not cause any statistically relevant deviation of the differential leukocyte count compared to the animals that received distilled water (Table 3). Likewise, there were no major variations observed in the white blood cells between blood samples from mice treated with APAP and APAP-v and those from the control batch (Table 3).

Table 3. The white blood count in CS, APAP, APAP-v groups. The data are displayed as arithmetic mean ± S.D. of the average values for 6 animals per group.

Groups.	Time Elapsed	White Blood Count %				
		PMN	Ly	E	M	B
Control	24 h	19.57 ± 3.23	78.20 ± 4.59	0.35 ± 0.17	1.67 ± 1.13	0.21 ± 0.03
	7 days	19.96 ± 2.59	77.31 ± 3.72	0.41 ± 0.26	2.11 ± 1.42	0.21 ± 0.07
CS	24 h	18.35 ± 2.57	79.45 ± 4.42	0.45 ± 0.35	1.55 ± 1.59	0.20 ± 0.05
	7 days	19.17 ± 1.71	78.41 ± 1.82	0.53 ± 0.42	1.68 ± 1.67	0.21 ± 0.07
APAP	24 h	18.70 ± 3.03	78.47 ± 2.87	0.55 ± 0.23	2.07 ± 1.03	0.21 ± 0.11
	7 days	19.67 ± 2.01	77.43 ± 2.23	0.47 ± 0.59	2.15 ± 1.26	0.22 ± 0.13
APAP-v	24 h	20.84 ± 3.81	75.96 ± 3.71	0.58 ± 0.87	2.41 ± 0.42	0.21 ± 0.07
	7 days	21.42 ± 4.07	76.10 ± 3.23	0.57 ± 0.76	2.35 ± 0.70	0.20 ± 0.18

Determining the activity of certain liver enzymes helps to diagnose and monitor liver disease or damage. The tests measure the levels of certain enzymes and proteins in the blood. These liver function tests quantify the activity of specific enzymes that liver cells release in response to injury or disease [56].

The administration of CS and APAP did not exhibit any statistically significant alterations in ALT and AST values when compared to the group receiving distilled water after 7 days (Figure 7). Furthermore, the hematological assessments did not show noteworthy deviations in ALT and AST values in the groups that were administered APAP-v versus the control group at any of the time points during the evaluations (Figure 7).

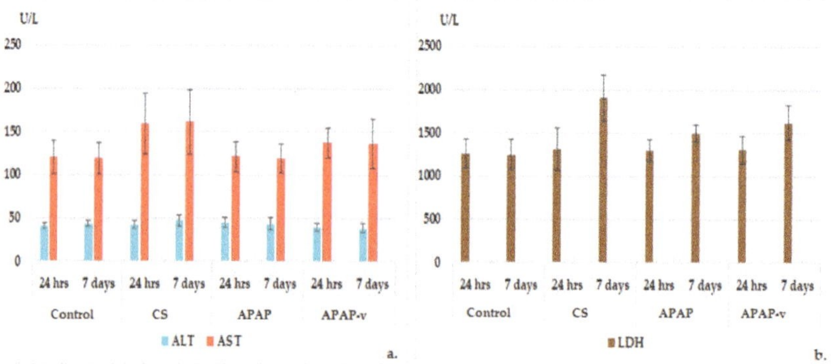

Figure 7. Liver enzymes: ALT, AST (**a**) and LDH (**b**) profile in CS, APAP, APAP-v groups. The data are displayed as arithmetic mean ± S.D. of the average values for 6 animals per group.

These observations indicate that the use of these substances does not impact liver function.

The measurement of blood creatinine and urea levels was performed to assess kidney function because their values reflect the glomerular filtration rate; a parameter that defines kidney function. Regardless of its cause, kidney disease is associated with decreased glomerular filtration rate, and the severity of kidney disease correlates closely but indirectly with glomerular filtration rate [57]. There were no significant alterations observed in the plasma urea and creatinine profile in the animals treated with polymer and APAP in comparison to those receiving distilled water, either at 24 h or 7 days, in the experiment (Figure 8). Similarly, the use of APAP-v did not exhibit any apparent changes in blood urea

0.2 mL of extraction medium from each sample was placed in a water bath at 37 °C for one hour. Subsequently, each tube underwent centrifugation for 10 min at 1000 RCF (relative centrifugal force). The absorbance of the resulting supernatant was measured at a wavelength of 545 nm using a Hewlett Packard 8453 UV–VIS spectrophotometer (Waldbronn, Germany).

The hemolysis ratio (hemolysis %) was determined as opposed to a completely hemolyzed solution, employing the following formula [71]:

$$\text{Hemolysis (\%)} = (\text{Tested substance absorbance} - \text{Negative control absorbance}) \times 100/(\text{Positive control absorbance} - \text{Negative control absorbance})$$

3.7. *Evaluation of the In Vivo Biocompatibility of Vesicles Entrapping APAP*

To assess the biocompatibility of substances in the nano- or micro-range, in vivo testing involved evaluating the biochemical and hematological profiles of animals exposed to the test substances, including double-distilled water, APAP, and APAP-v-.

Before the experiment, the mice were placed on an elevated wire mesh within a transparent plastic box and allowed to acclimate to the test chamber for a period of 2 h. The testing was conducted between 8 and 12 a.m.

The mice were acclimatized to the laboratory environment for 3 days and were used for only one procedure to ensure consistent conditions. Random assignment was performed, resulting in four groups of six animals each. The animals were administered a single oral dose of substances following the specified treatment regimen:

Group 1 (control): double-distilled water 0.1 mL/10 g body weight;

Group 2 (CS): 0.1 mL/10 body weight CS;

Group 3 (APAP): 150 mg/kg body weight APAP;

Group 4 (APAP-v): CS-based vesicles incorporating APAP, 150 mg/kg body weight;

In the experiment, hematological and biochemical assessments were conducted at two specific time points: 24 h and 7 days following the administration of the substances. Blood samples (0.3 mL) were collected from one of the lateral caudal veins of the mice for analysis. This collection procedure, performed under general anesthesia using 5% isoflurane, is swift (lasting 10–15 s) and does not cause stress to the animals [71,72].

The evaluation of the biocompatibility of APAP-v focused on investigating their impact on various parameters. These included the assessment of full blood count, serum biochemical markers such as liver enzyme activity (alanine aminotransferase—ALT, aspartate aminotransferase—AST, lactate dehydrogenase—LDH), as well as serum urea and creatinine levels. Additionally, certain immunological parameters were examined, namely the phagocytic capacity of peripheral blood polymorphonuclear neutrophils (PC), serum opsonic capacity (OC), and bactericidal capacity of peritoneal macrophages (BC). For the biochemical assessments, fasting venous blood samples collected on heparin were used. The determination of full blood count, as well as the levels of GTP, GOT, and LDH, was performed using a Hematology Analyzer 5 DIFF model BF-5180 (DIRUI, Istanbul, Turkey). To evaluate the PC and BC capacity of peritoneal macrophages, the animals were sacrificed under anesthesia with 5% isoflurane. Peritoneal macrophages were obtained through peritoneal lavage using HANKS solution maintained at a temperature of 37 °C. Thereafter, the samples were centrifuged at 1000 rpm for 10 min. Subsequently, they were exposed to various cultures of *Staphylococcus aureus 94* in a 0.2% glucose broth solution, diluted 1:1000 with saline solution. The specimens were cultured for a 48 h period at a temperature of 37 °C. Afterwards, they were re-plated onto culture media, and subsequent identification of colony formation on plates was conducted.

The animals were sacrificed and liver and kidney fragments were harvested for histopathological evaluation. These tissue specimens were preserved in a 10% formalin solution, subsequently embedded in paraffin, sectioned at 5 μm intervals, and stained with hematoxylin-eosin (H&E). The microscopic analysis of the slides was conducted using a Nikon TI Eclipse optical microscope (Tokyo, Japan), and the images were captured using a

Nikon Coolpix 950 digital camera with a resolution of 1600 × 1200 (1.92 Mpx) and optical zoom of x3.

3.8. Data Analysis

The results are reported as the mean values ± standard deviation (S.D.) of the mean. Data analysis and statistical processing were conducted using the SPSS 17.0 software, using the ANOVA one-way method for multivariable data analysis. To assess the significance of differences between groups, Newman–Keuls and Tukey post hoc tests were used for multiple comparisons. Statistical significance was indicated for p (probability) values below 0.05 compared to the control group.

4. Conclusions

We developed a new approach to create CS-coated lipid vesicles loaded with APAP and thoroughly characterized their physicochemical and structural properties. We successfully obtained unique lipid vesicles coated with CS, capable of efficiently encapsulating APAP, while maintaining moderate stability in colloidal dispersion and displaying a prolonged drug in vitro and in vivo release. Initial assessments, including a blood hemolysis assay, indicated promising in vitro biocompatibility of the APAP-v.

The APAP-v were shown to be biocompatible, showing no significant disturbances in hematological, biochemical and immune parameters, after oral administration in mice. Additionally, the histological assessment revealed no notable changes in the configuration of liver and kidney tissues in mice treated with APAP-v, as compared to control animals. These investigations suggest that APAP-v are promising in vivo drug delivery vehicles, holding great potential for future medical applications.

Author Contributions: Conceptualization and methodology, L.M.-T., E.G.P., L.L.P. and L.N.H.; data analysis and investigation, L.N.H., L.L.P., B.R.B., A.-M.P., D.A.P., I.L.G., P.A.F. and L.M.-T.; writing—original draft preparation, L.M.-T., L.N.H., E.G.P., L.L.P., I.L.G., B.R.B., A.-M.P. and D.A.P.; writing—review and editing, L.M.-T., L.N.H. and E.G.P., supervision and project administration, L.M.-T., L.L.P., E.G.P., I.L.G. and L.N.H. All authors have read and agreed to the published version of the manuscript.

Funding: This work is part of Loredana Nicoleta Hilitanu's PhD studies and was financially supported by the Doctoral School of "Grigore T. Popa" University of Medicine and Pharmacy, Iasi, Romania (Doctoral Grant financing approval No. 05/20 January 2021) and the European Social Fund—the Human Capital Operational Programme, Project/Grant No: POCU/993/6/13/154722.

Institutional Review Board Statement: The protocol of the experimental research was approved by the Ethics Committee from the University of Medicine and Pharmacy "Grigore T. Popa" Iasi, Romania (Certificate No. 24/14 July 2020).

Informed Consent Statement: Not applicable.

Data Availability Statement: Data are contained within the article.

Conflicts of Interest: The authors declare no conflict of interest.

References

1. Mallet, C.; Desmeules, J.; Pegahi, R.; Eschalier, A. An updated review on the metabolite (AM404)-mediated central mechanism of action of paracetamol (acetaminophen): Experimental evidence and potential clinical impact. *J. Pain Res.* **2023**, *16*, 1081–1094. [CrossRef] [PubMed]
2. Przybyła, G.W.; Szychowski, K.A.; Gmiński, J. Paracetamol—An old drug with new mechanisms of action. *Clin. Exp. Pharmacol. Physiol.* **2020**, *4891*, 3–19. [CrossRef] [PubMed]
3. Esh, C.J.; Chrismas, B.C.R.; Mauger, A.R.; Taylor, L. Pharmacological hypotheses: Is acetaminophen selective in its cyclooxygenase inhibition? *Pharmacol. Res. Perspect.* **2021**, *9*, e00835. [CrossRef] [PubMed]
4. Franco, F.W.; Malonn, M.C. acetaminophen pharmacokinetic and toxicological aspects: A review. *Rev. Científica Multidiscip. Núcleo Do Conhecimento* **2021**, *12*, 18–27. [CrossRef]

5. Graham, G.G.; Davies, M.J.; Day, R.O.; Mohamudally, A.; Scott, K.F. The modern pharmacology of paracetamol: Therapeutic actions, mechanism of action, metabolism, toxicity and recent pharmacological findings. *Inflammopharmacology* **2013**, *21*, 201–232. [CrossRef] [PubMed]
6. McGill, M.R.; Jaeschke, H. Metabolism and disposition of acetaminophen: Recent advances in relation to hepatotoxicity and diagnosis. *Pharm. Res.* **2013**, *30*, 2174–2187. [CrossRef] [PubMed]
7. Morse, J.D.; Stanescu, I.; Atkinson, H.C.; Anderson, B.J. Population pharmacokinetic modelling of acetaminophen and ibuprofen: The influence of body composition, formulation and feeding in healthy adult volunteers. *Eur. J. Drug Metab. Pharmacokinet.* **2022**, *47*, 497–507. [CrossRef] [PubMed]
8. Mazaleuskaya, L.L.; Sangkuhl, K.; Thorn, C.F.; FitzGerald, G.A.; Altman, R.B.; Klein, T.E. PharmGKB summary: Pathways of acetaminophen metabolism at the therapeutic versus toxic doses. *Pharmacogenet. Genom.* **2015**, *25*, 416–426. [CrossRef]
9. Jafernik, K.; Ładniak, A.; Blicharska, E.; Czarnek, K.; Ekiert, H.; Wiącek, A.E.; Szopa, A. Chitosan-based nanoparticles as effective drug delivery systems—A review. *Molecules* **2023**, *28*, 1963. [CrossRef]
10. Herdiana, Y.; Wathoni, N.; Shamsuddin, S.; Muchtaridi, M. Drug release study of the chitosan-based nanoparticles. *Heliyon* **2022**, *8*, e08674. [CrossRef]
11. Kumar, R.; Sirvi, A.; Kaur, S.; Samal, S.K.; Roy, S.; Sangamwar, A.T. Polymeric micelles based on amphiphilic oleic acid modified carboxymethyl chitosan for oral drug delivery of bcs class iv compound: Intestinal permeability and pharmacokinetic evaluation. *Eur. J. Pharm. Sci.* **2020**, *153*, 105466. [CrossRef] [PubMed]
12. Shoueir, K.R.; El-Desouky, N.; Rashad, M.M.; Ahmed, M.K.; Janowska, I.; El-Kemary, M. Chitosan based-nanoparticles and nanocapsules: Overview, physicochemical features, applications of a nanofibrous scaffold, and bioprinting. *Int. J. Biol. Macromol.* **2021**, *167*, 1176–1197. [CrossRef] [PubMed]
13. Seidi, F.; Khodadadi Yazdi, M.; Jouyandeh, M.; Dominic, M.; Naeim, H.; Nezhad, M.N.; Bagheri, B.; Habibzadeh, S.; Zarrintaj, P.; Saeb, M.R.; et al. Chitosan-based blends for biomedical applications. *Int. J. Biol. Macromol.* **2021**, *183*, 1818–1850. [CrossRef] [PubMed]
14. Sangnim, T.; Dheer, D.; Jangra, N.; Huanbutta, K.; Puri, V.; Sharma, A. Chitosan in oral drug delivery formulations: A review. *Pharmaceutics* **2023**, *15*, 2361. [CrossRef] [PubMed]
15. Barange, S.; Singh, S.; Ahmad, S.; Ahmad, S.W.; Manjhi, S.K.; Thakur, A.S.; Rathi, J.C.; Sharma, R. Preparation and optimization of paracetamol loaded niosomes. *JETIR* **2022**, *9*, d452–d464, eISSN: 2349-5162.
16. Shariare, M.H.; Masum, A.-A.; Alshehri, S.; Alanazi, F.K.; Uddin, J.; Kazi, M. Preparation and optimization of pegylated nano graphene oxide-based delivery system for drugs with different molecular structures using design of experiment (DoE). *Molecules* **2021**, *26*, 1457. [CrossRef] [PubMed]
17. Wang, W.; Zhou, S.; Sun, L.; Huang, C. Controlled delivery of paracetamol and protein at different stages from core–shell biodegradable microspheres. *Carbohydr. Polym.* **2010**, *79*, 437–444. [CrossRef]
18. Al-Nemrawi, N.K.; Dave, R.H. Formulation and characterization of acetaminophen nanoparticles in orally disintegrating films. *Drug Deliv.* **2016**, *23*, 540–549. [CrossRef]
19. Hallaj-Nezhadi, S.; Hassan, M. Nanoliposome-based antibacterial drug delivery. *Drug Deliv.* **2015**, *22*, 581–589. [CrossRef]
20. Gomez, G.A.; Hosseinidoust, Z. Liposomes for antibiotic encapsulation and delivery. *ACS Infect. Dis.* **2020**, *6*, 896–908. [CrossRef]
21. Yusuf, A.; Almotairy, A.R.Z.; Henidi, H.; Alshehri, O.Y.; Aldughaim, M.S. Nanoparticles as drug delivery systems: A review of the implication of nanoparticles' physicochemical properties on responses in biological systems. *Polymers* **2023**, *15*, 1596. [CrossRef] [PubMed]
22. Liu, P.; Chen, G.; Zhang, J. A review of liposomes as a drug delivery system: Current status of approved products, regulatory environments, and future perspectives. *Molecules* **2022**, *27*, 1372. [CrossRef]
23. Inglut, C.T.; Sorrin, A.J.; Kuruppu, T.; Vig, S.; Cicalo, J.; Ahmad, H.; Huang, H.-C. Immunological and toxicological considerations for the design of liposomes. *Nanomaterials* **2020**, *10*, 190. [CrossRef]
24. Barba, A.A.; Bochicchio, S.; Bertoncin, P.; Lamberti, G.; Dalmoro, A. Coating of nanolipid structures by a novel simil-microfluidic technique: Experimental and theoretical approaches. *Coatings* **2019**, *9*, 491. [CrossRef]
25. Sercombe, L.; Veerati, T.; Moheimani, F.; Wu, S.Y.; Sood, A.K.; Hua, S. Advances and challenges of liposome assisted drug delivery. *Front. Pharmacol.* **2015**, *6*, 286. [CrossRef]
26. Li, M.; Du, C.; Guo, N.; Teng, Y.; Meng, X.; Sun, H.; Li, S.; Yu, P.; Galons, H. Composition design and medical application of liposomes. *Eur. J. Med. Chem.* **2019**, *164*, 640–653. [CrossRef]
27. Daraee, H.; Etemadi, A.; Kouhi, M.; Alimirzalu, S.; Akbarzadeh, A. Application of liposomes in medicine and drug delivery. *Artif. Cells Nanomed. Biotechnol.* **2016**, *44*, 381–391. [CrossRef] [PubMed]
28. Basu, B.; Prajapati, B.; Dutta, A.; Paliwal, H. Medical application of liposomes. *J. Explor. Res. Pharmacol.* **2023**, *8*, 1–10. [CrossRef]
29. Garcia-Fuentes, M.; Prego, C.; Torres, D.; Alonso, M.J. A comparative study of the potential of solid triglyceride nanostructures coated with chitosan or poly(ethylene glycol) as carriers for oral calcitonin delivery. *Eur. J. Pharm. Sci.* **2005**, *25*, 133–143. [CrossRef]
30. Mohite, P.; Singh, S.; Pawar, A.; Sangale, A.; Prajapati, B.G. Lipid-based oral formulation in capsules to improve the delivery of poorly water-soluble drugs. *Front. Drug Deliv.* **2023**, 3, 1232012. [CrossRef]

31. Sharifi-Rad, J.; Quispe, C.; Butnariu, M.; Rotariu, L.S.; Sytar, O.; Sestito, S.; Rapposelli, S.; Akram, M.; Iqbal, M.; Krishna, A.; et al. Chitosan nanoparticles as a promising tool in nanomedicine with particular emphasis on oncological treatment. *Cancer Cell Int.* **2021**, *21*, 318. [CrossRef]
32. Kukushkina, E.A.; Hossain, S.I.; Sportelli, M.C.; Ditaranto, N.; Picca, R.A.; Cioffi, N. Ag-based synergistic antimicrobial composites. A critical review. *Nanomaterials* **2021**, *11*, 1687. [CrossRef]
33. Pathak, K.; Misra, S.K.; Sehgal, A.; Singh, S.; Bungau, S.; Najda, A.; Gruszecki, R.; Behl, T. Biomedical applications of quaternized chitosan. *Polymers* **2021**, *13*, 2514. [CrossRef]
34. Confederat, L.G.; Tuchilus, C.G.; Dragan, M.; Sha'at, M.; Dragostin, O.M. Preparation and antimicrobial activity of chitsan and its derivatives: A concise review. *Molecules* **2021**, *26*, 3694. [CrossRef]
35. Boroumand, H.; Badie, F.; Mazaheri, S.; Seyedi, Z.S.; Nahand, J.S.; Nejati, M.; Baghi, H.B.; Abbasi-Kolli, M.; Badehnoosh, B.; Ghandali, M.; et al. Chitosan-based nanoparticles against viral infections. *Front. Cell. Infect. Microbiol.* **2021**, *11*, 643953. [CrossRef]
36. Herdiana, Y.; Wathoni, N.; Shamsuddin, S.; Joni, I.M.; Muchtaridi, M. Chitosan-based nanoparticles of targeted drug delivery system in breast cancer treatment. *Polymers* **2021**, *13*, 1717. [CrossRef]
37. Abd-Elhakeem, A.; Radwan, M.S.; Rady, A.H. Chitosan nanoparticles as hepato-protective agent against alcohol and fatty diet stress in rats. *J. Biochem. Int.* **2017**, *4*, 5–10. Available online: https://www.ikprress.org/index.php/JOBI/article/view/3368 (accessed on 23 October 2023).
38. Midekessa, G.; Godakumara, K.; Ord, J.; Viil, J.; Lättekivi, F.; Dissanayake, K.; Kopanchuk, S.; Rinken, A.; Andronowska, A.; Bhattacharjee, S.; et al. Zeta potential of extracellular vesicles: Toward understanding the attributes that determine colloidal stability. *ACS Omega* **2020**, *5*, 16701–16710. [CrossRef]
39. Sebaaly, C.; Trifan, A.; Sieniawska, E.; Greige-Gerges, H. Chitosan-coating effect on the characteristics of liposomes: A focus on bioactive compounds and essential oils: A review. *Processes* **2021**, *9*, 445. [CrossRef]
40. Ladiè, R.; Cosentino, C.; Tagliaro, I.; Antonini, C.; Bianchini, G.; Bertini, S. Supramolecular structuring of hyaluronan-lactose-modified Chitosan matrix: Towards high-performance biopolymers with excellent biodegradation. *Biomolecules* **2021**, *11*, 389. [CrossRef]
41. Harugade, A.; Sherje, A.P.; Pethe, A. Chitosan: A review on properties, biological activities and recent progress in biomedical applications. *React. Funct. Polym.* **2023**, *191*, 105634. [CrossRef]
42. Yan, L.; Crayton, S.H.; Thawani, J.P.; Amirshaghaghi, A.; Tsourkas, A.; Cheng, Z. A pH-responsive drug-delivery platform based on glycol chitosan-coated liposomes. *Small* **2015**, *11*, 4870–4874. [CrossRef] [PubMed]
43. Rusu, A.G.; Chiriac, A.P.; Nita, L.E.; Rosca, I.; Pinteala, M.; Mititelu-Tartau, L. Chitosan derivatives in macromolecular co-assembly nanogels with potential for biomedical applications. *Biomacromolecules* **2020**, *21*, 4231–4243. [CrossRef] [PubMed]
44. Lunardi, C.N.; Gomes, A.J.; Rocha, F.S.; De Tommaso, J.; Patience, G.S. Experimental methods in chemical engineering: Zeta potential. *Can. J. Chem. Eng.* **2021**, *99*, 627–639. [CrossRef]
45. Taqvi, S.; Bassioni, G. Understanding wettability through zeta potential measurements [Internet]. In *Wettability and Interfacial Phenomena—Implications for Material Processing*; IntechOpen: London, UK, 2019. [CrossRef]
46. Carrasco-Sandoval, J.; Aranda, M.; Henríquez-Aedo, K.; Fernández, M.; López-Rubio, A.; Fabra, M.J. Impact of molecular weight and deacetylation degree of chitosan on the bioaccessibility of quercetin encapsulated in alginate/chitosan-coated zein nanoparticles. *Int. J. Biol. Macromol.* **2023**, *242 Pt 2*, 124876. [CrossRef] [PubMed]
47. Silva, H.D.; Beldíkova, E.; Poejo, J.; Abrunhosa, L.; Serra, A.T.; Duarte, C.M.M.; Br'anyik, T.; Cerqueira, M.A.; Pinheiro, A.C.; Vicente, A.A. Evaluating the effect of chitosan layer on bioaccessibility and cellular uptake of curcumin nanoemulsions. *J. Food Eng.* **2019**, *243*, 89–100. [CrossRef]
48. Nguyen, T.X.; Huang, L.; Liu, L.; Elamin Abdalla, A.M.; Gauthier, M.; Yang, G. Chitosan-coated nano-liposomes for the oral delivery of berberine hydrochloride. *J. Mater. Chem. B* **2014**, *2*, 7149–7159. [CrossRef] [PubMed]
49. Chen, M.X.; Li, B.K.; Yin, D.K.; Liang, J.; Li, S.S.; Peng, D.Y. Layer-by-layer assembly of chitosan stabilized multilayered liposomes for paclitaxel delivery. *Carbohydr. Polym.* **2014**, *111*, 298–304. [CrossRef]
50. Alshraim, M.O.; Sangi, S.; Harisa, G.I.; Alomrani, A.H.; Yusuf, O.; Badran, M.M. Chitosan-Coated Flexible Liposomes Magnify the Anticancer Activity and Bioavailability of Docetaxel: Impact on Composition. *Molecules* **2019**, *24*, 250. [CrossRef]
51. Goyal, T.; Schmotzer, C.L. Validation of hemolysis index thresholds optimizes detection of clinically significant hemolysis. *Am. J. Clin. Pathol.* **2015**, *143*, 579–583. [CrossRef]
52. de Lima, J.M.; Sarmento, R.R.; de Souza, J.R.; Brayner, F.A.; Feitosa, A.P.; Padilha, R.; Alves, L.C.; Porto, I.J.; Batista, R.F.; de Oliveira, J.E.; et al. Evaluation of hemagglutination activity of chitosan nanoparticles using human erythrocytes. *BioMed Res. Int.* **2015**, *2015*, 247965. [CrossRef]
53. Jesus, S.; Marques, A.P.; Duarte, A.; Soares, E.; Costa, J.P.; Colaço, M.; Schmutz, M.; Som, C.; Borchard, G.; Wick, P.; et al. Chitosan nanoparticles: Shedding light on immunotoxicity and hemocompatibility. *Front. Bioeng. Biotechnol.* **2020**, *8*, 100. [CrossRef] [PubMed]
54. Guo, X.; Sun, T.; Zhong, R.; Ma, L.; You, C.; Tian, M.; Li, H.; Wang, C. Effects of chitosan oligosaccharides on human blood components. *Front. Pharmacol.* **2018**, *9*, 1412. [CrossRef] [PubMed]
55. Fan, P.; Zeng, Y.; Zaldivar-Silva, D.; Agüero, L.; Wang, S. Chitosan-based hemostatic hydrogels: The concept, mechanism, application, and prospects. *Molecules* **2023**, *28*, 1473. [CrossRef] [PubMed]

56. Lala, V.; Zubair, M.; Minter, D.A. Liver function tests. [Updated 2023 Jul 30]. In *StatPearls [Internet]*; StatPearls Publishing: Treasure Island, FL, USA, 2023. Available online: https://www.ncbi.nlm.nih.gov/books/NBK482489/ (accessed on 17 October 2023).
57. Gounden, V.; Bhatt, H.; Jialal, I. Renal function tests. [Updated 2023 Jul 17]. In *StatPearls [Internet]*; StatPearls Publishing: Treasure Island, FL, USA, 2023. Available online: https://www.ncbi.nlm.nih.gov/books/NBK507821/ (accessed on 25 October 2023).
58. Chaplin, D.D. Overview of the immune response. *J. Allergy Clin. Immunol.* **2010**, *125* (Suppl. S2), S3–S23. [CrossRef] [PubMed]
59. Marshall, J.S.; Warrington, R.; Watson, W.; Kim, H.L. An introduction to immunology and immunopathology. *Allergy Asthma Clin. Immunol.* **2018**, *14* (Suppl. S2), 49. [CrossRef] [PubMed]
60. Legea nr. 206/27 mai 2004. Available online: http://legislatie.just.ro/Public/DetaliiDocument/52457 (accessed on 29 October 2023).
61. European Union. *DIRECTIVE 2010/63/EU of the European Parliament and of the Council of 22 September 2010 on the Protection of Animals Used for Scientific Purposes*; European Union: Brussels, Belgium, 2010; Available online: https://eur-lex.europa.eu/LexUriServ/LexUriServ.do?uri=OJ:L:2010:276:0033:0079:en:PDF (accessed on 23 October 2023).
62. Träger, J.; Meister, A.; Hause, G.; Harauz, G.; Hinderberger, D. Shaping membrane interfaces in lipid vesicles mimicking the cytoplasmic leaflet of myelin through variation of cholesterol and myelin basic protein contents. *Biochim. Biophys. Acta Biomembr.* **2023**, *1865*, 184179. [CrossRef]
63. Disalvo, A.; Frias, M.A. Surface characterization of lipid biomimetic systems. *Membranes* **2021**, *11*, 821. [CrossRef]
64. Nsairat, H.; Khater, D.; Sayed, U.; Odeh, F.; Al Bawab, A.; Alshaer, W. Liposomes: Structure, composition, types, and clinical applications. *Heliyon* **2022**, *8*, e09394. [CrossRef]
65. Guillot, A.J.; Martínez-Navarrete, M.; Garrigues, T.M.; Melero, A. Skin drug delivery using lipid vesicles: A starting guideline for their development. *J. Control. Release* **2023**, *355*, 624–654. [CrossRef]
66. Le, Q.V.; Lee, J.; Lee, H.; Shim, G.; Oh, Y.K. Cell membrane-derived vesicles for delivery of therapeutic agents. *Acta Pharm. Sin. B.* **2021**, *11*, 2096–2113. [CrossRef] [PubMed]
67. Lombardo, D.; Kiselev, M.A. Methods of liposomes preparation: Formation and control factors of versatile nanocarriers for biomedical and nanomedicine application. *Pharmaceutics* **2022**, *14*, 543. [CrossRef] [PubMed]
68. Tartau, L.; Cazacu, A.; Melnig, V. Ketoprofen-liposomes formulation for clinical therapy. *J. Mater. Sci. Mater. Med.* **2012**, *23*, 2499–2507. [CrossRef] [PubMed]
69. Pauna, A.R.; Mititelu-Tartau, L.; Bogdan, M.; Meca, A.D.; Popa, G.E.; Pelin, A.M.; Drochioi, C.I.; Pricop, D.A.; Pavel, L.L. Synthesis, characterization and biocompatibility evaluation of novel chitosan lipid micro-systems for modified release of diclofenac sodium. *Biomedicines* **2023**, *11*, 453. [CrossRef] [PubMed]
70. Sæbø, I.P.; Bjørås, M.; Franzyk, H.; Helgesen, E.; Booth, J.A. Optimization of the hemolysis assay for the assessment of cytotoxicity. *Int. J. Mol. Sci.* **2023**, *24*, 2914. [CrossRef]
71. Parasuraman, S.; Raveendran, R.; Kesavan, R. Blood sample collection in small laboratory animals. *J. Pharmacol. Pharmacother.* **2010**, *1*, 87–93. [CrossRef]
72. Marin, N.; Moragon, A.; Gil, D.; Garcia-Garcia, F.; Bisbal, V. Acclimation and blood sampling: Effects on stress markers in C57Bl/6J mice. *Animals* **2023**, *13*, 2816. [CrossRef]

molecules

Article

Characteristics of Hybrid Bioglass-Chitosan Coatings on the Plasma Activated PEEK Polymer

Kacper Przykaza [1,2,*], Małgorzata Jurak [1], Grzegorz Kalisz [3], Robert Mroczka [4] and Agnieszka Ewa Wiącek [1]

1 Department of Interfacial Phenomena, Institute of Chemical Sciences, Faculty of Chemistry, Maria Curie-Sklodowska University, Maria Curie-Sklodowska Sq. 3, 20-031 Lublin, Poland
2 Department of Bioanalytics, Faculty of Biomedicine, Medical University of Lublin, Jaczewskiego St. 8b, 20-090 Lublin, Poland
3 Independent Unit of Spectroscopy and Chemical Imaging, Medical University of Lublin, Chodzki St. 4a, 20-093 Lublin, Poland
4 Laboratory of X-ray Optics, Centre for Interdisciplinary Research, The John Paul II Catholic University of Lublin, Konstantynow St. 1J, 20-708 Lublin, Poland
* Correspondence: przykaza.kacper@poczta.umcs.lublin.pl

Abstract: Polyetheretherketone (PEEK) is a biocompatible, chemically and physically stable radiolucent polymer that exhibits a similar elastic modulus to the normal human bone, making it an attractive orthopedic implant material. However, PEEK is biologically inert, preventing strong enough bonding with the surrounding bone tissue when implanted in vivo. Surface modification and composite preparation are the two main strategies for the improvement of the bioactivity of PEEK. In this study, the plasma activated PEEK surfaces with the embedded bioglass, chitosan, and bioglass-chitosan mixed layers applying from the solution dip-coating technique were investigated. The most prominent factors affecting the coating biocompatibility are strictly connected with the composition of its outer surface (its charge and functional groups), hydrophilic-hydrophobic character, wettability and surface free energy, and topography (size of pores/substructures, roughness, stiffness), as well as the personal characteristics of the patient. The obtained surfaces were examined in terms of wettability and surface-free energy changes. Additionally, FTIR (Fourier Transformation Infrared Spectrometry) and SIMS (Secondary Ion Mass Spectrometry) were applied to establish and control the coating composition. Simultaneously the structure of coatings was visualized with the aid of SEM (Scanning Electron Microscopy). Finally, the obtained systems were incubated in SBF (Simulated Body Fluid) to verify the modifications' influence on the bioactivity/biocompatibility of the PEEK surface. Different structures with variable compositions, as well as changes of the wettability, were observed depending on the applied modification. In addition, the incubation in SBF suggested that the bioglass-chitosan ratio influenced the formation of apatite-like structures on the modified PEEK surfaces.

Keywords: PEEK polymer; chitosan; bioglass; biocompatible composite

check for updates

Citation: Przykaza, K.; Jurak, M.; Kalisz, G.; Mroczka, R.; Wiącek, A.E. Characteristics of Hybrid Bioglass-Chitosan Coatings on the Plasma Activated PEEK Polymer. *Molecules* **2023**, *28*, 1729. https://doi.org/10.3390/molecules28041729

Academic Editor: Alexandru Mihai Grumezescu

Received: 30 November 2022
Revised: 26 January 2023
Accepted: 31 January 2023
Published: 11 February 2023

1. Introduction

Biocompatibility is an extremely important issue in terms of modern science, particularly in medicine, biotechnology, material, and tissue engineering. To this day, a huge challenge for scientists is to obtain an ideal material that is highly compatible with appropriate applications, e.g., as a bone substitute. Synthetic polymers often exhibit mechanical properties that can allow them to be applied in a specific area of medicine, however, the lack of desired response from the body frequently prevents their use. This effect is influenced by many factors, such as surface composition, degradation, topography, surface wettability, and its free energy, as well as competitive protein binding [1]. For example, the presence of appropriate functional groups on the surface of the modified material can promote cell adhesion and proliferation, and thus the tissue will grow onto the foreign body as if it were a product of a given organism. One of the promising polymeric materials is polyetheretherketone (PEEK), owing to its excellent mechanical strength, biocompatibility, chemical resistance, and radiolucency. Moreover, PEEK reveals no toxicity to human tissues, shows elastic modulus very close to human bone, and does not interfere with the postoperative X-ray or magnetic resonance imaging (MRI) [2]. That is why PEEK was already used for the orthopedic implants, such as spinal cages, skull plates, and joints [3,4]. Unfortunately, its integration with the bone tissue is highly limited, most likely due to the insufficient bioactivity of PEEK induced by the small surface free energy, hydrophobic character, lack of active groups, and, thus, inappropriate protein binding to the surface [5]. To overcome this problem, various chemical and physical modifications of the top polymer layer are used. For example: sandblasting [6], plasma treatment [7], acid etching [8], laser processing [9], and the deposition of bioactive substances, such as polysaccharides, sterols, peptides, phospholipids, hydroxyapatite, or titanium layers [4,10–12]. One of the biocompatible, degradable, and non-toxic polysaccharides is chitosan (Chit). This derivative of natural chitin gained significant popularity in many branches of science, especially in tissue engineering. Chitosan is characterized by antibacterial properties and stimulates proliferation and cell adhesion, thus accelerating the wound healing process [13].

The Bioactivity Index (I_B) of a material is the time taken for more than half of the interface to bind:

$$(t_{0.5bb})\ I_B = 100/t_{0.5bb}.$$

The scale of I_B starts with 0, which means no bioactivity, and goes up to 13 so far. The materials characterized by the I_B index greater than 8 will bind to both the soft and hard tissues. These are called osteoinductive and are assigned to the class A biomaterials [14]. Their surface promotes the colonization of bone stem cells and induces their proliferation and differentiation. The examples of class A bioceramics are bioactive glasses and glass-ceramic. When the I_B index is smaller than 8, the materials can be included in class B and are called osteoconductive, which means they elicit only an extracellular tissue response. The substances from the hydroxyapatite family are considered as the class B biomaterials [15].

Still, one of the most biocompatible materials ever discovered is bioglass (BG) (the in vivo bioactivity index I_B is around 12.5). The most common BG material is the bioglass named 45S5. It was developed in 1969 by Larry Hench and its major components are: SiO_2 (45% wt); Na_2O (24.5% wt); CaO (24.5% wt); and P_2O_5 (6% wt) [14]. Similar to chitosan, this material is established as degradable in the body and what is most important is that it promotes osteoblasts adhesion, migration, proliferation, and mineralization of extra cellular matrix (ECM) into the bone tissue. Thus, it is suitable for applications involving direct contact with the bone, however, due to its insufficient mechanical properties, only as a graft or coating. The combination of bioglass and chitosan as a coating on the PEEK surface was already the main topic of other studies [6,16,17]. For example, Hong et al. tested the BG-Chit coatings deposited on the sandblasted and acid etched PEEK surfaces obtained by the dip-coating technique and observed a potential increase in the material biocompatibility based on the in vitro tests [6]. The other group developed multilayered PEEK/BG containing chitosan/gelatin/Ag-Mn with the aid of electrophoretic deposition (EPD). They established a great bioactivity, applying immersion in SBF, as well as the antibacterial activity, against the Gram-positive and Gram-negative bacteria [18]. Rehman et al. described extensively the preparation of PEEK/BG coatings on the stainless steel and its further coverage with a layer of chitosan-drug applying the EPD technique, which improved the corrosion resistance, wetting properties, and allowed to release the drug from the coating [17]. The coatings based on bioglass and chitosan exhibited good physicochemical and mechanical properties, indicating their possible use in regenerative medicine. However, the substrate had to be previously modified to ensure the proper adhesion to the coating. These modifications are usually time-consuming and can lead to excessive surface smoothing or deterioration of its adhesive properties [6]. Considering the unique ability of the plasma treatment, which can largely enhance the adhesion properties of the treated surface, in our experiment we applied cold plasma activation of PEEK surface to develop the bioglass-chitosan hybrid coating. Combining this approach with the dip-coating technique, we deposited stable bioglass, chitosan, and bioglass-chitosan mixed layers successfully on the activated PEEK surface. The surface properties (morphology, chemistry, wetting) of the PEEK were significantly altered by these modifications. Additionally, we proposed a faster and less complicated approach compared with the EPD technique. Moreover, the impact of the amount of the chitosan on the surface properties in combination with the bioglass was determined. Indeed, the presence of bioglass and chitosan in the coating should improve the biocompatibility and induce antibacterial properties [17,19]. We believe that precise combination in terms of the composition and adhesive strength of bioglass and chitosan can lead to the invention of new hybrid materials, which can merge the biocompatibility and physicochemical properties of both components.

2. Results and Discussion

2.1. FTIR Analysis

The full spectra acquired from the investigated samples confirm the identification of chitosan and bioglass within the samples and are presented in Figure 1A. The control sample of the cold-plasma activated PEEK is consistent with the literature data [20,21].

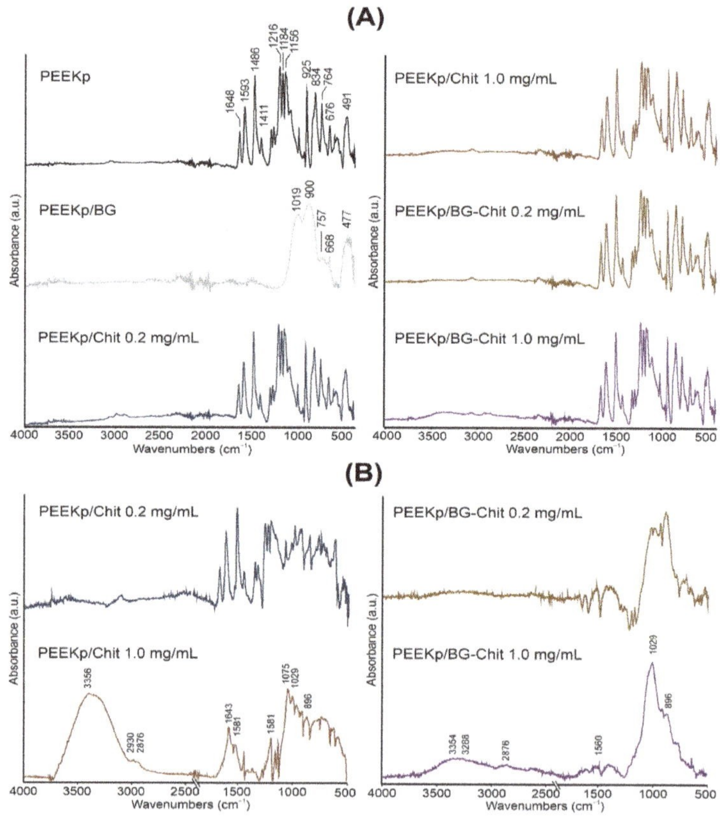

Figure 1. (**A**). FTIR spectra of the modified PEEK surfaces, normalized to the highest absorbance band. (**B**). FTIR subtraction spectra of PEEKp control material and chitosan-containing PEEKp surfaces.

The PEEK spectra reveal typical bands at 951, 834, 764, 676 cm^{-1} from the CH out- of-plane bending vibrations of an aromatic ring. The band at 926 cm^{-1} belongs to the $Aryl - (C = O) - Aryl$ symmetric stretching vibrations, and the maxima of bands at 1593, 1486, 1411, and 1156 cm^{-1} are assigned to the benzene skeleton vibrations, asymmetric stretching and rocking. The band at 1648 cm^{-1} indicates $C = O$ stretching and asymmetric stretching of $C - O - C$ at 1216 and 1184 cm^{-1}. The spectrum derived from bioglass is presented in Figure 1A. The maxima of bands at the 477 cm^{-1} peak are corresponding to $Si - O - Si$ bending vibration in the SiO_4^{4-} group, indicating the presence of amorphous silicate [22,23] and $Si - O - Si$ asymmetric stretching at 1019 cm^{-1} [24]. The band located at 900 cm^{-1} come from the stretching of the non-bridging oxygen atoms in the $Si - O$ bond [24]. The absorption of $P - O$ bending vibrations is located at 757 cm^{-1}, combined with the 668 cm^{-1} value, is attributed to the stretching vibrations of phosphate groups [25]. No peaks that can be assigned to the organic matter are observed, confirming the bioglass purity. To remove the strongly overlapping signal of PEEK in the experimental samples, the spectrum of the PEEKp is subtracted from the others. The results of the subtraction are presented in Figure 1B. The characteristic bands of chitosan can be found in PEEKp/Chit 1.0 mg/mL and PEEKp/BG−Chit 1.0 mg/mL with the visible bands at 3354 cm^{-1} and 3288 cm^{-1}, and originate from the $N - H$ stretching of primary amines and $O - H$ stretching. The bands at 2876 cm^{-1} are due to the $C - H$ symmetric stretching. The band at 1560 cm^{-1} was attributed to amide II ($N - H$), however, there were no observable bands from the amide I and amide III bonds [26,27]. Nevertheless, the amide I, II, and III bands not corresponding to those observed in PEEKp were detected in PEEKp/BG-Chit 1.0 mg/mL. The band around 1029 cm^{-1} is from the $C - N$ stretching and the $C - O$ stretching [28]. The bands at 896 and 1075 cm^{-1} were corresponding to the glycosidic $C - O - C$ stretching of the cross-linked chitosan [27,29]. The above description correlates with the conclusions drawn by other authors. Applying the XPS technique, Wiącek et al. observed a significant increase in the $C - O$, $C - N$, and $C - O - C$ groups, and a simultaneous decrease in the $C - C$ group on the plasma activated PEEK surface covered with the chitosan coating [30].

At this point, despite the rich spectrum of the PEEK substrate and the additional signals from bioglass, many signals characteristic of the chitosan coating were successfully located and identified, but only for the high concentration of chitosan (1.0 mg/mL). In similar research by other authors, when PEEK was activated by the nitrogen plasma and the 0.1 mg/mL chitosan solution was spread, they observed only higher signal intensity in the range of 3500–3200 cm^{-1}, without identification of specific peaks [31]. This is probably due to the insufficient sensitivity of the FTIR technique for the samples prepared from a solution with a lower concentration of chitosan (0.2 mg/mL), since its ions were detected when applying the much more sensitive TOF-SIMS analysis [11]. Nevertheless, a significant reduction in the intensity (depletion of the spectrum) of the chitosan signals after deposition with bioglass for the 1.0 mg/mL chitosan samples was observed and can be associated with the specific co-adsorption of both components, or the attenuation of one signal.

2.2. SEM Imaging

Scanning electron microscopy was used to evaluate the structures obtained during the modification of the PEEK surface (Figure 2A). The first crystal-like structures were observed after the adhesion of bioglass to the activated PEEK. These structures were scattered regularly and were characterized by irregular shapes. Their sizes were different, but they did not exceed the size of 10 × 10 µm^2. This constancy can be the effect of the concentration of polar functional groups on the PEEK surface, which supports the adhesion of bioglass to its surface. When the chitosan layer was present on the activated PEEK, no specific structures were observed, but rather the presence of a smoothing layer compared to the PEEKp control sample (Figure 2A). Considering the experiments by the other authors, where the chitosan layers on PEEK substrate were investigated, the structures varied depending on the deposition technique, chitosan molecular properties, and concentration, as well as prefabrication of the PEEK surface [31,32].

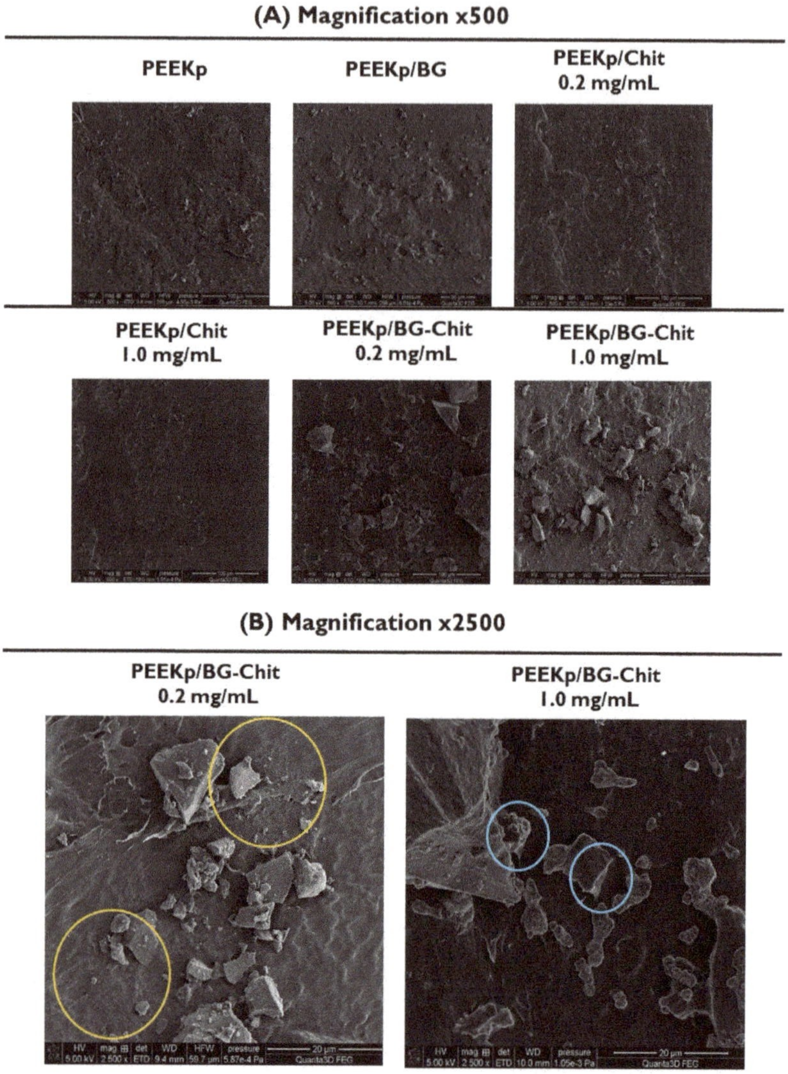

Figure 2. (**A**). SEM images of the modified PEEK surfaces with ×500 magnitude. The yellow circles show the structure of chitosan films. (**B**). SEM images of PEEKp/BG-Chit 0.2 and 1.0 mg/mL with ×2500 magnitude for better visualization of the BG–Chit mixed coatings. The blue circles denote bioglass fusion sites with the solid support.

Finally, for both BG-Chit mixed coatings deposited on the PEEKp substrate, significantly larger crystal structures of bioglass compared with the PEEKp/BG were observed (Figure 2A). Additionally, on the crystals surface, the presence of an additional lattice-like coating was noticed (in particular, for the chitosan concentration of 1mg/mL) connecting the crystals with the surface of the PEEK polymer. At this point, the structure of chitosan appeared and, for the better visualization of the BG-Chit mixed coating, larger magnitude images (×2500) were presented in Figure 2B. For the lower concentration of chitosan (0.2 mg/mL), the presence of chitosan structures was observed on the bioglass crystallites and near them (marked in yellow), thus the PEEK surface was partially covered with chitosan, or the chitosan molecules accumulated to form visible aggregates. When the BG-

Chit mixed coating was deposited from the larger concentration of chitosan (1.0 mg/mL), the chitosan-like structure was present on both BG and PEEK materials, making the coating more homogeneous. Similar results were described by Ravarian et al., when after drying the chitosan-bioglass sol-gel hybrids, they observed rigid monolith surfaces with visible bioglass crystals [33]. The other studies based on the EPD technique to produce the BG-Chit coating on the modified PEEK surfaces reported a more homogeneous coating layer but still with clearly visible BG crystals, very well bound to the substrate [17]. In our case, probable binding with the surface was observed at the point of contact between the crystallites and the PEEK surface (marked with blue circles). This can suggest that at different concentrations chitosan with the bioglass adhere to the activated PEEK surface differently, revealing inconsistent structures and coating properties. However, Hong et al. observed insufficient BG-Chit adhesion to the sandblasted PEEK surface, which prevented the preparation of the samples [6]. Thus, the sample wettability and/or energetic properties of PEEK are necessary to obtain a robust adhesion of the BG-Chit coatings. In our experiment, this was successfully achieved by the plasma activation of the PEEK polymer.

2.3. TOF-SIMS Analysis

The secondary ion mass spectrometry coupled with the time-of-flight detector is a novel technique allowing for the determination of chemical composition of solids, and thus was exploited to develop the chemical composition of the modified PEEK surfaces. Figure 3A depicts the intensity distribution of the calcium, chitosan, and phosphate ions for the multi-step modification of the PEEK substrate. In the TOF-SIMS spectra, the signals of Ca^+ and PO_2^- fragments are typical artifact peaks identified for most substrates that are not covered with a specific layer or very thin layers. Therefore, small artifact signals were observed for the PEEKp and PEEKp/Chit surfaces (Figure 3A).

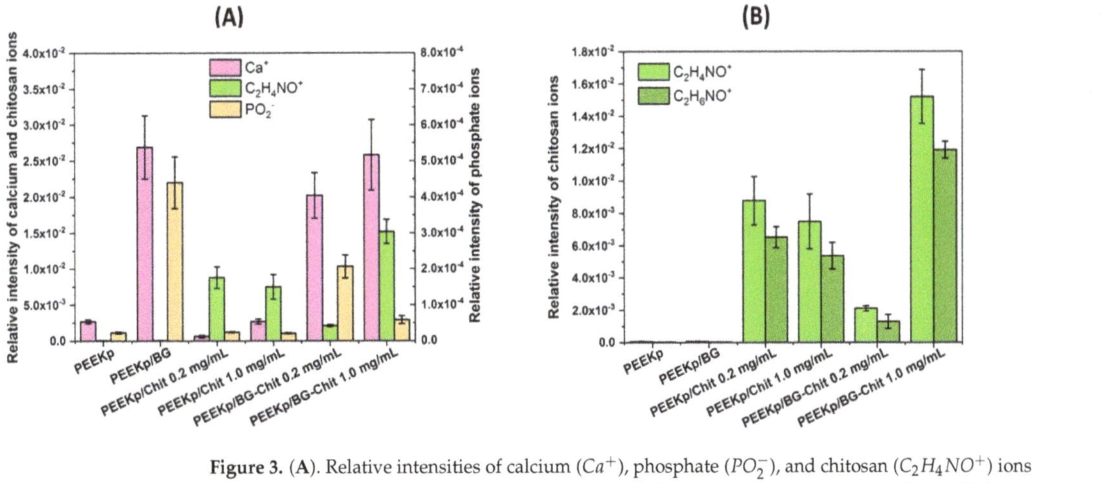

Figure 3. (**A**). Relative intensities of calcium (Ca^+), phosphate (PO_2^-), and chitosan ($C_2H_4NO^+$) ions determined on the modified PEEK surfaces. (**B**). Relative intensities of chitosan ($C_2H_4NO^+$ and $C_2H_6NO^+$) characteristic ions determined on the modified PEEK surfaces.

The largest abundance of Ca^+ fragment is observed for the layers with the bioglass, which proves its presence on the modified PEEK. Slightly larger intensities of calcium were observed for BG-Chit 1.0 mg/mL, compared to 0.2 mg/mL. This can be related to the enhancement of the excitation efficiency of calcium ions in the presence of larger amounts of chitosan and by the different distribution of the Ca^{2+} ions within the BG-Chit layer. At the lower concentration of chitosan, the calcium ions Ca^{2+} reside in closer proximity to each other, whilst at higher chitosan concentration, the calcium ions are more isolated. On the other hand, the distribution of intensity of PO_2^- is proportional to the amount of bioglass within the BG-Chit layer. The phosphate groups are incorporated into the SiO_2 network by chemical bonds (one phosphorous atom is connected to two oxygen atoms), while calcium ions are trapped by interactions with the hydroxyl groups [34]. Some Ca^+ ions were observed for the chitosan layers deposited on the PEEK without the bioglass, as well as for the PEEKp substrate, however, this was a result of the artifact signals. For the chitosan identification, two of the most prominent chitosan fragments, $C_2H_4NO^+$ and $C_2H_6NO^+$, were identified for all chitosan-containing layers (Figure 3B).

These ions were already described for the chitosan identification in our previous papers [11,35] and were found to be complementary; thus, we decided to use the $C_2H_4NO^+$ ion for the chitosan description (Figure 3A). The slightly smaller intensity of the $C_2H_4NO^+$ chitosan fragment is identified for the layer deposited from the more concentrated solution (PEEKp/Chit 1.0 mg/mL). In this case, the chitosan layer is thicker, thus the yield of the chitosan fragment received from the denser and thicker layer is diminished. This is opposite to the observed yield of the phospholipid monolayers, examined using the TOF-SIMS analyser, where the intensity of secondary ions is proportional to the concentration of the components in the monolayer [35,36]. However, the intensity of the chitosan fragments is roughly proportional to the concentration of chitosan in the BG-Chit layer. This is very likely determined by the chitosan molecular rearrangement within this layer. The chitosan molecules might not be cross-linked as in the pure chitosan layer due to the bioglass environment. Consequently, the intensity of the chitosan fragment in the BG-Chit coating reflects the layer composition. On the other hand, the SEM images revealed a specific location of chitosan when co-adsorbed with the bioglass (Figure 2A,B). Thus, at the low concentration of chitosan when both components attach to the polymer surface, the chitosan can remain between the PEEK and the bioglass surface connecting each other. In this case, more bioglass-related ions and fewer chitosan ions would appear. The effect would be different when more chitosan is present. Therefore, chitosan could create more cross-linked structures and cover more bioglass, thus less bioglass ions and more chitosan ions will appear (Figure 3A).

Figure 4 presents the chemical maps of chitosan and calcium ions for the investigated PEEK samples. Their intensities are in agreement with the quantitative analysis (Figure 3A) and reveal specific bioglass and bioglass-chitosan structures and distribution (Figure 4).

It was particularly visible for the PEEKp/BG-Chit 0.2 mg/mL, where the chitosan ions only partially covered the modified surface, while for the larger chitosan concentration, the chitosan coverage looked more homogeneous. The chemical maps for the calcium ions revealed bioglass related structures, notably evident for the PEEKp/BG and PEEKp/BG-Chit 0.2 mg/mL surfaces (Figure 4). Since the calcium ion intensities and distribution for PEEKp/BG-Chit 1.0 mg/mL were unusual, we compared the P^- and PO_2^- chemical maps for both BG-Chit mixed coatings (Figure 5).

The results were consistent with the ion distribution presented in Figures 3 and 4. Thus, there could be a certain mechanism that enhances the calcium ionization and distribution for higher concentrations of chitosan in the BG-Chit hybrid coatings obtained on the PEEKp surfaces. Apart from the PO_2^- fragment that was shown in Figure 3A, phosphorous (P^-) and phosphate ions (PO^-, PO_3^-), corresponding to the bioglass were identified (Figure 6).

The distribution of all phosphate fragments in the bioglass-containing samples was very similar. Nevertheless, the differences of their intensities were significant. Again, this can be correlated with the presence of chitosan on the BG crystals (or the opposite effect), and the specific concentration/grouping process of the BG-Chit suspension on the PEEK surfaces. On the other hand, bioglass can penetrate the chitosan structures and the unique swelling properties of chitosan can have an additional effect for the greater concentrations of chitosan that are present in the coating.

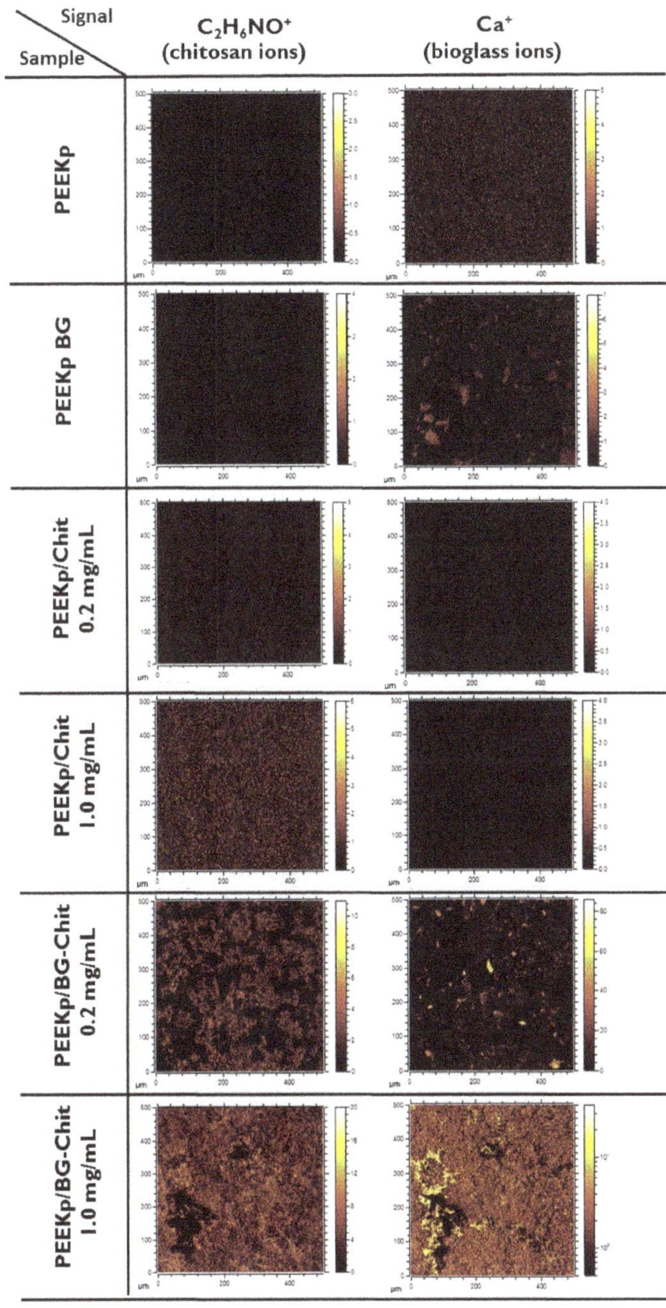

Figure 4. Lateral distribution of chitosan ($C_2H_4NO^+$) and calcium ions (Ca^+) obtained for the modified PEEK surfaces.

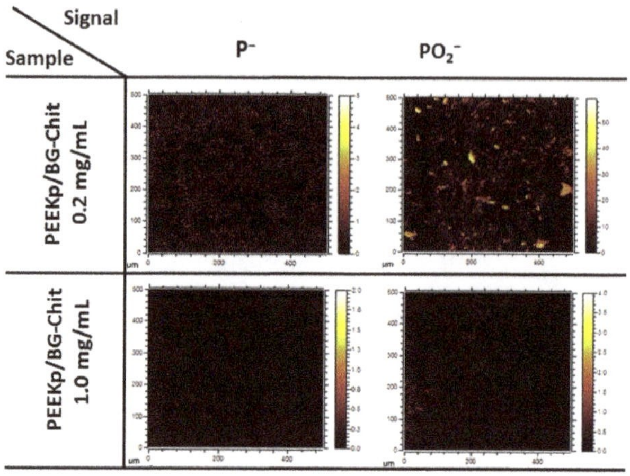

Figure 5. Lateral distribution of phosphorus P^- and phosphate PO_2^- ions for PEEKp/BG-Chit 0.2 and 1.0 mg/mL mixed coatings.

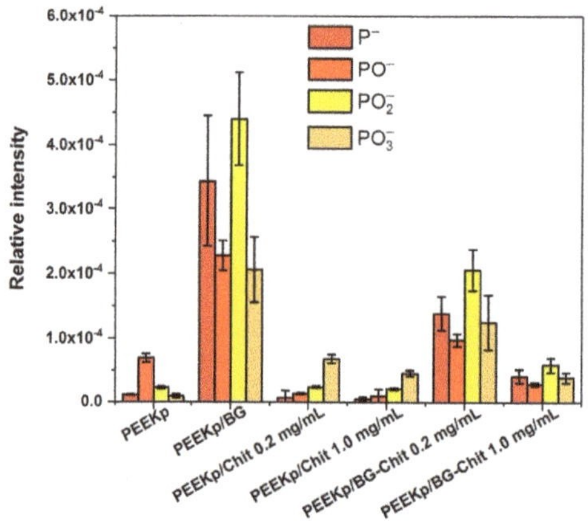

Figure 6. Relative intensities of phosphorous (P^-) and phosphate (PO^-, PO_2^-, PO_3^-) moieties determined on the modified PEEK surfaces.

2.4. Wettability and Surface Free Energy of the Modified PEEK Surfaces

Wetting properties of solids are one of the key surface parameters which have a significant effect on the host response at the cellular level. This phenomenon is mostly associated with the adsorption of the proteins and their conformation. Based on numerous experiments, the scientists established that the optimal water contact angle which best induces the cell proliferation, attachment, and spreading process occurs between 55.0–70.0° [37,38]. Figure 7A presents the contact angles of two polar (water, formamide) and one nonpolar (diiodomethane) test liquids, measured on differently modified PEEK surfaces.

The commercial PEEK revealed a hydrophobic character (contact angle of water $\Theta_W = 87.5°$), which is consistent with the findings of other authors and our previous studies [39,40]. The contact angle for more polar formamide was estimated to be $\Theta_F = 70.0°$ and $\Theta_D = 23.7°$ for the non-polar diiodomethane. The hydrophilic-hydrophobic character of the PEEK surface changed significantly after the plasma treatment ($\Theta_W = 36.4°$). This was due to the surface activation and creation of multiple polar groups on the PEEK surface rich in oxygen and nitrogen. The observed effect was even larger for formamide ($\Theta_F = 0.0°$—total wetting), but not noticeable in the case of diiodomethane ($\Theta_D = 23.1°$).

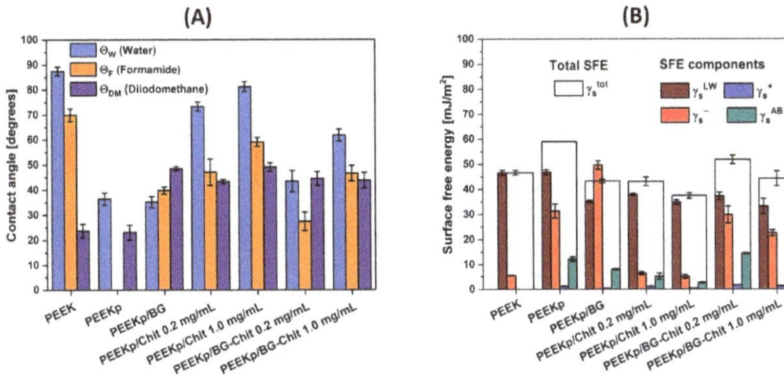

Figure 7. (**A**). Mean contact angle of water, formamide, and diiodomethane, and (**B**) surface free energy, as well as its components determined for the modified PEEK surfaces. The error bars in (**A**) denote standard deviations from 10–15 contact angle values of single test liquid of each surface (twice). The error bars in (**B**) denote the maximum and minimum values of surface free energy and its components calculated for 2 samples of each studied surface from the 10–15 contact angle values of each liquid (their averages).

The contact angles of water did not change significantly after the deposition of BG (Θ_W = 35.2°), nevertheless, it was possible to measure the contact angle of formamide (Θ_F = 39.8°) (Figure 7A). Additionally, the contact angle of diiodomethane changed significantly and was equal to Θ_D = 48.5°. Considering the SEM images, the activated PEEK surface was found to be covered evenly but partially with the BG. Those contact angles were the result of the wettability of the activated PEEK surface and bioglass. When the plasma-activated PEEK surfaces were covered with the chitosan layers obtained from two different chitosan concentrations (0.2 mg/mL or 1.0 mg/mL), the contact angles of water and formamide decreased compared to the PEEK control. For 0.2 mg/mL, the contact angle of water was Θ_W = 73.4°, while that of formamide was equal to Θ_F = 47.0° (Figure 7A). This observation can suggest that chitosan binds to the highly polar (activated) PEEK surface via its polar groups (–OH; –NH_2), and orients its non-polar residues towards the air phase. In the case of concentration of chitosan that is five-times greater (1.0 mg/mL), the PEEK surface was less polar, revealing contact angles of water and formamide equal to Θ_W = 81.3° and Θ_F = 59.0°. Since the polar character of PEEK covered with chitosan decreased at a greater concentration of chitosan, it can be concluded that the coverage of PEEK increased, and the probe liquids had less space to contact/interact with the activated substrate.

The most interesting observations were made when the BG-Chit mixed coatings were deposited on the activated PEEK surface. For the smaller chitosan concentration (0.2 mg/mL), significant decreases in the water (Θ_W = 43.5°) and formamide (Θ_F = 27.4°) contact angles were observed (Figure 7A). This was probably due to the preferential binding of chitosan with the bioglass crystallites. Thus, the chitosan can cover the PEEK surfaces only slightly and remain concentrated on the BG crystallites. This can result in a significant increase in the accessibility of the polar PEEK surface to the test liquids. For the latter examined system with the largest amount of chitosan, an increase in water (Θ_W = 61.9°) and formamide (Θ_F = 46.5°) contact angles compared with the system with a smaller chitosan concentration was found. This phenomenon can also be explained by the preferential adsorption of chitosan on the BG crystallites, followed by adsorption on the PEEK surface. Nevertheless, at the higher concentration of chitosan with bioglass, the separation of the BG-Chit suspension in the solution was observed. Hence, a creation of something that resembles the bonds between chitosan and bioglass and their specific adsorption on the activated PEEK surface cannot be ruled-out. Considering the optimal wettability for the biocompatibility assessment (Θ_W = 55.0–70.0°), the PEEKp/BG-Chit

1 mg/mL system achieved the best value of water contact angle ($\Theta_W = 61.9°$) [38]. The other studies investigating the BG-Chit materials in terms of their wetting properties reported water contact angle values of 45.0–55.0°, which were also considered to be suitable for osteoblast cell adhesion [17,41].

The measurements of contact angles with a wide range of well characterized probe liquids on the solid surfaces are the most common method for the determination of the surface free energy (SFE) and its components. Generally, SFE is the rise in energy strictly connected with moving the atom of the material from its bulk to its surface. When the new surface is created, these atoms reveal coordinative unsaturation of the bonds because now they are exposed to the different environment (air phase) [42]. Hence, the new surface atoms are characterized by different energies compared to the bulk atoms and SFE is the measure of this difference in the type and number of outer bonds. Primarily, when these bonds are mostly Lifshitz-van der Waals in type, SFE will exhibit non-polar nature. On the other hand, when the outer bonds are mostly the ionic/covalent type, there will be a greater contribution of the Lewis acid-base to the total SFE of material.

PEEK itself exhibits relatively low values of total SFE ($\gamma_s^{tot} = 46.6$ mJ/m^2) and mainly consists of the non-polar Lifshitz-van der Waals (γ_s^{LW}) component and minor electron-donor (γ_s^-) parameter (Figure 7B). The lack of the Lewis acid-base component (γ_s^{AB}) can suggest why PEEK is strongly inert to the biological environment. The cold air plasma treatment of the PEEK surface affects γ_s^{tot} largely, and particularly γ_s^- (Figure 7B). Moreover, this step causes the appearance of the γ_s^{AB} component, which is equal to 12.1 mJ/m^2. Unfortunately, this effect weakens over time and the PEEK surface returns to its original state in about 10 days. Similar observations were described by the other authors when the PEEK substrate was treated with cold oxygen plasma [40]. The adsorption of the bioglass crystallites on the plasma activated PEEK surfaces resulted in a significant increase in the electron-donor parameter (γ_s^-) compared with the PEEK control and the plasma activated PEEK (Figure 7B). This was undoubtedly the result of the contact of the measuring liquids with the bioglass structures rich in oxygen and silicon, as well as sodium and calcium ions. When the chitosan layers from the two different solution concentrations were adsorbed on the activated PEEK surface, its ionic nature of bonds was clearly suppressed as a result of blocking the polar groups by the chitosan layer. The more chitosan, the more suppressive effect was observed, however, only in the value of the total surface free energy $\gamma_s^{tot} = 43.1$ and 37.4 mJ/m^2 for PEEKp/Chit 0.2 mg/mL and PEEKp/Chit 1.0 mg/mL, respectively. Nevertheless, for both chitosan coatings, the Lewis acid-base component remained and was determined at the similar level (Figure 7B).

For the mixed BG-Chit layers deposited on the activated PEEK surface, relatively high values of the surface free energy were observed, which were different depending on the amount of chitosan. At the smaller ratio of chitosan (0.2 mg/mL) γ_s^{tot} increased to 51.8 mJ/m^2 and the highest value of γ_s^{AB} of 14.5 mJ/m^2 was found. Simultaneously, this surface revealed the electron-donor ($\gamma_s^- = 29.9$ mJ/m^2) and electron-acceptor ($\gamma_s^+ = 1.76$ mJ/m^2) parameters which were close to the value obtained for the plasma activated PEEK ($\gamma_s^- = 31.4$ mJ/m^2, $\gamma_s^+ = 1.17$ mJ/m^2). For the twin sample with a larger chitosan content (1.0 mg/mL), the obtained parameters of the surface free energy were very close, but γ_s^- decreased slightly and, in consequence, reduced γ_s^{tot} by 7.5 mJ/m^2 (Figure 7B). The described values point to the formation of cation and anion sites from the chitosan and bioglass moieties on the PEEK surface modified by the BG-Chit coating.

Additionally, for PEEKp/BG-Chit 0.2 and 1.0 mg/mL, the energetic surface parameters were determined after 7 days and no significant differences in the surface free energy parameters were found (Figure 8).

At this point, this is a very promising observation because for the long lasting charged surface sites, it is possible for the different molecules (from ions to proteins) to interact with the surface through the ion-dipole attractions and orbital overlap. Considering the complexity of the tissue regeneration process in this stage, it is important that specific modifications introduce permanent and pre-planned changes regarding the chemistry and surface properties of the potential bone replacement materials.

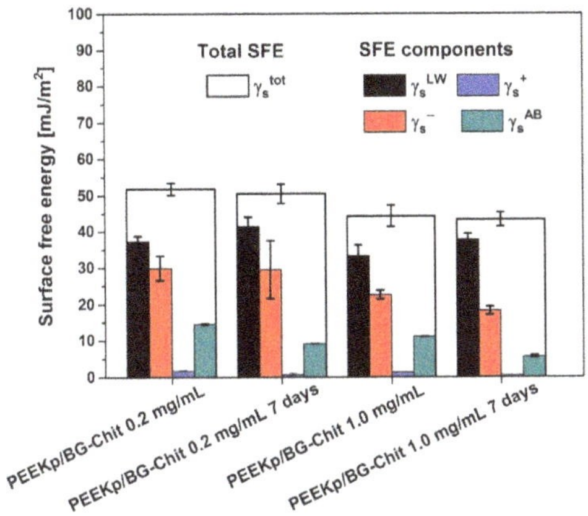

Figure 8. Surface free energy and its components, determined for modified PEEK surfaces, fresh and after 7 days. The error bars denote the maximum and minimum values of surface free energy and its components calculated for 2 samples of each studied surface from the 10–15 contact angle values of each liquid (their averages).

2.5. Biocompatibility Assessment Applying Incubation in the Simulated Body Fluid

The in vitro bioactivity studies include research based on biomaterial incubation in the simulated body fluid (SBF). The SBF solution contains all the necessary inorganic components of human blood plasma. The incubation of materials is made at a temperature close to the physiological temperature of the human body (37 °C). The pH value of the SBF solution (7.25 or 7.40) is also close to the physiological pH value of human blood (7.45). After the sample incubation in the artificial blood plasma, changes of its surface morphology and surface chemistry take place. Indicating the formation of the phosphate-calcium forms related to the material bioactivity, it is possible to predict and estimate the material usefulness. In optimal conditions when the new structures appear, the examined surface is covered with cauliflower-like forms of calcium phosphates. The resulting material, in terms of morphology, is characteristic ofcrystalline hydroxyapatite (HAP) $(Ca_{10}PO_4)_6(OH)_2$ [43].

In our experiment, after the incubation in SBF, a strong intensity increase in the Ca^+ fragment is observed for the BG-containing samples (Figure 9A). For the chitosan-containing PEEKp, no difference is visible, while the activated PEEK polymer substrate demonstrated only a slight increase in the Ca^+ amount. The additional information about the hydroxyapatite layer is provided by the P^-, PO^-, PO_2^-, PO_3^- fragments (Figure 9B–E). There is a similar correlation between the different types of ions identified by TOF-SIMS and the formation of hydroxyapatite. They exhibit significantly greater intensities for the BG-Chit mixed layers compared with the single BG or the chitosan layers. However, greater intensities were always determined for the smaller chitosan ratio (PEEKp/BG-Chit 0.2 mg/mL), except P^- ions, which are single atoms and, thus, cannot reflect the current structure sufficiently. Finally, the new ions, $Ca_2PO_3^+$ and $Ca_2PO_4^+$, were identified (Figure 9F). These ions were already established as corresponding to the HAP structures in the studies applying the TOF-SIMS technique [44]. Their intensities were larger and similar for the pure BG and BG-Chit 0.2 mg/mL and half smaller compared with BG-Chit 1.0 mg/mL. Moreover, the lack of these signals was observed for the PEEK/Chit 0.2 and 1.0 mg/mL samples, while for PEEKp control sample, the noticeable presence of calcium phosphate ions with their proper correlation was observed (Figure 9F). This suggests that the HAP formation on the PEEK surface was suppressed by the presence of chitosan. Similar results were obtained by Fraga et al. when the combination of the sodium silicate with chitosan allowed obtaining the HAP on its surface during the incubation in SBF [45]. Nevertheless, in our experiment this can be associated with blocking the most active PEEKp and bioglass sites by chitosan and, thus, affecting the HAP formation. Moreover, for the PEEKp/BG-Chit 1.0 mg/mL, the smaller intensity ratios of $Ca_2PO_3^+$ and $Ca_2PO_4^+$ suggest a slightly different molecular arrangement of the hydroxyapatite molecules on the surface, compared with the smaller amount of chitosan. The intensity distribution of the PO^-, PO_2^-, PO_3^- fragments for BG-Chit 1.0 mg/mL and BG-Chit 0.2 mg/mL correlates properly with the distribution of $Ca_2PO_4^+$, which exhibits a greater amount of HAP for BG-Chit 0.2 mg/mL. Surprisingly, the intensity of the negative phosphate ions corresponding to the HAP deposited for the pure bioglass (PEEKp/BG) was significantly smaller than that observed for BG/Chit 0.2 mg/mL and BG/Chit 1.0 mg/mL.

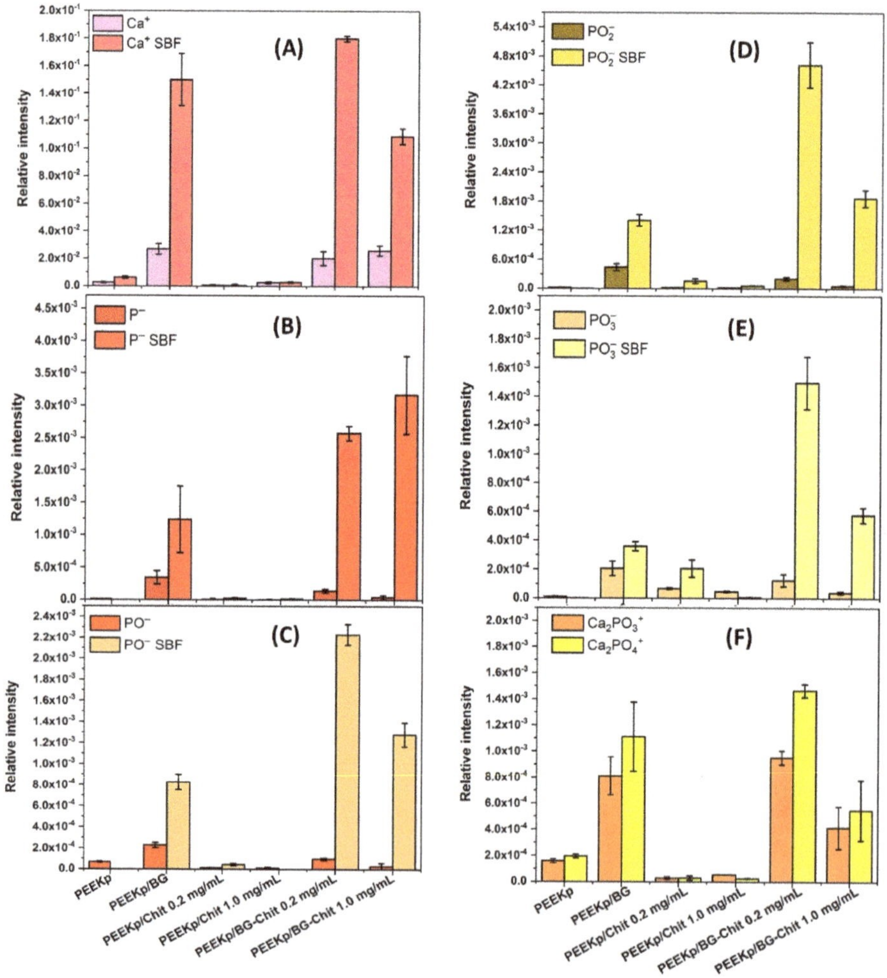

Figure 9. (**A**) Relative intensities of calcium (Ca^+), (**B**) phosphorus (P^-), phosphate fragments (**C**) PO^-, (**D**) PO_2^-, and (**E**) PO_3^- ions determined on the modified PEEK surfaces before and after the incubation in SBF. Figure (**F**) calcium phosphate $Ca_2PO_3^+$ and $Ca_2PO_4^+$ ions appeared after incubation process.

This can be determined by the different yield of the negative ions from hydroxyapatite bound on the pure bioglass. This suggests that the chemical bonds and molecular arrangements of HAP layer on the single bioglass are different compared with the HAP deposited on bioglass-chitosan coating. It can be hypothesized that, in such systems, it is important to select the optimal concentration of chitosan. Too high a concentration of this polysaccharide can lead to a strong swelling process and, thus, block the formation of hydroxyapatite on the bioglass crystallites. The additional insight into these questions can be provided by the chemical maps that show the distributions of individual components within the layer with the lateral resolution below 0.5 μm (Figure 10).

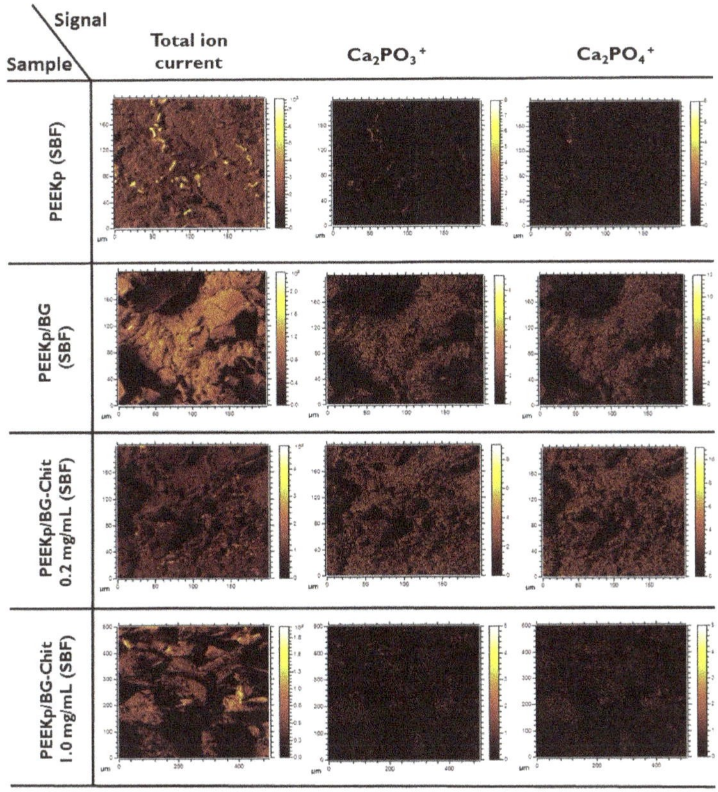

Figure 10. Total ion current and lateral distribution of calcium phosphates ($Ca_2PO_3^+$ and $Ca_2PO_4^+$) ions corresponding to hydroxyapatite.

The distribution of the most prominent hydroxyapatite $Ca_2PO_3^+$ and $Ca_2PO_4^+$ fragments is presented in Figure 10. On the PEEKp substrate only, several HAP related island-like areas are visible that correspond to the high intensity points in the total ion current image (Figure 9). For the pure bioglass, larger crystalline structures were observed. After mixing BG with chitosan (PEEKp/BG-Chit 0.2 mg/mL), a fine-crystalline structure of HAP layer was present. When the concentration of chitosan increased (PEEKp/BG-Chit 1.0 mg/mL), significantly bigger crystals were observed. Simultaneously, the absolute value of HAP decreased notably (about 10-times) together with the total ion current (3-times). In consequence, the relative intensity decreases about 3-times that corresponds to the distribution shown in Figure 9F. Figure 11 depicts the SIMS and SEM images of the modified PEEK structures after incubation in SBF.

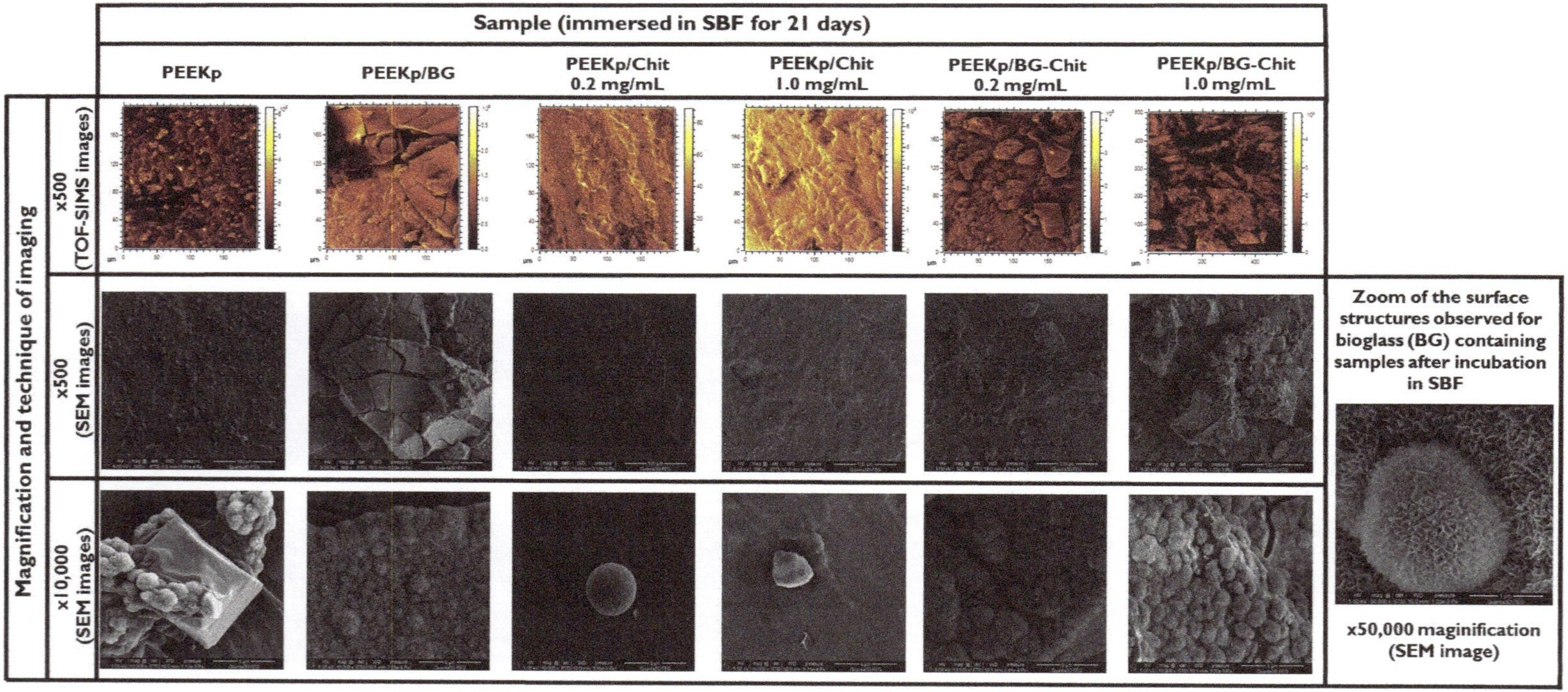

Figure 11. TOF-SIMS and SEM images of the modified PEEK surfaces after the immersion in SBF for 21 days.

After the incubation in the SBF process, all of the investigated PEEK surfaces revealed the creation of new structures. On the plasma activated PEEK, small (max. 20 × 20 μm^2) apatite-like structures were observed as the small islands scattered over the whole surface. Interestingly, the cauliflower structures grew not directly from PEEK surface, but from the ice crystal-like structures created on it (Figure 11). This can be related to the most active PEEK sites after the plasma activation ($-CO, -COO, -CNH_x, -CN <$), which induced the crystallization and mineralization of new structures. On the other hand, for the pure bioglass deposited on the PEEKp polymer, very intense crystal growth was observed, and the new HAP layer mineralized not only on the BG crystals, but also on the PEEKp surface. Additionally, the structures visible on both the SIMS and SEM images are partially cracked. At the larger magnitude (×10,000), the coverage revealed sponge-like mineral structures (Figure 11). The situation was different for the surfaces that contained only chitosan. No HAP-related structures were present. Only small spherical crystals (2.5 μm diameter) were unevenly scattered on both the PEEKp/Chit 0.2 and 1.0 mg/mL surfaces. It can be hypothesized that the chitosan deposited onto the plasma activated PEEK surface inhibits the formation of apatite-like structures. For the BG-Chit mixed coatings, fine crystals covered with newly crystallized structures were observed. In this case, no cracked sites were present. This can be a result of the incorporation of the chitosan to its structure and enhancing its flexibility. Considering the micro-structure of the obtained surfaces, it can be concluded that at higher concentration of chitosan (1.0 mg/mL), less homogenous structures appeared, and more isolated islands were present compared with the smaller concentration of chitosan (0.2 mg/mL) (Figure 11). Our results are in agreement with the other studies where the presence of bioglass in the chitosan coating always enhanced the formation of the porous apatite structures [16,18]. These structures appeared even after 3 days in SBF and were less porous with the more time spent in SBF. However, in our experiment new apatite structures maintained their porous structure after 21 days of immersing in SBF.

For the in-depth structure analysis of newly obtained surfaces, the distribution of chitosan and HAP related ions was performed applying depth profiling (Figure 12).

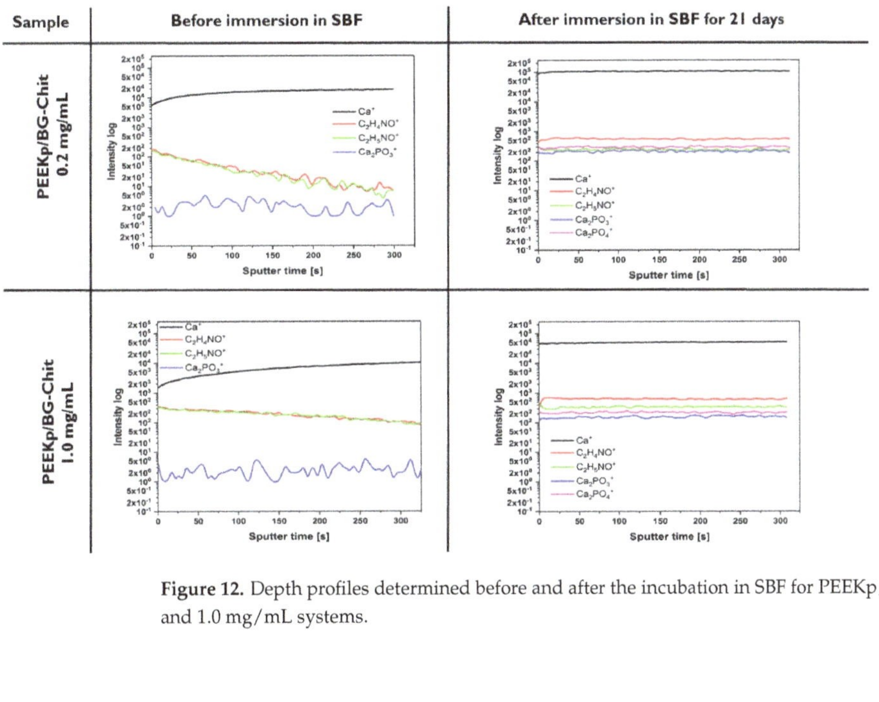

Figure 12. Depth profiles determined before and after the incubation in SBF for PEEKp/BG-Chit 0.2 and 1.0 mg/mL systems.

Before the incubation in SBF, no HAP related signals ($Ca_2PO_3^+$ and $Ca_2PO_4^+$) were observed. On the other hand, the intensity of the Ca^+ ions increased as a function of the sputtering time (depth of sample), while the intensity of chitosan ions decreased. A similar effect was observed for both BG-Chit mixed layers deposited on the PEEKp surface. This confirms the hypothesis that chitosan is adsorbed on the surface of the bioglass crystallites and remains in the topmost layer of the coating. A significantly different behavior was observed for the PEEK/BG-Chit samples after the incubation. All ions related to the bioglass, chitosan, and hydroxyapatite demonstrated constant intensity as a function of the sputtering time (Figure 12). This could indicate the homogeneous distribution of each individual component as a function of depth of the investigated samples. The HAP formation during the incubation in SBF is a complex process and involves a few steps containing K^+ and Na^+ exchanging with H^+ or H_3O^+; the formation of silica layer and then the breaking of $O-Si-O$ connections, condensation, and polymerization of silanol groups on the surface of BG; the deposition of Ca^{2+} ions; and finally the crystallization of hydroxyapatite. In the investigated systems when the whole process is induced by the presence of bioglass on the PEEKp surface, the chitosan molecules can possibly migrate from the coating and create a new forming structure. However, the other studies reported the degradation of chitosan when the BG-Chit-lawsone hybrid coatings formed by the EPD technique were immersed in SBF [17]. This discrepancy can be a result of different coating and/or different substrate preparation procedure. During this process, positively charged chitosan molecules in suspension interacted with the hydroxyl groups on the bioactive glass particles surface to form hydrogen bonds, which leads to the co-adsorption of chitosan and glass on the support [46].

The described experiments are very important from the point of view of designing biomaterials with specific properties. For example, tests of wettability connected with the surface free energy estimation allow the selection of the optimal surface in terms of its energy for the best cell adhesion and proliferation. This goes along with the surface topography which can be examined by different imaging techniques (e.g., SEM) to ensure the most appropriate conditions for cell attachment. Moreover, the surface chemistry measurement techniques, such as FTIR, XRD, and TOF-SIMS, are usually applied for the coatings investigations to establish their composition, structure, stability, or the presence of additional bioactive substances, e.g., drugs. Finally, the incubation tests in SBF provide very useful information about if and how fast apatite structure appears on the material, which helps predict its biocompatibility.

From the implantology point of view, the presented material (PEEK) shows the beneficial properties such as: mechanical properties (modulus of elasticity and tensile as well as yield strength, hardness, and toughness), which are very similar to those of human bone, corrosion resistivity, and bioinertness [47]. Due to this, it was already introduced for application as the spinal cage, dental materials, knee joints, and many other bone replacements [3,48,49]. The enrichment of PEEK with the bioglass-chitosan coating improved its surface properties, mainly wettability and roughness, which now are in the more optimal range for osteoblasts adhesion (water contact angle of 55.0–70.0° and roughness 1–3 μm) [37,38]. Nevertheless, the proposed material is not without its limitations. Firstly, it can be considered as unknown for the long-term stability of the coating. Secondly, it is associated with the high geometry of dental implants, so more studies are required to examine plasma activation and film deposition for such materials (bone screws). Finally, a serious limitation occurs with the implant placement. However, the degradation of the coating during the strong friction associated with the installation of the implant in the osteon or the trabeculae of the bone should be ruled out. In fact, for the spinal cages implants there would be beneficial fusion of two vertebrae with a bone, unless there is a pressure on the spinal nerves and the spinal cord itself. The PEEK based implants would fit better where the metal ones generate mechanical overstress (spinal cages, knee joints). Moreover, due to the 3D printing technology, they can be superior to the metallic implants where complicated shapes are required (tooth screws, skulls, and face bones implants).

Finally, PEEK does not induce allergy, compared to the metal alloys; thus, it can be in contact with a large surface area of living tissue and also its color is more appropriate, taking the esthetic implants into account.

3. Materials and Methods

3.1. Materials

The 20 × 30 × 5 mm^3 PEEK plates were cut from the commercially available TECAP EEK natural (1000 × 500 × 5 mm^3, PROFILEX) and were used as a substrate. The chitosan (molecular weight of 100,000–300,000 and the deacetylation degree (DD) of 82.0–84.0%) was purchased from Acrōs Organics (Geel, Belgium). The bioglass (45S5) was obtained from XL Sci-Tech, Inc. (Richland, WA, USA) in the form of granules (45–300 μm size, IsoSpheres, Richland, WA, USA). Before use, the bioglass powder was additionally ground in an agate mortar. Both materials were applied for PEEK modification. Methanol, acetic, and hydrochloric acid were purchased from Avantor Performance Materials Poland S.A. with a purity above 99.0%. Each time, the Milli-Q system water (resistivity ~18.2 MΩ*cm) was used. For the preparation of the simulated body fluid (SBF), a set of the following chemicals was used: NaCl, Na_2CO_3, $Na_2HPO_4 \times 12\,H_2O$, $MgCl_2 \times 6\,H_2O$, and $CaCl_2$ obtained from POCh (Gliwice, Poland); KCl, Na_2SO_4 from Chempur (Piekary Śląskie, Poland); and Tris(hydroxymethyl)aminomethane ($NH_2C(CH_2OH)_3$) purchased from Sigma-Aldrich (Burlington, MA, USA). All of them had a minimum purity of 99.0%. For the contact angle measurements, a set of the following liquids was applied: Milli-Q water, formamide (Acrōs Organics, 99.5%), and diiodomethane (Sigma-Aldrich, 99.0%).

3.2. PEEK Substrate Purification

Commercial PEEK plates were put into a 1L beaker and purified with a neutral extran (1.0 mL/100 mL of Milli-Q water), methanol, and Milli-Q water 2–3 times, until its conductivity did not exceed 5.0 μS/cm. After each liquid filling, the samples remained for 15 min in the ultrasonic bath. After this procedure, the plates were dried in the vacuum oven and then stored in desiccators.

3.3. Air Plasma Treatment of PEEK Surface

The air plasma (20 °C and 0.2 mbar) was applied to increase the adhesion of the PEEK surface. The optimal plasma parameters were established based on our previous research [20,30]. The plasma treatment was conducted with the aid of the microwave plasma generator (Diener Electronic, Germany). The duration of the plasma process was 60 s, the power was 460 W, and the continuous air flow was 22 sccm (standard cubic centimeters per min). The activated PEEK surfaces were further modified immediately after the plasma treatment.

3.4. Preparation of Chitosan (Chit), Bioglass (BG), and Bioglass-Chitosan (BG-Chit) Mixed Layers on the Plasma Activated PEEK Surface

The proper amount of chitosan was dissolved in 0.1% acetic acid (AA) to obtain the concentration of 1.0 mg/mL. The 0.2 mg/mL solution of the Chit was obtained by mixing the appropriate amount of 1.0 mg/mL Chit with 0.1% AA. The bioglass solution was prepared by mixing 0.125 g of BG with 25 mL of 0.1% AA. The mixed BG-Chit suspensions were prepared by mixing 0.125 g of BG with 25 mL of chitosan solution (1.0 mg/mL or 0.2 mg/mL). Immediately after the activation, the PEEK plates were submerged in 25 mL of the proper solution for 5 min, with continuous mixing of the solution/suspension with a magnetic stirrer. Next, the plates were dried in the vacuum oven (24 h, 20 °C) and subjected to further testing.

3.5. Wettability Tests and Determination of the Surface Free Energy

To study the changes of the PEEK surfaces wettability after the multi-step modification, the contact angle measurements were performed. The set of probe liquids consisted of Milli-

Q water, formamide, and diiodomethane. The contact angle measuring system DGD ADR from GBX S.A.R.L was exploited. An amount of 6 µL of the test liquid was carefully placed on the PEEK surface using a microsyringe. The advancing contact angle was measured with the aid of a digital camera and the WinDrop++ software. For each sample, these measurements were conducted twice and for each tested liquid, the contact angle was examined for not less than 10 droplets on the surface. The values of the average contact angles of Milli-Q water (W), formamide (F), and diiodomethane (DM) were further used for the surface free energy calculations.

Applying the Lifshitz-van der Waals-acid-base (LWAB) approach it was possible to estimate changes of the surface free energy of the investigated PEEK surfaces. This approach involves the values of the advancing contact angles of three liquids of different polarity and the well-established surface tensions contributions. As a result, the values of the total surface free energy (γ_s^{tot}) and its components, apolar Lifshitz-van der Waals (γ_s^{LW}), electron-donor (γ_s^-), electron-acceptor (γ_s^+), and acid-base (γ_s^{AB}), were obtained [50]. For the probe liquids, the following values of the surface tension and its components were used for calculations (Table 1):

Table 1. Values of surface tension and its components of the used testing liquids.

Probe Liquid	γ^{tot} [mJ/m^2]	γ^{LW} [mJ/m^2]	γ^{AB} [mJ/m^2]	γ^+ [mJ/m^2]	γ^- [mJ/m^2]
Water	72.8	21.8	51.0	25.5	25.5
Formamide	58.0	39.0	19.0	2.3	39.6
Diiodomethane	50.8	50.8	0	0	0

3.6. Fourier Transform Infrared Spectroscopy (FTIR) Analysis

The FTIR analysis was used for the initial assessment of the surface chemistry of the modified PEEK plates. The samples were studied under the same conditions with the FTIR spectroscope equipped with the attenuated total reflectance (ATR) accessory. The deuterated triglycine sulfate (DTGS) KBr detector was used for obtaining the IR spectral data. The spectra were recorded in the range of 4000–400 cm^{-1} with 256 scans with a resolution of 4 cm^{-1} and the optimal signal-to-noise ratio. There were at least two samples of each surface and 3 measurements on each sample. The data collection and analysis were performed with Omnic 12 (Thermo Fisher Scientific, Madison, WI, USA). For the analysis, three single spectra were recorded at different points of the investigated samples, averaged into one spectrum representative for a given sample, and normalized to the band at 2918 cm^{-1} assigned to the $C-H$ stretching. The baseline corrections were multipoint and applied at 4000, 1260, 560, and 400 cm^{-1}.

3.7. Scanning Electron Microscopy (SEM) Analysis

Surface imaging was conducted by means of the high-resolution scanning electron-ion microscope (SEM) Quanta3D FEG (USA). The detection was made using the Everhart-Thornley detector (ETD) with a voltage of 5 kV. The pressure in the chamber was always below 7×10^{-3} Pa. There were at least two samples of each surface measured and the magnification varied from 100× to 200,000×. The best images with the magnifications of 500–50,000× were selected for presentation. Since the PEEK polymer is a weak conductor, the samples were sputtered with a thin layer of silver and palladium just before the measurement with the Leica EM SCD 500 (Vienna, Austria) sputtering machine.

3.8. Time of Flight Secondary Ion Mass Spectrometry (TOF-SIMS) Analysis

The TOF-SIMS spectra were obtained by means of the TOF-SIMS.5 instrument (ION-TOF GmbH, Münster, Germany). The primary ion source of Bi^+ was used as a primary beam for the analysis (30 kV, cyclic time 100 µs, primary beam current 1.2 pA) in the positive and negative Spectrometry mode. The analysis area was 200 µm × 200 µm, with

256 × 256 resolution and 1 shot/pixel. All the measurements were performed under static mode (dose no larger than 1 × 10^{12} ions/cm^2). The surface potential equal to −430 V was applied to eliminate sample charging. Moreover, for depth profiles, a gun cluster ion beam (GCIB) at 7.5 keV was used for sputtering. For this purpose, the measurements were taken in the not-interlaced mode, including one analysis scan (Bi^+) and, subsequently, three sputtering scans (GCIB). The scanned area for the in-depth profile measurements was 100 μm × 100 μm, while the sputtering area was 300 μm × 300 μm. For the acquisition of chemical maps, the Fast Imaging mode was applied (cyclic time 100 μs, primary beam current 0.4 pA) that allowed the obtaining of a lateral resolution below 0.15 μm. The analysis area was 200 μm × 200 μm or 400 μm × 400 μm with a resolution 512 × 512 pixels. For the analysis, there were at least two samples of each surface and 3 measurements of each sample.

The post-processing data analysis was conducted using the SurfaceLab 6.7 software (ION-TOF) and Origin 2019 (OriginLab, Northampton, MA, USA). The negative spectra were recorded and calibrated using the positions of CH^-, CH_2^-, and CH_3^-, while the positive mode used the CH^+, $C_2H_3^+$, and $C_2H_5^+$ fragments. All intensities were normalized to the total intensity.

3.9. Bioactivity Tests Applying Incubation in the Simulated Body Fluid (SBF)

The simulated body fluid was prepared using a widely known procedure established by Kokubo and Takadama [43]. After preparation, the SBF was poured into the plastic bottle and stored in the fridge (8 °C). For the incubation purposes, 50 mL of SBF was poured to the plastic containers (7 cm height and 3 cm diameter) with a tight closure. Next, the dried PEEK samples were put into the containers with SBF and placed in the laboratory dryer at 37 °C for 21 days. Two samples of each surface were simultaneously incubated (in different bottles). After, the PEEK plates were gently rinsed with the Milli-Q water, dried, and then subjected for further analyses.

4. Conclusions

The activation of the PEEK polymer surface with the air plasma allowed the immobilization of bioglass, chitosan, and bioglass-chitosan mixed coatings on its surface, which was confirmed by the FTIR, SEM, and TOF-SIMS analyses. Chitosan was accumulated partially on both the PEEK and bioglass surfaces depending on its concentration. This affected the PEEK wetting properties, as well as changed the proportions of the surface free energy components, mainly related to the interactions through polar groups (γ_s^{AB} and γ_s^-). The creation of apatite structures was observed for the PEEKp and bioglass-containing samples after immersing in SBF. The amount of apatite was also dependent on the chitosan concentration.

We believe that the discussed creation of the hybrid coatings based on chitosan and bioglass on the activated PEEK surface can contribute definitely to the development of a largely biocompatible material that combines various features, such as antibacterial or anti-inflammatory properties. For further perspectives, more specialized experiments involving the cell lines (e.g., MG63), as well as bacteria, should be conducted to confirm the predicted biocompatibility and antibacterial properties.

Author Contributions: Conceptualization, A.E.W. and M.J.; methodology, A.E.W., M.J and K.P; software, K.P, R.M and G.K.; validation, K.P, R.M and G.K.; formal analysis, K.P.; investigation, K.P.; resources, K.P.; data curation, K.P., R.M. and G.K.; writing—original draft preparation, K.P.; writing—review and editing, A.E.W. and M.J.; visualization, K.P.; supervision, A.E.W. and M.J. All authors have read and agreed to the published version of the manuscript.

Funding: This research received no external funding.

Institutional Review Board Statement: Not applicable.

Informed Consent Statement: Not applicable.

Data Availability Statement: Data available on request.

Conflicts of Interest: The authors declare no conflict of interest.

Sample Availability: Not applicable.

References

1. Jurak, M.; Wiącek, A.E.; Ładniak, A.; Przykaza, K.; Szafran, K. What Affects the Biocompatibility of Polymers? *Adv. Colloid Interface Sci.* **2021**, *294*, 102451. [CrossRef] [PubMed]
2. Ning, L.; Deqiang, C.; Xiyan, G.; Lirong, L.; Weizeng, C. Biological Tribology Properties of the Modified Polyether Ether Ketone Composite Materials. *Rev. Adv. Mater. Sci.* **2020**, *59*, 399–405. [CrossRef]
3. Kurtz, S.M.; Devine, J.N. PEEK Biomaterials in Trauma, Orthopedic, and Spinal Implants. *Biomaterials* **2007**, *28*, 4845–4869. [CrossRef] [PubMed]
4. Ma, R.; Tang, S.; Tan, H.; Qian, J.; Lin, W.; Wang, Y.; Liu, C.; Wei, J.; Tang, T. Preparation, Characterization, in vitro Bioactivity, and Cellular Responses to a Polyetheretherketone Bioactive Composite Containing Nanocalcium Silicate for Bone Repair. *ACS Appl. Mater. Interfaces* **2014**, *6*, 12214–12225. [CrossRef] [PubMed]
5. Lim, J.Y.; Shaughnessy, M.C.; Zhou, Z.; Noh, H.; Vogler, E.A.; Donahue, H.J. Surface Energy Effects on Osteoblast Spatial Growth and Mineralization. *Biomaterials* **2008**, *29*, 1776–1784. [CrossRef] [PubMed]
6. Hong, W.; Guo, F.; Chen, J.; Wang, X.; Zhao, X.; Xiao, P. Bioactive Glass–Chitosan Composite Coatings on PEEK: Effects of Surface Wettability and Roughness on the Interfacial Fracture Resistance and in vitro Cell Response. *Appl. Surf. Sci.* **2018**, *440*, 514–523. [CrossRef]
7. Przykaza, K.; Woźniak, K.; Jurak, M.; Wiącek, A.E. Characteristics of Polypeptide/Phospholipid Monolayers on Water and the Plasma-Activated Polyetheretherketone Support. *J. Surfactants Deterg.* **2019**, *22*, 1213–1228. [CrossRef]
8. Chaijareenont, P.; Prakhamsai, S.; Silthampitag, P.; Takahashi, H.; Arksornnukit, M. Effects of Different Sulfuric Acid Etching Concentrations on PEEK Surface Bonding to Resin Composite. *Dent. Mater. J.* **2018**, *37*, 385–392. [CrossRef]
9. Czwartos, J.; Budner, B.; Bartnik, A.; Wachulak, P.; Fiedorowicz, H.; Mierczyk, Z. Physico-Chemical Surface Modifications of Polyetheretherketone (PEEK) Using Extreme Ultraviolet (EUV) Radiation and EUV-Induced Nitrogen Plasma. *Materials* **2020**, *13*, 4466. [CrossRef]
10. Schulze, K.; Hausmann, J.; Wielage, B. The Stability of Different Titanium-PEEK Interfaces against Water. *Procedia Mater. Sci.* **2013**, *2*, 92–102. [CrossRef]
11. Przykaza, K.; Jurak, M.; Wiącek, A.E.; Mroczka, R. Characteristics of Hybrid Chitosan/Phospholipid-Sterol, Peptide Coatings on Plasma Activated PEEK Polymer. *Mater. Sci. Eng. C* **2021**, *120*. [CrossRef]
12. Przykaza, K.; Woźniak, K.; Jurak, M.; Wiącek, A.E.; Mroczka, R. Properties of the Langmuir and Langmuir–Blodgett Monolayers of Cholesterol-Cyclosporine A on Water and Polymer Support. *Adsorption* **2019**, *25*, 111658. [CrossRef]
13. Aguilar, A.; Zein, N.; Harmouch, E.; Hafdi, B.; Bornert, F.; Offner, D.; Clauss, F.; Fioretti, F.; Huck, O.; Benkirane-Jessel, N.; et al. Application of Chitosan in Bone and Dental Engineering. *Molecules* **2019**, *24*, 3009. [CrossRef]
14. Hench, L.L. The Story of Bioglass®. *J. Mater. Sci. Mater. Med.* **2006**, *17*, 967–978. [CrossRef]
15. De Aza, P.N.; De Aza, A.H.; Pena, P.; De Aza, S. Bioactive Glasses and Glass-Ceramics. *Bol. La Soc. Esp. Ceram. Y Vidr.* **2007**, *46*, 45–55. [CrossRef]
16. Atiq Ur Rehman, M.; Bastan, F.E.; Haider, B.; Boccaccini, A.R. Electrophoretic Deposition of PEEK/Bioactive Glass Composite Coatings for Orthopedic Implants: A Design of Experiments (DoE) Study. *Mater. Des.* **2017**, *130*, 223–230. [CrossRef]
17. Ur Rehman, M.A.; Bastan, F.E.; Nawaz, Q.; Goldmann, W.H.; Maqbool, M.; Virtanen, S.; Boccaccini, A.R. Electrophoretic Deposition of Lawsone Loaded Bioactive Glass (BG)/Chitosan Composite on Polyetheretherketone (PEEK)/BG Layers as Antibacterial and Bioactive Coating. *J. Biomed. Mater. Res.—Part A* **2018**, *106*, 3111–3122. [CrossRef]
18. Nawaz, A.; Bano, S.; Yasir, M.; Wadood, A.; Ur Rehman, M.A. Ag and Mn-Doped Mesoporous Bioactive Glass Nanoparticles Incorporated into the Chitosan/Gelatin Coatings Deposited on PEEK/Bioactive Glass Layers for Favorable Osteogenic Differentiation and Antibacterial Activity. *Mater. Adv.* **2020**, *1*, 1273–1284. [CrossRef]
19. Ładniak, A.; Jurak, M.; Palusińska-Szysz, M.; Wiącek, A.E. The Influence of Polysaccharides/TiO_2 on the Model Membranes of Dipalmitoylphosphatidylglycerol and Bacterial Lipids. *Molecules* **2022**, *27*, 343. [CrossRef]
20. Rymuszka, D.; Terpiłowski, K.; Borowski, P.; Holysz, L. Time-Dependent Changes of Surface Properties of Polyether Ether Ketone Caused by Air Plasma Treatment. *Polym. Int.* **2016**, *65*, 827–834. [CrossRef]
21. Al Lafi, A.G. FTIR Spectroscopic Analysis of Ion Irradiated Poly (Ether Ether Ketone). *Polym. Degrad. Stab.* **2014**, *105*, 122–133. [CrossRef]
22. Adams, L.A.; Essien, E.R.; Shaibu, R.O.; Oki, A. Sol-Gel Synthesis of SiO_2-CaO-Na_2O-P_2O_5 Bioactive Glass Ceramic from Sodium Metasilicate. *New J. Glas. Ceram.* **2013**, *03*, 11–15. [CrossRef]
23. Crovace, M.C.; Souza, M.T.; Chinaglia, C.R.; Peitl, O.; Zanotto, E.D. Biosilicate®—A Multipurpose, Highly Bioactive Glass-Ceramic. in vitro, in vivo and Clinical Trials. *J. Non. Cryst. Solids* **2016**, *432*, 90–110. [CrossRef]

24. Herradi, S.; Bouhazma, S.; Chajri, S.; Khaldi, M.; El Hachadi, A.; El Bali, B.; Lachkar, M. The Effect of Strontium and Silver on the Bioactivity of a Quaternary Bioglass in the System SiO_2-CaO-Na_2O-P_2O_5. *J. Phys. Conf. Ser.* **2018**, *984*, 012011. [CrossRef]
25. Terzopoulou, Z.; Baciu, D.; Gounari, E.; Steriotis, T.; Charalambopoulou, G.; Tzetzis, D.; Bikiaris, D. Composite Membranes of Poly(ε-Caprolactone) with Bisphosphonate-Loaded Bioactive Glasses for Potential Bone Tissue Engineering Applications. *Molecules* **2019**, *24*, 3067. [CrossRef] [PubMed]
26. Queiroz, M.F.; Melo, K.R.T.; Sabry, D.A.; Sassaki, G.L.; Rocha, H.A.O. Does the Use of Chitosan Contribute to Oxalate Kidney Stone Formation? *Mar. Drugs* **2015**, *13*, 141–158. [CrossRef] [PubMed]
27. Gieroba, B.; Sroka-Bartnicka, A.; Kazimierczak, P.; Kalisz, G.; Lewalska-Graczyk, A.; Vivcharenko, V.; Nowakowski, R.; Pieta, I.S.; Przekora, A. Surface Chemical and Morphological Analysis of Chitosan/1,3-β-d-Glucan Polysaccharide Films Cross-Linked at 90 °C. *Int. J. Mol. Sci.* **2022**, *23*, 5953. [CrossRef]
28. Mincke, S.; Asere, T.G.; Verheye, I.; Folens, K.; Bussche, F.V.; Lapeire, L.; Verbeken, K.; van der Voort, P.; Tessema, D.A.; Fufa, F.; et al. Functionalized Chitosan Adsorbents Allow Recovery of Palladium and Platinum from Acidic Aqueous Solutions. *Green Chem.* **2019**, *21*. [CrossRef]
29. Lv, M.-Z.; Wang, L.-F.; Fang, L.; Li, P.-W.; Li, S.-D. Preparation and Properties of Natural Rubber/Chitosan Microsphere Blends. *Micro Nano Lett.* **2017**, *12*, 386–390. [CrossRef]
30. Wiacek, A.E.; Terpiłowski, K.; Jurak, M.; Worzakowska, M. Effect of Low-Temperature Plasma on Chitosan-Coated PEEK Polymer Characteristics. *Eur. Polym. J.* **2016**, *78*, 1–13. [CrossRef]
31. Terpiłowski, K.; Wiącek, A.E.; Jurak, M. Influence of Nitrogen Plasma Treatment on the Wettability of Polyetheretherketone and Deposited Chitosan Layers. *Adv. Polym. Technol.* **2018**, *37*, 1557–1569. [CrossRef]
32. Qiu, P.; Feng, L.; Fu, Q.; Dai, T.; Liu, M.; Wang, P.; Lan, Y. Dual-Functional Polyetheretherketone Surface with an Enhanced Osteogenic Capability and an Antibacterial Adhesion Property in vitro by Chitosan Modification. *Langmuir* **2022**, *38*, 14712–14724. [CrossRef] [PubMed]
33. Ravarian, R.; Craft, M.; Dehghani, F. Enhancing the Biological Activity of Chitosan and Controlling the Degradation by Nanoscale Interaction with Bioglass. *J. Biomed. Mater. Res.—Part A* **2015**, *103*, 2898–2908. [CrossRef] [PubMed]
34. Pajares-Chamorro, N.; Chatzistavrou, X. Bioactive Glass Nanoparticles for Tissue Regeneration. *ACS Omega* **2020**, *5*, 12716–12726. [CrossRef]
35. Jurak, M.; Wiącek, A.E.; Mroczka, R.; Łopucki, R. Chitosan/Phospholipid Coated Polyethylene Terephthalate (PET) Polymer Surfaces Activated by Air Plasma. *Colloids Surf. A Physicochem. Eng. Asp.* **2017**, *532*, 155–164. [CrossRef]
36. Jurak, M.; Mroczka, R.; Łopucki, R. Properties of Artificial Phospholipid Membranes Containing Lauryl Gallate or Cholesterol. *J. Membr. Biol.* **2018**, *251*, 277–294. [CrossRef]
37. Groth, T.; Altankov, G. Studies on Cell-Biomaterial Interaction: Role of Tyrosine Phosphorylation during Fibroblast Spreading on Surfaces Varying in Wettability. *Biomaterials* **1996**, *17*, 1227–1234. [CrossRef] [PubMed]
38. Menzies, K.L.; Jones, L. The Impact of Contact Angle on the Biocompatibility of Biomaterials. *Optom. Vis. Sci.* **2010**, *87*, 387–399. [CrossRef]
39. Przykaza, K.; Woźniak, K.; Jurak, M.; Wiącek, A.E. Wettability of Chitosan-Modified and Lipid/Polypeptide-Coated PEEK Surfaces. *Prog. Chem. Appl. Chitin Its Deriv.* **2019**, *24*. [CrossRef]
40. Sundriyal, P.; Sahu, M.; Prakash, O.; Bhattacharya, S. Long-Term Surface Modification of PEEK Polymer Using Plasma and PEG Silane Treatment. *Surf. Interfaces* **2021**, *25*, 101253. [CrossRef]
41. Akhtar, M.A.; Mariotti, C.E.; Conti, B.; Boccaccini, A.R. Electrophoretic Deposition of Ferulic Acid Loaded Bioactive Glass/Chitosan as Antibacterial and Bioactive Composite Coatings. *Surf. Coat. Technol.* **2021**, *405*, 126657. [CrossRef]
42. Chibowski, E. Surface Free Energy of a Solid from Contact Angle Hysteresis. *Adv. Colloid Interface Sci.* **2003**, *103*, 149–172. [CrossRef] [PubMed]
43. Kokubo, T.; Takadama, H. How Useful Is SBF in Predicting in vivo Bone Bioactivity? *Biomaterials* **2006**, *27*, 2907–2915. [CrossRef] [PubMed]
44. Morozowich, N.L.; Lerach, J.O.; Modzelewski, T.; Jackson, L.; Winograd, N.; Allcock, H.R. Characterization of Hydroxyapatite Deposition on Biomimetic Polyphosphazenes by Time-of-Flight Secondary Ion Mass Spectrometry (ToF-SIMS). *RSC Adv.* **2014**, *4*, 19680–19689. [CrossRef]
45. Fraga, A.F.; Filho, E.D.A.; Rigo, E.C.D.S.; Boschi, A.O. Synthesis of Chitosan/Hydroxyapatite Membranes Coated with Hydroxycarbonate Apatite for Guided Tissue Regeneration Purposes. *Appl. Surf. Sci.* **2011**, *257*, 3888–3892. [CrossRef]
46. Pishbin, F.; Mouriño, V.; Flor, S.; Kreppel, S.; Salih, V.; Ryan, M.P.; Boccaccini, A.R. Electrophoretic Deposition of Gentamicin-Loaded Bioactive Glass/Chitosan Composite Coatings for Orthopaedic Implants. *ACS Appl. Mater. Interfaces* **2014**, *6*, 8796–8806. [CrossRef]
47. Saini, M. Implant Biomaterials: A Comprehensive Review. *World J. Clin. Cases* **2015**, *3*, 52. [CrossRef]
48. Cowie, R.M.; Briscoe, A.; Fisher, J.; Jennings, L.M. PEEK-OPTIMA^{TM} as an Alternative to Cobalt Chrome in the Femoral Component of Total Knee Replacement: A Preliminary Study. *Proc. Inst. Mech. Eng. Part H J. Eng. Med.* **2016**, *230*, 1008–1015. [CrossRef]

49. Schwitalla, A.D.; Spintig, T.; Kallage, I.; Müller, W.D. Flexural Behavior of PEEK Materials for Dental Application. *Dent. Mater.* **2015**, *31*, 1377–1384. [CrossRef]
50. van Oss, C.J.; Chaudhury, M.K.; Good, R.J. Interfacial Lifshitz—van der Waals and Polar Interactions in Macroscopic Systems. *Chem. Rev.* **1988**, *88*, 927–941. [CrossRef]

molecules

MDPI

Communication

Novel Chitosan-Based Schiff Base Compounds: Chemical Characterization and Antimicrobial Activity

Riccardo Fontana [1], Peggy Carla Raffaella Marconi [1], Antonella Caputo [1,*] and Vasak B. Gavalyan [2,*]

[1] Department of Chemical, Pharmaceutical and Agricultural Sciences, University of Ferrara, Via Fossato di Mortara 64b, 44121 Ferrara, Italy; fntrcr1@unife.it (R.F.); mcy@unife.it (P.C.R.M.)

[2] A.I. Alikhanyan National Science Laboratory, Yerevan Physics Institute, 2-Alikhanyan Brothers St., Yerevan 0036, Armenia

* Correspondence: cpa@unife.it (A.C.); vasak@yerphi.am (V.B.G.)

Abstract: Chitosan (CS) and its derivatives are receiving considerable attention for their great biocompatibility and broad-spectrum activities in many fields. In this work, we aimed to characterize the antimicrobial activity of novel chitosan Schiff bases (CSSB). CS was synthesized by double deacetylation of chitin (Cn) after its extraction from the armors of crustaceans *Astacus leptodactylus*, and CSSB-1 and CSSB-2 were synthesized by interaction of CS with 4-(2-chloroethyl) benzaldehyde (aldehyde-1) and 4-(bromoethyl) benzaldehyde (aldehyde-2), respectively, at room temperature. The synthesized compounds were characterized by elemental analysis, gel permeation chromatography (GPC), infrared spectroscopy (FTIR), thermogravimetry (TG), and differential scanning calorimetry (DSC). The antimicrobial activity against Gram-positive (*Staphylococcus aureus*) and Gram-negative (*Pseudomonas aeruginosa*) bacteria and against yeasts (*Candida albicans*) was significantly increased due to their higher solubility as compared to unmodified CS opening perspectives for the use of these compounds for antimicrobial prevention in different fields as, for example, food industry, cosmetics, or restoration.

Keywords: chitosan; chitosan-based Schiff base; antibacterial activity; antifungal activity

check for updates

Citation: Fontana, R.; Marconi, P.C.R.; Caputo, A.; Gavalyan, V.B. Novel Chitosan-Based Schiff Base Compounds: Chemical Characterization and Antimicrobial Activity. *Molecules* **2022**, *27*, 2740. https://doi.org/10.3390/molecules27092740

Academic Editor: Agnieszka Ewa Wiącek

Received: 6 April 2022
Accepted: 20 April 2022
Published: 24 April 2022

1. Introduction

The importance of finding new antimicrobial platforms stems from the study of antimicrobial resistance and the fact that microorganisms are widespread and can be responsible for major infections. Antibiotic resistance is currently the main threat to infectious diseases due to the epidemiological and economic impact of this phenomenon [1]. In an attempt to find therapeutic alternatives to combat microorganisms, no alternative is precluded: from synthetic molecules to plant extracts, from food processing of by-products to biofermentative processes, and even to urban waste. All these are potentially rich sources of bioactive molecules, such as antimicrobials, or molecules capable of altering the mechanisms of biofilm formation or capable of dissolving the matrix; in this perspective, chitosan (CS) and natural antimicrobial compounds could play a key role, so much so that CS is currently at the center of great international research interest. The use of natural compounds and medicinal plants dates back to ancient times, and chitosan, thanks to its multiple properties in different fields, could be a stimulating starting point for the discovery of new pharmacological/antimicrobial platforms. Chitosan has been studied as an antimicrobial material against a wide range of target organisms such as algae, bacteria, yeasts, and fungi in experiments involving in vivo and in vitro interactions with chitosan in different forms (solutions, films, and composites) [2]. Since the broad-spectrum antibacterial activity of this material was first proposed by Allan and Hardwiger [3], together with the great commercial potential, the antimicrobial property of chitosan and its derivatives has attracted great attention from researchers. With this in mind, we investigated the effect of

CS and CSSB since the CSSBs reported so far exhibit better antimicrobial properties than bare chitosan.

Chitosan (CS) (β-(1 → 4)-2-amino-2-deoxy-D-glucopyranose) and its various derivatives are used in many fields due to their unique characteristics [4,5]. CS is prepared by deacetylation of chitin (β-(1 → 4)-2-acetamido-2-deoxy-D-glucopyranose), which is one of the most common natural polysaccharide biopolymers, inferior in nature only to cellulose. Among derivatives of CS, in our opinion, of particular interest are Schiff bases of chitosan (CSSB). Their synthesis is carried out via the reaction of the amino groups of CS with aldehydes or ketones. CS and some CSSB derivatives are inexpensive, non-toxic, highly biodegradable, biocompatible, and biologically active substances and are being used wildly in biomedicine, cosmetics, restoration, packaging of food, and environmental remediation. The biological activities of CSSB, including the antimicrobial properties, are important characteristics in terms of their specific use. Key physicochemical characteristics, such as average molecular weight (Mw), degree of N-acetylation (DA), polydispersity (Mw/Mn), crystallinity, and pattern of acetylation, are important determinants of CS and CSSB activities since they influence solubility in water and organic solvents [6–10].

The biological characteristics of CSSB synthesized on the basis of some aromatic aldehydes were recently investigated by several groups for their antimicrobial activity. CSSB was synthetized by reaction of CS with salicylaldehyde and showed that both the nature of the metal (Zn (II), Pd (II), and Pt (II)) complexes with the Schiff bases and the molecular weight of the starting CS (223 and 64 kDa) have a significant effect on the cytotoxicity, and in general on the antimicrobial activity of the compounds [11,12]. Based on 2-hydroxy-3-metoxybenzaldehyde, CSSB and its complex compounds with metal ions Cu (II), Ni (II), and Zn (II) were synthetized and shown to have better anticancer activity in vitro than pure CS [13]. In addition, acetophenone derivatives (4-aminoacetophenone enone and 4-bromoacetophenone) were used to synthesize CSSBs, and the antimicrobial activity of these chitosan derivatives was higher than unmodified CS, while their anticancer activity was inferior to CS [14]. In this paper, we present the results of our studies of CSSBs synthesized using aromatic aldehydes, namely 4-(2-chloroethyl) benzaldehyde (aldehyde-1) and 4-(2-bromoethyl) benzaldehyde (aldehyde-2) according to a previously described procedure [15]. In the reaction products (CSSB-1 and CSSB-2), there is an XCH_2CH_2 group capable of HX elimination reactions (HX = HCl, HBr) and the formation of pending vinyl groups [15]. By the reacting pending vinyl groups of CSSBs, novel polymer and copolymer systems with substantially valuable properties can be synthesized. Vinyl groups of CSSBs can also be considered as new potential centers for complexation (π complexes) with metal ions.

The starting CS was a double de-acetylated product with a degree of deacetylation of 81.7% and Mw of 84.5 kDa. By reacting CS with aldehyde-1 and aldehyde-2, CSSB-1 and CSSB-2 were synthesized, respectively, and physicochemically characterized by means of TG, FTIR, and GPC, as well as characterized for their antibacterial and antifungal activity.

2. Results and Discussion

2.1. *Characteristics of Chitosan-Based Schiff Base Compounds*

The CS used for the synthesis of CSSB-1 and CSSB-2 had a degree of deacetylation (DDA) of 81.7% and an average molecular weight value of Mn-48.8 kDa, Mw-84.5 kDa, and coefficient of polydispersity Mw/Mn 1.73. We believe that such a relatively low value of average molecular weight (AMW) was achieved by double deacetylation of the CS product. The scheme (Supplementary Materials, Figure S1) describes the synthesis of CSSB-1 and CSSB-2 by means of reaction of CS with aldehydes-1 and -2 (condensation reaction) to obtain the final products CSSB-1 and CSSB-2, respectively. Rather soft conditions of synthesis of CSSB-1 and CSSB-2 (temperature of the reaction environment or of condensation reaction 22 °C) suggest that values of AMW-s of CSSB-1 and CSSB-2 products do not undergo serious changes staying within limits Mn-48.8 kDa, Mw-84.5 kDa. This is important since the DDA and AMW values of CS and chitosan derivatives have a significant effect on many properties, including their biological activity.

The results of elemental analysis of the CS, CSSB-1, and CSSB-2 samples are shown in Table 1. The data from the elemental analysis show that in the tested samples, the percentages of carbon and nitrogen change noticeably. This is evident when comparing CS data with CSSB-1 and CSSB-2. Note that the fully de-acetylated CS contains 44.66% and 8.68% of carbon and nitrogen, respectively. The degree of substitution (DS) is an important characteristic for CS derivatives, including CSSB. The degree of substitution of hydrogen of the CS amino groups by the corresponding radical of aldehyde-1 or -2 is shown in the reaction of the scheme. The DS calculated by equation previously described [16] for CSSB-1 and CSSB-2 are 14.2% and 15.7%, respectively.

Table 1. Elemental analysis data of Cn, CS, CSSB-1, and CSSB-2.

Sample	Obtained				Calculated [a]							Yield %
	%C	%N	%H	C/N	%C	%N	%H	%O	% Cl	% Br	C/N	
Cn, DA 90.5%	46.33	6.98	6.36	6.64	47.09	7.03	6.48	39.40	-	-	6.70	-
CS, DA 18.3%	45.37	8.31	6.73	5.46	45.28	8.29	6.80	39.63	-	-	5.46	-
CSSB-1	48.13	7.54	6.37	6.38	47.92	7.47	6.59	35.68	2.35	-	6.41	17.3
CSSB-2	47.00	6.86	6.25	6.85	46.80	7.05	6.32	33.68	-	6.16	6.77	19.4

[a] In calculations, the content of water is not considered.

The data of thermogravimetry (TG) and differential scanning calorimetry (DSC) of CS, CSSB-1, and CSSB-2 samples are shown in Figure 1. It is noteworthy that intense weight loss (TG curves) for samples CSSB-1 and CSSB-2 (Figure 1b,c) starts at a lower temperature (200 °C) in comparison with sample CS (240 °C) (Figure 1a), while for all samples, CS, CSSB-1, and CSSB-2, it ends at temperatures of about 420 °C. The process of intense weight loss is accompanied by a single exothermic effect that clearly manifests itself on DSC curves. The exothermic effect for the CS sample (Figure 1a) begins at 287 °C and reaches the maximum temperature value (the minimum point on the DSC curve) at 322 °C. The starting temperatures and the minimum points on the DSC curve for samples CSSB-1 and CSSB-2 have values of 243, 270 °C and 239, 271 °C, respectively. Our results suggest that the thermal stability of CSSB relative to CS is significantly lower in agreement with other results [11,13,17]. We believe that in the synthesis of CS derivatives, the supramolecular structure of CS is destroyed, mainly due to numerous hydrogen bonds between macromolecules. The mobility of CS derivative macromolecules increases; thus, significantly less thermal energy is required to break down the weakest bonds compared to CS. In our opinion, it is this that can explain the fact that in terms of thermal stability, CSSB-1 and CSSB-2 are inferior to CS.

The main results of absorption spectra of the CS, CSSB-1, and CSSB-2 samples are summarized in Table 2 and Supplementary Figure S2. Attention is drawn to the lack of absorption in the region around 1651 cm^{-1} for samples of CSSB-1 and CSSB-2. We suggest that the absorption of the imine bond (C=N stretch, imine), which takes place with sufficient intensity in the region around 1640 cm^{-1}, is inferior to the absorbance characteristic of the OH and NH2 (OH; NH2-stretch) and C-O-C and C-N (asymmetric stretch C-O-C and C-N stretch) groups. In fact, there is complete overlap in the absorption of the C=O bond (C=O stretch amide 1) in the region of about 1651 cm^{-1}. As a result, in the ranges of absorption of samples CSSB-1 and CSSB-2, there is no absorption in the region near 1651 cm^{-1}. The CS, CSSB-1, and CSSB-2 absorption spectra are in agreement with the data discussed by other groups [11,18–21].

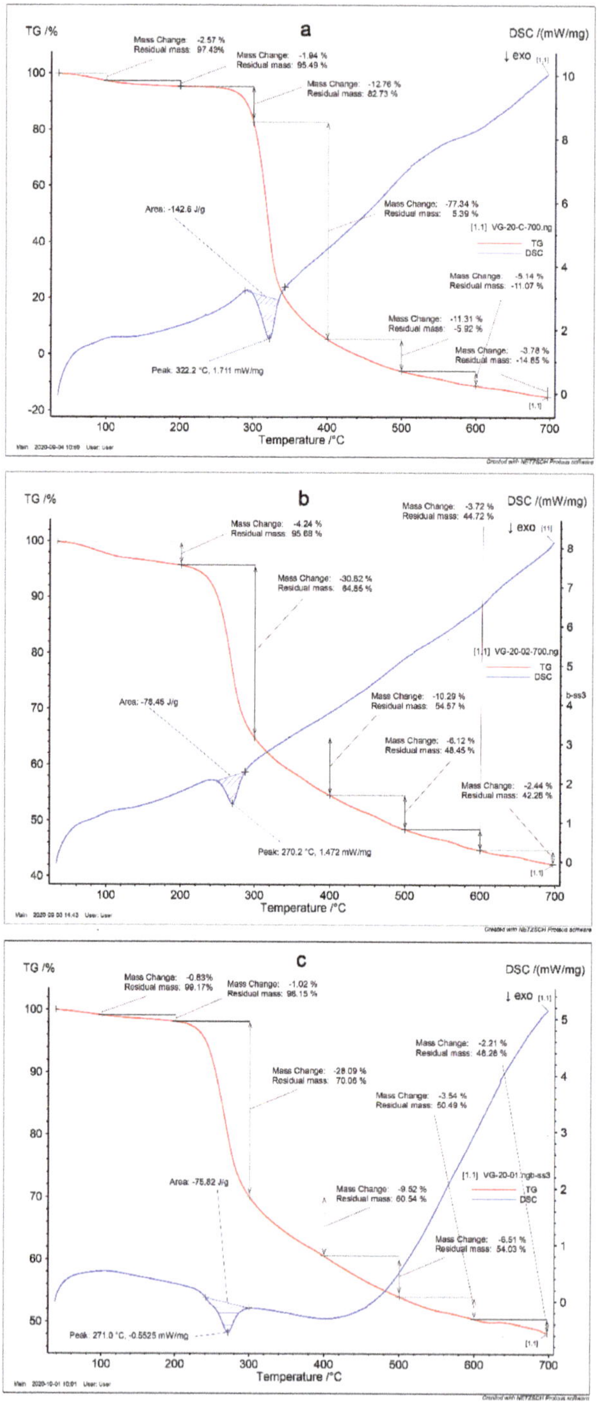

Figure 1. Characterization of the compounds. TG and DSC data of CS (**a**), CSSB−1 (**b**), and CSSB−2 (**c**) samples.

Table 2. FTIR data of CS, CSSB-1, and CSSB-2.

Sample	Vibration Type in Group and Its Wavenumber, cm^{-1}						
	OH; NH_2 Stretch	C=N Stretch (Imine)	C-H Stretch (CH_2)	C=O Stretch Amide 1	C=C; C-H Stretch Aromatic Ring	Asymmetric Stretch C-O-C and C-N Stretch	O-bridge Stretch (Glucosamine)
CS	3362.5 3296.2	-	2874.5	1651.9	-	1150.7 1081.7	1028.7
CSSB-1	3367.7 3296.3	1639.4	2871.5	-	1602.1; 1522.1 1454.5	1150.4 1064.3	1030.1
CSSB-2	3364.6 3306.8	1641.0	2875.0	-	1609.2; 1518.2 1447.8	1150.3 1067.9	1025.5

2.2. Antimicrobial Properties of CS, CSSB-1, and CSSB-2

In order to assess and compare the antimicrobial activity of CS and its Schiff bases derivatives, CSSB-1 and CSSB-2, two bacterial strains (i.e., the Gram-negative *P. aeruginosa* and the Gram-positive *S. aureus*) and one yeast strain (*C. albicans*) were used as model microorganisms. They were incubated with three different concentrations (0.006%, 0.012%, and 0.024%) of each compound for 6, 18, and 24 h to determine both the lower effective dose and a time-dependent effect. The antimicrobial activity of CS, CSSB-1, and CSSB-2 was measured as the reduction in growth (i.e., reduction in colony-forming units/mL (CFU/mL) and the reduction in the optical density at 600 nm compared to each untreated microorganism).

As shown in Figures 2–4 and Supplementary Tables S1–S3, as observed by the numbers of alive cells (CFU/mL) and the absorbance values of the cultures, in general, the antimicrobial activity of the CS, CSSB-1, and CSSB-2 compounds was inversely related to their concentration at each time point and for each microorganism. Of note, for *S. aureus*, the effect was also inversely related to the time of contact since at the lower dose of 0.006%, the reduction in bacterial growth was higher after 24 h of contact (for all compounds, 100% of growth reduction compared to untreated cells) than at shorter time points of 6 h (CS: 36%; CSSB-1: 47%; CSSB-2: 29%) and 18 h (CS: 67%; CSSB-1: 92%; CSSB-2: 90%) (Figures 5 and S3 and Table S4). A slightly different result was observed for *P. aeruginosa* as, in general, the lower dose was less effective. In fact, although the antibacterial effect of the CS, CSSB-1, and CSSB-2 compounds was inversely related to their concentration at each time point, as for *S. aureus*, however, all three compounds at 0.006% were less effective as compared to *S. aureus*, in controlling the growth of *P. aeruginosa* at all-time points, whereas at the higher concentration (0.024%) 100% inhibition of *P. aeruginosa* growth was observed in a fashion similar to *S. aureus* (Figures 6 and S4 and Table S5). These results suggest a different behavior of CS and CSSBs depending on the nature and structure of the outer surface of the Gram-positive and Gram-negative bacterial cells. The lower antibacterial activity at dose of 0.006% observed on *P. aeruginosa* suggest that the outer membrane of Gram-negative cells creates a barrier, likely reducing the entry of the molecules and thus their effect on cell growth. That the chemical structure of the cell surface may have a role in modulating the antimicrobial activity of CS, CSSB-1, and CSSB-2 is also suggested by the observation that all three compounds showed significantly greater inhibitory activity on yeast than bacterial cells. In fact, almost 100% reduction in *C. albicans* growth was observed even at the lowest concentration of 0.006% already after 6 h of contact. This may depend on the different chemical compositions of the walls of yeasts and bacteria, made of chitin and peptidoglycan, respectively (Figures 7 and S5, and Table S6). It is plausible to hypothesize that these compounds (CS, CSSB-1, and CSSB-2) compete with chitin precursors and interfere with the synthesis of the yeast wall, inhibiting the growth of yeast cells very efficiently.

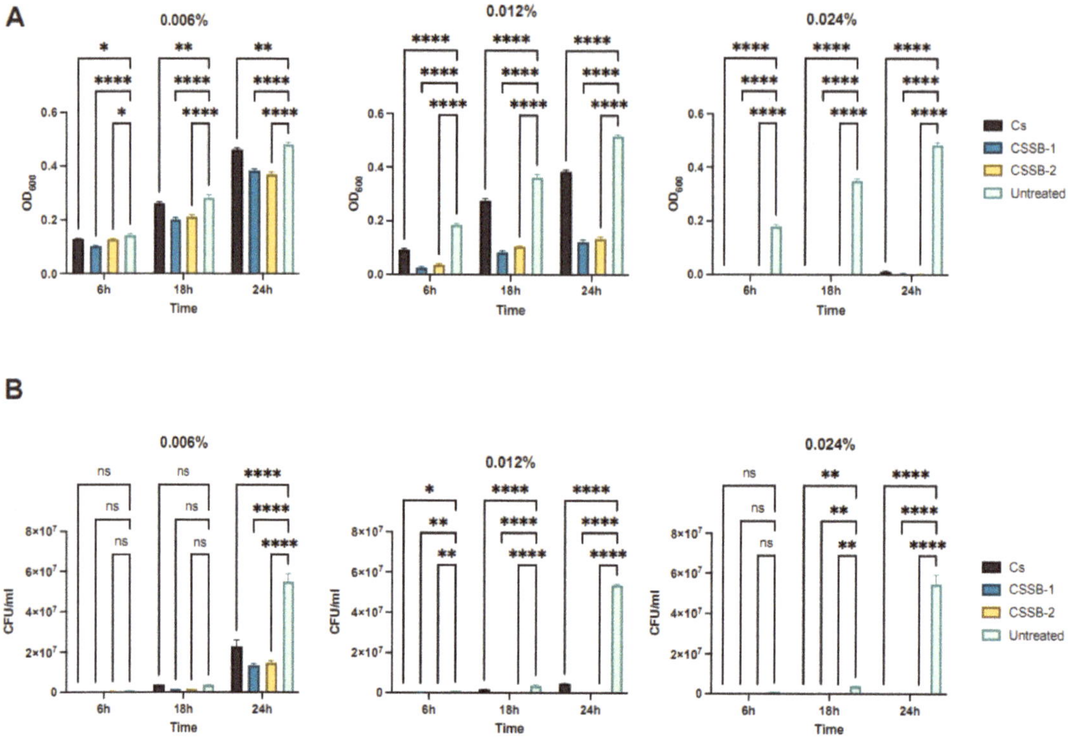

Figure 2. Antimicrobial activity of CS, CSSB-1, and CSSB-2 compounds on *S. Aureus*. Fresh cultures of *S. aureus* were incubated in liquid cultures with three different concentrations (0.006%, 0.012%, and 0.024%) of CS, CSSB-1, and CSSB-2 for 6, 18, and 24 h or without the compounds (untreated cells). After incubation, the antimicrobial activity was evaluated by measuring the optical density at 600 nm of each liquid culture (**A**) and by determining the number of colony-forming units/mL (CFU/mL) (**B**) after plating the cultures onto agar plates and incubation to allow colony formation. *, $p = 0.01$; **, $p = 0.001$; ****, $p < 0.0001$.

Finally, our results indicate that the CS Schiff base derivatives, CSSB-1 and CSSB-2, exerted higher antimicrobial activity than unmodified CS as compared to the untreated cultures. The action of these compounds may be related to their positive charged amino groups binding to the surface of bacterial wall or membrane, which are negatively charged, leading to alteration of cell permeability and cell death. This is particularly evident against *S. aureus* and *P. aeruginosa* at all time points with doses of 0.006% and 0.012%. The greater activity might be due to the higher water solubility of CSSB-1 and CSSB-2 compounds, allowing for their almost complete suspension, and to the substitutions themselves (Figures 5–7 and S3–S5, and Table S4–S6).

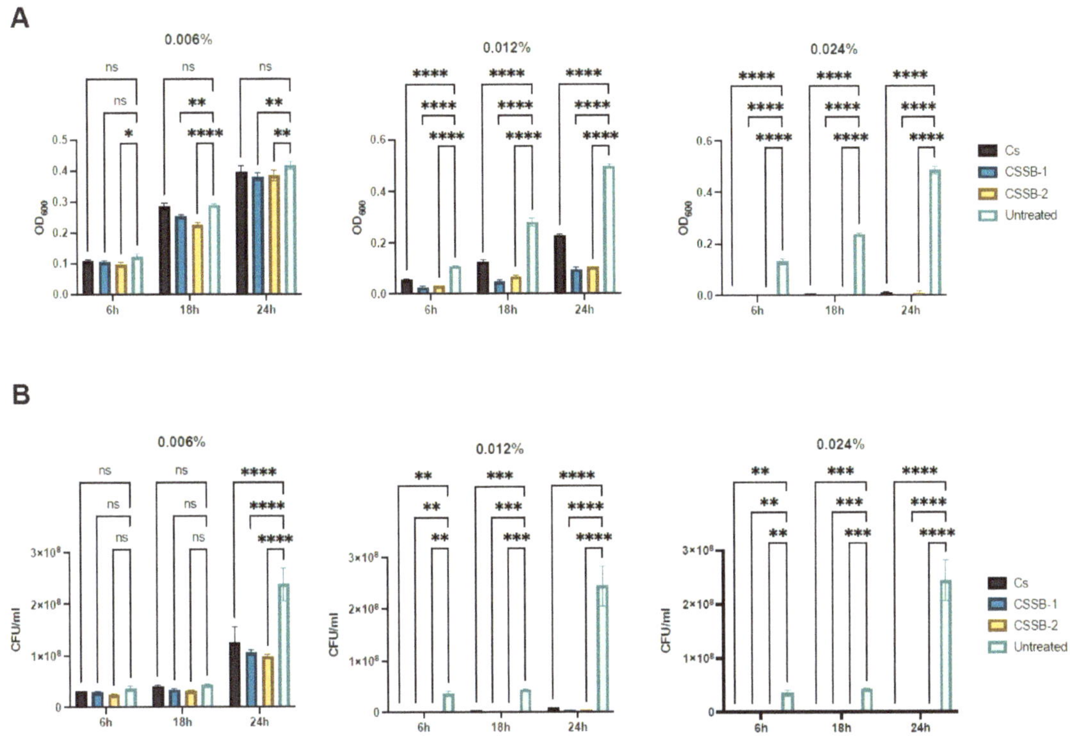

Figure 3. Antimicrobial activity of CS, CSSB-1, and CSSB-2 compounds on *P. aeurginosa*. Fresh cultures of *P. aeruginosa* were incubated with three different concentrations (0.006%, 0.012%, and 0.024%) of CS, CSSB-1, and CSSB-2 for 6, 18, and 24 h or without the compounds (untreated cells). After incubation, the antimicrobial activity was evaluated by measuring the optical density at 600 nm of each liquid culture (**A**) and by determining the number of colony-forming units/mL (CFU/mL) (**B**) after plating the cultures onto agar plates and incubation to allow colony formation. *, $p = 0.01$; **, $p = 0.001$; ***, $p = 0.0001$; ****, $p < 0.0001$.

Other studies showed that modification of CS can increase its antimicrobial activity. CS modified with acetophenone derivatives through Schiff base reaction possesses increased efficacy due to its electron-donating nature [14]. Recently, it was also reported that a new CSSB (3EtO-4OH/Chit) and its 3EtO-4OH/Chit/Fe_2O_3 nanocomposite have greater antimicrobial activity than CS itself with 3EtO-4OH/Chit/Fe_2O_3 showing the highest activity due to the presence of Fe_2O_3 nanoparticles, which increase the total antimicrobial property of 3EtO-4OH/Chit [20]. Yin et al. tried an approach similar to ours [22] and showed that CSSBs synthetized from an aldehyde showed better antimicrobial properties, thanks to these electron-withdrawing substituents. Our findings and data from the literature allow us to state that these substitutions enhance the antibacterial activity of CS-based products.

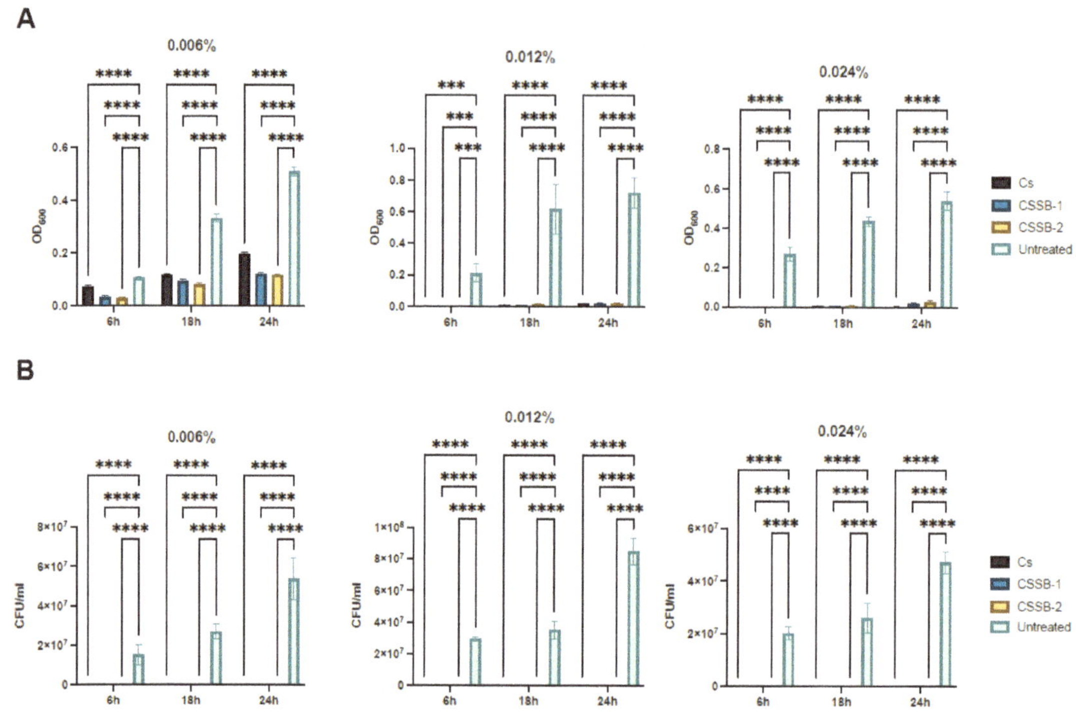

Figure 4. Antimicrobial activity of CS, CSSB-1, and CSSB-2 compounds on *C. albicans*. Fresh cultures of *C. albicans* were incubated with three different concentrations (0.006%, 0.012%, and 0.024%) of CS, CSSB-1, and CSSB-2 for 6, 18, and 24 h or without the compounds (untreated cells). After incubation, the antimicrobial activity was evaluated by measuring the optical density at 600 nm of each liquid culture (**A**) and by determining the number of colony-forming units/mL (CFU/mL) (**B**) after plating the cultures onto agar plates and incubation to allow colony formation. ***, $p = 0.0001$; ****, $p < 0.0001$.

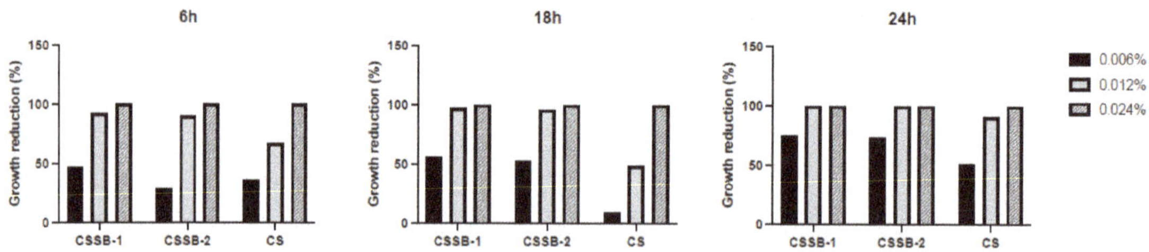

Figure 5. Growth reduction in *S. Aureus* after treatment with CS, CSSB-1, and CSSB-2 compounds. Fresh cultures of *S. aureus* were incubated in liquid cultures with three different concentrations (0.006%, 0.012%, and 0.024%) of CS, CSSB-1, and CSSB-2 for 6, 18, and 24 h or without the compounds (untreated cells). After incubation, each culture was plated onto agar plates and incubated to allow colony formation (CFU/mL, as illustrated in Figure 2). The results are represented as percentage (%) of cell growth reduction after each treatment as compared to control, untreated cells.

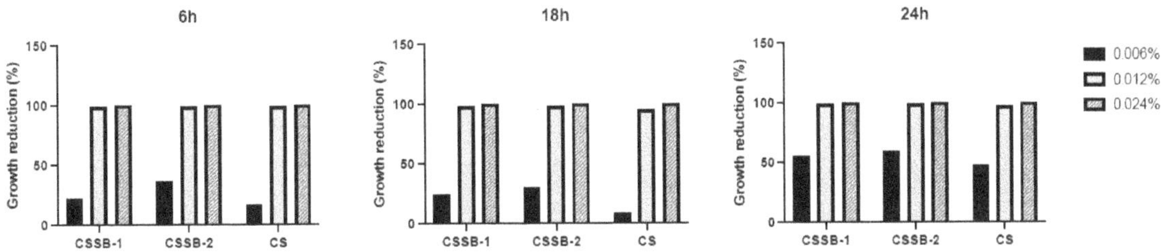

Figure 6. Growth reduction in *P. aeruginosa* after treatment with CS, CSSB-1, and CSSB-2 compounds. Fresh cultures of *P. aeruginosa* were incubated in liquid cultures with three different concentrations (0.006%, 0.012%, and 0.024%) of CS, CSSB-1, and CSSB-2 for 6, 18, and 24 h or without the compounds (untreated cells). After incubation, each culture was plated onto agar plates and incubated to allow colony formation (CFU/mL, as illustrated in Figure 3). The results are represented as percentage (%) of cell growth reduction after each treatment as compared to control, untreated cells.

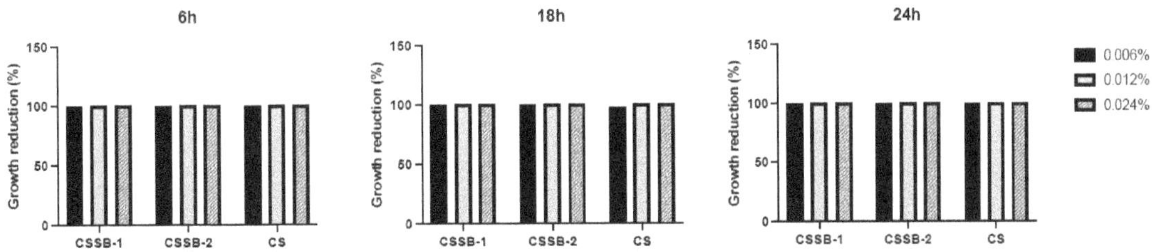

Figure 7. Growth reduction in *C. albicans* after treatment with CS, CSSB-1, and CSSB-2 compounds. Fresh cultures of *C. albicans* were incubated in liquid cultures with three different concentrations (0.006%, 0.012%, and 0.024%) of CS, CSSB-1, and CSSB-2 for 6, 18, and 24 h or without the compounds (untreated cells). After incubation, each culture was plated onto agar plates and incubated to allow colony formation (CFU/mL, as illustrated in Figure 4). The results are represented as percentage (%) of cell growth reduction after each treatment as compared to control, untreated cells.

We are proceeding with work on the synthesis and characterization of completely new O-carboxymethyl derivatives of CSSB-1 and CSSB-2 and with the study of their antimicrobial properties.

3. Materials and Methods

3.1. Reagents, Instruments, and Products

The materials and methods used are described by Gavalyan [15]. The molecular weights of the CS samples were obtained by Gel Penetration Chromatograph (GPC) Agilent 1200 (Agilent Technologies, Santa Clara, CA, USA) columns U-Hydrogel 500 and 250 (Waters), T = 25 °C. Elemental analyses (C, H, N) were performed on Euro EA 3000 analyzer (Euro Vector, Pavia, Italy). Attenuated total reflection Fourier-transform infrared spectra (FTIR ATR) was registered by an FTIR Microscope LUMOS (Bruker, Billerica, MA, USA) spectrometer (Ge prism, Happ-Genzel apodization, ATR distortion is corrected, number of scans 32; resolution: 4 cm^{-1}). The thermogravimetry (TG) was carried out on STA 449 F3 JUPITER from Netzsch (Selb, Germany), helium atmosphere (the flow rate is 20 mL/s), temperature gradient from room temperature to 700 °C/10 °C/min, the initial weight of each sample was 10 mg.

The extraction of Cn from armors of crayfish *Astacus leptodactylus* (Lake Sevan, Armenia), the synthesis of CS, aromatic aldehydes 4-(2-chloroethyl) benzaldehyde (aldehyde-1), 4-(2-bromoethyl) benzaldehyde (aldehyde-2) and chitosan Schiff bases (CSSB-1 and CSSB-2) were carried out as described [15]. The elemental analysis of CS after the second round of deacetylation was C, 45.37%, N, 8.31%, and H, 6.73%. The degree of acetylation (DA) of synthesized CS (18.3%) was calculated by using elemental analysis data and an equation reported by Kasaai et al. [23]. The yields of CSSB-1 and CSSB-2 synthesized at 22 °C (RT) were 17.3% and 19.4%, respectively. CSSB-1 and CSSB-2 were lyophilized and sent to the University of Ferrara for antimicrobial activity testing. For this purpose, they were dissolved in sterile water at 10 mg/mL for 24 h at 37 °C, filtered through 0.22 µm filters (Starlab, Milan, Italy), aliquoted, and stored at RT until use.

3.2. Bacterial Strains and Culture Conditions

Bacteria and yeast isolates were purchased from ATCC (*S. aureus* (ATCC 25923), *P. aeruginosa* (ATCC 89033) and *C. albicans* (ATCC 90028)) and stored at −80 °C in Luria-Bertani (LB) broth (Liofilchem, Roseto degli Abruzzi, TE, Italy) with 50% glycerol. The experiments were carried out with bacteria cultured in Tryptic Soy Broth (TSB) whereas yeasts in Sabouraud Dextrose Broth (SDB) (Liofilchem) in agitation at 150 rpm (MaxQ 4000, Thermo Scientific, Milan, Italy) or plated on Tryptic Soy Agar (TSA) (Scharlab Italia, Milan, Italy) or Sabouraud Dextrose Agar (SDA) (Liofilchem), respectively, and incubation at 35 °C/37 °C (bacteria) and 25 °C (yeasts).

3.3. Determination of the Antimicrobial Activity of CSSB

To determine the antimicrobial activity of the CSSB products, microorganisms were cultured in 5 mL of TSB or SDB broth for 18 h and, after incubation, the absorbance values of the cultures were read at 600 nm (OD_{600}) with a spectrophotometer to calculate the number of cells/mL using the formula "sample $OD_{600} \times 10^6/0,1$". For *P. aeruginosa* and *S. aureus* cultures, the values were further multiplied by 70 and 100, respectively, to obtain the real concentration of cells/mL. The optical density, in fact, underestimates the number of *P. aeruginosa* and *S. aureus* bacterial cells in liquid media. Thus, each bacterial strain was titrated on agar plates, and in parallel, the bacterial inocula were read with the spectrophotometer to determine the conversion factors (70 and 100, respectively, for *P. aeruginosa* and *S. aureus*). Microorganisms (5×10^2 cells/mL) were then diluted 10^{-3} and 10^{-4} in a liquid medium and incubated (3 mL of each dilution in 10 mL tubes) with CS, CSSB-1, and CSSB-2 at three different final concentrations (0.006%, 0.012%, and 0.024%) or without any compound (untreated control) for three different time points (6, 18 and 24 h). Following incubation, the tubes were observed for the presence of turbidity and the absorbance (OD_{600}) of each tube measured. Each sample (1 mL) was then plated in duplicate on the surface of TSA or SDA plates (undiluted or previously diluted 10^{-1} up to 10^{-8}) and incubated for 24 h (bacteria) and 2–5 days (yeasts). Where present, colonies were counted and reported as colony-forming units (CFU) per mL. For each compound concentration and for each time point, the analyses were performed on the results of three independent experiments and are represented as the mean ± SD.

3.4. Statistical Analysis

Statistical analysis was performed using one-way ANOVA followed by Dunnett's multiple comparisons test with GraphPad Prism version 9.0.0 for MacOS (GraphPad Software, San Diego, CA, USA), with $p \leq 0.05$ to identify significant differences. *, $p = 0.01$; **, $p = 0.001$; ***, $p = 0.0001$; ****, $p < 0.0001$.

4. Conclusions

CS was synthesized from Cn, and then by its interaction with aromatic aldehydes, novel chitosan Schiff bases (CSSB) were produced and structurally characterized. We have shown that the thermal stability of CSSB is inferior, and the supramolecular structure of CS is destroyed, mainly due to numerous hydrogen bonds between macromolecules. The mobility of CS derivative macromolecules increases; thus, significantly less thermal energy is required to break down the weakest bonds compared to CS. The antibacterial and antifungal activity of the novel compounds (CSSB-1 and CSSB-2) have significantly improved due to their higher solubility as compared to CS.

Supplementary Materials: The following supporting information can be downloaded at: https://www.mdpi.com/article/10.3390/molecules27092740/s1, Tables S1–S6 and Figures S1–S5.

Author Contributions: Contribution of the authors are as follows: Conceptualization, preparation, and physicochemical characterization of the CS-derived compounds, V.B.G.; Methodology standardization and testing of antimicrobial activity, R.F., P.C.R.M. and A.C.; Formal Analysis of all data, R.F., A.C. and V.B.G.; Writing—Original Draft Preparation, R.F., A.C. and V.B.G.; Review and Editing, R.F. and A.C.; Discussion of the results and of the final manuscript, R.F., A.C., P.C.R.M. and V.B.G.; Funding Acquisition, V.B.G., P.C.R.M. and A.C. All authors have read and agreed to the published version of the manuscript.

Funding: This work stems from a collaboration agreement between the A.I. Alikhanyan National Science laboratory (Yerevan Physics Institute) Foundation and the University of Ferrara. The synthesis and physicochemical characterization of the compounds was performed within the basic funding program from the AANL-YerPhI Foundation of the Government of Armenia, whereas no specific grants from funding agencies in the public, commercial or not-for-profit sectors granted the studies of biological activities. We thank the University of Ferrara for providing the funding for the project. R.F. was supported by the Ph.D. program of the University of Ferrara.

Institutional Review Board Statement: Not applicable.

Informed Consent Statement: Not applicable.

Data Availability Statement: The data are presented in the text and in supplementary material.

Acknowledgments: We thank Inesa V. Blagodatskikh (Russian Federation, Russian Academy of Sciences, INEOS, Vavilova 28, Moscow) for gel permeation chromatografic analysis of CS samples, Vardan Tadevosyan for assistance in drafting the article, and Gabriele Perrone (University of Bologna, Bologna, Italy) for assistance in the statistical analysis.

Conflicts of Interest: The authors declare no conflict of interest.

Sample Availability: Samples of the compounds are not available from the authors.

References

1. La Fauci, V.; Costa, G.B.; Genovese, C.; Palamara, M.A.R.; Alessi, V.; Squeri, R. Drug-resistant bacteria on hands of healthcare workers and in the patient area: An environmental survey in Southern Italy's hospital. *Rev. Esp. Quimioter. Publ. Soc. Esp. Quimioter.* **2019**, *32*, 303–310.
2. Kong, M.; Chen, X.G.; Xing, K.; Park, H.J. Antimicrobial properties of chitosan and mode of action: A state of the art review. *Int. J. Food Microbiol.* **2010**, *144*, 51–63. [CrossRef] [PubMed]
3. Allan, C.R.; Hadwiger, L.A. The fungicidal effect of chitosan on fungi of varying cell wall composition. *Exp. Mycol.* **1979**, *3*, 285–287. [CrossRef]
4. Hassan, M.A.; Omer, A.M.; Abbas, E.; Baset, W.M.A.; Tamer, T.M. Preparation, physicochemical characterization and antimicrobial activities of novel two phenolic chitosan Schiff base derivatives. *Sci. Rep.* **2018**, *8*, 11416. [CrossRef]
5. Aranaz, I.; Alcántara, A.R.; Civera, M.C.; Arias, C.; Elorza, B.; Caballero, A.H.; Acosta, N. Chitosan: An Overview of Its Properties and Applications. *Polymers* **2021**, *13*, 3256. [CrossRef]
6. Kumirska, J.; Weinhold, M.X.; Thöming, J.; Stepnowski, P. Biomedical Activity of Chitin/Chitosan Based Materials—Influence of Physicochemical Properties Apart from Molecular Weight and Degree of N-Acetylation. *Polymers* **2011**, *3*, 1875–1901. [CrossRef]
7. Pal, P.; Pal, A.; Nakashima, K.; Yadav, B.K. Applications of chitosan in environmental remediation: A review. *Chemosphere* **2020**, *266*, 128934. [CrossRef]
8. Pal, K.; Bharti, D.; Sarkar, P.; Anis, A.; Kim, D.; Chałas, R.; Maksymiuk, P.; Stachurski, P.; Jarzębski, M. Selected Applications of Chitosan Composites. *Int. J. Mol. Sci.* **2021**, *22*, 10968. [CrossRef]
9. Suteu, D.; Rusu, L.; Zaharia, C.; Badeanu, M.; Daraban, G.M. Challenge of Utilization Vegetal Extracts as Natural Plant Protection Products. *Appl. Sci.* **2020**, *10*, 8913. [CrossRef]
10. Antony, R.; Arun, T.; Manickam, S.T.D. A review on applications of chitosan-based Schiff bases. *Int. J. Biol. Macromol.* **2019**, *129*, 615–633. [CrossRef]
11. Barbosa, H.F.; Attjioui, M.; Leitao, A.; Moerschbacher, B.M.; Cavalheiro, T. Characterization, solubility and biological activity of amphihilic biopolymeric Schiff bases synthesized using chitosans. *Carbohydr. Polym.* **2019**, *220*, 1–11. [CrossRef] [PubMed]

12. Barbosa, H.F.G.; Attjioui, M.; Ferreira, A.P.G.; Moerschbacher, B.M.; Cavalheiro, T. New series of metal complexes by amphiphilic biopolymeric Schiff bases from modified chitosans: Preparation, characterization and effect of molecular weight on its biological applications. *Int. J. Biol. Macromol.* **2019**, *145*, 417–428. [CrossRef] [PubMed]
13. Malekshah, R.E.; Shakeri, F.; Khaleghian, A.; Salehi, M. Developing a biopolymeric chitosan supported Schiff-base and Cu(II), Ni(II) and Zn(II) complexes and biological evaluation as pro-drug. *Int. J. Biol. Macromol.* **2020**, *152*, 846–861. [CrossRef] [PubMed]
14. Raouf, O.H.; Selim, S.; Mohamed, H.; Abdel-Gawad, O.F.; Elzanaty, A.M.; Ahmed, S.A.-K. Synthesis, Characterization and Biological Activity of Schiff Bases Based on Chitosan and Acetophenone Derivatives. *Adv. J. Chem. A* **2020**, *3*, 274–282. [CrossRef]
15. Gavalyan, V.B. Synthesis and characterization of new chitosan-based Schiff base compounds. *Carbohydr. Polym.* **2016**, *145*, 37–47. [CrossRef]
16. dos Santos, J.E.; Dockal, E.R.; Cavalheiro, T. Synthesis and characterization of Schiff bases from chitosan and salicylaldehyde derivatives. *Carbohydr. Polym.* **2005**, *60*, 277–282. [CrossRef]
17. El Hamdaoui, L.; El Marouani, M.; El Bouchti, M.; Kifani-Sahban, F.; El Moussaouiti, M. Thermal Stability, Kinetic Degradation and Lifetime Prediction of Chitosan Schiff Bases Derived from Aromatic Aldehydes. *ChemistrySelect* **2021**, *6*, 306–317. [CrossRef]
18. Sadeghi, A.; Moztarzadeh, F.; Mohandesi, J.A. Investigating the effect of chitosan on hydrophilicity and bioactivity of conductive electrospun composite scaffold for neural tissue engineering. *Int. J. Biol. Macromol.* **2018**, *121*, 625–632. [CrossRef]
19. Brugnerotto, J.; Lizardi, J.; Goycoolea, F.M.; Argüelles-Monal, W.; Desbrières, J.; Rinaudo, M. An infrared investigation in relation with chitin and chitosan characterization. *Polymer* **2001**, *42*, 3569–3580. [CrossRef]
20. Foroughnia, A.; Khalaji, A.D.; Kolvari, E.; Koukabi, N. Synthesis of new chitosan Schiff base and its Fe2O3 nanocomposite: Evaluation of methyl orange removal and antibacterial activity. *Int. J. Biol. Macromol.* **2021**, *177*, 83–91. [CrossRef]
21. Rasweefali, M.; Sabu, S.; Sunooj, K.; Sasidharan, A.; Xavier, K.M. Consequences of chemical deacetylation on physicochemical, structural and functional characteristics of chitosan extracted from deep-sea mud shrimp. *Carbohydr. Polym. Technol. Appl.* **2021**, *2*, 100032. [CrossRef]
22. Yin, X.; Chen, J.; Yuan, W.; Lin, Q.; Ji, L.; Liu, F. Preparation and antibacterial activity of Schiff bases from O-carboxymethyl chitosan and para-substituted benzaldehydes. *Polym. Bull.* **2011**, *68*, 1215–1226. [CrossRef]
23. Kasaai, M.R.; Arul, J.; Charlet, G. Intrinsic viscosity-molecular weight relationship for chitosan. *J. Polym. Sci. Part B Polym. Phys.* **2000**, *38*, 2591–2598. [CrossRef]

molecules MDPI

Article

Chitosan and Its Carboxymethyl-Based Membranes Produced by Crosslinking with Magnesium Phytate

Adam Zając [1,*], Wojciech Sąsiadek [1], Lucyna Dymińska [1,*], Paulina Ropuszyńska-Robak [1], Jerzy Hanuza [2], Maciej Ptak [2], Szymon Smółka [2], Radosław Lisiecki [2] and Katarzyna Skrzypczak [3]

1 Department of Bioorganic Chemistry, Faculty of Production Engineering, Wroclaw University of Economics and Business, 118-120 Komandorska Str., 53-345 Wrocław, Poland
2 Institute of Low Temperature and Structure Research, 2 Okólna Str., 50-422 Wrocław, Poland
3 Faculty of Chemistry, Wrocław University of Science and Technology, 27 Wybrzeże Wyspiańskiego Str., 50-370 Wrocław, Poland
* Correspondence: adam.zajac@ue.wroc.pl (A.Z.); lucyna.dyminska@ue.wroc.pl (L.D.); Tel.: +48-71-368-0299 (A.Z. & L.D.)

Abstract: Membranes produced by crosslinking chitosan with magnesium phytate were prepared using highly deacetylated chitosan and its *N*-carboxymethyl, *O*-carboxymethyl and *N,O*-carboxymethyl derivatives. The conditions of the membrane production were described. IR, Raman, electron absorption and emission spectra were measured and analyzed for all the substrates. It was found that *O*-carboxymethyl chitosan derivative is the most effectively crosslinked by magnesium phytate, and the films formed on this substrate exhibit good mechanical parameters of strength, resistance and stability. Strong O–H···O hydrogen bonds proved to be responsible for an effective crosslinking process. Newly discovered membrane types produced from chitosan and magnesium phytate were characterized as morphologically homogenous and uniform by scanning electron microscopy (SEM) and IR measurements. Due to their good covering properties, they do not have pores or channels and are proposed as packaging materials.

Keywords: phytate crosslinked chitosan; synthesis and spectroscopic properties; crosslinking mechanism; new packaging material; mechanical characterization

Citation: Zając, A.; Sąsiadek, W.; Dymińska, L.; Ropuszyńska-Robak, P.; Hanuza, J.; Ptak, M.; Smółka, S.; Lisiecki, R.; Skrzypczak, K. Chitosan and Its Carboxymethyl-Based Membranes Produced by Crosslinking with Magnesium Phytate. *Molecules* **2023**, *28*, 5987. https://doi.org/10.3390/molecules28165987

Academic Editor: Agnieszka Ewa Wiącek

Received: 29 May 2023
Revised: 6 August 2023
Accepted: 8 August 2023
Published: 10 August 2023

1. Introduction

Polysaccharide-based membranes have been used for years in medicine, pharmacy, the food industry and technical areas. Polysaccharides are mainly extracted from plants, algae and animal sources. Chitosan is an example of a polysaccharide obtained by deacetylation of chitin, the second most abundant organic compound in nature after cellulose. Due to its biocompatibility, biodegradability, non-toxicity and chemical properties, it was widely used for producing films, gels, nano- and micro-particles and hydrogel beads [1–9]. For these reasons, chitosan is considered an attractive material for the preparation of membranes for various applications [10–20]. However, it has been found that chitosan-based hydrogel membranes exhibit low mechanical stability, high water content and a loose three-dimensional arrangement. Bulk and surface crosslinking was proposed as a new method for improving the chemical and mechanical stability of chitosan and prepared from their products [15,16,19–31]. Different types of crosslinkers are used, both covalently and ionically coupled to chitosan. In these processes glutaraldehyde and glyoxal [22,23,26,32], genipine, citrate, tripolyphosphate and sulfate were applied [32,33]. In other studies, starch [25], montmorillonite [29] and vanillin [30] were used as crosslinkers for chitosan. It was found that the properties of the final products, namely, mechanical stretching, swelling, permeation of low molecular weight molecules and drug releasing [34] depend on the pH of the used media and charge density [35] as well as the degree of deacetylation of chitosan [32]. The effect of the preparation conditions and molecular weight of chitosan on the structure, crystallinity, morphology, hydrophilic properties and water state were studied using several methods [21]. The obtained chitosan-based membranes with different ionic crosslinking densities were proposed for pharmaceutical and industrial applications [21].

In the present work, we propose a new technique of chitosan crosslinking using magnesium phytate. We believe that membranes produced by this process will be applied in the medicine and food industries. The materials we obtained were studied by infrared and Raman spectroscopy methods, including using a confocal Raman microscope for mapping their surface. Electron absorption and emission measurements were also taken. These methods are very sensitive to the presence of structural changes in materials and changes in their chemical composition. In addition, they register the presence of various functional groups in their composite. They play a special role in determining how to link biopolymers with a component in the form of a composite that bonds them by a system of chemical bonds, ionic interactions and hydrogen bridges.

2. Results and Discussion

Three types of carboxymethyl derivatives of chitosan (Section 3.3), namely, *N*-carboxymethyl chitosan (**NCC**), *O*-carboxymethyl chitosan (**OCC**) and *N,O*-carboxymethyl chitosan (**NOCC**), were obtained as the results of the below-described syntheses. The structures of these derivatives are shown in Figure 1. All the obtained products were identified by measuring their IR and Raman spectra and comparing them with the spectra of free chitosan.

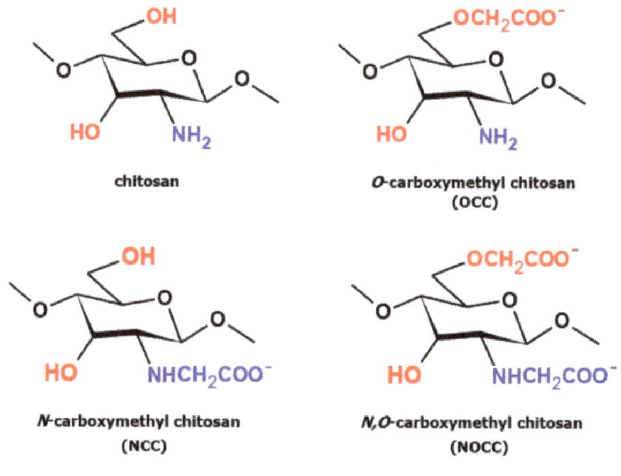

Figure 1. Structures of chitosan and three types of its carboxymethyl derivatives.

2.1. Vibrational Spectra

The MIR and FIR spectra as well as Raman spectra of the studied chitosan and its derivatives are shown in Figures 2 and 3, respectively.

Analyzing the IR and Raman spectra of the studied materials it should be noted that all spectral patterns are very similar because all the studied compounds contain the same basic skeleton—the pyranoid ring. The positions of vibrational bands corresponding to this unit are practically stable for all the studied samples. The differences between them appears because carboxymethyl derivatives contain an additional –CH_2-COO– unit substituted at the hydroxyl or amine group. The bands corresponding to this chromophore are observed in the following ranges of the IR spectra: 1590–1593, 1381–1414, 1262–1283, 773–822, 694–605 and 512–575 cm^{-1}. The substitution of the carboxymethyl unit at hydroxyl or amine groups provokes specific changes in typical ranges of the IR and Raman spectra. Such characteristic vibrations are observed in the following ranges:

(a) Very strong bands observed at 1080 cm^{-1} in the Raman spectra of **NOCC** and **OCC** derivatives correspond to the symmetric υ_s(C-O-C) vibration; they originate from the φ-CH_2-OCH_2COO fragment of these compounds.

(b) The carboxymethylation of the pyranoid ring at the C-CH_2-OH substituent also brings about changes in the IR spectra of **OCC** and **NOCC** derivatives—medium intensity bands are observed at 890 and 879 cm^{-1}. They vibrations correspond to coupled $\upsilon_{as}(\varphi)$ + δ(CH) + ρ(CH_2) vibrations of the pyranoid ring and methyl group of the φ-CH_2-O- substituent.

(c) Other changes are observed in the Raman spectra of the **NCC** and **NOCC** derivatives—new bands observed at 960 and 955 cm^{-1} correspond to the vibrations of the pyranoid ring coupled with the γ(NH) motion.

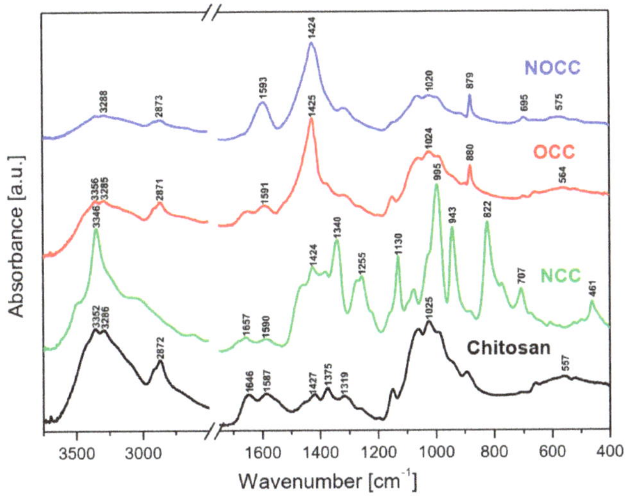

Figure 2. FT-IR spectra of the studied chitosan and its *N*- and *O*-carboxymethyl derivatives.

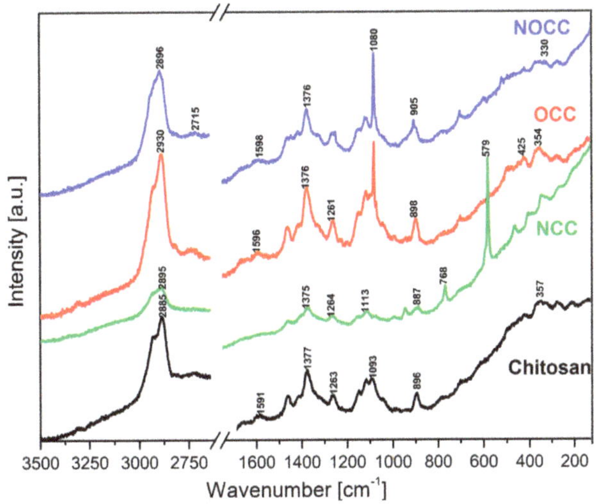

Figure 3. Raman spectra of the studied chitosan and its *N*- and *O*-carboxymethyl derivatives.

The abovementioned changes in the IR and Raman spectra should be considered diagnostic effects of the carboxymethylation of the pyranoid ring in chitosan derivatives.

The assignment of the observed wavenumbers to the respective vibrations is proposed on the basis of quantum chemical PED calculations performed for *N*-acetylated chitosan reported in our previous paper [36]. Supplementary Table S1A,B presents these data.

2.2. UV–Vis Spectra

The diffuse reflectance spectra of chitosan and its **NCC**, **OCC** and **NOCC** derivatives measured in the solid state are shown in Figure 4. The spectral patterns observed in the 200–600 nm range exhibit very broad bands with the maxima in the 250–340 nm range. The spectrum of chitosan contains five components observed as shoulders at 255, 286, 312, 340 and 370 nm. The spectra of the **OCC** and **NOCC** derivatives are somewhat different containing a broad band at 286 nm with four shoulders of its slope at 255, 281, 295, 340 and 370 nm. The spectrum of the **NCC** derivative is very weak, probably due to amorphization of this sample. The assignment of the observed bands to the respective electron transitions could be made using the same approximation as for other heterocyclic rings. The singlet states of the pyranoid ring in chitosan should be assigned to the following transitions: S_1 = 27,027 cm^{-1}, S_2 = 29,411 cm^{-1}, S_3 =32,051 cm^{-1}, S_4 = 34,965 cm^{-1} and S_5 = 39,215 cm^{-1}. The carboxymethylation of chitosan shifts some transitions into the following wavelengths: S_1 = 27,027 cm^{-1}, S_2 = 29,411 cm^{-1}, S_3 = 33,898 cm^{-1}, S_4 = 35,587 cm^{-1} and S_5 = 39,215 cm^{-1}.

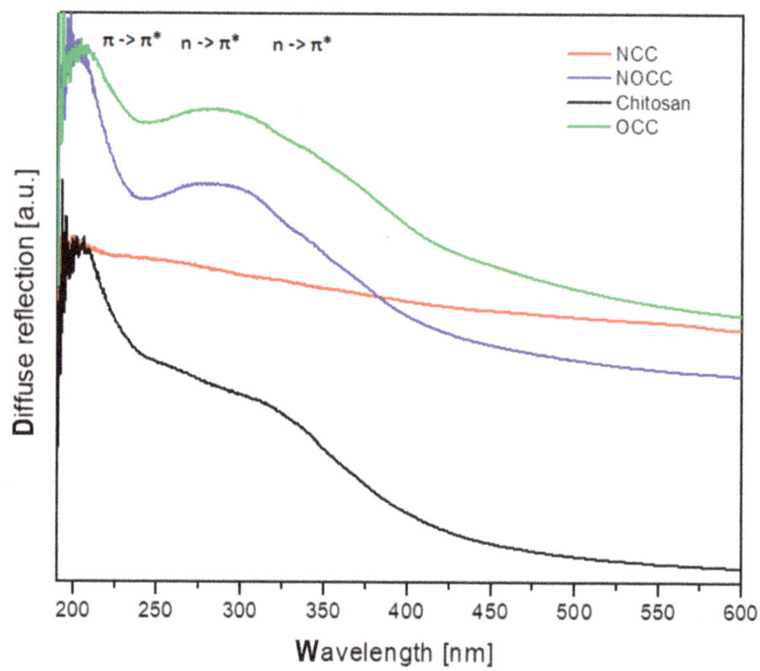

Figure 4. Diffuse reflectance spectra of chitosan and its studied derivatives.

The electron transitions observed in the UV–Vis spectra should be assigned to the $\pi \rightarrow \pi^*$ at 200–255 nm and n(Opyranoid) $\rightarrow \pi^*$ at 255–300 nm, n(Namine) π^* at 300–400 nm. These energy states will be compared to those obtained from the emission studies.

2.3. Emission Spectra

The emission spectra of the studied chitosan and its **OCC**, **NCC** and **NOCC** derivatives were measured as solids using excitation at 350 nm (28,571 cm^{-1}); they are shown in Figure 5. All the spectra contain a very broad band observed in the 400–600 nm range with the maxima at 458 nm (21,834 cm^{-1}) for chitosan, 429 nm (23,310 cm^{-1}) for **NCC**, 470 nm (21,276 cm^{-1}) for **OCC** and 461 nm (21,691 cm^{-1}) for the **NOCC** derivative. On the slope of these bands 2–3 shoulders are observed showing that they have a triplet structure. They are observed at the following wavenumbers: 23,410 and 19,607 cm^{-1} for chitosan, 22,573 and 20,000 cm^{-1} for **NCC**, 33,471 and 20,408 cm^{-1} for **OCC** and 23,255 and 20,408 cm^{-1} for the **NOCC** derivative.

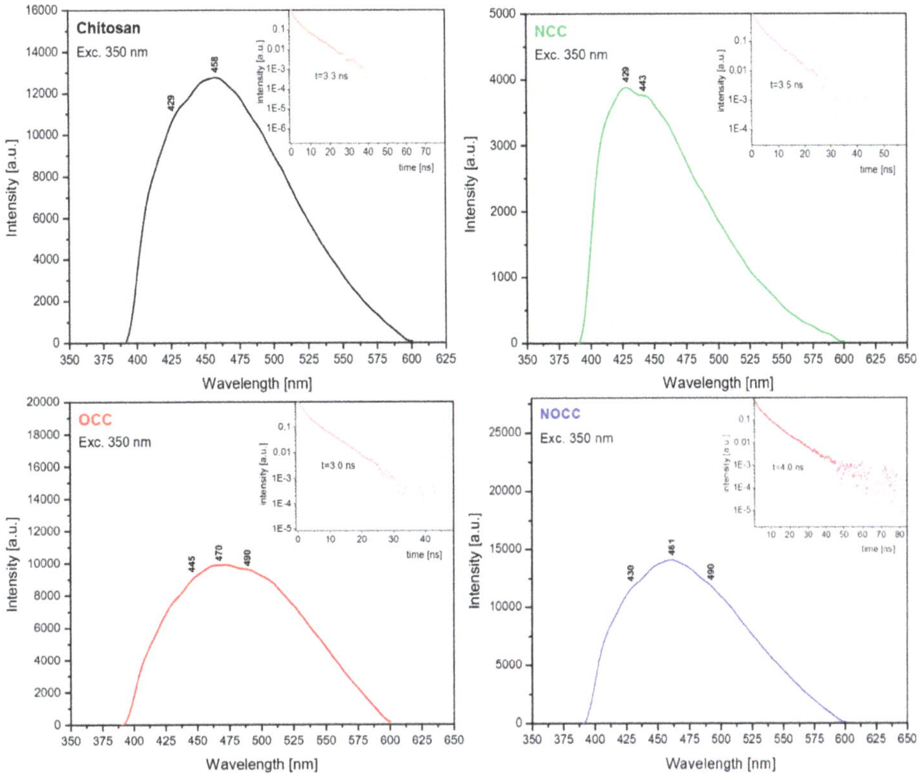

Figure 5. Comparison of the emission spectra of chitosan and its studied derivatives.

The energy gap between the electron absorption and emission studies (Stokes shift) can be calculated for each studied compound. It is equal to 12,648 cm^{-1} for chitosan, 13,689 cm^{-1} for the **OCC** derivative and 13,896 cm^{-1} for the **NOCC** one. Such energy gaps of the 12,600–13,900 cm^{-1} order could not be explained as having derived from simple $S_1 \rightarrow S_0$ emission. The emission transitions are excited at 28,571 cm^{-1}, i.e., slightly above the S_1 level, but the emissions occur from the levels located in the 19,400–23,260 cm^{-1} range. The explanation of this behavior is that the depopulation mechanism of the excited states in chitosan derivatives proceeds by the intersystem crossing of the singlet states to triplet levels $S_0 \rightarrow S_1 \rightarrow T_{1,2,3}$. On the basis of the measured emission spectra, the following succession of the triplet states in the studied compounds should be proposed (Table 1).

The small energetic splitting between the T_1, T_2 and T_3 states causes overlapping of these levels in the form of a broad spectral band in the emission spectra. The participation of the triplet levels in the depopulation of the excited states of the chitosan derivatives is confirmed by the measurements of their lifetimes. They are 3.3 ns for chitosan, 3.6 ns for **NCC**, 3.0 ns for **OCC** and 4.0 ns for **NOCC**. They are very short, characteristic for these types of electron processes.

Table 1. The triplet states in the studied compounds.

	T_1	T_2	T_3
Chitosan	19,607	21,834	23,310 cm^{-1}
NCC	20,000	22,573	23,310 cm^{-1}
OCC	20,408	21,276	22,471 cm^{-1}
NOCC	20,408	21,691	23,255 cm^{-1}

2.4. Spectral and Microscopic Studies of the Crosslinked Chitosan

To characterize the interaction between chitosan and phytate in crosslinked products the IR spectra of the synthesized films were measured. They are presented in Figure 6. They revealed bands characteristic for both these components [36–38]. The fundamental differences between the substrates and final products are observed in the 2000–3700 cm^{-1} range. A very strong and broad spectral pattern is observed in this range built of a clear triplet with maxima at about 3175–3234, 2870–2921 and 2330–2390 cm^{-1}. Such a complex band is characteristic for strong hydrogen bonds O–H···O formed between the O–H group of phosphate from the phytate unit and C=O chromophore of chitosan, or, inversely, between the C=O group of phytate and H–O bond from chitosan. Such a triplet structure in this vibrational range is typical for Fermi resonance of the υ(O-H) and δ(OH) vibrations [39]. It should be noted that the HB formed in the films obtained from the **OCC** derivative are weaker that those from the **NOCC** derivative observed at 3234, 2917–2921 and 2334–2386 cm^{-1}.

The bands originating from the vibrations of the phytate component are observed at 2873, 2324–2385, 2125–2139, 1627–1632, 1110–1120, 1020, 938–939 and 476–494 cm^{-1} in the IR spectra of films [37,38]. On the other hand, the bands typical for chitosan and its derivatives are observed in the ranges listed in Table S1A.

Very strong and broad bands observed for the films appear in the 600–1250 cm^{-1} range; such a band is also observed in the spectra of **OCC** and **NOCC** derivatives. These bands and those originating from the HB interactions as well as υ(C=O) vibrations at 1720 cm^{-1} should be considered as proof of the crosslinking process in the chitosan/phytate system.

The IR studies of the obtained membranes allowed to define the interaction between the crosslinked components. The strong hydrogen bonds O–H···O formed between the O–H group of phosphate from phytate unit and C=O chromophore of chitosan as well as between the C=O group of phytate and H–O bond from chitosan are responsible for the crosslinking process in the chitosan/phytate system. Such interactions lead to the aggregation of the adjacent chitosan chains formed the assembled structures as building blocks similar to those described by Lei et al. [40].

The surface morphology and tightness of the obtained membranes for **OCC** chitosan derivatives were studied using field emission scanning microscopy. Both sides of the films were studied to compare their properties. Figure 7 presents the results of these measurements.

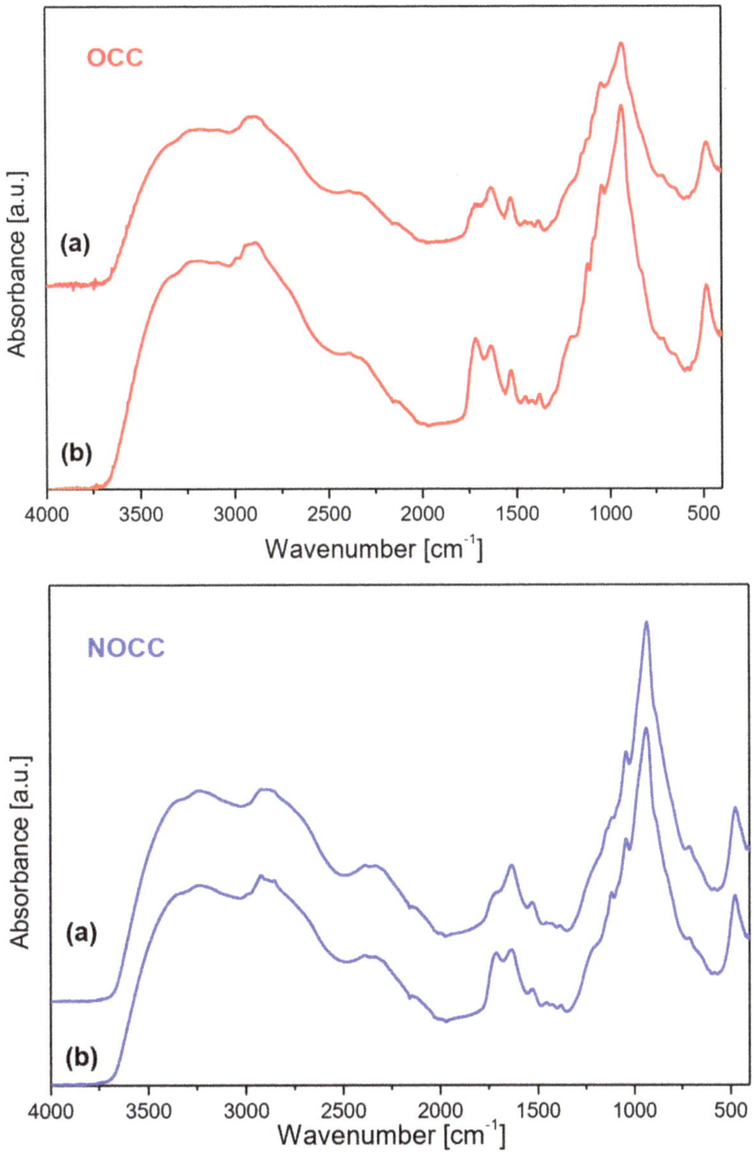

Figure 6. FTIR spectra of films produced by crosslinking of magnesium phytate with **OCC** and **NOCC** chitosan derivatives. **Top** layer (**a**) and **bottom** layer (**b**) of membrane.

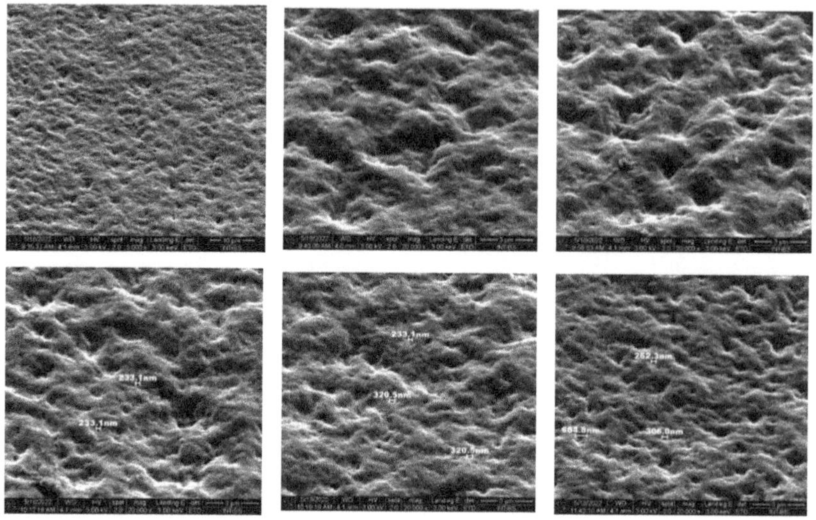

Top layer of OCC membrane

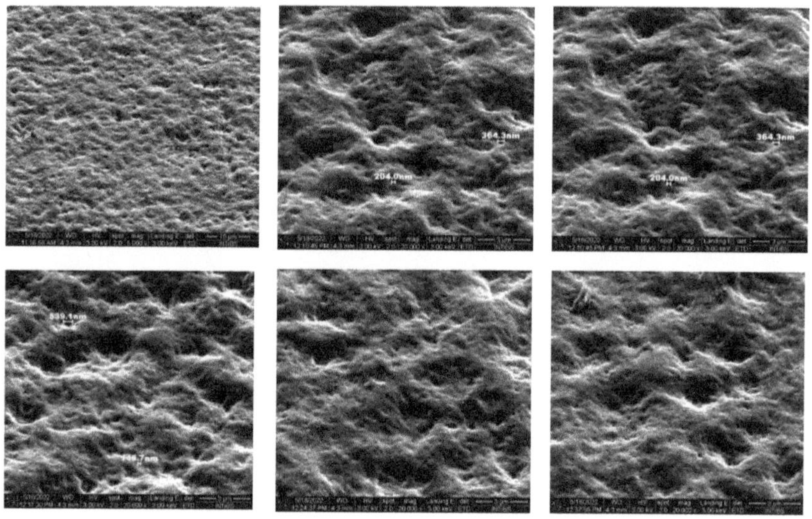

Bottom layer of OCC membrane

Figure 7. Photograph of the films recorded for **OCC** by electron microscope.

Comparing the surface morphology of both sides of the membranes it is seen that they are similar. Regular unevenness is visible: depressions and elevations of similar depth (height). No pores or channels are extended into the bulk. This means that the components used for crosslinking interact effectively giving stable and uniform membranes. Both **OCC** and **NOCC** chitosan derivatives could be used for the production of good quality membranes. The final product exhibits good covering properties and could be used as a packaging material. The membrane is uniform because the water drop located on its surface does not penetrate the second side.

3. Materials and Methods

3.1. Materials

Chitosan of low viscosity (<200 mPa.s, 1% in acetic acid, CAS 9012-76-4) was obtained from Merck KGaA (Darmstadt, Germany). This product was previously characterized by the degree of deacetylation [36]. The degree of deacetylation (DD) for chitosan was 76%.

Phytic acid (50 wt. % solution in water, d = 1.432 g/cm^3), chloroacetic acid (99%, CAS 79-11-8), magnesium carbonate basic (tested according to Ph. Eur., heavy, CAS 39409-82-0), glyoxylic acid monohydrate (98%, CAS 563-96-2), sodium borohydride (≥98.0% powder, CAS 16940-66-2) and acetic acid (100%, glacial, anhydrous for analysis, CAS 16940-66-2) were purchased from Merck KGaA (Darmstadt, Germany). Sodium hydroxide (for analysis, CAS 1310-73-2), ethanol (96.0%, for analysis, CAS 64-17-5) and isopropanol (min. 99%, for analysis, CAS 67-63-0) were obtained from Chempur (Piekary Śląskie, Poland).

3.2. Synthesis of Magnesium Phytate

The phytate complexes were synthesized by changing phytic acid to a metal mole ratio. About 10 mL solution magnesium phytate was prepared dissolving 14.32 g (50%) IP6 and 0.91 g $MgCO_3$ at room temperature for 2–3 h. After complexation, the content of the beaker was freeze-dried. The freeze-drying process was conducted by treating reaction products at −80 °C for 24 h, followed by drying them at pressure 0.02 mbar for 48 h using a M. Christ GmbH Alpha freeze-dryer.

3.3. Synthesis of Chitosan Derivatives

N-carboxymethyl chitosan (**NCC**) was synthesized using 1 g of chitosan (CC) dissolved in 100 mL of 0.15 mol aqueous acetic acid. It was treated with 4 g of solid glyoxylic acid and the obtained solution was mixed with magnetic stirrer. After storing for one hour, 3 g of solid sodium borohydride $NaBH_4$ dissolved in 5 mL water was added to this solution. This mixture was treated by pure ethanol in a 1:1 volume ratio, which produced the precipitate which was filtered giving an **NCC** residue. It was washed with a small amount of water and three times with 70% ethanol. White powder was dried at 60 °C for 12 h.

O-carboxymethyl chitosan (**OCC**) was synthesized by soaking 2 g of chitosan in 30 mL NaOH solution for 3 h. Such a pre-treated chitosan was filtered and dissolved in 30 mL of 50% isopropanol by stirring it for 0.5 h. Independently, the second solution was prepared with 5 g of monochloroacetic acid dissolved in 7 mL of 50% isopropanol and added to the formed mixture. The obtained solution was stirred at 4 °C for 5 h, kept frozen for 24 h and filtered by washing with 70% ethanol. Separated powdered **OCC** was dried at 60 °C for 12 h.

N,O-carboxymethyl chitosan (**NOCC**) was synthesized from initially pre-soaked 2 g of chitosan with 20 mL of 20% solution prepared from sodium hydroxide. After 12 h this soaked chitosan was separated by filtration and placed in a two-necked flask in which 15 mL isopropanol was introduced. The resulting mixture was stirred using a magnetic stirrer for 0.5 h at room temperature. The separately-prepared 10 mL of isopropanolic solution with 1.43g of solid monochloroacetic acid was added to the flask with chitosan and stirred for 0.5 h. The content of the flask was refluxed for 3 h at 60 °C, filtered after 3 h and washed with 70% ethanol three times—the last time with isopropanol. The final **NOCC** residue was dried at 60 °C for 12 h.

The carboxymethylated chitosan derivatives were used for their crosslinking with magnesium phytate. This solution was used for crosslinking carboxymethylated chitosan derivatives which were prepared by dissolving of 0.04 g in 1 mL water. The **NCC** solution obtained in this way was fully transparent with pH = 8.73, but **OCC** and **NOCC** solutions were turbid showing pH: **OCC**—10.14 and **NOCC**—11.94, respectively. Their transparency grew gradually while adding some amount of 80% lactic acid. Adding 250 µL of the solution changed the pH of **OCC** solution to pH 5.02, but adding 300 µL of lactic acid to **NOCC** solution changed its pH to 5.14.

The phytate/chitosan crosslinking chitosan-based membranes were produced in the form of thin films. The above-described mixtures of magnesium phytate and **OCC** and **NOCC** chitosan derivatives were cast in a Petri dish obtaining thin transparent films that were sprinkled using a diluted solution of magnesium phytate and kept in a desiccator containing silica gel. The photographs of the obtained films are shown in Figure 8.

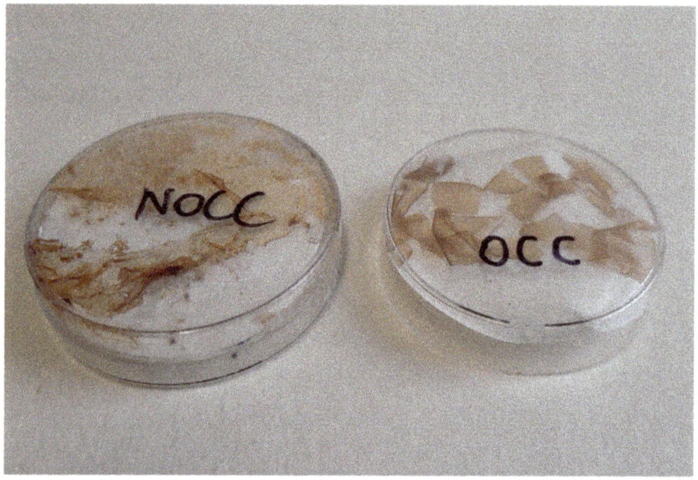

Figure 8. Photographs of the films obtained by crosslinking of the **OCC** and **NOCC** solutions by magnesium phytate.

The following conclusions can be drawn on the production process used for crosslinking of the chitosan derivatives: (1) the crosslinking for the **NCC** derivative does not occur; (2) among the studied chitosan derivatives, **OCC** is the most effectively crosslinked by magnesium phytate–the film formed by this substrate exhibits good mechanical parameters of strength, resistance and stability.

The obtained films were studied using scanning electron microscopy (SEM), IR and Raman measurements, as well as electron UV–Vis and emission spectra.

3.4. Spectroscopic Studies

IR spectra were measured using a Nicolet iS50 FT-IR (Thermo Fisher Scientific Inc., Waltham, MA, USA) spectrometer equipped with an Automated Beamsplitter exchange system (iS50 ABX containing DLaTGS KBr detector and DLaTGS Solid Substrate detector for mid-IR and far-IR regions, respectively) and a built-in all-reflective diamond ATR module (iS50 ATR). Polycrystalline mid-IR spectra were collected in the 4000–400 cm^{-1} range in KBr pellets and far-IR spectra in 600–50 cm^{-1} in Nujol mull. Spectral resolution was set to 2 cm^{-1}.

Raman spectra in the 4000–80 cm^{-1} range were measured in back scattering geometry with a FT Bruker 110/S spectrometer (Bruker Corporation, Billerica, MA, USA). The resolution was 2.0 cm^{-1}. The YAG:Nd (excitation wavelength 1064 nm) laser was used as an excitation source.

Room temperature electron absorption spectra were measured in the 200–1500 nm spectral range using a Cary-Varian 5E UV–Vis-near-IR spectrophotometer (Agilent Technologies, Inc., Santa Clara, CA, USA). Diffuse reflectance spectra were recorded with a Praying Mantis diffuse reflectance accessory (Harrick Scientific Products, Inc., Pleasantville, NY, USA). In these measurements the baseline was first recorded for Al_2O_3 powder, and next this line was subtracted from that obtained for particular powder sample spectra.

Decay profiles and emission spectra were recorded with a grating spectrograph (Princeton Instr. Model Acton 2500i, Princeton Instruments, Trenton, NJ, USA) coupled to a streak camera (Hamamatsu Model C5680, Hamamatsu Photonics Europe GmbH, Herrsching, Germany). For excitation, a femtosecond laser (Coherent Model "Libra") was used. The laser delivers a train of 89 fs pulses at a wavelength of 800 nm and a pulse energy of 1 mJ, with repetition rates regulated up to 1 kHz. To attain light pulses at different wavelengths the laser was coupled to an optical parametric amplifier (Light Conversion Model OPerA) that can operate in the range of 230–2800 nm.

The surface morphology of the foil samples (both sides) was studied with a field emission scanning (FE-SEM) microscope (FEI NovaNanoSEM 230, FEI Company, Hillsboro, OR, USA). A secondary electron (SE) detector was applied for imaging, and low beam energy (3 keV) was used to avoid charging.

4. Conclusions

Novel chitosan/magnesium phytate membranes were obtained using highly deacetylated chitosan and its *N*-carboxymethyl (**NCC**), *O*-carboxymethyl (**OCC**) and *N,O*-carboxymethyl (**NOCC**) derivatives. **OCC** crosslinked by magnesium phytate proved to be the most effective. The films that form on this substrate exhibit good mechanical parameters of strength, resistance and stability. The spectroscopic studies of the obtained membranes showed that the strong O–H···O hydrogen bonds formed between the OH and C=O groups of phytate and chitosan are responsible for the crosslinking process and strong interactions between these components. It was found that new type membranes produced from chitosan and magnesium phytate are mechanically stable and morphologically homogenous and uniform. They do not have pores or channels and, due to their good covering properties, are proposed as packaging materials. The membrane is impermeable to water, which does not penetrate the second side of the foil.

Supplementary Materials: The following supporting information can be downloaded at: https://www.mdpi.com/article/10.3390/molecules28165987/s1, Table S1. A. Assignment of the bands observed in the MIR spectra. B. Assignment of the bands observed in the Raman spectra.

Author Contributions: Conceptualization, J.H. and A.Z.; methodology, A.Z.,W.S., L.D., M.P. and R.L.; software, P.R.-R., K.S. and S.S.; validation, A.Z. and W.S.; formal analysis, A.Z. and W.S.; investigation, J.H., A.Z., W.S., M.P., K.S. and R.L.; resources, A.Z. and W.S.; data curation, A.Z.; writing—original draft preparation, J.H., A.Z., W.S. and L.D.; writing—review and editing, J.H.; visualization, A.Z., W.S. and L.D.; supervision, J.H.; project administration, A.Z.; funding acquisition, A.Z. All authors have read and agreed to the published version of the manuscript.

Funding: This research received no external funding.

Institutional Review Board Statement: Not applicable.

Informed Consent Statement: Not applicable.

Data Availability Statement: Not applicable.

Conflicts of Interest: The authors declare no conflict of interest.

Sample Availability: Samples of the compounds (*N*-carboxymethyl, *O*-carboxymethyl, and *N,O*-carboxymethyl) are available from the authors.

References

1. Rinaudo, M. Chitin and Chitosan: Properties and Applications. *Prog. Polym. Sci.* **2006**, *31*, 603–632. [CrossRef]
2. Dash, M.; Chiellini, F.; Ottenbrite, R.M.; Chiellini, E. Chitosan—A Versatile Semi-Synthetic Polymer in Biomedical Applications. *Prog. Polym. Sci.* **2011**, *36*, 981–1014. [CrossRef]
3. Bernkop-Schnürch, A.; Dünnhaupt, S. Chitosan-Based Drug Delivery Systems. *Eur. J. Pharm. Biopharm.* **2012**, *81*, 463–469. [CrossRef] [PubMed]
4. Yogeshkumar, N.G.; GuravAtul, S.; Adhikrao, V.Y. Chitosan and Its Applications: A Review of Literature. *Int. J. Res. Pharm. Biomed. Sci.* **2013**, *4*, 312–331.

5. Vimal, S.; Abdul Majeed, S.; Taju, G.; Nambi, K.S.N.; Sundar Raj, N.; Madan, N.; Farook, M.A.; Rajkumar, T.; Gopinath, D.; Sahul Hameed, A.S. Chitosan Tripolyphosphate (CS/TPP) Nanoparticles: Preparation, Characterization and Application for Gene Delivery in Shrimp. *Acta Trop.* **2013**, *128*, 486–493. [CrossRef]
6. Cheung, R.; Ng, T.; Wong, J.; Chan, W. Chitosan: An Update on Potential Biomedical and Pharmaceutical Applications. *Mar. Drugs* **2015**, *13*, 5156–5186. [CrossRef]
7. Hamed, I.; Özogul, F.; Regenstein, J.M. Industrial Applications of Crustacean By-Products (Chitin, Chitosan, and Chitooligosaccharides): A Review. *Trends Food Sci. Technol.* **2016**, *48*, 40–50. [CrossRef]
8. Oladzadabbasabadi, N.; Mohammadi Nafchi, A.; Ariffin, F.; Wijekoon, M.M.J.O.; Al-Hassan, A.A.; Dheyab, M.A.; Ghasemlou, M. Recent Advances in Extraction, Modification, and Application of Chitosan in Packaging Industry. *Carbohydr. Polym.* **2022**, *277*, 118876. [CrossRef]
9. Liu, Z.; Fan, B.; Zhao, J.; Yang, B.; Zheng, X. Benzothiazole Derivatives-Based Supramolecular Assemblies as Efficient Corrosion Inhibitors for Copper in Artificial Seawater: Formation, Interfacial Release and Protective Mechanisms. *Corros. Sci.* **2023**, *212*, 110957. [CrossRef]
10. Higuchi, A.; Komiyama, J.; Iijima, T. The States of Water in Gel Cellophane Membranes. *Polym. Bull.* **1984**, *11*, 203–208. [CrossRef]
11. Amaral, I.F.; Granja, P.L.; Melo, L.V.; Saramago, B.; Barbosa, M.A. Functionalization of Chitosan Membranes through Phosphorylation: Atomic Force Microscopy, Wettability, and Cytotoxicity Studies. *J. Appl. Polym. Sci.* **2006**, *102*, 276–284. [CrossRef]
12. López-Pérez, P.M.; Marques, A.P.; da Silva, R.M.P.; Pashkuleva, I.; Reis, R.L. Effect of Chitosan Membrane Surface Modification via Plasma Induced Polymerization on the Adhesion of Osteoblast-like Cells. *J. Mater. Chem.* **2007**, *17*, 4064. [CrossRef]
13. Xu, D.; Hein, S.; Wang, K. Chitosan Membrane in Separation Applications. *Mater. Sci. Technol.* **2008**, *24*, 1076–1087. [CrossRef]
14. Lue, S.J.; Shieh, S.-J. Water States in Perfluorosulfonic Acid Membranes Using Differential Scanning Calorimetry. *J. Macromol. Sci. Part B* **2009**, *48*, 114–127. [CrossRef]
15. Ostrowska-Czubenko, J.; Gierszewska-Drużyńska, M. Effect of Ionic Crosslinking on the Water State in Hydrogel Chitosan Membranes. *Carbohydr. Polym.* **2009**, *77*, 590–598. [CrossRef]
16. Mengatto, L.; Luna, J.A.; Cabrera, M.I. Influence of Cross-Linking Density on Swelling and Estradiol Permeation of Chitosan Membranes. *J. Mater. Sci.* **2010**, *45*, 1046–1051. [CrossRef]
17. Chakrabarty, T.; Kumar, M.; Shahi, V.K. Chitosan Based Membranes for Separation, Pervaporation and Fuel Cell Applications: Recent Developments. In *Biopolymers;* Sciyo: Rijeka, Croatia, 2010.
18. Wu, X.; He, G.; Gu, S.; Hu, Z.; Yan, X. The State of Water in the Series of Sulfonated Poly (Phthalazinone Ether Sulfone Ketone) (SPPESK) Proton Exchange Membranes. *Chem. Eng. J.* **2010**, *156*, 578–581. [CrossRef]
19. Ostrowska-Czubenko, J.; Pieróg, M.; Gierszewska-Drużyńska, M. Water State in Chemically and Physically Crosslinked Chitosan Membranes. *J. Appl. Polym. Sci.* **2013**, *130*, 1707–1715. [CrossRef]
20. Lima, H.A.; Lia, F.M.V.; Ramdayal, S. Preparation and Characterization of Chitosan-Insulin-Tripolyphosphate Membrane for Controlled Drug Release: Effect of Cross Linking Agent. *J. Biomater. Nanobiotechnol.* **2014**, *05*, 211–219. [CrossRef]
21. Gierszewska, M.; Ostrowska-Czubenko, J. Chitosan-Based Membranes with Different Ionic Crosslinking Density for Pharmaceutical and Industrial Applications. *Carbohydr. Polym.* **2016**, *153*, 501–511. [CrossRef]
22. Monteiro, O.A.; Airoldi, C. Some Studies of Crosslinking Chitosan–Glutaraldehyde Interaction in a Homogeneous System. *Int. J. Biol. Macromol.* **1999**, *26*, 119–128. [CrossRef] [PubMed]
23. Crescenzi, V.; Francescangeli, A.; Taglienti, A.; Capitani, D.; Mannina, L. Synthesis and Partial Characterization of Hydrogels Obtained via Glutaraldehyde Crosslinking of Acetylated Chitosan and of Hyaluronan Derivatives. *Biomacromolecules* **2003**, *4*, 1045–1054. [CrossRef] [PubMed]
24. Harish Prashanth, K.V.; Tharanathan, R.N. Crosslinked Chitosan—Preparation and Characterization. *Carbohydr. Res.* **2006**, *341*, 169–173. [CrossRef]
25. Li, H.; Gao, X.; Wang, Y.; Zhang, X.; Tong, Z. Comparison of Chitosan/Starch Composite Film Properties before and after Cross-Linking. *Int. J. Biol. Macromol.* **2013**, *52*, 275–279. [CrossRef] [PubMed]
26. Nagireddi, S.; Katiyar, V.; Uppaluri, R. Pd(II) Adsorption Characteristics of Glutaraldehyde Cross-Linked Chitosan Copolymer Resin. *Int. J. Biol. Macromol.* **2017**, *94*, 72–84. [CrossRef]
27. Guerrero, P.; Muxika, A.; Zarandona, I.; de la Caba, K. Crosslinking of Chitosan Films Processed by Compression Molding. *Carbohydr. Polym.* **2019**, *206*, 820–826. [CrossRef]
28. Yeamsuksawat, T.; Liang, J. Characterization and Release Kinetic of Crosslinked Chitosan Film Incorporated with α-Tocopherol. *Food Packag. Shelf Life* **2019**, *22*, 100415. [CrossRef]
29. Gierszewska, M.; Jakubowska, E.; Olewnik-Kruszkowska, E. Effect of Chemical Crosslinking on Properties of Chitosan-Montmorillonite Composites. *Polym. Test.* **2019**, *77*, 105872. [CrossRef]
30. Hunger, M.; Domalik-Pyzik, P.; Reczyńska, K.; Chłopek, J. Double Crosslinking of Chitosan/Vanillin Hydrogels as a Basis for Mechanically Strong Gradient Scaffolds for Tissue Engineering. *Eng. Biomater.* **2020**, *23*, 2–11. [CrossRef]
31. Yang, Y.; Chen, G.; Murray, P.; Zhang, H. Porous Chitosan by Crosslinking with Tricarboxylic Acid and Tuneable Release. *SN Appl. Sci.* **2020**, *2*, 435. [CrossRef]
32. Berger, J.; Reist, M.; Mayer, J.M.; Felt, O.; Peppas, N.A.; Gurny, R. Structure and Interactions in Covalently and Ionically Crosslinked Chitosan Hydrogels for Biomedical Applications. *Eur. J. Pharm. Biopharm.* **2004**, *57*, 19–34. [CrossRef] [PubMed]

33. Mi, F.-L.; Sung, H.-W.; Shyu, S.-S.; Su, C.-C.; Peng, C.-K. Synthesis and Characterization of Biodegradable TPP/Genipin Co-Crosslinked Chitosan Gel Beads. *Polymer* **2003**, *44*, 6521–6530. [CrossRef]
34. Peppas, N.A.; Mikos, A.G. Preparation Methods and Structure of Hydrogels. In *Hydrogels in Medicine and Pharmacy*; Peppas, N.A., Ed.; CRC Press: Boca Raton, FL, USA, 1986; pp. 1–26. ISBN 9780429285097.
35. Shu, X.; Zhu, K. The Influence of Multivalent Phosphate Structure on the Properties of Ionically Cross-Linked Chitosan Films for Controlled Drug Release. *Eur. J. Pharm. Biopharm.* **2002**, *54*, 235–243. [CrossRef] [PubMed]
36. Zając, A.; Hanuza, J.; Wandas, M.; Dymińska, L. Determination of N-Acetylation Degree in Chitosan Using Raman Spectroscopy. *Spectrochim. Acta Part A Mol. Biomol. Spectrosc.* **2015**, *134*, 114–120. [CrossRef] [PubMed]
37. Zając, A.; Dymińska, L.; Lorenc, J.; Kaczmarek, S.M.; Leniec, G.; Ptak, M.; Hanuza, J. Spectroscopic Properties and Molecular Structure of Copper Phytate Complexes: IR, Raman, UV–Vis, EPR Studies and DFT Calculations. *JBIC J. Biol. Inorg. Chem.* **2019**, *24*, 11–20. [CrossRef]
38. Zając, A.; Solarz, P.; Ptak, M.; Lorenc, J.; Kaczmarek, S.M.; Leniec, G.; Hermanowicz, K.; Hanuza, J. Synthesis, Optical and Magnetic Studies of Cerium and Europium Phytate Complexes—New Microporous Materials. *J. Mol. Struct.* **2021**, *1233*, 130114. [CrossRef]
39. Novak, A. Hydrogen Bonding in Solids Correlation of Spectroscopic and Crystallographic Data. In *Large Molecules*; Springer: Berlin/Heidelberg, Germany, 1974; pp. 177–216. ISBN 978-3-540-37932-4.
40. Lei, M.; Huang, W.; Jin, Z.; Sun, J.; Zhang, M.; Zhao, S. Effect of Molecular Structure and Ionization State on Aggregation of Carboxymethyl Chitosan: A Molecular Dynamics Study. *Carbohydr. Polym.* **2022**, *297*, 119993. [CrossRef]

molecules

MDPI

Article

Effects of Chitosan Molecular Weight and Degree of Deacetylation on Chitosan—Cellulose Nanocrystal Complexes and Their Formation

Hezhong Wang [1,†] and Maren Roman [1,2,*]

1 Department of Sustainable Biomaterials, Virginia Tech, Blacksburg, VA 24061, USA
2 Macromolecules Innovation Institute, Virginia Tech, Blacksburg, VA 24061, USA
* Correspondence: maren.roman@vt.edu
† Current address: Department of Pesticide Science, NanoAgro Center, Henan Agricultural University, Zhengzhou 450002, China.

Abstract: This study was conducted to determine the effects of chitosan molecular weight and degree of deacetylation (DD) on chitosan–cellulose nanocrystal (CNC) polyelectrolyte–macroion complexes (PMCs) and their formation. Chitosan samples with three different molecular weights (81, $3 \cdot 10^3$, $6 \cdot 10^3$ kDa) and four different DDs (77, 80, 85, 89%) were used. The effects on PMC formation were determined by turbidimetric titration. An effect of the molecular weight of chitosan was not observed in turbidimetric titrations. Turbidity levels were higher for CNCs with lower sulfate group density and larger hydrodynamic diameter than for CNCs with higher sulfate group density and smaller hydrodynamic diameter. Conversely, turbidity levels were higher for chitosans with higher DD (higher charge density) than for chitosans with lower DD (lower charge density). PMC particles from chitosans with different molecular weights were characterized by scanning electron microscopy, laser Doppler electrophoresis, and dynamic light scattering. PMCs from high-molecular-weight chitosan were more spherical and those from medium-molecular-weight chitosan had a slightly larger hydrodynamic diameter than PMCs from the respective other two chitosans. The molecular weight of the chitosan was concluded to have no effect on the formation of chitosan–CNC PMC particles and only a minor effect on the shape and size of the particles. The higher turbidity levels for CNCs with lower sulfate group density and larger hydrodynamic diameter and for chitosans with higher DD were attributed to a larger number of CNCs being required for charge compensation.

Keywords: cellulose; chitosan; polysaccharide; polyelectrolyte; complexation; nanocrystals

Citation: Wang, H.; Roman, M. Effects of Chitosan Molecular Weight and Degree of Deacetylation on Chitosan—Cellulose Nanocrystal Complexes and Their Formation. *Molecules* **2023**, *28*, 1361. https://doi.org/10.3390/molecules28031361

Academic Editor: Agnieszka Ewa Wiącek

Received: 20 December 2022
Revised: 19 January 2023
Accepted: 20 January 2023
Published: 31 January 2023

1. Introduction

Chitosan is one of very few known polysaccharides that bear exclusively positive charges. A positive charge in polysaccharides is generally related to the amino group of 2-amino-2-deoxy-D-glucopyranosyl (GlcN) residues, which is protonated in acidic aqueous media. In addition to these residues, chitosan contains 2-acetamido-2-deoxy-D-glucopyranosyl (GlcNAc) residues, which under most conditions bear no charge. The proportions of GlcN and GlcNAc residues in a chitosan sample depend on the processing conditions during manufacture or preparation, involving the chemical or enzymatic deacetylation of the parent polysaccharide chitin. The fraction of GlcN residues in a chitosan sample is termed the degree of deacetylation (DD), generally expressed in percent. The polycationic nature and availability of chitosan render it attractive for applications involving ionic interactions. For such applications, the DD is a crucial parameter as it is related to the charge density along the polymer chain.

An area of intense research, into which chitosan has been absorbed, is the area of polyelectrolyte complexes (PECs). PECs are intermolecular complexes of oppositely charged polyelectrolytes based on attractive Coulomb interactions [1]. The interest in PECs is motivated by their numerous current and potential applications, such as the encapsulation of sensitive ingredients in food products [2], the delivery of drugs [3] and genes [4], the entrapment and delivery of proteins and immobilization of enzymes [5], and the encapsulation of cells [6]. The properties of PECs are governed by many factors, primarily the molecular weights and densities of ionizable groups of the two polyelectrolytes, the polyelectrolyte concentrations after mixing, the ratio of mixing, and the ionic strength and pH of the reaction medium.

The effect of molecular weight on the properties of PECs is incompletely understood. Several groups have reported the composition of PECs to be independent of the molecular weight of the polyelectrolytes [7,8]. Dautzenberg [9] and his collaborators have observed no effect of the molecular weight of sodium poly(styrene sulfate) on the structural parameters of its PECs with poly(diallyldimethylammonium chloride). De Vasconcelos et al. [10], on the other hand, have reported an increase and shift in the turbidity maximum in turbidimetric titrations of poly(methacrylic acid) (PMMA) with chitosan toward a higher chitosan/PMMA mass ratio with increasing molecular weight of PMMA. The authors concluded that PECs from PMMA of higher molecular weight were more water-soluble and larger than PECs from PMMA of lower molecular weight.

The effect of the density of ionizable groups, generally equated with the charge density, on the properties of PECs is much better understood as it has been more widely studied [7,11–16]. The results of these studies indicate that densely structured PECs are formed when the charge densities of the two polyelectrolytes are similar whereas highly swellable PECs with lower structural densities are formed in the case of a mismatch in the charge densities of the polyelectrolytes [13,17].

In previous studies [18–20], we investigated a new type of PEC between chitosan and cellulose nanocrystals (CNCs), which are anionic, rod-like nanoparticles. We referred to these complexes as polyelectrolyte–macroion complexes (PMCs) to emphasize the particulate nature of one of the two components. The properties and formation of these PMCs were governed by the large mismatch in density of the ionizable groups. A decrease in pH of the surrounding medium was found to increase the mismatch and cause PMC particle swelling whereas an increase in pH resulted in particle shrinking. Moreover, an increase in ionic strength caused a slight decrease in particles size, which was attributed to charge screening. Here we study the effects of chitosan molecular weight and DD on the preparation of chitosan–CNC PMC particles to test the hypothesis that their properties and formation can be controlled through these parameters. The impact of CNC properties on PMC formation was also assessed.

2. Results

Three commercial chitosan samples with different molecular weight specifications (high, medium, and low) were characterized according to molecular weight, degree of deacetylation, hydrodynamic diameter, electrophoretic mobility, and amino group density. The characteristics of the three chitosan samples are listed in Table 1.

CNCs with two different sulfate group densities, denoted CNC-18 and CNC-33, were prepared by sulfuric acid hydrolysis of a dissolving-grade softwood sulfite pulp and characterized according to hydrodynamic diameter, electrophoretic mobility, and sulfate group density. The characteristics of the two CNC samples are listed in Table 2.

Table 1. Characteristics of the high, medium, and low molecular weight chitosans.

Chitosan	Molecular Weight [a] (kDa)	DD [b] (%)	Hydrodynamic Diameter [c] (nm)	Electrophoretic Mobility [d] (μm·cm/V·s)	Amino Group Density [e] (mol/kg)
High Mw [f]	$6 \cdot 10^3$	87	319 ± 18	2.6 ± 0.2	6.1
Medium Mw [f]	$3 \cdot 10^3$	88	265 ± 14	3.2 ± 0.1	5.8
Low Mw [f]	81	88	238 ± 52	1.7 ± 0.9	5.3

[a] Calculated from the intrinsic viscosity; [b] degree of deacetylation determined by ^{1}H NMR; [c] Determined by dynamic light scattering; [d] Measured with a Malvern Zetasizer Nano ZS 90; [e] Determined by conductometric titration; [f] molecular weight.

Table 2. Characteristics of the cellulose nanocrystal (CNC) samples.

CNC Sample	Hydrodynamic Diameter [a] (nm)	Electrophoretic Mobility [b] (μm·cm/V·s)	Sulfate Group Density [c] (mol/kg)
CNC-18	104 ± 4	−2.8 ± 0.2	0.18
CNC-33	68 ± 5	−3.1 ± 0.4	0.33

[a] Determined by dynamic light scattering; [b] Measured with a Malvern Zetasizer Nano ZS 90; [c] Determined by conductometric titration.

2.1. Effect of Chitosan Molecular Weight

The complexation of CNCs with chitosans of different molecular weights was analyzed by turbidimetric titration. The turbidity curves for titrations of chitosan solutions with suspensions of CNC-18 and CNC-33 are shown in Figure 1. In all titrations, on addition of the CNC suspension to the chitosan solution, the turbidity increased rapidly initially and then leveled off or decreased slightly at higher S/N ratios. The titration curves for different molecular weights of chitosan differed slightly. Because of the large standard deviations of the measured turbidity values (error bars in Figure 1), the significance of the observed differences in the titration curves was assessed by one-way ANOVA of the turbidity data. The ANOVA test results are listed in Table 3. Statistical analysis of the turbidity data revealed that there was no statistically significant difference between the data from different molecular weight chitosans in both cases, titration with CNC-18 and with CNC-33. Thus, an effect of the molecular weight of the chitosan was not observed in the turbidimetric titrations experiments.

Our results are in accordance with those studies that have found that the molecular weight to have no effect on the parameters of PMCs [7–9].

Averaging of the turbidity levels observed with different chitosan molecular weights (inset in Figure 1a), showed that on average, CNC-18 caused higher turbidity levels than CNC-33 at any given S/N ratio. The higher turbidity levels are consistent with a higher cellulose/chitosan mass ratio required for charge stoichiometry in the case of CNC-18, with a lower sulfate group density. At a pH of 2.6, used in these titrations, the degree of ionization of the CNCs was 0.58 (pK_a of 2.46 [19]), i.e., 58% of the available sulfate groups were deprotonated and negatively charged. Charge stoichiometry, or an NH_3^+/SO_3^- ratio of unity, required a cellulose/chitosan mass ratio of about 57:1 for CNC-18 and of about 30:1 for the CNC-33. As a result, PMC particles prepared with CNC-18 probably contained more (and larger) CNCs and were bulkier than PMC particles prepared with CNC-33. The more rapid formation and bulkier nature of the PMC particles formed by CNC-18 might be the reason for the observed faster increase in turbidity.

The effect of chitosan molecular weight on the morphology of chitosan–CNC PMC particles was studied by SEM. FE-SEM images of PMC particles prepared with CNC-18 and different molecular weight chitosans are shown in Figure 2. The particles from high-molecular-weight chitosan appeared to be more spherical than the particles from medium and low molecular weight chitosans. This finding is in accordance with the results of Ko et al. [21], who reported more spherical particle shapes at higher molecular weights for tripolyphosphate-crosslinked chitosan microparticles.

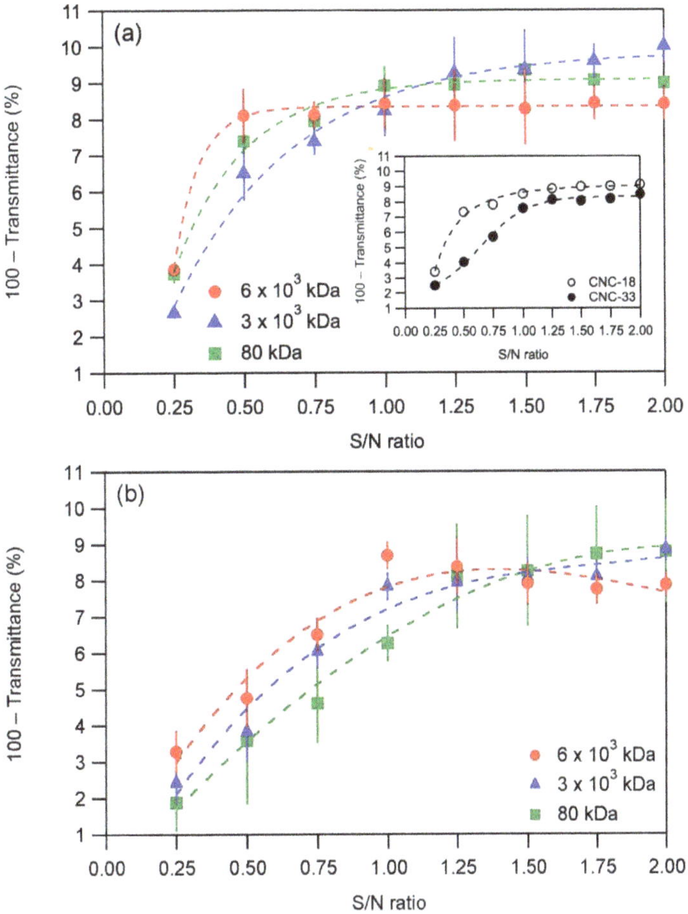

Figure 1. Turbidity curves for titrations of solutions of chitosans with different molecular weights with (**a**) CNC-18 suspensions and (**b**) CNC-33 suspensions. (The dashed lines are merely visual guides. Each data point is a mean of three measurements. The error bars correspond to ± one standard deviation.) The inset shows the averaged data for CNC-18 and CNC-33. S/N ratio = number of CNC sulfate groups divided by number of chitosan amino groups, 100 − Transmittance of deionized water = 0%.

Table 3. One-way ANOVA test results (α = 0.05) for the effect of molecular weight of the commercial chitosan samples on the turbidity.

Source of CNCs [a]	DF [b]	F Ratio [c]	Prob > F [d]
Batch 1	2	0.1250	0.8827
Batch 2	2	0.4208	0.6580

[a] Cellulose nanocrystals; [b] degrees of freedom; [c] mean of squares between groups divided by mean of squares within groups; [d] null hypothesis: all groups for the study are the same, rejected for Prob > F values larger than α.

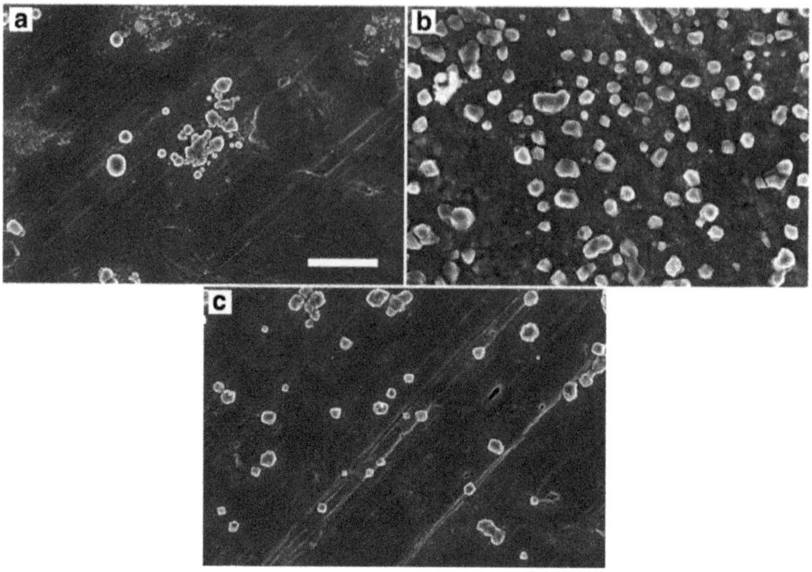

Figure 2. Field emission scanning electron microscopy images of chitosan–cellulose nanocrystal (CNC) polyelectrolyte–macroion complex particles prepared by addition of a CNC-18 suspension to solutions of chitosan with different molecular weights: (**a**) $6 \cdot 10^3$ kDa, (**b**) $3 \cdot 10^3$ kDa, (**c**) 80 kDa. Scale bar (applying to all images): 3 μm.

Table 4 lists the electrophoretic mobilities and hydrodynamic diameters of chitosan–CNC PMC particles from different molecular weight chitosans. The particles were formed at an N/S ratio of 0.3 to yield sizes within the analytical range, i.e., particles formed at a higher N/S ratio exceeded the size limit of the instrument. The particles from medium-molecular-weight chitosan were slightly larger than those from high and low-molecular weight chitosan. PMC particles from the latter two types were of comparable size. The slightly larger size of the PMC particles from medium-molecular-weight chitosan might indicate an optimum length ratio of the PMC components. The estimated length ratios (contour length of the chitosan molecules/length of the CNCs) were 150, 80, and 2:1 for the high, medium, and low-molecular weight chitosan, respectively, using 120 nm as an average CNC length. The electrophoretic mobilities of the PMC particles were positive for all chitosan types, indicating a non-stoichiometric composition and excess of ammonium groups in the particles.

Table 4. Hydrodynamic diameter and electrophoretic mobility of polyelectrolyte–macroion comples (PMC) particles from different molecular weight chitosans.

Sample	Hydrodynamic Diameter (nm)	Electrophoretic Mobility (μm·cm/V·s)
PMC with high-molecular-weight chitosan	373 ± 9	3.0 ± 0.1
PMC with medium-molecular-weight chitosan	539 ± 32	3.4 ± 0.1
PMC with low-molecular-weight chitosan	363 ± 37	2.9 ± 0.2

2.2. *Effect of Chitosan Degree of Deacetylation*

The effect of the DD of chitosan on chitosan–CNC PMC particles was studied by turbidimetric titration using chitosan samples subjected to a chitin deacetylation procedure.

The characteristics of these chitosan samples are listed in Table 5. The data in Table 5 shows that the deacetylation procedure caused a decrease in the molecular weight in addition to an increase in the DD.

Table 5. Characteristics of the chitosan samples prepared in-house by chitin deacetylation.

Number of Deacetylation Cycles	Molecular Weight [a] (kDa)	DD [b] (%)
1	1.7	77
2	1.9	80
3	1.4	85
4	1.0	89

[a] Calculated from intrinsic viscosity; [b] Degree of deacetylation determined by FTIR spectroscopy.

The turbidity curves for titrations of CNC-18 suspensions with chitosan solutions of different DD are shown in Figure 3. Upon addition of the chitosan solution to the CNC suspension, the turbidity increased rapidly initially and then leveled off. At higher volumes of chitosan solution added, the turbidity decreased slightly in accordance with the results reported in Wang and Roman [18] for the effect of mixing sequence.

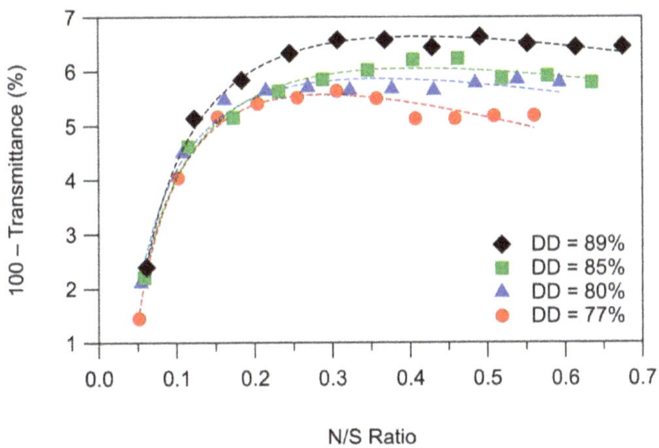

Figure 3. Titration curves for turbidimetric titrations of CNC-18 suspensions with solutions of chitosans with DDs ranging from 77 to 89%. (The dashed lines are merely visual guides. Each data point is a mean of three measurements. Error bars are omitted for clarity.) N/S ratio = number of chitosan amino groups divided by number of CNC sulfate groups, 100 − Transmittance of deionized water = 0%.

Overall, the turbidity at any given N/S ratio was higher for higher DD values. To establish whether the observed effect was related to the sample differences in DD or molecular weight, we analyzed the turbidity data by one-way ANOVA. The ANOVA test results are shown in Table 6.

Table 6. One-way ANOVA test results ($\alpha = 0.05$) for the effects of DD and molecular weight of the deacetylated chitin samples on the turbidity.

Effect	DF [b]	F Ratio [c]	Prob > F [d]
Molecular weight	2	1.7644	0.2497
DD [a]	3	8.7221	<0.0001 *

[a] degree of deacetylation; [b] degrees of freedom; [c] means of squares between groups divided by mean of squares within groups; [d] null hypothesis: all groups for the study are the same, rejected for Prob > F values larger than α; * indicates rejection of the null hypothesis.

Analysis of the data by one-way ANOVA showed that the molecular weight did not have a statistically significant effect on the turbidity and that the differences in the turbidity curves were due to the differences in DD. Post hoc comparisons using the Tukey HSD (Honestly Significant Difference) test (Table 7) indicated that the turbidity levels for a DD of 80 and 77% were statistically significantly lower than for a DD of 89 and 85%.

Table 7. Significance levels and least squares mean (Tukey HSD, $\alpha = 0.05$) for the effect of degree of deacetylation (DD) of the deacetylated chitin samples on the turbidity.

Number of Deacetylation Cycles	Level (DD %)	Significance [a]	Least Squares Mean
1	77	A	4.87
2	80	A	5.27
3	85	B	5.42
4	89	B	5.98

[a] Levels not connected by same letter are statistically significantly different.

A higher DD signifies a higher charge density, requiring a larger number of CNCs for charge compensation. Thus, PMC particles from chitosans with higher DDs might contain a larger number of CNCs and therefore be larger than PMC particles from chitosans with lower DDs.

3. Discussion

Chitosan and cellulose are both natural polysaccharides with good biodegradation and biocompatibility profiles. CNCs prepared by sulfuric acid hydrolysis are negatively charged because they undergo partial surface esterification during the hydrolysis process, resulting in highly acidic sulfate groups [22]. Chitosan carries amino groups, resulting from the deacetylation of acetamido groups that are protonated and positively charged at acidic pH levels [23]. Upon mixing, the negatively charged CNCs and positively charged chitosan molecules aggregate and form well-defined PMC particles. These PMC particles have a strong potential for applications in agriculture and oral drug delivery, where their biodegradability and biocompatibility are important attributes. Different applications will require different PMC properties, making a detailed understanding of the factors that govern their formation and properties essential.

The present study is the third in a series of studies investigating the effects of process parameters and chitosan properties on the formation and properties of chitosan–CNC PMC particles [20]. The first study [18] focused on the effects of the sequence of mixing of a chitosan solution and a CNC suspension as well as the CNC concentration. The study revealed that when a CNC suspension is added to a chitosan solution, the turbidity increases and then levels off, whereas when a chitosan solution is added to a CNC suspension, the turbidity reaches a maximum and then decreases on further chitosan addition, suggesting a partial dissociation of aggregates due to charge overcompensation. The study also showed that chitosan–CNC PMC particles are composed primarily of CNCs, due to the lower charge density of CNCs (0.18–0.33 versus 5.3–6.1 mol/kg), and range in size from a few hundred nanometers to several micrometers, depending on the cellulose/chitosan ratio. Interestingly, particles formed at amino/sulfate group molar ratios >1, carrying a positive charge, had more or less spherical shapes, whereas particles formed at ratios <1, carrying a negative charge, had well-defined polygonal shapes.

The second study [19] investigated the effects of pH and salt concentration on the formation and properties of chitosan–CNC PMC particles. The pK_b of chitosan was measured by potentiometric titration as 6.40 ± 0.03. The pK_a of the CNCs was determined through extrapolation of a degree of ionization vs. pH curve as 2.46 ± 0.12. The pH range in which both the CNCs and chitosan were completely ionized was found to be narrow and centered at 4.5. A pH increase from this range was observed to cause a turbidity decrease due to shrinking of the PMC particles on a decrease in the degree of ionization of the chitosan chains. A pH decrease, on the other hand, resulted in a steep increase in turbidity, which was attributed to an increase in particles size as a result of a decrease in the degree of ionization of the CNCs. The size changes in the PMC particles were confirmed by SEM. A decrease in turbidity was also observed on an increase in salt concentration and was shown to be attributable to charge-screening-related shrinking of the PMC particles.

This study investigated two additional variables that may be expected to affect the formation and properties of chitosan–CNC PMC particles, namely chitosan molecular weight and DD. The results of this study complete the picture by revealing that chitosan molecular weight has no apparent effect on the formation of the PMC particles and only a minor, if any, effect on their shape and size. The observed effect of DD, specifically higher DDs resulting in higher turbidity levels, is consistent with the previously reported increase in PMC hydrodynamic diameter with increasing amino/sulfate group molar ratio, observed when titrating a CNC suspension with a chitosan solution [18], as well as the previously observed increase in turbidity after lowering of the pH from 4.5, causing a decrease of CNC degree of ionization [19].

In summary, the three studies allow the following conclusions:

- Chitosan–CNC PMC particles consist primarily of CNCs held together by (less bulky) chitosan molecules.
- PMC particle size can be adjusted from a few hundred nanometers to several micrometers and net particle charge from negative to positive by controlling the amino/sulfate group molar ratio through selection of chitosan DD and chitosan-to-CNC mass ratio. (The charge density of CNCs is low, compared to the charge density of chitosan, and controllable only within limits).
- In environments with a pH below 2, chitosan–CNC PMC particles will be larger, and in environments with a pH above 7, smaller, than in environments with a pH between 2 and 7. This behavior is particularly interesting for the oral delivery of therapeutics and nutraceuticals.
- In environments with a salt concentration above 0.1 M, chitosan–CNC PMC particles will be smaller and more compact than in environments with a salt concentration below 0.1 M.

4. Materials and Methods

4.1. Materials

High, medium, and low molecular weight chitosans (Fluka BioChemika) were purchased from Sigma-Aldrich (St. Louis, MO, USA). Chitin was purchased from ACROS Organics (Morris Plains, New Jersey, USA). NaCl (certified), HCl (0.1 N, both certified), H_2SO_4 (>95%), and NaOH (0.1 N and 1.0 N, certified) were obtained from Fisher Scientific (Waltham, MA, USA). 50% NaOH was ACS reagent grade and purchased from Ricca Chemical Company (Arlington, TX, USA). All experiments were conducted with deionized water (resistivity at 25 °C: 18.2 MΩ·cm) prepared with a Direct-Q 5 Ultrapure Water System (Millipore, Burlington, MA, USA).

4.2. CNC Preparation

The method of preparation of the CNCs was described in Wang and Roman [18]. Briefly, a 50 g portion of softwood dissolving pulp (Temalfa 93 A-A, sulfite pulp), generously donated by Tembec, Inc. (Témiscaming, QC, Canada) was ground to pass a 60-mesh screen. The obtained powder was added to 500 mL of 64 wt. % H_2SO_4, preheated to 45 °C, and hydrolyzed under stirring at that temperature. After 45 min, the reaction medium was poured into 4.5 L of deionized water after which the CNCs were isolated by centrifugation and purified by dialysis with deionized water. When the pH of the dialysis water remained unchanged, the dialysis was discontinued. The suspension was then cooled with an ice-bath for sonication and filtered sequentially through 1, 0.45, and 0.22 μm Millipore polyvinylidene fluoride (PVDF) syringe filters. After filtration, the CNC stock suspension generally had a concentration between 0.6 and 0.9% (w/v). To determine the effects of CNC properties, a second CNC sample was prepared with a hydrolysis time of 60 min instead of 45 min. The CNCs were characterized as described in Wang and Roman [18].

4.3. Purification of Commerical Chitosans

For purification, 1 g of commercial chitosan was added to 250 mL 0.1 N HCl and allowed to dissolve overnight. Next, the solution was filtered sequentially through 1, 0.45, and 0.22 μm PVDF syringe filters. Following filtration, 1 N NaOH was added to the solution until its pH was between 9 and 10. After isolation by centrifugation at 4 °C and 4900 rpm for 15 min, the precipitated chitosan was further purified by washing thrice with deionized water, followed by freeze-drying.

4.4. Preparation of Chitosan Samples with Different DDs

Chitosan samples with different DDs were prepared by deacetylation of chitin according to literature procedures [24,25]. In detail, 20 g chitin was heated for 1 h under stirring in 400 mL 47% NaOH at 110 °C in a nitrogen atmosphere. The reaction mixture was allowed to cool to 80 °C and the reaction product was washed with deionized water and dried for 2 h in an oven set to 105 °C. Approximately 5 g of the reaction product was set aside, and the rest was subjected again to the procedure above. This process was repeated twice more to yield a total of four samples with different DDs. Finally, the chitosan samples were purified as described above by dissolution, filtration, precipitation, washing, and freeze-drying.

4.5. Chitosan Characterization

4.5.1. Molecular Weight

The molecular weights of the four prepared chitosan samples were determined as described in Wang and Roman [18] using the viscosity method that is based on the Mark–Houwink equation:

$$[\eta] = KM^a \tag{1}$$

where $[\eta]$ is the intrinsic viscosity, M is the viscosity-average molecular weight, and K and a are the Mark–Houwink parameters. Viscosity measurements were performed at 25 °C using an Ubbelohde capillary viscometer with an inner capillary diameter of 0.53 mm and chitosan solutions in the binary solvent system 0.3 M acetic acid/0.2 M sodium acetate [26], ranging in concentration from 0.1 to 1 mg/mL. The values used for the Mark–Houwink parameters were $a = -1.02 \cdot 10^{-2} \cdot \mathrm{DD} + 1.82$ and $K = 1.64 \cdot 10^{-30} \cdot \mathrm{DD}^{14.0}$ [27].

4.5.2. Degree of Deacetylation

The DDs of the commercial chitosans were determined by ^{1}H-NMR spectroscopy as described in Wang and Roman [18]. Briefly, chitosan was dissolved in a mixture of D_2O and DCl, and spectra were acquired at 90 ± 1 °C using a Varian Unity 400 MHz NMR spectrometer (Varian, Inc., Palo Alto, CA, USA). The DD was calculated as

$$\mathrm{DD} = \left[\frac{H1D}{(H1D + HAc/3)}\right] \times 100\% \tag{2}$$

where $H1D$ is the integral of the H1 GlcN peak and HAc is the integral of the peak corresponding to the three protons of the GlcNAc acetyl group [28].

The DDs of the four chitosan samples obtained by the deacetylation of chitin were determined by FTIR spectroscopy according to the method by Baxter et al. [29], which was chosen for its greater convenience in combination with an equivalent accuracy, compared to the NMR-based method above. The method is based on the intensity ratio of the amide I band at 1650 cm^{-1} and the OH band at 3450 cm^{-1}. FTIR spectra were recorded from KBr pellets, prepared as follows. Ninety-eight milligrams of dry KBr were ground in a mortar with 2 mg of sample. The mixture was transferred into a deep stainless-steel nut, with a stainless-steel bolt inserted from one end. The KBr pellet was generated by inserting a second bolt from the opposite end and compressing the powder for a period of about 10 min. Before data collection, the nut with the KBr pellet was placed in a vacuum oven for sample drying. FTIR spectra were recorded with a Thermo Nicolet Nexus 470 FTIR spectrometer (Thermo Fisher Scientific, Inc., Waltham, MA, USA) using 128 as the number of scans and 4 cm^{-1} as the spectral resolution.

4.5.3. Hydrodynamic Diameter and Electrophoretic Mobility

The z-average hydrodynamic diameter (cumulants mean) and electrophoretic mobility of the chitosan samples were measured in triplicate at 25 °C using a 0.001% (w/v) chitosan solution and Malvern Zetasizer Nano ZS with Malvern DTS1060 folded capillary cells (Malvern Instruments, Inc., Westborough, MA, USA) without adjustment of the pH or ionic strength.

4.5.4. Amino Group Density

The amino group densities of the commercial chitosans were determined through conductometric titration using a Mettler Toledo SevenMulti S47 pH/conductivity meter equipped with an InLab 730 conductivity probe (Mettler-Toledo Inc., Columbus, OH, USA). A 0.1N NaOH solution was added dropwise under stirring and N_2 to 25 mL of a 0.1% (w/v) chitosan solution with an NaCl concentration of 0.1 M. The conductivity was recorded after every 10 drops and the amino group density calculated from the titrant volume between the two equivalence points. Measurements were performed in triplicate.

4.6. Preparation of Dilute Chitosan Solutions

After drying in an oven for 2 h at 105 °C, a 0.1 g amount of purified chitosan was added to 100 mL of 0.1 N HCl to give a ~0.1% (w/v) chitosan stock solution. The stock solution's concentration was measured by thermogravimetric analysis and calculated by averaging three measurements. Further details are provided in Wang and Roman [18]. Starting solutions for the complexation experiments were obtained by dilution of the stock solution with deionized water. Prior to the complexation experiments, 0.1 N HCl or 0.1 N NaOH and NaCl were added to the solutions until the desired pH and ionic strength, respectively, were reached.

4.7. Preparation of Dilute CNC Suspensions

Starting suspensions for the complexation experiments were obtained by dilution of the filtered stock suspension with deionized water. Prior to the complexation experiments, 0.1 N HCl or 0.1 N NaOH and NaCl were added to the suspensions until the desired pH and ionic strength, respectively, were reached.

4.8. Turbidimetric Titrations

In a turbidimetric titration, a chitosan solution (0.001% (w/v)) was added to a CNC suspension (0.02% (w/v)), or vice versa. Prior to the experiment, the ionic strength and pH of the chitosan solution and CNC suspension were adjusted to 1 mM and 2.6, respectively. Titrations were performed by drop-wise addition under vigorous agitations with a magnetic stir bar. A probe colorimeter (PC 950, Brinkmann Instruments, Inc., Westbury, NY, USA), operating at a wavelength of 420 nm and with an optical cell with a 2 cm path length, was used for measuring the transmittance of the reaction mixture. Turbidity values are reported as 100 − transmittance (%). Each experiment was performed after zeroing of the probe colorimeter in deionized water and was repeated two times.

4.9. Characterization of Chitosan–CNC PMC Particles

4.9.1. Morphology

The morphology of chitosan–CNC PMC particles was investigated with a LEO 1550 field emission scanning electron microscope (LEO Electron Microscopy, Inc., Thornwood,

NY, USA). A working distance of 4 mm and an accelerating voltage of 5 kV were used for FE-SEM imaging. Samples were prepared by mixing of a CNC suspension (0.02% (w/v)) and a chitosan solution (0.001% (w/v)), each with an ionic strength of 1 mM and a pH of 3.4, under strong agitation with a magnetic stir bar. The volumes of the two liquids were chosen to give an amino/sulfate group molar ratio (N/S ratio) of 1. Drops of 10 µL of the suspensions, containing PMC particles, were placed onto conductive Ni–Cu tape (Ted Pella) on standard SEM stubs (Ted Pella). After drying of the droplets under ambient conditions, a thin (~6 nm) layer of carbon was deposited onto the samples.

4.9.2. Hydrodynamic Diameter and Electrophoretic Mobility

The electrophoretic mobility and z-average hydrodynamic diameter of chitosan–CNC PMC particles were measured by laser Doppler electrophoresis in a Malvern Zetasizer Nano ZS90 (Malvern Instruments, Inc., Westborough, MA, USA). Samples were prepared by mixing of a CNC suspension (0.02% (w/v)) and a chitosan solution (0.001% (w/v)), each with an ionic strength of 1 mM and a pH of 2.6, under strong agitation with a magnetic stir bar. The volumes of the two liquids were chosen to give sulfate/amino group molar ratio (S/N ratio) of 0.3. The obtained PMC particle suspensions were used without further dilution or filtration. Each measurement was performed at 25 °C in a Malvern DTS1060 folded capillary cell and was repeated two times.

4.10. Statistical Data Analysis

The turbidity data from the turbidimetric titrations were analyzed by one-way analysis of variance (ANOVA) with a 95% confidence interval (α = 0.05) and Tukey's HSD test. Statistical analysis of the data was carried out with the SAS software JMP 7.0.2.

5. Conclusions

The effects of chitosan molecular weight and DD on the formation and properties of chitosan–CNC PMC particles were investigated to assess the impact of these parameters. The molecular weight of the chitosan appeared to have no apparent effect on the formation of chitosan–CNC PMC particles and only a minor, if any, effect on the shape and size of the particles. The sulfate group density of CNCs showed an inverse relationship with turbidity levels, likely due to a larger number of CNCs required for charge compensation at lower sulfate group densities. Similarly, chitosans with higher DDs gave higher turbidity levels, attributable to the higher chitosan charge density requiring a larger number of CNC for charge compensation.

Author Contributions: Conceptualization, H.W. and M.R.; formal analysis, H.W. and M.R.; funding acquisition, M.R.; methodology, H.W.; project administration, M.R.; supervision, M.R.; visualization, M.R.; writing—original draft, H.W.; writing—review and editing, M.R. All authors have read and agreed to the published version of the manuscript.

Funding: This material is based on work supported in part by the USDA/CSREES under Grant No. 2005-35504-16088, the National Science Foundation under Grant No. CHE-0724126, and the Institute for Critical Technology and Applied Science at Virginia Tech. M.R. acknowledges financial support by the Virginia Agricultural Experiment Station and the Hatch Multistate Program of the National Institute of Food and Agriculture, U.S. Department of Agriculture.

Institutional Review Board Statement: Not applicable.

Informed Consent Statement: Not applicable.

Data Availability Statement: Reported data are available upon request. Inquiries should be directed to the corresponding author.

Acknowledgments: The authors also acknowledge generous support from Omnova, Inc. and Tembec, Inc., H.W. thanks the staff of the Nanoscale Characterization and Fabrication Laboratory for assistance with the SEM images.

Conflicts of Interest: The authors declare no conflict of interest.

Sample Availability: There are no available samples.

References

1. Thünemann, A.F.; Müller, M.; Dautzenberg, H.; Joanny, J.F.O.; Löwen, H. Polyelectrolyte complexes. In *Polyelectrolytes with Defined Molecular Architecture II*; Schmidt, M., Ed.; Advances in Polymer Science; Springer: Berlin/Heidelberg, Germany, 2004; Volume 166, pp. 113–171.
2. De Kruif, C.G.; Weinbreck, F.; de Vries, R. Complex coacervation of proteins and anionic polysaccharides. *Curr. Opin. Colloid Interface Sci.* **2004**, *9*, 340–349. [CrossRef]
3. Liu, Z.H.; Jiao, Y.P.; Wang, Y.F.; Zhou, C.R.; Zhang, Z.Y. Polysaccharides-based nanoparticles as drug delivery systems. *Adv. Drug Deliv. Rev.* **2008**, *60*, 1650–1662.
4. Midoux, P.; Breuzard, G.; Gomez, J.P.; Pichon, C. Polymer-based gene delivery: A current review on the uptake and intracellular trafficking of polyplexes. *Curr. Gene Ther.* **2008**, *8*, 335–352. [CrossRef]
5. Cooper, C.L.; Dubin, P.L.; Kayitmazer, A.B.; Turksen, S. Polyelectrolyte-protein complexes. *Curr. Opin. Colloid Interface Sci.* **2005**, *10*, 52–78. [CrossRef]
6. Bhatia, S.R.; Khattak, S.F.; Roberts, S.C. Polyelectrolytes for cell encapsulation. *Curr. Opin. Colloid Interface Sci.* **2005**, *10*, 45–51. [CrossRef]
7. Vishalakshi, B.; Ghosh, S.; Kalpagam, V. The effects of charge-density and concentration on the composition of polyelectrolyte complexes. *Polymer* **1993**, *34*, 3270–3275. [CrossRef]
8. Becherán-Marón, L.; Peniche, C.; Argüelles-Monal, W. Study of the interpolyelectrolyte reaction between chitosan and alginate: Influence of alginate composition and chitosan molecular weight. *Int. J. Biol. Macromol.* **2004**, *34*, 127–133. [CrossRef]
9. Dautzenberg, H. Polyelectrolyte complex formation in highly aggregating systems: Methodical aspects and general tendencies. In *Physical Chemistry of Polyelectrolytes*; Radeva, T., Ed.; Marcel Dekker: New York City, NY, USA, 2001; pp. 743–792.
10. de Vasconcelos, C.L.; Bezerril, P.M.; dos Santos, D.E.S.; Dantas, T.N.C.; Pereira, M.R.; Fonseca, J.L.C. Effect of molecular weight and ionic strength on the formation of polyelectrolyte complexes based on poly (methacrylic acid) and chitosan. *Biomacromolecules* **2006**, *7*, 1245–1252. [CrossRef]
11. Shovsky, A.V.; Varga, I.; Makuska, R.; Claesson, P.M. Formation and stability of soluble stochiometric polyelectrolyte complexes: Effects of charge density and polyelectrolyte concentration. *J. Dispers. Sci. Technol.* **2009**, *30*, 980–988. [CrossRef]
12. Shovsky, A.; Varga, I.; Makuska, R.; Claesson, P.M. Formation and stability of water-soluble, molecular polyelectrolyte complexes: Effects of charge density, mixing ratio, and polyelectrolyte concentration. *Langmuir* **2009**, *25*, 6113–6121. [CrossRef] [PubMed]
13. Dautzenberg, H.; Jaeger, W. Effect of charge density on the formation and salt stability of polyelectrolyte complexes. *Macromol. Chem. Phys.* **2002**, *203*, 2095–2102. [CrossRef]
14. Hugerth, A.; Caram-Lelham, N.; Sundelöf, L.O. The effect of charge density and conformation on the polyelectrolyte complex formation between carrageenan and chitosan. *Carbohydr. Polym.* **1997**, *34*, 149–156. [CrossRef]
15. Koetz, J.; Linow, K.J.; Philipp, B.; Li, P.H.; Vogl, O. Effects of charge-density and structure of side-chain branching on the composition of polyanion polycation complexes. *Polymer* **1986**, *27*, 1574–1580. [CrossRef]
16. Dautzenberg, H.; Linow, K.J.; Philipp, B. The formation of water-soluble polysalts (simplexes) of anionic and cationic co-polymers of acrylamide. 2. Effect of charge-density and conversion on the structure of the simplexes. *Acta Polym.* **1982**, *33*, 619–625. [CrossRef]
17. Dautzenberg, H.; Hartmann, J.; Grunewald, S.; Brand, F. Stoichiometry and structure of polyelectrolyte complex particles in diluted solution. *Ber. Bunsenges. Phys. Chem.* **1996**, *100*, 1024–1032. [CrossRef]
18. Wang, H.; Roman, M. Formation and properties of chitosan-cellulose nanocrystal polyelectrolyte-macroion complexes for drug delivery applications. *Biomacromolecules* **2011**, *12*, 1585–1593. [CrossRef] [PubMed]
19. Wang, H.Z.; Qian, C.; Roman, M. Effects of pH and salt concentration on the formation and properties of chitosan–cellulose nanocrystal polyelectrolyte–macroion complexes. *Biomacromolecules* **2011**, *12*, 3708–3714. [CrossRef] [PubMed]
20. Wang, H. Chitosan–Cellulose Nanocrystal Polyelectrolyte Complex Particles: Preparation, Characterization, and In Vitro Drug Release Properties. Ph.D. Thesis, Virginia Tech, Blacksburg, VA, USA, 22 October 2009.
21. Ko, J.A.; Park, H.J.; Hwang, S.J.; Park, J.B.; Lee, J.S. Preparation and characterization of chitosan microparticles intended for controlled drug delivery. *Int. J. Pharm.* **2002**, *249*, 165–174. [CrossRef] [PubMed]
22. Rånby, B.G. Aqueous colloidal solutions of cellulose micelles. *Acta Chem. Scand.* **1949**, *3*, 649–650. [CrossRef]
23. Sabu, S.; Sasidharan, A.; Venugopal, V. Influence of isolation conditions on the physicochemical and biological properties of chitosan and chitosan oligosaccharides from marine crustacean shell wastes. In *Chitooligosaccharides: Prevention and Control of Diseases*; Kim, S.-K., Ed.; Springer International Publishing: Cham, Switzerland, 2022; pp. 333–352.
24. No, H.K.; Meyers, S.P. Preparation and characterization of chitin and chitosan—A review. *J. Aquat. Food Prod. Technol.* **1995**, *4*, 27–52. [CrossRef]
25. Mima, S.; Miya, M.; Iwamoto, R.; Yoshikawa, S. Highly deacetylated chitosan and its properties. *J. Appl. Polym. Sci.* **1983**, *28*, 1909–1917. [CrossRef]
26. Rinaudo, M.; Milas, M.; Dung, P.L. Characterization of chitosan. Influence of ionic strength and degree of acetylation on chain expansion. *Int. J. Biol. Macromol.* **1993**, *15*, 281–285. [PubMed]

27. Wang, W.; Bo, S.; Li, S.; Qin, W. Determination of the Mark-Houwink equation for chitosans with different degrees of deacetylation. *Int. J. Biol. Macromol.* **1991**, *13*, 281–285. [CrossRef] [PubMed]
28. Lavertu, M.; Xia, Z.; Serreqi, A.N.; Berrada, M.; Rodrigues, A.; Wang, D.; Buschmann, M.D.; Gupta, A. A validated H-1 NMR method for the determination of the degree of deacetylation of chitosan. *J. Pharm. Biomed. Anal.* **2003**, *32*, 1149–1158. [CrossRef]
29. Baxter, A.; Dillon, M.; Taylor, K.D.A.; Roberts, G.A.F. Improved method for I.R. Determination of the degree of N-acetylation of chitosan. *Int. J. Biol. Macromol.* **1992**, *14*, 166–169.

molecules

Article

Flexible and Wearable Strain–Temperature Sensors Based on Chitosan/Ink Sponges

Xiaoying Lin, Feng Wu, Yunqing He and Mingxian Liu *

Department of Materials Science and Engineering, College of Chemistry and Materials Science, Jinan University, Guangzhou 511443, China; linxiaoying0508@163.com (X.L.)
* Correspondence: liumx@jnu.edu.cn

Abstract: A simple and economic strategy to construct a chitosan-ink carbon nanoparticle sponge sensor was proposed by freeze-drying of chitosan and Chinese ink mixture solution. The microstructure and physical properties of the composite sponges with different ratios are characterized. The interfacial compatibility of chitosan and carbon nanoparticles in ink is satisfied, and the mechanical property and porosity of chitosan was increased by the incorporation of carbon nanoparticles. Due to excellent conductivity and good photothermal conversion effect of the carbon nanoparticles in ink, the constructed flexible sponge sensor has satisfactory strain and temperature sensing performance and high sensitivity (133.05 ms). In addition, these sensors can be successfully applied to monitor the large joint movement of the human body and the movement of muscle groups near the esophagus. Dual functionally integrated sponge sensors show great potential for strain and temperature detection in real time. The prepared chitosan-ink carbon nanoparticle composite shows promising applications in wearable smart sensors.

Keywords: chitosan; ink; photothermal; strain sensor; temperature sensor

check for updates

Citation: Lin, X.; Wu, F.; He, Y.; Liu, M. Flexible and Wearable Strain–Temperature Sensors Based on Chitosan/Ink Sponges. *Molecules* **2023**, *28*, 4083. https://doi.org/10.3390/molecules28104083

Academic Editor: Agnieszka Ewa Wiącek

Received: 6 April 2023
Revised: 12 May 2023
Accepted: 12 May 2023
Published: 14 May 2023

1. Introduction

With the continuous development of modern science and technology, the development and application of electronic equipment has increased rapidly. Various sensors have been used in human motion detection, human–computer interaction, artificial intelligence medical care, human health detection, electronic skin, and other fields [1–4]. Sensors can sensitively monitor changes in pressure [5], temperature [6], humidity [7,8], gas [9,10], etc. Due to their excellent flexibility and reliability, flexible sensors have attracted extensive and growing interest. Flexible sensors can be directly attached to human skin or curved surfaces, and the obtained signals can accurately detect human physiological information. They are generally consistent with flexible substrates and conductive active substances. Many kinds of construction materials, including fiber, sponge, hydrogel, aerogel, rubber, film, and so on, are used for the design of the sensor [11–17]. Hydrogel and rubber materials have good stretchability and deformation recoverability, but the compactness of the materials leads to poor air permeability. In addition, adhering these materials directly to the skin easily causes inflammation, which greatly affects their development and application in the wearable field [18]. Conductive materials used for the sensor include carbon materials (carbon black, carbon nanotubes, and graphene) [19–21], metal nanowires [22], MXene [23], polypyrrole [24], etc. Most of the conductive materials are expensive, increasing the cost of flexible sensors. Therefore, cheap sensor materials are widely sought-after.

Chinese ink, as a traditional material that has been widely used in calligraphy and painting in Asian countries for thousands of years, is a mixture of carbon nanoparticles and adhesive [25,26]. Carbon nanoparticles in the ink are the products of incomplete combustion of pinewood or burning oil at a wick [27]. The ink has good water dispersion stability and fluidity, and it has strong adhesion to the surface of the materials. Chinese ink has a simple preparation process with a low price, which has been used in many fields. In modern science and technology, Chinese ink has been used in solar steam systems [28], photothermal therapy (PTT) of tumors [29], strain sensors [30,31], etc. For example, Wang et al. developed a PTT technique for tumor metastatic lymph nodes using Chinese ink [32]. Zhou et al. prepared a flexible 3D electrode with a high-density current (16.3 ± 0.5 mA cm^3) by coating a natural loofah sponge with Chinese ink [33]. Zhang et al. used ink to prepare polydimethylhydrosiloxane-nickel fabric with excellent anti-icing and good photothermal properties, which was used in the anti-icing/icephobic project of outdoor clothing [34]. Our previous work also reported the advanced application of Chinese ink. Melamine foam was coated with Chinese ink by the dip-coating method, which exhibited excellent conductivity, Joule heat characteristics, and desalination ability [30]. In our other work, we used ink and water-based glue to prepare a flexible rubber sensor with multiple response functions, which can quickly detect temperature, humidity, and strain signals [31].

Chitosan is a kind of natural biopolymer, which is widely available and has the advantages of biocompatibility, environmental friendliness, regeneration, biodegradability, and antibacterial properties [35–37]. Chitosan sponge is a three-dimensional porous material with the advantages of being lightweight, having high porosity, and having a high strain range. However, the chitosan sponge has low compression resistance and a very unstable structure. Therefore, the combination of sponge and conductive material can improve mechanical properties. The conductive materials in the three-dimensional structure of the sponge contact each other under pressure, and the increased compactness drives the change of conductivity, which can be used as material for biological intelligence detection. Zheng et al. successfully prepared ultra-light aerogels containing chitosan, polyimide, and $FeCl_3$ by freeze-drying and carbonization, which showed a wide range of excellent sensing sensitivity (10.28 kPa^{-1}) and remarkable long-term compressive stability (1000 cycles of compression under 50% strain without deformation) [38]. Similarly, Wu et al. developed a carbon nanotube/chitosan aerogel sensor with hydrophobicity, high sensitivity, fast response, and repeatability. The sensor was prepared by freeze-drying and dip-coating methods, being immersed in graphene oxide and 1H, 1H, 2H, 2H-perfluoroctyliethoxysilane [39].

In this study, we proposed a strategy for preparing a flexible chitosan-ink carbon nanoparticle (CS-ink) composite sponge using the solution blending and freeze-drying method. The positively charged chitosan is closely connected with the negatively charged ink through electrostatic interaction, which enhances the mechanical properties of chitosan. Due to the excellent photothermal conversion effect of carbon nanoparticles, the temperature of the composite sponge can be rapidly increased to 254.7 °C within 5 s under the irradiation of a near-infrared laser (0.9 W/cm^2). Moreover, CS-ink exhibits a negative temperature coefficient behavior and an outstanding temperature sensitivity, which can be used as a temperature sensor. Since the carbon nanoparticles in ink are conductive materials, the prepared CS-ink composite sponge can be used as a strain sensor. The sensor can sensitively detect human movement with a response time of 133.4 ms. This simple and low-cost flexible strain–temperature sensor can be applied to human motion and temperature detection, which shows promising potential in wearable devices.

2. Results and Discussion

2.1. Preparation of CS-Ink Sponge

Figure 1a shows the schematic of the preparation process of the chitosan-ink carbon nanoparticle sponge. Composite sponges were obtained by blending chitosan solution with ink solution and subsequently employing the freeze-drying method. Carbon nanoparticles in ink can enter the chitosan molecular network. The sponge displayed a porous three-dimensional structure, as shown in Figure 1b. With the addition of ink, the pore size of the sponge decreased. According to the SEM, the pore size of the CS-ink sponge was 34.1 ± 11.1 μm, while the size of CS was 79 ± 28.2 μm. The magnified SEM image displayed that the surface of pure chitosan was smooth, while the surface of the CS-ink sponge was covered with carbon nanoparticles uniformly. The large number of carbon nanoparticles on the surface of the chitosan suggested that the carbon nanoparticles adhered well to the surface of the chitosan, and chitosan and carbon nanoparticles have good interfacial compatibility.

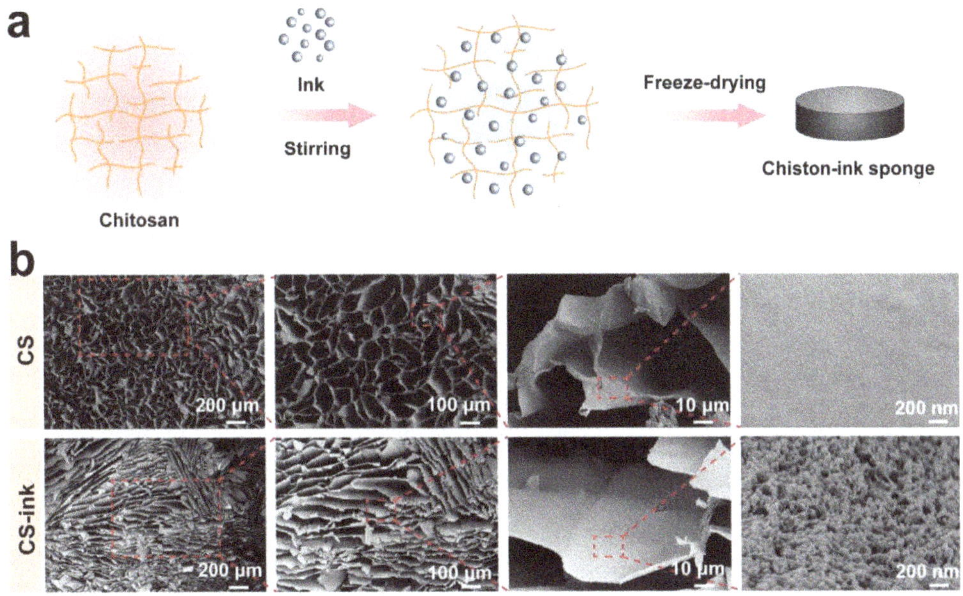

Figure 1. Schematic illustration of CS-ink sponge (**a**); SEM images of CS and CS-ink sponge (**b**).

2.2. Characterization of CS-Ink Sponge

CS-ink sponges with different shapes were shown in Figure 2a. CS-ink had a low density and can be placed on the plant leaves without bending them. The CS sponge was insulating, while the CS-ink was conductive. A circuit with a blue light-emitting diode (LED) light was designed to verify the conductivity of CS-ink. The LED lamp emitted light when the CS-ink sponge was connected to the circuit. The FTIR spectral measurement was used to explore further the composition of the CS-ink sponge (Figure 2b). The stretching vibration of the hydroxyl group (3258 cm^{-1}), bending vibration of the protonated amino group (1542 cm^{-1}), stretching vibration of the C–N bond (1405 cm^{-1}), and the stretching vibration of the C–O bond (1018 cm^{-1}) were attributed to the functional groups of pure chitosan [40]. These characteristic peaks decreased in intensity or even disappeared with the addition of ink, while the ink showed no strong absorption peaks, which were attributed to the fact that carbon nanoparticles can shield the structure information since it has a strong absorption towards the light. These results revealed that the carbon nanoparticles in ink were encapsulated in chitosan.

To further verify the combination of carbon nanoparticles and chitosan, Raman spectrum detection was performed in Figure 2c. Carbon nanoparticles in ink had a strong D band (1339 cm^{-1}) and G band (1584 cm^{-1}), which were typical Raman peaks of graphite-like carbon [41]. G band refers to E_2g phonons of carbon atoms, while D band derives from in-plane defects associated with graphite carbon. These two peaks also appeared in the CS-ink sponge, proving that the characteristic bonds of the carbon nanoparticles in the composite material were not destroyed despite the addition of chitosan solution to the ink. The XRD patterns of the CS, CS-ink, and ink were illustrated in Figure 2f. There are diffraction peaks of CS-ink at 24° and 43.5°. Carbon nanoparticles broadened the peak of CS at 21.1° [42], which was influenced by (002) crystal plane of carbon. Simultaneously, CS-ink had (100) crystal plane at 43.5° [30], which was unique to the carbon nanoparticles. There was no secondary phase with any new structure in the CS-ink composite sponge. These results also suggested the successful incorporation of carbon nanoparticles with chitosan.

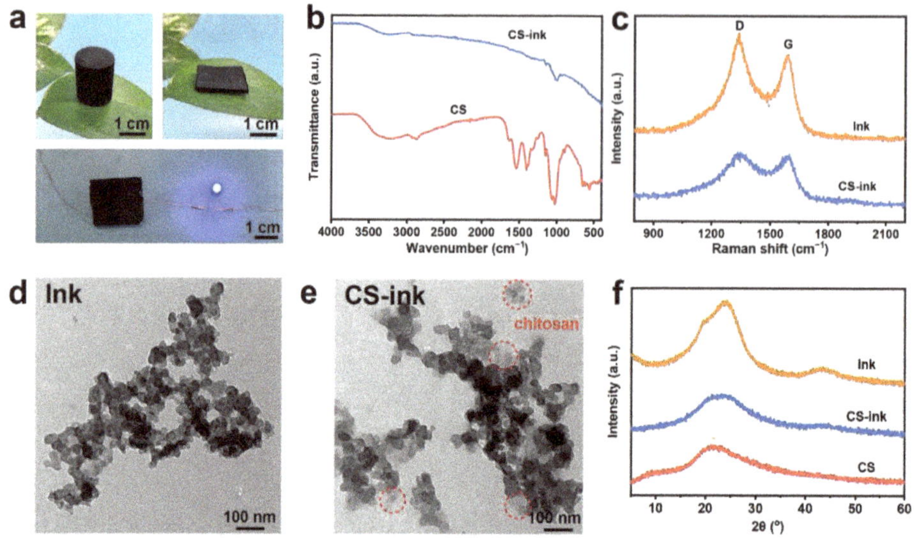

Figure 2. The appearance of the CS-ink sponge and the illumination of the LED bulb when linked with the CS-ink sponge in the circuit (**a**), FIRT spectrum of CS and CS-ink (**b**), Raman spectrogram of ink and CS-ink (**c**), TEM image of ink (**d**) and CS-ink (**e**), XRD patterns of CS, ink, and CS-ink (**f**).

The binding of carbon nanoparticles and chitosan can be visualized in the TEM image. The morphology of carbon nanoparticles in Figure 2d was spherical, and the carbon nanoparticles in CS-ink (Figure 2e) were connected by the chitosan layer. This indicated that chitosan and carbon nanoparticles were successfully combined. Moreover, there existed electrostatic interactions between chitosan and ink. The zeta potential of ink and chitosan was −18.7 mV and +48.4 mV, respectively (Figure 3a). As expected, the zeta potential increased to +64.8 mV after the combination of ink and chitosan. The increase in zeta potential meant that the carbon nanoparticles could firmly adhere to the chitosan.

As shown in Figure 3b,c, the ink content greatly influenced the apparent density and porosity. The density of the pure chitosan sponge was 31.1 mg/cm^3, and with the increase in ink content, the density of the sponge increased to 33.8–61.5 mg/cm^3. The porosity of the sponge rose from 66.8% to 83.4–88.9%. It can also be seen from Figure 3d that the higher the content of ink, the greater the conductivity. The electrical conductivity of the sponges increased to 0.0047–0.2869 S/m, with the ink content increasing to 10%–50%, meaning that the addition of carbon nanoparticles can significantly improve the conductivity of the sponge.

The mechanical robustness of the sponges was investigated by the compressive stress–strain test. Figure 3e shows the representative stress–strain curve. The compressive strength of CS-ink at 80% strain was 0.44 MPa, which was 2.1 times higher than that of CS (0.21 MPa). The compressive strength of the sponge with 50% ink content reached 1.02 MPa. However, the sponge with high compressive strength was less flexible and more brittle, while those with low compressive strength tended to deform during use. Both of them were unfavorable for being used as wearable devices in sensors. Considering the flexibility and electrical conductivity of the sponge, a 1:0.25 weight ratio of chitosan solution and ink was selected as the sensor sponge in this work.

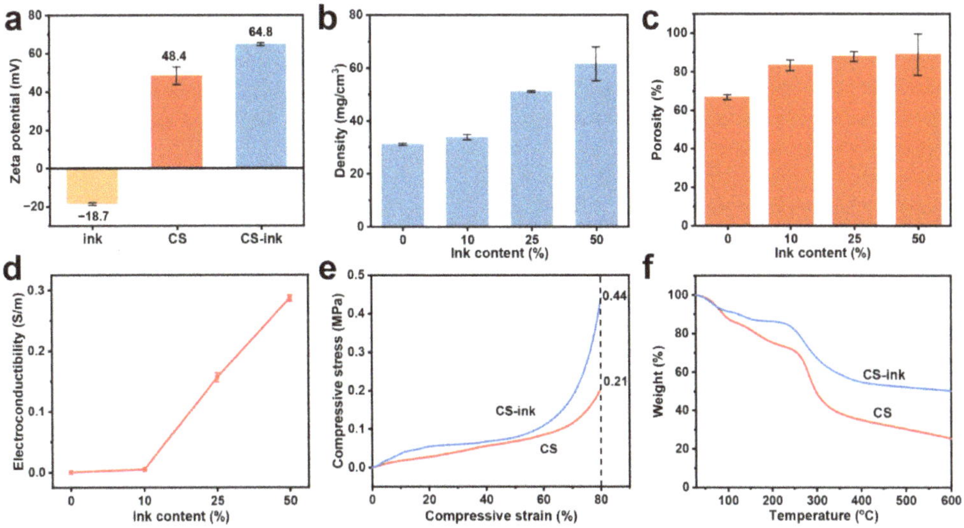

Figure 3. Zeta potential of ink, CS, and CS-ink (**a**); density (**b**) and porosity (**c**) of the CS-ink sponge; conductivity of the CS-ink sponge (**d**); the compressive stress–strain curves (**e**); and TG curves of CS and CS-ink sponges (**f**).

The thermal stability of the pure chitosan and the CS-ink sponge was determined by TGA, and the representative curve is shown in Figure 3f. The residual amount of CS-ink sponge was higher than that of pure chitosan in the range of 30 to 600 °C. The weight loss of CS was 74.7%, while that of the CS-ink sponge was 49.8%. Less weight loss of the sponge was attributed to the addition of ink, which also demonstrated that carbon nanoparticles facilitated the thermal stability of the sponge. Therefore, the decomposition of CS might be delayed by the incorporation of carbon nanoparticles. The loading of carbon nanoparticles was calculated to be as high as 24.9%, which was close to the ink content of 25%, indicating an effective binding of carbon nanoparticles to chitosan.

2.3. Photothermal Property of CS-Ink Sponge

The incorporation of carbon nanoparticles into the sponges not only endows the sponges with electrical conductivity but also with excellent photothermal conversion performance. The carbon nanoparticles in ink can effectively absorb light and convert it into heat, so the temperature of the sponge will rise with exposure to irradiation, which is known as photothermal behavior.

Figure 4a showed the surface temperature variation of the sponges with different ink contents under 808 nm laser irradiation (0.7 W/cm^2) for 5 min. At room temperature around 25 °C, the surface temperature of CS only increased by 12.6 °C. The photothermal conversion performance of the sponge was effectively improved by 10% ink. In particular, the sponge with 25% ink content had the highest photothermal effect, which was chosen as the temperature sensor. Figure 4b shows the temperature changes of the CS sponge with 25% ink contents under laser irradiation in detail. Obviously, the temperature rise was positively correlated with the laser power density, and the temperature could reach a basic equilibrium value within 60 s. Most importantly, the surface temperature could maintain a slight fluctuation at the equilibrium value with a further extension of the illumination time. Figure 4c reflected the infrared thermography of the sponge under laser irradiation of different power densities. The temperature rise was limited when the power density was 0.5 W/cm^2. The surface temperature of the sponge can be increased to 89.3 °C within 5 min. Excitingly, the surface temperatures of sponges can sharply rise to 180.9 °C and 254.7 °C within 5 s at the power of 0.7 and 0.9 W/cm^2, respectively. Moreover, the maximum temperature of the CS-ink was 200.2 °C and 267.8 °C under irradiation for 1 min, respectively. The photothermal stability of the CS-ink sponge was tested in Figure 4d. The photothermal effect of the CS-ink sponge can be reproduced several times when the 808 nm laser was turned on or off. Moreover, the shape of the chitosan sponge was not collapsed or deformed at the end of the cycling test. Therefore, it has been proved that CS-ink has the ultra-fast near-infrared response, efficient photothermal conversion performance, and excellent photothermal stability.

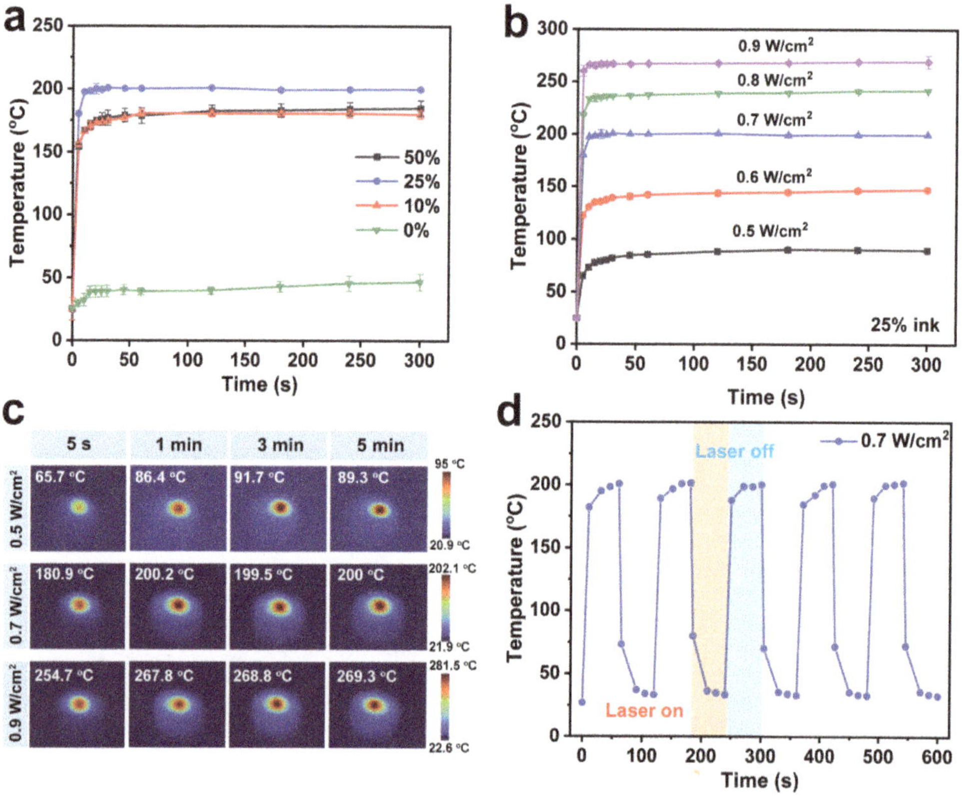

Figure 4. Temperature curve of the chitosan sponge with different ink content under 808 nm laser irradiation (0.7 W/cm^2) for 5 min (**a**); temperature change (**b**) and IR thermal images (**c**) of CS-ink sponge under 808 nm laser irradiation with different power densities for 5 min; and temperature change of CS-ink sponge with an 808 nm laser switch-on/off five times (**d**).

2.4. Application of CS-Ink Sponge as Strain Sensor

Based on the superior conductivity and high porosity, CS-ink can be employed as a strain sensor. First, a weight of 250 N was pressed against the sensor (Figure 5a), and the output signal was rapid and repeatable with a response time of 133.3 ± 4.7 ms, demonstrating that the sensor can provide stable signal monitoring in real time. To evaluate the application of the sponge as a flexible strain sensor for human motion detection, the sponges were attached to the human body to sense the signal change. The sensor was attached to the cervical vertebra. As volunteers nodded their heads, the cervical vertebra drove the sponge to be stretched (Figure 5b). As a result, the relative resistance decreased. The sponge was slightly squeezed when the cervical vertebra returned to its original position. Consequently, a large peak followed by a small peak in the signal appeared throughout the monitoring process. CS-ink sensors could detect not only large movements of the human joints but also small movements of the body. As shown in Figure 5c, when the tester swallowed water, the R/R_0 decreased sharply because the sensor was subjected to pressure from the muscle movement near the esophagus. This characteristic can be applicable to the initial diagnosis and real-time therapeutic monitoring of patients suffering from swallowing difficulties. The strain sensor was applied to the finger joint to detect the bending of the finger (Figure 5d). When the finger was bent, the conductive network in the sensor was squeezed, causing the R/R_0 to decrease rapidly while remaining balanced. When the finger was straightened, the R/R_0 returned to its original position. Similarly, the R/R_0 of the sponge attached to the elbow decreased when the elbow was bent, which was also caused by the sponge being squeezed. There will be regular and periodic signal responses when bending and straightening the wrist (Figure 5e) and elbow (Figure 5f) at the same angle and frequency. As shown in Figure 5g, the response time was rapid and stable. In summary, CS-ink sensors as wearable flexible devices have the ability to monitor human movement in real time, which shows great potential for applications in medical health monitoring and motion detection.

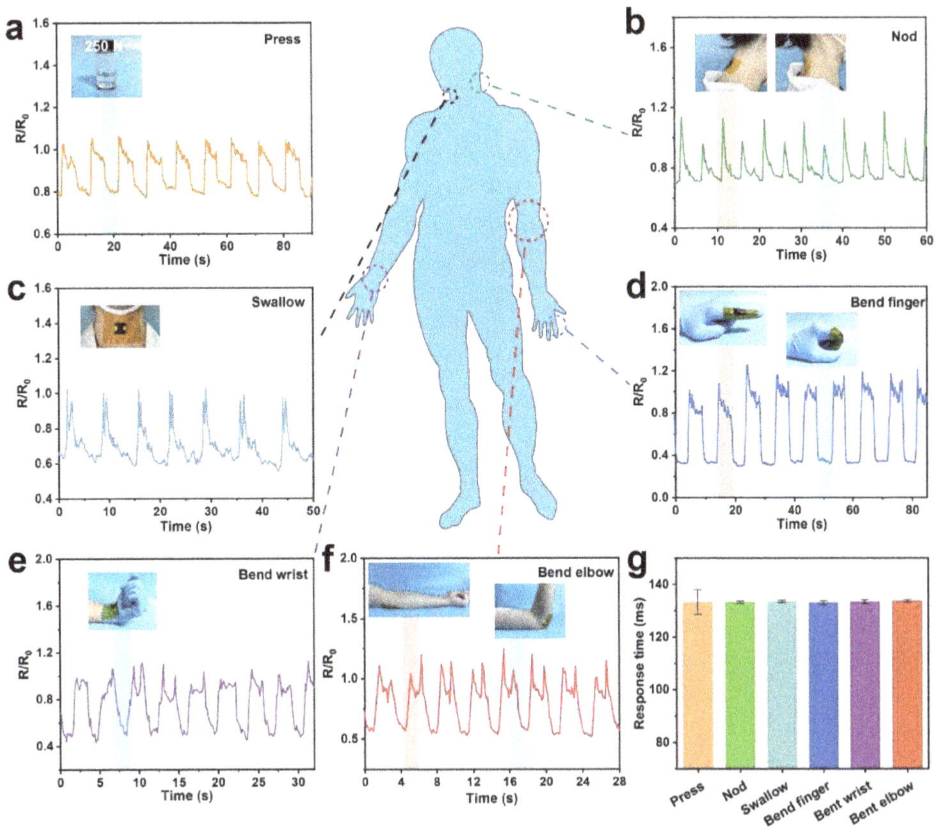

Figure 5. Representative resistant signal of the CS-ink sensor during pressing (**a**), nodding (**b**), swallowing (**c**), bending finger (**d**), wrist (**e**), and elbow (**f**), and response time of different movements (**g**).

2.5. Application of CS-Ink as Temperature Detection Sensor

The CS-ink sponge with excellent photothermal conversion performance makes the sponge as a sensor that can be used for temperature monitoring. Firstly, the relationship of sponge resistance changing with temperature was tested. Figure 6a was a schematic diagram for detecting the temperature resistance change of the sponge from 25 to 50 °C. As can be seen from Figure 6b, $\Delta R/R_0$ decreased when the temperature increased, exhibiting negative temperature coefficient (NTC) behavior [43]. The NTC effect is caused by the fact that an increase in temperature promotes the migration of charge carriers, resulting in a decrease in resistance. The temperature coefficient of resistance (TCR) was 0.321% according to the slope of the fitting curve, which showed outstanding temperature sensitivity [44]. $\Delta R/R_0$ is monotonic and linear in the temperature range of 25–50 °C. It was also found that the experimental data were in agreement with the theoretical curve (R^2 = 0.999).

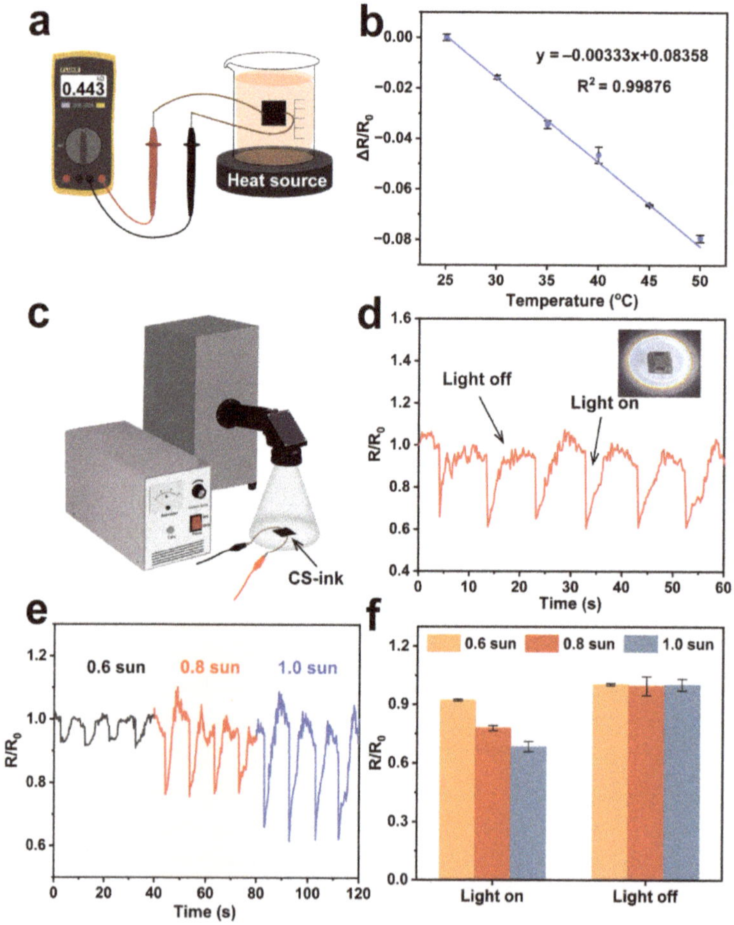

Figure 6. Schematic diagram of the CS-ink sensor used for temperature detection (**a**), resistance changes of the sensor during the temperature rise (**b**), schematic diagram of the test device of temperature response sensor (**c**), detection of temperature changes caused by light irradiation (**d**), resistance changes of the sensor under the irradiation of different xenon lamps (**e**), and resistance changes of the sensor when the xenon lamp was lighted on or off (**f**).

Furthermore, the temperature change of the sponge was detected by the resistance change of the sensor. Figure 6c shows a schematic diagram of detecting the resistance change of a sample irradiated by simulated sunlight. When irradiated by the xenon lamp, the sensor temperature quickly rose to 71 °C within 5 s. The carbon nanoparticle network was capable of converting light energy into heat energy. Some of the heat was used to raise the temperature. As a result, the resistance of the sensor was reduced. When the xenon lamp irradiation was removed, the temperature of the sensor gradually decreased, and the resistance returned to the initial value (Figure 6d). The temperature sensor responded quickly, with a response time of only 132.7 ± 2.1 ms. In addition, the intensity of sunlight also affected the change in sensor resistance. As shown in Figure 6e, the resistance changes of the sensor increased as the irradiation intensity increased, and the resistance change was stable (Figure 6f). Therefore, CS-ink can be utilized as a wearable temperature sensor to track temperature changes in the environment.

3. Materials and Methods

3.1. Materials

Zhujiang brand Chinese ink was purchased from Guangzhou Jinjian office manufacturing plant, China. Chitosan (deacetylation degree ≥ 95%, viscosity-average molecular weight of 600,000 g/mol, viscosity (10 g/L, 20 °C) of 200 mPa·s) was purchased from Jinan Haidebei Marine Bioengineering Co., Ltd., Jinan, China. Glacial acetic acid was obtained from Guangdong Guangshi Reagent Technology Co., Ltd., Shenzhen, China.

3.2. Preparation of Chitosan-Ink Carbon Nanoparticles Sponge

A 2 wt% chitosan solution was prepared in a 2 wt% glacial acetic acid solution and stirred for 12 h. Chinese ink (10%, 25%, and 50% relative to the mass of the chitosan solution) was added to the chitosan solution. The chitosan-ink solution was obtained after stirring for 1 h and being subjected to ultrasonic shock treatment. Then, the solution was poured into a mold and frozen at −70 °C. After freeze-drying for 48 h in a vacuum freeze-dryer (Ningbo Scientz Biotechnology Co., Ltd., Ningbo, China), chitosan-ink carbon nanoparticle sponges with a length, width, and height of 20 × 20 × 4 mm and a diameter of 15.6 mm and a height of 20 mm were obtained. Chitosan sponges with 25% ink content were named as CS-ink, and this sample was used for the following measurement. The preparation method of the pure chitosan sponge (named CS) was the same as above. Finally, all samples were placed in an oven at 60 °C for a period of time to remove the acid from the sponge.

3.3. Characterization Methods

The cross-section morphology of the sponges was examined using a scanning electrospinning microscopy (SEM, Zeiss Sigma 500, Oberkochen, Germany). Before observation, the surface of the samples was coated with a thin gold layer. The microstructure of carbon nanoparticles and CS-ink was observed with a transmission electron microscope (TEM, JEOL/JEM-1400 Flash, Tokyo, Japan). The chemical structure of chitosan and CS-ink was characterized using a Fourier transform infrared spectrometer (Nicolet iS50, Thermo Fisher Scientific Ltd., Waltham, MA, USA) in a scope of wavenumber from 4000 to 400 cm^{-1}. The XRD patterns of CS and CS-ink were determined by an X-ray diffractometer (MiniFlex-600, Rigaku Corporation, Tokyo, Japan) at an accelerating voltage of 40 kV and a current of 40 mA. The diffraction angles were 5 to 60° with a 15°/min scan rate. From the measurements of the electrophoretic mobility, the zeta potentials of ink, CS, and CS-ink dispersions were calculated from the Henry equation using zetameter software via a Nano ZS zeta potential analyzer (Malvern Instruments Co., Malvern, UK). The thermolysis curves of CS and CS-ink sponges were analyzed by a thermogravimetric analyzer (Mettler Toledo, Co., Ltd., Zürich, Switzerland). About 10 mg of the sponge was placed in an Al_2O_3 crucible and heated from 30 to 600 °C at a heating rate of 10 °C/min under a nitrogen atmosphere. The first derivative of TG (DTG) was also analyzed by the software. The experimental data were processed and expressed by the mean values ± standard deviation (SD).

3.4. Mechanical Properties

A universal testing machine (AGS-X, Shimadzu, Japan) was conducted to measure the compression resistance of CS and CS-ink sponges at 25 °C and 55% RH. The samples (diameter of 15.6 mm, height of 7 mm) were compressed to 80% of their original volume by a longitudinal pressure at a 5 mm/min compression rate.

3.5. Porosity

The porosity of the sponges was analyzed by the absolute alcohol displacement method. The weight and size of the samples were measured first. Then, the dried samples were soaked in alcohol. After 6 h, the samples were taken out and weighed. Each group should be tested three times. The porosity was calculated using the following formula (Equation (1)),

$$Porosity\ (\%) = \frac{m_2 - m_1}{\rho \times V} \times 100\% \tag{1}$$

where m_2 is the weight of the samples after immersing in alcohol, ρ is the density of alcohol (0.79 g/cm^3 at room temperature), and m_1 and V are the mass and volume of the samples at the initial state, respectively.

3.6. Electrical Conductivity

The conductivity of the cylindrical samples was measured using a multimeter (Fluke 17B+, Fluke Electronic Instrument Ltd., Everett, WA, USA). The sample's left and right ends were wrapped with copper wires, and the other copper wires were connected to a multimeter. The conductivity was calculated as follows:

$$G\ (S/m) = \frac{L}{R \times S} \tag{2}$$

where G is the conductivity, L is the length of the sample in the current direction, R is the resistance, and S is the cross-sectional area of the sample perpendicular to the current direction.

3.7. Photothermal Performance

The samples (diameter of 15.6 mm and height of 20 mm) were irradiated by an 808 nm fiber-coupled laser (MD-III-808, Changchun New Industry Optoelectronic Technology Ltd., Changchun, China). The temperature changes of the chitosan sponges with different ink contents were measured by irradiating them at 0.7 W/cm^2 power density for 5 min. The surface temperatures and images of the CS-ink sponge were recorded by an infrared imager (TIS 55, Fluke Electronic Instrument Ltd., Everett, WA, USA) in real time. The power density of the laser was adjusted in the range of 0.5–0.9 W/cm^2, and the measuring distance between the laser and the sample was 1 cm. The parallel test was done three times.

3.8. The Application of CS-Ink Sponge as a Strain Sensor

The sponge with a size of 20 × 20 × 4 mm was attached to the fingers, wrists, elbow, spine, and throat of the human body with polyimide tape. One end of the copper wire was connected to the sponge and the other to the electrometer (Keithley 6514, Tektronix Inc., Beaverton, OR, USA). Subsequently, the signal response was recorded using an electrometer by pressing the sample; bending the finger, wrist, and elbow; nodding; and swallowing. Three groups of parallel samples were tested for each action, and each test cycle was repeated several times.

3.9. Temperature Monitoring

The CS-ink sponge adhered to the outer wall of a beaker filled with 450 mL of silicone oil. The sponge was connected to the multimeter by copper wire. The temperature of the sensor was changed by heating the silicone oil, and the resistance change was measured by a multimeter.

The temperature of the sensor was varied by changing the power of simulated sunlight (Oriel Class A solar simulator, Oriel 91195A, Newport Co., Irvine, CA, USA). The sponge under simulated sunlight was connected to the electrometer through wires. The wires were taped to the table to reduce errors caused by vibration. The xenon lamp was turned on to irradiate the sample for 5 s and then turned off for 5 s, cycling the process several times. The relationship between the signal response from the sensor and temperature was measured. The measurement distance between the xenon lamp and the sample was 16 cm. The parallel test was done three times.

4. Conclusions

In conclusion, a flexible temperature–pressure response sensor with excellent photothermal properties and good electrical conductivity was developed. The sensor consists of Chinese ink and chitosan, obtained by a simple solution blending and freeze-drying method. The carbon nanoparticles in ink impart good conductivity and superior photothermal conversion properties to the chitosan sponge, with a conductivity of 0.1565 S/m and a rapid temperature rise to 254.7 °C under NIR irradiation at low power density (0.9 W/cm^2) within 5 s. Additionally, the composite sponge was used as a strain–temperature sensor for the detection of human motion and temperature changes in real time. The average response time of the sensor is 133.05 ms, which provides a fast output of stable and repeatable signals. Therefore, CS-ink sensors have a wide range of application prospects in wearable electronics, electronic skin, health monitoring, environmental temperature detection, and other fields.

Author Contributions: Conceptualization, X.L. and M.L.; methodology, X.L. and F.W.; validation, X.L. and M.L.; formal analysis, X.L.; resources, M.L.; data curation, Y.H.; writing—original draft preparation, X.L.; writing—review and editing, X.L. and F.W.; visualization, X.L. and F.W.; supervision, M.L.; project administration, M.L.; funding acquisition, M.L. All authors have read and agreed to the published version of the manuscript.

Funding: We acknowledge the financial support received from the National Natural Science Foundation China (52073121), Science and Technology Program of Guangzhou (202102010117), Fundamental Research Funds for the Central Universities (21622406), and Project Team of Foshan National Hi-tech Industrial Development Zone Industrialization Entrepreneurial Teams Program (2220197000129).

Institutional Review Board Statement: Not applicable.

Informed Consent Statement: Informed consent was obtained from all subjects involved in the study.

Data Availability Statement: All the data have been included in the study.

Acknowledgments: The authors would like to thank the technicians of Jinan University's test platform for their help with the SEM analysis.

Conflicts of Interest: The authors declare no conflict of interest.

Sample Availability: Not applicable.

References

1. Chen, J.; Zhu, Y.; Chang, X.; Pan, D.; Song, G.; Guo, Z.; Naik, N. Recent progress in essential functions of soft electronic skin. *Adv. Funct. Mater.* **2021**, *31*, 2104686. [CrossRef]
2. Wang, J.; Lu, C.; Zhang, K. Textile-based strain sensor for human motion detection. *Energy Environ. Mater.* **2020**, *3*, 80–100. [CrossRef]
3. Zang, Y.; Zhang, F.; Di, C.-A.; Zhu, D. Advances of flexible pressure sensors toward artificial intelligence and health care applications. *Mater. Horiz.* **2015**, *2*, 140–156. [CrossRef]
4. Zhou, Y.; Shen, M.; Cui, X.; Shao, Y.; Li, L.; Zhang, Y. Triboelectric nanogenerator based self-powered sensor for artificial intelligence. *Nano Energy* **2021**, *84*, 105887. [CrossRef]
5. Sun, H.; Bu, Y.; Liu, H.; Wang, J.; Yang, W.; Li, Q.; Guo, Z.; Liu, C.; Shen, C. Superhydrophobic conductive rubber band with synergistic dual conductive layer for wide-range sensitive strain sensor. *Sci. Bull.* **2022**, *67*, 1669–1678. [CrossRef]

6. Li, M.; Chen, J.; Zhong, W.; Luo, M.; Wang, W.; Qing, X.; Lu, Y.; Liu, Q.; Liu, K.; Wang, Y. Large-area, wearable, self-powered pressure–temperature sensor based on 3D thermoelectric spacer fabric. *ACS Sens.* **2020**, *5*, 2545–2554. [CrossRef]
7. Sang, M.; Kang, K.; Zhang, Y.; Zhang, H.; Kim, K.; Cho, M.; Shin, J.; Hong, J.H.; Kim, T.; Lee, S.K. Ultrahigh sensitive Au-doped silicon nanomembrane based wearable sensor arrays for continuous skin temperature monitoring with high precision. *Adv. Mater.* **2022**, *34*, 2105865. [CrossRef]
8. Cuana, R.; Nasution, T.; Agusnar, H.; Susilowati, A.; Lubis, N.; Pradana, I. Humidity Detection Based on Chitosan/PEO Film Sensor. *J. Phys. Conf. Ser.* **2023**, *2421*, 012038. [CrossRef]
9. Choi, S.-J.; Kim, I.-D. Recent developments in 2D nanomaterials for chemiresistive-type gas sensors. *Electron. Mater. Lett.* **2018**, *14*, 221–260. [CrossRef]
10. Park, H.; Kim, W.; Lee, S.W.; Park, J.; Lee, G.; Yoon, D.S.; Lee, W.; Park, J. Flexible and disposable paper-based gas sensor using reduced graphene oxide/chitosan composite. *J. Mater. Sci. Technol.* **2022**, *101*, 165–172. [CrossRef]
11. Shen, Z.; Zhang, Z.; Zhang, N.; Li, J.; Zhou, P.; Hu, F.; Rong, Y.; Lu, B.; Gu, G. High-Stretchability, Ultralow-Hysteresis ConductingPolymer Hydrogel Strain Sensors for Soft Machines. *Adv. Mater.* **2022**, *34*, 2203650. [CrossRef] [PubMed]
12. Xu, Z.; Zhang, D.; Liu, X.; Yang, Y.; Wang, X.; Xue, Q. Self-powered multifunctional monitoring and analysis system based on dual-triboelectric nanogenerator and chitosan/activated carbon film humidity sensor. *Nano Energy* **2022**, *94*, 106881. [CrossRef]
13. Liu, H.; Xu, T.; Cai, C.; Liu, K.; Liu, W.; Zhang, M.; Du, H.; Si, C.; Zhang, K. Multifunctional superelastic, superhydrophilic, and ultralight nanocellulose-based composite carbon aerogels for compressive supercapacitor and strain sensor. *Adv. Funct. Mater.* **2022**, *32*, 2113082. [CrossRef]
14. Zeng, R.; Lu, S.; Qi, C.; Jin, L.; Xu, J.; Dong, Z.; Lei, C. Polyacrylamide/carboxymethyl chitosan double-network hydrogels with high conductivity and mechanical toughness for flexible sensors. *J. Appl. Polym. Sci.* **2022**, *139*, 51993. [CrossRef]
15. Quan, L.; Tie, J.; Wang, Y.; Mao, Z.; Zhang, L.; Zhong, Y.; Sui, X.; Feng, X.; Xu, H. Mussel-inspired chitosan-based hydrogel sensor with pH-responsive and adjustable adhesion, toughness and self-healing capability. *Polym. Adv. Technol.* **2022**, *33*, 1867–1880. [CrossRef]
16. Cai, J.; Du, M.; Li, Z. Flexible temperature sensors constructed with fiber materials. *Adv. Mater. Technol.* **2022**, *7*, 2101182. [CrossRef]
17. Wei, L.; Wu, Z.; Tang, S.; Qin, X.; Xiong, Y.; Li, J.; Ruiz-Hitzky, E.; Wang, X. Tracheid-inspired nanoarchitectured carbon-based aerogels with ultra-compressibility for wearable piezoresistive sensors. *Carbon* **2023**, *203*, 386–396. [CrossRef]
18. Miyamoto, A.; Lee, S.; Cooray, N.F.; Lee, S.; Mori, M.; Matsuhisa, N.; Jin, H.; Yoda, L.; Yokota, T.; Itoh, A. Inflammation-free, gas-permeable, lightweight, stretchable on-skin electronics with nanomeshes. *Nat. Nanotechnol.* **2017**, *12*, 907–913. [CrossRef]
19. He, Y.; Zhou, M.; Mahmoud, M.; Lu, X.; He, G.; Zhang, L.; Huang, M.; Elnaggar, A.Y.; Lei, Q.; Liu, H. Multifunctional wearable strain/pressure sensor based on conductive carbon nanotubes/silk nonwoven fabric with high durability and low detection limit. *Adv. Compos. Hybrid Mater.* **2022**, *5*, 1939–1950. [CrossRef]
20. Liu, C.; Tan, Q.; Deng, Y.; Ye, P.; Kong, L.; Ma, X.; Xu, L.; Xu, Y.; Qiang, Q.; Chen, W. Highly sensitive and stable 3D flexible pressure sensor based on carbon black and multi-walled carbon nanotubes prepared by hydrothermal method. *Compos. Commun.* **2022**, *32*, 101178. [CrossRef]
21. Chen, Y.; Wang, L.; Wu, Z.; Luo, J.; Li, B.; Huang, X.; Xue, H.; Gao, J. Super-hydrophobic, durable and cost-effective carbon black/rubber composites for high performance strain sensors. *Compos. Part B Eng.* **2019**, *176*, 107358. [CrossRef]
22. Araki, T.; Uemura, T.; Yoshimoto, S.; Takemoto, A.; Noda, Y.; Izumi, S.; Sekitani, T. Wireless monitoring using a stretchable and transparent sensor sheet containing metal nanowires. *Adv. Mater.* **2020**, *32*, 1902684. [CrossRef] [PubMed]
23. Chen, T.; Liu, Z.; Zhao, G.; Qin, Z.; Zheng, P.; Aladejana, J.T.; Tang, Z.; Weng, M.; Peng, X.; Chang, J. Piezoresistive Sensor Containing Lamellar MXene-Plant Fiber Sponge Obtained with Aqueous MXene Ink. *ACS Appl. Mater. Interfaces* **2022**, *14*, 51361–51372. [CrossRef]
24. An, K.H.; Jeong, S.Y.; Hwang, H.R.; Lee, Y.H. Enhanced sensitivity of a gas sensor incorporating single-walled carbon nanotube–polypyrrole nanocomposites. *Adv. Mater.* **2004**, *16*, 1005–1009. [CrossRef]
25. Lin, J.; Zhou, P.; Chen, Q.; Zhang, W.; Luo, Z.; Chen, L. Reprogrammable, light-driven and sensing actuators based on Chinese ink composite: A synergetic use of shape-memory and self-healing strategies. *Sens. Actuators B Chem.* **2022**, *362*, 131776. [CrossRef]
26. Wei, S.; Fang, X.; Yang, J.; Cao, X.; Pintus, V.; Schreiner, M.; Song, G. Identification of the materials used in an Eastern Jin Chinese ink stick. *J. Cult. Herit.* **2012**, *13*, 448–452. [CrossRef]
27. Swider, J.R.; Hackley, V.A.; Winter, J. Characterization of Chinese ink in size and surface. *J. Cult. Herit.* **2003**, *4*, 175–186. [CrossRef]
28. Zhang, X.-F.; Wang, Z.; Song, L.; Feng, Y.; Yao, J. Chinese ink enabled wood evaporator for continuous water desalination. *Desalination* **2020**, *496*, 114727. [CrossRef]
29. Chen, D.; Chen, C.; Huang, C.; Chen, T.; Liu, Z. Injectable hydrogel for NIR-II photo-thermal tumor therapy and dihydroartemisinin-mediated chemodynamic therapy. *Front. Chem.* **2020**, *8*, 251. [CrossRef]
30. Cao, X.; Liu, H.; Cai, J.; Chen, L.; Yang, X.; Liu, M. Chinese ink coated melamine foam with Joule heating and photothermal effect for strain sensor and seawater desalination. *Compos. Part A Appl. Sci. Manuf.* **2021**, *149*, 106535. [CrossRef]
31. Liu, B.; Zhou, Y.; Chen, L.; Feng, Y.; Liu, M. High-performance sensors based on chinese ink and water-based glue for detection of strain, temperature, and humidity. *ACS Sustain. Chem. Eng.* **2022**, *10*, 1847–1856. [CrossRef]
32. Wang, S.; Cao, Y.; Zhang, Q.; Peng, H.; Liang, L.; Li, Q.; Shen, S.; Tuerdi, A.; Xu, Y.; Cai, S. New application of old material: Chinese traditional ink for photothermal therapy of metastatic lymph nodes. *ACS Omega* **2017**, *2*, 5170–5178. [CrossRef] [PubMed]

33. Zhou, L.; Sun, L.; Fu, P.; Yang, C.; Yuan, Y. Carbon nanoparticles of Chinese ink-wrapped natural loofah sponge: A low-cost three-dimensional electrode for high-performance microbial energy harvesting. *J. Mater. Chem. A* **2017**, *5*, 14741–14747. [CrossRef]
34. Zhang, S.; Zhang, F.; Zhang, Z.; Li, G.; Fu, H.; Huang, J.; Wang, Y.; Lei, Z.; Qian, X.; Lai, Y. An electroless nickel plating fabric coated with photothermal Chinese ink for powerful passive anti-icing/icephobic and fast active deicing. *Chem. Eng. J.* **2022**, *450*, 138328. [CrossRef]
35. Ong, S.-Y.; Wu, J.; Moochhala, S.M.; Tan, M.-H.; Lu, J. Development of a chitosan-based wound dressing with improved hemostatic and antimicrobial properties. *Biomaterials* **2008**, *29*, 4323–4332. [CrossRef]
36. Hu, Z.; Zhang, D.-Y.; Lu, S.-T.; Li, P.-W.; Li, S.-D. Chitosan-based composite materials for prospective hemostatic applications. *Mar. Drugs* **2018**, *16*, 273. [CrossRef]
37. Li, Q.; Dunn, E.; Grandmaison, E.; Goosen, M.F. Applications and properties of chitosan. In *Applications of Chitin and Chitosan*; CRC Press: Boca Raton, FL, USA, 2020; pp. 3–29.
38. Zheng, H.; Chen, G.; Gan, L.; Liu, J.; Lin, L. Iron-Adjustable Compressible Elastic Chitosan-Derived Carbon Aerogel with Wide-Range Linear Sensitivity and Super Sensing Performances for Wearable Piezoresistive Sensors. *ACS Sustain. Chem. Eng.* **2022**, *10*, 10604–10614. [CrossRef]
39. Wu, J.; Li, H.; Lai, X.; Chen, Z.; Zeng, X. Conductive and superhydrophobic F-rGO@ CNTs/chitosan aerogel for piezoresistive pressure sensor. *Chem. Eng. J.* **2020**, *386*, 123998. [CrossRef]
40. Lawrie, G.; Keen, I.; Drew, B.; Chandler-Temple, A.; Rintoul, L.; Fredericks, P.; Grøndahl, L. Interactions between alginate and chitosan biopolymers characterized using FTIR and XPS. *Biomacromolecules* **2007**, *8*, 2533–2541. [CrossRef]
41. Nagyte, V.; Kelly, D.J.; Felten, A.; Picardi, G.; Shin, Y.; Alieva, A.; Worsley, R.E.; Parvez, K.; Dehm, S.; Krupke, R. Raman fingerprints of graphene produced by anodic electrochemical exfoliation. *Nano Lett.* **2020**, *20*, 3411–3419. [CrossRef]
42. Zhang, Y.; Xue, C.; Xue, Y.; Gao, R.; Zhang, X. Determination of the degree of deacetylation of chitin and chitosan by X-ray powder diffraction. *Carbohydr. Res.* **2005**, *340*, 1914–1917. [CrossRef] [PubMed]
43. Zhang, Y.; Chen, K.; Li, Y.; Lan, J.; Yan, B.; Shi, L.; Ran, R. High-strength, self-healable, temperature-sensitive, MXene-containing composite hydrogel as a smart compression sensor. *ACS Appl. Mater. Interfaces* **2019**, *11*, 47350–47357. [CrossRef] [PubMed]
44. Wang, Q.; Huang, X.; Han, F.; Wu, Y.; Wang, L.; Dai, H.; Song, P.; Tang, L.; Gao, J. Superhydrophobic, biocompatible and durable nanofiber composite with an asymmetric structure for anisotropic strain sensing and body motion detection. *Chem. Eng. J.* **2022**, *450*, 137899. [CrossRef]

molecules

MDPI

Article

A Simple Method for Synthesis of Chitosan Nanoparticles with Ionic Gelation and Homogenization

Nicolas Van Bavel [1], Travis Issler [1], Liping Pang [2,*], Max Anikovskiy [3,*] and Elmar J. Prenner [1,*]

1 Department of Biological Sciences, University of Calgary, Calgary, AB T2N 1N4, Canada; nicolas.vanbavel@ucalgary.ca (N.V.B.); travis.issler@ucalgary.ca (T.I.)
2 Institute of Environmental Science and Research, P.O. Box 29181, Christchurch 8540, New Zealand
3 Department of Chemistry, University of Calgary, Calgary, AB T2N 1N4, Canada
* Correspondence: liping.pang@esr.cri.nz (L.P.); m.anikovskiy@ucalgary.ca (M.A.); eprenner@ucalgary.ca (E.J.P.)

Abstract: Chitosan nanoparticles (CNPs) are known to have great utility in many fields (pharmaceutical, agricultural, food industry, wastewater treatment, etc.). In this study we aimed to synthesize sub-100 nm CNPs as a precursor of new biopolymer-based virus surrogates for water applications. We present a simple yet efficient synthesis procedure for obtaining high yield, monodisperse CNPs with size 68–77 nm. The CNPs were synthesized by ionic gelation using low molecular weight chitosan (deacetylation 75–85%) and tripolyphosphate as crosslinker, under rigorous homogenization to decrease size and increase uniformity, and purified by passing through 0.1 μm polyethersulfone syringe filters. The CNPs were characterized using dynamic light scattering, tunable resistive pulse sensing, and scanning electron microscopy. We demonstrate reproducibility of this method at two separate facilities. The effects of pH, ionic strength and three different purification methods on the size and polydispersity of CNP formation were examined. Larger CNPs (95–219) were produced under ionic strength and pH controls, and when purified using ultracentrifugation or size exclusion chromatography. Smaller CNPs (68–77 nm) were formulated using homogenization and filtration, and could readily interact with negatively charge proteins and DNA, making them an ideal precursor for the development of DNA-labelled, protein-coated virus surrogates for environmental water applications.

Keywords: nanoparticle synthesis; chitosan; biopolymers; ionic gelation

Citation: Van Bavel, N.; Issler, T.; Pang, L.; Anikovskiy, M.; Prenner, E.J. A Simple Method for Synthesis of Chitosan Nanoparticles with Ionic Gelation and Homogenization. *Molecules* **2023**, *28*, 4328. https://doi.org/10.3390/molecules28114328

Academic Editors: Eric Guibal and Agnieszka Ewa Wiącek

Received: 15 April 2023
Revised: 10 May 2023
Accepted: 22 May 2023
Published: 25 May 2023

1. Introduction

Chitosan is an abundant natural cationic biopolymer derived from the partial deacetylation of chitin [1]. The biopolymer is made up of linear β-(1→4) glycosidic linkages of 2-acetamido-d-glucose and 2-amino-d-glucose units (Figure 1). The amine functional groups of chitosan have a pKb value of ~6.5, which increases its solubility in neutral and acidic media [2]. Additionally, the amine groups contribute to the physiochemical properties of the biopolymer and allow for chitosan-based materials to be developed via synthetic processes such as gelation.

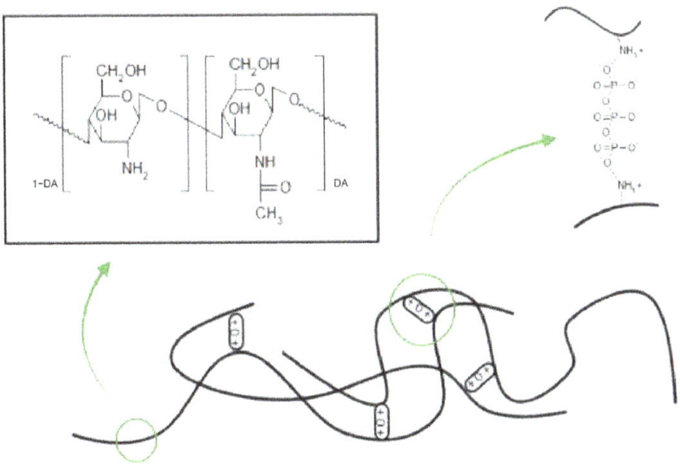

Figure 1. Chitosan crosslinking with tripolyphosphate, showcasing the structure of chitosan polymers and the anionic crosslinker tripolyphosphate. Created with BioRender.com.

Chitosan is biodegradable, biocompatible, nontoxic and has notable antimicrobial properties. Chitosan nanoparticles (CNPs), microparticles (CMPs) and membranes have attracted considerable interests in applications in food industry, water and wastewater treatment, drug delivery and pharmaceutical industry [3–7] (Figure 2).

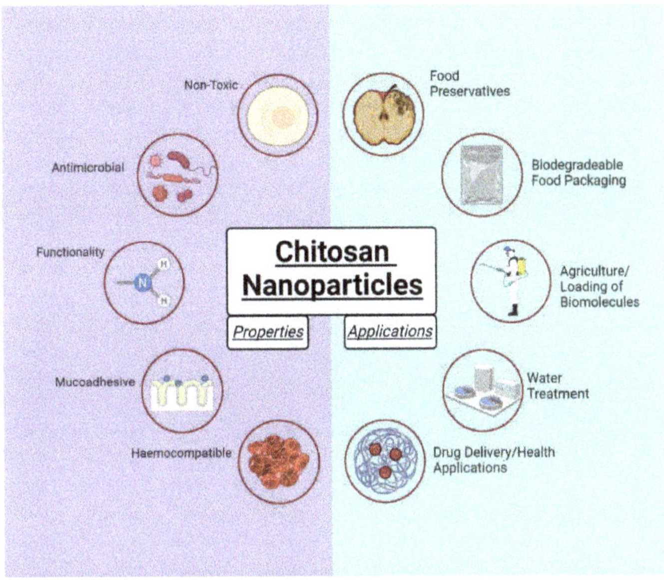

Figure 2. Properties and applications of chitosan nanoparticles. Adapted from "Increased Risk of Cancers with Obesity." Created with BioRender.com.

Chitosan-based materials have been used as antimicrobial agents in the food industry as a food preservative [8]. Chitosan-coated foods were found to substantially reduce surface microbial colonies [9]. Moreover, the application of CNPs to reinforce the biodegradable thermoplastic polylactic acid may become a key method in the production of biodegradable food packaging [10,11]. The antimicrobial property of chitosan is largely attributed to its amine groups, which enable the polymer to bind the negatively charged bacterial cell walls, ultimately leading to changes in membrane permeability [12,13]. Chitosan disrupts biofilm formation, is mucoadhesive and capable of opening tight junctions in epithelial cells [14–17].

CNPs have demonstrated great utility as a drug delivery vehicle, acting as scaffolding for the loading of drugs and other biomolecules. Specifically, CNPs have been reported to improve drug release, permeability, and stability [18]. Furthermore, the ability for CNPs to be functionalized with specific markers may allow for targeted delivery with a range of therapeutics. For its pharmaceutical uses, chitosan has been implemented to enhance the uptake of beneficial chemical compounds, such as flavonoids. For example, incorporation of tea polyphenols into chitosan nanoparticles has been shown to increase the absorption and bioavailability of the phenols [19].

Applications in the agricultural industry include herbicide-loaded CNPs as a more environmentally friendly option compared to traditional weed control methods [20], which may reduce the impact of herbicides on the environmental and human health.

In the field of wastewater treatment, the hydroxyl groups associated with chitosan make the biopolymer a promising candidate for the removal of pesticides, dyes, and heavy metals. Magnetic CNPs prepared with iron oxide (Fe_3O_4) have been demonstrated as an effective tool for removing heavy metals and dyes from wastewater [21,22].

CNPs are routinely formulated with the ionic gelation method, in which polymer units of interest are cross-linked through electrostatic interactions with an oppositely charged molecule [23]. For CNP synthesis, an anionic cross linker such as tripolyphosphate (TPP) may be used (Figure 1). While setup and execution of this synthesis method is relatively simple, the desired outcome with a target size and uniformity may not be so readily achieved. Synthesis parameters including homogenization time and speed must be carefully tuned, while parameters such as pH and ionic strength may need to be considered to avoid aggregation and excess growth of the particles [24,25].

For most applications, it is essential to control the size and size distribution of NPs, which play a critical role in determining material properties and performance [26–28]. Small CNPs (30–60 nm) are highly efficient for DNA encapsulation and are an effective vehicle for delivery of DNA vaccines [29].

For water and wastewater treatment, smaller particles will allow for a greater degree of adsorption thus the large surface area-volume ratio of CNPs is of great interest. The formulation of sub-100 nm CNPs with a high degree of uniformity is particularly applicable in water and wastewater treatment as a potential surrogate for waterborne viruses. In this way, CNPs may be labelled with a marker, and used to determine the fate of viruses within a treatment facility.

Previously, protein-coated, DNA-labelled 70 nm silica nanoparticles have been used as virus surrogates for investigating virus removal and transport in groundwater and soils [30–32]. However, silica NPs-based virus surrogates are neither biocompatible nor biodegradable thus have limited uses in the nature environments. To address a lack of appropriate virus substitutes made of biocompatible biopolymers, we aimed to produce uniform sub-100 nm CNPs as a precursor for the development of biopolymer-based virus surrogates. The effects of pH, ionic strength and purification methods on the size and polydispersity of CNPs formation during ionic gelation were examined. Finally, future work was discussed.

2. Results and Discussion

2.1. Chitosan Nanoparticle Synthesis

In this study we successfully synthesized high yield (3.57×10^9 particles/mL) monodispersed (PDI 0.2–0.3) CNPs with size 68–77 nm using the ionic gelation method. This size range is very similar to some enteric viruses for example rotavirus and adenovirus. Thus, the produced CNPs can be used as a precursor for developing novel biopolymer-based virus surrogates, which will be DNA-labelled and protein-coated to mimic the physicochemical properties of the target viruses [30]. The new virus surrogates can be used to predict water contamination risks in freshwaters and help to design improved water treatment systems. Such a biopolymer-based abiotic virus surrogate has not been reported in the literature.

The virus-sized CNPs were obtained under the following optimized conditions: initial concentrations: 1 mg/mL (final 0.7 mg/mL) LMW chitosan (50–190 kDa, deacetylation 75–85%) (pH 3.2), 1 mg/mL TPP (final 0.3 mg/mL). Samples were homogenized using a handheld homogenizer motor (7000 rpm for 2 min), and purified by passing through a 0.1 μm syringe filter at flow rate 0.15 mL/min. Larger CNPs (95–219 nm) were produced under TPP pH 2–4, NaCl 100–200 mM, and when using ultracentrifugation and size exclusion chromatography. Particles reached >1000 nm at 200 mM NaCl.

The first demonstration of CNP synthesis was performed by Calvo et al. who focused on the ratio of chitosan:TPP as a control of particle size and dispersity [33]. However, this widely used method is prone to uncontrollable intra- and intermolecular crosslinking between chitosan and TPP, resulting in aggregation, excess growth, and high polydispersity of CNPs [34–36].

The size and dispersity (reflected in PDI) of CNPs formation can be influenced by many processing factors, such as concentrations of chitosan and TPP, ionic strength, pH, reaction time, temperature, homogenization speed and duration, purification methods, etc. Kunjachan et al. found that chitosan and TPP concentrations had significant effects on particle size and PDI, while homogenization time only influenced the PDI [23]. However, Zu et al. observed that a time of 1–3 min decreased particle size, while 3–9 min introduced mechanical energy that necessitated higher particle density, thereby lowering specific surface area and increasing particle size [37]. Budi et al. (2020) observed a reduced CNPs size with lower TPP concentration due to a decrease in cross-link bonding and the formation of aggregate particles [38].

Sreekumar et al. reported that size of the CNPs is strongly dependent on the initial chitosan concentration, degree of chitosan acetylation, molecular weight, and salt concentration in the medium [39]. Using initial chitosan 0.1–5 mg/mL, 20–50% acetylation, 125–450 kDa, and NaCl 50–150 mM, CNPs with size 100–1200 nm and PDI 0.1–0.4 were produced [39]. Many of these experimental parameters were comparable to our study, but smaller size CNPs were achieved with our protocol.

Due to limited resources, it was not feasible to investigate the influence of many processing factors on the size of CNPs formation. Thus, we had focused on assessing the influence of ionic strength, pH and purification methods on CNPs size and dispersity. These are described blow.

2.1.1. Characteristics of Chitosan Nanoparticles under Ionic Strength Control

Control of CNPs size by altering ionic strength has been reported previously [24,40,41]. Increased ionic strength is proposed to limit electrostatic interactions between chitosan and TPP resulting in smaller monodisperse CNPs [24]. In particular, Cl^- ions compete with TPP and reduce bridging between CNPs, thereby increasing colloidal stability [40] (Figure 3). Although, at high ionic strengths, this effect will be offset by increased particle collisions induced by electrostatic screening. We found that the addition of increasing concentrations of NaCl (0–200 mM) correlated with a decrease in polydispersity index (PDI) of the CNPs, yet expectedly increased their size into the micro range (Figure 4).

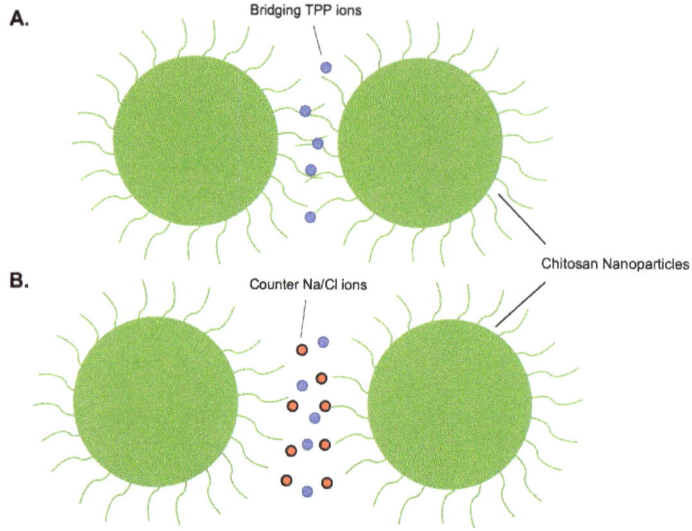

Figure 3. Effect of ionic strength on colloidal stability. (**A**) Bridging of chitosan nanoparticles by tripolyphosphate interactions with chitosan chains. (**B**) Disruption of tripolyphosphate-chitosan interactions by the presence of counter ions.

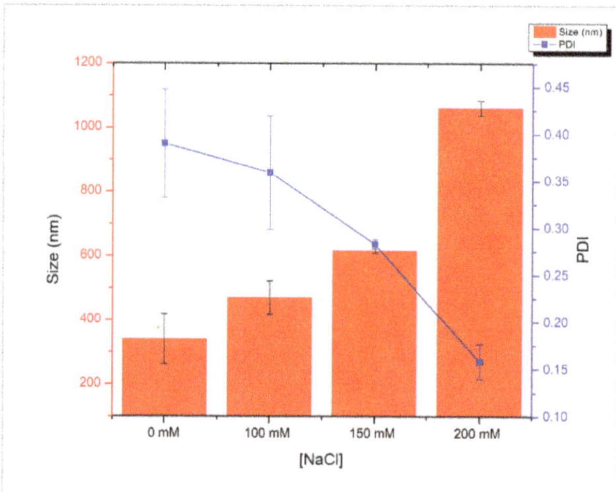

Figure 4. Size (red bars) and polydispersity index (blue line) of chitosan nanoparticles prepared with 1 mg/mL LMW chitosan and 1 mg/mL tripolyphosphate with increasing concentration of NaCl (0–200 mM). Measurements performed with dynamic light scattering in triplicate.

At 100, 150 and 200 mM NaCl, PDI values decreased from 0.38 to 0.36, 0.27, and 0.14 respectively (Figure 4). This is in agreement with previous reports that found a decrease in PDI at all NaCl concentrations between 100 and 500 mM [24]. Likewise, Jonassen et al., found narrower CNPs size distributions with increasing ionic strengths [41]. We also found that particle size dramatically increased at higher concentrations, reaching >1000 nm at 200 mM NaCl (Figure 4). A similar phenomenon has been reported previously and is attributed to the screening of electrostatic repulsion, while the low PDI values are attributed to slower kinetics at high salt concentrations, thereby leading to a large population of monodisperse CNPs overtime [24].

2.1.2. Characteristics of Chitosan Nanoparticles under pH Control

Alteration of pH has also been proposed as an effective means of avoiding aggregation and minimizing particle growth [25]. The kinetics of CNP formation are strongly governed by the protonation state of chitosan amine groups. Increasing the pH of the chitosan solution nearer to the amine group's pKb value of 6.5 increases the number of internal hydrogen bonds, which restricts crosslinking with TPP ions. Consequently, the chitosan chains become more condensed within the particle, leading to a decrease in size, while the reduced availability of protonated amines do not allow for aggregation via bridging TPP ions.

The pH of the TPP solution has been shown to influence CNP size and polydispersity [25,42,43]. Mattu et al. reported that CNPs prepared with basic TPP solutions (pH 9.5) had a larger size compared with those obtained with acidic TPP solutions (pH 5.5). It is proposed that lower pH values limit TPP interaction with chitosan, due to the presence of H_3O^+ and H^+ counter ions. At TPP pH >5, both the size and PDI of CNPs has been reported to increase dramatically [25], therefore, we examined the pH effect by fixing the chitosan solution to pH 5 and altering the TPP solution pH between 2 to 4.

Consistent with previous findings, we also observed that smaller CNPs were formed at lower TPP pH [25,42,43]. CNP size was the largest at pH 4, with a size of 142 nm, compared to 117 nm, and 121 nm for pH 3 and 2 respectively (Figure 5). PDI values fluctuated with varying pH but were lowest at 0.15 for pH 2 (Figure 5). These changes are minor and suggest that in this TPP pH range, the pH of the chitosan solution is the determining factor for particle growth and polydispersity. Therefore, altering the pH of the TPP solution may be used for fine-tuning of size and PDI after an appropriate pH for the chitosan solution is found.

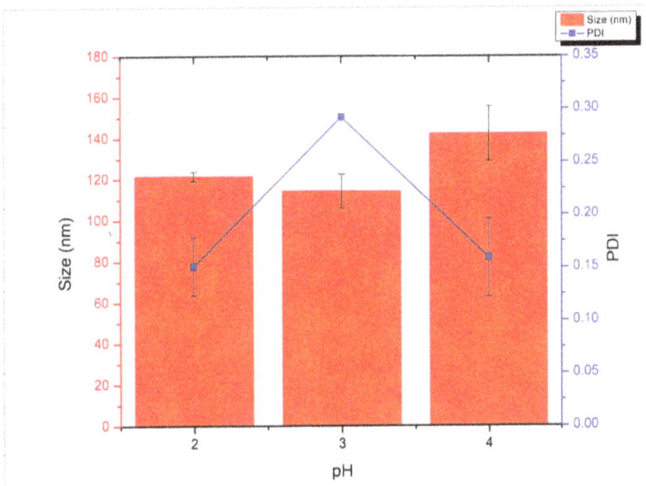

Figure 5. Size (red bars) and polydispersity index (blue line) of chitosan nanoparticles prepared with 1 mg/mL LMW chitosan and 1 mg/mL tripolyphosphate with increasing tripolyphosphate solution pH (2–4). Measurements performed with dynamic light scattering in triplicate.

Our experimental results are in agreement with previous reports [25], and the method we used can serve as a more efficient control for particle size and dispersity than ionic strength. However, our desired goal of <100 nm CNPs could not be achieved through this control alone.

2.1.3. Characteristics of Chitosan Nanoparticles with a Homogenizer

In order to synthesize smaller CNPs, an alternate homogenization method was investigated. A high number of procedures in the literature employ magnetic stirring throughout the growth phase and rely on stir speed and time to control for particle size and polydispersity. This is not a new idea, as the increased shear forces introduced at higher speeds disrupts particle formation, resulting in smaller sizes. This idea is commonly used in the sol-gel technique, for formulating silica NPs of a desired size.

We have found that the use of magnetic stirring is unable to formulate <100 nm CNPs with low dispersity, even when chitosan and TPP ratios are selected appropriately, or other factors such as ionic strength and pH are controlled, as discussed previously. However, implementing rigorous homogenization with a handheld homogenizer motor at 7000 rpm for 2 min consistently resulted in <80 nm CNPs. The ability to formulate small CNPs with low polydispersity, without the use of additional reagents and timely procedures, is an important step in developing this efficient synthesis procedure. Similar resulted have been reported in which applying higher homogenization speeds (1000 to 7000 rpm) decreased particle size [44]. However, our protocol is the first to achieve sub-100 nm sizes using this mixing method.

2.2. *Purification and Collection of Chitosan Nanoparticles*

2.2.1. Purification of Chitosan Nanoparticles with Ultracentrifugation

The next barrier to overcome in this procedure was the collection of CNPs from the homogenized solution. The small size of the CNPs restricted our ability to wash and collect via centrifugation, a staple method in the purification step of most NP syntheses. Ultracentrifugation is necessary for the isolation of particles <100 nm, however, the high centrifugal forces consequently disrupted the structure of the CNPs, resulting in a rigid pellet that was not susceptible to resuspension.

Attempts to resuspend the pellet via vortexing and bath sonication resulted in a slightly cloudy solution with visible aggregates from the pellet. The measured size and PDI of this sample were 219 nm and 0.7 respectively. These DLS results suggest that any CNPs that had resuspended had lost their original structure and uniformity. Decreases to ultracentrifugation speed (149,000 RCF to 118,000 RCF) and time (60 min to 40 min) were assessed, however similar results were observed.

2.2.2. Purification of Chitosan Nanoparticles with Size Exclusion Chromatography

Size exclusion chromatography was evaluated next as a means to separate the CNPs from any precursor material retained in solution, while maintaining their size and uniformity. However, attempts to purify in this manner presented other challenges.

Light scattering and DLS measurements of initial samples suggested that our CNPs of interest primarily eluted over two fractions, while other fractions contribute to a polydisperse sample (Figure 6A, Table 1). The collected fractions were pooled and subjected to a second pass through the column. In comparison to the scattering profile of CNPs, unlinked chitosan polymers display similar scattering values across all 5 eluted fractions, with the exception of fraction 2 (Figure 6B). Furthermore, these scattering values are significantly larger than those recorded for the CNPs (OD 550 nm: ~40,000 to ~24,000). This supports our assumption that CNPs are being collected and that little unlinked chitosan polymer remains after particle formation.

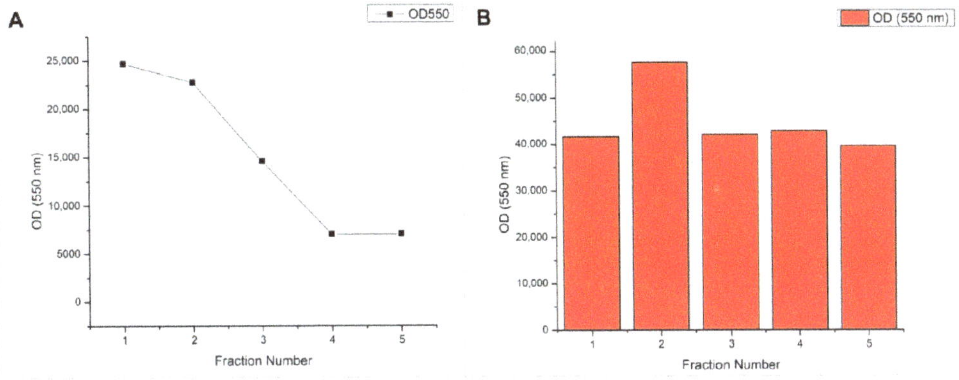

Figure 6. Light scattering results of chitosan fractions collected with size exclusion chromatography. (**A**) Chitosan nanoparticles synthesized under homogenization. (**B**) LMW chitosan dissolved in 1% acetic acid to a concentration of 1 mg/mL.

Table 1. Results of Size Exclusion Chromatography of CNPs synthesized using 1 mg/mL chitosan and 1 mg/mL tripolyphosphate.

SEC Fraction	Major Species Hydrodynamic Diameter (nm)	Z-Average (nm)
Fraction 1	62.12	146.7
Fraction 2	75.42	60.15
SEC Pooled Fractions	**Z-Average (nm)**	**PDI**
1, 2 Pooled	103	0.218
3, 4, 5, 6, 7 Pooled	95.18	0.399
Not Subjected	83.36	0.301

The resulting fractions expected to contain CNPs of the desired size range were pooled and measured using DLS, while the subsequent fractions were pooled separately (Table 1). Unfortunately, this process of pooling fractions to avoid high polydispersity greatly diluted the CNPs. This in turn makes DLS measurements unreliable, and even more importantly, limits the use of these CNPs for relevant applications.

Table 1 summarizes the sizes of CNPs in fractions eluted with size exclusion chromatography (SEC). The initial exclusion may be used to obtain a certain number of particles in this range. However, this method may not be efficient as evident in the changes between Z-average as well as in the major species present. Ultimately, this method diluted CNP concentrations to a point where their characterization became unreliable.

This method of purification greatly limits the use for CNP synthesis on large scales, where high concentrations and monodispersity are required.

2.2.3. Purification of Chitosan Nanoparticles Using Syringe Filtration

To effectively purify CNPs out of the homogenized solution and maintain high concentration values and monodispersity, we opted to filter the solution through a 0.1 μm polyethersulfone filter. This would remove any larger particles from solution that were not within the desired size range, increasing monodispersity.

This method proved the most effective of the three for collection of <80 nm CNPs with relatively low PDI values of ~0.3 (Table 2). Following homogenization, CNPs are drawn into a syringe, which is then loaded into a syringe pump (Harvard Instruments). CNPs are then infused through the filter into a collection vial at a rate of 0.15 mL/min. Large nanoparticles and unreacted chitosan may cause clogging of the filter, and the infusion can be paused to facilitate filter replacement.

Table 2. Characteristics of chitosan nanoparticles synthesized with a 1 mg/mL chitosan solution and 1 mg/mL tripolyphosphate solution and subsequently purified using syringe filtration.

Facility/Equipment	Size (nm)	PDI	Zeta Potential (mV)	Particle Concentration (Particle/mL)
University of Calgary/Malvern ZetaSizer Nano ZS	73	0.344	10.17 ± 3.35	-
University of Calgary/Malvern ZetaSizer Nano ZS	68	0.330	21.00 ± 0.70	-
University of Calgary/Malvern ZetaSizer Nano ZS	77	0.319	21.10 ± 1.98	-
IZON Science (Christchurch, New Zealand) TRPS with Exoid	72	0.227 *	-	3.57×10^9

* calculated from the square of the standard deviation divided by the mean particle diameter.

The above procedure was replicated four times and samples were measured across two facilities with different characterization instruments, including DLS at University of Calgary and Tunable Resistive Pulse Sensing (TRPS) with Exoid-034 at IZON Science (Christchurch, New Zealand). Results were strikingly similar.

Zeta potential measurements of these CNPs revealed a value of ~+20 mV. This is an indication of adequate particle stability, as electrostatic repulsion between one another will prevent aggregation over time.

Considering the three approaches used in this study to purify CNPs following their synthesis using homogenization, purification by syringe filtration has proved to be the most effective way to isolate these nanoparticles within the <80 nm size range and to have adequate monodispersity (Figure 7). Ultracentrifugation applies significant force resulting in the formation of gels, while SEC may be capable of isolating for particle size the wide elution range and indirect dilution of samples introduces unreliability into measurements and hampers scalability for application.

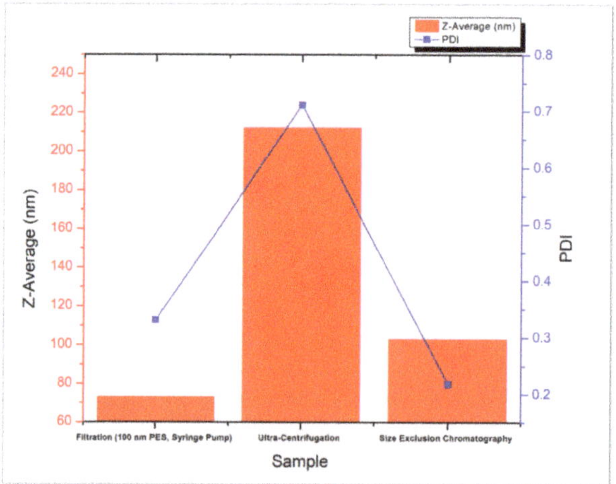

Figure 7. Summary of three chitosan nanoparticle purification methods and their ability to isolate nanoparticles of small size and low polydispersity.

2.3. Characterization of Chitosan Nanoparticles with Scanning Electron Microscopy

CNPs were analyzed via SEM as a secondary size characterization method in addition to DLS. SEM images were taken for CNPs synthesized with a homogenizer and purified with syringe filtration, as this procedure proved effective for our goals of formulating <100 nm CNPs with low polydispersity. The CNPs displayed a circular morphology, similar to literature, and were split into two size populations, with average sizes of 75 $\pm$ 8 nm and 33 $\pm$ 11 nm (Figure 8).

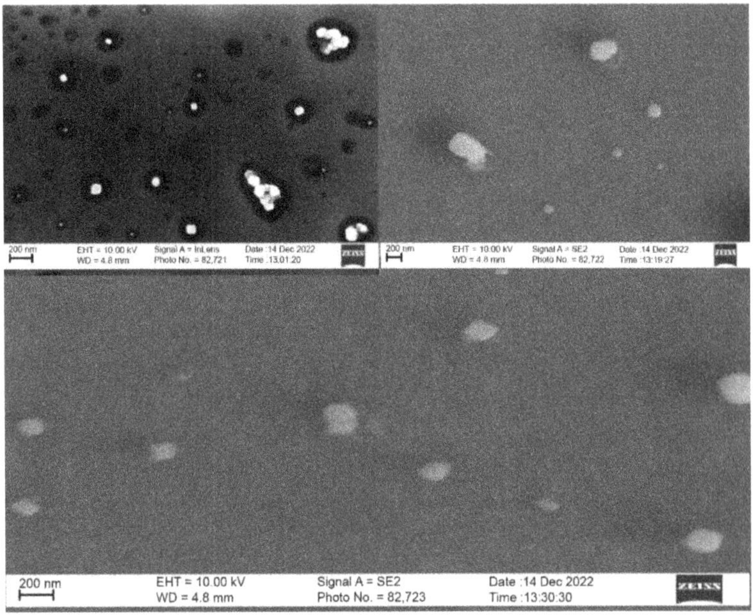

Figure 8. SEM characterization of CNPs following synthesis using homogenization and purification using filtration. Images obtained using both InLens and SE2 Signal A (Zeiss Σigma VP Scanning Electron Microscope).

The larger population is in strong agreement with DLS measurements, while the presence of a second smaller population is reflective of the somewhat poor PDI value (Table 2). These smaller particles may be the result of an incomplete growth phase. Therefore, increasing the homogenization time may allow for these particles to equilibrate with the average size.

3. Materials and Methods

3.1. Materials

Low molecular weight chitosan (MW 50–190 kDa, degree of deacetylation 75–85%), sodium tripolyphosphate, and sodium chloride were purchased from Sigma Aldrich (Markham, ON, Canada). Acetic acid was obtained from EMD chemicals (Mississauga, ON, Canada).

3.2. Synthesis of Chitosan Nanoparticles

The procedure was adapted from Iswanti et al. [45]. Briefly, the chitosan solution was prepared by dissolving chitosan in 1% acetic acid to a concentration of 1 mg/mL. The chitosan solution (21 mL) was set on a magnetic stir plate (1000 rpm) and 9 mL of 1 mg/mL aqueous TPP solution was added dropwise at a rate of 1 mL/min. The mixture was homogenized at 7000 rpm over 2 min with a handheld homogenizer motor (Fisherbrand™ 150) prior to purification and collection (Figure 9).

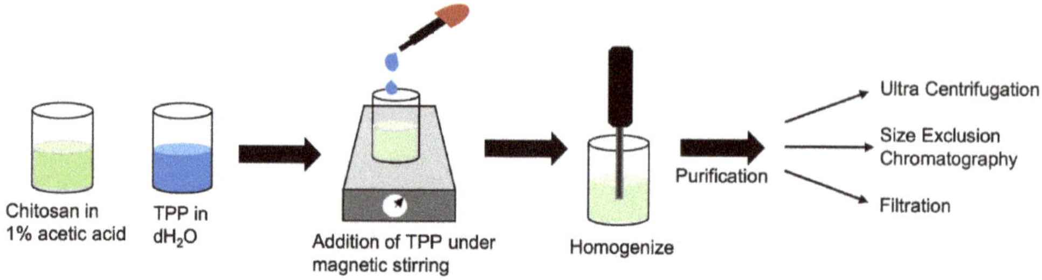

Figure 9. Schematic of chitosan nanoparticle synthesis.

3.3. Purification of Chitosan Nanoparticles

CNPs were washed and collected using three different methods: (1) ultracentrifugation and resuspension in a 1% acetic acid solution; (2) size exclusion chromatography and pooling of collected fractions; and (3) filtration through a 0.1 μm polyethersulfone filter.

Ultracentrifugation was carried out with an Optima L-1000K centrifuge (Beckman Coulter, Indianapolis, IN, USA) at 149,000 RCF for 60 min (18 °C). The pellet was resuspended in 1% acetic acid (10 mL) via vortexing and bath sonication. Alternatively, CNPs were added to a Sephadex G25 size exclusion column. Scattering spectra (500 nm excitation and emission wavelengths) were recorded with a spectrofluorometer (Shimadzu, Columbia, MD, USA) for collected fractions (10 drops/fraction) to determine the elution profile of CNPs. Lastly, the CNP suspension was filtered through a 0.1 μm PES syringe filter with a syringe pump (Harvard Apparatus, Holliston, MA, USA) at a flow rate of 0.3 mL/min.

3.4. Synthesis of Chitosan Nanoparticles with Ionic Strength Control

Adapting methods by Sawtarie et al. the chitosan and TPP solutions were prepared to a concentration of 2 mg/mL in 6 mL of 2% acetic and 2 mL of ddH_2O, respectively [24]. Three NaCl solutions were made to concentrations of 200, 300, and 400 mM. The NaCl solutions were then mixed with the chitosan and TPP solutions at a 1:1 volume ratio to bring the final precursor concentrations to 1 mg/mL and final NaCl concentrations to 100, 150, and 200 mM. CNPs were then synthesized by dropwise addition of the TPP-NaCl solution (4 mL; 1 mL/min) to the chitosan-NaCl (12 mL) solution set on a magnetic stir plate at 1000 rpm. After 5 min, the CNP suspension was subject to 2 rounds of microcentrifugation (18,000 RCF, 40 min) and resuspended in 10 mL ddH_2O.

3.5. Synthesis of Chitosan Nanoparticles with pH Control

Adapting methods by Masarudin et al. the chitosan and TPP solutions were prepared to a concentration of 1 mg/mL in 12 mL of 1% acetic and 4 mL of ddH_2O respectively [25]. The chitosan solution was adjusted to pH 5 with 1 M HCl and three TPP solutions were adjusted to pH values of 2, 3, and 4 with 1 M HCl. CNP synthesis was carried out by dropwise addition of the TPP solution (1 mL/min) to the chitosan solution under magnetic stirring at 1000 rpm. After 5 min of homogenization, the CNPs were collected via centrifugation (18,000 RCF, 40 min) and resuspended in 2 mL ddH_2O.

3.6. Characterization of Chitosan Nanoparticles

The concentration of CNP samples was determined by drying and weighing of a 50 μL aliquot. The size, PDI, and zeta potential of CNPs were measured using a Zetasizer Nano ZS (Malvern, Worcestershire, UK). Measurements were performed in triplicate at 25 °C with 10 runs per measurement and 10 s per run. Size and PDI of CNPs were measured in disposable polystyrene cuvettes and zeta potential was measured in folded capillary zeta cells.

Imaging of the CNPs was done with a scanning electron microscope (Zeiss Sigma VP, Field Emission, Zeiss, Heidenheim, Germany) using an InLens and SE2 Signal A. Samples (50 μL) were deposited on silicon wafers and dried under nitrogen prior to imaging.

4. Conclusions and Future Work

Synthesis methods of CNPs are many, however, reproducible methods for achieving sub-100 nm particles are lacking in the literature. Certain groups employ controls on salt concentration and pH to facilitate the growth phase in producing these desired sizes, however, these methods can be time consuming and costly for large-scale productions. Our approach utilizes rigorous homogenization followed by filtration of the CNP solution and requires no additional reagents, which greatly simplifies the synthesis procedure. We have demonstrated reproducible results in synthesis of high yield, monodispersed, 68–77 nm CNPs.

The positively charged sub-100 nm CNPs could readily interact with negatively charge proteins and DNA, making them an ideal precursor for the development of DNA-labelled, protein-coated virus surrogates [30]. We do not anticipate any difficulty in encapsulating biomolecules within the sub-100 nm CNPs developed in this study. We had previously encapsulated DNA tracers within 310 nm chitosan microparticles with a thin alginate outer layer [46]; encapsulation changed very little on the size of chitosan microparticles. Furthermore, the amino groups present in chitosan enable CNPs to be easily functionalized, leading to many potential applications such as water and wastewater treatment, food-packaging, biomedical and agricultural products.

Author Contributions: Conceptualization, methodology: E.J.P., M.A. and L.P.; Formal analysis, investigation, data curation: N.V.B., T.I., M.A. and L.P.; Writing–original draft preparation: N.V.B., T.I. and E.J.P.; Writing–review and editing: N.V.B., T.I., E.J.P., L.P. and M.A.; Visualization: N.V.B. and T.I.; Supervision, project administration: E.J.P., M.A. and L.P.; Funding acquisition: L.P. and E.J.P. All authors have read and agreed to the published version of the manuscript.

Funding: This research was funded through the Institute of Environmental Science and Research's 2022 Strategic Science Investment Fund from the New Zealand Ministry of Business, Innovation & Employment (to LP). We also acknowledge Natural Sciences and Engineering Research Council of Canada (Discovery grant to EJP). RGPIN/03911-2018.

Institutional Review Board Statement: Not applicable.

Informed Consent Statement: Not applicable.

Data Availability Statement: Data will be made available upon request.

Conflicts of Interest: The authors declare no conflict of interest.

References

1. Shukla, S.K.; Mishra, A.K.; Arotiba, O.A.; Mamba, B.B. Chitosan-Based Nanomaterials: A State-of-the-Art Review. *Int. J. Biol. Macromol.* **2013**, *59*, 46–58. [CrossRef]
2. Mohammed, M.A.; Syeda, J.T.M.; Wasan, K.M.; Wasan, E.K. An Overview of Chitosan Nanoparticles and Its Application in Non-Parenteral Drug Delivery. *Pharmaceutics* **2017**, *9*, 53. [CrossRef]
3. Liu, T.; Li, J.; Tang, Q.; Qiu, P.; Gou, D.; Zhao, J. Chitosan-Based Materials: An Overview of Potential Applications in Food Packaging. *J. Foods* **2022**, *11*, 1490. [CrossRef]
4. Al-Manhel, A.J.; Al-Hilphy, A.R.S.; Niamah, A.K. Extraction of Chitosan, Characterisation and Its Use for Water Purification. *J. Saudi Soc. Agric. Sci.* **2018**, *17*, 186–190. [CrossRef]
5. Garg, U.; Chauhan, S.; Nagaich, U.; Jain, N. Current Advances in Chitosan Nanoparticles Based Drug Delivery and Targeting. *Adv. Pharm. Bull.* **2019**, *9*, 195–204. [CrossRef] [PubMed]
6. Bhatnagar, A.; Sillanpää, M. Applications of Chitin- and Chitosan-Derivatives for the Detoxification of Water and Wastewater—A Short Review. *Adv. Colloid Interface Sci.* **2009**, *152*, 26–38. [CrossRef]
7. Doshi, B.; Sillanpää, M.; Kalliola, S. A Review of Bio-Based Materials for Oil Spill Treatment. *Water Res.* **2018**, *135*, 262–277. [CrossRef] [PubMed]

8. Kong, M.; Chen, X.G.; Xing, K.; Park, H.J. Antimicrobial Properties of Chitosan and Mode of Action: A State of the Art Review. *Int. J. Food Microbiol.* **2010**, *144*, 51–63. [CrossRef]
9. Wei, C.; Liao, X.; Huang, Y.; Zhou, F. Effects of Chitosan and Licorice Antioxidants on Skin Color of Duck Meat Refrigerated and the Number of Micro-Organisms on the Surface. *J. Food Sci. Technol.* **2020**, *45*, 280–285.
10. Yanat, M.; Schroën, K. Preparation Methods and Applications of Chitosan Nanoparticles; with an Outlook toward Reinforcement of Biodegradable Packaging. *React. Funct. Polym.* **2021**, *161*, 104849. [CrossRef]
11. Elsawy, M.A.; Saad, G.R.; Sayed, A.M. Mechanical, Thermal, and Dielectric Properties of Poly(Lactic Acid)/Chitosan Nanocomposites. *Polym. Eng. Sci.* **2016**, *56*, 987–994. [CrossRef]
12. Yilmaz Atay, H. Antibacterial Activity of Chitosan-Based Systems. In *Functional Chitosan: Drug Delivery and Biomedical Applications*; Jana, S., Ed.; Springer: Singapore, 2019. [CrossRef]
13. Másson, M. Antimicrobial Properties of Chitosan and Its Derivatives. In *Chitosan for Biomaterials III. Advances in Polymer Science*; Jayakumar, R., Prabaharan, M., Eds.; Springer International Publishing: Cham, Switzerland, 2021; pp. 131–168. [CrossRef]
14. Duan, C.; Meng, X.; Meng, J.; Khan, M.I.H.; Dai, L.; Khan, A.; An, X.; Zhang, J.; Huq, T.; Ni, Y. Chitosan as A Preservative for Fruits and Vegetables: A Review on Chemistry and Antimicrobial Properties. *J. Bioresour. Bioprod.* **2019**, *4*, 11–21. [CrossRef]
15. Al-Zahrani, S.S.; Bora, R.S.; Al-Garni, S.M. Antimicrobial Activity of Chitosan Nanoparticles. *Biotechnol. Biotechnol. Equip.* **2021**, *35*, 1874–1880. [CrossRef]
16. Felipe, V.; Breser, M.L.; Bohl, L.P.; Rodrigues da Silva, E.; Morgante, C.A.; Correa, S.G.; Porporatto, C. Chitosan Disrupts Biofilm Formation and Promotes Biofilm Eradication in Staphylococcus Species Isolated from Bovine Mastitis. *Int. J. Biol. Macromol.* **2019**, *126*, 60–67. [CrossRef]
17. Ways, T.M.M.; Lau, W.M.; Khutoryanskiy, V.V. Chitosan and Its Derivatives for Application in Mucoadhesive Drug Delivery Systems. *Polymers* **2018**, *10*, 267. [CrossRef]
18. Deshmukh, P.R.; Joshi, A.; Vikhar, C.; Khdabadi, S.S.; Tawar, M. Current Applications of Chitosan Nanoparticles. *Syst. Rev. Pharm.* **2022**, *13*, 685–693. [CrossRef]
19. Liang, J.; Yan, H.; Puligundla, P.; Gao, X.; Zhou, Y.; Wan, X. Applications of Chitosan Nanoparticles to Enhance Absorption and Bioavailability of Tea Polyphenols: A Review. *Food Hydrocoll.* **2017**, *69*, 286–292. [CrossRef]
20. Grillo, R.; Pereira, A.E.S.; Nishisaka, C.S.; De Lima, R.; Oehlke, K.; Greiner, R.; Fraceto, L.F. Chitosan/Tripolyphosphate Nanoparticles Loaded with Paraquat Herbicide: An Environmentally Safer Alternative for Weed Control. *J. Hazard. Mater.* **2014**, *278*, 163–171. [CrossRef]
21. Thinh, N.N.; Hanh, P.T.B.; Ha, L.T.T.; Anh, L.N.; Hoang, T.V.; Hoang, V.D.; Dang, L.H.; Van Khoi, N.; Lam, T.D. Magnetic Chitosan Nanoparticles for Removal of Cr(VI) from Aqueous Solution. *Mater. Sci. Eng. C* **2013**, *33*, 1214–1218. [CrossRef]
22. Cao, C.; Xiao, L.; Chen, C.; Shi, X.; Cao, Q.; Gao, L. In Situ Preparation of Magnetic Fe_3O_4/Chitosan Nanoparticles via a Novel Reduction-Precipitation Method and Their Application in Adsorption of Reactive Azo Dye. *Powder Technol.* **2014**, *260*, 90–97. [CrossRef]
23. Kunjachan, S.; Jose, S. Understanding the Mechanism of Ionic Gelation for Synthesis of Chitosan Nanoparticles Using Qualitative Techniques. *Asian J. Pharm.* **2010**, *4*, 148–153. [CrossRef]
24. Sawtarie, N.; Cai, Y.; Lapitsky, Y. Preparation of Chitosan/Tripolyphosphate Nanoparticles with Highly Tunable Size and Low Polydispersity. *Colloids Surf. B Biointerfaces* **2017**, *157*, 110–117. [CrossRef]
25. Masarudin, M.J.; Cutts, S.M.; Evison, B.J.; Phillips, D.R.; Pigram, P.J. Factors Determining the Stability, Size Distribution, and Cellular Accumulation of Small, Monodisperse Chitosan Nanoparticles as Candidate Vectors for Anticancer Drug Delivery: Application to the Passive Encapsulation of [14 C]-Doxorubicin. *Nanotechnol. Sci. Appl.* **2015**, *8*, 67–80. [CrossRef]
26. Dolai, J.; Mandal, K.; Jana, N.R. Nanoparticle Size Effects in Biomedical Applications. *ACS Appl. Nano Mater.* **2021**, *4*, 6471–6496. [CrossRef]
27. Mozaffari, S.; Li, W.; Dixit, M.; Seifert, S.; Lee, B.; Kovarik, L.; Mpourmpakis, G.; Karim, A.M. The Role of Nanoparticle Size and Ligand Coverage in Size Focusing of Colloidal Metal Nanoparticles. *Nanoscale Adv.* **2019**, *1*, 4052–4066. [CrossRef] [PubMed]
28. Hoshyar, N.; Gray, S.; Han, H.; Bao, G. The Effect of Nanoparticle Size on in Vivo Pharmacokinetics and Cellular Interaction. *Nanomedicine* **2016**, *11*, 673–692. [CrossRef] [PubMed]
29. Nunes, R.; Sousa, Â.; Simaite, A.; Aido, A.; Buzgo, M. Sub-100 Nm Chitosan-Triphosphate-DNA Nanoparticles for Delivery of DNA Vaccines. *Proceedings* **2021**, *78*, 12. [CrossRef]
30. Pang, L.; Farkas, K.; Bennett, G.; Varsani, A.; Easingwood, R.; Tilley, R.; Nowostawska, U.; Lin, S. Mimicking Filtration and Transport of Rotavirus and Adenovirus in Sand Media Using DNA-Labeled, Protein-Coated Silica Nanoparticles. *Water Res.* **2014**, *62*, 167–179. [CrossRef]
31. Farkas, K.; Varsani, A.; Pang, L. Adsorption of Rotavirus, MS2 Bacteriophage and Surface-Modified Silica Nanoparticles to Hydrophobic Matter. *Food Environ. Virol.* **2015**, *7*, 261–268. [CrossRef]
32. Clemens, H.; Pang, L.; Morgan, L.K.; Weaver, L. Attenuation of Rotavirus, MS2 Bacteriophage and Biomolecule-Modified Silica Nanoparticles in Undisturbed Silt Loam over Gravels Dosed with Onsite Wastewater. *Water Res.* **2020**, *169*, 115272. [CrossRef]
33. Alonso, M.J.; Calvo, P.; Remuñán-López, C.; Vila-Jato, J.L. Novel Hydrophilic Chitosan-Polyethylene Oxide Nanoparticles as Protein Carriers. *J. Appl. Polym. Sci.* **1997**, *63*, 125–132. [CrossRef]
34. Rampino, A.; Borgogna, M.; Bellich, B.; Blasi, P.; Virgilio, F.; Cesàro, A. Chitosan-Pectin Hybrid Nanoparticles Prepared by Coating and Blending Techniques. *Eur. J. Pharm. Sci.* **2016**, *84*, 37–45. [CrossRef] [PubMed]

35. Tammam, S.N.; Azzazy, H.M.E.; Breitinger, H.G.; Lamprecht, A. Chitosan Nanoparticles for Nuclear Targeting: The Effect of Nanoparticle Size and Nuclear Localization Sequence Density. *Mol. Pharm.* **2015**, *12*, 4277–4289. [CrossRef] [PubMed]
36. Kamat, V.; Bodas, D.; Paknikar, K. Chitosan Nanoparticles Synthesis Caught in Action Using Microdroplet Reactions. *Sci. Rep.* **2016**, *6*, 8–11. [CrossRef] [PubMed]
37. Zu, Y.; Zhang, Y.; Wang, W.; Zhao, X.; Han, X.; Wang, K.; Ge, Y. Preparation and in Vitro/in Vivo Evaluation of Resveratrol-Loaded Carboxymethyl Chitosan Nanoparticles. *Drug Deliv.* **2016**, *23*, 981–991. [CrossRef]
38. Budi, S.; Asih Suliasih, B.; Rahmawati, I.; Erdawati. Size-Controlled Chitosan Nanoparticles Prepared Using Ionotropic Gelation. *ScienceAsia* **2020**, *46*, 457–461. [CrossRef]
39. Sreekumar, S.; Goycoolea, F.M.; Moerschbacher, B.M.; Rivera-Rodriguez, G.R. Parameters Influencing the Size of Chitosan-TPP Nano- and Microparticles. *Sci. Rep.* **2018**, *8*, 4695. [CrossRef]
40. Huang, Y.; Lapitsky, Y. On the Kinetics of Chitosan/Tripolyphosphate Micro- and Nanogel Aggregation and Their Effects on Particle Polydispersity. *J. Colloid Interface Sci.* **2017**, *486*, 27–37. [CrossRef]
41. Jonassen, H.; Kjøniksen, A.L.; Hiorth, M. Effects of Ionic Strength on the Size and Compactness of Chitosan Nanoparticles. *Colloid Polym. Sci.* **2012**, *290*, 919–929. [CrossRef]
42. Mattu, C.; Li, R.; Ciardelli, G. Chitosan Nanoparticles as Therapeutic Protein Nanocarriers: The Effect of Ph on Particle Formation and Encapsulation Efficiency. *Polym. Compos.* **2013**, *34*, 1538–1545. [CrossRef]
43. Katas, H.; Alpar, H.O. Development and Characterisation of Chitosan Nanoparticles for SiRNA Delivery. *J. Control. Release* **2006**, *115*, 216–225. [CrossRef] [PubMed]
44. Amin, M.K.; Boateng, J.S. Comparison and Process Optimization of PLGA, Chitosan and Silica Nanoparticles for Potential Oral Vaccine Delivery. *Ther. Deliv.* **2019**, *10*, 493–514. [CrossRef] [PubMed]
45. Iswanti, F.C.; Nurulita, I.; Djauzi, S.; Sadikin, M.; Witarto, A.B.; Yamazaki, T. Preparation, Characterization, and Evaluation of Chitosan-Based Nanoparticles as CpG ODN Carriers. *Biotechnol. Biotechnol. Equip.* **2019**, *33*, 390–396. [CrossRef]
46. Pang, L.; Abeysekera, G.; Hanning, K.; Premaratne, A.; Robson, B.; Abraham, P.; Sutton, R.; Hanson, C.; Hadfield, J.; Heiligenthal, L.; et al. Water Tracking in Surface Water, Groundwater and Soils Using Free and Alginate-Chitosan Encapsulated Synthetic DNA Tracers. *Water Res.* **2020**, *184*, 116192. [CrossRef] [PubMed]

molecules

MDPI

Article

Fabrication of PEGylated Chitosan Nanoparticles Containing Tenofovir Alafenamide: Synthesis and Characterization

Muhammad Zaman [1,*], Muhammad Hammad Butt [2,*], Waqar Siddique [3,4], Muhammad Omer Iqbal [5,6], Naveed Nisar [7,8], Asma Mumtaz [9], Hafiza Yusra Nazeer [8], Abdulrahman Alshammari [10] and Muhammad Shahid Riaz [11]

1 Faculty of Pharmacy, University of Central Punjab, Lahore 54000, Pakistan
2 Department of Medicinal Chemistry, Faculty of Pharmacy, Uppsala University, 75123 Uppsala, Sweden
3 Department of Pharmacy, University of South Asia (USA), Lahore 54000, Pakistan
4 College of Pharmacy, University of Sargodha, Sargodha 40100, Pakistan
5 Shandong Provincial Key Laboratory of Glycoscience and Glycoengineering, School of Medicine and Pharmacy, Ocean University of China, Qingdao 266003, China
6 Royal Institute of Medical Sciences, Multan 59300, Pakistan
7 Institute of Research and Advanced Studies of Pharmacy (IRASP), Multan 59300, Pakistan
8 Faculty of Pharmacy, Bahauddin Zakariya University Multan, Multan 59300, Pakistan
9 Multan Medical and Dental College, Multan 59300, Pakistan
10 Department of Pharmacology and Toxicology, College of Pharmacy, King Saud University, Post Box 2455, Riyadh 11451, Saudi Arabia
11 Nutrition and Food Science Area, Preventive Medicine and Public Health, Food Science, Toxicology and Forensic Medicine Department, Faculty of Pharmacy, Universitat de València, Avda, Vicent Andrés Estellés, s/n, Burjassot, 46100 València, Spain
* Correspondence: m.zaman2157@gmail.com (M.Z.); muhammad-hammad.butt.7587@student.uu.se (M.H.B.)

check for updates

Citation: Zaman, M.; Butt, M.H.; Siddique, W.; Iqbal, M.O.; Nisar, N.; Mumtaz, A.; Nazeer, H.Y.; Alshammari, A.; Riaz, M.S. Fabrication of PEGylated Chitosan Nanoparticles Containing Tenofovir Alafenamide: Synthesis and Characterization. *Molecules* **2022**, *27*, 8401. https://doi.org/10.3390/molecules27238401

Academic Editor: Agnieszka Ewa Wiącek

Received: 2 November 2022
Accepted: 17 November 2022
Published: 1 December 2022

Publisher's Note: MDPI stays neutral with regard to jurisdictional claims in published maps and institutional affiliations.

Abstract: Tenofovir alafenamide (TAF) is an antiretroviral (ARV) drug that is used for the management and prevention of human immunodeficiency virus (HIV). The clinical availability of ARV delivery systems that provide long-lasting protection against HIV transmission is lacking. There is a dire need to formulate nanocarrier systems that can help in revolutionizing the way to fight against HIV/AIDS. Here, we aimed to synthesize a polymer using chitosan and polyethylene glycol (PEG) by the PEGylation of chitosan at the hydroxyl group. After successful modification and confirmation by FTIR, XRD, and SEM, TAF-loaded PEGylated chitosan nanoparticles were prepared and analyzed for their particle size, zeta potential, morphology, crystallinity, chemical interactions, entrapment efficacy, drug loading, in vitro drug release, and release kinetic modeling. The fabricated nanoparticles were found to be in a nanosized range (219.6 nm), with ~90% entrapment efficacy, ~14% drug loading, and a spherical uniform distribution. The FTIR analysis confirmed the successful synthesis of PEGylated chitosan and nanoparticles. The in vitro analysis showed ~60% of the drug was released from the PEGylated polymeric reservoir system within 48 h at pH 7.4. The drug release kinetics were depicted by the Korsmeyer–Peppas release model with thermodynamically nonspontaneous drug release. Conclusively, PEGylated chitosan has the potential to deliver TAF from a nanocarrier system, and in the future, cytotoxicity and in vivo studies can be performed to further authenticate the synthesized polymer.

Keywords: tenofovir alafenamide; antiretroviral drugs; polymer synthesis; chitosan; polyethylene glycol; nanocarrier system; nanoparticles

1. Introduction

Antiretroviral (ARV) drugs are the cornerstone of HIV prevention and management. Since the advent of combination ARV therapy in 1995, ARV drugs have proven to be effective treatment options for the prevention of HIV infection [1]. Tenofovir is an ARV drug that belongs to the class of acyclic nucleoside phosphonates. Tenofovir has the potential to inhibit HIV, but an active metabolite (tenofovir diphosphate) has a strong inhibitory activity against HIV reverse transcriptase. In spite of these favorable properties, tenofovir in its parent form cannot be administered orally owing to its poor membrane permeability, poor in vitro drug release, and low oral bioavailability. Moreover, tenofovir has free 'OH' groups that easily ionize at intestinal pH, leading to limited permeability of the drug across the intestinal wall [2].

Tenofovir alafenamide (TAF) is a caspase-activated prodrug of tenofovir (TFV). Based on an in vitro analysis, the 50% effective concentration (EC50) of TAF is in the low nanomolar range (5 to 11.2 nM). TAF is intracellularly converted by kinases into tenofovir diphosphate, an active form of the drug that competitively inhibits HIV reverse transcriptase and the generation of viral transcripts [3]. The diphosphate, due to its charge and pKa, is highly impermeable to cellular membranes and is trapped in the intracellular volume compared to the parent species. TAF is a more potent prodrug than tenofovir diphosphate fumarate (TDF), resulting in low drug exposures and reduced side effects compared to TDF. Together, these characteristics make TAF one of the leading drug molecules for long-acting ARV delivery because it is potent and cellularly long-acting [4].

Biodegradable nanoparticle-based systems are capable of improving the therapeutic effects of bioactive molecules by enhancing their bioavailability, solubility, and retention time in the body. The encapsulation of drugs in polymeric nanoparticulate systems helps to improve drug efficacy, specificity, and tolerability as well their therapeutic indices. Traditional drug delivery systems and routes have numerous drawbacks, including the need to administer large drug doses to achieve the desired therapeutic response, the inability to reach target sites, and the failure to achieve desired therapeutic effects due to the interactions of drugs with biological molecules [5]. To overcome these problems, innovation has shifted towards nanoformulations for the purpose of achieving targeted drug delivery, the controlled or sustained release of drugs, the maintenance of therapeutic efficacy, and the avoidance of unwanted interactions. In this regard, nanoparticles can serve as efficient drug delivery systems to improve drug targetability, absorption, and bioavailability. Nanoparticles are colloidal particles based on solid materials, and they have sizes within the range of 1–1000 nm [6]. They allow control of the release pattern of drugs, and more than one drug can be incorporated in them to accomplish the aim of combination therapy [7]. Nanoparticles can be functionalized on the basis of different polymers, functional groups, ligands, and protein molecules to customize them according to the type of drug delivery required. Polymeric nanoparticles are the rising stars of novel drug delivery systems. Polymers are macromolecules formed by the union of a large number of monomers by covalent bonding in a linear or branched form. Many polymers exhibit the characteristics of biocompatibility, biodegradability, and nonimmunogenicity, which make them ideal candidates for use in the preparation of nanoparticles [8].

Chitosan is a water-insoluble linear amino polysaccharide obtained by the deacetylation of chitin. The biocompatible, biodegradable, nontoxic, and mucoadhesive properties of chitosan make it an ideal candidate for the formulation of polymeric nanoparticles. It has the ability to load both lipophilic and hydrophilic drugs and is quite inexpensive and readily available in the market. Chitosan also possesses a positive charge, which allows to react with negatively charged drugs and biomolecules, improving the absorption of drugs [9]. Moreover, chitosan has remarkable bioadhesive properties that make it suitable for use in mucoadhesive drug delivery systems aimed at enhancing the retention times of drugs by adhering to mucosae for longer periods of time. These properties render chitosan a favorable water-soluble natural polymer that is ideal for use in the preparation of polymeric nanoparticles [10].

Despite its amazing properties, the use of chitosan in drug delivery systems comes with certain problems. One of them is the limited solubility of chitosan at a pH value that is higher than its pKa value, i.e., pH 5.5–6.5. Chitosan exhibits good solubility in acidic conditions as a result of the protonation of the D-glucosamine units in its structure. In contrast, it shows poor or negligible solubility at neutral or alkaline pH values. Moreover, chitosan-based polymeric nanoparticles often suffer from the unwanted attacks of serum nucleases and proteins and undergo rapid clearance from blood circulation via the reticuloendothelial system (RES) [11]. This issue has again drawn attention from scientists hoping to modify the polymer and formulate long-acting polymeric nanoparticles. The literature reported that biocompatible polymers, such as polyethylene glycol, have the capacity to overcome such issues. The modification of chitosan with PEG prevents interaction with blood components and plasma proteins, reduces interaction with opsonins, improves the solubility of chitosan, and ultimately prevents the removal of nanoparticles from the RES. PEG is an ideal polymer that is highly soluble in water as well as inorganic solvents. It also shows good biocompatibility and biodegradability and low cytotoxicity [12].

PEG is known for its 'stealth behavior' and has the ability to be modified with a large number of functional groups on its terminal end groups. It is an FDA-approved linear polyether compound that is nonimmunogenic, nontoxic, and non-antigenic and exhibits very low protein adsorption compared to any other polymer used in drug delivery systems. PEGylated chitosan is obtained by attaching PEG molecules to various reactive groups present on the backbone of chitosan. The modification of chitosan with PEG improves the pharmacokinetic and pharmacodynamic properties of the drug delivery system prepared with PEGylated chitosan derivatives, specifically enhancing body residence time, bioavailability, and water solubility while reducing renal clearance and cytotoxicity [13].

The aforementioned problems and reasoning associated with the choice of drug, polymer, delivery systems, and functional groups has led to the decision to formulate nanoparticles with this particular combination of materials and methods, i.e., PEGylated chitosan nanoparticles of TAF, to achieve the best possible therapeutic effect of the drug backed by a strong, biocompatible, and nontoxic drug delivery system. This study aims to synthesize PEGylate chitosan by the initial protection of its amine groups with phthalic anhydride, followed by the deprotection and chemoselective conjugation of PEG to the hydroxyl groups. The drug-loaded PEGylated chitosan nanoparticles were subjected to characterization tests to assess their potential for the successful drug delivery of TAF.

2. Materials and Methods

2.1. Materials

Tenofovir Alafemaide was received as gift from GENIX PHARMA PVT. LTD., Pakistan. High-molecular-weight chitosan (>75% deacetylated, MW: 310,000–375,000 Da, 800–2000 cP, CAS number: 9012-76-4) was purchased from SigmaAldrich, Burlington, MA, USA. Polyethylene glycol (PEG) 6000 was purchased from Fluka Chemie GmbH, Germany. Thionyl chloride ($SOCl_2$, $\geq$99%), anhydrous tetrahydrofuran (THF, $\geq$99%, CAS number: 109-99-9), sodium tripolyphosphate (TPP, 85%, CAS number: 7758-29-4), phthalic anhydride ($\geq$99%, CAS number: 85-44-9), sodium hydride (NaH, 60% dispersion in mineral oil, CAS number: 7646-69-7), hydrazine monohydrate (98%, contains 64% hydrazine, CAS number: 7803-57-8), anhydrous N,N-dimethylformamide (DMF, 99.8%, CAS number: 68-12-2), and anhydrous sodium hydroxide pellets (NaOH, $\geq$98%, CAS number: 1310-73-2) were purchased from Merck & Co., Inc., New York, USA, and pyridine (99.5%) was purchased from Riedel-de Haen. All reagents and solvents used in the synthesis process were of analytical grade. Deionized and double-distilled water was taken from a research laboratory of the University of Central Pharmacy, Pakistan.

2.2. Method for Synthesis of PEGylated Chitosan

The PEGylated chitosan was synthesized using the modification of chitosan through hydroxyl groups. In the first step, the deacetylation of chitosan was performed using NaOH, followed by the phthylation of deacetylated chitosan using DMF (step 2). In the third step, the chlorination of phthylated chitosan was achieved by using $SOCL_2$, and PEGylated phthylated chitosan was achieved in the PEG activation step (step 4). In the last step, PEGylated chitosan was synthesized for further use in the fabrication of nanoparticles.

The deacetylation of chitosan was achieved by preparing a 40% (w/v) solution of NaOH and added the required amount of chitosan powder to it. The chitosan-containing solution was stirred for 4 h at 110 °C in a nitrogenous environment. The uniform solution was filtered and a 40% NaOH solution was added again under same conditions. The final solution was freeze-dried at −50 °C in a lyophilizer.

After the deacetylation of chitosan, the phthylation of chitosan was achieved by taking deacetylated chitosan in a beaker and adding a freshly prepared phthalic anhydride solution. The phthalic anhydride solution was prepared by adding 2.76 g of phthalic anhydride to 20 mL of DMF. The solution was stirred for 8 h at 120 °C in a nitrogenous environment. The resulting product was then cooled at room temperature (25 °C), and ice-cold water was added in the resultant to ensure the precipitation of phthylated chitosan. The precipitates were filtered, washed with methanol to remove unreacted chitosan, and dried overnight.

The PEGylated chitosan was synthesized chemoselectively, where phthylated chitosan was attached to PEG after the activation of PEG with NaH (used as a catalyst to make a conjugate of chitosan and PEG). The synthesis of PEGylated chitosan was completed in substeps. First, phthylated chitosan was added to 20 mL of pyridine, and $SOCL_2$ was added to the resultant solution in a volume ten times greater than that of chitosan. The reaction was stirred for 30 min at 80 °C in a nitrogenous environment. After stirring, the resultant solution was cooled at room temperature, and chilled water was added to the solution for precipitation. The solution was filtered and dried to yield chlorinated phthaloyl chitosan. In the next step, the desired amount of PEG was taken, added to the already prepared NaH solution after mixing with anhydrous THF, and stirred for two hours at 60 °C in a nitrogenous environment. The previously prepared chlorinated phthylated chitosan was added to the reaction mixture and again stirred for 16 h. The reaction mixture was allowed to cool, and methanol was added to make precipitates. The mixture was filtered and dried to obtain PEGylated chitosan.

After the deacetylation of chitosan, the protection of the amine group was ensured, and in the last step, the deprotection of PEGylated phthylated chitosan was achieved by taking 15 mL of hydrazine monohydrate along with 30 mL of distilled water and mixing for 16 h at 100 °C under constant magnetic stirring until a viscous solution was left in the beaker. The same procedure was repeated three times by reconstituting the beaker contents with distilled water until a solid residue remained. The solid residue was vacuum-dried to obtain the desired PEGylated chitosan and was stored for further use. The complete synthesis process is presented in Figure 1.

2.3. Method for Preparation of PEGylated Chitosan TAF Nanoparticles

An ionic gelation method was used to prepare PEGylated chitosan nanoparticles because it was extensively used in the literature for the preparation of polymeric nanoparticles. First, the polymer was dissolved in a 1% (v/v) acetic acid solution, and a pH of 5 was maintained using NaOH or HCL as required. The TPP solution was prepared after dissolving TPP in double-distilled water to obtain a concentration of 0.7 mg/mL and a pH of 3 was maintained using NaOH or HCL as required. The polymer solution was stirred on a magnetic stirrer for one hour at room temperature, and a dropwise TPP solution was added at a constant speed to make polymeric nanoparticles. The prepared nanoparticles were centrifuged at 14,000 rpm and 5 °C for 30 min. Then, the supernatant and sediment were separated. The sediment was redispersed using deionized water and characterized by a particle size analyzer for the particle size, zeta potential, and polydispersity index. A graphical representation of the preparation of nanoparticles is presented in Figure 2.

2.4. Characterization

The physicochemical properties of the synthesized polymer and fabricated nanoparticles were checked to confirm the successful synthesis of the polymer and the potential of the nanocarrier system for drug delivery.

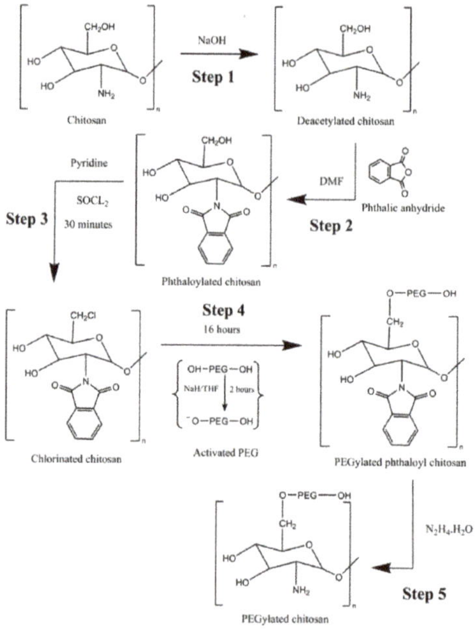

Figure 1. Reaction scheme of PEGylated chitosan.

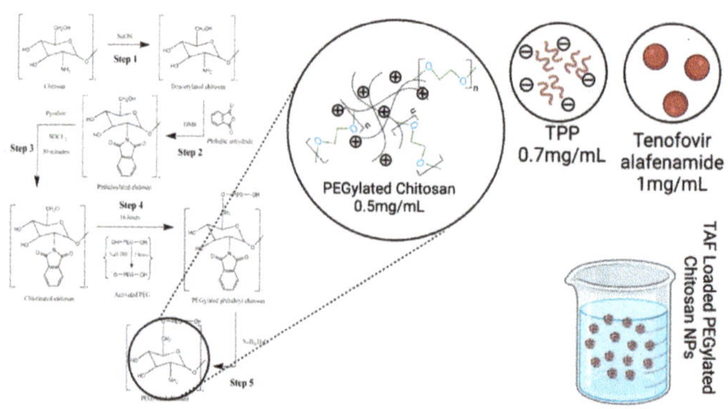

Figure 2. Graphical representation of the preparation of TAF nanoparticles.

2.4.1. Determination of λ_{max} and Calibration Curve of TAF

The wavelength of a substance that shows the maximum absorption is called the λ_{max}. This value is usually calculated with a UV-visible spectrophotometer after preparing drug solutions. A solution of TAF was prepared by dissolving the drug in distilled water and making up the volume, and wavelength scans were carried out on a spectrophotometer. The calibration curve of TAF was determined by making serial dilutions of the drug from a stock solution in a distilled water and PBS buffer solution. The stock solution was prepared by dissolving 10 mg of drug in 100 mL of distilled water to obtain a concentration of 0.1 mg/mL. The first dilution was prepared by taking 2 mL of stock solution and adding distilled water up to 10 mL. The second dilution was prepared by taking 4 mL from the stock solution and making up to 10 mL with distilled water. The next dilutions were prepared by taking 6 mL from the stock solution, and distilled water was added up to

10 mL. The absorbance values of all prepared dilutions were checked with a UV-visible spectrophotometer at the wavelength determined in the previous step.

2.4.2. Fourier-Transformed Infrared (FTIR) Spectroscopy

FTIR studies were conducted to determine the compatibility of the drug with other formulation excipients. FTIR spectroscopy is sensitive to the chemical surface of nanoparticles and helps in the identification of the functional groups and bonds present in the drug and other excipients. [14]. FTIR was performed on all steps during the synthesis of PEGylated chitosan, the pure drug, polyethylene glycol, chitosan, and drug-loaded nanoparticle formulation using an Agilent Cary 630 FTIR spectrometer (California, USA).

2.4.3. Scanning Electron Microscopy (SEM)

SEM studies were conducted to check the surface morphology of the polymer and the size of the nanoparticles [15]. The SEM studies were carried out using a Zeiss EVO LS10 (Carl Zeiss, Oberkochen, Germany).

2.4.4. X-ray Diffraction (XRD) Analysis

An XRD analysis was conducted to analyze degree of crystallinity of the structures of the polymers and prepared formulations. It helped to evaluate the impact of changes on the crystallinity of formulations [16]. The XRD analysis was carried out with a Bruker D8 Discover diffractometer (Massachusetts, USA).

2.4.5. Particle Size Analysis

The nanoparticles were characterized for physical parameters such as the particle size, zeta potential, and polydispersity index (PDI) using a Zetasizer Nano ZS (Malvern, UK). The nanoparticles were checked in a standard range for nanoparticles, which is 1–1000 nm size. The zeta potential is the measure of the surface charge of nanoparticles. It is measured to assess the degree of stability of nanoparticles in a liquid formulation, and the magnitude of the zeta potential indicates the degree of repulsion between adjacent nanoparticles with similar charges within a liquid. The PDI is a measure of the standard deviation from the average particle diameter within a particle solution. It is measured to check the uniformity of a solution in terms of particle size [17]. Physical characterization was carried out for all prepared nanoparticle formulations in order to determine the above-mentioned physical parameters.

2.4.6. Drug-Loading Capacity

The drug-loading capacity (DLC) is a measure of the amount of drug that is loaded compared to the total weight of the nanoparticles. It determines what percent of the nanoparticle mass comprises the encapsulated drug [18]. The percent DLC was determined for drug-loaded nanoparticles. To determine the % DLC, the nanoparticle formulations were subjected to centrifugation at 14,000 rpm for 30 min. After centrifugation, the supernatant liquid was used to check the absorbance value in a UV spectrophotometer at a wavelength of 262 nm. This value was put into the calibration curve regression equation to find the amount of drug present in the supernatant liquid, from which the drug-loading capacity was determined using the following formula:

$$Drug\ loading\ capacity = \frac{Total\ drug - unentrapped\ drug}{Total\ weight\ of\ nanoparticles} \times 100$$

2.4.7. Entrapment Efficacy

The entrapment efficiency is the measure of the amount of drug entrapped in the nanoparticles compared to the total amount of drug subjected to loading into nanoparticles. It determines what percentage of the total added drug was entrapped into the nanoparticles, thus determining the efficiency of the nanoparticles in terms of entrapment [19].

The %EE was determined for drug-loaded nanoparticles. For this purpose, the amount of unentrapped drug in the nanoparticles was determined by separating the nanoparticles from their aqueous medium by centrifugation at 14,000 rpm for 30 min. The supernatant obtained after centrifugation was evaluated in a UV-visible spectrophotometer. The resulting absorbance value was put into the regression equation obtained from the calibration curve of TAF to determine the amount of unentrapped drug in the nanoparticles. Then, the %EE of the nanoparticles was determined using the following formula:

$$Entrapment\ efficacy = \frac{Total\ drug - unentraped\ drug}{Total\ drug} \times 100$$

2.4.8. In Vitro Drug Release

For in vitro drug release studies, a 12 kD dialysis membrane was used to assess the release of TAF from PEGylated chitosan nanoparticles and non-PEGylated nanoparticles using a USP drug dissolution apparatus II (Paddle). A 1 mL nanoparticle solution was mixed with a 4 mL pH 7.4 PBS solution of in dialysis membrane bag. The membrane bag was tied with the paddle of the dissolution apparatus. The paddle was immersed in 500 mL of dissolution medium with the temperature set at 37 °C and a speed of 50 rpm. Then, 5 mL of sample solution was removed every hour for 9 h and at 24 h, 47 h, and 48 h, with 5 mL of fresh dissolution medium added after every sample. The samples were run in a UV-visible spectrophotometer at 262 nm. Absorbance was noted, and the % drug release was calculated using the following equation [20]:

$$\%\ Drug\ Release = \frac{Absorbance\ of\ sample}{Absorbance\ of\ standard} \times 100$$

2.4.9. Drug Release Kinetic Modeling

Drug release kinetic studies were performed to check the pattern of TAF release from the nanoparticles. To check this, zero-order, first-order, Higuchi's, Hixson–Crowell, and Korsmeyer–Peppas models were generated in DD Solver software. In the Korsmeyer–Peppas model, n was also calculated to check the diffusion pattern of the drug from the carrier. The drug followed a Fickian diffusion mechanism if the n value was less than 0.45; otherwise, the mechanism was non-Fickian ($n > 0.45$). The drug release data from all formulations of the nanoparticles were subjected to the following equations:

Zero-order equation:

$$F = k_o \times t$$

where t is time and k_o is the release rate constant for zero-order kinetics;

First-order equation:

$$F = 100 \times [1 - Exp(-k_1 \times t)]$$

where k_1 is the release rate constant for the first-order kinetics;

Higuchi's equation:

$$F = k_H \times t^{0.5}$$

where k_H represents the Higuchi release rate constant;

Hixson–Crowell equation:

$$F = 100 \times \left[1 - (1 - k_{HC} \times t)^3\right]$$

where k_{HC} is the Hixson–Crowell rate constant;

Korsmeyer–Peppas equation:

$$F = k_{KP} \times t^n$$

where k_{KP} is a constant corresponding to the structural and geometric characteristics of the device and n is the release exponent, which is indicative of the mechanism of the drug release [21]. The R^2 values of all the models and the n value from the Korsmeyer–Peppas model obtained from the results of DD solver were assessed to determine which model best fit the mechanism of drug release from the nanoparticles.

3. Results and Discussions

3.1. Determination of Absorption Spectra of TAF

The absorption spectra of TAF were calculated from UV-visible spectrophotometer via a wavelength scan. The figure showed the maximum absorbance at 260 nm, and in the literature it was reported as 262 nm. The obtained wavelength is presented in Figure 3.

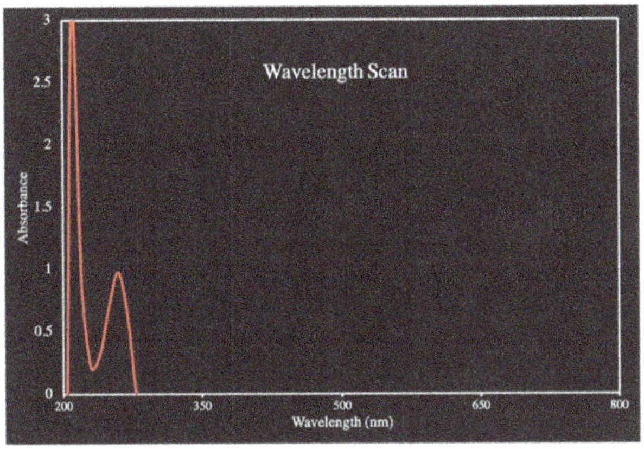

Figure 3. The absorption spectra of TAF.

3.2. Determination of Calibration Curve of TAF

The calibration curve of TAF was determined according to the method mentioned in the Materials and Methods Section. The absorbance was determined for all dilutions at a λ_{max} of 262nm. The absorbance values of all dilutions with different concentrations in distilled water as well as a PBS buffer of 7.4 pH are presented in Figure 4.

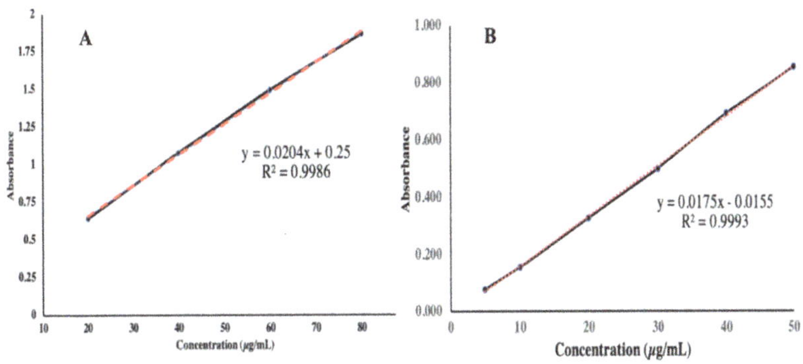

Figure 4. Calibration curve of TAF in (**A**) distilled water and (**B**) PBS buffer solution.

3.3. Chemical Compatibility Studies

3.3.1. FTIR of Synthesized Polymer

The FTIR analysis was conducted on all steps of polymer synthesis. The first step was the deacetylation of chitosan. In Figure 5A, the FTIR of deacetylated chitosan shows a prominent band in the range of 3000–3500 cm^{-1}, which corresponds to the N-H and O-H stretching vibrations. The peaks at 2950.18 cm^{-1} and 2847.86 cm^{-1} represent C-H symmetric and asymmetric vibrations, respectively. These bands are characteristic bands of polysaccharides that represent the high linkage prevalence of C-H bonds and are exhibited in the spectra of many other polysaccharides such as carrageenans, xylan, and glucans [22]. C=O amide I stretching, the N-H bending of amide II, and C-N amide III stretching are represented by the bands at 1727.62 cm^{-1}, 1574.80 cm^{-1}, and 1366.06 cm^{-1}, respectively, which are indicative of the residual N-acetyl groups that were left after deacetylation. The band at 1149.88 cm−1 represents the asymmetric stretching vibrations of the C-O-C bridge. CH_2 bending vibrations are depicted by the peak at 1433.16 cm^{-1}. The peaks at 1060.42 cm^{-1} and 1026.88 cm^{-1} exhibit the C-O stretching and bending vibrations. The FTIR of chitosan showed the absence of the acetyl (CH_3) peaks at 1375 cm^{-1}, which were reported by Queiroz et al. in their study [23]. This confirms the successful deacetylation of chitosan by treating it with a NaOH solution. However, the FTIR spectrum of deacetylated MMWC showed some bands at 1718.30 cm^{-1}, 1578.52 cm^{-1}, and 1423.84 cm^{-1}, which are indicative of the residual N-acetyl groups that were left after deacetylation. Similar results were reported by Silva et al., who prepared multifunctional chitosan and gold nanoparticles and observed bands of residual n-acetyl groups in the FTIR spectrum of deacetylated chitosan [24]. In Figure 5B, the FTIR of phthaloylated chitosan shows a prominent band in the range of 3200–3400 cm^{-1}, which corresponds to the N-H and O-H stretching vibrations. The successful phthaloylation of deacetylated chitosan was confirmed by the appearance of the peaks at 1772.34 cm^{-1} and 1705.25 cm^{-1}, corresponding to the C=O imide bond of the phthaloyl group in the FTIR spectrum of phthaloylated MMWC, and at 1774.21 cm^{-1} and 1701.52 cm^{-1} in the FTIR spectrum of phthaloylated HMWC. A study conducted by Malhotra et al. reported similar results, where the imide bond could be observed at 1774 cm^{-1} and 1702 cm^{-1} in the FTIR spectrum of phthaloylated chitosan [25]. In Figure 5C, phthaloylated chitosan was chlorinated, and the successful completion of the chlorination process was confirmed by the appearance of a peak at 719.37 cm^{-1}, which represented the formation of an organic halogen bond within the structure of chitosan, which is in line with the results reported by Najafabadi et al. [26]. Furthermore, the reduction in the magnitude of the peaks from 3200–3400 cm^{-1} shows the reduction in O-H groups in chlorinated phthaloyl chitosan. In Figure 5D, the peak at 3410.14 cm^{-1} exhibits C-H stretching vibrations of PEG. The PEGylation of chitosan was confirmed by the appearance of characteristic peaks of PEG at 1386.56 cm^{-1}, 1194.61 cm^{-1}, 989.60 cm^{-1}, and 874.06 cm^{-1}. C-O stretching vibrations are presented by the peak at 1011.97 cm^{-1}, and the peak at 1105.55 cm^{-1} shows the C-O-C bridge of chitosan. In Figure 5E, the FTIR spectrum of deprotected PEGylated chitosan shows O-H stretching vibrations at 1062.29 cm^{-1}, which represent the characteristic peaks of PEG, confirming the PEGylation of chitosan. The peaks at 1608.34 cm^{-1}, 1576.66 cm^{-1}, and 892.69 cm^{-1} correspond to the amide I, amide II, and pyranose structures, respectively, which are the main peaks of the chitosan molecule. The deprotection of the sample was confirmed by the presence of a C=O imide peak at 1710 cm^{-1}, which came from the hydrazine used during deprotection. The results were similar to the study reported by Malhotra et al., where they prepared peptide-tagged PEGylated chitosan nanoparticles for siRNA delivery [27].

3.3.2. FTIR of Synthesized NPs

In Figure 6A, the FTIR spectrum of TPP shows a peak at 1470.43cm−1 corresponding to P=O stretching vibrations. PO_3 symmetric and asymmetric stretching vibrations are shown by the peak at 1455.52 cm^{-1}. The antisymmetric stretching vibrations of P-O-P bridge are exhibited by the peak at 1092.10 cm^{-1}. In Figure 6B, the FTIR spectrum of TAF shows a weak N-H band at 3330.38 cm^{-1} and a peak of aromatic stretching vibrations at 3170.10 cm^{-1}. The C=O and P=O stretching vibrations are depicted by the peaks at 1744.39 cm^{-1} and 1658.66 cm^{-1}, respectively. The NH_3 scissoring band stretching vibrations are exhibited by a sharp peak at 1604.61 cm^{-1}. The peaks at 1420.11 cm^{-1} and 1487.20 cm^{-1} correspond to the aromatic CN stretching vibrations in pairs. The deformation of C-N is represented by the peak at 1263.56 cm^{-1}. The band from 600 cm^{-1} to 900 cm^{-1} corresponds to the C-H out-of-plane deformation, and lastly the peak at 918.78 cm^{-1} exhibits the N-H wagging band. In Figure 6C, the deprotected PEGylated chitosan was further treated with TPP to formulate nanoparticles. The FTIR spectrum of nanoparticles depicts very sharp peaks for N-H and O-H stretching vibrations within the range of 3200–3500 cm^{-1}, which are suggestive of the stronger hydrogen bonding within the nanoparticles. Moreover, a hypochromic shift in the peaks of the $CONH_2$ and NH_2 vibrations is observed in the nanoparticles compared to those of simple deacetylated chitosan, which is a result of the interaction between the NH_3 of chitosan and the phosphate groups of the TPP used during nanoparticle formulation, which coincides with the results of a study reported by Qureshi et al. [28]. In Figure 6D, the FTIR spectrum of TAF-loaded PEGylated chitosan nanoparticles shows the same characteristic peaks as that of chitosan nanoparticles. The absence of characteristic peaks of TAF shows that TAF was fully encapsulated without any chemical interaction. The spectrum shows all the peaks of PEGylated chitosan nanoparticles. However, none of the peaks of TAF appear in the FTIR of the drug-loaded nanoparticles. Similar results were reported by Ulu et al. in their study where they proposed that the absence of peaks of TAF suggests that the drug was fully encapsulated within the nanoparticles and no possible interaction of the drug with the components of the nanoparticles was observed, which is why only the peaks of PEGylated chitosan nanoparticles were observed and the peaks of TAF were masked within the nanoparticle structure [29].

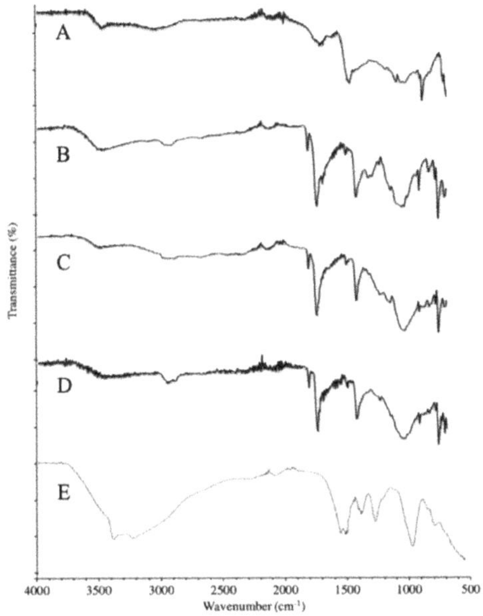

Figure 5. FTIR analysis of synthesized PEGylated chitosan. (**A**) Deacetylated chitosan. (**B**) Phthaloylated chitosan. (**C**) Chlorinated chitosan. (**D**) PEGylated chitosan. (**E**) Deprotected PEGylated chitosan.

3.4. XRD Studies

X-ray diffraction studies were performed on pure PEG, PEGylated chitosan, TAF, and the TAF-loaded NP formulation. The obtained diffractogram of PEG showed high-intensity sharp peaks at 19° and 24° as well as minor peaks at 27° and 28°, which confirmed the crystalline structure of PEG (Figure 7A). The PEGylated chitosan polymers of different molecular weights showed amorphous structures with zero peaks (Figure 7B). The pure TAF was a crystalline powder with many high-intensity sharp peaks at different points as well as small peaks indicating the various functional groups of the drug (Figure 7C). The diffractograms of the drug-loaded nanoparticles (Figure 7D) showed no prominent peaks. However, diffused peaks were observed, which depicted the amorphous form of the prepared formulation. The absence of sharp peaks of the drug in the XRD of formulated nanoparticles can be attributed to the use of various chemicals during the formation of the nanoparticles, which resulted in the amorphization process, causing the disappearance of characteristic peaks of the crosslinked polymer and drug. Similar results were reported by Melo et al. [30] during the preparation of chitosan/PEG nanoparticles loaded with indole-3-carbinol and by Maity et al. [31] during the study of naringenin-loaded alginate-coated chitosan core–shell nanoparticles, where they observed the disappearance of characteristic peaks in the drug-loaded nanoparticles. Thus, it is possible to consider that the disappearance of the drug peaks in the nanoparticles is a clear indication of the dispersion, at the molecular level, of the drug within the nanoparticle matrix.

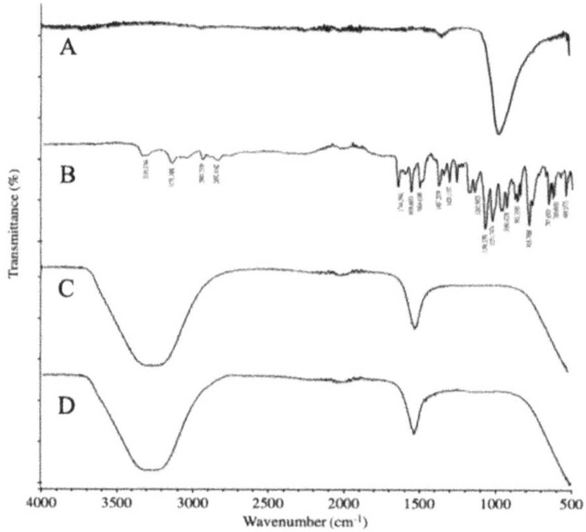

Figure 6. FTIR analysis of synthesized PEGylated chitosan. (**A**) TPP. (**B**) TAF. (**C**) Deprotected PEGylated chitosan nanoparticles. (**D**) TAF-loaded PEGylated chitosan nanoparticles.

3.5. SEM Analysis

Scanning electron microscopy was performed on the PEGylated polymer (Figure 8) and the nanoparticle formulation (Figure 9). The SEM images showed that the polymer particles were rectangular to cylindrical in shape, along with spherical and irregularly shaped particles. With such different results, it was difficult to claim the best structural morphology of the synthesized polymeric particles. In the SEM images, it was also revealed that the particles had slightly spongy as well as amorphous outer surfaces. In the XRD diffractogram, the synthesized polymer had an amorphous nature, and the SEM results were authenticated with XRD where distinct or sharp peaks were observed. Ulu et al. developed simple and drug-loaded chitosan nanoparticles and reported that the nanoparticles were roughly spherical in shape and evenly distributed [29]. The spherical shape of the nanoparticles was due to the tendency of chitosan molecules to fuse with each other with extensive hydrogen bonds in the structure [32]. The current results correlate with previous studies where similar results were reported [33,34]. Additionally, Ulu et al. also reported that TAF loaded in nanoparticles did not change the morphology of fabricated nanoparticles [29]. The PEGylation of chitosan and drug-loaded PEGylated nanoparticles was performed, and the results of the FTIR and SEM analyses authenticated the successful synthesis of the polymer and nanoparticles. For further identification to confirm the drug loading, XRD was performed on the pure drug, chitosan, PEG, PEGylated chitosan, and prepared nanoparticles. The XRD spectra of PEGylated chitosan nanoparticles showed an amorphous nature, which confirmed that the crystal structure of the drug was tapered.

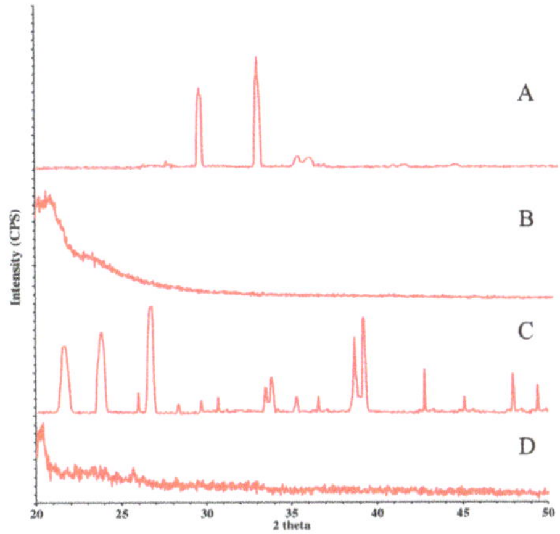

Figure 7. Illustration of XRD of PEG (**A**), PEGylated chitosan (**B**), the drug (**C**), and TAF PEGylated nanoparticles (**D**).

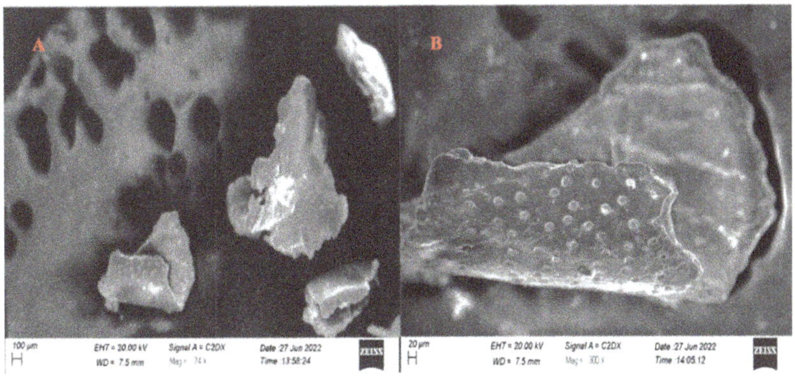

Figure 8. SEM image of synthesized PEGylated chitosan at (**A**) 74× and (**B**) 300×.

Figure 9. SEM image of PEGylated chitosan nanoparticles at 5000×.

The SEM images of nanoparticles showed well-segregated particles, as authenticated by the particle size analyzer results. The particles in the SEM images showed the successful fabrication of nanoparticles with PEGylated chitosan and advocate for the stability of nanocarriers. The figure illustrates the spherical shape of nanoparticles and the uniform distribution. From the SEM images, it can be observed that the particles were of a nanosized range that was also studied or reflected by the zeta sizer outcomes. The well-segregated particles might be due to a considerable charge on the particle surfaces that was observed in the zeta potential analysis.

3.6. Drug-Loading Capacity and Entrapment Efficacy of NPs

The %EE of TAF loaded in PEGylated chitosan nanoparticles was calculated to be ~90% and ~15%. Ulu et al. aimed to develop TAF loaded in simple chitosan nanoparticles and found an EE of approximately 50% [29]. In another study, Shailender et al. found an EE of 48.2%, but they prepared nanoparticles of chitosan loaded with a drug that was similar to that used in the current project [35]. The literature reported the effect of the molecular weight of the polymer and found that the EE and molecular weight have a direct relation with each other [36]. In this study, we also reported the same results, where EE increased with an increase in the molecular weight of the polymer. These findings can be attributed to the idea that an increased molecular weight of the polymers results in a more porous or expanded nanoparticle matrix, allowing more drug to enter and encapsulate. The results were authenticated in the literature, where Meng et al. prepared tenofovir nanoparticles using different molecular weights of chitosan (low, medium, and high) and reported less EE, but with an increase in the molecular weight, the EE surged [37]. Therefore, the % of the drug loading into the chitosan-based nanoparticles might involve different mechanisms, such as adsorption, encapsulation, and electrostatic interaction [36]. Therefore, it can be hypothesized that encapsulation occurs due to the physical confinement of the drug in the interstices of the nanostructured material during the gelation process by various interactions or due to hydrogen bonding between the drug and the polymers [38]. Ideally, an efficient drug delivery system would possess an excellent degree of drug association. A high EE was possible due to the hydrophobic nature of drugs, which helps the drug to enter the nanocarrier system. In this study, TAF showed good EE and indicated ionic interactions between the polymer and TPP. This interaction ultimately resulted in the preparation of highly entrapped drug nanoparticles. In a previous study conducted by Shahid et al., they prepared ticagrelor-loaded simple chitosan and ticagrelor-loaded thiolated chitosan nanoparticles and reported EE values of 84.1% and 94.3%, respectively [39]. In another study, Bo Fan et al. reported an EE of 98% in thiolated chitosan nanoparticles [40].

3.7. Particle Size Determination

The particle size of the drug-loaded PEGylated chitosan was analyzed, and the observed particle size distribution was 219.4 nm with a PDI of 0.369 (Figure 10), which indicated the stability of the nanoparticles (Figure 11). Ulu et al. checked the hydrodynamic diameters of drug-unloaded and drug-loaded nanoparticles and found average particle sizes of 200 to 340 nm [29]. The increase in the average particle size of nanoparticles might be due to drug incorporation between the chitosan network [32]. The zeta potential was 23.4mV (Figure 11). The results were similar to the study reported by Ulu et al., where a higher zeta potential of 42.8 mV was observed for chitosan due to its cationic nature [29,41]. In other studies, Yien et al. [42] and Shailender et al. [35] prepared drug-loaded chitosan nanoparticles and reported higher zeta potential values. Similar results were reported by Machado et al., where they fabricated composites of nanoparticles (tenofovir-loaded poly(lactic-co-glycolic acid)/stearylamine) [43]. The PDI was also checked for the prepared nanoparticles, where a PDI of 0.369 was observed in the prepared formulation. In a previous study, a low PDI was observed (0.44) [29], and a lower PDI indicated the fabrication of stable nanoparticles [35]. However, a higher PDI was also observed in a previous study where they reported a PDI of 0.65 for TAF-loaded chitosan nanoparticles [44]. The PDI in the current study showed a uniform distribution. However, an increased PDI supports the good colloidal nature and long-term stability of nanoparticles [29]. Our results were comparable with many studies, such as Mariadoss et al. [45], Zhang et al., [46] and Ulu et al. [29].

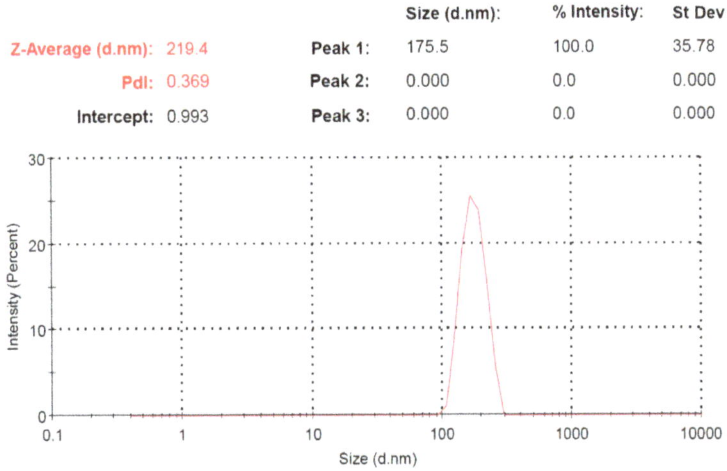

Figure 10. Particle size distribution of TAF PEGylated chitosan nanoparticles.

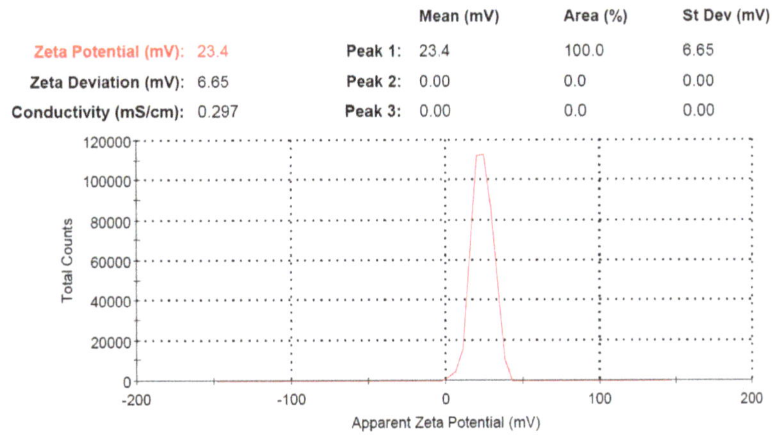

Figure 11. Zeta potential of TAF PEGylated chitosan nanoparticles.

3.8. *In Vitro Drug Release*

The in vitro % drug release of all formulations was carried out in USP Dissolution apparatus II using a dissolution medium (PBS) with a pH of 7.4. The % of the drug released from the nanoparticles was calculated for 48 h (Figure 12). Approximately 60% of the drug was released from the PEGylated nanoparticle formulation. The TAF was released from the prepared biodegradable polymeric nanocarrier system from the surface because of desorption, erosion, degradation, reabsorption, and diffusion mechanisms of the polymeric network [45,47]. A study conducted by Ulu et al. reported a burst release of approximately 53% in a day [29], but in our study almost 35% of the drug was released in first 24 h. The burst release was due to the higher amount of drug dispersed on the surface or near the surface of the nanocarrier system [36]. After two days, approximately 60% of the drug was released from the nanoparticles. These results were in agreement with the previous study where Meng et al. reported almost complete drug release from nanoparticles using low-molecular-weight chitosan within 48 h and 50–70% from high- and medium-molecular-weight chitosan within 120 h [37]. In addition, 80–90% release was reported by Ulu et al. [29], and 80% release within 72 h was reported by Zhang et al. [48]. Another study prepared thiol-modified chitosan nanofibers to check the release of tenofovir [49]. They reported an 80–100% release from the carrier within 50 h. To summarize, we can say that our release outcomes corresponded with the literature.

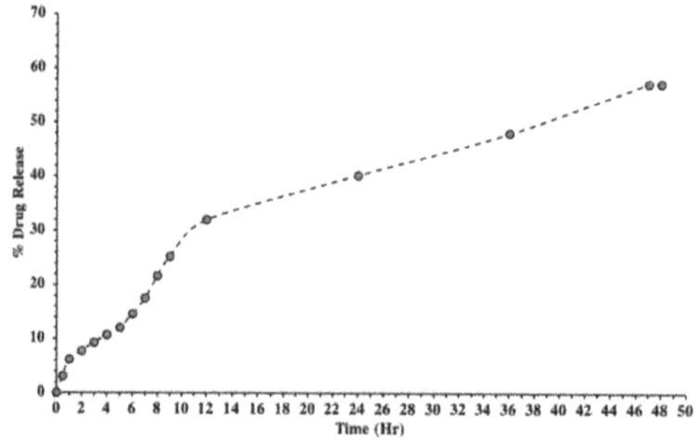

Figure 12. In vitro percent of drug release from PEGylated chitosan nanoparticle formulations.

Polymers such as PEG have the capacity to solubilize in water as well as organic solvents, which is extensively used in grafting many polymers. It is biocompatible and biodegradable and has no immunogenicity, antigenicity, or toxicity [25]. In 2004, Gorochovceva and Makuska proposed a novel method for the synthesis of PEGylated chitosan by modifications at hydroxyl groups [50]. Previously, PEG was extensively used as a cross-linker for forming graft polymers to enable drug release [51]. Different techniques were used by changing parameters to prepare a nanocarrier system of modified chitosan. Those parameters were the concentration of the polymer, the crosslinker, the pH, the molecular weight, the drug-to-polymer ratio, the surfactants, and the stabilizers. However, all those parameters directly or indirectly alter the structure as well as the morphology of prepared chitosan nanoparticles and the release patterns of drugs [52]. Furthermore, drug release was controlled by PEG, and the same pattern was reported in our study, where drug release was decreased and approximately 60% of the drug was released in 2 days. Our results were supported by a similar study conducted by Moradkhannejhad et al., where they checked the curcumin drug release from a nanocarrier system. They reported that with an increase in the PEG content the drug release was improved because of the hydrophilic nature of PEG. They also studied the drug release pattern by varying the molecular weight of PEG and found that the formulation with PEG400 showed more drug release compared to the formulation with PEG6000. This change was due to the hydrophilicity of PEG and its consequent swelling capacity, which ultimately improved drug release. Therefore, we can say that the lower the molecular weight, the higher the hydrophilicity and vice versa [53].

3.9. Kinetic Model Studies

For kinetics studies, the results of the % of drug release were added in the DD Solver software to check all the R2 values for kinetic models such as first-order, zero-order, Hixson–Crowell, Korsmeyer–Peppas, and Higuchi (Table 1). The drug-loaded PEGylated chitosan formulation showed first-order kinetics. First-order release indicates that the drug release was dependent on the initial concentration of the formulation. Moreover, the analysis of the R2 values of all kinetics models depicted that our formulation followed the Korsmeyer–Peppas model (R^2 = 0.9885). The Korsmeyer–Peppas model represents a one-dimensional drug release relative to the drug dosage form [54]. A study conducted by Jusu et al. checked the drug release on all five models that we used in our study and found that all PEGylated drug-loaded microspheres followed the Korsmeyer–Peppas model [55]. In the Korsmeyer–Peppas model, the n value represents the diffusion type. The release may follow Fickian diffusion if the n value is less than or equal to 0.5. Values ranging from 0.5 to 0.85 represent anomalous or non-Fickian diffusion, and values greater than 0.85 refer to a case II transport mechanism [26]. Our formulation showed non-Fickian diffusion with an n value greater than 0.45. A non-Fickian-type release mechanism means that the phenomena responsible for drug release are drug diffusion processes from the NPs coupled to the relaxation of the polymeric chains [56]. Our results corresponded with the study of Jusu et al., where the n value was in range of 0.446 to 0.889, consistent with non-Fickian diffusion or the anomalous transport of drug release that involves two phenomena: drug diffusion as well as drug relaxation from the matrix [55,57]. A similar study was conducted on tenofovir-loaded chitosan nanoparticles and found the best fitting model to be Higuchi (0.9794). This model depicts that liquid penetrates into the nanoparticles and dissolves the drug molecules, and ultimately the release of the drug is predominately controlled by the diffusion mechanism [29].

Table 1. Drug release kinetic models.

Formulation		Formulation
Zero-Order	K_0	1.329
	R^2	0.8693
First-Order	K_1	0.021
	R^2	0.9557
Higuchi Model	K_H	7.757
	R^2	0.9588
Korsmeyer–Peppas Model	K_{KP}	5.066
	N	0.632
	R^2	0.9885
Hixson–Crowell Model	K_{HC}	0.006
	R^2	0.9316
Diffusion mechanism		Non-Fickian

4. Conclusions

We concluded that the successful synthesis of PEGylated chitosan was achieved using a novel approach to form PEG alkoxide using NaH as a catalyst. The synthesized PEGylated polymer has the ability to deliver TAF. The formulated nanoparticles showed a nanosize range for their particle sizes with a polydispersity index of 0.369 and a relatively high entrapment efficiency. The SEM analysis showed that the polymer particles have a rectangular to cylindrical shape along with some spherical and irregularly shaped particles due to which the determination of the exact structural morphology of the particles was complex. The findings of the XRD analysis showed the masking of the sharp peaks of the drug in the XRD of the formulated nanoparticles, indicating amorphization and dispersion, at the molecular level, of the drug in the nanoparticle matrix. The drug release studies showed that approximately 60% of the drug was released from the formulation within 48 h, following the Korsmeyer–Peppas model of drug release. All the findings suggest that the developed nanocarrier system can effectively deliver TAF and in the future can help in providing promising treatment and management of hepatitis B. However, cellular cytotoxicity studies and in vivo analyses are further required to authenticate the formulated nanoparticle system. Therefore, the proposed formulation is a major breakthrough in this field, and a nonviral vector can be further prepared for biomedical applications that involve the target-mediated delivery of drugs.

Author Contributions: Conceptualization, M.Z. and M.H.B.; Data curation, W.S.; Formal analysis, M.Z., M.H.B., N.N., A.M. and H.Y.N.; Funding acquisition, M.Z., W.S., A.A. and M.S.R.; Investigation, M.H.B., W.S., N.N., H.Y.N. and A.A.; Methodology, M.H.B., M.O.I. and M.S.R.; Project administration, M.Z. and M.S.R.; Resources, M.O.I., N.N., A.M. and H.Y.N.; Software, M.H.B. and N.N.; Supervision, A.A. and M.S.R.; Validation, M.O.I.; Visualization, A.M.; Writing—original draft, M.Z., M.H.B., W.S., N.N. and A.M.; Writing—review and editing, M.O.I., H.Y.N., A.A. and M.S.R. All authors have read and agreed to the published version of the manuscript.

Funding: This research was supported Project number RSP2022R491, King Saud University, Riyadh, Saudi Arabia.

Institutional Review Board Statement: Not applicable.

Informed Consent Statement: Not applicable.

Acknowledgments: The authors are thankful to King Saud University, Riyadh, Saudi Arabia.

Conflicts of Interest: The authors declare no conflict of interest.

Sample Availability: Samples of the compounds are not available from the authors.

References

1. Saag, M.S.; Benson, C.A.; Gandhi, R.T.; Hoy, J.F.; Landovitz, R.J.; Mugavero, M.J.; Sax, P.E.; Smith, D.M.; Thompson, M.A.; Buchbinder, S.P.; et al. Antiretroviral drugs for treatment and prevention of HIV infection in adults: 2018 recommendations of the International Antiviral Society–USA Panel. *JAMA* **2018**, *320*, 379–396. [CrossRef]
2. Hawkins, T.; Veikley, W.; Claire, R.L.S., III; Guyer, B.; Kearney, N.C.B.P. Intracellular pharmacokinetics of tenofovir diphosphate, carbovir triphosphate, and lamivudine triphosphate in patients receiving triple-nucleoside regimens. *JAIDS J. Acquir. Immune Defic. Syndr.* **2005**, *39*, 406–411. [CrossRef] [PubMed]
3. Agarwal, K.; Brunetto, M.; Seto, W.K.; Lim, Y.-S.; Fung, S.; Marcellin, P.; Ahn, S.H.; Izumi, N.; Chuang, W.; Bae, H.; et al. 96 weeks treatment of tenofovir alafenamide vs. tenofovir disoproxil fumarate for hepatitis B virus infection. *J. Hepatol.* **2018**, *68*, 672–681. [CrossRef] [PubMed]
4. Wassner, C.; Bradleyand, N.; Lee, Y. A review and clinical understanding of tenofovir: Tenofovir disoproxil fumarate versus tenofovir alafenamide. *J. Int. Assoc. Provid. AIDS Care (JIAPAC)* **2020**, *19*, 2325958220919231. [CrossRef] [PubMed]
5. Jain, K.K. An overview of drug delivery systems. In *Drug Delivery Systems*; Humana: New York, NY, USA, 2020; pp. 1–54.
6. Mohanraj, V.; Chen, Y. Nanoparticles—A review. *Trop. J. Pharm. Res.* **2006**, *5*, 561–573. [CrossRef]
7. Begines, B.; Ortiz, T.; Pérez-Aranda, M.; Martínez, G.; Merinero, M.; Argüelles-Arias, F.; Alcudia, A. Polymeric nanoparticles for drug delivery: Recent developments and future prospects. *Nanomaterials* **2020**, *10*, 1403. [CrossRef]
8. Essa, D.; Kondiah, P.P.; Choonaraand, Y.E.; Pillay, V. The design of poly (lactide-co-glycolide) nanocarriers for medical applications. *Front. Bioeng. Biotechnol.* **2020**, *8*, 48. [CrossRef]
9. Bernkop-Schnürch, A.; Dünnhaupt, S. Chitosan-based drug delivery systems. *Eur. J. Pharm. Biopharm.* **2012**, *81*, 463–469. [CrossRef]
10. Tiyaboonchai, W. Chitosan nanoparticles: A promising system for drug delivery. *Naresuan Univ. J. Sci. Technol. (NUJST)* **2013**, *11*, 51–66.
11. Fan, M.; Huand, Q.; Shen, K. Preparation and structure of chitosan soluble in wide pH range. *Carbohydr. Polym.* **2009**, *78*, 66–71. [CrossRef]
12. Du, J.; Hsieh, Y.-L. PEGylation of chitosan for improved solubility and fiber formation via electrospinning. *Cellulose* **2007**, *14*, 543–552. [CrossRef]
13. Veronese, F.M.; Pasut, G. PEGylation, successful approach to drug delivery. *Drug Discov. Today* **2005**, *10*, 1451–1458. [CrossRef] [PubMed]
14. Praseptiangga, D.; Zahara, H.L.; Widjanarko, P.I.; Joniand, I.M.; Panatarani, C. Preparation and FTIR spectroscopic studies of SiO_2-ZnO nanoparticles suspension for the development of carrageenan-based bio-nanocomposite film. In *AIP Conference Proceedings*; AIP Publishing LLC: Melville, NY, USA, 2020.
15. Sagadevan, S.; Koteeswari, P. Analysis of structure, surface morphology, optical and electrical properties of copper nanoparticles. *J. Nanomed. Res.* **2015**, *2*, 00040–00048. [CrossRef]
16. Vorontsov, A.V.; Tsybulya, S.V. Influence of nanoparticles size on XRD patterns for small monodisperse nanoparticles of Cu0 and TiO_2 anatase. *Ind. Eng. Chem. Res.* **2018**, *57*, 2526–2536. [CrossRef]
17. Clayton, K.N.; Salameh, J.W.; Wereleyand, S.T.; Kinzer-Ursem, T.L. Physical characterization of nanoparticle size and surface modification using particle scattering diffusometry. *Biomicrofluidics* **2016**, *10*, 054107. [CrossRef]
18. Rosenblatt, K.M.; Bunjes, H. Evaluation of the drug loading capacity of different lipid nanoparticle dispersions by passive drug loading. *Eur. J. Pharm. Biopharm.* **2017**, *117*, 49–59. [CrossRef] [PubMed]
19. Ajun, W.; Yan, S.; Liand, G.; Huili, L. Preparation of aspirin and probucol in combination loaded chitosan nanoparticles and in vitro release study. *Carbohydr. Polym.* **2009**, *75*, 566–574. [CrossRef]

20. Nawaz, S.; Khan, S.; Farooq, U.; Haider, M.S.; Ranjha, N.M.; Rasul, A.; Nawaz, A.; Arshad, N.; Hameed, R. Biocompatible hydrogels for the controlled delivery of anti-hypertensive agent: Development, characterization and in vitro evaluation. *Des. Monomers Polym.* **2018**, *21*, 18–32. [CrossRef]
21. Mircioiu, C.; Voicu, V.; Anuta, V.; Tudose, A.; Celia, C.; Paolino, D.; Fresta, M.; Sandulovici, R.; Mircioiu, I. Mathematical modeling of release kinetics from supramolecular drug delivery systems. *Pharmaceutics* **2019**, *11*, 140. [CrossRef]
22. Kouchak, M.; Avadi, M.; Abbaspour, M.; Jahangiriand, A.; Boldaji, S.K. Effect of different molecular weights of chitosan on preparation and characterization of insulin loaded nanoparticles by ion gelation method. *Int. J. Drug Dev. Res.* **2012**, *4*, 271–277.
23. Queiroz, M.F.; Melo, K.R.T.; Sabry, D.A.; Sassakiand, G.L.; Rocha, H.A.O. Does the use of chitosan contribute to oxalate kidney stone formation? *Mar. Drugs* **2014**, *13*, 141–158. [CrossRef]
24. Silva, I.O.; Ladchumananandasivam, R.; Nascimento, J.H.O.; Silva, K.K.O.S.; Oliveira, F.R.; Souto, A.P.; Felgueiras, H.P.; Zille, A. Multifunctional chitosan/gold nanoparticles coatings for biomedical textiles. *Nanomaterials* **2019**, *9*, 1064. [CrossRef] [PubMed]
25. Malhotra, M.; Lane, C.; Tomaro-Duchesneau, C.; Sahaand, S.; Prakash, S. A novel method for synthesizing PEGylated chitosan nanoparticles: Strategy, preparation, and in vitro analysis. *Int. J. Nanomed.* **2011**, *6*, 485.
26. Najafabadi, A.H.; Abdoussand, M.; Faghihi, S. Synthesis and evaluation of PEG-O-chitosan nanoparticles for delivery of poor water soluble drugs: Ibuprofen. *Mater. Sci. Eng. C* **2014**, *41*, 91–99. [CrossRef]
27. Malhotra, M.; Tomaro-Duchesneauand, C.; Prakash, S. Synthesis of TAT peptide-tagged PEGylated chitosan nanoparticles for siRNA delivery targeting neurodegenerative diseases. *Biomaterials* **2013**, *34*, 1270–1280. [CrossRef]
28. Qureshi, J.; Iqbal, F.M.; Danish, Z.; Shah, S.N.H.; Umar, H.; Zaman, M.; Alvi, M.N.; Islam, N. Synthesis and characterization of self-assembling chitosan-based nanoparticles. *Pak. J. Pharm. Sci.* **2022**, *35*, 227–231.
29. Ulu, A.; Sezer, S.K.; Yüksel, Ş.; Koçand, A.; Ateş, B. Preparation, Controlled Drug Release, and Cell Viability Evaluation of Tenofovir Alafenamide-Loaded Chitosan Nanoparticles. *Starch-Stärke* **2021**, 2100144. [CrossRef]
30. Melo, M.N.; Pereira, F.M.; Rocha, M.A.; Ribeiro, J.G.; Junges, A.; Monteiro, W.F.; Diz, F.M.; Ligabue, R.A.; Morrone, F.B.; Severino, P.; et al. Chitosan and chitosan/PEG nanoparticles loaded with indole-3-carbinol: Characterization, computational study and potential effect on human bladder cancer cells. *Mater. Sci. Eng. C* **2021**, *124*, 112089. [CrossRef]
31. Maity, S.; Mukhopadhyay, P.; Kunduand, P.P.; Chakraborti, A.S. Alginate coated chitosan core-shell nanoparticles for efficient oral delivery of naringenin in diabetic animals—An in vitro and in vivo approach. *Carbohydr. Polym.* **2017**, *170*, 124–132. [CrossRef] [PubMed]
32. Gopalakrishnan, L.; Ramana, L.N.; Sethuramanand, S.; Krishnan, U.M. Ellagic acid encapsulated chitosan nanoparticles as anti-hemorrhagic agent. *Carbohydr. Polym.* **2014**, *111*, 215–221. [CrossRef] [PubMed]
33. Kahdestani, S.A.; Shahriariand, M.H.; Abdouss, M. Synthesis and characterization of chitosan nanoparticles containing teicoplanin using sol–gel. *Polym. Bull.* **2021**, *78*, 1133–1148. [CrossRef]
34. Van, S.N.; Minh, H.D.; Anh, D.N. Study on chitosan nanoparticles on biophysical characteristics and growth of Robusta coffee in green house. *Biocatal. Agric. Biotechnol.* **2013**, *2*, 289–294.
35. Shailender, J.; Ravi, P.R.; Sirukuri, M.R.; Dalviand, A.; Priya, O.K. Chitosan nanoparticles for the oral delivery of tenofovir disoproxil fumarate: Formulation optimization, characterization and ex vivo and in vivo evaluation for uptake mechanism in rats. *Drug Dev. Ind. Pharm.* **2018**, *44*, 1109–1119. [CrossRef]
36. Nallamuthu, I.; Deviand, A.; Khanum, F. Chlorogenic acid loaded chitosan nanoparticles with sustained release property, retained antioxidant activity and enhanced bioavailability. *Asian J. Pharm. Sci.* **2015**, *10*, 203–211. [CrossRef]
37. Meng, J.; Sturgisand, T.F.; Youan, B.-B.C. Engineering tenofovir loaded chitosan nanoparticles to maximize microbicide mucoadhesion. *Eur. J. Pharm. Sci.* **2011**, *44*, 57–67. [CrossRef] [PubMed]
38. Liu, Q.; Jing, Y.; Han, C.; Zhangand, H.; Tian, Y. Encapsulation of curcumin in zein/caseinate/sodium alginate nanoparticles with improved physicochemical and controlled release properties. *Food Hydrocoll.* **2019**, *93*, 432–442. [CrossRef]
39. Shahid, N.; Erum, A.; Zaman, M.; Iqbal, M.O.; Riaz, R.; Tulain, R.; Hussain, T.; Amjad, M.W.; Raja, M.A.; Farooq, U.; et al. Fabrication of thiolated chitosan based biodegradable nanoparticles of ticagrelor and their pharmacokinetics. *Polym. Polym. Compos.* **2022**, *30*, 09673911221108742. [CrossRef]
40. Fan, B.; Xing, Y.; Zheng, Y.; Sunand, C.; Liang, G. pH-responsive thiolated chitosan nanoparticles for oral low-molecular weight heparin delivery: In vitro and in vivo evaluation. *Drug Deliv.* **2016**, *23*, 238–247. [CrossRef]
41. Chowdhury, P.; Roy, B.; Mukherjee, N.; Mukherjee, S.; Joardar, N.; Mondal, M.K.; Roy, D.; Babu, S.P.S. Chitosan biopolymer functionalized gold nanoparticles with controlled cytotoxicity and improved antifilarial efficacy. *Adv. Compos. Hybrid Mater.* **2018**, *1*, 577–590. [CrossRef]
42. Yien, L.; Zin, N.M.; Sarwarand, A.; Katas, H. Antifungal activity of chitosan nanoparticles and correlation with their physical properties. *Int. J. Biomater.* **2012**, *2012*, 632698.
43. Machado, A.; Cunha-Reis, C.; Araújo, F.; Nunes, R.; Seabra, V.; Ferreira, D.; das Neves, J.; Sarmento, B. Development and in vivo safety assessment of tenofovir-loaded nanoparticles-in-film as a novel vaginal microbicide delivery system. *Acta Biomater.* **2016**, *44*, 332–340. [CrossRef]
44. Shailender, J.; Ravi, P.R.; Saha, P.; Dalviand, A.; Myneni, S. Tenofovir disoproxil fumarate loaded PLGA nanoparticles for enhanced oral absorption: Effect of experimental variables and in vitro, ex vivo and in vivo evaluation. *Colloids Surf. B Biointerfaces* **2017**, *158*, 610–619. [CrossRef]

45. Mariadoss, A.V.A.; Vinayagam, R.; Senthilkumar, V.; Paulpandi, M.; Murugan, K.; Xu, B.; Gothandam, K.M.; Kotakadi, V.S.; David, E. Phloretin loaded chitosan nanoparticles augments the pH-dependent mitochondrial-mediated intrinsic apoptosis in human oral cancer cells. *Int. J. Biol. Macromol.* **2019**, *130*, 997–1008. [CrossRef]
46. Zhang, J.; Wang, Y.; Li, J.; Zhao, W.; Yangand, Z.; Feng, Y. α-Santalol functionalized chitosan nanoparticles as efficient inhibitors of polo-like kinase in triple negative breast cancer. *RSC Adv.* **2020**, *10*, 5487–5501. [CrossRef] [PubMed]
47. Jain, A.; Agarwal, A.; Majumder, S.; Lariya, N.; Khaya, A.; Agrawal, H.; Majumdar, S.; Agrawal, G.P. Mannosylated solid lipid nanoparticles as vectors for site-specific delivery of an anti-cancer drug. *J. Control. Release* **2010**, *148*, 359–367. [CrossRef] [PubMed]
48. Zhang, T.; Sturgisand, T.F.; Youan, B.-B.C. pH-responsive nanoparticles releasing tenofovir intended for the prevention of HIV transmission. *Eur. J. Pharm. Biopharm.* **2011**, *79*, 526–536. [CrossRef]
49. Meng, J.; Agrahari, V.; Ezoulin, M.J.; Zhang, C.; Purohit, S.S.; Molteni, A.; Dim, D.; Oyler, N.A.; Youan, B.-B.C. Tenofovir containing thiolated chitosan core/shell nanofibers: In vitro and in vivo evaluations. *Mol. Pharm.* **2016**, *13*, 4129–4140. [CrossRef]
50. Gorochovceva, N.; Makuška, R. Synthesis and study of water-soluble chitosan-O-poly (ethylene glycol) graft copolymers. *Eur. Polym. J.* **2004**, *40*, 685–691. [CrossRef]
51. Lin, W.; Leeand, H.; Wang, D. The influence of plasticizers on the release of theophylline from microporous-controlled tablets. *J. Control. Release* **2004**, *99*, 415–421. [CrossRef]
52. Liu, X.; Howard, K.A.; Dong, M.Ø.; Andersen, M.; Rahbek, U.L.; Johnsen, M.G.; Hansen, O.C.; Besenbacher, F.; Kjems, J. The influence of polymeric properties on chitosan/siRNA nanoparticle formulation and gene silencing. *Biomaterials* **2007**, *28*, 1280–1288. [CrossRef]
53. Moradkhannejhad, L.; Abdouss, M.; Nikfarjam, N.; Shahriariand, M.H.; Heidary, V. The effect of molecular weight and content of PEG on in vitro drug release of electrospun curcumin loaded PLA/PEG nanofibers. *J. Drug Deliv. Sci. Technol.* **2020**, *56*, 101554. [CrossRef]
54. Helmi, O.; Elshishinyand, F.; Mamdouh, W. Targeted doxorubicin delivery and release within breast cancer environment using PEGylated chitosan nanoparticles labeled with monoclonal antibodies. *Int. J. Biol. Macromol.* **2021**, *184*, 325–338. [CrossRef] [PubMed]
55. Jusu, S.M.; Obayemi, J.D.; Salifu, A.A.; Nwazojie, C.C.; Uzonwanne, V.; Odusanya, O.S.; Soboyejo, W.O. Drug-encapsulated blend of PLGA-PEG microspheres: In vitro and in vivo study of the effects of localized/targeted drug delivery on the treatment of triple-negative breast cancer. *Sci. Rep.* **2020**, *10*, 1–23. [CrossRef] [PubMed]
56. Scheeren, L.E.; Nogueira, D.R.; Macedo, L.B.; Vinardell, M.P.; Mitjans, M.; Infante, M.R.; Rolim, C.M. PEGylated and poloxamer-modified chitosan nanoparticles incorporating a lysine-based surfactant for pH-triggered doxorubicin release. *Colloids Surf. B Biointerfaces* **2016**, *138*, 117–127. [CrossRef] [PubMed]
57. Wan, L.S.; Hengand, P.W.; Wong, L.F. Relationship between swelling and drug release in a hydrophilic matrix. *Drug Dev. Ind. Pharm.* **1993**, *19*, 1201–1210. [CrossRef]

MDPI
St. Alban-Anlage 66
4052 Basel
Switzerland
www.mdpi.com

Molecules Editorial Office
E-mail: molecules@mdpi.com
www.mdpi.com/journal/molecules

www.ingramcontent.com/pod-product-compliance
Lightning Source LLC
Chambersburg PA
CBHW041940260926
51850CB00018B/312